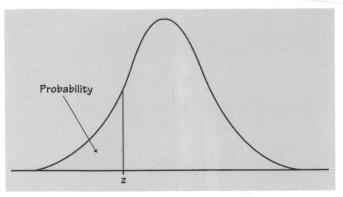

Table entry for z is the
probability lying below z.

TABLE A	Standard normal probabilities								
z	.00	.01	.02	.03	.04	.05	.06	.07	.08	.09
−3.4	.0003	.0003	.0003	.0003	.0003	.0003	.0003	.0003	.0003	.0002
−3.3	.0005	.0005	.0005	.0004	.0004	.0004	.0004	.0004	.0004	.0003
−3.2	.0007	.0007	.0006	.0006	.0006	.0006	.0006	.0005	.0005	.0005
−3.1	.0010	.0009	.0009	.0009	.0008	.0008	.0008	.0008	.0007	.0007
−3.0	.0013	.0013	.0013	.0012	.0012	.0011	.0011	.0011	.0010	.0010
−2.9	.0019	.0018	.0018	.0017	.0016	.0016	.0015	.0015	.0014	.0014
−2.8	.0026	.0025	.0024	.0023	.0023	.0022	.0021	.0021	.0020	.0019
−2.7	.0035	.0034	.0033	.0032	.0031	.0030	.0029	.0028	.0027	.0026
2.6	.0047	.0045	.0044	.0043	.0041	.0040	.0039	.0038	.0037	.0036
−2.5	.0062	.0060	.0059	.0057	.0055	.0054	.0052	.0051	.0049	.0048
−2.4	.0082	.0080	.0078	.0075	.0073	.0071	.0069	.0068	.0066	.0064
−2.3	.0107	.0104	.0102	.0099	.0096	.0094	.0091	.0089	.0087	.0084
−2.2	.0139	.0136	.0132	.0129	.0125	.0122	.0119	.0116	.0113	.0110
−2.1	.0179	.0174	.0170	.0166	.0162	.0158	.0154	.0150	.0146	.0143
−2.0	.0228	.0222	.0217	.0212	.0207	.0202	.0197	.0192	.0188	.0183
−1.9	.0287	.0281	.0274	.0268	.0262	.0256	.0250	.0244	.0239	.0233
−1.8	.0359	.0351	.0344	.0336	.0329	.0322	.0314	.0307	.0301	.0294
−1.7	.0446	.0436	.0427	.0418	.0409	.0401	.0392	.0384	.0375	.0367
−1.6	.0548	.0537	.0526	.0516	.0505	.0495	.0485	.0475	.0465	.0455
−1.5	.0668	.0655	.0643	.0630	.0618	.0606	.0594	.0582	.0571	.0559
−1.4	.0808	.0793	.0778	.0764	.0749	.0735	.0721	.0708	.0694	.0681
−1.3	.0968	.0951	.0934	.0918	.0901	.0885	.0869	.0853	.0838	.0823
−1.2	.1151	.1131	.1112	.1093	.1075	.1056	.1038	.1020	.1003	.0985
−1.1	.1357	.1335	.1314	.1292	.1271	.1251	.1230	.1210	.1190	.1170
−1.0	.1587	.1562	.1539	.1515	.1492	.1469	.1446	.1423	.1401	.1379
−0.9	.1841	.1814	.1788	.1762	.1736	.1711	.1685	.1660	.1635	.1611
−0.8	.2119	.2090	.2061	.2033	.2005	.1977	.1949	.1922	.1894	.1867
−0.7	.2420	.2389	.2358	.2327	.2296	.2266	.2236	.2206	.2177	.2148
−0.6	.2743	.2709	.2676	.2643	.2611	.2578	.2546	.2514	.2483	.2451
−0.5	.3085	.3050	.3015	.2981	.2946	.2912	.2877	.2843	.2810	.2776
−0.4	.3446	.3409	.3372	.3336	.3300	.3264	.3228	.3192	.3156	.3121
−0.3	.3821	.3783	.3745	.3707	.3669	.3632	.3594	.3557	.3520	.3483
−0.2	.4207	.4168	.4129	.4090	.4052	.4013	.3974	.3936	.3897	.3859
−0.1	.4602	.4562	.4522	.4483	.4443	.4404	.4364	.4325	.4286	.4247
−0.0	.5000	.4960	.4920	.4880	.4840	.4801	.4761	.4721	.4681	.4641

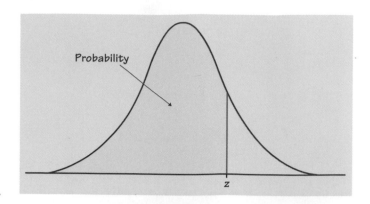

Table entry for _z_ is the probability lying below _z_.

z	.00	.01	.02	.03	.04	.05	.06	.07	.08	.09
0.0	.5000	.5040	.5080	.5120	.5160	.5199	.5239	.5279	.5319	.5359
0.1	.5398	.5438	.5478	.5517	.5557	.5596	.5636	.5675	.5714	.5753
0.2	.5793	.5832	.5871	.5910	.5948	.5987	.6026	.6064	.6103	.6141
0.3	.6179	.6217	.6255	.6293	.6331	.6368	.6406	.6443	.6480	.6517
0.4	.6554	.6591	.6628	.6664	.6700	.6736	.6772	.6808	.6844	.6879
0.5	.6915	.6950	.6985	.7019	.7054	.7088	.7123	.7157	.7190	.7224
0.6	.7257	.7291	.7324	.7357	.7389	.7422	.7454	.7486	.7517	.7549
0.7	.7580	.7611	.7642	.7673	.7704	.7734	.7764	.7794	.7823	.7852
0.8	.7881	.7910	.7939	.7967	.7995	.8023	.8051	.8078	.8106	.8133
0.9	.8159	.8186	.8212	.8238	.8264	.8289	.8315	.8340	.8365	.8389
1.0	.8413	.8438	.8461	.8485	.8508	.8531	.8554	.8577	.8599	.8621
1.1	.8643	.8665	.8686	.8708	.8729	.8749	.8770	.8790	.8810	.8830
1.2	.8849	.8869	.8888	.8907	.8925	.8944	.8962	.8980	.8997	.9015
1.3	.9032	.9049	.9066	.9082	.9099	.9115	.9131	.9147	.9162	.9177
1.4	.9192	.9207	.9222	.9236	.9251	.9265	.9279	.9292	.9306	.9319
1.5	.9332	.9345	.9357	.9370	.9382	.9394	.9406	.9418	.9429	.9441
1.6	.9452	.9463	.9474	.9484	.9495	.9505	.9515	.9525	.9535	.9545
1.7	.9554	.9564	.9573	.9582	.9591	.9599	.9608	.9616	.9625	.9633
1.8	.9641	.9649	.9656	.9664	.9671	.9678	.9686	.9693	.9699	.9706
1.9	.9713	.9719	.9726	.9732	.9738	.9744	.9750	.9756	.9761	.9767
2.0	.9772	.9778	.9783	.9788	.9793	.9798	.9803	.9808	.9812	.9817
2.1	.9821	.9826	.9830	.9834	.9838	.9842	.9846	.9850	.9854	.9857
2.2	.9861	.9864	.9868	.9871	.9875	.9878	.9881	.9884	.9887	.9890
2.3	.9893	.9896	.9898	.9901	.9904	.9906	.9909	.9911	.9913	.9916
2.4	.9918	.9920	.9922	.9925	.9927	.9929	.9931	.9932	.9934	.9936
2.5	.9938	.9940	.9941	.9943	.9945	.9946	.9948	.9949	.9951	.9952
2.6	.9953	.9955	.9956	.9957	.9959	.9960	.9961	.9962	.9963	.9964
2.7	.9965	.9966	.9967	.9968	.9969	.9970	.9971	.9972	.9973	.9974
2.8	.9974	.9975	.9976	.9977	.9977	.9978	.9979	.9979	.9980	.9981
2.9	.9981	.9982	.9982	.9983	.9984	.9984	.9985	.9985	.9986	.9986
3.0	.9987	.9987	.9987	.9988	.9988	.9989	.9989	.9989	.9990	.9990
3.1	.9990	.9991	.9991	.9991	.9992	.9992	.9992	.9992	.9993	.9993
3.2	.9993	.9993	.9994	.9994	.9994	.9994	.9994	.9995	.9995	.9995
3.3	.9995	.9995	.9995	.9996	.9996	.9996	.9996	.9996	.9996	.9997
3.4	.9997	.9997	.9997	.9997	.9997	.9997	.9997	.9997	.9997	.9998

Introduction to the Practice of Statistics

INTRODUCTION TO THE PRACTICE OF STATISTICS

THIRD EDITION

David S. Moore
George P. McCabe

Purdue University

W. H. Freeman and Company
New York

Acquisitions editor: Holly Hodder/Michelle Russel Julet
Sponsoring editor: Patrick Farace
Development editor: Don Gecewicz
Project editor: Diane Cimino Maass
Director of media: Patrick Shriner
Cover and text designer: Diana Blume
Illustration coordinator: Bill Page
Illustrations: Publication Services
Production coordinator: Maura Studley / Susan Wein
Composition: Publication Services
Manufacturing: Von Hoffman Press, Inc.
Marketing manager: Kimberly Manzi

Cover image: Andy Warhol, *200 Campbell's Soup Cans,* 1962. Oil on canvas, 72 x 100 inches. © 1998 Andy Warhol Foundation for the Visual Arts/ARS, New York. Photo courtesy Leo Castelli Photo Archives.

TI-83 screens are used with permission of the publisher. Copyright ©1996, Texas Instruments, Inc.
TI-83 Graphics Calculator is a registered trademark of Texas Instruments, Inc.
Minitab is a registered trademark of Minitab, Inc.
SAS is a registered trademark of SAS Institute, Inc.
Microsoft and Windows are registered trademarks of the Microsoft Corporation in the U.S.A. and other countries.
Excel screen shots reprinted with permission from the Microsoft Corporation.

Library of Congress Cataloging-in-Publication Data

Moore, David S.
   Introduction to the practice of statistics / David S. Moore,
George P. McCabe. — 3rd ed.
      p.   cm.
   Includes index.
   ISBN 0-7167-3502-4. — ISBN 0-7167-3286-6 (Minibook)
   1. Mathematical statistics.   I. McCabe, George P.   II. Title.
QA276.12.M65   1998
519.5—dc21   97-45727
                                                                   CIP

Printed in the United States of America

Third printing, 1999

CONTENTS

*This section contains optional material.

*This section contains optional material.

*This section contains optional material.

*This section contains optional material.

*This section contains optional material.

_____

*This section contains optional material.

CHAPTER **12**

One–Way Analysis of Variance   742

CHAPTER **13**

Two–Way Analysis of Variance   798

*This section contains optional material.

Additional chapters are on CD-ROM and available in a separate print format:

CHAPTER **14**
Nonparametric Tests

Introduction

14.1 Wilcoxon Rank Sum Test

The rank transformation
The Wilcoxon rank sum test
The normal approximation
What hypothesis does Wilcoxon test?
Ties
Limitations of nonparametric tests
Section 14.1 Exercises

14.2 The Wilcoxon Signed Rank Test

The normal approximation
Ties
Section 14.2 Exercises

14.3 The Kruskal–Wallis Test

Hypotheses and assumptions
The Kruskal-Wallis test
Section 14.3 Exercises

CHAPTER **15**
Logistic Regression

Binomial distributions and odds
The logistic regression model
Fitting and interpreting the logistic regression model
Inference for logistic regression
Multiple logistic regression
Chapter 15 Exercises

PREFACE

Introduction to the Practice of Statistics (IPS) is an introductory text that focuses on data and on statistical reasoning. It is elementary in mathematical level, but conceptually rich in statistical ideas and serious in its aim to help students think about data and use statistical methods with understanding. Although the first edition of *IPS* was a somewhat radical departure from the then-standard course, which emphasized probability and inference, this third edition now represents the current standard, in which data analysis, design of data production, and the demands of statistical practice join probability-based inference as foci for study.

Statisticians have, in fact, reached a general consensus on the nature of first courses for general college audiences. As Richard Scheaffer says in discussing a survey paper by one of us, "With regard to the content of an introductory statistics course, statisticians are in closer agreement today than at any previous time in my career."[1] *IPS* is one expression of that consensus, the principles of which have been summarized by a joint committee of the American Statistical Association and the Mathematical Association of America as follows:

- Emphasize the elements of statistical thinking.
- Incorporate more data and concepts, fewer recipes and derivations. Wherever possible, automate computations and graphics.[2]

In this preface we first briefly describe our philosophy and then discuss new features of the third edition. The title of the book expresses our intent to introduce readers to statistics as it is used in practice. Statistics in practice is concerned with gaining understanding from data; it focuses on problem solving rather than on methods that may be useful in specific settings. A text cannot fully imitate practice, because it must teach specific methods in a logical order and must use data that are not the reader's own. Nonetheless, our interest and experience in applying statistics have influenced the nature of *IPS* in several ways.

Statistical Thinking   Statistics is interesting and useful because it provides strategies and tools for using data to gain insight into real problems. As the continuing revolution in computing automates most of the tiresome details, an emphasis on statistical concepts and on insight from data becomes both more practical for students and teachers and more important for users who must supply what is not automated. No student should complete a first statistics course, for example, without a firm grasp of the distinction between observational studies and experiments and of why randomized comparative experiments are the gold standard for evidence of causation. We have seen many statistical mistakes, but few that simply involved getting a calculation wrong. We therefore ask students to learn to explore data, always starting with plots, to think about the context of the data and the design of the study that produced the data, the possible influence of wild observations on conclusions, and the reasoning that lies behind standard methods of inference. Users of statistics who form these habits from the beginning are well prepared to learn and use more advanced methods.

Data   Data are numbers with a context. The number 10.3 alone is meaningless. Hearing that a friend's new baby weighed 10.3 pounds at birth, however, engages our background knowledge and brings immediate meaning. Note that the weight could not plausibly be 10.3 ounces or 10.3 kilograms. Because context makes numbers meaningful, our examples and exercises use real data with real contexts that we briefly describe. Calculating the mean of five numbers is arithmetic, not statistics. We hope that the presence of background information, even in exercises intended for routine drill, will encourage students to always consider the meaning of their calculations as well as the calculations themselves. Note in this connection that a calculation or a graph is rarely "the answer" to a statistical problem. We strongly encourage requiring students to always state a brief conclusion in the context of the problem. This helps build data sense as well as the communications skills that employers value.

Mathematics   Although statistics is a mathematical science, it is not a field of mathematics and should not be taught as if it were. A fruitful mathematical theory (based on probability, which *is* a field of mathematics) underlies some parts of basic statistics, but by no means all. The distinction between observation and experiment, for example, is a core statistical idea that is ignored by the theory.[3] Mathematically-trained teachers, rightly resisting a formula-based approach, sometimes identify conceptual understanding with mathematical understanding. When teaching statistics, we must emphasize statistical ideas and recognize that mathematics is not the only vehicle for conceptual understanding. ***Introduction to the Practice of Statistics*** requires only the ability to read and use equations without having each step parsed. We require no algebraic derivations, let alone calculus. Because this is a *statistics* text, it is richer in ideas and requires more thought than the low mathematical level suggests.

Calculators and Computers   Statistical calculations and graphics are, in practice, automated by software. We encourage instructors to use software of their choice or a graphing calculator that includes functions for both data analysis and basic inference. ***IPS*** includes some topics that reflect the dominance of software in practice, such as normal quantile plots and the version of the two-sample t procedures that does not require equal variances. *All* *students should have at least a "two-variable statistics" calculator* with functions for correlation and the least-squares regression line as well as for the mean and standard deviation. Although not all exercises are feasible for students with such a calculator, we have taken care to make the book usable by students without access to computing.

Judgment Statistics in practice requires judgment. It is easy to list the mathematical assumptions that justify use of a particular procedure, but not so easy to decide when the procedure can be safely used in practice. Because judgment develops through experience, an introductory course should present clear guidelines and not make unreasonable demands on the judgment of students. We have given guidelines—for example, on using the *t* procedures for comparing two means but avoiding the *F* procedures for comparing two variances—that we follow ourselves. Similarly, many exercises require students to use some judgment and (equally important) to justify their choices in words. Many students would prefer to stick to calculating, and many statistics texts allow them to. Requiring more will do them much good in the longer run.

Teaching Experiences We have successfully used **IPS** in courses taught to quite diverse student audiences. For general undergraduates from mixed disciplines, we cover Chapters 1 to 8 and one of Chapters 9, 10, and 12, omitting all optional material. For sophomores planning to major in actuarial science or statistics, we add Chapters 10 and 11 to the core material in Chapters 1 to 8 and include most optional content. We deemphasize Chapter 4 (probability) because these students will take a probability course later in their program, and we make very intensive use of software. The third group we teach contains beginning graduate students in such fields as education, family studies, and retailing. These mature, but sometimes quantitatively unprepared, students read the entire text (Chapters 11 and 13 lightly) with little emphasis on Chapter 4 and some parts of Chapter 5. In all cases, beginning with data analysis and data production (Part I) helps students overcome their fear of statistics and builds a sound base for studying inference.

The Third Edition

In revising ***Introduction to the Practice of Statistics*** after two successful editions, we have concentrated on helping students read the book and on maintaining close ties to statistical practice.

Organization

The **table of contents** for the latter part of the text has been revised. The new contents makes it clearer that instructors can choose from among the material after Chapter 8 by making each natural unit into a separate, shorter, chapter. This also allows a three-part structure that makes the major divisions of the book more apparent.

There are some changes in **scope and sequence.** For example, correlation now precedes regression in Chapter 2. Correlation is a somewhat more general topic, because it does not require the distinction between explanatory and response variables. More important, the new order allows the least-squares regression line to be described in terms of the means and standard deviations of the two variables and the correlation between them. A completely new treatment of conditional probability (Chapter 4) and of inference for two-way tables (Chapter 9) are other examples of major changes.

Accessibility

We have carefully rewritten the text to improve **readability** without losing the conceptual richness and real-problem settings that characterize *IPS.* In particular, we have provided more structure for readers via **more frequent subsection headings** and by bringing more material out of the flow of the text into **boxes** and other displays.

Technology

We assume better technology—at least a "two-variable statistics" calculator. Calculation recipes intended for basic calculators have disappeared. Our principle is that if a formula is "just a rule," it should be automated. For example, equations for the slope of the least-squares regression line based on sums of squares provide no insight to students and are just rules. No such formula appears in this edition: they are replaced by a button on a calculator or a menu item in software. On the other hand, the equation $b = r s_y / s_x$ for the regression slope does provide insight. We discuss it carefully and expect students to know the equation and its implications. Some exercises in Chapters 4 and 5 now suggest simulations that help students understand probability.

Statistical Output

We recognize the diversity of technology and prevalence of graphical interfaces. There are no longer any Minitab commands in the text, because these are now menu items in the software. We include **examples of output from a wider range of software,** adding the TI-83 graphing calculator and the Excel spreadsheet to statistical software such as Data Desk, Minitab, and SAS. The purpose of displaying output is primarily to convince students that they can read and use the information provided by any of these platforms once they have mastered the ideas and terminology in the text.

"Beyond the basics" Sections

We have added short subsections entitled "Beyond the basics" to most chapters. These briefly introduce somewhat more advanced material that is now becoming standard in statistical practice. Although this material is considered "enrichment" rather than required learning, instructors using software that implements these methods may wish to ask students to use them regularly. It is often easy, for example, to ask for a density estimate rather than a histogram or to employ a scatterplot smoother to summarize the overall pattern of a plot.

New Exercises and Examples

Many new exercises and examples based on new data sets have been added. Weaker or outdated exercises have been replaced and the overall exercise count has modestly increased. More exercises assume automated calculations and graphics, so that the separate "Computer Exercises" of previous editions are no longer needed. There are more exercises and examples from **business and finance,** so that the content reflects the full range of statistical applications.

Data Sets

The data from exercises and tables are available in several formats on a dual-platform CD-ROM included with each copy of *IPS*. Data sets can also be downloaded from the *IPS* website at **www.whfreeman.com/statistics.**

Additional Chapters on CD-ROM

In writing *IPS* we have concentrated on the material we think most valuable in a first course. Our criteria were direct usefulness in practice and the extent to which studying this material prepares students to go on to more advanced statistical methods. We think that encyclopedic introductory texts intimidate students and that very brief coverage of many methods reinforces students' perception that statistics consists of recipes. The new CD-ROM that accompanies each book allows us to offer additional (optional) material without violating these principles. The CD-ROM contains two new chapters, complete with exercises. **Chapter 14, Nonparametric Tests,** presents the most common rank tests: Wilcoxon-Mann-Whitney, Wilcoxon signed rank, and Kruskal-Wallis, along with comments on their use in practice. **Chapter 15, Logistic Regression,** introduces a topic that is very prominent in statistical practice and that we would like to see become more common in beginning instruction. These chapters are also available as a printed supplement.

Supplements

A full range of supplements is available to help teachers and students better teach and learn from *IPS*. Several **complimentary supplements** are available for adopters of the third edition of *IPS:*

■ **Instructor's Guide with Solutions** prepared by Darryl Nester of Bluffton College includes worked out solutions to all exercises and teaching suggestions.

- **Test Bank** prepared by William Notz and Michael Fligner of Ohio State University is available in both computerized and printed versions for generating quizzes and exams.

- **Instructor's CD-ROM** contains all material that appears on the Student CD-ROM and additional instructor-only features including all graphs and figures from *IPS* in an exportable presentation format.

And for Students . . .

- **Student CD-ROM** is included with every copy of *IPS*. It contains the Electronic Encyclopedia of Statistical Examples and Exercises (EESEE) application, which is a rich repository of case studies applying the concepts of *IPS* to various real-world venues such as the mass media, sports, natural sciences, social sciences, and medicine. Each case study is accompanied by practice problems, and most include full data sets which are exportable to various statistical software packages. EESEE was developed at Ohio State University by Professors William I. Notz, Dennis K. Pearl, and Elizabeth A. Stasny.

  In addition to the EESEE application, the Student CD also contains a "Q&A" interactive quiz with built-in feedback for each chapter, the additional chapters on nonparametric tests and logistical regression in a printable format, video clips showing real-world applications of statistics, and some special electronic statistical tools.

- **Study Guide** prepared by William Notz and Michael Fligner of Ohio State University offers students explanations of the crucial concepts in each section of *IPS,* detailed solutions to key text problems, step-through models of important statistical techniques, and useful case studies.

- **Statistical Software Manuals** will guide students in the use of particular statistical software with *IPS.* The chapters of each manual correlate to those of *IPS* and include a set of exercises specific to that chapter's concepts. These manuals are:
  - **Minitab Manual** prepared by Michael Evans of the University of Toronto
  - **Excel Manual** prepared by Fred Hoppe of McMaster University
  - **SPSS Manual** prepared by Paul Stephenson, Neal Rogness, Justine Ritchie, and Patricia Stephenson of Grand Valley State University.
- **TI-83 Graphing Calculator Manual** prepared by David Neal of Western Kentucky University demonstrates the use of the TI-83 in solving problems and analyzing data in *IPS.*

For Students *and* Instructors:

- *IPS* **Website** contains additional and continually-updated materials for learning and teaching from *IPS.* These offerings include the data sets from the text and additional EESEE case studies, practice problems, and data. The Website can be accessed at:

www.whfreeman.com/statistics

Acknowledgments

We are pleased that the first two editions of *Introduction to the Practice of Statistics* have helped move the teaching of introductory statistics in a direction supported by most statisticians. We are grateful to the many colleagues and students who have provided helpful comments, and we hope that they will find this new edition another step forward. In particular, we would like to thank the following colleagues who offered specific comments on the new edition:

John Barnard, Harvard University
Tim Brown, University of Melbourne
Patricia Buchanan, Pennsylvania State University
Smiley Cheng, University of Manitoba
Brian Corbitt, The Open Polytechnic of New Zealand
Constance Eshbach Cutchins, University of Central Florida
Roy V. Erickson, Michigan State University
Kenneth Fairbanks, Murray State University
Lloyd Gavin, California State, Sacramento
Mark E. Glickman, Boston University
Joel B. Greenhouse, Carnegie Mellon University
Miriam Schapiro Grosof, Stern College
Katherine Halvorson, Smith College
Robert Hannum, University of Denver
W. L. Harkness, Pennsylvania State University
I. Hibschweiler, Daemen College
Elizabeth A. Houseworth, University of Oregon
John Z. Imbrie, University of Virginia
T. Henry Jablonski Jr., East Tennessee State University
Dave Jobson, University of Alberta
Gerald Kaminski, University of St. Thomas
Douglas Kelker, University of Alberta
Eugene Klimko, Binghamton University
Marlene J. Kovaly, Florida Community College at Jacksonville
Don O. Loftsgaarden, University of Montana
George Marrah, James Madison University
Brendan McCabe, University of British Columbia
Kip Murray, University of Houston
Tom O'Bryan, University of Wisconsin, Milwaukee
Raman Patel, California State University, Chico
Richard Pulskamp, Xavier University
Robert L. Raymond, University of Minnesota
Larry J. Ringer, Texas A & M University
Neal Rogness, Grand Valley State University
Kenneth A. Ross, University of Oregon
Ronald M. Schrader, University of New Mexico
Al Schainblatt, San Francisco State University
Robert K. Smidt, California Polytechnic University
Patty Solomon, University of Adelaide

Elizabeth A. Stasny, The Ohio State University
Bert Steece, University of Southern California
Paul Stephenson, Grand Valley State University
Linn M. Stranak, Union University
Thaddeus Tarpey, Wright State University
David van Dyk, Harvard University
Chamont Wang, The College of New Jersey
Peter Westfall, Texas Tech University
William H. Woodall, University of Alabama
Alan Zaslavsky, Harvard University

Most of all, we are grateful to the many people in varied disciplines and occupations with whom we have worked to gain understanding from data. They have provided both material for this book and the experience that enabled us to write it. What the eminent statistician John Tukey called "the real-problems experience and the real-data experience" has shaped our view of statistics. It has convinced us of the need for beginning instruction to focus on data and concepts, building intellectual skills that transfer to more elaborate settings and remain essential when all details are automated. We hope that users and potential users of statistical techniques will find this emphasis helpful.

Notes

1. D. S. Moore and discussants, "New pedagogy and new content: the case of statistics," *International Statistical Review,* 65 (1997), pp.123–165. Richard Scheaffer's comment appears on page 156.

2. These are main heads in a brief version of the committee's report endorsed by the Board of Directors of the American Statistical Association. The full text appears on page 127 of the article cited in Note 1.

3. A detailed discussion appears in G. Cobb and D. S. Moore, "Mathematics, statistics, and teaching," *American Mathematical Monthly,* 104 (1997), pp. 801–823.

INTRODUCTION:
What Is Statistics?

Statistics is the science of collecting, organizing, and interpreting numerical facts, which we call *data*. We are bombarded by data in our everyday lives. Most of us associate "statistics" with the bits of data that appear in news reports: baseball batting averages, imported car sales, the latest poll of the president's popularity, and the average high temperature for today's date. Advertisements often claim that data show the superiority of the advertiser's product. All sides in public debates about economics, education, and social policy argue from data. Yet the usefulness of statistics goes far beyond these everyday examples.

The study and collection of data are important in the work of many professions, so that training in the science of statistics is valuable preparation for a variety of careers. Each month, for example, government statistical offices release the latest numerical information on unemployment and inflation. Economists and financial advisors, as well as policy makers in government and business study these data in order to make informed decisions. Doctors must understand the origin and trustworthiness of the data that appear in medical journals if they are to offer their patients the most effective treatments. Politicians rely on data from polls of public opinion. Business decisions are based on market research data that reveal consumer tastes. Farmers study data from field trials of new crop varieties. Engineers gather data on the quality and reliability of manufactured products. Most areas of academic study make use of numbers, and therefore also make use of the methods of statistics.

We can no more escape data than we can avoid the use of words. Just as words on a page are meaningless to the illiterate or confusing to the partially educated, so data do not interpret themselves but must be read with understanding. Just as a writer can arrange words into convincing arguments or incoherent nonsense, so data can be compelling, misleading, or simply irrelevant. Numerical literacy, the ability to follow and understand numerical arguments, is important for everyone. The ability to express yourself numerically, to be an author rather than just a reader, is a vital skill in many professions and areas of study. The study of statistics is therefore essential to a sound education. We must learn how to read data, critically and with comprehension. We must learn how to produce data that provide clear answers to important questions. And we must learn sound methods for drawing trustworthy conclusions based on data.

The Rise of Statistics

Historically, the ideas and methods of statistics developed gradually as society grew interested in collecting and using data for a variety of applications. The earliest origins of statistics lie in the desire of rulers to count the number of inhabitants or measure the value of taxable land in their domains. As the physical sciences developed in the seventeenth and eighteenth centuries, the importance of careful measurements of weights, distances, and other physical quantities grew. Astronomers and surveyors striving for exactness had to deal with variation in their measurements. Many measurements should be better than a single measurement, even though they vary among themselves. How can we best combine many varying observations? Statistical methods that are still important were invented in order to analyze scientific measurements.

By the nineteenth century, the agricultural, life, and behavioral sciences also began to rely on data to answer fundamental questions. How are the heights of parents and children related? Does a new variety of wheat produce higher yields than the old, and under what conditions of rainfall and fertilizer? Can a person's mental ability and behavior be measured just as we measure height and reaction time? Effective methods for dealing with such questions developed slowly and with much debate.[1]

As methods for producing and understanding data grew in number and sophistication, the new discipline of statistics took shape in the twentieth century. Ideas and techniques that originated in the collection of government data, in the study of astronomical or biological measurements, and in the attempt to understand heredity or intelligence came together to form a unified "science of data." That science of data—statistics—is the topic of this text.

The Organization of this Book

Part I of this book, called simply "Data," concerns data analysis and data production. The first two chapters deal with statistical methods for organizing and describing data. These chapters progress from simpler to more complex data. Chapter 1 examines data on a single variable, Chapter 2 is devoted to relationships among two or more variables. You will learn both how to examine data produced by others and how to organize and summarize your own data. These summaries will be first graphical, then numerical, then when appropriate in the form of a mathematical model that gives a compact description of the overall pattern of the data. Chapter 3 outlines arrangements (called designs) for producing data that answer specific questions. The principles presented in this chapter will help you to design proper samples and experiments, and to evaluate such investigations in your field of study.

Part II, consisting of Chapters 4 to 8, introduces statistical inference—formal methods for drawing conclusions from properly produced data. Statistical inference uses the language of probability to describe how reliable its conclusions are, so some basic facts about probability are needed to understand inference. Probability is the subject of Chapters 4 and 5. Chapter 6, perhaps the most important chapter in the text, introduces the reasoning of statistical inference. Effective inference is based on good procedures for producing data (Chapter 3), careful examination of the data (Chapters 1 and 2), and an understanding of the nature of statistical inference as discussed in Chapter 6. Chapters 7 and 8 describe some of the most common specific methods of inference, for drawing conclusions about means and proportions from one and two samples.

The five shorter chapters in Part III and two short chapters on CD-ROM introduce somewhat more advanced methods of inference, dealing with relations in categorical data, regression and correlation, and analysis of variance.

Understanding from Data

The practice of statistics involves the use of many recipes for numerical calculation, some quite simple and some very complex. As you are learning how to use these recipes, remember that the goal of statistics is not calculation for its own sake, but gaining understanding from numbers. Many of the calculations can be automated by a calculator or computer, but you must supply the understanding. Chapters 7 to 13 present only a few of the many specific procedures for inference. The more complex procedures are always carried out by computers using specialized software. A thorough grasp of the principles of statistics will enable you to quickly learn more advanced methods as needed. On the other hand, a fancy computer analysis carried out without attention to basic principles will often produce elaborate nonsense. As you read, seek to understand the principles as well as the necessary details of methods and recipes.

Note

1. The rise of statistics from the physical, life, and behavioral sciences is described in detail by S. M. Stigler, *The History of Statistics: The Measurement of Uncertainty Before 1900*, Harvard-Belknap, Cambridge, Mass., 1986. Much of the information in the brief historical notes appearing throughout the text is drawn from this book.

ABOUT THE AUTHORS

David S. Moore is Shanti S. Gupta Distinguished Professor of Statistics at Purdue University and 1998 President of the American Statistical Association. He received his A.B. from Princeton and the Ph.D. from Cornell, both in mathematics. He has written many research papers in statistical theory and served on the editorial boards of several major journals. Professor Moore is an elected fellow of the American Statistical Association and of the Institute of Mathematical Statistics, and an elected member of the International Statistical Institute. He has served as program director for statistics and probability at the National Science Foundation.

In recent years, Professor Moore has devoted his attention to the teaching of statistics. He was the content developer for the Annenberg/Corporation for Public Broadcasting college-level telecourse *Against All Odds: Inside Statistics* and for the series of video modules *Statistics: Decisions Through Data*, intended to aid the teaching of statistics in schools. He is the author of influential articles on statistics education and of several leading texts. Professor Moore has served as president of the International Association for Statistical Education, and has received the Mathematical Association of America's national award for distinguished college or university teaching of mathematics. He is currently a member of the National Research Council's Mathematical Sciences Education Board.

George P. McCabe is Professor of Statistics and Head of the Statistical Consulting Service at Purdue University. In 1966 he received a B.S. degree in mathematics from Providence College and in 1970 a Ph.D. in mathematical statistics from Columbia University. His entire professional career has been spent at Purdue with sabbaticals at Princeton, the Commonwealth Scientific and Industrial Research Organization (CSIRO) in Melbourne, Australia, the University of Berne (Switzerland), and the National Institute of Standards and Technology (NIST) in Boulder, Colorado. Professor McCabe is an elected fellow of the American Statistical Association and is 1998 Chair of its section on Statistical Consulting. He has served on the editorial boards of several statistics journals. He has consulted with many major corporations and has testified as an expert witness on the use of statistics in several cases.

Professor McCabe's research interests have focused on applications of statistics. He is author or co-author of over 100 publications in many different journals. Since working on a US Agency for International Development project on functional implications of malnutrition in Egypt, his collaborations on problems in human nutrition have increased. Other recent interests include problems associated with metal fatigue in aging aircraft structures, epidemiology, and industrial applications of statistics.

Andy Warhol, *Campbell's Soup Can*, 1964

Data

Prelude

The goal of statistics is to gain information from data. In this chapter you will learn statistical tools for describing data that consist of observations on just one variable, such as the grade point averages of a group of students, the list prices of new cars, or the daily numbers of customers at a theme park.

The first step is to display the data in a graph so that our eyes can take in the overall pattern and spot unusual observations. To see the shape of data on the grade point averages of college students, for example, we might use a histogram, a boxplot, a density curve, or a normal quantile plot. We will discuss all of these graphs in this chapter.

We supplement graphs with numbers that summarize specific aspects of the data, such as the average of these students' grades. As we study graphs and numerical summaries, keep firmly in mind where the data come from and what we hope to learn from them. Good data analysis requires not only that we be able to make the graphs and calculate the numerical summaries, but that we know which to choose and how to interpret in plain language the results of our work. Graphs and numbers are not ends in themselves, but aids to understanding.

Here are some questions we will meet, and answer from data, in this chapter:

- Government regulations group hot dogs into three types, beef, meat, and poultry. Are there systematic differences in how many calories hot dogs of these types contain?

- Is there a regular pattern in the changing price of fresh oranges over time? What about the changing value of stock-market indexes?

- A publication for investors says that in the long run the mean return on stocks has been 13.8% per year and that the standard deviation of returns has been 16.4%. How do these numbers warn us that stocks are profitable but risky?

- Eleanor scores 680 on the SAT college entrance examination. Gerald scores 27 on the ACT examination. How can we compare their scores?

Rebeca Mendoza, The Golden Structure, 1996.

Looking at Data— Distributions

Introduction

Statistics uses data to gain understanding. Understanding comes from combining knowledge of the background of the data with the ability to use graphs and calculations to see what the data tell us. The numbers in a medical study, for example, mean little without some knowledge of the goals of the study and of what blood pressure, heart rate, and other measurements contribute to those goals. On the other hand, measurements from the study's several hundred subjects are of little value even to the most knowledgeable medical expert until the tools of statistics organize, display, and summarize them. We begin our study of statistics by mastering the art of examining data.

Variables

Any set of data contains information about some group of *individuals*. The information is organized in *variables*.

Individuals and Variables

Individuals are the objects described by a set of data. Individuals may be people, but they may also be animals or things.

A **variable** is any characteristic of an individual. A variable can take different values for different individuals.

A college's student data base, for example, includes data about every currently enrolled student. The students are the individuals described by the data set. For each individual, the data contain the values of variables such as date of birth, gender (female or male), choice of major, and grade point average. In practice, any set of data is accompanied by background information that helps us understand the data. When you plan a statistical study or explore data from someone else's work, ask yourself Why? Who? What?

1. **Why? What purpose** do the data have? Do we hope to answer some specific questions? Do we want to draw conclusions about individuals other than the ones we actually have data for?
2. **Who?** What **individuals** do the data describe? **How many** individuals appear in the data?
3. **What?** How many **variables** do the data contain? What are the **exact definitions** of these variables? In what **units of measurement** is each variable recorded? Weights, for example, might be recorded in pounds, in thousands of pounds, or in kilograms.

Some variables, like gender and college major, simply place individuals into categories. Others, like height and grade point average, take numerical values with which we can do arithmetic. It makes sense to give an average income for a company's employees, but it does not make sense to give an "average" gender. We can, however, count the numbers of female and male employees and do arithmetic with these counts.

Categorical and Quantitative Variables

A **categorical variable** places an individual into one of several groups or categories.

A **quantitative variable** takes numerical values for which arithmetic operations such as adding and averaging make sense.

The **distribution** of a variable tells us what values it takes and how often it takes these values.

EXAMPLE 1.1

Here is a small part of the data set in which CyberStat Corporation records information about its employees:

	A	B	C	D	E	F
1	Name	Age	Gender	Race	Salary	Job Type
2	Fleetwood, Delores	39	Female	White	62,100	Management
3	Perez, Juan	27	Male	White	47,350	Technical
4	Wang, Lin	22	Female	Asian	18,250	Clerical
5	Johnson, LaVerne	48	Male	Black	77,600	Management

Enter   NUM

case

Each row records data on one individual. You will often see each row of data called a **case.** Each column contains the values of one variable for all the individuals. In addition to the person's name, there are 5 variables. Gender, race, and job type are categorical variables; age in years and salary in dollars are quantitative variables. Most statistical software uses this format to enter data—each row is an individual, and each column is a variable. This data set appears in a **spreadsheet** program that has rows and columns ready for your use. Spreadsheets are commonly used to enter and transmit data. Most statistical software can read data from the major spreadsheet programs.

spreadsheet

1.1 Displaying Distributions with Graphs

exploratory data analysis

Statistical tools and ideas help us examine data in order to describe their main features. This examination is called **exploratory data analysis.** Like an explorer crossing unknown lands, we want first to simply describe what we see. Here are two basic strategies that help us organize our exploration of a set of data:

- Begin by examining each variable by itself. Then move on to study the relationships among the variables.

- Begin with a graph or graphs. Then add numerical summaries of specific aspects of the data.

We will follow these principles in organizing our learning. This chapter presents methods for describing a single variable. We study relationships among several variables in Chapter 2. Within each chapter, we begin with graphical displays, then add numerical summaries for more complete description.

Graphs for categorical variables

The values of a categorical variable are labels for the categories, such as "male" and "female." The distribution of a categorical variable lists the categories and gives either the **count** or the **percent** of individuals who fall in each category. For example, here is the distribution of marital status for all Americans age 18 and over:[1]

Marital status	Count (millions)	Percent
Never married	43.9	22.9
Married	116.7	60.9
Widowed	13.4	7.0
Divorced	17.6	9.2

bar graph

pie chart

The graphs in Figure 1.1 display these data. The **bar graph** in Figure 1.1(a) quickly compares the sizes of the four marital status groups. The heights of the four bars show the counts in the four categories. The **pie chart** in Figure 1.1(b) helps us see what part of the whole each group forms. For example, the "married" slice makes up 61% of the pie because 61% of adults

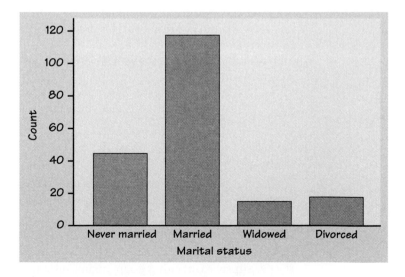

FIGURE 1.1(a)   Bar graph of the marital status of U.S. adults.

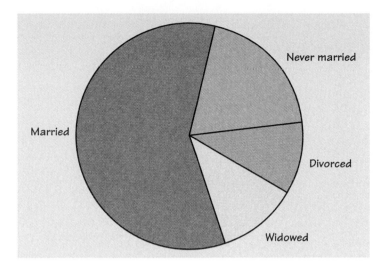

FIGURE 1.1(b) Pie chart of the same data.

are married. Pie charts require that you include all the categories that make up a whole. Bar graphs are more flexible. For example, you can use a bar graph to compare the numbers of students at your college majoring in biology, business, and political science. A pie chart cannot make this comparison because not all students fall into one of these three majors.

Bar graphs and pie charts help an audience grasp the distribution quickly. They are, however, of limited use for data analysis because it is easy to understand categorical data on a single variable such as marital status without a graph. We can move on to quantitative variables, where graphs are essential tools.

Measuring the speed of light

How do we begin to examine the distribution of a single quantitative variable? Let's look at an important scientific study.

EXAMPLE 1.2

Light travels fast, but it is not transmitted instantaneously. Light takes over a second to reach us from the moon and over 10 billion years to reach us from the most distant objects yet observed in the expanding universe. Because radio and radar also travel at the speed of light, an accurate value for that speed is important in communicating with astronauts and orbiting satellites. An accurate value for the speed of light is also important to computer designers because electrical signals travel at light speed. The first reasonably accurate measurements of the speed of light were made over 100 years ago by A. A. Michelson and Simon Newcomb. Table 1.1 contains 66 measurements made by Newcomb between July and September 1882.[2]

TABLE 1.1	Newcomb's measurements of the passage time of light				
28	22	36	26	28	28
26	24	32	30	27	24
33	21	36	32	31	25
24	25	28	36	27	32
34	30	25	26	26	25
−44	23	21	30	33	29
27	29	28	22	26	27
16	31	29	36	32	28
40	19	37	23	32	29
−2	24	25	27	24	16
29	20	28	27	39	23

A set of numbers such as those in Table 1.1 is meaningless without some background information. The individuals (or cases) here are Newcomb's 66 repetitions of his experiment. We need to know exactly *what variable* he measured, and in *what units*. Newcomb measured the time in seconds that a light signal took to pass from his laboratory on the Potomac River to a mirror at the base of the Washington Monument and back, a total distance of about 7400 meters. Just as you can compute the speed of a car from the time required to drive a mile, Newcomb could compute the speed of light from the passage time. Newcomb's first measurement of the passage time of light was 0.000024828 second, or 24,828 nanoseconds. (There are 10^9 nanoseconds in a second.) The entries in Table 1.1 record only the deviation from 24,800 nanoseconds. The table entry 28 is short for the original 0.000024828. The entry −44 stands for 24,756 nanoseconds.

Measurement

A detailed answer to the question "What variable is being measured?" requires a description of the **instrument** used to make the measurement. Then we must judge whether the variable measured is appropriate for our purpose. This judgment often requires expert knowledge of the particular field of study. For example, Newcomb invented a novel and complicated apparatus to measure the passage time of light. We accept the judgment of physicists that this instrument is appropriate for its intended task and more accurate than earlier instruments.

Newcomb measured a clearly defined and easily understood variable, the time light takes to travel a fixed distance. Questions about measurement are often harder to answer in the social and behavioral sciences than in the physical sciences. We can agree that a tape measure is an appropriate way to measure a person's height. But how shall we measure her intelligence?

Questionnaires and interview forms as well as tape measures are measuring instruments. A psychologist wishing to measure "general intelligence" might use the Wechsler Adult Intelligence Scale (WAIS). The WAIS is a standard "IQ test" that asks subjects to solve a large number of problems. The appropriateness of this instrument as a measure of intelligence is not accepted by all psychologists, many of whom doubt that there is such a thing as "general intelligence." Because questions about measurement usually require knowledge of the particular field of study, we will say little about them.

rate

You should nonetheless always ask yourself if a particular variable really does measure what you want it to. Often, for example, the **rate** at which something occurs is a more meaningful measure than a simple **count** of occurrences.

EXAMPLE 1.3

The 1970s saw the completion of most of the interstate highway system, an emphasis on better safety features in new cars, and a reduction in the national speed limit to 55 miles per hour. Did driving become safer? Here's one measure: the number of motor vehicle deaths was 52,600 in 1970 and 51,091 in 1980. That's a small change. We might think that driving was no safer in 1980 than in 1970.

However, the number of motor vehicles registered in the United States grew from 108 million in 1970 to 156 million in 1980. The count of deaths does not take account of the fact that more people did more driving in 1980 than in 1970.

A better measure of the dangers of driving is a *rate*, the number of deaths divided by the number of miles driven. This rate takes into account the total miles driven by all vehicles. In 1980, vehicles drove 1,527,000,000,000 miles in the United States. Because this number is so large, it is usual to give the death rate per 100 million miles driven. For 1980, this rate is

$$\frac{\text{motor vehicle deaths}}{\text{hundreds of millions of miles driven}} = \frac{51{,}091}{15{,}270}$$
$$= 3.3$$

The death rate fell from 4.7 deaths per 100 million miles in 1970 to 3.3 in 1980. That 30% drop tells us that driving did get safer.

Variation

Look again at Newcomb's observations in Table 1.1. They are not all the same. Why should this be? After all, each observation records the travel time of light over the same path, measured by a skilled observer using the same apparatus each time. Newcomb knew that careful measurements almost always vary. The environment of every measurement is slightly different. The apparatus changes a bit with the temperature, the density of the atmosphere changes from day to day, and so on. Newcomb did his best to eliminate the sources of variation that he could anticipate, but even the most careful experiments produce variable results. Newcomb took many measurements rather than just one because the average of 66 passage times should be less variable than the result of a single measurement. The average does not depend on the specific conditions at just one time.

Putting ourselves in Newcomb's place, we might rush to compute the average passage time, convert this time to a new and better estimate of the speed of light, and publish the result. The temptation to do a routine calculation and announce an answer arises whenever data have been collected to answer a specific question. Because computers are very good at calculations, succumbing to temptation is the easy road. The first step toward statistical sophistication is to resist the temptation to calculate without thinking. Since variation is surely present in any set of data, we must first examine the nature of the variation. We may be surprised at what we find.

Distribution of a Quantitative Variable

The pattern of variation of a variable is called its **distribution.** The distribution of a quantitative variable records its numerical values and how often each value occurs.

The distribution of a variable is best displayed graphically. We introduce some graphical tools for displaying distributions in the next few pages. With a better-stocked tool kit, we will then return to Newcomb's measurements.

Stemplots

A *stemplot* (also called a stem-and-leaf plot) gives a quick picture of the shape of a distribution while including the actual numerical values in the graph. Stemplots work best for small numbers of observations that are all greater than 0.

Stemplot

To make a **stemplot**:

1. Separate each observation into a **stem** consisting of all but the final (rightmost) digit and a **leaf**, the final digit. Stems may have as many digits as needed, but each leaf contains only a single digit.
2. Write the stems in a vertical column with the smallest at the top, and draw a vertical line at the right of this column.
3. Write each leaf in the row to the right of its stem, in increasing order out from the stem.

EXAMPLE 1.4

Here are the number of home runs that Babe Ruth hit in each of his 15 years with the New York Yankees, 1920 to 1934:[3]

| 54 | 59 | 35 | 41 | 46 | 25 | 47 | 60 | 54 | 46 | 49 | 46 | 41 | 34 | 22 |

1.1   Displaying Distributions with Graphs   **11**

```
2 |                2 | 5 2              2 | 2 5
3 |                3 | 5 4              3 | 4 5
4 |                4 | 1 6 7 6 9 6 1    4 | 1 1 6 6 6 7 9
5 |                5 | 4 9 4            5 | 4 4 9
6 |                6 | 0                6 | 0
   (a)                    (b)                    (c)
```

FIGURE 1.2   Making a stemplot of the data in Example 1.4. (a) Write the stems. (b) Go through the data writing each leaf on the proper stem. (c) Arrange the leaves on each stem in order out from the stem.

To make a stemplot for these data, we use the first digits as stems and the second digits as leaves. Thus 54, for example, appears as a leaf 4 on the stem 5. Figure 1.2 shows the the steps in making the plot. We see from the completed stemplot that Ruth hit 54 home runs in two different years.

back-to-back stemplot

When you wish to compare two related distributions, a **back-to-back stemplot** with common stems is useful. The leaves on each side are ordered out from the common stem. Ruth's record of 60 home runs in one season was broken by another Yankee, Roger Maris, who hit 61 home runs in 1961. Here is a back-to-back stemplot comparing the home run data for Ruth and Maris. Ruth's superiority as a home run hitter is clearly evident.

```
    Ruth |   | Maris
         | 0 | 8
         | 1 | 3 4 6
     5 2 | 2 | 3 6 8
     5 4 | 3 | 3 8
 9 7 6 6 6 1 1 | 4 |
     9 4 4 | 5 |
       0 | 6 | 1
```

splitting stems

rounding

Stemplots do not work well for large data sets, where each stem must hold a large number of leaves. Fortunately, there are several modifications of the basic stemplot that are helpful when plotting the distribution of a moderate number of observations. You can increase the number of stems in a plot by **splitting each stem** into two, one with leaves 0 to 4 and the other with leaves 5 through 9. When the observed values have many digits, it is often best to **round** the numbers to just a few digits before making a stemplot. You must use your judgment in deciding whether to split stems or round, though statistical software will often make these choices for you. Remember that the purpose of a stemplot is to display the shape of a distribution. If a stemplot has fewer than about five stems, you should usually split the stems unless there are few observations. If many stems have no leaves or only one leaf, rounding may help. Here is an example that makes use of both of these modifications.

EXAMPLE 1.5

A marketing consultant observed 50 consecutive shoppers at a supermarket. One variable of interest was how much each shopper spent in the store. Here are the data (in dollars), arranged in increasing order:

3.11	8.88	9.26	10.81	12.69	13.78	15.23	15.62	17.00	17.39
18.36	18.43	19.27	19.50	19.54	20.16	20.59	22.22	23.04	24.47
24.58	25.13	26.24	26.26	27.65	28.06	28.08	28.38	32.03	34.98
36.37	38.64	39.16	41.02	42.97	44.08	44.67	45.40	46.69	48.65
50.39	52.75	54.80	59.07	61.22	70.32	82.70	85.76	86.37	93.34

To make a stemplot of this distribution, we first round the purchases to the nearest dollar. We can then use tens of dollars as our stems and dollars as leaves. This gives us the single-digit leaves that a stemplot requires. For example, we round 12.69 to 13 and write it as a leaf of 3 on the 1 stem. Figure 1.3(a) displays the stemplot. For comparison, we split each stem to make the stemplot in Figure 1.3(b). Both graphs show the distribution shape, but the split stems provide more detail.

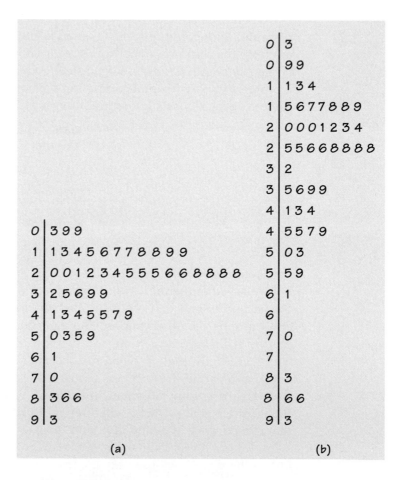

FIGURE 1.3   Stemplots of the amounts spent by 50 supermarket shoppers (in dollars), with and without splitting the stems.

Examining distributions

Making a statistical graph is not an end in itself. The purpose of the graph is to help us understand the data. After you make a graph, always ask, "What do I see?" Once you have displayed a distribution, you can see its important features as follows.

Examining a Distribution

In any graph of data, look for the **overall pattern** and for striking **deviations** from that pattern.

You can describe the overall pattern of a distribution by its **shape, center,** and **spread.**

An important kind of deviation is an **outlier,** an individual value that falls outside the overall pattern.

We will learn how to describe center and spread numerically in Section 1.2. For now, we can describe the center of a distribution by its *midpoint*, the value with roughly half the observations taking smaller values and half taking larger values. We can describe the spread of a distribution by giving the range between the *smallest and largest values*. To see the shape of a distribution more clearly, turn the stemplot on its side so that the larger values lie to the right. Some things to look for in describing shape are:

modes
unimodal

- Does the distribution have one or several major peaks, called **modes?** A distribution with one major peak is called **unimodal.**

symmetric
skewed

- Is it approximately symmetric or is it skewed in one direction? A distribution is **symmetric** if the values smaller and larger than its midpoint are mirror images of each other. It is **skewed** to the right if the right tail (larger values) is much longer than the left tail (smaller values).

EXAMPLE 1.6

The distribution of Babe Ruth's home run counts (Figure 1.2) is symmetric and unimodal. Note that when describing data we do not insist on exact symmetry. Our goal is an approximate overall description, not mathematical exactness. The midpoint is 46 home runs (count up to the eighth of the 15 values), and the range is from 22 to 60. There are no outliers. In particular, Ruth's record 60 home runs in one season does not stand outside his overall pattern.

EXAMPLE 1.7

Maris's 61 home run season, on the other hand, is an outlier that stands far above his other years. You should search for an explanation for any outlier. Sometimes outliers point to errors made in recording the data. In other cases, the outlying observation may be caused by equipment failure or other unusual circumstances. The outlier in Roger Maris's home run data marks an exceptional individual achievement. Maris's home run counts, like many distributions with only a few observations, have an irregular shape. We might describe Maris's performance, omitting the outlier, as roughly symmetric with midpoint 23 and range 8 to 39.

EXAMPLE 1.8 | The distribution of supermarket spending (Figure 1.3) is skewed to the right. There are many moderate values and a few quite large dollar amounts. The direction of the long tail (to the right) determines the direction of the skewness. Because the largest amounts do not fall outside the overall skewed pattern, we do not call them outliers. The midpoint of the distribution (count up 25 entries in the stemplot) is $28. The range is $3 to $93. The shape revealed by the stemplot will interest the store management. If the big spenders can be studied in more detail, steps to attract them may be profitable. Finally, this distribution is unimodal. Although there are several minor bulges in the stemplot, there is only one major peak, containing the 22 shoppers who spent between $15 and $28. The right-skewed shape in Figure 1.3 is common among distributions of money amounts. There are many moderately priced houses, for example, but the few very expensive mansions give the distribution of house prices a strong right skew.

Histograms

Stemplots display the actual values of the observations. This feature makes stemplots awkward for large data sets. Moreover, the picture presented by a stemplot divides the observations into groups (stems) determined by the number system rather than by judgment. Histograms do not have these lim-

histogram

itations. A **histogram** breaks the range of values of a variable into intervals and displays only the count or percent of the observations that fall into each interval. You can choose any convenient number of intervals, but you should always choose intervals of equal width. Histograms are slower to construct by hand than stemplots and do not display the actual values observed. For these reasons we prefer stemplots for small data sets. The construction of a histogram is best shown by example. Any statistical software package will of course make a histogram for you.

EXAMPLE 1.9 | A study of the teaching of reading in the public schools of Gary, Indiana, began by examining the performance of seventh graders on the reading portion of the Iowa Test of Basic Skills.[4] In all, 947 Gary seventh graders took this test. Their vocabulary scores, expressed as an equivalent grade level, ranged from 2.0 (that is, equivalent to a beginning second grader) to 12.1. The scores of the first few of the 947 students are

| 5.4 | 6.8 | 2.0 | 7.6 | 6.6 | 7.8 | 8.1 | 5.6 | 6.0 | 7.9 | ... |

To make a *histogram* of the distribution of scores, proceed as follows:

1. *Divide the range of the data into classes of equal width.* In this case, it is natural to use grade levels, so the classes are

$$2.0 \leq \text{score} < 3.0$$
$$3.0 \leq \text{score} < 4.0$$
$$\vdots$$
$$12.0 \leq \text{score} < 13.0$$

We have specified the classes precisely so that each score falls in exactly one class. A score of 2.9 falls in the first class, for example, and a score of 3.0 falls in the second.

TABLE 1.2 Vocabulary scores for seventh graders		
Class	Number of students	Percent
2.0–2.9	9	0.95
3.0–3.9	28	2.96
4.0–4.9	59	6.23
5.0–5.9	165	17.42
6.0–6.9	244	25.77
7.0–7.9	206	21.75
8.0–8.9	146	15.42
9.0–9.9	60	6.34
10.0–10.9	24	2.53
11.0–11.9	5	0.53
12.0–12.9	1	0.11
Total	947	100.01

frequency

2. *Count the number of observations in each class.* These counts are called **frequencies,** and a table of frequencies for all classes is a **frequency table.** Table 1.2 is a frequency table of the Gary vocabulary scores. A frequency table is a convenient summary of a large set of data.

3. *Draw the histogram.* In the histogram in Figure 1.4, the vocabulary score scale is horizontal and the frequency scale vertical. Each bar represents a class. The base of the bar covers the class, and the bar height is the class frequency. The graph is drawn with no horizontal space between the bars (unless a class is empty, so that its bar has 0 height).

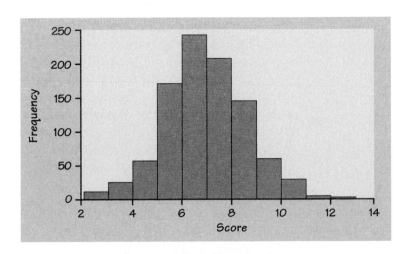

FIGURE 1.4 Histogram of the vocabulary scores of 947 seventh graders in Gary, Indiana.

The distribution in Figure 1.4 is centered in the 6.0 to 6.9 class. It is roughly symmetric and unimodal in shape. The choice of classes can slightly affect the appearance of a histogram, as can the automatic choice of classes forced by a stemplot. This is another reason why we do not require exact symmetry in order to describe a distribution as "symmetric." The distribution in Figure 1.4 is quite regular. The histogram shows no large gaps or obvious outliers, and both tails fall off quite smoothly from a single center peak.

Large sets of data are often reported in the form of frequency tables, because it is not practical to publish the individual observations. In addition to the frequency (count) for each class, the fraction or percentage of the observations that fall in each class can be reported. These fractions are sometimes called **relative frequencies.** Table 1.2 gives both frequencies and relative frequencies for the Gary reading scores. When you report statistical results, however, it is clearer to use the nontechnical terms "number" and "percent." Expressed as percents, the relative frequencies of all of the classes should add to 100%. The percents in Table 1.2 total 100.01% because of **roundoff error.** Each is rounded to the nearest 0.01%, so that the rounded numbers do not have a sum exactly equal to 100%.

relative frequency

roundoff error

A histogram of relative frequencies looks just like a frequency histogram such as Figure 1.4. Simply relabel the vertical scale to read in percents. Use histograms of relative frequencies for comparing several distributions with different numbers of observations. The bars of a frequency histogram for reading scores in Los Angeles would tower over a similar histogram from Gary, because there are many more seventh graders in Los Angeles. All relative frequency histograms have the same vertical scale (0% to 100%), so we can more easily compare the shapes of the two distributions.

Our eyes respond to the *area* of the bars in a histogram. Because the classes are all the same width, area is determined by height and all classes are fairly represented. There is no one right choice of the classes in a histogram. Too few classes will give a "skyscraper" graph, with all values in a few classes with tall bars. Too many will produce a "pancake" graph, with most classes having one or no observations. Neither choice will give a good picture of the shape of the distribution. You must use your judgment in choosing classes to display the shape. Statistical software will choose the classes for you. The software's choice is usually a good one, but you can change it if you want.

Although histograms resemble bar graphs, their details and uses are distinct. A histogram shows the distribution of frequencies or relative frequencies among the values of a single variable, and a bar graph compares the size of different items. The horizontal axis of a bar graph need not have any measurement scale but simply identifies the items being compared. Draw bar graphs with blank space between the bars to separate the items being compared, and histograms with no space to indicate that all values of the variable are covered. Some software not primarily intended for statistics, such as spreadsheet programs, will draw histograms as if they were bar graphs, with space between the bars. You can tell the software to eliminate the space in order to produce a proper histogram.

Looking at data

What of Simon Newcomb and the speed of light? Scientists know that careful measurements will vary, but they hope to see a symmetric unimodal distribution like that in Figure 1.4. If there is no systematic bias in the measurements, the best estimate of the true value of the measured quantity is the center of the distribution. The histogram in Figure 1.5 shows that Newcomb's data do have a symmetric unimodal distribution—but there are two low outliers that stand outside this pattern. What can be done?

The handling of outliers is another matter for judgment. Sometimes outliers are of special interest as evidence of an extraordinary event. A high outlier in the distribution of brightness seen from a surveillance satellite may represent a missile launch. A low outlier in the scores of a machine-graded examination may point to a student who misread the instructions. Newcomb, however, hoped for a well-behaved distribution with a clear center, and the two outliers (−44 and −2 in Table 1.1) were disturbing.

When outliers are surprising and unwanted, you should first search for a clear cause for each outlier, such as equipment failure during the experiment or an error in typing the data. Almost all large data sets have errors, often caused by mistakes in recording the data. Outliers sometimes point out these errors and allow us to correct them by looking at the original data records. If equipment failure or some other abnormal condition caused the outlier, we can delete it from the data with a clear conscience. When no cause is found, the decision is difficult. The outlier may be evidence of an extraordinary occurrence or of unexpected variability in the data. Newcomb decided to drop the worse outlier (−44), but he kept the

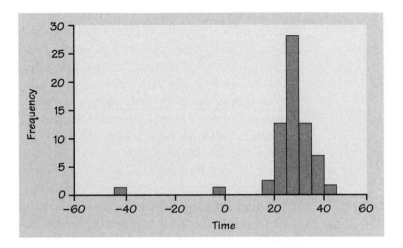

FIGURE 1.5 Histogram of Simon Newcomb's 66 measurements of the passage time of light, from Table 1.1.

other. He based his estimate of the speed of light on the average (the mean, in the language of Section 1.2) of his observations. The mean of all 66 observations is 26.21. The mean of the 65 observations he kept is 27.29. The strong effect of the single value −44 on the mean is one motivation for discarding it when we are interested in the center of the distribution as a whole.

Adjustments based on judgment alone, as Newcomb's was, are risky. As the following example illustrates, it is always best to find the reason for an outlier, if possible.

EXAMPLE 1.10

In 1985 British scientists reported a hole in the ozone layer of the earth's atmosphere over the South Pole. This news was disturbing, because ozone protects us from cancer-causing ultraviolet radiation. The British report was at first disregarded, because it was based on ground instruments looking up. More comprehensive observations from satellite instruments looking down had shown nothing unusual. Examination of the satellite data revealed that the South Pole ozone readings were so low that the computer software used to analyze the data had automatically set these values aside as suspicious outliers. Readings dating back to 1979 were reanalyzed and showed a large and growing hole in the ozone layer that is now thought to result from burning fossil fuels such as coal and oil. International efforts to reduce use of these fuels are under way.[5]

Computers analyzing large volumes of data are often programmed to suppress outliers as protection against errors in the data. As the example of the hole in the ozone illustrates, suppressing an outlier without investigating it can conceal valuable information.

Time plots

We can gain more insight into Newcomb's data from a different kind of graph. When data represent similar observations made over time, it is wise to plot them against either time or against the order in which the observations were taken. Figure 1.6 plots Newcomb's passage times for light in the order that they were recorded. There is some suggestion in this plot that the variability (the vertical spread in the plot) is decreasing over time. In particular, both outlying observations were made early on. Perhaps Newcomb became more adept at using his equipment as he gained experience. Learning effects like this are quite common; the experienced data analyst will check for them routinely. If we allow Newcomb 20 observations for learning, the mean of the remaining 46 measurements is 28.15. The best modern measurements suggest that the "true value" for the passage time in Newcomb's experiment is 33.02. Allowing for learning does move the average result closer to the true value.

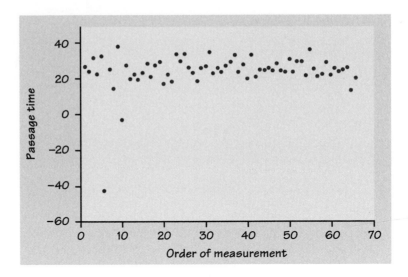

FIGURE 1.6   Plot of Newcomb's measurements against the time order in which they were made.

Whenever data are collected over time, it is a good idea to plot the observations in time order. Always put time on the horizontal scale of your plot and the variable you are measuring on the vertical scale. Connecting the data points by lines helps emphasize any change over time. Patterns in a time plot can help us to better understand the data. Summaries of the distribution of a variable that ignore time order, such as stemplots and histograms, can be misleading when there is systematic change over time.

time series

Many interesting data sets are **time series,** measurements of a variable taken at regular intervals over time. Government economic and social data are often published as time series. Some examples are the monthly unemployment rate and the quarterly gross domestic product. Weather records, the demand for electricity, and measurements on the items produced by a manufacturing process are other examples of time series.

Plots against time can reveal the main features of a time series. As in our examination of distributions, we look first for overall patterns and then for striking deviations from those patterns. Here are some types of overall patterns to look for in a time series.

Seasonal Variation and Trend

A pattern in a time series that repeats itself at known regular intervals of time is called **seasonal variation.**

A **trend** in a time series is a persistent, long-term rise or fall.

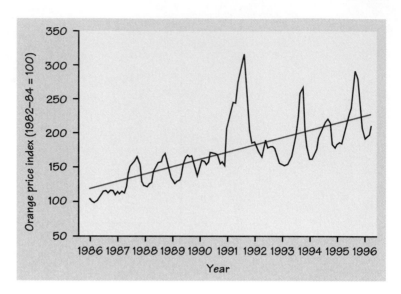

FIGURE 1.7 Time plot of the index number (1982–84 = 100) for the price of fresh oranges, 1986 to mid-1996. The line shows the increasing trend.

EXAMPLE 1.11

index number

Figure 1.7 is a time plot of the average price of fresh oranges over the period 1986 to mid-1996.[6] This information is collected each month as part of the government's reporting of retail prices. The monthly Consumer Price Index is the most publicized product of this effort. The price is presented as an **index number.** That is, the price scale gives the price as a percent of the average price of oranges in the years 1982 to 1984. This is indicated by the legend "1982–84 = 100." The first value is 103.8 for January 1986, so at that time oranges cost about 104% of their 1982 to 1984 average price. The index number is based on the retail price of oranges at many stores in all parts of the country. The price of oranges at a single store will drop when the manager decides to advertise a sale on oranges, and it will rise when the sale ends. The index number is a nationwide average price that is less variable than the price at one store.

Figure 1.7 shows a clear *trend* of increasing price. Statistical software can draw a line on the time plot that represents the trend. Such a line appears in Figure 1.7. Superimposed on this trend is a strong *seasonal variation,* a regular rise and fall that recurs each year. Orange prices are usually highest in August or September, when the supply is lowest. Prices then fall in anticipation of the harvest and are lowest in January or February, when the harvest is complete and oranges are plentiful.

seasonally adjusted

Because many economic time series show strong seasonal variation, government agencies often adjust for this variation before releasing economic data. The data are then said to be **seasonally adjusted.** Seasonal adjustment helps avoid misinterpretation. A rise in the unemployment rate from December to January, for example, does not mean that the economy is slipping. Unemployment almost always rises in January as temporary holiday help is laid off and outdoor employment in the north drops because of bad weather. The seasonally adjusted unemployment rate reports an increase only if unemployment rises more than normal from December to January.

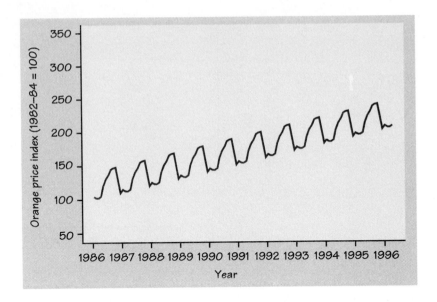

FIGURE 1.8 The estimated trend and seasonal variation in orange prices.

Beyond the basics ▶ decomposing time series*

Statistical software can help us examine a time series by "decomposing" the data into systematic patterns such as trends and seasonal variation and the "residuals" that remain after we remove these patterns. Figure 1.8 shows the combined trend and seasonal variation for the time series of orange prices in Figure 1.7. The seasonal variation represents an average of the seasonal pattern for all the years in the original data, automatically extracted by software. In terms of our strategy for looking at graphs, Figure 1.8 displays the overall pattern of the time plot.

Can we use the overall pattern in Figure 1.8 to predict next year's orange prices? Software also subtracts the trend and the seasonal variation from the original orange prices, giving the residuals in Figure 1.9. This is a graph of the deviations of orange prices from the overall pattern of Figure 1.8. Figure 1.9 shows quite a bit of erratic fluctuation and several major deviations from the pattern set by the trend and seasonal variation. The large jump in orange prices in 1991, for example, resulted from a freeze in Florida and could not be predicted. It is clear that trend and seasonal variation do not allow us to predict orange prices well.

---

*The "Beyond the basics" sections briefly discuss supplementary topics. Your software may make some of these available to you. For example, the results plotted in Figures 1.8 and 1.9 come from the Minitab statistical software.

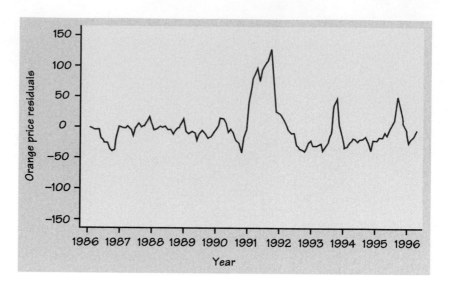

FIGURE 1.9   The erratic fluctuations that remain when both the trend and the seasonal variation are removed from orange prices.

SUMMARY

A data set contains information on a collection of **individuals.** Individuals may be people, animals, or things. The data for one individual make up a **case.** For each individual, the data give values for one or more **variables.** A variable describes some characteristic of an individual, such as a person's height, gender, or salary.

Some variables are **categorical** and others are **quantitative.** A categorical variable places each individual into a category, such as male or female. A quantitative variable has numerical values that measure some characteristic of each individual, such as height in centimeters or annual salary in dollars.

Exploratory data analysis uses graphs and numerical summaries to describe the variables in a data set and the relations among them.

A large number of observations on a single variable can be summarized in a table of **frequencies** (counts) or **relative frequencies** (percents or fractions).

Bar graphs and **pie charts** display the distributions of categorical variables. These graphs use the frequencies or relative frequencies of the categories.

Stemplots and **histograms** display the distributions of quantitative variables. Stemplots separate each observation into a **stem** and a one-digit **leaf.** Histograms plot the frequencies or relative frequencies of classes of values.

When examining a distribution, look for **shape, center,** and **spread** and for clear **deviations** from the overall shape.

Some distributions have simple shapes, such as **symmetric** and **skewed.** The number of **modes** (major peaks) is another aspect of overall shape. Not all distributions have a simple overall shape, especially when there are few observations.

Outliers are observations that lie outside the overall pattern of a distribution. Always look for outliers and try to explain them.

When observations on a variable are taken over time, make a **time plot** that graphs time horizontally and the values of the variable vertically. A time plot can reveal **trends** or other changes over time.

SECTION 1.1 EXERCISES

1.1 Data from a medical study contain values of many variables for each of the people who were the subjects of the study. Which of the following variables are categorical and which are quantitative?

(a) Gender (female or male)

(b) Age (years)

(c) Race (Asian, black, white, or other)

(d) Smoker (yes or no)

(e) Systolic blood pressure (millimeters of mercury)

(f) Level of calcium in the blood (micrograms per milliliter)

1.2 Political party preference in the United States depends in part on the age, income, and gender of the voter. A political scientist selects a large sample of registered voters. For each voter, she records gender, age, household income, and whether they voted for the Democratic or for the Republican candidate in the last congressional election. Which of these variables are categorical and which are quantitative?

1.3 Here is part of a data set that describes 1997 model motor vehicles:

Vehicle	Type	Where made	City MPG	Highway MPG
Acura 2.5TL	Compact	Foreign	20	25
Buick Skylark	Compact	Domestic	22	32
Audi A8 Quattro	Midsize	Foreign	17	25
Chrysler Concorde	Large	Domestic	19	27

Identify the individuals. Then list the variables recorded for each individual and classify each variable as categorical or quantitative.

1.4 You want to compare the "size" of several statistics textbooks. Describe at least three possible numerical variables that describe the "size" of a book.

In what *unit* would you measure each variable? What *measuring instrument* does each require? Describe a variable that is appropriate for estimating how long it would take you to read the book. Describe a variable that helps decide whether the book will fit easily into your book bag.

1.5   You are studying the relationship between political attitudes and length of hair among male students. You will measure political attitudes with a standard questionnaire. How will you measure length of hair? Give precise instructions that an assistant could follow. Include a statement of the *unit* and the *measuring instrument* that your assistant is to use.

1.6   Popular magazines often rank cities in terms of how desirable it is to live and work in each city. Describe five variables that you would measure for each city if you were designing such a study. Give reasons for each of your choices.

1.7   Congress wants the medical establishment to show that progress is being made in fighting cancer. Some variables that might be used are:

   (a) Total deaths from cancer. These have risen sharply over time, from 331,000 in 1970 to 505,000 in 1980 to 537,000 in 1994.

   (b) The percent of all Americans who die from cancer. The percent of deaths due to cancer has also risen steadily, from 17.2% in 1970 to 20.9% in 1980 to 23.4% in 1994.

   (c) The percent of cancer patients who survive for five years from the time the disease was discovered. These rates are rising slowly. For whites, the five-year survival rate was 52.2% in the 1974 to 1977 period and 60.6% in the 1986 to 1991 period.

   Discuss the usefulness of each of these variables as a measure of the effectiveness of cancer treatment. In particular, explain why both (a) and (b) could increase even if treatment is getting more effective, and why (c) could increase even if treatment is getting less effective.

1.8   Bicycle riding has become more popular. Is it also getting safer? During 1988, 24,800,000 people rode a bicycle at least six times and 949 people were killed in bicycle accidents. In 1992, there were 54,632,000 riders and 903 bicycle fatalities.

   (a) Compare the death rates for the two years. What do you conclude?

   (b) Although the data come from the same government publication, we suspect that some change in measurement took place between 1988 and 1992 that makes the data for the two years not directly comparable. Why should we be suspicious?

1.9   You are writing an article for a consumer magazine based on a survey of the magazine's readers on the reliability of their household appliances. Of 13,376 readers who reported owning Brand A dishwashers, 2942 required a service call during the past year. Only 192 service calls were reported by the 480 readers who owned Brand B dishwashers. Describe an appropriate variable to measure the reliability of a make of dishwasher, and compute the values of this variable for Brand A and for Brand B.

1.10 All the members of a physical education class are asked to measure their pulse rate as they sit in the classroom. The students use a variety of methods. Method 1: count heart beats for 6 seconds and multiply by 10 to get beats per minute. Method 2: count heart beats for 30 seconds and multiply by 2 to get beats per minute.

(a) Which method do you prefer? Why?

(b) One student proposes a third method: starting exactly on a heart beat, measure the time needed for 50 beats and convert this time into beats per minute. This method is more accurate than either method in (a). Why?

1.11 Each year *Fortune* magazine lists the top 500 companies in the United States, ranked according to their total annual sales in dollars. Describe three other variables that could reasonably be used to measure the "size" of a company.

1.12 In 1994, there were 12,263,000 undergraduate students in U.S. colleges. According to the U.S. Department of Education, there were 117,000 American Indian students, 674,000 Asian, 1,317,000 non-Hispanic black, 968,000 Hispanic, and 8,916,000 non-Hispanic white students. In addition, 269,000 foreign undergraduates were enrolled in U.S. colleges.

(a) Each number is rounded to the nearest thousand. Has this resulted in roundoff errors?

(b) Present these data in a graph.

1.13 Here are data on the percent of females among people earning doctorates in 1993 in several fields of study (from the 1996 *Statistical Abstract of the United States*):

Computer science	14.4%
Life sciences	39.9%
Education	59.2%
Engineering	9.6%
Physical sciences	21.9%
Psychology	61.2%

(a) Present these data in a well-labeled bar graph.

(b) Would it also be correct to use a pie chart to display these data? Explain your answer.

1.14 In 1993 there were 90,523 deaths from accidents in the United States. Among these were 41,893 deaths from motor vehicle accidents, 13,141 from falls, 3807 from drowning, 3900 from fires, and 7382 from poisoning. (Data from the 1996 *Statistical Abstract of the United States*.)

(a) Find the percent of accidental deaths from each of these causes, rounded to the nearest percent. What percent of accidental deaths were due to other causes?

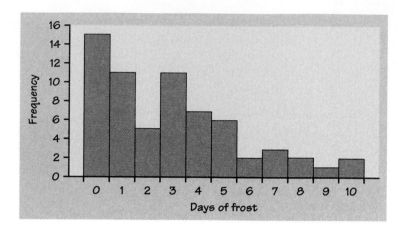

FIGURE 1.10(a) Histogram of the number of frost days during April at Greenwich, England, over a 65-year period, for Exercise 1.15.

(b) Make a well-labeled bar graph of the distribution of causes of accidental deaths. Be sure to include an "other causes" bar.

(c) Would it also be correct to use a pie chart to display these data? Explain your answer.

1.15 The histograms in Figure 1.10 display the distributions of two weather-related variables. Figure 1.10(a) shows the number of days of frost (minimum temperature below freezing) in Greenwich, England, in the month of April over a 65-year period. Figure 1.10(b) shows the number of hurricanes reaching the east coast of the United States each year over a 70-year period. Give a brief description of the overall shape of each of these distributions. What is the most important difference between

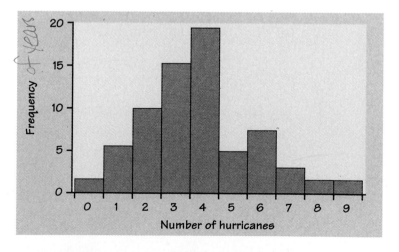

FIGURE 1.10(b) Histogram of the number of hurricanes reaching the U.S. east coast over a 70-year period, for Exercise 1.15.

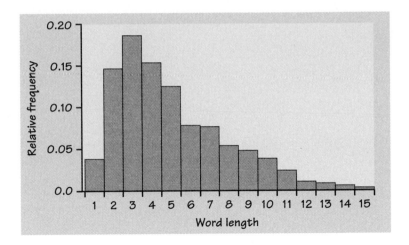

FIGURE 1.11 The distribution of lengths of words used in articles of *Popular Science* magazine, for Exercise 1.16.

the two shapes? About where does the center of each distribution lie? (Data on frosts from C. E. Brooks and N. Carruthers, *Handbook of Statistical Methods in Meteorology*, H. M. Stationery Office, 1953. Hurricane data from H. C. S. Thom, *Some Methods of Climatological Analysis*, World Meteorological Organization, Geneva, Switzerland, 1966.)

1.16 The number of letters in a word is a measure of its length. Figure 1.11 displays the distribution of the lengths of words used in articles in *Popular Science* magazine. These data were collected by students who opened the magazine arbitrarily and recorded the lengths of all words in the first complete paragraph on the page. Describe the main features of this distribution. Is it symmetric, right-skewed, or left-skewed? Unimodal or double-peaked? Are there gaps or outliers? (Statistical methods have been used to distinguish between works by different authors. Although the distribution of word lengths is a rather crude measure, *Popular Science* uses more long words than do authors of serious fiction.)

1.17 Figure 1.12 is a histogram of the average scores on the verbal part of the SAT college admissions test in each of the states and the District of Columbia in 1996. In some states most college-bound students take the SAT, while in others most students take the ACT examination instead. In the ACT states, only students who hope to enter selective colleges take the SAT. Describe the overall shape of this distribution. How are the two groups of states just described visible in the histogram?

1.18 Outliers are sometimes the most interesting feature of a distribution. Figure 1.13 displays the distribution of batting averages for all 167 American League baseball players who batted at least 200 times in the 1980 season. The outlier is the .390 batting average of George Brett, the highest batting average in the major leagues since Ted Williams hit .406 in 1941. (See Exercise 1.80 on page 87 for a comparison of Brett and Williams.) Is the

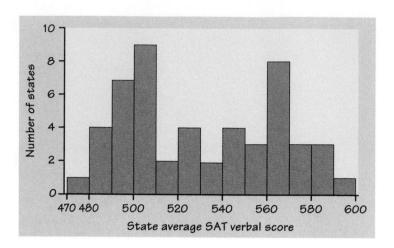

FIGURE 1.12 Histogram of the mean scores on the verbal part of the SAT for the 50 states and the District of Columbia, for Exercise 1.17.

overall shape (ignoring the outlier) roughly symmetric or clearly skewed? What is the approximate midpoint of American League batting averages? What is the range if we ignore the outlier?

1.19   Sketch a histogram for a distribution that is skewed to the left. Suppose that you and your friends emptied your pockets of coins and recorded the year marked on each coin. The distribution of dates would be skewed to the left. Explain why.

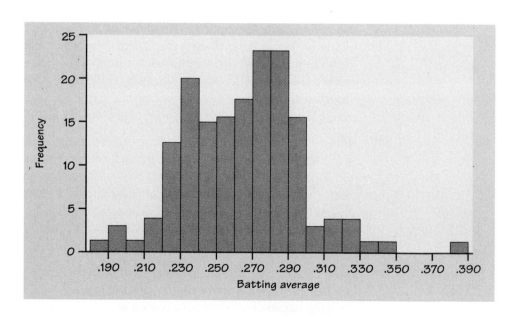

FIGURE 1.13 The distribution of batting averages of American League players in 1980, for Exercise 1.18.

1.20 The Survey of Study Habits and Attitudes (SSHA) is a psychological test that evaluates college students' motivation, study habits, and attitudes toward school. A selective private college gives the SSHA to a sample of 18 of its incoming first-year college women. Their scores are

| 154 | 109 | 137 | 115 | 152 | 140 | 154 | 178 | 101 |
| 103 | 126 | 126 | 137 | 165 | 165 | 129 | 200 | 148 |

The college also administers the test to a sample of 20 first-year college men. Their scores are

| 108 | 140 | 114 | 91 | 180 | 115 | 126 | 92 | 169 | 146 |
| 109 | 132 | 75 | 88 | 113 | 151 | 70 | 115 | 187 | 104 |

(a) Make a back-to-back stemplot of the men's and women's scores. The overall shapes of the distributions are indistinct, as often happens when only a few observations are available. Are there any outliers?

(b) Compare the midpoints and the ranges of the two distributions. What is the most noticeable contrast between the female and male scores?

1.21 There is some evidence that increasing the amount of calcium in the diet can lower blood pressure. In a medical experiment one group of men was given a daily calcium supplement, while a control group received a placebo (a dummy pill). The seated systolic blood pressure of all the men was measured before the treatments began and again after 12 weeks. The blood pressure distributions in the two groups should have been similar at the beginning of the experiment. Here are the initial blood pressure readings for the two groups:

Calcium group

| 107 | 110 | 123 | 129 | 112 | 111 | 107 | 112 | 136 | 102 |

Placebo group

| 123 | 109 | 112 | 102 | 98 | 114 | 119 | 112 | 110 | 117 | 130 |

Make a back-to-back stemplot of these data. Does your plot show any major differences in the two groups before the treatments began? In particular, are the centers of the two blood pressure distributions close together?

1.22 Plant scientists have developed varieties of corn that have increased amounts of the essential amino acid lysine. In a test of the protein quality of this corn, an experimental group of 20 one-day-old male chicks was fed a ration containing the new corn. A control group of another 20 chicks was fed a ration that was identical except that it contained normal corn. Here are the weight gains (in grams) after 21 days:

Control				Experimental			
380	321	366	356	361	447	401	375
283	349	402	462	434	403	393	426
356	410	329	399	406	318	467	407
350	384	316	272	427	420	477	392
345	455	360	431	430	339	410	326

Make a back-to-back stemplot of these data. Report the approximate midpoints of both groups. Does it appear that the chicks fed high-lysine corn grew faster? Are there any outliers or other problems? (Based on G. L. Cromwell et al., "A comparison of the nutritive value of *opaque-2, floury-2* and normal corn for the chick," *Poultry Science*, 47 (1968), pp. 840–847.)

1.23   The Degree of Reading Power (DRP) test is often used to measure the reading ability of children. Here are the DRP scores of 44 third-grade students, measured during research on ways to improve reading performance:

40	26	39	14	42	18	25	43	46	27	19
47	19	26	35	34	15	44	40	38	31	46
52	25	35	35	33	29	34	41	49	28	52
47	35	48	22	33	41	51	27	14	54	45

Make a stemplot of these data. Then make a histogram. Which display do you prefer, and why? Describe the main features of the distribution. (Data provided by Maribeth Cassidy Schmitt, from her Ph.D. dissertation, "The effects of an elaborated directed reading activity on the metacomprehension skills of third graders," Purdue University, 1987.)

1.24   In 1798 the English scientist Henry Cavendish measured the density of the earth by careful work with a torsion balance. The variable recorded was the density of the earth as a multiple of the density of water. Here are Cavendish's 29 measurements:

5.50	5.61	4.88	5.07	5.26	5.55	5.36	5.29	5.58	5.65
5.57	5.53	5.62	5.29	5.44	5.34	5.79	5.10	5.27	5.39
5.42	5.47	5.63	5.34	5.46	5.30	5.75	5.68	5.85	

Present these measurements graphically by either a stemplot or a histogram and explain the reason for your choice. Then briefly discuss the main features of the distribution. In particular, what is your estimate of the density of the earth based on these measurements? (Cavendish's data are reported by S. M. Stigler in the article cited in Note 2.)

TABLE 1.3　Percent of population 65 years old and over, by state (1995)

State	Percent	State	Percent	State	Percent
Alabama	13.0	Louisiana	11.4	Ohio	13.4
Alaska	4.9	Maine	13.9	Oklahoma	13.5
Arizona	13.3	Maryland	11.3	Oregon	13.6
Arkansas	14.5	Massachusetts	14.2	Pennsylvania	15.9
California	11.0	Michigan	12.4	Rhode Island	15.7
Colorado	10.0	Minnesota	12.4	South Carolina	12.0
Connecticut	14.3	Mississippi	12.3	South Dakota	14.4
Delaware	12.6	Missouri	13.9	Tennessee	12.5
Florida	18.6	Montana	13.1	Texas	10.2
Georgia	10.0	Nebraska	13.9	Utah	8.8
Hawaii	12.6	Nevada	11.4	Vermont	12.0
Idaho	11.4	New Hampshire	11.9	Virginia	11.1
Illinois	12.5	New Jersey	13.7	Washington	11.6
Indiana	12.6	New Mexico	10.9	West Virginia	15.3
Iowa	15.2	New York	13.4	Wisconsin	13.3
Kansas	13.7	North Carolina	12.5	Wyoming	11.1
Kentucky	12.6	North Dakota	14.5		

SOURCE: *Statistical Abstract of the United States*, 1996.

Exercises 1.25 to 1.28 present somewhat larger data sets with background information. We will use these data sets repeatedly as we master the tools of data analysis. We begin by making and interpreting a graph of the data.

1.25 Table 1.3 contains the percent of residents 65 years of age and over in each of the 50 states.

(a) Make a stemplot of these data using percent as stems and tenths of a percent as leaves. Then make a second stemplot splitting each stem in two (for leaves 0 to 4 and leaves 5 to 9). Which display do you prefer?

(b) Describe the shape, center, and spread of the distribution. Is it roughly symmetric or distinctly skewed? Which states are clear outliers? Use what you know about the states to explain these outliers.

1.26 Table 1.4 gives the monthly percent return on Philip Morris stock for the period from July 1990 to May 1997. (The return on an investment consists of the change in its price plus any cash payments made, given here as a percent of its price at the start of each month.)

(a) Make either a histogram or a stemplot of these data. How did you decide which graph to make?

(b) There is one clear outlier. What is the value of this observation? (It is explained by news of action against smoking, which depressed this tobacco company stock.) Describe the shape, center, and spread of the data after you omit the outlier.

TABLE 1.4 Monthly percent returns on Philip Morris stock from July 1990 to May 1997

−5.7	1.2	4.1	3.2	7.3	7.5	18.6	3.7	−1.8	2.4
−6.5	6.7	9.4	−2.0	−2.8	−3.4	19.2	−4.8	0.5	−0.6
2.8	−0.5	−4.5	8.7	2.7	4.1	−10.3	4.8	−2.3	−3.1
−10.2	−3.7	−26.6	7.2	−2.9	−2.3	3.5	−4.6	17.2	4.2
0.5	8.3	−7.1	−8.4	7.7	−9.6	6.0	6.8	10.9	1.6
0.2	−2.4	−2.4	3.9	1.7	9.0	3.6	7.6	3.2	−3.7
4.2	13.2	0.9	4.2	4.0	2.8	6.7	−10.4	2.7	10.3
5.7	0.6	−14.2	1.3	2.9	11.8	10.6	5.2	13.8	−14.7
3.5	11.7	1.3							

(c) The data appear in time order reading from left to right across each row in turn, beginning with the −5.7% return in July 1990. Make a time plot of the data. This was a period of increasing action against smoking, so we might expect a trend toward lower returns. But it was also a period in which stocks in general rose sharply, which would produce an increasing trend. What does your time plot show?

1.27   Table 1.5 gives the survival times in days of 72 guinea pigs after they were injected with tubercle bacilli in a medical experiment. Make a suitable graph and describe the shape, center, and spread of the distribution of survival times. Are there any outliers? (Data from T. Bjerkedal, "Acquisition of resistance in guinea pigs infected with different doses of virulent tubercle bacilli," *American Journal of Hygiene*, 72 (1960), pp. 130–148.)

1.28   Table 1.6 presents data on 78 seventh-grade students in a rural midwestern school. The researcher was interested in the relationship between the students' "self-concept" and their academic performance. The data we give here include each student's grade point average (GPA), score on a standard IQ test, and gender, taken from school records. Gender is coded as F for female and M for male. The students are identified only by an

TABLE 1.5 Survival times (days) of guinea pigs in a medical experiment

43	45	53	56	56	57	58	66	67	73
74	79	80	80	81	81	81	82	83	83
84	88	89	91	91	92	92	97	99	99
100	100	101	102	102	102	103	104	107	108
109	113	114	118	121	123	126	128	137	138
139	144	145	147	156	162	174	178	179	184
191	198	211	214	243	249	329	380	403	511
522	598								

TABLE 1.6 Educational data for 78 seventh-grade students

OBS	GPA	IQ	Gender	Self-concept	OBS	GPA	IQ	Gender	Self-concept
001	7.940	111	M	67	043	10.760	123	M	64
002	8.292	107	M	43	044	9.763	124	M	58
003	4.643	100	M	52	045	9.410	126	M	70
004	7.470	107	M	66	046	9.167	116	M	72
005	8.882	114	F	58	047	9.348	127	M	70
006	7.585	115	M	51	048	8.167	119	M	47
007	7.650	111	M	71	050	3.647	97	M	52
008	2.412	97	M	51	051	3.408	86	F	46
009	6.000	100	F	49	052	3.936	102	M	66
010	8.833	112	M	51	053	7.167	110	M	67
011	7.470	104	F	35	054	7.647	120	M	63
012	5.528	89	F	54	055	0.530	103	M	53
013	7.167	104	M	54	056	6.173	115	M	67
014	7.571	102	F	64	057	7.295	93	M	61
015	4.700	91	F	56	058	7.295	72	F	54
016	8.167	114	F	69	059	8.938	111	F	60
017	7.822	114	F	55	060	7.882	103	F	60
018	7.598	103	F	65	061	8.353	123	M	63
019	4.000	106	M	40	062	5.062	79	M	30
020	6.231	105	F	66	063	8.175	119	M	54
021	7.643	113	M	55	064	8.235	110	M	66
022	1.760	109	M	20	065	7.588	110	M	44
024	6.419	108	F	56	068	7.647	107	M	49
026	9.648	113	M	68	069	5.237	74	F	44
027	10.700	130	F	69	071	7.825	105	M	67
028	10.580	128	M	70	072	7.333	112	F	64
029	9.429	128	M	80	074	9.167	105	M	73
030	8.000	118	M	53	076	7.996	110	M	59
031	9.585	113	M	65	077	8.714	107	F	37
032	9.571	120	F	67	078	7.833	103	F	63
033	8.998	132	F	62	079	4.885	77	M	36
034	8.333	111	F	39	080	7.998	98	F	64
035	8.175	124	M	71	083	3.820	90	M	42
036	8.000	127	M	59	084	5.936	96	F	28
037	9.333	128	F	60	085	9.000	112	F	60
038	9.500	136	M	64	086	9.500	112	F	70
039	9.167	106	M	71	087	6.057	114	M	51
040	10.140	118	F	72	088	6.057	93	F	21
041	9.999	119	F	54	089	6.938	106	M	56

observation number (OBS). The missing OBS numbers show that some students dropped out of the study. The final variable is each student's score on the Piers-Harris Children's Self-Concept Scale, a psychological test administered by the researcher. (Data provided by Darlene Gordon.)

(a) How many variables does this data set contain? Which are categorical variables and which are quantitative variables?

(b) Make a stemplot of the distribution of GPA, after rounding to the nearest tenth of a point.

(c) Describe the shape, center, and spread of the GPA distribution. Identify any suspected outliers from the overall pattern.

(d) Make a back-to-back stemplot of the rounded GPAs for female and male students. Write a brief comparison of the two distributions.

1.29   Make a graph of the distribution of IQ scores for the seventh-grade students in Table 1.6. Describe the shape, center, and spread of the distribution, as well as any outliers. IQ scores are usually said to be centered at 100. Is the midpoint for these students close to 100, clearly above, or clearly below?

1.30   Based on a suitable graph, briefly describe the distribution of self-concept scores for the students in Table 1.6. Be sure to identify any suspected outliers.

1.31   The distribution of the ages of a nation's population has a strong influence on economic and social conditions. The following table shows the age distribution of U.S. residents in 1950 and 2075, in millions of people. The 1950 data come from that year's census, while the 2075 data are projections made by the Census Bureau.

Age group	1950	2075
Under 10 years	29.3	34.9
10 to 19 years	21.8	35.7
20 to 29 years	24.0	36.8
30 to 39 years	22.8	38.1
40 to 49 years	19.3	37.8
50 to 59 years	15.5	37.5
60 to 69 years	11.0	34.5
70 to 79 years	5.5	27.2
80 to 89 years	1.6	18.8
90 to 99 years	0.1	7.7
100 to 109 years	—	1.7
Total	151.1	310.6

(a) Because the total population in 2075 is much larger than the 1950 population, comparing percents (relative frequencies) in each age group is clearer than comparing counts. Make a table of the percent of the total population in each age group for both 1950 and 2075.

(b) Make a histogram of the 1950 age distribution (in percents). Describe the main features of the distribution. In particular, look at the percent of children relative to the rest of the population.

(c) Make a histogram of the projected age distribution for the year 2075. Use the same scales as in (b) for easy comparison. What are the most important changes in the U.S. age distribution projected for the 125-year period between 1950 and 2075?

1.32   A manufacturing company is reviewing the salaries of its full-time employees below the executive level at a large plant. The clerical staff is almost entirely female, while a majority of the production workers and technical staff are male. As a result, the distributions of salaries for male and female employees may be quite different. Table 1.7 gives the frequencies and relative frequencies for women and men. Make histograms from these data, choosing the type that is most appropriate for comparing the two distributions. Then describe the overall shape of the two salary distributions and the chief differences between them.

1.33   **(Optional)** Sometimes you want to make a histogram from data that are already grouped into classes of unequal width. A report on the recent graduates of a large state university includes the following relative frequency table of the first-year salaries of last year's graduates. Salaries are in $1000 units, and it is understood that each class includes its left endpoint but not its right endpoint—for example, a salary of exactly $20,000 belongs in the second class.

TABLE 1.7   Salary distributions of female and male workers in a large factory

Salary ($1000)	Women Number	Women %	Men Number	Men %
10–15	89	11.8	26	1.1
15–20	192	25.4	221	9.0
20–25	236	31.2	677	27.9
25–30	111	14.7	823	33.6
30–35	86	11.4	365	14.9
35–40	25	3.3	182	7.4
40–45	11	1.5	91	3.7
45–50	3	0.4	33	1.4
50–55	2	0.3	19	0.8
55–60	0	0.0	11	0.4
60–65	0	0.0	0	0.0
65–70	1	0.1	3	0.1
Total	756	100.1	2451	100.0

Salary	15–20	20–25	25–30	30–35	35–40	40–50	50–60	60–80
Percent	4	10	21	26	18	13	5	3

The last three classes are wider than the others. An accurate histogram must take this into account. If the base of each bar in the histogram covers a class and the height is the percent of graduates with salaries in that class, the areas of the three rightmost bars will overstate the percent who have salaries in these classes. To make a correct histogram, the area of each bar must be proportional to the percent in that class. Most classes are $5000 wide. A class *twice* as wide ($10,000) should have a bar *half* as tall as the percent in that class. This keeps the area proportional to the percent. How should you treat the height of the bar for a class $20,000 wide? Make a correct histogram with the heights of the bars for the last three classes adjusted so that the areas of the bars reflect the percent in each class.

1.34   The years around 1970 brought unrest to many U.S. cities. Here are data on the number of civil disturbances in each 3-month period during the years 1968 to 1972:

	Period	Count		Period	Count
1968	Jan.–Mar.	6	1970	July–Sept.	20
	Apr.–June	46		Oct.–Dec.	6
	July–Sept.	25	1971	Jan.–Mar.	12
	Oct.–Dec.	3		Apr.–June	21
1969	Jan.–Mar.	5		July–Sept.	5
	Apr.–June	27		Oct.–Dec.	1
	July–Sept.	19	1972	Jan.–Mar.	3
	Oct.–Dec.	6		Apr.–June	8
1970	Jan.–Mar.	26		July–Sept.	5
	Apr.–June	24		Oct.–Dec.	5

(a) Make a time plot of these counts. Connect the points in your plot by straight line segments to make the pattern clearer.

(b) Describe the trend and seasonal variation in this time series. Can you suggest an explanation for the seasonal variation in civil disorders?

1.35   We often look at time series data to see the effect of a social change or new policy. Here are data on motor vehicle deaths in the United States. Because the *count* of deaths will tend to rise as motorists drive more miles, we look instead at the *rate* of deaths, which is the number of deaths per 100 million miles driven.

Year	Rate	Year	Rate	Year	Rate
1960	5.1	1972	4.3	1984	2.6
1962	5.1	1974	3.5	1986	2.5
1964	5.4	1976	3.3	1988	2.4
1966	5.5	1978	3.3	1990	2.2
1968	5.2	1980	3.3	1992	1.8
1970	4.7	1982	2.8	1994	1.7

(a) Make a time plot of these death rates. Describe the overall pattern of the data.

(b) In 1974 the national speed limit was lowered to 55 miles per hour in an attempt to conserve gasoline after the 1973 Mideast war. In the mid-1980s most states raised speed limits on interstate highways to 65 miles per hour. Some said that the lower speed limit saved lives. Is the effect of lower speed limits between 1974 and the mid-1980s visible in your plot?

(c) Does it make sense to make a histogram of these 18 death rates? Explain your answer.

1.36 The impression that a time plot gives depends on the scales you use on the two axes. If you stretch the vertical axis and compress the time axis, change appears to be more rapid. Compressing the vertical axis and stretching the time axis make change appear slower. Make two more time plots of the data in Exercise 1.35, one that makes motor vehicle death rates appear to decrease very rapidly and one that shows only a slow decrease. The moral of this exercise is: pay close attention to the scales when you look at a time plot.

1.37 Babe Ruth was a pitcher for the Boston Red Sox in the years 1914 to 1917. In 1918 and 1919 he played some games as a pitcher and some as an outfielder. From 1920 to 1934 Ruth was an outfielder for the New York Yankees. He ended his career in 1935 with the Boston Braves. Here are the number of home runs Ruth hit in each year:

Year	HRs	Year	HRs	Year	HRs
1914	0	1921	59	1928	54
1915	4	1922	35	1929	46
1916	3	1923	41	1930	49
1917	2	1924	46	1931	46
1918	11	1925	25	1932	41
1919	29	1926	47	1933	34
1920	54	1927	60	1934	22
				1935	6

Earlier we examined the distribution of Ruth's home run totals during his Yankee years. Now make a time plot and describe its main features.

1.38   The following table gives the times (in minutes, rounded to the nearest minute) for the winning man in the Boston Marathon in the years 1959 to 1997:

Year	Time	Year	Time	Year	Time	Year	Time
1959	143	1970	131	1981	129	1992	128
1960	141	1971	139	1982	129	1993	130
1961	144	1972	136	1983	129	1994	127
1962	144	1973	136	1984	131	1995	129
1963	139	1974	134	1985	134	1996	129
1964	140	1975	130	1986	128	1997	131
1965	137	1976	140	1987	132		
1966	137	1977	135	1988	129		
1967	136	1978	130	1989	129		
1968	142	1979	129	1990	128		
1969	134	1980	132	1991	131		

Women were allowed to enter the race in 1972. Here are the times for the winning woman from 1972 to 1997:

Year	Time	Year	Time	Year	Time
1972	190	1981	147	1990	145
1973	186	1982	150	1991	144
1974	167	1983	143	1992	144
1975	162	1984	149	1993	145
1976	167	1985	154	1994	142
1977	168	1986	145	1995	145
1978	165	1987	146	1996	147
1979	155	1988	145	1997	146
1980	154	1989	144		

Make time plots of the men's and women's winning times on the same graph for easy comparison. Give a brief description of the pattern of Boston Marathon winning times over these years. Have times stopped improving in recent years? If so, when did improvement end for men and for women?

1.39   There are seasonal patterns in the growth of children. We can examine the pattern for some children using data from the Egyptian village of Kalama. The following tables give the average weights and heights of Egyptian toddlers who reach the age of 24 months in each month of the year.[7] (Data courtesy of Linda D. McCabe and the Agency for International Development.)

Month	Weight (kg)	Month	Weight (kg)	Month	Weight (kg)
Jan.	11.26	May	10.87	Sept.	10.87
Feb.	11.39	June	10.80	Oct.	10.74
Mar.	11.37	July	10.77	Nov.	10.89
Apr.	11.10	Aug.	10.79	Dec.	11.25

Month	Height (cm)	Month	Height (cm)	Month	Height (cm)
Jan.	79.80	May	80.10	Sept.	79.94
Feb.	79.66	June	80.23	Oct.	79.80
Mar.	79.91	July	80.07	Nov.	79.39
Apr.	80.00	Aug.	79.79	Dec.	79.51

(a) Make a time plot of the weight data. Describe any seasonal variation that appears in the plot. In particular, which months correspond to the highest weights? Which correspond to the lowest?

(b) Repeat the analysis of (a) for the height data.

(c) Describe the similarities and differences between the weight plot and the height plot.

1.40 In villages such as Kalama, illness may slow the growth of children. In particular, diarrhea is a serious problem and a major cause of death for small children. The following table gives the average percent of days ill with diarrhea per month for the children whose weight and height information appears in the previous exercise:

Month	Percent	Month	Percent	Month	Percent
Jan.	2.54	May	8.42	Sept.	1.20
Feb.	1.82	June	6.87	Oct.	2.51
Mar.	2.28	July	4.51	Nov.	2.34
Apr.	6.86	Aug.	3.42	Dec.	1.27

(a) Make a time plot. Describe any seasonal variation that appears in the plot. Which months have the highest average percent of days ill with diarrhea? Which months have the lowest?

(b) Egyptian physicians say there are two peaks in the frequency of diarrhea, a primary peak and a secondary, lower peak. Is there evidence in these data to support this claim?

(c) Compare your plot with those of the previous exercise. Can you formulate any theories concerning the relationship between diarrhea and the growth of these children? Note that illness in one month may not have an effect on growth until subsequent months. On the other hand, a child who is not growing well may be more susceptible to illness.

1.2 Describing Distributions with Numbers

"Face it. A hot dog isn't a carrot stick." So said *Consumer Reports*, commenting on the low nutritional quality of the all-American frank. Table 1.8 shows the magazine's laboratory test results for calories and milligrams of sodium (mostly due to salt) in a number of major brands of hot dogs. There are three types: all beef, "meat" (mainly pork and beef, but government regulations allow up to 15% poultry meat), and poultry. Because people concerned about health may prefer low-calorie, low-salt hot dogs, we ask: "Are there any systematic differences among the three types in these two variables?" We can begin to answer this question by comparing stemplots of the three distributions of calorie content, one for each type, and by making a similar comparison for sodium content. We can make the comparisons more explicit by using some numerical tools for describing distributions. This section will develop these tools.

A brief description of a distribution should include its *shape* and numbers describing its *center* and *spread*. We describe the shape of a distribution based

TABLE 1.8 Calories and sodium in three types of hot dogs

Beef hot dogs		Meat hot dogs		Poultry hot dogs	
Calories	Sodium	Calories	Sodium	Calories	Sodium
186	495	173	458	129	430
181	477	191	506	132	375
176	425	182	473	102	396
149	322	190	545	106	383
184	482	172	496	94	387
190	587	147	360	102	542
158	370	146	387	87	359
139	322	139	386	99	357
175	479	175	507	170	528
148	375	136	393	113	513
152	330	179	405	135	426
111	300	153	372	142	513
141	386	107	144	86	358
153	401	195	511	143	581
190	645	135	405	152	588
157	440	140	428	146	522
131	317	138	339	144	545
149	319				
135	298				
132	253				

SOURCE: *Consumer Reports*, June 1986, pp. 366–367.

on inspection of a histogram or a stemplot. Now we will learn specific ways to use numbers to measure the center and spread of a distribution. We can calculate these numerical measures for any quantitative variable. But to interpret measures of center and spread, and to choose among the several measures we will learn, you must think about the shape of the distribution and the meaning of the data. The numbers, like graphs, are aids to understanding, not "the answer" in themselves.

Measuring center: the mean

A description of a distribution almost always includes a measure of its center or average. The two common measures of center are the *mean* and the *median*. The mean is the "average value" and the median is the "middle value." These are two different ideas for "center," and the two measures behave differently. We need precise recipes for the mean and the median.

The Mean $\bar{x}$

To find the **mean $\bar{x}$** of a set of observations, add their values and divide by the number of observations. If the n observations are $x_1, x_2, \ldots, x_n$, their mean is

$$\bar{x} = \frac{x_1 + x_2 + \cdots + x_n}{n}$$

or, in more compact notation,

$$\bar{x} = \frac{1}{n} \sum x_i$$

The $\sum$ (capital Greek sigma) in the formula for the mean is short for "add them all up." The bar over the x indicates the mean of all the x-values. Pronounce the mean $\bar{x}$ as "x-bar." This notation is so common that writers who are discussing data use $\bar{x}$, $\bar{y}$, etc. without additional explanation. The subscripts on the observations x_i are just a way of keeping the n observations distinct. They do not necessarily indicate order or any other special facts about the data.

EXAMPLE 1.12 | Here are the number of home runs hit by Babe Ruth in each of his seasons as a Yankee:

| 54 | 59 | 35 | 41 | 46 | 25 | 47 | 60 | 54 | 46 | 49 | 46 | 41 | 34 | 22 |

The Babe's mean number of home runs hit in a year is

$$\bar{x} = \frac{1}{15}(54 + 59 + \cdots + 22) = \frac{659}{15} = 43.9$$

Roger Maris's mean, from the data on page 11, is

$$\bar{x} = \frac{1}{10}(8 + 13 + \cdots + 61) = \frac{261}{10} = 26.1$$

Ruth's superiority is evident from these averages.

The distribution of Maris's yearly home run production shows a clear outlier, his record 61 homers in 1961. If we exclude this as atypical, the mean number of home runs in the other 9 years of his American League career is 22.2. The single outlier has increased the 10-year average by about 4 homers per year. This illustrates an important weakness of the mean as a measure of center: it is sensitive to the influence of a few extreme observations. These may be outliers, but a skewed distribution that has no outliers will also pull the mean toward its long tail. Because the mean cannot resist the influence of extreme observations, we say that it is not a **resistant measure** of center. A measure that is resistant does more than limit the influence of outliers. Its value does not respond strongly to changes in a few observations, no matter how large those changes may be. The mean fails this requirement, because we can make the mean as large as we wish by making a large enough increase in just one observation.

resistant measure

Measuring center: the median

We used the midpoint of a distribution as an informal measure of center in the previous section. The *median* is the formal version of the midpoint, with a specific rule for calculation.

The Median M

The **median** M is the midpoint of a distribution, the number such that half the observations are smaller and the other half are larger. To find the median of a distribution:

1. Arrange all observations in order of size, from smallest to largest.
2. If the number of observations n is odd, the median M is the center observation in the ordered list. Find the location of the median by counting $(n + 1)/2$ observations up from the bottom of the list.
3. If the number of observations n is even, the median M is the mean of the two center observations in the ordered list. The location of the median is again $(n + 1)/2$ from the bottom of the list.

Note that the formula $(n + 1)/2$ does *not* give the median, but only the location of the median in the ordered list. Medians require little arithmetic, so they are easy to find by hand for small sets of data. Arranging even a moderate number of observations in order is very tedious, however, so that finding the median by hand for larger sets of data is unpleasant. Even simple calculators have an $\bar{x}$ button, but you will need computer software or a graphing calculator to automate finding the median.

EXAMPLE 1.13

To find the median number of home runs hit in a year by Babe Ruth, we arrange the data in increasing order:

22	25	34	35	41	41	46	**46**	46	47	49	54	54	59	60

The median is the bold 46, the eighth observation in the ordered list. You can find the median by eye—there are seven observations to the left and seven to the right. Or you can use the recipe $(n + 1)/2 = 16/2 = 8$ to locate the median in the list.

What about Roger Maris? His yearly home run totals in increasing order are

8	13	14	16	**23**	**26**	28	33	39	61

Because the number of observations $n = 10$ is even, there is a center pair of numbers rather than a single center point. This pair of numbers is shown in bold in the list. The median M is their average:

$$M = \frac{23 + 26}{2} = \frac{49}{2} = 24.5$$

The recipe $(n + 1)/2 = 11/2 = 5.5$ for the position of the median in the list means that the median is at location "five and one-half," that is, halfway between the fifth and sixth observations.

Notice that because a stemplot arranges the observations in increasing order you can find the median from a stemplot without rewriting the data. Notice also that the median, unlike the mean, is resistant. Maris's record 61 home runs is simply one observation falling above the median. Even if this observation were erroneously entered as 610, the median would not change.

Mean versus median

The median and mean are the most common measures of the center of a distribution. The mean and median of a symmetric distribution are close together. If the distribution is exactly symmetric, the mean and median are exactly the same. In a skewed distribution, the mean is farther out in the long tail than is the median. For example, the distribution of house prices is strongly skewed to the right. There are many moderately priced houses and a few very expensive mansions. The few expensive houses pull the mean up but do not affect the median. The mean price of new houses sold in the first quarter of 1996 was $170,800, but the median price for these same houses was only $143,000. Reports about house prices, incomes, and other strongly skewed distributions usually give the median ("middle value") rather than the mean ("arithmetic average value").

We now have two general strategies for dealing with outliers and other irregularities in data. The first strategy asks us to detect outliers and investigate their causes. We can then correct outliers if they are wrongly recorded, delete them for good reason, or otherwise give them individual attention. The second strategy is to use resistant methods, so that outliers have little influence over our conclusions. In general, we prefer to examine data carefully and consider any irregularity in the light of the individual situation. This will sometimes lead to a decision to employ resistant methods.

EXAMPLE 1.14

In the case of Roger Maris, the median is more typical of his performance than is the mean, but it obscures the great event of his career. A better summary might be: "Maris hit 61 home runs in 1961 and averaged 22.2 homers in his other 9 years in the league."

Newcomb's data, on the other hand, are supposed to be repeated measurements of the same quantity. The histogram on page 17 shows unexplained outliers. Even though we may not be able to explain the outlying values, we should not give them much influence over our estimate of the velocity of light. We can delete the outliers from a computation of the mean, or we can report the median. If the two outliers are omitted, the mean of the remaining 64 measurements is $\bar{x} = 27.75$. The median of 66 observations has position $(66 + 1)/2 = 33.5$ in the ordered list of passage times from Table 1.1. Its value is $M = 27$. Either of these values is preferable to the overall mean ($\bar{x} = 26.21$) as a statement of Newcomb's value for the travel time of light.

Measuring spread: the quartiles

A measure of center alone can be misleading. Two nations with the same median family income are very different if one has extremes of wealth and poverty and the other has little variation among families. A drug with the correct mean concentration of active ingredient is dangerous if some batches are much too high and others much too low. We are interested in the *spread* or *variability* of incomes and drug potencies as well as their centers. **The simplest useful numerical description of a distribution consists of both a measure of center and a measure of spread.**

percentile

We can describe the spread or variability of a distribution by giving several percentiles. The *p*th **percentile** of a distribution is the value such that p percent of the observations fall at or below it. The median is just the 50th percentile, so the use of percentiles to report spread is particularly appropriate when the median is our measure of center. The most commonly used percentiles other than the median are the *quartiles*. The first quartile is the 25th percentile, and the third quartile is the 75th percentile. (The second quartile is the median itself.) To calculate a percentile, arrange the observations in increasing order and count up the required percent from the bottom of the list. Our definition of percentiles is a bit inexact, because there is not always a value with exactly p percent of the data at or below it. We will be content to take the nearest observation for most percentiles, but the quartiles are important enough to require an exact recipe.

The Quartiles Q_1 and Q_3

To calculate the quartiles:

1. Arrange the observations in increasing order and locate the median M in the ordered list of observations.

2. The **first quartile Q_1** is the median of the observations whose position in the ordered list is to the left of the location of the overall median.

3. The **third quartile Q_3** is the median of the observations whose position in the ordered list is to the right of the location of the overall median.

EXAMPLE 1.15

In Example 1.5 on page 12, we arranged the shopping data in order in a stemplot, after rounding to eliminate the cents. The rounded data are adequate to give us a quick numerical summary of the distribution. Here they are, rewritten for convenience from the stemplot:

3	9	9	11	13	14	15	16	17	17	18	18	19	20	20	20	21
22	23	24	25	25	26	26	28	‖	28	28	28	32	35	36	39	39
41	43	44	45	45	47	49	50	53	55	59	61	70	83	86	86	93

There are $n = 50$ observations, so the position of the median is $(50 + 1)/2 = 25.5$, or midway between the 25th and 26th observations. This location is marked by ‖ in the list. Both center observations are $28, so the median is $28. To find the first quartile, consider the 25 observations falling to the left of the location ‖ of the median. Note carefully that the leftmost of the two center observations is included in this group, because the location ‖ falls between two data points. The median of these 25 observations is the thirteenth in order, or $19. This is the first quartile. You can find the third quartile similarly as the thirteenth value above ‖, or $45. We can summarize these results in compact form as

$$Q_1 = \$19 \qquad M = \$28 \qquad Q_3 = \$45$$

We find other percentiles more informally. For example, we take the 95th percentile of the 50 observations to be the 48th in the ordered list, because $0.95 \times 50 = 47.5$, which we round to 48. The 95th percentile is therefore $86.

When there is an odd number of observations, the median is the unique center observation, and the rule for finding the quartiles excludes this center value. For example, the median of Ruth's 15 home run totals

---

22	25	34	35	41	41	46	**46**	46	47	49	54	54	59	60

---

is the bold 46 (the eighth entry). The first quartile is the median of the seven observations falling to the left of this point in the list, $Q_1 = 35$. Similarly, $Q_3 = 54$.

Data Desk

```
Summary of        spending
No Selector

Percentile     25

         Count      50
          Mean      34.7022
        Median      27.8550
        StdDev      21.6974
           Min       3.11000
           Max      93.3400
Lower ith %tile     19.2700
Upper ith %tile     45.4000
```

Minitab

```
Descriptive Statistics

Variable        N      Mean    Median    TrMean     StDev    SEMean
spending       50     34.70     27.85     32.92     21.70      3.07

Variable      Min       Max        Q1        Q3
spending     3.11     93.34     19.06     45.72
```

FIGURE 1.14 Numerical descriptions of the unrounded shopping data from the Data Desk and Minitab software.

EXAMPLE 1.16 | Statistical software often provides several numerical measures in response to a single command. Figure 1.14 displays such output from two software systems for the unrounded shopping data from Example 1.5. Both give us the number of observations and the mean, median, and quartiles, as well as other measures. Notice that the values for the quartiles differ slightly—for example, the first quartile is $19.27 in one output and $19.06 in the other. You can check that applying our rules to Example 1.5 gives $19.27. Some statistical software uses slightly different rules to compute the quartiles, so the results given by a computer may not agree exactly with the results found by using our rules. However, any differences will be small. Just use the values that your software gives you.

Measuring spread: the interquartile range

The quartiles together with the median give some indication of the center, spread, and shape of a distribution. In Example 1.15, the fact that the third quartile $Q_3 = 45$ lies farther above the median $M = 28$ than the first quartile $Q_1 = 19$ lies below it reflects the right skewness of the data. The distance between the quartiles is a simple measure of spread that gives the range covered by the middle half of the data. This distance is called the *interquartile range*.

The Interquartile Range _IQR_

The **interquartile range _IQR_** is the distance between the first and third quartiles,

$$IQR = Q_3 - Q_1$$

In Example 1.15, $IQR = \$45 - \$19 = \$26$. The quartiles and the IQR are not affected by changes in either tail of the distribution. They are therefore resistant, because changes in a few data points have no further effect once these points move outside the quartiles. You should be aware that no single numerical measure of spread, such as IQR, is very useful for describing skewed distributions. The spreads of the two sides of a skewed distribution are unequal, so that the two quartiles Q_1 and Q_3 (given with the median M) are more informative.

The interquartile range is the basis of a rule of thumb for identifying suspected outliers.

The 1.5 × _IQR_ Criterion for Outliers

Call an observation a suspected outlier if it falls more than $1.5 \times IQR$ above the third quartile or below the first quartile.

In Example 1.15,

$$1.5 \times IQR = 1.5 \times \$26 = \$39$$

Any values below $\$19 - \$39 = -\$20$ or above $\$45 + \$39 = \$84$ are flagged as possible outliers. In the case of the strongly skewed distribution of supermarket purchases, the flagged values (\$86 and \$93) do not appear to be outliers in the sense of deviating from the overall pattern of the distribution. Using the $1.5 \times IQR$ rule to spot outliers does not replace using judgment, but it is helpful when the process must be automated.

The five-number summary and boxplots

In Section 1.1, we used the smallest and largest observations to indicate the spread of a distribution. These single observations tell us little about the distribution as a whole, but they give information about the tails of the distribution that is missing if we know only Q_1, M, and Q_3. To get a quick summary of both center and spread, combine all five numbers.

> ### The Five-Number Summary
>
> The **five-number summary** of a set of observations consists of the smallest observation, the first quartile, the median, the third quartile, and the largest observation, written in order from smallest to largest. In symbols, the five-number summary is
>
> $$\text{Minimum} \quad Q_1 \quad M \quad Q_3 \quad \text{Maximum}$$

The five-number summary for the supermarket spending data of Example 1.15 is

$$\$3 \quad \$19 \quad \$28 \quad \$45 \quad \$93$$

The median describes the center of the distribution; the quartiles show the spread of the center half of the data; the minimum and maximum show the full spread of the data. The IQR can be computed from the five-number summary, as can other helpful facts such as the distances of the quartiles from the median.

The five-number summary leads to another visual representation of a distribution, the *boxplot*. Figure 1.15 is a boxplot of the distribution of supermarket purchases from Example 1.15.

> ### Boxplot
>
> A **boxplot** is a graph of the five-number summary, with suspected outliers plotted individually.
>
> - A central box spans the quartiles.
> - A line in the box marks the median.
> - Observations more than $1.5 \times IQR$ outside the central box are plotted individually as possible outliers.
> - Lines extend from the box out to the smallest and largest observations that are not suspected outliers.

The center, spread, and overall range of the distribution are immediately apparent from a boxplot. When we plot suspected outliers individually, they are also immediately apparent—in fact, using software to make a boxplot is a quick way to apply the $1.5 \times IQR$ criterion. The details of a boxplot may vary a bit depending on your choice of software. Some software will not plot suspected outliers individually, and some may use a rule other than the $1.5 \times IQR$ criterion to identify outliers.

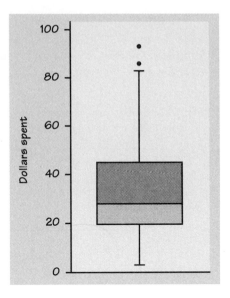

FIGURE 1.15 Boxplot of the amount spent by 50 shoppers in a supermarket, from Example 1.15.

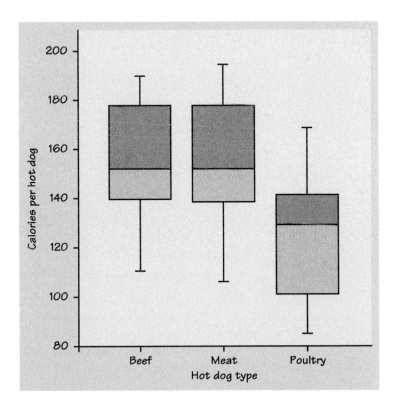

FIGURE 1.16 Boxplots of the calorie content of three types of hot dogs, from Table 1.8.

Comparing distributions

Boxplots are most useful for comparing distributions. Let us therefore return to the calorie content of hot dogs.

EXAMPLE 1.17 | The five-number summaries of the distributions of calories for the three types of hot dogs in Table 1.8 on page 40 are

Type	Min.	Q_1	M	Q_3	Max.
Beef	111	140	152.5	178.5	190
Meat	107	138.5	153	180.5	195
Poultry	86	100.5	129	143.5	170

Calculate the IQR for each type, and check that no observations fall more than $1.5 \times IQR$ outside the quartiles. Figure 1.16 presents boxplots based on these calculations. We see at once that poultry franks as a group contain fewer calories than beef or meat hot dogs. The median number of calories in a poultry hot dog is less than the first quartile of either of the other distributions. But each type shows quite a large spread among brands, so that simply buying a poultry frank does not guarantee a low-calorie food.

Figure 1.16 also illustrates why boxplots are generally inferior to stemplots and histograms as displays of a single distribution. Here is a stemplot of the calorie content of the 17 brands of meat hot dogs:

```
10 | 7
11 |
12 |
13 | 5 6 8 9
14 | 0 6 7
15 | 3
16 |
17 | 2 3 5 9
18 | 2
19 | 0 1 5
```

There are two distinct clusters of brands (the distribution has two peaks with a gap between them) and one outlier in the lower tail. The boxplot hid the clusters. The quartiles $Q_1 = 139$ and $Q_3 = 179$ are roughly at the centers of the clusters, so that much of the IQR is the spread between the two clusters. Because of this, the $1.5 \times IQR$ rule used in drawing the boxplot did not call attention to the outlier. Returning to the *Consumer Reports* article from which the data were taken supplies an explanation for the shape of the distribution. Not all brands of hot dogs have the same weight. It turns out that all of the brands in the high-calorie cluster weigh 2 ounces each, while 7 of the 8 in the lower cluster weigh 1.6 ounces each. The heavier franks naturally tend to have more calories. The outlier is unusual in two ways. It is the calorie content of "Eat Slim Veal Hot Dogs," the only frank in the group that is not composed of some mix of pork, beef, and poultry. But this brand also weighs only 1.4 ounces. If *Consumer Reports* had reported calories per ounce, "Eat Slim" would not be an outlier.[8]

As this example illustrates, boxplots are best used for comparing several distributions. A stemplot (or, for large data sets, a histogram) provides a clearer display of a single distribution, especially when accompanied by the median and quartiles as numerical signposts.

The distribution of calorie content of meat hot dogs also illustrates the limitations of numerical summaries of distributions. Neither the mean nor the median conveys much useful information about this distribution, because a clear "center" is not present. The median falls near the gap between the two clusters and is not a typical calorie value. The main message about a double-peaked distribution is simply its shape, and shape is best portrayed visually. The routine methods of statistics compute numerical measures and draw conclusions based on their values. These methods are extremely useful, and we will study them carefully in later chapters. But they cannot be applied blindly, by feeding data to a computer program, because **statistical measures and methods based on them are generally meaningful only for distributions of sufficiently regular shape**. This principle will become clearer as we progress, but it is good to be aware at the beginning that quickly resorting to fancy calculations is the mark of a statistical amateur. The trained practitioner looks, thinks, and chooses calculations selectively.

Measuring spread: the standard deviation

The five-number summary is not the most common numerical description of a distribution. That distinction belongs to the combination of the mean to measure center and the *standard deviation* to measure spread. The standard deviation measures spread by looking at how far the observations are from their mean.

The Standard Deviation s

The **variance s^2** of a set of observations is the average of the squares of the deviations of the observations from their mean. In symbols, the variance of n observations $x_1, x_2, \ldots, x_n$ is

$$s^2 = \frac{(x_1 - \bar{x})^2 + (x_2 - \bar{x})^2 + \cdots + (x_n - x)^2}{n - 1}$$

or, in more compact notation,

$$s^2 = \frac{1}{n-1} \sum (x_i - \bar{x})^2$$

The **standard deviation s** is the square root of the variance s^2:

$$s = \sqrt{\frac{1}{n-1} \sum (x_i - \bar{x})^2}$$

The idea behind the variance and the standard deviation as measures of spread is as follows: The deviations $x_i - \overline{x}$ display the spread of the values x_i about their mean $\overline{x}$. Some of these deviations will be positive and some negative because some of the observations fall on each side of the mean. In fact, *the sum of the deviations of the observations from their mean will always be zero.* Squaring the deviations makes them all positive, so that observations far from the mean in either direction have large positive squared deviations. The variance is the average squared deviation. Therefore, s^2 and s will be large if the observations are widely spread about their mean, and small if the observations are all close to the mean.

EXAMPLE 1.18

A person's metabolic rate is the rate at which the body consumes energy. Metabolic rate is important in studies of weight gain, dieting, and exercise. Here are the metabolic rates of 7 men who took part in a study of dieting. (The units are calories per 24 hours. These are the same calories used to describe the energy content of foods.)

$$1792 \quad 1666 \quad 1362 \quad 1614 \quad 1460 \quad 1867 \quad 1439$$

Enter these data into your calculator or software and verify that

$$\overline{x} = 1600 \text{ calories} \quad s = 189.24 \text{ calories}$$

Figure 1.17 plots these data as dots on the calorie scale, with their mean marked by an asterisk (*). The arrows mark two of the deviations from the mean. If you were calculating s by hand, you would find the first deviation as

$$x_1 - \overline{x} = 1792 - 1600 = 192$$

Exercise 1.54 asks you to calculate the seven deviations, square them, and find s^2 and s directly from the deviations. Working one or two short examples by hand helps you understand how the standard deviation is obtained. In practice you will always use either software or a calculator that will find s from keyed-in data.

The idea of the variance is straightforward: it is the average of the squares of the deviations of the observations from their mean. The details we have just presented, however, raise some questions.

Why do we square the deviations? Why not just average the distances of the observations from their mean? There are two reasons, neither of them obvious. First, the sum of the squared deviations of any set of observations from

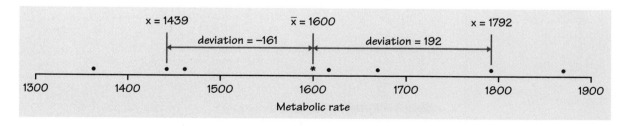

FIGURE 1.17 Metabolic rates for seven men, with the mean (*) and the deviations of two observations from the mean.

their mean is the smallest that the sum of squared deviations from any number can possibly be. This is not true of the unsquared distances. So squared deviations point to the mean as center in a way that distances do not. Second, the standard deviation turns out to be the natural measure of spread for a particularly important class of symmetric unimodal distributions, the *normal distributions*. We will meet the normal distributions in the next section. We commented earlier that the usefulness of many statistical procedures is tied to distributions of particular shapes. This is distinctly true of the standard deviation.

Why do we emphasize the standard deviation rather than the variance? One reason is that s, not s^2, is the natural measure of spread for normal distributions. There is also a more general reason to prefer s to s^2. Because the variance involves squaring the deviations, it does not have the same unit of measurement as the original observations. The variance of the metabolic rates, for example, is measured in squared calories. Taking the square root remedies this. The standard deviation s measures spread about the mean in the original scale.

Why do we average by dividing by $n - 1$ rather than n in calculating the variance? Because the sum of the deviations is always zero, the last deviation can be found once we know the other $n - 1$. So we are not averaging n unrelated numbers. Only $n - 1$ of the squared deviations can vary freely, and we average by dividing the total by $n - 1$. The number $n - 1$ is called the **degrees of freedom** of the variance or standard deviation. Many calculators offer a choice between dividing by n and dividing by $n - 1$, so be sure to use $n - 1$.

degrees of freedom

Properties of the standard deviation

Here are the basic properties of the standard deviation s as a measure of spread.

Properties of the Standard Deviation

■ s measures spread about the mean and should be used only when the mean is chosen as the measure of center.

■ $s = 0$ only when there is *no spread*. This happens only when all observations have the same value. Otherwise, $s > 0$. As the observations become more spread out about their mean, s gets larger.

■ s, like the mean $\bar{x}$, is not resistant. A few outliers can make s very large.

The use of squared deviations renders s even more sensitive than $\bar{x}$ to a few extreme observations. For example, Roger Maris's 61 home run year raises his mean homers per season from 22.2 for the remaining 9 years to 26.1, and it increases the standard deviation from 10.2 for 9 years to 15.6 for all 10 years. A skewed distribution with a few observations in the single long tail will have a large standard deviation. The number s does not give much helpful information about such a distribution.

Choosing measures of center and spread

How do we choose between the five-number summary and $\bar{x}$ and s to describe the center and spread of a distribution? Because the two sides of a strongly skewed distribution have different spreads, no single number such as s describes the spread well. The five-number summary, with its two quartiles and two extremes, does a better job.

Choosing a Summary

The five-number summary is usually better than the mean and standard deviation for describing a skewed distribution or a distribution with strong outliers. Use $\bar{x}$ and s only for reasonably symmetric distributions that are free of outliers.

EXAMPLE 1.19

A central principle in the study of investments is that taking bigger risks is rewarded by higher returns, at least on the average over long periods of time. It is usual in finance to measure risk by the standard deviation of returns, on the grounds that investments whose returns show a large spread from year to year are less predictable and therefore more risky than those whose returns have small spread. Compare, for example, the approximate mean and standard deviation of the annual percent returns on American common stocks and U.S. Treasury bills over the period from 1950 to 1996:[9]

Investment	Mean return	Standard deviation
Common stocks	13.8%	16.4%
Treasury bills	5.4%	3.1%

Stocks are risky. They went up almost 14% per year on the average during this period, but they dropped more than 25% in the worst year. The large standard deviation reflects the fact that stocks have produced both large gains and large losses. When you buy a Treasury bill, on the other hand, you are lending money to the government for one year. You know that the government will pay you back with interest. That is much less risky than buying stocks, so (on the average) you get a smaller return.

Are $\bar{x}$ and s good summaries for distributions of investment returns? Figure 1.18 displays stemplots of the annual returns for both investments. (Because stock returns are so much more spread out, a back-to-back stemplot does not work well. The stems in the stock stemplot are tens of percents; the stems for bills are percents. The lowest returns are -26% for stocks and 0.9% for bills.) The distribution of returns on stocks is roughly symmetric. But returns on Treasury bills, which have never had negative returns because the U.S. government has always paid its debts, have a right-skewed distribution. Custom in the financial world calls for $\bar{x}$ and s because some parts of investment theory use them. For describing distributions of returns, however, the five-number summary would be more informative.

```
-2 | 6                        0  | 9
-1 | 5 0 0                    1  | 2 5 7 7 8 9
-0 | 9 8 7 5 3                2  | 4 7 8 8
 0 | 0 1 1 4 5 6 7 7 8        3  | 0 2 3 5 5 6
 1 | 0 1 2 2 4 6 6 7 9 9 9    4  | 0 2 4 4 4 9
 2 | 0 2 2 3 3 4 4 7 9        5  | 1 1 4 5 5 6 9 9
 3 | 0 2 2 2 2 7 8            6  | 1 4 9
 4 | 3                        7  | 0 2 5 7 7
 5 | 0                        8  | 0 4
                              9  | 0 9
                             10  | 6 8
                             11  |
                             12  | 2
                             13  |
                             14  | 7
           (a)                          (b)
```

FIGURE 1.18 Stemplots of annual returns for stocks and Treasury bills, 1950 to 1996. (a) Stock returns, in whole percents. (b) Treasury bill returns, in percents and tenths of a percent.

Remember that a graph gives the best overall picture of a distribution. Numerical measures of center and spread report specific facts about a distribution, but they do not describe its entire shape. Numerical summaries do not disclose the presence of multiple modes or gaps, for example. Our examination of calories in meat hot dogs provides an example of a distribution for which numerical summaries alone are misleading. **Always plot your data**.

Changing the unit of measurement

The same variable can be recorded in different units of measurement. Americans commonly record distances in miles and temperatures in degrees Fahrenheit, while the rest of the world measures distances in kilometers and temperatures in degrees Celsius. Fortunately, it is easy to convert numerical descriptions of a distribution from one unit of measurement to another. This is true because a change in the measurement unit is a *linear transformation* of the measurements.

> **Linear Transformations**
>
> A **linear transformation** changes the original variable x into the new variable x_{new} given by an equation of the form
>
> $$x_{\text{new}} = a + bx$$
>
> Adding the constant a shifts all values of x upward or downward by the same amount. In particular, such a shift changes the origin (zero point) of the variable. Multiplying by the positive constant b changes the size of the unit of measurement.

EXAMPLE 1.20

(a) If a distance x is measured in kilometers, the same distance in miles is

$$x_{\text{new}} = 0.62x$$

For example, a 10-kilometer race covers 6.2 miles. This transformation changes the units without changing the origin—a distance of 0 kilometers is the same as a distance of 0 miles.

(b) A temperature x measured in degrees Fahrenheit must be reexpressed in degrees Celsius to be easily understood by the rest of the world. The transformation is

$$x_{\text{new}} = \frac{5}{9}(x - 32) = -\frac{160}{9} + \frac{5}{9}x$$

Thus the high of 95°F on a hot American summer day translates into 35°C. In this case

$$a = -\frac{160}{9} \quad \text{and} \quad b = \frac{5}{9}$$

This linear transformation changes both the unit size and the origin of the measurements. The origin in the Celsius scale (0°C, the temperature at which water freezes) is 32° in the Fahrenheit scale.

Linear transformations do not change the shape of a distribution. If measurements on a variable x have a right-skewed distribution, any new variable x_{new} obtained by a linear transformation $x_{\text{new}} = a + bx$ (for $b > 0$) will also have a right-skewed distribution. If the distribution of x is symmetric and unimodal, the distribution of x_{new} remains symmetric and unimodal.

Although a linear transformation preserves the basic shape of a distribution, the center and spread will change. Because linear changes of measurement scale are common, we must be aware of their effect on numerical descriptive measures of center and spread. Fortunately, the changes follow a simple pattern.

EXAMPLE 1.21 | Julie and John both measure the lengths of the same five cockroaches, but Julie measures in centimeters and John uses millimeters. There are 10 millimeters in a centimeter, so each of John's measurements is 10 times as large as Julie's measurement of the same roach. Here are their results:

Roach	1	2	3	4	5
Julie	1.4 cm	2.2 cm	1.1 cm	1.6 cm	1.2 cm
John	14 mm	22 mm	11 mm	16 mm	12 mm

These two sets of numbers measure the same five lengths in different units. Their means represent the same average length in different units, so John's mean (in millimeters) must be 10 times as large as Julie's mean (in centimeters). The spread of John's numbers is likewise 10 times the spread of Julie's, so John's standard deviation (again in millimeters) is 10 times Julie's standard deviation (in centimeters). You can verify these facts with your calculator.

Here is a summary of such common-sense relationships in more formal language.

Effect of a Linear Transformation

To see the effect of a linear transformation on measures of center and spread, apply these rules:

- Multiplying each observation by a positive number b multiplies both measures of center (mean and median) and measures of spread (interquartile range and standard deviation) by b.

- Adding the same number a (either positive or negative) to each observation adds a to measures of center and to quartiles and other percentiles but does not change measures of spread.

The measures of spread IQR and s do not change when we add the same number a to all of the observations because adding a constant changes the location of the distribution but leaves the spread unaltered. You can find the effect of a linear transformation $x_{\text{new}} = a + bx$ by combining these rules. For example, if x has mean $\bar{x}$ the transformed variable x_{new} has mean $a + b\bar{x}$.

SUMMARY

A numerical summary of a distribution should report its **center** and its **spread** or **variability.**

The **mean** $\bar{x}$ and the **median** M describe the center of a distribution in different ways. The mean is the arithmetic average of the observations, and the median is their midpoint.

When you use the median to describe the center of the distribution, describe its spread by giving the **quartiles**. The **first quartile Q_1** has one-fourth of the observations below it, and the **third quartile Q_3** has three-fourths of the observations below it.

The **interquartile range** is the difference between the quartiles. It is the spread of the center half of the data. The **1.5 × *IQR* criterion** flags observations more than $1.5 \times IQR$ beyond the quartiles as possible outliers.

The **five-number summary** consisting of the median, the quartiles, and the smallest and largest individual observations provides a quick overall description of a distribution. The median describes the center, and the quartiles and extremes show the spread.

Boxplots based on the five-number summary are useful for comparing several distributions. The box spans the quartiles and shows the spread of the central half of the distribution. The median is marked within the box. Lines extend from the box to the extremes and show the full spread of the data, except that points identified by the $1.5 \times IQR$ criterion are often plotted individually.

The **variance s^2** and especially its square root, the **standard deviation s**, are common measures of spread about the mean as center. The standard deviation s is zero when there is no spread and gets larger as the spread increases.

A **resistant measure** of any aspect of a distribution is relatively unaffected by changes in the numerical value of a small proportion of the total number of observations, no matter how large these changes are. The median and quartiles are resistant, but the mean and the standard deviation are not.

The mean and standard deviation are good descriptions for symmetric distributions without outliers. They are most useful for the normal distributions introduced in the next section. The five-number summary is a better exploratory summary for skewed distributions.

Linear transformations have the form $x_{\text{new}} = a + bx$. A linear transformation changes the origin if $a \neq 0$ and changes the size of the unit of measurement if $b > 0$. Linear transformations do not change the overall shape of a distribution. A linear transformation multiplies a measure of spread by b, and changes a percentile or measure of center m into $a + bm$.

Numerical measures of particular aspects of a distribution, such as center and spread, do not report the entire shape of most distributions. In some cases, particularly distributions with multiple peaks and gaps, these measures may not be very informative.

SECTION 1.2 EXERCISES

1.41 A college rowing coach tests the 10 members of the women's varsity rowing team on a Stanford Rowing Ergometer (a stationary rowing machine). The variable measured is revolutions of the ergometer's flywheel in a 1-minute session. The data are

| 446 | 552 | 527 | 504 | 450 | 583 | 501 | 545 | 549 | 506 |

(a) Make a stemplot of these data after rounding to two digits. Then find the mean and the median of the original, unrounded ergometer scores. Explain the similarity or difference in these two measures in terms of the symmetry or skewness of the distribution.

(b) The coach used $\bar{x}$ and s to summarize these data. Find the standard deviation s. Do you agree that this is a suitable summary?

1.42 Here are the percents of the popular vote won by the successful candidate in each U.S. presidential election from 1948 to 1996:

Year	1948	1952	1956	1960	1964	1968	1972
Percent	49.6	55.1	57.4	49.7	61.1	43.4	60.7

Year	1976	1980	1984	1988	1992	1996	
Percent	50.1	50.7	58.8	53.9	43.2	49.2	

(a) Make a graph to display this distribution. What are the main features of the distribution?

(b) What is the median percent of the vote won by the successful candidate in presidential elections?

(c) Call an election a landslide if the winner's percent falls at or above the third quartile. Which elections were landslides?

1.43 Here are the scores on the Survey of Study Habits and Attitudes (SSHA) for 18 first-year college women:

| 154 | 109 | 137 | 115 | 152 | 140 | 154 | 178 | 101 |
| 103 | 126 | 126 | 137 | 165 | 165 | 129 | 200 | 148 |

and for 20 first-year college men:

| 108 | 140 | 114 | 91 | 180 | 115 | 126 | 92 | 169 | 146 |
| 109 | 132 | 75 | 88 | 113 | 151 | 70 | 115 | 187 | 104 |

(a) Make a back-to-back stemplot of these data, or use your result from Exercise 1.20.

(b) Find the mean $\bar{x}$ and the median M for both sets of SSHA scores. What feature of each distribution explains the fact that $\bar{x} > M$?

(c) Find the five-number summaries for both sets of SSHA scores. Your plot in (a) suggests that there is an outlier among the women's scores.

Does the $1.5 \times IQR$ criterion flag this observation? Make side-by-side boxplots for the two distributions. (Be sure to use a common scale to make comparison easy.)

(d) Use your results to write a brief comparison of the two groups. Do women as a group score higher than men? Which of your descriptions (stemplots, boxplots, numerical measures) show this? Which group of scores is more spread out when we ignore outliers? Which of your descriptions shows this most clearly?

1.44 The SSHA data for women given in the previous exercise contain one high outlier. Calculate the mean $\bar{x}$ and the median M for these data with and without the outlier. How does removing the outlier affect $\bar{x}$? How does it affect M? Your results illustrate the greater resistance of the median.

1.45 Exercise 1.22 presented data on the growth of chicks fed normal corn (the control group) and a new variety with better protein quality (the experimental group). Here are the weight gains in grams:

Control				Experimental			
380	321	366	356	361	447	401	375
283	349	402	462	434	403	393	426
356	410	329	399	406	318	467	407
350	384	316	272	427	420	477	392
345	455	360	431	430	339	410	326

(a) The researchers used $\bar{x}$ and s to summarize the data and as a basis for further statistical analysis. Find these measures for both groups.

(b) What kinds of distributions are best summarized by $\bar{x}$ and s? Do these distributions seem to fit the criteria? (Use your graph from Exercise 1.22; if you did not do that exercise, make a back-to-back stemplot now. It is often hard to decide if $\bar{x}$ and s are good descriptions for small sets of data because the shapes of such distributions are frequently irregular. We will learn a better tool for making this decision in the next section.)

1.46 The weights in the previous exercise are given in grams. There are 28.35 grams in an ounce. Use the results of part (a) of the previous exercise to find the mean and standard deviation of the weight gains measured in ounces.

1.47 The business magazine *Forbes* estimates (November 6, 1995) that the "average" household wealth of its readers is either about $800,000 or about $2.2 million, depending on which "average" it reports. Which of these numbers is the mean wealth and which is the median wealth? Explain your answer.

1.48 The NASDAQ Composite Index describes the average price of common stock traded over the counter, that is, not on one of the stock exchanges. In 1991, the mean capitalization of the companies in the NASDAQ index was

$80 million and the median capitalization was $20 million. (A company's capitalization is the total market value of its stock.) Explain why the mean capitalization is much higher than the median.

1.49   Last year a small accounting firm paid each of its five clerks $25,000, two junior accountants $60,000 each, and the firm's owner $255,000. What is the mean salary paid at this firm? How many of the employees earn less than the mean? What is the median salary?

1.50   In computing the median income of any group, some federal agencies omit all members of the group who had no income. Give an example to show that the reported median income of a group can go down even though the group becomes economically better off. Is this also true of the mean income?

1.51   This year, the firm in Exercise 1.49 gives no raises to the clerks and junior accountants, while the owner's take increases to $455,000. How does this change affect the mean? How does it affect the median?

1.52   Henry Cavendish (see Exercise 1.24 on page 30) used $\bar{x}$ to summarize his 29 measurements of the density of the earth.

(a) Find $\bar{x}$ and s for his data.

(b) Cavendish recorded the density of the earth as a multiple of the density of water. The density of water is almost exactly 1 gram per cubic centimeter, so his measurements have these units. In American units, the density of water is 62.43 pounds per cubic foot. This is the weight of a cube of water measuring 1 foot on each side. Express Cavendish's first result for the earth (5.50) in pounds per cubic foot. Then find $\bar{x}$ and s in pounds per cubic foot.

1.53   The following are the golf scores of 12 members of a women's golf team in tournament play:

$$89 \quad 90 \quad 87 \quad 95 \quad 86 \quad 81 \quad 102 \quad 105 \quad 83 \quad 88 \quad 91 \quad 79$$

(a) Present the distribution by a stemplot and describe its main features.

(b) Compute the mean, variance, and standard deviation of these golf scores.

(c) Compute the median, the quartiles, and the IQR. Are there any suspected outliers by the $1.5 \times IQR$ criterion?

(d) Based on the shape of the distribution, would you report the standard deviation or the quartiles as a measure of spread?

1.54   Calculate the mean and standard deviation of the metabolic rates in Example 1.18, showing each step in detail. First find the mean $\bar{x}$ by summing the 7 observations and dividing by 7. Then find each of the deviations $x_i - \bar{x}$ and their squares. Check that the deviations have sum 0. Calculate the variance as an average of the squared deviations (remember to divide by $n - 1$). Finally, obtain s as the square root of the variance.

1.55   This is a standard deviation contest. You must choose four numbers from the whole numbers 0 to 10, with repeats allowed.

(a) Choose four numbers that have the smallest possible standard deviation.

(b) Choose four numbers that have the largest possible standard deviation.

(c) Is more than one choice possible in either (a) or (b)? Explain.

1.56 This exercise requires a calculator with a standard deviation button, or statistical software on a computer. The observations

$$10,001 \quad 10,002 \quad 10,003$$

have mean $\bar{x} = 10,002$ and standard deviation $s = 1$. Adding a 0 in the center of each number, the next set becomes

$$100,001 \quad 100,002 \quad 100,003$$

The standard deviation remains $s = 1$ as more 0s are added. Use your calculator or computer to calculate the standard deviation of these numbers, adding extra 0s until you get an incorrect answer. How soon did you go wrong? This demonstrates that calculators and computers cannot handle an arbitrary number of digits correctly.

1.57 Table 1.3 (page 31) gives the percent of each state's residents who are at least 65 years old. Give a brief graphical and numerical description of the distribution. Explain your choice of numerical measures.

1.58 We saw in Example 1.19 that it is usual in the study of investments to use the mean and standard deviation to summarize and compare investment returns. Table 1.4 gives the monthly returns on one company's stock for 82 consecutive months.

(a) Find the mean monthly return and the standard deviation of the returns. If you invested $100 in this stock at the beginning of a month and got the mean return, how much would you have at the end of the month?

(b) The distribution can be described as "symmetric and unimodal, with one low outlier." If you invested $100 in this stock at the beginning of the worst month in the data (the outlier), how much would you have at the end of the month? Find the mean and standard deviation again, this time leaving out the low outlier. How much did this one observation affect the summary measures? Would leaving out this one observation change the median? The quartiles? How do you know, without actual calculation? (Returns over longer periods of time, or returns on portfolios containing several investments, tend to follow a normal distribution more closely than these monthly returns. So use of the mean and standard deviation is better justified for such data.)

1.59 Table 1.5 (page 32) gives the survival times of 72 guinea pigs in a medical study. Survival times—whether of cancer patients after treatment or of car batteries in everyday use—are almost always right-skewed. Make a graph to verify that this is true of these survival times. Then give a numerical summary that is appropriate for such data. Explain in simple language, to someone who knows no statistics, what your summary tells us about the guinea pigs.

1.60 IQ scores are usually said to follow a normal distribution closely. In particular, they are symmetric and unimodal. Is this true of the IQ scores for the 78 seventh-grade students in Table 1.6 (page 33)? Find the mean, the median, and the standard deviation of these IQ scores. Are the mean and median close in value, as they should be in a symmetric distribution?

1.61 Find the 10th and 90th percentiles of the distribution of percent of residents age 65 and over in the states, from Table 1.3 (page 31). Which states are in the top 10%? In the bottom 10%?

1.62 Find the *quintiles* (the 20th, 40th, 60th, and 80th percentiles) of the IQ scores in Table 1.6 (page 33). For quite large sets of data, the quintiles or the *deciles* (10th, 20th, 30th, etc. percentiles) give a more detailed summary than the quartiles.

1.63 We have already compared the distribution of calories in the three types of hot dogs represented in Table 1.8 on page 40. Make a similar comparison of the sodium content of the three types. Write a summary of your findings suitable for readers who know no statistics. Can we hold down our sodium intake by buying poultry hot dogs?

1.64 Figure 1.12 (page 28) shows the distribution of the mean SAT verbal scores in the 50 states and the District of Columbia for 1996. In 1995, the SATs were "recentered" to locate the mean scores for individuals close to the center of the 200 to 800 range of possible scores. To see the effect of recentering, we enter the state results for the mathematics and verbal SATs for both 1990 and 1996 into a statistical software package. Here are the results. (This output is from the Minitab statistical software; other software produces similar results.)

```
SATV-1990
       N      MEAN    MEDIAN    STDEV      MIN      MAX       Q1       Q3
      51    448.16    443.00    30.82   397.00   511.00   422.00   476.00

SATM-1990
       N      MEAN    MEDIAN    STDEV      MIN      MAX       Q1       Q3
      51    497.39    490.00    34.57   437.00   577.00   470.00   523.00

SATV-1996
       N      MEAN    MEDIAN    STDEV      MIN      MAX       Q1       Q3
      51    531.90    525.00    33.76   480.00   596.00   501.00   565.00

SATM-1996
       N      MEAN    MEDIAN    STDEV      MIN      MAX       Q1       Q3
      51    529.30    521.00    34.83   473.00   600.00   500.00   557.00
```

(a) Use the output to make side-by-side boxplots of the four distributions. What was the effect of recentering on the distribution of verbal SAT scores in the states? How did the distributions of state verbal and mathematics scores compare before recentering? How do they compare after recentering?

(b) Look again at Figure 1.12. What important feature of the distribution of 1996 verbal scores does the boxplot fail to display?

1.65 In each of the following settings, give the values of a and b for the linear transformation $x_{new} = a + bx$ that expresses the change in units of measurement.

(a) You are writing a report on the power of car engines. Your sources use horsepower x. Your professor requires you to reexpress power in watts x_{new}. One horsepower is 746 watts. What is the power in watts of a 140-horsepower engine?

(b) You must change a speed x measured in miles per hour into the metric system value x_{new} in kilometers per hour. (A mile is 0.62 kilometers.) What is 65 miles per hour in metric units?

(c) You want to restate water temperature x in a swimming pool (measured in degrees Fahrenheit) as the difference x_{new} between x and the "normal" body temperature of 98.6 degrees.

(d) The recommended daily allowance (RDA) for vitamin C is 30 milligrams. You measure milligrams of vitamin C in foods and must convert your results to percent of RDA.

1.66 **(Optional)** Give an approximate five-number summary of the salary data in Exercise 1.33 (page 35) by pretending that all salaries fall at the midpoint of their class. You see that you can find approximate numerical descriptions from grouped data when the actual data are not available.

1.67 **(Optional)** A change of units that multiplies each unit by b, such as the change $x_{new} = 2.54x$ from inches x to centimeters x_{new}, multiplies our usual measures of spread by b. This is true of IQR and the standard deviation. What happens to the variance when we change units in this way?

1.68 **(Optional)** The *trimmed mean* is a measure of center that is more resistant than the mean but uses more of the available information than the median. To compute the 10% trimmed mean, discard the highest 10% and the lowest 10% of the observations and compute the mean of the remaining 80%. Trimming eliminates the effect of a small number of outliers. Compute the 10% trimmed mean of the guinea pig survival time data in Table 1.5 (page 32). Then compute the 20% trimmed mean. Compare the values of these measures with the median and the ordinary untrimmed mean.

1.3   The Normal Distributions

We now have a kit of graphical and numerical tools for describing distributions. What is more, we have a clear strategy for exploring data on a single quantitative variable:

1. **Always plot your data:** make a graph, usually a stemplot or a histogram.
2. Look for the overall **pattern** and for striking **deviations** such as outliers.
3. Calculate a numerical summary to briefly describe **center** and **spread.**

Here is one more step to add to this strategy:

4. Sometimes the overall pattern of a large number of observations is so regular that we can describe it by a smooth curve.

Figure 1.19 is a histogram of the vocabulary scores of 947 seventh-grade students in Gary, Indiana. The smooth curve drawn through the histogram gives a compact description of the overall pattern. Because this curve describes the entire distribution in a single expression, it is actually easier to work with than our familiar graphs and numerical summaries. The curve **mathematical model** is a **mathematical model**—an idealized mathematical description—for the distribution.

The use of mathematical models is a common and powerful tool in statistics, as in other sciences. Suppose, for example, that we are watching a caterpillar crawl forward. Once each minute we measure the distance it has moved.

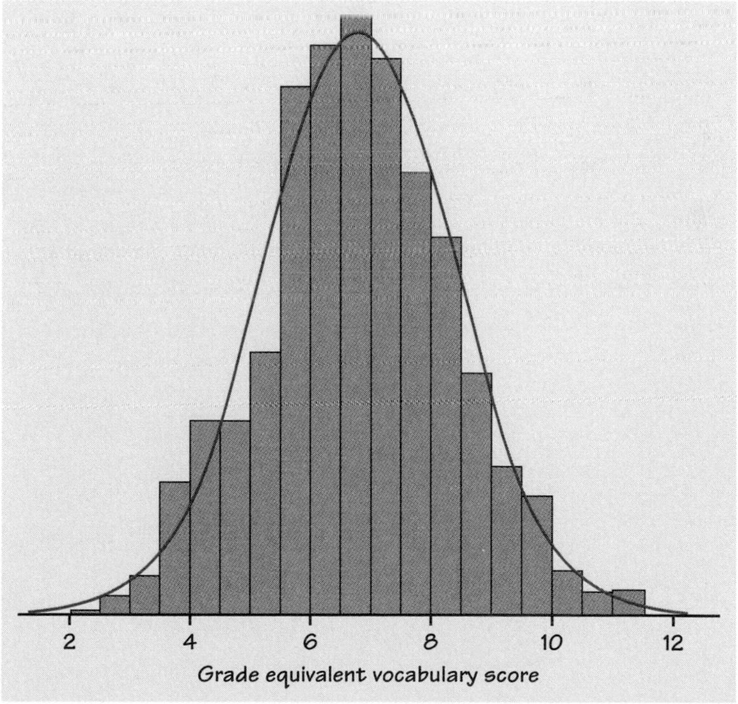

FIGURE 1.19 Histogram of 947 vocabulary scores, showing the approximation of the distribution by a normal curve.

We plot the observed values against time, and see that the points fall close to a straight line. It is natural to draw a straight line on our graph as a compact description of these data. The straight line, with its equation of the form $y = a + bt$ (where y is distance and t is time), is a mathematical model of the caterpillar's progress. It is an idealized description because the points on the graph do not fall *exactly* on the line. We want to give a similarly compact description of the distribution of a quantitative variable, to replace stemplots and histograms. Just as a straight line is one of many types of curves that can be used to describe a plot of distance traveled against time, there are many types of mathematical distributions that can be used to describe a set of single-variable data. This section will concentrate on one type, the normal distributions.

Density curves

The idea for a mathematical model to describe a distribution is illustrated by Figure 1.19. We draw a smooth curve through a histogram to describe the shape of the distribution without the lumpiness of the histogram.

The areas of the bars in a histogram represent either counts or proportions of the observations. For example, Figure 1.20(a) is a copy of Figure 1.19 with the leftmost bars shaded. The area of the shaded bars in Figure

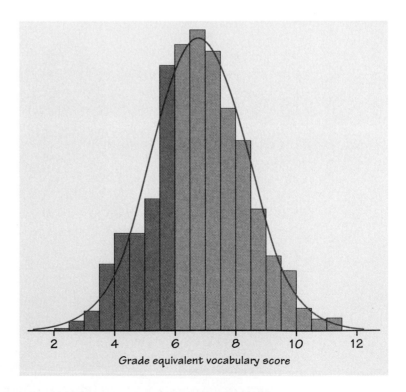

FIGURE 1.20(a)   The relative frequency of scores less than or equal to 6.0 from the histogram is 0.303.

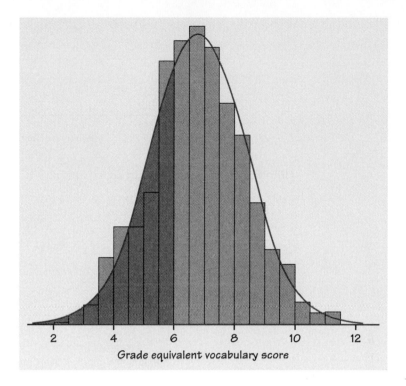

Grade equivalent vocabulary score

FIGURE 1.20(b) The relative frequency of scores less than or equal to 6.0 from the density curve is 0.293.

1.20(a) represents the students with vocabulary scores 6.0 or lower. There are 287 such students, who make up the proportion 287/947 = 0.303 of all Gary seventh graders. In Figure 1.20(b), the area under the smooth curve to the left of 6.0 is shaded. To make such curves easier to compare for different distributions, we adjust the scale so that the total area under the curve is exactly 1. Areas under the curve then represent proportions of the observations. That is, *area = relative frequency*. The curve is then a *density curve*. The shaded area under the density curve in Figure 1.20(b) represents the proportion of students with score 6.0 or lower. This area is 0.293, only 0.010 away from the histogram result. You can see that areas under the density curve give quite good approximations of areas given by the histogram.

Density Curve

A **density curve** is a curve that

- is always on or above the horizontal axis, and
- has area exactly 1 underneath it.

A density curve describes the overall pattern of a distribution. The area under the curve and above any range of values is the relative frequency of all observations that fall in that range.

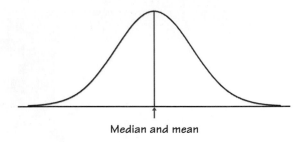

Median and mean

FIGURE 1.21(a)   A symmetric density curve with its mean and median marked.

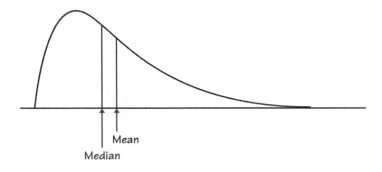

Mean

Median

FIGURE 1.21(b)   A right-skewed density curve with its mean and median marked.

The density curve in Figures 1.19 and 1.20 is a *normal curve*. Density curves, like distributions, come in many shapes. Figure 1.21 shows two density curves, a symmetric normal density curve and a right-skewed curve. A density curve of the appropriate shape is often an adequate description of the overall pattern of a distribution. Outliers, which are deviations from the overall pattern, are not described by the curve.

Measuring center and spread for density curves

Our measures of center and spread apply to density curves as well as to actual sets of observations, but only some of these measures are easily seen from the curve. **A mode** of a distribution described by a density curve is a peak point of the curve, the location where the curve is highest. Because areas under a density curve represent proportions of the observations, the **median** is the point with half the total area on each side. You can roughly locate the **quartiles** by dividing the area under the curve into quarters as accurately as possible by eye. The IQR is then the distance between the first and third quartiles. There are mathematical ways of calculating areas under curves. These allow us to locate the median and quartiles exactly on any density curve.

What about the mean and standard deviation? The mean of a set of observations is their arithmetic average. If we think of the observations as weights strung out along a thin rod, the mean is the point at which the rod would balance. This fact is also true of density curves. The mean is the point at which

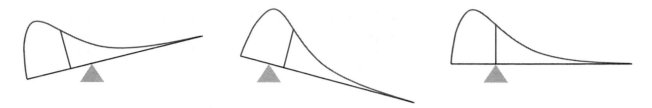

FIGURE 1.22 The mean of a density curve is the point at which it would balance.

the curve would balance if it were made out of solid material. Figure 1.22 illustrates this interpretation of the mean. We have marked the mean and median on the density curves in Figure 1.21. A symmetric curve, such as the normal curve in Figure 1.21(a), balances at its center of symmetry. Half the area under a symmetric curve lies on either side of its center, so this is also the median. For a right-skewed curve, such as that shown in Figure 1.21(b), the small area in the long right tail tips the curve more than the same area near the center. The mean (the balance point) therefore lies to the right of the median. It is hard to locate the balance point by eye on a skewed curve. There are mathematical ways of calculating the mean for any density curve, so we are able to mark the mean as well as the median in Figure 1.21(b). The standard deviation can also be calculated mathematically, but it can't be located by eye on most density curves.

Median and Mean of a Density Curve

The **median** of a density curve is the equal-areas point, the point that divides the area under the curve in half.

The **mean** of a density curve is the balance point, at which the curve would balance if made of solid material.

The median and mean are the same for a symmetric density curve. They both lie at the center of the curve. The mean of a skewed curve is pulled away from the median in the direction of the long tail.

A density curve is an idealized mathematical model for a distribution of data. For example, the symmetric density curve in Figures 1.19 and 1.20 is exactly symmetric, but the histogram of vocabulary scores is only approximately symmetric. We therefore need to distinguish between the mean and standard deviation of the density curve and the numbers $\bar{x}$ and s computed from the actual observations. The usual notation for the mean of an idealized distribution is μ (the Greek letter mu). We write the standard deviation of a density curve as σ (the Greek letter sigma).

mean μ

standard deviation σ

Our usual strategy in examining data is to look first for an overall pattern and then for deviations from that pattern. We can describe the overall pattern by pointing to the major features of a stemplot or histogram. Sometimes, as in the discussion of the calorie content of hot dogs that followed

Example 1.17 on page 50, no more compact description is adequate. When the pattern of data is quite regular, statisticians use a mathematical model to give a compact description of the overall pattern. When the data are observations on one quantitative variable, the models employed are density curves for distributions.

Normal distributions

normal curves

One particularly important class of density curves has already appeared in Figures 1.20 and 1.21(a). These density curves are symmetric, unimodal, and bell-shaped. They are called **normal curves,** and they describe *normal distributions*. All normal distributions have the same overall shape. The exact density curve for a particular normal distribution is specified by giving its mean μ and its standard deviation σ. The mean is located at the center of the symmetric curve and is the same as the median. Changing μ without changing σ moves the normal curve along the horizontal axis without changing its spread. The standard deviation σ controls the spread of a normal curve. Figure 1.23 shows two normal curves with different values of σ. The curve with the larger standard deviation is more spread out.

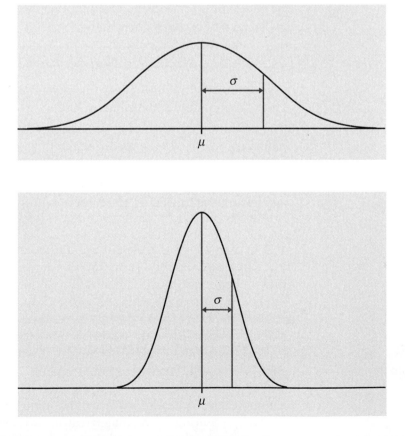

FIGURE 1.23   Two normal curves, showing the mean μ and standard deviation σ.

The standard deviation σ is the natural measure of spread for normal distributions. Not only do μ and σ completely determine the shape of a normal curve, but we can locate σ by eye on the curve. Here's how. As we move out in either direction from the center μ, the curve changes from falling ever more steeply

to falling ever less steeply

The points at which this change of curvature takes place are located at distance σ on either side of the mean μ. You can feel the change as you run a pencil along a normal curve, and so find the standard deviation. Remember that μ and σ alone do not specify the shape of most distributions, and that the shape of density curves in general does not reveal σ. These are special properties of normal distributions.

There are other symmetric bell-shaped density curves that are not normal. The normal density curves are specified by a particular equation. The height of the density curve at any point x is given by

$$\frac{1}{\sigma\sqrt{2\pi}} e^{-\frac{1}{2}\left(\frac{x-\mu}{\sigma}\right)^2}$$

We will not make direct use of this fact, although it is the basis of mathematical work with normal distributions. Note that the equation of the curve is completely determined by μ and σ.

Why are the normal distributions important in statistics? Here are three reasons. First, normal distributions are good descriptions for some distributions of *real data*. Distributions that are often close to normal include scores on tests taken by many people (such as SAT exams and many psychological tests), repeated careful measurements of the same quantity, and characteristics of biological populations (such as lengths of cockroaches and yields of corn). Second, normal distributions are good approximations to the results of many kinds of *chance outcomes,* such as tossing a coin many times. Third, and most important, we will see that many *statistical inference* procedures based on normal distributions work well for other roughly symmetric distributions. HOWEVER . . . even though many sets of data follow a normal distribution, many do not. Most income distributions, for example, are skewed to the right and so are not normal. Nonnormal data, like nonnormal people, not only are common but are sometimes more interesting than their normal counterparts.

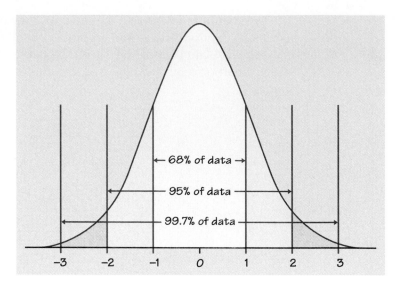

FIGURE 1.24   The 68–95–99.7 rule for normal distributions.

The 68–95–99.7 rule

Although there are many normal curves, they all have common properties. Here are some of the most important.

The 68–95–99.7 Rule

In the normal distribution with mean μ and standard deviation σ:

- ■ **68%** of the observations fall within σ of the mean μ.
- ■ **95%** of the observations fall within 2σ of μ.
- ■ **99.7%** of the observations fall within 3σ of μ.

Figure 1.24 illustrates the 68–95–99.7 rule. By remembering these three numbers, you can think about normal distributions without constantly making detailed calculations.

EXAMPLE 1.22   The distribution of heights of young women aged 18 to 24 is approximately normal with mean $\mu = 64.5$ inches and standard deviation $\sigma = 2.5$ inches. Figure 1.25 shows what the 68–95–99.7 rule says about this distribution.

Two standard deviations is 5 inches for this distribution. The 95 part of the 68–95–99.7 rule says that the middle 95% of young women are between $64.5 - 5$ and $64.5 + 5$ inches tall, that is, between 59.5 inches and 69.5 inches. This fact is exactly true for an exactly normal distribution. It is approximately true for the heights of young women because the distribution of heights is approximately normal.

The other 5% of young women have heights outside the range from 59.5 to 69.5 inches. Because the normal distributions are symmetric, half of these women are on the tall side. So the tallest 2.5% of young women are taller than 69.5 inches.

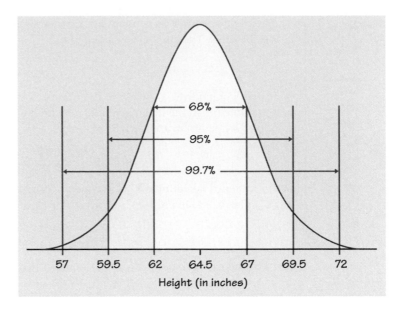

FIGURE 1.25 The 68–95–99.7 rule applied to the heights of young women.

Because we will mention normal distributions often, a short notation is helpful. We abbreviate the normal distribution with mean μ and standard deviation σ as $N(\mu, \sigma)$. For example, the distribution of young women's heights is $N(64.5, 2.5)$.

Normal distribution $N(\mu, \sigma)$

Standardizing observations

As the 68–95–99.7 rule suggests, all normal distributions share many properties. In fact, all normal distributions are the same if we measure in units of size σ about the mean μ as center. Changing to these units is called *standardizing*. To standardize a value, subtract the mean of the distribution and then divide by the standard deviation.

Standardizing and z-Scores

If x is an observation from a distribution that has mean μ and standard deviation σ, the **standardized value** of x is

$$z = \frac{x - \mu}{\sigma}$$

A standardized value is often called a z-**score.**

A z-score tells us how many standard deviations the original observation falls away from the mean, and in which direction. Observations larger than the mean are positive when standardized, and observations smaller than the mean are negative.

EXAMPLE 1.23   The heights of young women are approximately normal with $\mu = 64.5$ inches and $\sigma = 2.5$ inches. The standardized height is

$$z = \frac{\text{height} - 64.5}{2.5}$$

A woman's standardized height is the number of standard deviations by which her height differs from the mean height of all young women. A woman 68 inches tall, for example, has standardized height

$$z = \frac{68 - 64.5}{2.5} = 1.4$$

or 1.4 standard deviations above the mean. Similarly, a woman 5 feet (60 inches) tall has standardized height

$$z = \frac{60 - 64.5}{2.5} = -1.8$$

or 1.8 standard deviations less than the mean height.

We need a way to write variables, such as "height" in Example 1.23, that follow a theoretical distribution such as a normal distribution. We use capital letters near the end of the alphabet for such variables. If X is the height of a young woman, we can then shorten "the height of a young woman is less than 68 inches" to "$X < 68$." We will use lowercase x to stand for any specific value of the variable X.

We often standardize observations from symmetric distributions to express them in a common scale. We might, for example, compare the height of two children of different ages by calculating their z-scores. The standardized heights tell us where each child stands in the distribution for his or her age group.

Standardizing is a linear transformation that transforms the data into the standard scale of z-scores. We know that a linear transformation does not change the shape of a distribution, and that the mean and standard deviation change in a simple manner. In fact, *any variable obtained from a normal variable by a linear transformation remains normal*. Of course, the mean and standard deviation will change in the usual way. If X has the normal distribution with mean μ and standard deviation σ, then $X_{\text{new}} = a + bX$ for a positive b has the normal distribution with mean $a + b\mu$ and standard deviation $b\sigma$. In particular, the standardized values for any distribution always have mean 0 and standard deviation 1. If the original distribution was normal, the standardized values have the normal distribution with mean 0 and standard deviation 1.

The standard normal distribution

If the variable we standardize has a normal distribution, standardizing does more than give a common scale. It makes all normal distributions into a single distribution, and this distribution is still normal. Standardizing a variable that has any normal distribution produces a new variable that has the *standard normal distribution*.

> ### The Standard Normal Distribution
>
> The **standard normal distribution** is the normal distribution $N(0, 1)$ with mean 0 and standard deviation 1.
>
> If a variable X has any normal distribution $N(\mu, \sigma)$ with mean μ and standard deviation σ, then the standardized variable
>
> $$Z = \frac{X - \mu}{\sigma}$$
>
> has the standard normal distribution.

Table A in the back of the book gives areas under the standard normal curve. Table A also appears on the inside front cover. You can use Table A to do normal calculations, although software or a statistical calculator is usually more efficient.

The Standard Normal Table

Table A is a table of areas under the standard normal curve. The table entry for each value z is the area under the curve to the left of z.

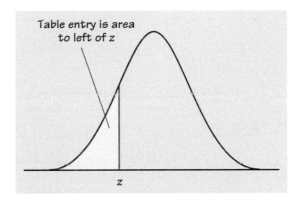

EXAMPLE 1.24

Problem: What proportion of observations on a standard normal variable Z take values less than 1.4? That is, what is the relative frequency of $Z < 1.4$?

Solution: To find the area to the left of 1.40, locate 1.4 in the left-hand column of Table A, then locate the remaining digit 0 as .00 in the top row. The entry opposite 1.4 and under .00 is 0.9192. This is the area we seek. Figure 1.26(a) illustrates this area.

Problem: Find the proportion of observations from the standard normal distribution that are greater than -2.15. This is the relative frequency of $Z > -2.15$.

Solution: Enter Table A under $z = -2.15$. That is, find -2.1 in the left-hand column and .05 in the top row. The table entry is 0.0158. This is the area to the *left* of -2.15. Because the total area under the curve is 1, the area lying to the *right* of -2.15 is $1 - 0.0158 = 0.9842$. Figure 1.26(b) shows these areas.

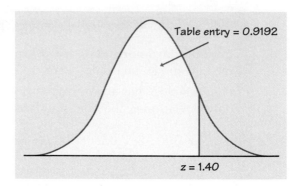

FIGURE 1.26(a)   The area under a standard normal curve to the left of the point $z = 1.4$ is 0.9192. Table A gives areas under the standard normal curve.

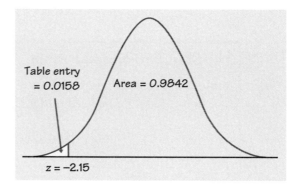

FIGURE 1.26(b)   Areas under the standard normal curve to the right and left of $z = -2.15$. Table A gives the area to the left.

Normal distribution calculations

We saw in Example 1.23 that standardizing the height $x = 68$ inches for a young woman gives the z-score $z = 1.4$. Example 1.24 then showed that the area under the standard normal curve to the left of $z = 1.4$ is 0.9192. This is the relative frequency of z-scores less than 1.4. But the z-score is less than 1.4 whenever the height is less than 68 inches. So the proportion of all young women who are less than 68 inches tall is 0.9192, or about 92%. We can find relative frequencies for *any* normal distribution by standardizing and using Table A. Of course, software or a statistical calculator often automates the calculation. Here is an example.

EXAMPLE 1.25   The level of cholesterol in the blood is important because high cholesterol levels increase the risk of heart disease. The distribution of blood cholesterol levels in a large population of people of the same age and sex is roughly normal. For 14-year-old boys, the mean is $\mu = 170$ milligrams of cholesterol per deciliter of blood (mg/dl) and the standard deviation is $\sigma = 30$ mg/dl.[10] Levels above 240 mg/dl may require medical attention. What percent of 14-year-old boys have more than 240 mg/dl of cholesterol?

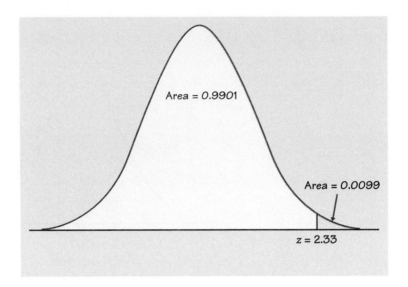

FIGURE 1.27 Areas under the standard normal curve for Example 1.25.

1. *State the problem.* Call the level of cholesterol in the blood X. The variable X has the $N(170, 30)$ distribution. We want the proportion of boys with $X > 240$.

2. *Standardize.* Subtract the mean, then divide by the standard deviation, to transform the problem about X into a problem about a standard normal Z:

$$X > 240$$
$$\frac{X - 170}{30} > \frac{240 - 170}{30}$$
$$Z > 2.33$$

Figure 1.27 shows the standard normal curve with the areas of interest marked.

3. *Use the table.* From Table A, we see that the proportion of observations less than 2.33 is 0.9901. About 99% of boys have cholesterol levels less than 240. The area to the right of 2.33 is therefore $1 - 0.9901 = 0.0099$. This is about 0.01, or 1%. Only about 1% of boys have high cholesterol.

There is *no* area under a smooth curve and exactly over the point 240. Consequently, the area to the right of 240 (the relative frequency of $X > 240$) is the same as the area at or to the right of this point (the relative frequency of $X \geq 240$). An actual data set may contain a boy with exactly 240 mg/dl of blood cholesterol. The fact that it is not possible for a normal variable to have $X = 240$ exactly is a consequence of the idealized smoothing of normal models for data.

Sometimes we encounter a value of z more extreme than those appearing in Table A. For example, the area to the left of $z = -4$ is not given directly in the table. The z-values in Table A leave only area 0.0002 in each tail unaccounted for. For practical purposes, we can act as if there is zero area outside the range of Table A.

The key to using either software or Table A to do a normal calculation is to sketch the area you want, then match that area with the areas that the table or software gives you. Here is another example.

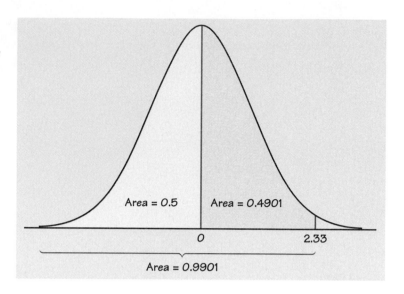

FIGURE 1.28 Areas under the standard normal curve for Example 1.26.

EXAMPLE 1.26

What percent of 14-year-old boys have blood cholesterol between 170 and 240 mg/dl?

1. *State the problem*. We want the proportion of boys with $170 \leq X \leq 240$.
2. *Standardize*.

$$170 \quad \leq \quad X \quad \leq \quad 240$$
$$\frac{170 - 170}{30} \leq \frac{X - 170}{30} \leq \frac{240 - 170}{30}$$
$$0 \quad \leq \quad Z \quad \leq \quad 2.33$$

    Figure 1.28 shows the area under the standard normal curve. This picture guides our use of the table.

3. *Use the table*. The area between 2.33 and 0 is the area below 2.33 *minus* the area below 0. Look at Figure 1.28 to check this. From Table A,

    area between 0 and 2.33 = area below 2.33 − area below 0.00
    $$= 0.9901 - 0.5000 = 0.4901$$

About 49% of boys have cholesterol levels between 170 and 240 mg/dl.

Examples 1.25 and 1.26 illustrate the use of Table A to find the relative frequency of a given event, such as "cholesterol level between 170 and 240 mg/dl." We may instead want to find the observed value corresponding to a given relative frequency. To do this, we use Table A backward, finding the desired relative frequency in the body of the table and then reading the corresponding z from the left column and top row.

EXAMPLE 1.27

Scores on the SAT verbal test in recent years follow approximately the $N(505, 110)$ distribution. How high must a student score in order to place in the top 10% of all students taking the SAT?

1. *State the problem*. We want to find the SAT score x with area 0.1 to its *right* under the normal curve with mean $\mu = 505$ and standard deviation $\sigma = 110$. That's the same as finding the SAT score x with area 0.9 to its *left*. Figure 1.29

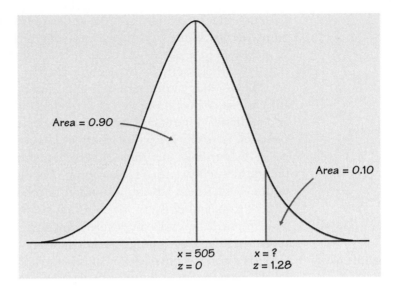

FIGURE 1.29 Locating the point on a normal curve with area 0.10 to its right.

poses the question in graphical form. Because Table A gives the areas to the left of z-values, always state the problem in terms of the area to the left of x.

2. *Use the table.* Look in the body of Table A for the entry closest to 0.9. It is 0.8997. This is the entry corresponding to $z = 1.28$. So $z = 1.28$ is the standardized value with area 0.9 to its left.

3. *Unstandardize* to transform the solution from the z back to the original x scale. We know that the standardized value of the unknown x is $z = 1.28$. So x itself satisfies

$$\frac{x - 505}{110} = 1.28$$

Solving this equation for x gives

$$x = 505 + (1.28)(110) = 645.8$$

This equation should make sense: it finds the x that lies 1.28 standard deviations above the mean on this particular normal curve. That is the "unstandardized" meaning of $z = 1.28$. We see that a student must score at least 646 to place in the highest 10%. The general rule for unstandardizing a z-score is

$$x = \mu + z\sigma$$

Normal quantile plots

The normal distributions provide good models for some distributions of real data. Examples include the Gary vocabulary scores and Newcomb's measurements of the passage time of light (after dropping the outliers). The distributions of some other common variables are usually skewed and therefore distinctly nonnormal. Examples include economic variables such as personal income and gross sales of business firms, the survival times of cancer patients after treatment, and the service lifetime of mechanical or electronic components. While experience can suggest whether or not a normal model is plausible in a particular case, it is risky to assume that a distribution is normal without actually inspecting the data.

The decision to describe a distribution by a normal model may determine the later steps in our analysis of the data. Both relative frequency calculations and statistical inference based on such calculations follow from the choice of a model. How can we judge whether data are approximately normal?

A histogram or stemplot can reveal distinctly nonnormal features of a distribution, such as outliers (Newcomb's histogram, Figure 1.5 on page 17), pronounced skewness (the supermarket purchase stemplot, Figure 1.3 on page 12), or gaps and clusters (the hot dog stemplot on page 50). If the stemplot or histogram appears roughly symmetric and unimodal, however, we need a more sensitive way to judge the adequacy of a normal model. The most useful tool for assessing normality is another graph, the **normal quantile plot.***

normal quantile plot

Here is the idea of a simple version of a normal quantile plot. It is not feasible to make normal quantile plots by hand, but software makes them for us, using more sophisticated versions of this basic idea.

1. Arrange the observed data values from smallest to largest. Record what percentile of the data each value occupies. For example, the smallest observation in a set of 20 is at the 5% point, the second smallest is at the 10% point, and so on.

2. Do normal distribution calculations to find the z-scores at these same percentiles. For example, $z = -1.645$ is the 5% point of the standard normal distribution, and $z = -1.282$ is the 10% point.

3. Plot each data point x against the corresponding z. If the data distribution is close to standard normal, the plotted points will lie close to the 45-degree line $x = z$. If the data distribution is close to any normal distribution, the plotted points will lie close to some straight line.

Any normal distribution produces a straight line on the plot because standardizing turns any normal distribution into a standard normal. Standardizing is a linear transformation that can change the slope and intercept of the line in our plot but cannot turn a line into a curved pattern.

Use of Normal Quantile Plots

If the points on a normal quantile plot lie close to a straight line, the plot indicates that the data are normal. Systematic deviations from a straight line indicate a nonnormal distribution. Outliers appear as points that are far away from the overall pattern of the plot.

---

*Some software calls these graphs *normal probability plots*. There is a technical distinction between the two types of graphs, but the terms are often used loosely.

Figures 1.30 to 1.33 are normal quantile plots for data we have met earlier. The data x are plotted vertically against the corresponding standard normal z-score plotted horizontally. The z-score scale extends from -3 to 3 because almost all of a standard normal curve lies between these values. These figures show how normal quantile plots behave.

EXAMPLE 1.28

Figure 1.30 is a normal quantile plot of Newcomb's passage time data. Most of the points lie close to a straight line, indicating that a normal model fits well. The two outliers deviate from the line and show how the plot responds to low outliers. These two points are below the line formed by the rest of the data—they are farther out in the low direction than we expect normal data to be. Compare the histogram of these data (Figure 1.5 on page 17).

Figure 1.31 is a normal quantile plot of Newcomb's measurements of the passage time of light with the two outliers omitted. The effect of omitting the outliers is to magnify the plot of the remaining data. As we saw from Figure 1.30, a normal distribution fits quite well. The only important deviation from normality is the "stairstep" appearance caused by numerous short horizontal runs of points. Each run represents repeated observations having the same value—there are six measurements at 27 and seven at 28, for example. This phenomenon is called **granularity.** It is caused by the limited precision of the measurements. The granularity would largely disappear if Newcomb had been able to measure one more decimal place to spread out the steps. This minor granularity does not prevent us from adopting a normal distribution as a model.

granularity

EXAMPLE 1.29

Figure 1.32 is a normal quantile plot of the supermarket spending data. The stemplots in Figure 1.3 (page 12) show that the distribution is right-skewed. To see the right skewness in the normal quantile plot, draw a line through the leftmost points, which correspond to the smaller observations. The larger observations fall systematically above this line. That is, the right-of-center observations have larger values than in a normal distribution. *In a right-skewed distribution, the largest observations fall distinctly above a line drawn through the main body of points.* Similarly, left skewness is evident when the smallest observations fall below the line. Unlike Figure 1.30, there are no individual outliers.

EXAMPLE 1.30

Figure 1.33 is a normal quantile plot of the 78 IQ scores from Table 1.6. We expect IQ scores to be roughly normal. The plot is roughly straight, showing good fit to a normal model except for some granularity due to the fact that all IQ scores are whole numbers.

As Figures 1.31 and 1.33 illustrate, real data almost always show some departure from the theoretical normal model. It is important to confine your examination of a normal quantile plot to searching for shapes that show clear departures from normality. Don't overreact to minor wiggles in the plot. When we discuss statistical methods that are based on the normal model, we will pay attention to the sensitivity of each method to departures from normality. Many common methods work well as long as the data are approximately normal and outliers are not present.

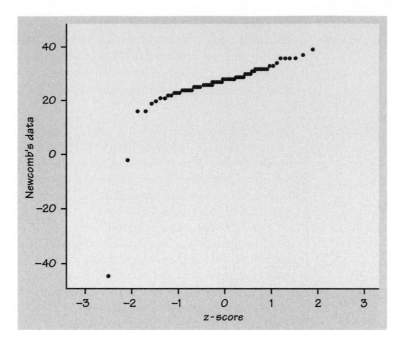

FIGURE 1.30   Normal probability plot for Newcomb's 66 measurements of the passage time of light. There are two low outliers.

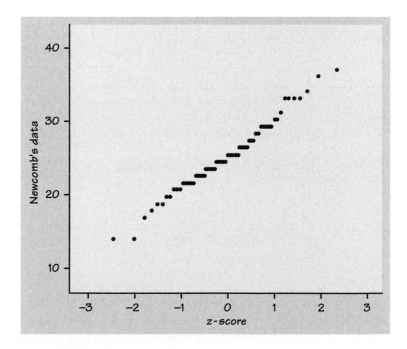

FIGURE 1.31   Newcomb's data without the two outliers. The data are close to normal except for slight granularity.

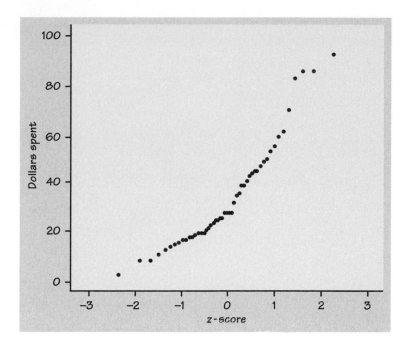

FIGURE 1.32   Normal probability plot of the supermarket spending data. The pattern bends up at the right, showing right skewness.

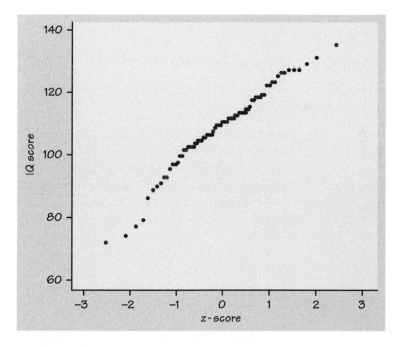

FIGURE 1.33   Normal probability plot of the IQ scores of 78 seventh-grade students. The straight pattern shows that this distribution is close to normal.

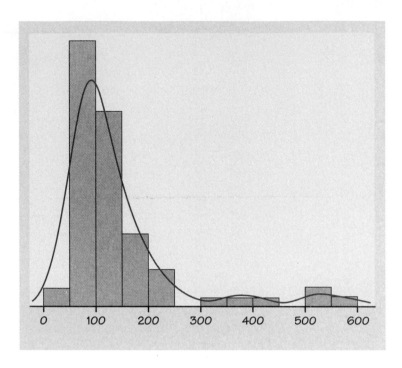

FIGURE 1.34 A density curve for the guinea pig survival data, calculated from the data by density estimation software.

Beyond the basics ▶ density estimation

A density curve gives a compact summary of the overall shape of a distribution. Figure 1.19 (page 65) shows a normal density curve that provides a good summary of the distribution of Gary vocabulary scores. Many distributions do not have the normal shape. There are other families of density curves that are used as mathematical models for various distribution shapes.

Modern software offers a more flexible option, **density estimation.** A density estimator does not start with any specific shape, such as the normal shape. It looks at the data and draws a density curve that describes the overall shape of the data. Figure 1.34 shows a histogram of the strongly right-skewed guinea pig survival data from Table 1.5. A density estimator produced the density curve drawn over the histogram. You can see that this curve does catch the overall pattern: the strong right skewness, the major peak near 100 days, and the two smaller clusters in the right tail. You can also see that the density estimator works blindly: in order to get a smooth curve, it starts below 0, even though survival times less than 0 days are not possible. Nonetheless, density estimation offers a quick way to picture the overall shapes of distributions. It is another useful tool for exploratory data analysis.[11]

SUMMARY

The overall pattern of a distribution can often be described compactly by a **density curve.** A density curve has total area 1 underneath it. Areas under a density curve give **relative frequencies** for the distribution.

The **mean** μ (balance point), the **median** (equal-areas point), and the **quartiles** can be approximately located by eye on a density curve. The **standard deviation** σ cannot be located by eye on most density curves. The mean and median are equal for symmetric density curves, but the mean of a skewed curve is located farther toward the long tail than is the median.

The **normal distributions** are described by bell-shaped, symmetric unimodal density curves. The mean μ and standard deviation σ completely specify the normal distribution $N(\mu, \sigma)$. The mean is the center of symmetry and σ is the distance from μ to the change-of-curvature points on either side.

To **standardize** any observation x, subtract the mean of the distribution and then divide by the standard deviation. The resulting **z-score** $z = (x - \mu)/\sigma$ says how many standard deviations x lies from the distribution mean. All normal distributions are the same when measurements are transformed to the standardized scale. In particular, all normal distributions satisfy the **68–95–99.7 rule.**

If X has the $N(\mu, \sigma)$ distribution, then the standardized variable $Z = (X - \mu)/\sigma$ has the **standard normal** distribution $N(0, 1)$. Relative frequencies for any normal distribution can be calculated from the **standard normal table** (Table A), which gives relative frequencies for the events $Z < z$ for many values of z.

The adequacy of a normal model for describing a distribution of data is best assessed by a **normal quantile plot,** which is available in most statistical software packages. A pattern on such a plot that deviates substantially from a straight line indicates that the data are not normal.

SECTION 1.3 EXERCISES

1.69 Many software packages have "random number generators" that produce numbers that are distributed uniformly between 0 and 1. Figure 1.35 graphs the density curve of the outcome of a random number generator.

(a) Check that the area under this curve is 1.

(b) What proportion of the outcomes are less than 0.25? (Sketch the density curve, shade the area that represents the proportion, then find that area. Do this for part (c) also.)

(c) What proportion of outcomes lie between 0.1 and 0.9?

1.70 Many random number generators allow users to specify the range of the random numbers to be produced. Suppose that you specify that the

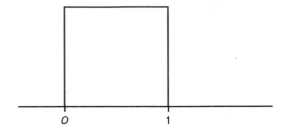

FIGURE 1.35 The density curve of a uniform distribution, for Exercise 1.69.

outcomes are to be distributed uniformly between 0 and 2. Then the density curve of the outcomes has constant height between 0 and 2, and height 0 elsewhere.

(a) What is the height of the density curve between 0 and 2? Draw a graph of the density curve.

(b) Use your graph from (a) and the fact that areas under the curve are relative frequencies of outcomes to find the proportion of outcomes that are less than 1.

(c) Find the proportion of outcomes that lie between 0.5 and 1.3.

1.71 What are the mean and the median of the distribution in Figure 1.35? What are the quartiles?

1.72 Figure 1.36 displays three density curves, each with three points marked on it. At which of these points on each curve do the mean and the median fall?

1.73 The Environmental Protection Agency requires that the exhaust of each model of motor vehicle be tested for the level of several pollutants. The level of oxides of nitrogen (NOX) in the exhaust of one light truck model was found to vary among individual trucks according to a normal distribution with mean $\mu = 1.45$ grams per mile driven and standard deviation $\sigma = 0.40$ grams per mile. Sketch the density curve of this normal distribution, with the scale of grams per mile marked on the horizontal axis.

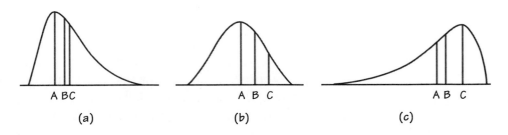

FIGURE 1.36 Three density curves, for Exercise 1.72.

1.74 A study of elite distance runners found a mean weight of 63.1 kilograms (kg), with a standard deviation of 4.8 kg. Assuming that the distribution of weights is normal, sketch the density curve of the weight distribution with the horizontal axis marked in kilograms. (Based on M. L. Pollock et al., "Body composition of elite class distance runners," in P. Milvy (ed.), *The Marathon: Physiological, Medical, Epidemiological, and Psychological Studies*, New York Academy of Sciences, 1977.)

1.75 Give an interval that contains the middle 95% of NOX levels in the exhaust of trucks using the model described in Exercise 1.73.

1.76 Use the 68–95–99.7 rule to find intervals centered at the mean that will include 68%, 95%, and 99.7% of the weights of the elite runners described in Exercise 1.74.

1.77 The length of human pregnancies from conception to birth varies according to a distribution that is approximately normal with mean 266 days and standard deviation 16 days. Use the 68–95–99.7 rule to answer the following questions.

(a) Between what values do the lengths of the middle 95% of all pregnancies fall?

(b) How short are the shortest 2.5% of all pregnancies? How long do the longest 2.5% last?

1.78 The IQ scores of the 78 seventh-grade students reported in Table 1.6 are approximately normal. How well do these scores satisfy the 68–95–99.7 rule? To find out, calculate the mean $\bar{x}$ and standard deviation s of the scores. Then calculate the percent of the 78 scores that fall between $\bar{x} - s$ and $\bar{x} + s$ and compare your result with 68%. Do the same for the intervals covering two and three standard deviations on either side of the mean. (The 68–95–99.7 rule is exact for any theoretical normal distribution. It will hold only approximately for actual data.)

1.79 Eleanor scores 680 on the mathematics part of the SAT examination. The distribution of SAT scores in a reference population is normal with mean 500 and standard deviation 100. Gerald takes the ACT mathematics test and scores 27. ACT scores are normally distributed with mean 18 and standard deviation 6. Find the z-scores for both students. Assuming that both tests measure the same kind of ability, who has the higher score?

1.80 Three landmarks of baseball achievement are Ty Cobb's batting average of .420 in 1911, Ted Williams's .406 in 1941, and George Brett's .390 in 1980. These batting averages cannot be compared directly because the distribution of major league batting averages has changed over the decades. The distributions are quite symmetric and (except for outliers such as Cobb, Williams, and Brett) reasonably normal. While the mean batting average has been held roughly constant by rule changes and the balance between hitting and pitching, the standard deviation has dropped over time. Here are the facts:

Decade	Mean	Std. Dev.
1910s	.266	.0371
1940s	.267	.0326
1970s	.261	.0317

Compute the standardized batting averages for Cobb, Williams, and Brett to compare how far each stood above his peers. (Data from Stephen Jay Gould, "Entropic homogeneity isn't why no one hits .400 anymore," *Discover*, August 1986, pp. 60–66.)

1.81 Using either Table A or your calculator or software, find the proportion of observations from a standard normal distribution that satisfies each of the following statements. In each case, sketch a standard normal curve and shade the area under the curve that is the answer to the question.

(a) $Z < 2.85$

(b) $Z > 2.85$

(c) $Z > -1.66$

(d) $-1.66 < Z < 2.85$

1.82 Using either Table A or your calculator or software, find the relative frequency of each of the following events in a standard normal distribution. In each case, sketch a standard normal curve with the area representing the relative frequency shaded.

(a) $Z \leq -2.25$

(b) $Z \geq -2.25$

(c) $Z > 1.77$

(d) $-2.25 < Z < 1.77$

1.83 Find the value z of a standard normal variable Z that satisfies each of the following conditions. (If you use Table A, report the value of z that comes closest to satisfying the condition.) In each case, sketch a standard normal curve with your value of z marked on the axis.

(a) The point z with 25% of the observations falling below it.

(b) The point z with 40% of the observations falling above it.

1.84 The variable Z has a standard normal distribution.

(a) Find the number z such that the event $Z < z$ has relative frequency 0.8.

(b) Find the number z such that the event $Z > z$ has relative frequency 0.35.

1.85 SAT scores are approximately normal with mean 500 and standard deviation 100. Scores of 800 or higher are reported as 800, so a perfect paper is not required to score 800 on the SAT. What percent of students who take the SAT score 800?

1.86 The rate of return on stock indexes (which combine many individual stocks) is approximately normal. Since 1945, the Standard & Poor's 500 index has had a mean yearly return of about 12%, with a standard deviation of about 16.5%. Take this normal distribution to be the distribution of yearly returns over a long period.

(a) In what range do the middle 95% of all yearly returns lie?

(b) The market is down for the year if the return on the index is less than zero. In what percent of years is the market down?

(c) In what percent of years does the index gain 25% or more?

1.87 The Graduate Record Examinations (GRE) are widely used to help predict the performance of applicants to graduate schools. The range of possible scores on a GRE is 200 to 900. The psychology department at a university finds that the scores of its applicants on the quantitative GRE are approximately normal with mean $\mu = 544$ and standard deviation $\sigma = 103$. Find the relative frequency of applicants whose score X satisfies each of the following conditions:

(a) $X > 700$

(b) $X < 500$

(c) $500 < X < 800$

1.88 The length of human pregnancies from conception to birth varies according to a distribution that is approximately normal with mean 266 days and standard deviation 16 days.

(a) What percent of pregnancies last less than 240 days (that's about 8 months)?

(b) What percent of pregnancies last between 240 and 270 days (roughly between 8 months and 9 months)?

(c) How long do the longest 20% of pregnancies last?

1.89 A patient is said to be hypokalemic (low potassium in the blood) if the measured level of potassium is 3.5 or less. (The units for this measure are meq/l, or milliequivalents per liter.) An individual's potassium level is not a constant, however, but varies from day to day. In addition, the measurement procedure itself has some variation. Suppose that the overall variation follows a normal distribution. Judy has a mean potassium level of 3.8 with a standard deviation of 0.2. If she is measured on many days, on what proportion of days will the measurement suggest that Judy is hypokalemic?

1.90 The Acculturation Rating Scale for Mexican Americans (ARSMA) is a psychological test that evaluates the degree to which Mexican Americans are adapted to Mexican/Spanish versus Anglo/English culture. The range of possible scores is 1.0 to 5.0, with higher scores showing more Anglo/English acculturation. The distribution of ARSMA scores in a population used to develop the test is approximately normal with mean 3.0 and standard deviation 0.8. A researcher believes that Mexicans will have an average score near 1.7, and that first-generation Mexican Americans will average about 2.1 on the ARSMA scale. What proportion of the population used to develop the test has scores below 1.7? Between 1.7 and 2.1?

1.91   Scores on the Wechsler Adult Intelligence Scale for the 20 to 34 age group are approximately normally distributed with mean 110 and standard deviation 25. Scores for the 60 to 64 age group are approximately normally distributed with mean 90 and standard deviation 25.

Sarah, who is 30, scores 135 on this test. Sarah's mother, who is 60, also takes the test and scores 120. Who scored higher relative to her age group, Sarah or her mother? Who has the higher absolute level of the variable measured by the test? At what percentile of their age groups are Sarah and her mother? (That is, what percent of the age group has lower scores?)

1.92   How high a score on the ARSMA test of Exercise 1.90 must a Mexican American obtain to be among the 30% of the population used to develop the test who are most Anglo/English in cultural orientation? What scores make up the 30% who are most Mexican/Spanish in their acculturation?

1.93   The scores of a reference population on the Wechsler Intelligence Scale for Children (WISC) are normally distributed with $\mu = 100$ and $\sigma = 15$.

(a) What percent of this population have WISC scores below 100?

(b) Below 80?

(c) Above 140?

(d) Between 100 and 120?

1.94   The distribution of scores on the WISC is described in the previous exercise. What score will place a child in the top 5% of the population? In the top 1%?

1.95   The median of any normal distribution is the same as its mean. We can use normal calculations to find the quartiles and related descriptive measures for normal distributions.

(a) What is the area under the standard normal curve to the left of the first quartile? Use this to find the value of the first quartile for a standard normal distribution. Find the third quartile similarly.

(b) Your work in (a) gives the z-scores for the quartiles of any normal distribution. Scores on the Wechsler Intelligence Scale for Children (WISC) are normally distributed with mean 100 and standard deviation 15. What are the quartiles of WISC scores?

(c) What is the value of the IQR for the standard normal distribution?

(d) What percent of the observations in the standard normal distribution are suspected outliers according to the $1.5 \times IQR$ criterion? (This percent is the same for any normal distribution.)

1.96 The lower and upper deciles of any distribution are the points that mark off the lowest 10% and the highest 10%. On a density curve, these are the points with area 0.1 and 0.9 to their left under the curve.

(a) What are the lower and upper deciles of the standard normal distribution?

(b) The length of human pregnancies is approximately normal with mean 266 days and standard deviation 16 days. What are the lower and upper deciles of this distribution?

1.97 Figure 1.37 is a normal quantile plot of the DRP scores from Exercise 1.23 (page 30). Are these scores approximately normally distributed? Discuss any major deviations from normality that appear in the plot.

1.98 Figure 1.38 is a normal quantile plot of the survival times of the guinea pigs in a medical experiment, from Table 1.5. Explain carefully how the strong right skewness of this distribution is seen in the plot.

1.99 The distance between two mounting holes is important to the performance of an electrical meter. The manufacturer measures this distance regularly for quality control purposes, recording the data as thousandths of an inch more than 0.600 inches. For example, 0.644 is recorded as 44. Figure 1.39 is a normal quantile plot of the distances for the last 27 electrical meters measured. Is the overall shape of the distribution approximately normal? Why does the plot have a "stairstep" appearance? (Data provided by Charles Hicks, Purdue University.)

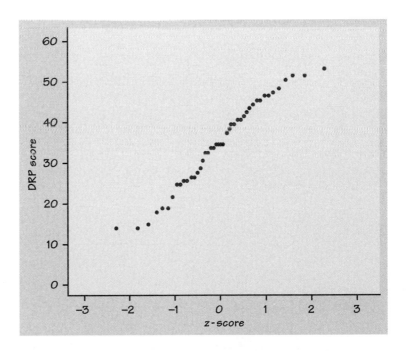

FIGURE 1.37 Normal probability plot of DRP scores, for Exercise 1.97.

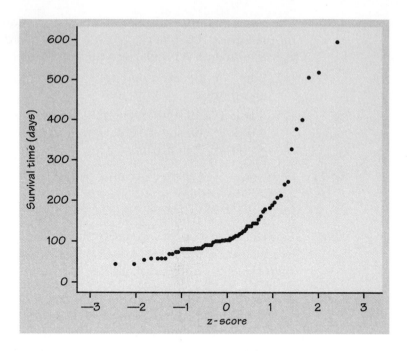

FIGURE 1.38   Normal probability plot of guinea pig survival times, for Exercise 1.98.

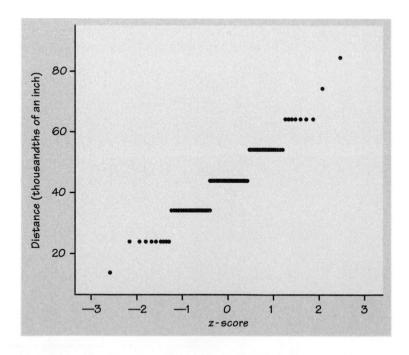

FIGURE 1.39   Normal probability plot of distance between mounting holes, for Exercise 1.99.

The remaining exercises for this section require the use of software that will make normal quantile plots.

1.100 We expect repeated careful measurements of the same quantity to be approximately normal. Make a normal quantile plot for Cavendish's measurements in Exercise 1.24 (page 30). Are the data approximately normal? If not, describe any clear deviations from normality.

1.101 Is the distribution of monthly returns on Philip Morris stock approximately normal with the exception of possible outliers? Make a normal quantile plot of the data in Table 1.4 and report your conclusions.

1.102 Table 1.6 reports the scores of 78 seventh-grade students on the Piers-Harris Children's Self-Concept Scale. Give a careful description of the distribution of self-concept scores using both graphs and numbers. Are there outliers? Is the distribution roughly normal if we ignore the outliers?

1.103 Table 1.6 gives the gender of each student, coded as F for females and M for males. Compare the distributions of self-concept for female and male students, using both graphs and numbers. Report your findings. Some people think that female students in mixed classes have lower self-concept than male students. Do you see any evidence that this is true for these students?

1.104 Use software to generate 100 observations from the standard normal distribution. Make a histogram of these observations. How does the shape of the histogram compare with a normal density curve? Make a normal quantile plot of the data. Does the plot suggest any important deviations from normality? (Repeating this exercise several times is a good way to become familiar with how normal quantile plots look when data actually are close to normal.)

1.105 Use software to generate 100 observations from the distribution described in Exercise 1.69. (The software will probably call this a "uniform distribution.") Make a histogram of these observations. How does the histogram compare with the density curve in Figure 1.35? Make a normal quantile plot of your data. According to this plot, how does the uniform distribution deviate from normality?

CHAPTER 1 EXERCISES

1.106 What type of graph or graphs would you plan to make in a study of each of the following issues?

(a) What makes of cars do students drive? How old are their cars?

(b) How many hours per week do students study? How does the number of study hours change during a semester?

(c) Which radio stations are most popular with students?

(d) When many students measure the concentration of the same solution for a chemistry course laboratory assignment, do their measurements follow a normal distribution?

1.107 The Canadian Province of Ontario carries out statistical studies to evaluate Canada's national health care system in the province. The bar graphs in Figure 1.40 come from a study of admissions and discharges from community hospitals in Ontario. They show the number of heart attack patients admitted and discharged on each day of the week during 1992 and 1993. (Based on Antoni Basinski, "Almost never on Sunday: implications of the patterns of admission and discharge for common conditions," Institute for Clinical Evaluative Sciences in Ontario, October 18, 1993.)

(a) Explain why you expect the number of patients admitted with heart attacks to be roughly the same for all days of the week. Do the data show that this is true?

(b) Describe how the distribution of the day on which patients are discharged from the hospital differs from that of the day on which they are admitted. What do you think explains the difference?

1.108 If two distributions have exactly the same mean and standard deviation, must their histograms have the same shape? If they have the same five-number summary, must their histograms have the same shape? Explain.

1.109 Voting patterns in the United States were affected by many social changes in the years between 1940 and 1980. Prior to the civil rights movement of

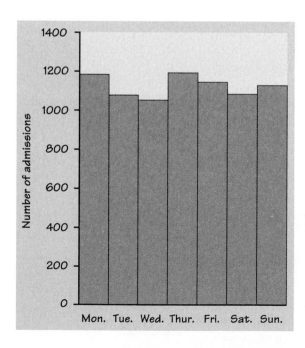
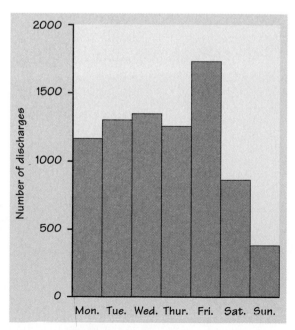

FIGURE 1.40 Bar graphs of the number of hospital patients admitted and discharged each day of the week in Ontario, Canada, for Exercise 1.107.

the 1960s, blacks were effectively prevented from voting in some southern states. Until 1971, citizens between the ages of 18 and 21 could not vote in national elections. Figure 1.41(a) is a histogram of the percent of the voting-age population who actually voted in the 1940 presidential election in each of the 48 states that existed at that time. Figure 1.41(b) shows the distribution of the percent of the voting-age population who voted in the 1980 presidential election in the 50 states and the District of Columbia. To allow a direct comparison, the horizontal and vertical scales and the class widths are the same for the two histograms. Describe the most important changes in the shape of the distribution between 1940 and 1980.

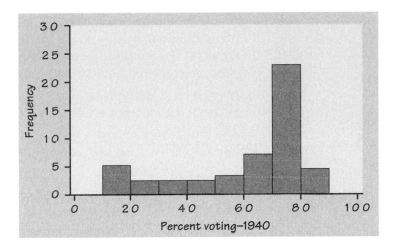

FIGURE 1.41(a)   Distribution of the percent of the voting-age population in each state who voted in the 1940 presidential election, for Exercise 1.109.

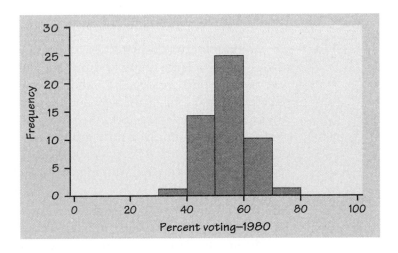

FIGURE 1.41(b)   Distribution of the percent of the voting-age population in each state who voted in the 1980 presidential election, for Exercise 1.109.

1.110 Deborah is a student at a midwestern college who lives off campus. She records the time she takes to drive to school each morning during the fall term. Here are the times (in minutes) for 42 consecutive weekdays, with the dates in order along the rows:

8.25	7.83	8.30	8.42	8.50	8.67	8.17	9.00	9.00	8.17	7.92
9.00	8.50	9.00	7.75	7.92	8.00	8.08	8.42	8.75	8.08	9.75
8.33	7.83	7.92	8.58	7.83	8.42	7.75	7.42	6.75	7.42	8.50
8.67	10.17	8.75	8.58	8.67	9.17	9.08	8.83	8.67		

(a) Make a graph of these drive times. Is the distribution roughly symmetric, clearly skewed, or neither? Are there any clear outliers?

(b) The data show three unusual situations: the day after Thanksgiving (no traffic on campus); a delay due to an accident; and a day with icy roads. Identify and remove these three observations, and give a numerical summary of the remaining 39 drive times.

(c) Are these data reasonably close to having a normal distribution when the unusual days are removed? How do you know?

(d) Make a time plot of the drive times. Do the data show any trend? For example, do drive times increase over time because the weather gets worse later in the term?

1.111 The 1996 *Statistical Abstract of the United States* reports FBI data on murders for 1994. In that year, 57.8% of all murders were committed with handguns, 12.2% with other firearms, 12.7% with knives, 5.3% with "personal weapons" (usually the hands or feet), and 4.1% with blunt objects. Make a graph to display these data. Do you need an "other methods" category?

1.112 The mean and median salaries paid to major league baseball players in 1993 were $490,000 and $1,160,000. Which of these numbers is the mean, and which is the median? Explain your answer.

1.113 Education researchers give a test designed to measure the reading ability of children to a large group of third- and sixth-grade children. The mean score for the third graders is 75 and the standard deviation is 10. For the sixth graders the mean is 82 and the standard deviation is 11. The researchers want to transform the scores to a scale that allows us to compare the standing of a third grader within her grade to the standing of a sixth grader within his grade. Although z-scores would work, the researchers prefer a scale with mean 100 and standard deviation 20 in each grade.

(a) What linear transformation will change third-grade scores x into new scores $x_{new} = a + bx$ that have the desired mean and standard deviation? (Use $b > 0$ to preserve the order of the scores.)

(b) Do the same for the sixth-grade scores.

(c) David is a third-grade student who scores 78 on the test. Find David's transformed score. Nancy is a sixth-grade student who scores 78. What is her transformed score? Who scores higher within his or her grade?

(d) Suppose that the distribution of scores in each grade is normal. Then both sets of transformed scores have the $N(100, 20)$ distribution. What percent of third graders have scores less than 78? What percent of sixth graders have scores less than 78?

1.114 The Chapin Social Insight Test evaluates how accurately the subject appraises other people. In the reference population used to develop the test, scores are approximately normally distributed with mean 25 and standard deviation 5. The range of possible scores is 0 to 41.

(a) What proportion of the population has scores below 20 on the Chapin test?

(b) What proportion has scores below 10?

(c) How high a score must you have in order to be in the top quarter of the population in social insight?

1.115 The scores of a reference population on the Wechsler Intelligence Scale for Children (WISC) are normally distributed with $\mu = 100$ and $\sigma = 15$. A school district classifies children as "gifted" if their WISC score exceeds 135. There are 1300 sixth graders in the school district. About how many of them are gifted?

1.116 The Chapin Social Insight Test described in Exercise 1.114 has a mean of 25 and a standard deviation of 5. You want to rescale the test using a linear transformation so that the mean is 100 and the standard deviation is 20. Let x denote the score in the original scale and x_{new} be the transformed score.

(a) Find the linear transformation required. That is, find the values of a and b in the equation $x_{\text{new}} = a + bx$.

(b) Give the rescaled score for someone who scores 30 in the original scale.

(c) What are the quartiles of the rescaled scores?

1.117 Explain why the normal quantile plot of the DRP reading scores (Figure 1.37 on page 91), suggests that the mean and standard deviation are satisfactory measures of center and spread for these data. Then calculate the mean and standard deviation from the scores given in Exercise 1.23 on page 30.

1.118 Exercise 1.22 (page 29) presents data on the weight gains of chicks fed two types of corn. The researchers use $\bar{x}$ and s to summarize each of the two distributions. Make a normal quantile plot for each group and report your findings. Is use of $\bar{x}$ and s justified?

1.119 Table 1.9 gives the weights in pounds and positions of the players on a major college football team. The positions are quarterback (QB), running back (RB), offensive line (OL), wide receiver (WR), tight end (TE), kicker (K), defensive back (DB), and defensive line (DL).

TABLE 1.9   Position and weight (pounds) for a major college football team

QB	205	QB	202	QB	210	QB	180	QB	190	RB	230
RB	190	RB	170	RB	190	RB	225	RB	207	RB	215
RB	227	RB	207	RB	225	OL	335	OL	284	OL	235
OL	295	OL	270	OL	295	OL	295	OL	270	OL	295
OL	295	OL	290	OL	300	OL	284	OL	275	OL	305
WR	155	WR	180	WR	188	WR	190	WR	202	WR	182
WR	188	WR	183	WR	170	WR	195	WR	180	WR	178
TE	256	TE	260	TE	230	TE	235	TE	240	TE	245
K	175	K	175	K	175	K	160	K	193	DB	180
DB	172	DB	190	DB	190	DB	193	DB	192	DB	170
DB	190	DB	193	DB	195	DB	180	DB	195	DB	176
DB	175	LB	205	LB	220	LB	225	LB	235	LB	237
LB	210	LB	220	LB	230	LB	215	LB	216	DL	240
DL	230	DL	245	DL	280	DL	250	DL	235	DL	235
DL	275	DL	245	DL	260	DL	250	DL	220	DL	265
DL	240	DL	245	DL	280	DL	285	DL	240		

(a) Make side-by-side boxplots of the weights for the eight player positions.

(b) Briefly compare the weight distributions. Which position has the heaviest players overall? Which has the lightest?

(c) Are any individual players outliers within their position?

1.120   Most statistical software packages have routines for generating values of variables having specified distributions. Use your statistical software to generate 25 observations from the $N(20, 5)$ distribution. Compute the mean and standard deviation $\bar{x}$ and s of the 25 values you obtain. How close are $\bar{x}$ and s to the μ and σ of the distribution from which the observations were drawn?

Repeat 20 times the process of generating 25 observations from the $N(20, 5)$ distribution and recording $\bar{x}$ and s. Make a stemplot of the 20 values of $\bar{x}$ and another stemplot of the 20 values of s. Make normal quantile plots of both sets of data. Briefly describe each of these distributions. Are they symmetric or skewed? Are they roughly normal? Where are their centers? (The distributions of measures like $\bar{x}$ and s when repeated sets of observations are made from the same theoretical distribution will be very important in later chapters.)

1.121   Table 1.10 shows the salaries paid to the members of the 1996 World Series champion New York Yankees baseball team (excluding bonuses). Display this distribution with a graph or graphs and appropriate numerical measures, then give a description of its main features. (Data released by Major League Baseball, reported in *USA Today*, November 15, 1996.)

TABLE 1.10   1996 salaries for the New York Yankees baseball team

Player	Salary	Player	Salary	Player	Salary
Cecil Fielder	$9,237,500	Charlie Hayes	$1,750,000	Darryl Strawberry	$560,000
Paul O'Neill	5,300,000	Tony Fernandez	1,500,000	Mike Aldrete	250,000
Kenny Rogers	5,000,000	Tim Raines	1,450,000	Graeme Lloyd	205,000
David Cone	4,666,666	Ricky Bones	1,425,000	Andy Pettitte	195,000
Melido Perez	4,650,000	Jim Leyritz	1,400,579	Mariano Rivera	131,125
John Wetteland	4,000,000	Scott Kamieniecki	1,100,000	Derek Jeter	130,000
Bernie Williams	3,000,000	Dwight Gooden	950,000	Wally Whitehurst	125,000
Joe Girardi	2,325,000	Pat Kelly	900,000	Dave Pavias	120,000
Tino Martinez	2,300,000	Jeff Nelson	860,000	Michael Figga	109,000
Pat Listach	2,200,000	Mariano Duncan	845,000	Dave Polley	109,000
Wade Boggs	2,050,000	Luis Sojo	625,000	Ruben Rivera	109,000
Jimmy Key	1,750,000				

1.122 At the time the salaries in Table 1.10 were announced, one dollar was worth 1.29 Swiss francs.

(a) Convert the Yankees' salaries into Swiss francs. (What arithmetic operation did you do?)

(b) Make a graph of the distribution in francs. How does this histogram compare with a graph of the salaries in dollars?

(c) Find the mean, median, and quartiles of the distribution in both dollars and francs. How are they related?

(d) Find the standard deviation and interquartile range of the distribution in both dollars and francs. How are these measures of spread related?

1.123 You are planning a sample survey of households in California. You decide to select households separately within each county and to choose more households from the more populous counties. To aid in the planning, Table 1.11 gives the Census Bureau estimates of the populations of California counties as of July 1, 1996. Examine the distribution of county populations both graphically and numerically, using whatever tools are most suitable. Then write a brief description of the main features of this distribution. Sample surveys often select households from all of the most populous counties but from only some of the less populous. How would you divide California counties into three groups according to population, with the intent of including all of the first group, half of the second, and a smaller fraction of the third in your survey?

1.124 The CHEESE data set described in the data appendix records measurements on 30 specimens of Australian cheddar cheese. Investigate the distributions of the variables H2S and LACTIC using graphical and

TABLE 1.11 Population of California counties

County	Population	County	Population	County	Population
Alameda	1,328,139	Marin	233,230	San Mateo	686,909
Alpine	1,232	Mariposa	15,869	Santa Barbara	385,573
Amador	33,315	Mendocino	83,298	Santa Clara	1,599,604
Butte	192,507	Merced	192,311	Santa Cruz	237,821
Calaveras	38,437	Modoc	9,693	Shasta	161,740
Colusa	18,223	Mono	10,497	Sierra	3,409
Contra Costa	881,490	Monterey	339,047	Siskiyou	44,193
Del Norte	26,947	Napa	116,512	Solano	365,536
El Dorado	151,706	Nevada	89,016	Sonoma	420,872
Fresno	751,272	Orange	2,636,888	Stanislaus	415,786
Glenn	26,202	Placer	213,227	Sutter	75,650
Humboldt	123,023	Plumas	20,597	Tehama	54,108
Imperial	142,651	Riverside	1,417,425	Trinity	13,418
Inyo	18,433	Sacramento	1,117,275	Tulare	349,922
Kern	622,729	San Benito	44,503	Tuolumne	52,196
Kings	113,351	San Bernardino	1,598,358	Ventura	714,733
Lake	55,261	San Diego	2,655,463	Yolo	149,925
Lassen	31,431	San Francisco	735,315	Yuba	60,905
Los Angeles	9,127,751	San Joaquin	533,392		
Madera	110,481	San Luis Obispo	229,437		

numerical summaries of your choice. Write a short description of the notable features for each distribution.

1.125 The CSDATA data set described in the data appendix contains information on 234 computer science students. We are interested in comparing the SAT mathematics scores and grade point averages of female students with those of male students. Make two sets of side-by-side boxplots to carry out these comparisons. Write a brief discussion of the male-female comparisons. Then make normal quantile plots of grade point averages and SAT math scores separately for men and women. Which of the four distributions are approximately normal?

NOTES

1. Data for 1995, collected by the Current Population Survey and reported in the 1996 *Statistical Abstract of the United States.* This annual publication is an essential source for many kinds of data.

2. Newcomb's data and the background information about his work appear in S. M. Stigler, "Do robust estimators work with real data?" *Annals of Statistics,* 5 (1977), pp. 1055–1078.

3. Data from *The Baseball Encyclopedia*, 3rd ed., Macmillan, New York, 1976. Maris's home run data are from the same source.

4. Data from Gary Community School Corporation, courtesy of Celeste Foster, Department of Education, Purdue University.

5. James Gleick, "Hole in ozone over South Pole worries scientists," *New York Times*, July 29, 1986. Note that the outliers, although omitted from the initial analysis, were not discarded. It was therefore possible to reanalyze the data later.

6. The data on which Figure 1.7 is based can be downloaded from the Consumer Price Index home page at the Bureau of Labor Statistics website, http://stats.bls.gov/cpihome.

7. Because there are only a few children who are exactly 24 months old in any one month, a fairly complicated statistical procedure has been used to obtain the table entries from the data for all children between 18 and 30 months of age. Changes in these standardized weights and heights indicate slower or faster average growth in the entire group of children.

8. We thank W. Robert Stephenson of Iowa State University for pointing out the importance of weight in explaining the distribution of calories.

9. The returns in Example 1.19 are calculated from data on securities prices compiled by Global Financial Data and made available at http://www.globalfindata.com.

10. Detailed data appear in P. S. Levy et al., "Total serum cholesterol values for youths 12–17 years," *Vital and Health Statistics Series 11*, No. 150 (1975), U.S. National Center for Health Statistics.

11. The density curve in Figure 1.34 was produced by a kernel density estimator. See B. W. Silverman, *Density Estimation for Statistics and Data Analysis*, Chapman and Hall, London, 1986.

Prelude

Statistical investigations only rarely focus on a single variable. We are more often interested in comparisons among several distributions, changes in a variable over time, or relationships among several variables. Chapter 1 provided graphical and numerical tools for examining the distribution of a single variable, for comparing several distributions, and for investigating change over time. In this chapter we will concentrate on methods for studying relationships among several variables.

Once again we begin with graphs and then move to numerical summaries of specific aspects of the data. Although we will meet a variety of numerical summaries in this chapter, we emphasize those that describe straight-line relations between two variables. These methods, called correlation and regression, are among the most common statistical tools. We will study data on issues such as these:

- Some people compare state school performance by comparing the mean scores of students taking the SAT exam in various states. Does this make sense?

- A report on investing in mutual funds says that the largest fund (Fidelity Magellan) has correlation 0.85 with the overall stock market. What does this tell us about the Magellan fund?

- The Florida manatee is endangered by careless users of powerboats. Is there a close connection between the number of boats and the number of manatees that they kill?

- Does taking the drug Bendectin to relieve nausea during pregnancy cause birth defects?

Guy Dill, *Pram*, 1996

Looking at Data— Relationships

Introduction

A medical study finds that short women are more likely to have heart attacks than women of average height, while tall women have fewer heart attacks. An insurance group reports that heavier cars have fewer deaths per 10,000 vehicles registered than do lighter cars. These and many other statistical studies look at the relationship between two variables. To understand such a relationship, we must often examine other variables as well. To conclude that shorter women have higher risk from heart attacks, for example, the researchers had to eliminate the effect of other variables such as weight and exercise habits. Our topic in this chapter is relationships between variables. One of our main themes is that the relationship between two variables can be strongly influenced by other variables that are lurking in the background.

To study the relationship between two variables, we measure both variables on the same individuals. If we measure both the height and the weight of each of a large group of people, we know which height goes with each weight. These data allow us to study the connection between height and weight. A list of the heights and a separate list of the weights, two sets of single-variable data, do not show the connection between the two variables. In fact, taller people also tend to be heavier. And people who smoke more cigarettes per day tend not to live as long as those who smoke fewer. We say that pairs of variables such as height and weight or smoking and life expectancy are *associated*.

Association Between Variables

Two variables measured on the same individuals are **associated** if some values of one variable tend to occur more often with some values of the second variable than with other values of that variable.

Statistical associations are overall tendencies, not ironclad rules. They allow individual exceptions. Although smokers on the average die earlier than nonsmokers, some people live to 90 while smoking three packs a day.

Getting started

When you examine the relationship between two or more variables, first ask the preliminary questions that are familiar from Chapter 1:

- What *individuals* do the data describe?
- What *variables* are present? How are they measured?
- Which variables are *quantitative* and which are *categorical*?

A medical study, for example, may record each subject's sex (female, male) and smoking status (current smoker, former smoker, never smoked) along with quantitative variables such as weight and blood pressure. We may be interested in possible associations between two quantitative variables (such as a person's weight and blood pressure), between a quantitative and a categorical variable (such as sex and blood pressure), or between two categorical variables (such as sex and smoking status).

When you examine the relationship between two variables, a new question becomes important:

■ Is your purpose simply to explore the nature of the relationship, or do you hope to show that one of the variables can explain variation in the other? That is, are some of the variables *response variables* and others *explanatory variables*?

> ### Response Variable, Explanatory Variable
>
> A **response variable** measures an outcome of a study. An **explanatory variable** explains or causes changes in the response variables.

It is easiest to identify explanatory and response variables when we actually set values of one variable in order to see how it affects another variable.

EXAMPLE 2.1

Alcohol has many effects on the body. One effect is a drop in body temperature. To study this effect, researchers give several different amounts of alcohol to mice, then measure the change in each mouse's body temperature in the 15 minutes after taking the alcohol. Amount of alcohol is the explanatory variable, and change in body temperature is the response variable.

When you don't set the values of either variable but just observe both variables, there may or may not be explanatory and response variables. Whether there are depends on how you plan to use the data.

EXAMPLE 2.2

A political scientist looks at the Democrats' share of the popular vote for president in each state in the two most recent presidential elections. She does not wish to explain one year's data by the other's, but rather to see a pattern that may shed light on political conditions. The political scientist has two related variables, but she does not use the explanatory-response distinction.

A political consultant looks at the same data with an eye to including the previous election result in a state as one of several explanatory variables in a model to predict how the state will vote in the next election. Now the earlier election result is an explanatory variable and the later result is a response variable.

In many studies, the goal is to show that changes in one or more explanatory variables actually *cause* changes in a response variable. But not all explanatory-response relationships involve direct causation. The SAT scores of high school students may help predict the students' future college grades, but high SAT scores certainly don't cause high college grades.

independent variable
dependent variable

Some of the statistical techniques in this chapter require us to distinguish explanatory from response variables; others make no use of this distinction. Explanatory variables are often called **independent variables,** and response variables are often referred to as **dependent variables.** The idea behind this language is that response variables depend on explanatory variables. Because the words "independent" and "dependent" have other meanings in statistics that are unrelated to the explanatory-response distinction, we prefer to avoid those words.

Most statistical studies examine data on more than one variable. Fortunately, statistical analysis of several-variable data builds on the tools used for examining individual variables. The principles that guide our work also remain the same:

- Start with graphical display, then add numerical summaries.
- Look for overall patterns and deviations from those patterns.
- When the overall pattern is quite regular, use a compact mathematical model to describe it.

2.1   Scatterplots

Relationships between two quantitative variables are best displayed graphically. The most useful graph for this purpose is a *scatterplot.*

Scatterplot

A **scatterplot** shows the relationship between two quantitative variables measured on the same individuals. The values of one variable appear on the horizontal axis, and the values of the other variable appear on the vertical axis. Each individual in the data appears as the point in the plot fixed by the values of both variables for that individual.

Always plot the explanatory variable, if there is one, on the horizontal axis (the x axis) of a scatterplot. As a reminder, we usually call the explanatory variable x and the response variable y. If there is no explanatory-response distinction, either variable can go on the horizontal axis.

EXAMPLE 2.3

Over a million American students take the SAT college entrance examination each year. The states differ greatly in the mean scores their students achieve on the SAT. A histogram of the SAT verbal scores (Figure 1.12 on page 28) shows that there are two modes, corresponding to two clusters of states. We suspect that the clusters reflect the percent of each state's students who take the SATs. Some states favor the ACT college entrance examination, so that only students applying to selective colleges take the SAT. In other states, most college-bound students take the SAT. The mean scores are lower in these states because a less selective group of students take the test.

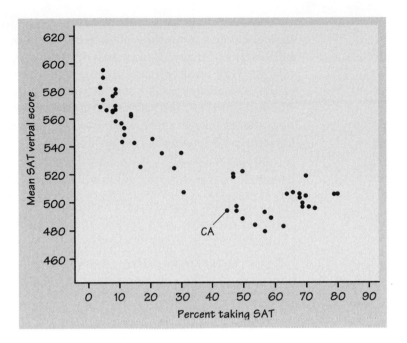

FIGURE 2.1 State mean SAT verbal scores plotted against the percent of high school graduates in each state who take the SAT exams.

A scatterplot allows us to see how state SAT performance is related to the percent of students who take the SAT. Figure 2.1 is a scatterplot of state mean SAT verbal score against the percent who take the exam. Each point on the plot represents a single individual—that is, a single state. Because we think that the percent taking the exam influences mean score, percent taking is the explanatory variable and we plot it horizontally. For example, 45% of California high school graduates take the SAT, and their mean verbal score is 495. California appears as the point (45, 495) in the scatterplot, above 45 on the *x* axis and to the right of 495 on the *y* axis. We have marked this point in Figure 2.1.

Interpreting scatterplots

To interpret a scatterplot, apply the strategies of exploratory analysis learned in Chapter 1.

Examining a Scatterplot

In any graph of data, look for the **overall pattern** and for striking **deviations** from that pattern.

You can describe the overall pattern of a scatterplot by the **form, direction, and strength** of the relationship.

An important kind of deviation is an **outlier,** an individual value that falls outside the overall pattern of the relationship.

---

clusters

Figure 2.1 shows a clear *form:* there are two distinct **clusters** of states. In one cluster, more than 40% of high school seniors take the SAT, and the mean scores are low. Fewer than 30% of seniors in states in the other cluster take the SAT, and these states have higher mean scores. Call the states in the under-30% cluster "ACT states" and those in the over-40% cluster "SAT states." The relationship also has a clear *direction:* states in which a higher percent of students take the SAT tend to have lower mean scores. This is a *negative association* between the two variables.

Positive Association, Negative Association

Two variables are **positively associated** when above-average values of one tend to accompany above-average values of the other and below-average values also tend to occur together.

Two variables are **negatively associated** when above-average values of one accompany below-average values of the other, and vice versa.

---

linear relationship

When a scatterplot shows distinct clusters, it is often useful to describe the overall pattern separately within each cluster. The *form* of the relationship in the ACT states is roughly **linear.** That is, the points roughly follow a straight line. The *strength* of a relationship in a scatterplot is determined by how closely the points follow a clear form. The linear pattern among the ACT states is moderately strong because the points show only modest scatter about the straight-line pattern. In summary, the ACT states in Figure 2.1 show a moderately strong negative linear relationship. The cluster of SAT states shows very little relationship between percent taking the SAT and mean SAT verbal score. Neither the form nor the direction within the cluster is clear.

Figure 2.1 does not show any striking outliers—none of the points lies clearly outside the two clusters that form the overall pattern, and none of the ACT states lies clearly outside the linear pattern of that cluster.

Adding categorical variables to scatterplots

The Census Bureau groups the states into four broad regions, named Midwest, Northeast, South, and West. We might ask about regional patterns in the SAT exam. Figure 2.2 repeats part of Figure 2.1, with an important addition. We have plotted only the Northeast and Midwest groups of states, using the plot symbol "e" for the northeastern states and the symbol "m" for the midwestern states.

The regional comparison is striking. The nine northeastern states are all SAT states—in fact, at least 68% of high school graduates in each of these states take the SAT. The twelve midwestern states are mostly ACT states. In ten of these states, the percent taking the SAT is between 5% and 14%. One

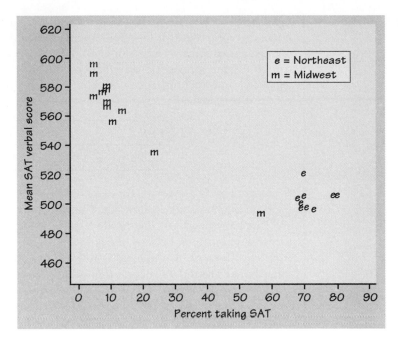

FIGURE 2.2 State mean SAT verbal scores and percent taking the SAT for the northeastern states (plot symbol "e") and the midwestern states (plot symbol "m").

midwestern state is clearly an outlier within the region. Indiana is an SAT state (57% take the SAT) that falls close to the northeastern cluster. Ohio, where 24% take the SAT, also lies slightly out of the midwestern cluster.

In dividing the states into regions, we introduced a third variable into the scatterplot. "Region" is a categorical variable that has four values, although we plotted data from only two of the four regions. The two regions are displayed by the two different plotting symbols.[1]

Categorical Variables in Scatterplots

To add a categorical variable to a scatterplot, use a different plot color or symbol for each category.

More examples of scatterplots

Experience in examining scatterplots is the foundation for more detailed study of relationships among quantitative variables. Here is an example with a stronger pattern than that in Figure 2.1.

EXAMPLE 2.4

Archaeopteryx is an extinct beast having feathers like a bird but teeth and a long bony tail like a reptile. Only six fossil specimens are known. Because these specimens differ greatly in size, some scientists think they are different species rather than individuals from the same species. Here are data on the lengths in centimeters of the femur (a leg bone) and the humerus (a bone in the upper arm) for the five specimens that preserve both bones:[2]

Femur	38	56	59	64	74
Humerus	41	63	70	72	84

Because there is no explanatory-response distinction, we can put either measurement on the x axis of a scatterplot. The plot appears in Figure 2.3.

The plot shows a strong positive linear association. The form is linear and the association is strong because the points lie close to a straight line. The association is positive because as the length of one bone increases, so does the length of the other bone. These data suggest that all five specimens belong to the same species and differ in size because some are younger than others. We expect that a specimen from a different species would have a different relationship between the lengths of the two bones, so that it would appear as an outlier.

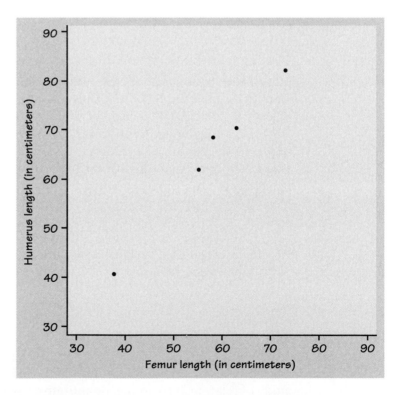

FIGURE 2.3   The lengths of two bones in the five surviving fossil specimens of the extinct beast *Archaeopteryx*.

Some scatterplots appear quite different from the cloud of points in Figure 2.1 and the strong linear pattern in Figure 2.3. This is particularly true in experiments in which measurements of a response variable are taken at a few selected levels of the explanatory variable. The following example illustrates the use of scatterplots in this setting.

EXAMPLE 2.5

How much corn per acre should a farmer plant to obtain the highest yield? Too few plants will produce a low yield. On the other hand, if there are too many plants, they will compete with each other for moisture and nutrients, and yields will fall. Table 2.1 shows the results of several years of field experiments in Nebraska.[3] Each entry is the mean yield of four small plots planted at the same rate per acre. All plots were irrigated, fertilized, and cultivated identically. The yield of each plot should therefore depend only on the planting rate—and of course on the uncontrolled aspects of each growing season, such as temperature and wind. The experiment lasted several years in order to avoid misleading conclusions due to the peculiarities of a single growing season.

The scatterplot in Figure 2.4 displays the results of this experiment. Because planting rate is the explanatory variable, we plot it horizontally as the x variable. The yield is the response variable y. The vertical spread of points over each planting rate shows the year-to-year variation in yield. We show the overall pattern by plotting the mean yield for each planting rate (averaged over all years). These means are marked by triangles and joined by line segments. As expected, the form of the relationship is not linear. Yields first increase with the planting rate, then decrease when too many plants are crowded in. Because there is no consistent direction, we cannot describe the association as either positive or negative. It appears that the best choice is about 20,000 plants per acre.

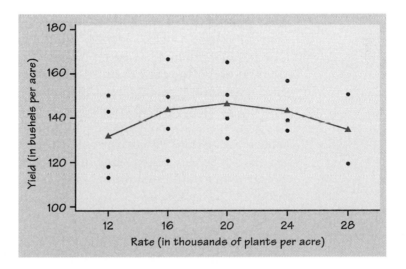

FIGURE 2.4 Yield of corn plotted against planting rate, from an agriculture experiment.

Plants per acre	1956	1958	1959	1960	Mean
12,000	150.1	113.0	118.4	142.6	131.0
16,000	166.9	120.7	135.2	149.8	143.2
20,000	165.3	130.1	139.6	149.9	146.2
24,000		134.7	138.4	156.1	143.1
28,000			119.0	150.5	134.8
Mean	160.8	124.6	130.1	149.8	

TABLE 2.1   Corn yields (bushels per acre) in an agricultural experiment

Plotting the means as in Figure 2.4 helps us see the overall pattern of results when responses are measured at a few fixed levels of an explanatory variable such as planting rate. The scatterplot also displays deviations from the pattern by showing how much variability lies behind each mean.

In this case, however, we must regard the means with caution. As the gaps in Table 2.1 show, the agronomists used only the three lowest planting rates in all four years. The means therefore cover different spans of time. In particular, 1956 was a year of very high yields, so the means that include 1956 are biased upward relative to those that do not. A closer look at Table 2.1 shows that 24,000 plants per acre was superior to 20,000 in two of the three years in which both rates were planted. The agronomists in fact concluded that yields continue to increase somewhat up to 24,000 plants per acre.

Incomplete data as in Example 2.5 are common in practice. The data would be more convincing if all five planting rates had been used in all four years—but it was only after the first results were in that the researchers realized that higher planting rates might give better results. Incomplete data often complicate a statistical analysis.

Example 2.5 illustrates a very important fact: **the relationship between two variables often cannot be fully understood without knowledge about other variables**. Figure 2.4 plots corn yield against planting rate, but the experiment spanned several growing seasons that differed from each other. If the agronomists had not carefully controlled many other variables (moisture, fertilizer, etc.), these variables would have confused the situation completely. You should be cautious in drawing conclusions from a strong relationship appearing in a scatterplot until you understand what other variables may be lurking in the background.

Beyond the basics ▶ scatterplot smoothers

A scatterplot provides a complete picture of the relationship between two quantitative variables. A complete picture is often too detailed for easy interpretation, so we try to describe the plot in terms of an overall pattern and deviations from that pattern. Though we can often do this by eye, more sys-

smoothing

tematic methods of extracting the overall pattern are helpful. This is called **smoothing** a scatterplot. Example 2.5 suggests how to proceed when we are plotting a response variable y against an explanatory variable x. We smoothed Figure 2.4 by averaging the y values separately for each x value. Though not all scatterplots have many y values at the same value of x, as did Figure 2.4, modern software provides scatterplot smoothers that form an overall pattern by looking at the y values for points in the neighborhood of each x value. Smoothers use *resistant* calculations, so they are not affected by outliers in the plot.

EXAMPLE 2.6

Crash a motorcycle into a wall. The rider, fortunately, is a dummy with an instrument to measure acceleration (change of velocity) mounted in its head. Figure 2.5 plots the acceleration of the dummy's head against time in milliseconds. Acceleration is measured in g's, or multiples of the acceleration due to gravity at the earth's surface. The motorcycle approaches the wall at a constant speed (acceleration near 0). As it hits, the dummy's head snaps forward and decelerates violently (negative acceleration reaching more than 100 g's), then snaps back again (up to 75 g's) and wobbles a bit before coming to rest.[4]

The scatterplot has a clear overall pattern, but it does not obey a simple form such as linear. Moreover, the strength of the pattern varies, from quite strong at the left of the plot to weaker (much more scatter) at the right. A scatterplot smoother deals with this complexity quite effectively and draws a line on the plot to represent the overall pattern.

EXAMPLE 2.7

During the Vietnam War, a random selection process—a draft lottery—was used to decide which young men would be drafted into the army. The first draft lottery was held in 1970. The 366 possible birth dates, from January 1 to December 31, were placed into identical plastic capsules, poured into a rotating drum, and picked out one by one. The first birth date drawn won draft number 1, the next 2, and so on. Men were drafted in order of their draft numbers, those with the lowest numbers first.

The outcome of the 1970 draft lottery appears in the scatterplot in Figure 2.6. Birth dates, numbered 1 to 366 beginning with January 1, are plotted on the x axis. Draft numbers are on the y axis, so that each point shows the draft number assigned to one birth date. A properly conducted lottery should produce no systematic relationship between these variables. It is hard to see any overall pattern in Figure 2.6. Yet it was charged that the lottery was biased against men born late in the year, that these men received systematically lower draft numbers than men born earlier.

A scatterplot smoother produces the curve in Figure 2.6 as the overall pattern of the plot. Sure enough, there is a clear negative association between birth date and draft number, with men born at the end of the year receiving particularly low draft numbers. An investigation showed that birth dates had been inserted into capsules and poured into a box one month at a time before being placed into the drum. The capsules for the later months were put in last and remained near the top because of inadequate mixing.[5]

Scatterplot smoothers help us see overall patterns, especially when these patterns are not simple. They can also, as in Example 2.7, reveal relationships that are not obvious from a scatterplot alone. The combination of graphing and calculating once again proves its effectiveness.

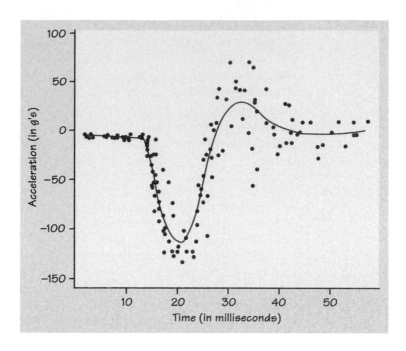

FIGURE 2.5   Time plot of the acceleration of the head of a crash test dummy as a motorcycle hits a wall, with the overall pattern calculated by a scatterplot smoother.

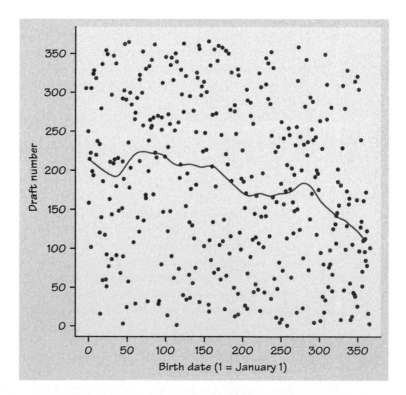

FIGURE 2.6   The 1970 draft lottery: draft selection number against birth date, with the overall pattern calculated by a scatterplot smoother.

Categorical explanatory variables

Scatterplots display the association between two quantitative variables. To display a relationship between a categorical explanatory variable and a quantitative response variable, make a side-by-side comparison of the distributions of the response for each category. We have already met some tools for such comparisons:

- A back-to-back stemplot compares two distributions. See the comparison of home run counts (the quantitative response) hit by Babe Ruth and Roger Maris (the categories) on page 11.
- Side-by-side boxplots compare any number of distributions. See the comparison of calories (the quantitative response) for beef, meat, and poultry hot dogs (the categories) in Figure 1.16 on page 49.

You can sometimes use a type of scatterplot to display the association between a categorical explanatory variable and a quantitative response. Suppose, for example, that the agronomists of Example 2.5 had compared the yields of five varieties of corn rather than five planting rates. The plot in Figure 2.4 remains helpful if the varieties A, B, C, D, and E are marked at equal intervals on the horizontal axis in place of the planting rate. In particular, a graph of the mean (or median) response for each category will show the overall nature of the relationship.

Many categorical variables, like corn variety or type of hot dog, have no natural order from smallest to largest. In such situations we cannot speak of a positive or negative association with the response variable. If the mean responses in our plot increase as we go from right to left, we could make them decrease by writing the categories in the opposite order. The plot simply presents a side-by-side comparison of several distributions. The categorical variable labels the distributions. Some categorical variables do have a least-to-most order, however. We can then speak of the direction of the association between the categorical explanatory variable and the quantitative response. Here is an example.

EXAMPLE 2.8

We suspect that there is a positive association between how much a person works and his or her earnings. The Census Bureau publishes relevant data, but "how much a person works" appears as a categorical variable with these values:

A = 26 weeks or less
B = 27 to 49 weeks
C = 50 weeks or more

The categories A to C are ordered from least work to most work. The side-by-side boxplots of Figure 2.7 show the distributions of annual earnings for men who worked at full-time jobs in 1995. We arrange the boxplots in order of amount worked to show the association. The lines in the plots extend out to the 5th and 95th percentiles of the distributions rather than to the single smallest and largest values. This variation on the boxplot idea is common when we have very many observations—more than 52 million men in this example.

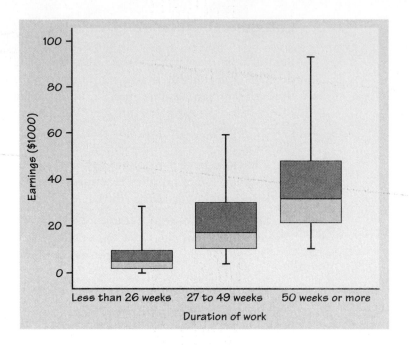

FIGURE 2.7   Boxplots for the earning of male full-time workers in three categories.

Figure 2.7 makes clear the strong positive association between work and earnings. The third quartile of earnings for men who held full-time jobs less than half the year lies below the first quartile for those who worked between 27 and 49 weeks, and below the 5% point for those who worked all year. The plot also shows that the earnings of men who work many weeks are much more spread out than the earnings of those who work few weeks.

SUMMARY

To study relationships between variables, we must measure the variables on the same group of individuals.

If we think that a variable x may explain or even cause changes in another variable y, we call x an **explanatory variable** and y a **response variable.**

A **scatterplot** displays the relationship between two quantitative variables. Mark values of one variable on the horizontal axis (x axis) and values of the other variable on the vertical axis (y axis). Plot each individual's data as a point on the graph.

Always plot the explanatory variable, if there is one, on the x axis of a scatterplot. Plot the response variable on the y axis.

Plot points with different colors or symbols to see the effect of a categorical variable in a scatterplot.

In examining a scatterplot, look for an overall pattern showing the **form, direction,** and **strength** of the relationship, and then for **outliers** or other deviations from this pattern.

Form: Linear relationships, where the points show a straight-line pattern, are an important form of relationship between two variables. Curved relationships and **clusters** are other forms to watch for.

Direction: If the relationship has a clear direction, we speak of either **positive association** (high values of the two variables tend to occur together) or **negative association** (high values of one variable tend to occur with low values of the other variable).

Strength: The **strength** of a relationship is determined by how close the points in the scatterplot lie to a simple form such as a line.

To display the relationship between a categorical explanatory variable and a quantitative response variable, make a graph that compares the distributions of the response for each category of the explanatory variable.

SECTION 2.1 EXERCISES

2.1 In each of the following situations, is it more reasonable to simply explore the relationship between the two variables or to view one of the variables as an explanatory variable and the other as a response variable? In the latter case, which is the explanatory variable and which is the response variable?

  (a) The amount of time spent studying for a statistics exam and the grade on the exam

  (b) The weight and height of a person

  (c) The amount of yearly rainfall and the yield of a crop

  (d) A student's scores on the SAT math exam and the SAT verbal exam

  (e) The occupational class of a father and of a son

2.2 How well does a child's height at age 6 predict height at age 16? To find out, measure the heights of a large group of children at age 6, wait until they reach age 16, then measure their heights again. What are the explanatory and response variables here? Are these variables categorical or quantitative?

2.3 Vehicle manufacturers are required to test their vehicles for the amount of each of several pollutants in the exhaust. Even among identical vehicles the amount of pollutant varies, so several vehicles must be tested. Figure 2.8 plots the amounts of two pollutants, carbon monoxide and nitrogen oxides, for 46 identical vehicles. Both variables are measured in grams of the pollutant per mile driven. (Data from T. J. Lorenzen, "Determining statistical characteristics of a vehicle emissions audit procedure," *Technometrics,* 22 (1980), pp. 483–493.)

  (a) Describe the nature of the relationship. Is the association positive or negative? Is it nearly linear or clearly curved? Are there any outliers?

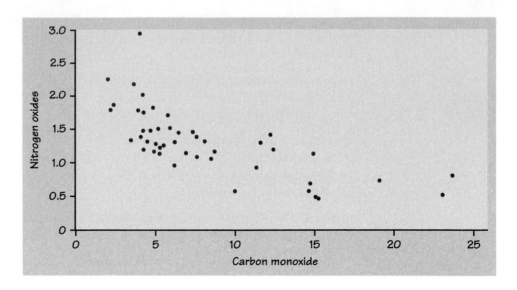

FIGURE 2.8   Nitrogen oxides versus carbon monoxide in the exhaust of 46 vehicles, for Exercise 2.3.

(b) A writer on automobiles says, "When an engine is properly built and properly tuned, it emits few pollutants. If the engine is out of tune, it emits more of all important pollutants. You can find out how badly a vehicle is polluting the air by measuring any one pollutant. If that value is acceptable, the other emissions will also be OK." Do the data in Figure 2.8 support this claim?

2.4   Figure 2.9 plots the city and highway fuel consumption of 1997 model midsize cars, from the Environmental Protection Agency's *Model Year 1997 Fuel Economy Guide*.

(a) There is one outlier, the Rolls-Royce Silver Spur. What are the city and highway gas mileages for this car?

(b) Describe the form, direction, and strength of the relationship. Explain why you might expect a relationship with this direction and strength.

2.5   There is some evidence that drinking moderate amounts of wine helps prevent heart attacks. Table 2.2 gives data on yearly wine consumption (liters of alcohol from drinking wine, per person) and yearly deaths from heart disease (deaths per 100,000 people) in 19 developed nations. (Data from M. H. Criqui, University of California, San Diego, reported in the *New York Times*, December 28, 1994.)

(a) Make a scatterplot that shows how national wine consumption helps explain heart disease death rates.

(b) Describe the form of the relationship. Is there a linear pattern? How strong is the relationship?

(c) Is the direction of the association positive or negative? Explain in simple language what this says about wine and heart disease. Do

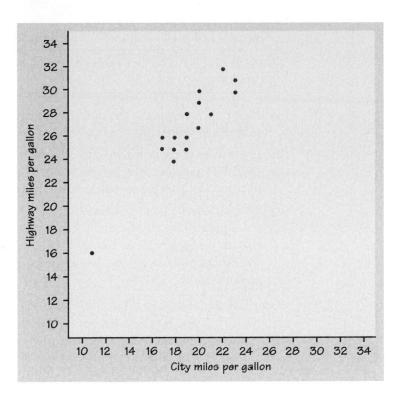

FIGURE 2.9 City and highway fuel consumption for 1997 model midsize car, for exercise 2.4

TABLE 2.2   Wine consumption and heart attacks

Country	Alcohol from wine	Heart disease deaths	Country	Alcohol from wine	Heart disease deaths
Australia	2.5	211	Netherlands	1.8	167
Austria	3.9	167	New Zealand	1.9	266
Belgium	2.9	131	Norway	0.8	227
Canada	2.4	191	Spain	6.5	86
Denmark	2.9	220	Sweden	1.6	207
Finland	0.8	297	Switzerland	5.8	115
France	9.1	71	United Kingdom	1.3	285
Iceland	0.8	211	United States	1.2	199
Ireland	0.7	300	West Germany	2.7	172
Italy	7.9	107			

you think these data give good evidence that drinking wine *causes* a reduction in heart disease deaths? Why?

2.6   Here are the golf scores of 12 members of a college women's golf team in two rounds of tournament play. (A golf score is the number of strokes required to complete the course, so low scores are better.)

Player	1	2	3	4	5	6	7	8	9	10	11	12
Round 1	89	90	87	95	86	81	102	105	83	88	91	79
Round 2	94	85	89	89	81	76	107	89	87	91	88	80

(a) Make a scatterplot of the data, taking the first-round score as the explanatory variable.

(b) Is there an association between the two scores? If so, is it positive or negative? Explain why you would expect scores in two rounds of a tournament to have an association like that you observed.

(c) The plot shows one outlier. Circle it. The outlier may occur because a good golfer had an unusually bad round or because a weaker golfer had an unusually good round. Can you tell from the data given whether the outlier is from a good player or from a poor player? Explain your answer.

2.7   Water flowing across farmland washes away soil. Researchers released water across a test bed at different flow rates and measured the amount of soil washed away. The following table gives the flow (in liters per second) and the weight (in kilograms) of eroded soil. (Data from G. R. Foster, W. R. Ostercamp, and L. J. Lane, "Effect of discharge rate on rill erosion," paper presented at the 1982 Winter Meeting of the American Society of Agricultural Engineers.)

Flow rate	0.31	0.85	1.26	2.47	3.75
Eroded soil	0.82	1.95	2.18	3.01	6.07

(a) Plot the data. Which is the explanatory variable?

(b) Describe the pattern that you see. Would it be reasonable to describe the overall pattern by a straight line? Is the association positive or negative?

2.8   How does the fuel consumption of a car change as its speed increases? Here are data for a British Ford Escort. Speed is measured in kilometers per hour, and fuel consumption is measured in liters of gasoline used per 100 kilometers traveled.[6]

Speed (km/h)	Fuel used (liters/100 km)	Speed (km/h)	Fuel used (liters/100 km)	Speed (km/h)	Fuel used (liters/100 km)
10	21.00	60	5.90	110	9.03
20	13.00	70	6.30	120	9.87
30	10.00	80	6.95	130	10.79
40	8.00	90	7.57	140	11.77
50	7.00	100	8.27	150	12.83

(a) Make a scatterplot. (Which variable should go on the x axis?)

(b) Describe the form of the relationship. In what way is it not linear? Explain why the form of the relationship makes sense.

(c) It does not make sense to describe the variables as either positively associated or negatively associated. Why not?

(d) Is the relationship reasonably strong or quite weak? Explain your answer.

2.9 In 1974, the Franklin National Bank failed. Franklin was one of the 20 largest banks in the nation, and the largest ever to fail. Could Franklin's weakened condition have been detected in advance by simple data analysis? The table below gives the total assets (in billions of dollars) and net income (in millions of dollars) for the 20 largest banks in 1973, the year before Franklin failed. Franklin is bank number 19. (Data from D. E. Booth, *Regression Methods and Problem Banks,* COMAP, Lexington, Mass., 1986.)

Bank	1	2	3	4	5	6	7	8	9	10
Assets	49.0	42.3	36.3	16.4	14.9	14.2	13.5	13.4	13.2	11.8
Income	218.8	265.6	170.9	85.9	88.1	63.6	96.9	60.9	144.2	53.6

Bank	11	12	13	14	15	16	17	18	19	20
Assets	11.6	9.5	9.4	7.5	7.2	6.7	6.0	4.6	3.8	3.4
Income	42.9	32.4	68.3	48.6	32.2	42.7	28.9	40.7	13.8	22.2

(a) We expect banks with more assets to earn higher income. Make a scatterplot of these data that displays the relation between assets and income. Mark Franklin (Bank 19) with a separate symbol.

(b) Describe the overall pattern of your plot. Are there any banks with unusually high or low income relative to their assets? Does Franklin stand out from other banks in your plot?

2.10 Metabolic rate, the rate at which the body consumes energy, is important in studies of weight gain, dieting, and exercise. The table below gives data on the lean body mass and resting metabolic rate for 12 women and 7 men who are subjects in a study of dieting. Lean body mass, given in kilograms, is a person's weight leaving out all fat. Metabolic rate is measured in calories burned per 24 hours, the same calories used to describe the energy content of foods. The researchers believe that lean body mass is an important influence on metabolic rate.

Subject	Sex	Mass	Rate	Subject	Sex	Mass	Rate
1	M	62.0	1792	11	F	40.3	1189
2	M	62.9	1666	12	F	33.1	913
3	F	36.1	995	13	M	51.9	1460
4	F	54.6	1425	14	F	42.4	1124
5	F	48.5	1396	15	F	34.5	1052
6	F	42.0	1418	16	F	51.1	1347
7	M	47.4	1362	17	F	41.2	1204
8	F	50.6	1502	18	M	51.9	1867
9	F	42.0	1256	19	M	46.9	1439
10	M	48.7	1614				

(a) Make a scatterplot of the data, using different symbols or colors for men and women.

(b) Is the association between these variables positive or negative? What is the form of the relationship? How strong is the relationship? Does the pattern of the relationship differ for women and men? How do the male subjects as a group differ from the female subjects as a group?

2.11   Here are data on a group of people who contracted botulism, a form of food poisoning that can be fatal. The variables recorded are the person's age in years, the incubation period (the time in hours between eating the infected food and the first signs of illness), and whether the person survived (S) or died (D). (Modified from data provided by Dana Quade, University of North Carolina.)

Person	1	2	3	4	5	6	7	8	9
Age	29	39	44	37	42	17	38	43	51
Incubation	13	46	43	34	20	20	18	72	19
Outcome	D	S	S	D	D	S	D	S	D

Person	10	11	12	13	14	15	16	17	18
Age	30	32	59	33	31	32	32	36	50
Incubation	36	48	44	21	32	86	48	28	16
Outcome	D	D	S	D	D	S	D	S	D

(a) Make a scatterplot of incubation period against age, using different symbols for people who survived and those who died.

(b) Is there an overall relationship between age and incubation period? If so, describe it.

(c) More important, is there a relationship between either age or incubation period and whether the victim survived? Describe any relations that seem important here.

(d) Are there any unusual observations that may require individual investigation?

2.12   Table 1.6 (page 33) reports data on 78 seventh-grade students. We wonder if IQ score helps explain grade point average (GPA). Make a scatterplot and describe the form, direction, and strength of the relationship between IQ and GPA. There are some unusual observations, in particular a student with IQ above 100 but very low GPA and another student with very low IQ but a respectable GPA. Write the observation numbers for these students next to the points on your plot. What are their genders?

2.13   Table 2.3 gives information about all single-family houses sold in a neighborhood in Boulder, Colorado, in the first part of 1995. (Data from the Boulder Area Board of Realtors.)

TABLE 2.3   Data on house sales in Boulder, Colorado

Selling price	Style	Square feet	Bedrooms	Selling price	Style	Square feet	Bedrooms
$113,000	TRI-LE	1350	3	$175,000	RANCH	2900	3
$113,542	BI-LEV	1700	4	$175,000	4-LEVE	2486	3
$120,000	TRI-LE	1350	4	$176,000	RANCH	2671	3
$120,000	TRI-LE	1350	3	$177,000	2STORY	3058	4
$122,100	RANCH	1700	4	$177,500	4-LEVE	2612	4
$123,000	TRI-LE	1350	4	$177,900	RANCH	2692	5
$128,500	RANCH	2457	3	$180,000	RANCH	3000	4
$137,000	RANCH	2100	3	$180,000	RANCH	2025	3
$139,900	RANCH	2811	3	$180,500	4-LEVE	1640	3
$142,000	RANCH	2920	3	$185,000	RANCH	3396	4
$142,900	4-LEVE	2034	3	$185,500	4-LEVE	2450	3
$146,000	BI-LEV	1700	4	$186,900	4-LEVE	2350	3
$146,221	TRI-LE	2342	3	$186,900	4-LEVE	2387	3
$148,000	RANCH	2962	4	$187,863	2STORY	2993	4
$149,000	TRI-LE	2342	3	$188,000	RANCH	2680	3
$152,000	RANCH	3244	3	$190,000	2STORY	3412	4
$153,000	4-LEVE	1808	3	$192,000	2STORY	3271	4
$156,500	RANCH	3244	3	$193,000	4-LEVE	2098	3
$157,250	RANCH	3101	3	$195,000	4-LEVE	2333	3
$159,900	2STORY	2976	4	$195,000	BI-LEV	2491	3
$162,500	2STORY	2439	3	$199,900	4-LEVE	2508	3
$167,900	4-LEVE	2447	3	$200,000	RANCH	3460	4
$167,900	4-LEVE	2674	4	$200,000	2STORY	3271	5
$168,500	4-LEVE	2616	4	$206,900	RANCH	3222	4
$169,000	RANCH	2822	3	$270,000	TRI-LE	2705	3
$170,100	4-LEVE	2508	3	$289,000	4-LEVE	3456	4
$171,000	RANCH	3044	3	$299,000	2STORY	3792	3
$172,000	2STORY	3077	4	$319,500	CONTEM	3746	4
$173,000	RANCH	2878	4	$327,500	2STORY	4167	5
$174,900	CAPECO	2400	3				

TABLE 2.4   Count and weight of seeds produced by common tree species

Tree species	Seed count	Seed weight (mg)	Tree species	Seed count	Seed weight (mg)
Paper birch	27,239	0.6	American beech	463	247
Yellow birch	12,158	1.6	American beech	1,892	247
White spruce	7,202	2.0	Black oak	93	1,851
Engelmann spruce	3,671	3.3	Scarlet oak	525	1,930
Red spruce	5,051	3.4	Red oak	411	2,475
Tulip tree	13,509	9.1	Red oak	253	2,475
Ponderosa pine	2,667	37.7	Pignut hickory	40	3,423
White fir	5,196	40.0	White oak	184	3,669
Sugar maple	1,751	48.0	Chestnut oak	107	4,535
Sugar pine	1,159	216.0			

(a) Describe the distribution of selling price with a graph and a numerical summary. What are the main features of this distribution?

(b) How is the selling price of a house related to its floor area in square feet? Make a scatterplot and describe the relationship.

(c) The neighborhood includes a group of new houses on a secluded street that are fancier than others in the area. How are these houses visible in your plot? How many of them are there?

2.14 Ecologists collect data to study nature's patterns. Table 2.4 gives data on the mean number of seeds produced in a year by several common tree species and the mean weight of the seeds produced. (Some species appear twice because their seeds were counted in two locations.) We might expect that trees with heavy seeds produce fewer of them, but what is the form of the relationship? (Data from many studies compiled in D. F. Greene and E. A. Johnson, "Estimating the mean annual seed production of trees," *Ecology*, 75 (1994), pp. 642–647.)

(a) Make a scatterplot showing how the weight of tree seeds helps explain how many seeds the tree produces. Describe the form, direction, and strength of the relationship.

(b) When dealing with sizes and counts, the logarithms of the original data are often the "natural" variables. Use your calculator or software to obtain the logarithms of both the seed weights and the seed counts in Table 2.4. Make a new scatterplot using these new variables. Now what are the form, direction, and strength of the relationship?

2.15 To demonstrate the effect of nematodes (microscopic worms) on plant growth, a botanist introduces different numbers of nematodes into 16 planting pots. He then transplants a tomato seedling into each pot. Here are data on the increase in height of the seedlings (in centimeters) 14 days after planting. (Data provided by Matthew Moore.)

Nematodes	Seedling growth
0	10.8 9.1 13.5 9.2
1,000	11.1 11.1 8.2 11.3
5,000	5.4 4.6 7.4 5.0
10,000	5.8 5.3 3.2 7.5

(a) Make a scatterplot of the response variable (growth) against the explanatory variable (nematode count). Then compute the mean growth for each group of seedlings, plot the means against the nematode counts, and connect these four points with line segments.

(b) Briefly describe the conclusions about the effects of nematodes on plant growth that these data suggest.

2.16 The presence of harmful insects in farm fields is detected by erecting boards covered with a sticky material and examining the insects trapped on the boards. Some colors are more attractive to insects than others. In an experiment aimed at determining the best color for attracting cereal leaf beetles, six boards of each of four colors were placed in a field of oats in July. The table below gives data on the number of cereal leaf beetles trapped. (Modified from M. C. Wilson and R. E. Shade, "Relative attractiveness of various luminescent colors to the cereal leaf beetle and the meadow spittlebug," *Journal of Economic Entomology*, 60 (1967), pp. 578–580.)

Board color	Insects trapped
Lemon yellow	45 59 48 46 38 47
White	21 12 14 17 13 17
Green	37 32 15 25 39 41
Blue	16 11 20 21 14 7

(a) Make a plot of the counts of insects trapped against board color (space the four colors equally on the horizontal axis). Compute the mean count for each color, add the means to your plot, and connect the means with line segments.

(b) Based on the data, state your conclusions about the attractiveness of these colors to the beetles.

(c) Does it make sense to speak of a positive or negative association between board color and insect count?

2.17 When animals of the same species live together, they often establish a clear pecking order. Lower-ranking individuals defer to higher-ranking animals, usually avoiding open conflict. A researcher on animal behavior wants to study the relationship between pecking order and physical characteristics such as weight. He confines four chickens in each of seven pens and observes the pecking order that emerges in each pen. Here is a table of the weights (in grams) of the chickens, arranged by pecking order. That is, the first row gives the weights of the dominant chickens in the seven pens, the second row the weights of the number 2 chicken in each pen, and so on. (Data collected by D. L. Cunningham, Cornell University.)

Pecking order	Weight (g)						
	Pen 1	Pen 2	Pen 3	Pen 4	Pen 5	Pen 6	Pen 7
1	1880	1300	1600	1380	1800	1000	1680
2	1920	1700	1830	1520	1780	1740	1460
3	1600	1500	1520	1520	1360	1520	1760
4	1830	1880	1820	1380	2000	2000	1800

(a) Make a plot of these data that is appropriate to study the effect of weight on pecking order. Include in your plot any means that might be helpful.

(b) We might expect that heavier chickens would tend to stand higher in the pecking order. Do these data give clear evidence for or against this expectation?

2.2 Correlation

A scatterplot displays the form, direction, and strength of the relationship between two quantitative variables. Straight-line (linear) relations are particularly important because a straight line is a simple pattern that is quite common. We say a linear relationship is strong if the points lie close to a straight line, and weak if they are widely scattered about a line. Our eyes are not good judges of how strong a relationship is. The two scatterplots in Figure 2.10 depict exactly the same data, but the lower plot is drawn smaller in a large field. The lower plot seems to show a stronger relationship. Our eyes can be fooled by changing the plotting scales or the amount of white space around the cloud of points in a scatterplot.[7] We need to follow our strategy for data analysis by using a numerical measure to supplement the graph. *Correlation* is the measure we use.*

The correlation *r*

We have data on variables x and y for n individuals. Think, for example, of measuring height and weight for n people. Then x_1 and y_1 are your height and your weight, x_2 and y_2 are my height and my weight, and so on. For the i th individual, height x_i goes with weight y_i. Here is the definition of correlation.

---

*Correlation was introduced by the English gentleman scientist Francis Galton (1822–1911) in 1888. Galton called r the "index of co-relation" and applied it to measurements such as the forearm lengths and the heights of a group of people.

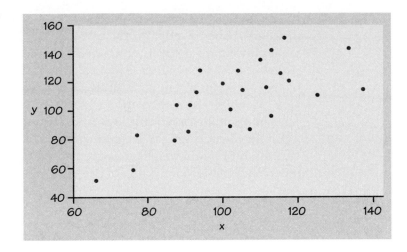

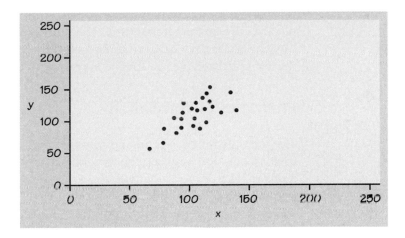

FIGURE 2.10 Two scatterplots of the same data; the linear pattern in the lower plot appears stronger because of the surrounding white space.

Correlation

The **correlation** measures the direction and strength of the linear relationship between two quantitative variables. Correlation is usually written as r.

Suppose that we have data on variables x and y for n individuals. The means and standard deviations of the two variables are $\bar{x}$ and s_x for the x-values, and $\bar{y}$ and s_y for the y-values. The correlation r between x and y is

$$r = \frac{1}{n-1} \sum \left(\frac{x_i - \bar{x}}{s_x} \right) \left(\frac{y_i - \bar{y}}{s_y} \right)$$

As always, the summation sign $\sum$ means "add these terms for all the individuals." The formula for the correlation r is a bit complex. It helps us see what correlation is but is not convenient for actually calculating r. In practice you should use software or a calculator that finds r from keyed-in values of two variables x and y. Exercise 2.18 asks you to calculate a correlation step-by-step from the definition to solidify its meaning.

The formula for r begins by standardizing the observations. Suppose, for example, that x is height in centimeters and y is weight in kilograms and that we have height and weight measurements for n people. Then $\bar{x}$ and s_x are the mean and standard deviation of the n heights, both in centimeters. The value

$$\frac{x_i - \bar{x}}{s_x}$$

is the standardized height of the i th person, familiar from Chapter 1. The standardized height says how many standard deviations above or below the mean a person's height lies. Standardized values have no units—in this example, they are no longer measured in centimeters. Standardize the weights also. The correlation r is an average of the products of the standardized height and the standardized weight for the n people.

Properties of correlation

The formula for correlation helps us see that r is positive when there is a positive association between the variables. Height and weight, for example, have a positive association. People who are above average in height tend to also be above average in weight. Both the standardized height and the standardized weight for such a person are positive. People who are below average in height tend also to have below-average weight. Then both standardized height and standardized weight are negative. In both cases, the products in the formula for r are mostly positive and so r is positive. In the same way, we can see that r is negative when the association between x and y is negative. More detailed study of the formula gives more detailed properties of r. Here is what you need to know in order to interpret correlation:

■ Correlation makes no use of the distinction between explanatory and response variables. It makes no difference which variable you call x and which you call y in calculating the correlation.

■ Correlation requires that both variables be quantitative, so that it makes sense to do the arithmetic indicated by the formula for r. We cannot calculate a correlation between the incomes of a group of people and what city they live in, because city is a categorical variable.

■ Because r uses the standardized values of the observations, r does not change when we change the units of measurement of x, y, or both. Measuring height in inches rather than centimeters and weight in pounds rather than kilograms does not change the correlation between height and weight. The correlation r itself has no unit of measurement; it is just a number.

■ Positive r indicates positive association between the variables, and negative r indicates negative association.

■ The correlation r is always a number between -1 and 1. Values of r near 0 indicate a very weak linear relationship. The strength of the relationship increases as r moves away from 0 toward either -1 or 1. Values of r close to -1 or 1 indicate that the points lie close to a straight line. The extreme values $r = -1$ and $r = 1$ occur only when the points in a scatterplot lie exactly along a straight line.

■ Correlation measures the strength of only the linear relationship between two variables. Correlation does not describe curved relationships between variables, no matter how strong they are.

■ Like the mean and standard deviation, the correlation is not resistant: r is strongly affected by a few outlying observations. Use r with caution when outliers appear in the scatterplot.

The scatterplots in Figure 2.11 illustrate how values of r closer to 1 or -1 correspond to stronger linear relationships. To make the essential meaning of r clear, the standard deviations of both variables in these plots are equal and the horizontal and vertical scales are the same. In general, it is not so easy to guess the value of r from the appearance of a scatterplot. Remember that changing the plotting scales in a scatterplot may mislead our eyes, but it does not change the standardized values of the variables and therefore cannot change the correlation.

EXAMPLE 2.9   The real data we have examined also illustrate the behavior of correlation.

Figure 2.9 (page 119) shows a strong positive linear association between city and highway gas mileage for midsize cars. The correlation is $r = 0.915$. If we remove the Rolls-Royce, the correlation drops to $r = 0.835$. The outlier extends the linear pattern and increases r. Remember that correlation is not resistant.

Figure 2.6 (page 114) shows a very weak relationship between birth date and draft lottery number. The scatterplot smoother suggests that the form is not too far from linear, so we can use correlation to measure the strength of the relationship. We expect a small negative r, and calculation gives $r = -0.226$.

The correlation between time and acceleration for the motorcycle crash data graphed in Figure 2.5 (page 114) is $r = 0.296$. Because the relationship is not at all linear, r provides no useful information. Always plot your data before calculating common statistical measures such as correlation.

Finally, remember that **correlation is not a complete description of two-variable data,** even when the relationship between the variables is linear. You should give the means and standard deviations of both x and y along with the correlation. (Because the formula for correlation uses the means and standard deviations, these measures are the proper choices to accompany a correlation.) Conclusions based on correlations alone may require rethinking in the light of a more complete description of the data.

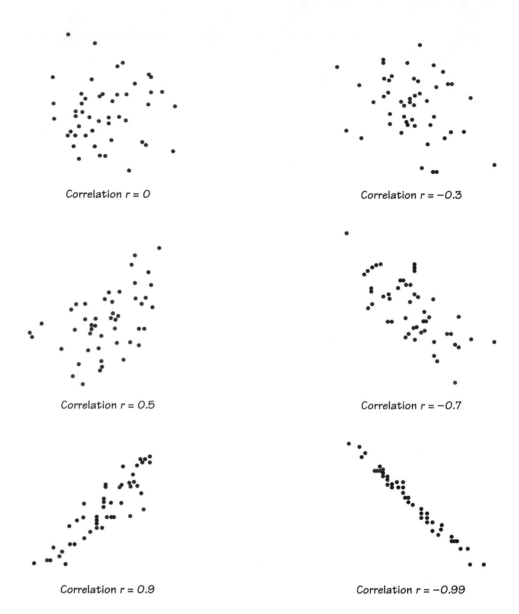

FIGURE 2.11 How the correlation r measures the strength of linear association.

EXAMPLE 2.10 | Competitive divers are scored on their form by a panel of judges who use a scale from 1 to 10. The subjective nature of the scoring often results in controversy. We have the scores awarded by two judges, Ivan and George, on a large number of dives. How well do they agree? We do some calculation and find that the correlation between their scores is $r = 0.9$. But the mean of Ivan's scores is 3 points lower than George's mean.

These facts do not contradict each other. They are simply different kinds of information. The mean scores show that Ivan awards much lower scores than George. But because Ivan gives *every* dive a score about 3 points lower than George, the correlation remains high. Adding the same number to all values of either x or y does not change the correlation. If Ivan and George both rate several divers, the contest is fairly scored because Ivan and George agree on which dives are better than others. The high r shows their agreement. But if Ivan scores one diver and George another, we must add 3 points to Ivan's scores to arrive at a fair comparison.

SUMMARY

The **correlation r** measures the direction and strength of the linear association between two quantitative variables x and y. Although you can calculate a correlation for any scatterplot, r measures only linear relationships.

Correlation indicates the direction of a linear relationship by its sign: $r > 0$ for a positive association and $r < 0$ for a negative association.

Correlation always satisfies $-1 \le r \le 1$ and indicates the strength of a relationship by how close it is to -1 or 1. Perfect correlation, $r = \pm 1$, occurs only when the points lie exactly on a straight line.

Correlation ignores the distinction between explanatory and response variables. The value of r is not affected by changes in the unit of measurement of either variable. Correlation is not resistant, so outliers can greatly change the value of r.

SECTION 2.2. EXERCISES

2.18   Example 2.4 (page 110) gives the lengths of two bones in five fossil specimens of the extinct beast *Archaeopteryx*. The scatterplot in Figure 2.3 shows a strong positive linear relationship.

    (a) Find the correlation r step-by-step. That is, find the mean and standard deviation of the femur lengths and of the humerus lengths. Then find the five standardized values for each variable and use the formula for r.

    (b) Now enter these data into your calculator or software and use the correlation function to find r. Check that you get the same result as in (a).

2.19   Exercise 2.10 (page 121) gives data on the lean body mass and metabolic rate for 12 women and 7 men.

    (a) If you did not do Exercise 2.10, make a scatterplot. Use different symbols or colors for women and men. Do you think the correlation will be about the same for men and women or quite different for the two groups? Why?

    (b) Find r for women alone and also for men alone.

    (c) Calculate the mean body mass for the women and for the men. Does the fact that the men are heavier than the women on the average influence the correlations? If so, in what way?

(d) Lean body mass was measured in kilograms. How would the correlations change if we measured body mass in pounds? (There are about 2.2 pounds in a kilogram.)

2.20   A student wonders if people of similar heights tend to date each other. She measures herself, her dormitory roommate, and the women in the adjoining rooms; then she measures the next man each woman dates. Here are the data (heights in inches):

Women	66	64	66	65	70	65
Men	72	68	70	68	71	65

(a) Make a scatterplot of these data. Based on the scatterplot, do you expect the correlation to be positive or negative? Near ± 1 or not?

(b) Find the correlation r between the heights of the men and women.

(c) How would r change if all the men were 6 inches shorter than the heights given in the table? Does the correlation help answer the question of whether women tend to date men taller than themselves?

(d) If heights were measured in centimeters rather than inches, how would the correlation change? (There are 2.54 centimeters in an inch.)

(e) If every woman dated a man exactly 3 inches taller than herself, what would be the correlation between male and female heights?

2.21   Here are the golf scores of 11 members of a women's golf team in two rounds of college tournament play:

Player	1	2	3	4	5	6	7	8	9	10	11
Round 1	89	90	87	95	86	81	105	83	88	91	79
Round 2	94	85	89	89	81	76	89	87	91	88	80

If you did not make a scatterplot in Exercise 2.6, do so now. Find the correlation between the Round 1 and Round 2 scores. Remove Player 7's scores and find the correlation for the remaining 10 players. Explain carefully why removing this single case substantially increases the correlation.

2.22   Exercise 2.8 (page 120) gives data on gas mileage against speed for a small car. Make a scatterplot if you have not already done so, then find the correlation r. Explain why r is close to zero despite a strong relationship between speed and gas used.

2.23   Changing the units of measurement can dramatically alter the appearance of a scatterplot. Consider the following data:

x	−4	−4	−3	3	4	4
y	0.5	−0.6	−0.5	0.5	0.5	−0.6

(a) Plot the data on x and y axes that extend from -6 to 6.

(b) Form new variables $x^* = x/10$ and $y^* = 10y$, starting from the values of x and y. Plot y^* against x^* on the same axes using a different plotting symbol. The two plots are very different in appearance.

(c) Find the correlation between x and y. Then find the correlation between x^* and y^*. How are the two correlations related? Explain why this is not surprising.

2.24 The British government conducts regular surveys of household spending. The following table shows the average weekly household spending on tobacco products and alcoholic beverages for each of the 11 regions of Britain. (Data from British official statistics, *Family Expenditure Survey*, Department of Employment, 1981.)

Region	Alcohol	Tobacco
North	£6.47	£4.03
Yorkshire	6.13	3.76
Northeast	6.19	3.77
East Midlands	4.89	3.34
West Midlands	5.63	3.47
East Anglia	4.52	2.92
Southeast	5.89	3.20
Southwest	4.79	2.71
Wales	5.27	3.53
Scotland	6.08	4.51
Northern Ireland	4.02	4.56

(a) Make a scatterplot of spending on tobacco y against spending on alcohol x.

(b) Describe the pattern of the plot and any important deviations.

(c) Find the correlation. Then compute the correlation for the 10 regions omitting Northern Ireland. Explain why this r differs so greatly from the r for all 11 cases.

2.25 Figure 2.1 (page 107) is a scatterplot of each state's mean SAT verbal score against the percent of the state's high school graduates who take the exam. The correlation between these variables is $r = -0.884$. Explain why r is negative and quite strong. What important features of the relationship does r fail to describe?

2.26 Financial experts use statistical measures to describe the performance of investments such as mutual funds. In the past, fund managers feared that investors would not understand statistical descriptions, but mounting pressure to give better information is moving standard deviations and correlations into the public eye.

(a) The T. Rowe Price mutual fund group reports the standard deviation of yearly percent returns for its funds. Recently, Equity Income Fund had standard deviation 9.94%, and Science & Technology Fund had standard deviation 23.77%. Explain to someone who knows no statistics how these standard deviations help investors compare the two funds.

(b) Some mutual funds act much like the stock market as a whole, as measured by a market index such as the Standard & Poor's 500 stock index (S&P 500). Others are very different from the overall market. We can use correlation to describe the association. Monthly returns from Fidelity Magellan Fund, the largest mutual fund, have correlation $r = 0.85$ with the S&P 500. Fidelity Small Cap Stock Fund has correlation $r = 0.55$ with the S&P 500. Explain to someone who knows no statistics how these correlations help investors compare the two funds.

2.27   Consider again the correlation r between the speed of a car and its gas consumption, from the data in Exercise 2.8 (page 120).

(a) Transform the data so that speed is measured in miles per hour and fuel consumption in gallons per mile. (There are 1.609 kilometers in a mile and 3.785 liters in a gallon.) Make a scatterplot and find the correlation for both the original and the transformed data. How did the change of units affect your results?

(b) Now express fuel consumption in miles per gallon. (So each value is $1/x$ if x is gallons per mile.) Again make a scatterplot and find the correlation. How did this change of units affect your results? (*Lesson*: The effects of a linear transformation of the form $x_{new} = a + bx$ are simple. The effects of a nonlinear transformation are more complex.)

2.28   Table 2.4 (page 124) gives data on the mean number of seeds produced in a year by several common tree species and the mean weight of the seeds produced. Exercise 2.14 examines these data and also the results of transforming both variables by taking logarithms. If you did not do Exercise 2.14, do it now. Then find the correlation between seed production and seed weight for both the original and the transformed data. What does r tell you in each case?

2.29   Table 1.6 (page 33) reports data on 78 seventh-grade students. Some people think that self-concept is related to school performance. Make a scatterplot of school GPA against self-concept score. Describe the overall pattern that you see. Does it make sense to use the correlation as a summary? If so, find r and interpret its value in words.

2.30   If women always married men who were 2 years older than themselves, what would be the correlation between the ages of husband and wife? (*Hint*: Draw a scatterplot for several ages.)

2.31   A college newspaper interviews a psychologist about student ratings of the teaching of faculty members. The psychologist says, "The evidence indicates that the correlation between the research productivity and teaching rating of faculty members is close to zero." The paper reports this as "Professor

McDaniel said that good researchers tend to be poor teachers, and vice versa." Explain why the paper's report is wrong. Write a statement in plain language (don't use the word "correlation") to explain the psychologist's meaning.

2.32 Each of the following statements contains a blunder. Explain in each case what is wrong.

(a) "There is a high correlation between the gender of American workers and their income."

(b) "We found a high correlation ($r = 1.09$) between students' ratings of faculty teaching and ratings made by other faculty members."

(c) "The correlation between planting rate and yield of corn was found to be $r = 0.23$ bushel."

2.3 Least-Squares Regression

Correlation measures the direction and strength of the straight-line (linear) relationship between two quantitative variables. If a scatterplot shows a linear relationship, we would like to summarize this overall pattern by drawing a line on the scatterplot. A *regression line* summarizes the relationship between two variables, but only in a specific setting: when one of the variables helps explain or predict the other. That is, regression describes a relationship between an explanatory variable and a response variable.*

Regression Line

A **regression line** is a straight line that describes how a response variable y changes as an explanatory variable x changes. We often use a regression line to **predict** the value of y for a given value of x. Regression, unlike correlation, requires that we have an explanatory variable and a response variable.

EXAMPLE 2.11

How do children grow? The pattern of growth varies from child to child, so we can best understand the general pattern by following the average height of a number of children. Table 2.5 presents the mean heights of a group of children in Kalama, an Egyptian village that was the site of a study of nutrition in developing countries.[8] The data were obtained by measuring the heights of 161 children from the village each month from 18 to 29 months of age.

Figure 2.12 is a scatterplot of the data in Table 2.5. Age is the explanatory variable, which we plot on the x axis. The plot shows a strong positive linear association with no outliers. The correlation is $r = 0.994$, close to the $r = 1$ of points that lie exactly on a line. A line drawn through the points will describe these data very well.

---

*The term "regression" and the general methods for studying relationships now included under this term were introduced by Francis Galton. Galton was engaged in the study of heredity. One of his observations was that the children of tall parents tended to be taller than average but not as tall as their parents. This "regression toward mediocrity" gave these statistical methods their name.

TABLE 2.5	Mean height of Kalama children
Age x in months	Height y in centimeters
18	76.1
19	77.0
20	78.1
21	78.2
22	78.8
23	79.7
24	79.9
25	81.1
26	81.2
27	81.8
28	82.8
29	83.5

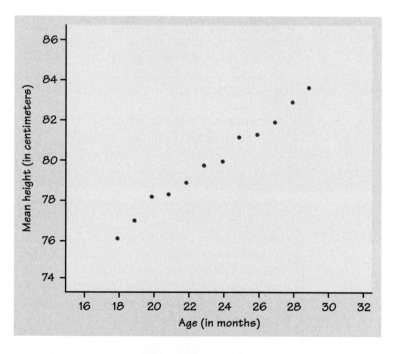

FIGURE 2.12   Mean height of children in Kalama, Egypt, plotted against age from 18 to 29 months, from Table 2.5.

Fitting a line to data

fitting a line

When a scatterplot displays a linear pattern, we can describe the overall pattern by drawing a straight line through the points. Of course, no straight line passes exactly through all of the points. **Fitting a line** to data means drawing a line that comes as close as possible to the points. The equation of a line fitted to the data gives a compact description of the dependence of the response variable y on the explanatory variable x. It is a mathematical model for the straight-line relationship.

Straight Lines

Suppose that y is a response variable (plotted on the vertical axis) and x is an explanatory variable (plotted on the horizontal axis). A straight line relating y to x has an equation of the form

$$y = a + bx$$

In this equation, b is the **slope,** the amount by which y changes when x increases by one unit. The number a is the **intercept,** the value of y when $x = 0$.

Any straight line describing the Kalama data has the form

$$\text{height} = a + (b \times \text{age})$$

A regression line for predicting mean height from age has the equation

$$\text{height} = 64.93 + (0.635 \times \text{age})$$

Figure 2.13 shows that this line fits the data well. The slope $b = 0.635$ tells us that the height of Kalama children increases by about 0.6 centimeter for each month of age. The slope b of a line $y = a + bx$ is the *rate of change* in the response y as the explanatory variable x changes. The slope of a regression line is an important numerical description of the relationship between the two variables. The intercept, $a = 64.93$ centimeters, would be the mean height at birth (age = 0) if the straight-line pattern of growth were true starting at birth. Children don't grow at a fixed rate from birth, so the intercept a is not important in our situation except as part of the equation of the line.

Prediction

prediction

We can use a regression line to **predict** the response y for a specific value of the explanatory variable x.

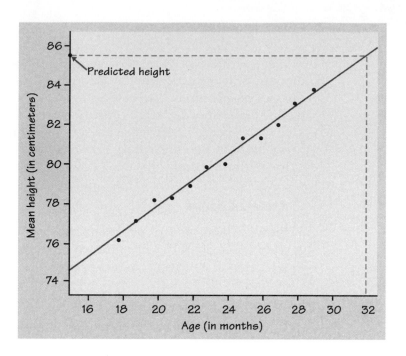

FIGURE 2.13   The least-squares line fitted to the Kalama data and used to predict height at age 32 months.

EXAMPLE 2.12 | Based on the linear pattern in Figure 2.12, we want to predict the mean height of Kalama children at 32 months of age. To use our fitted line to predict height, go "up and over" on the graph in Figure 2.13. From age 32 months on the x axis, go up to the fitted line and over to the y axis. It appears that the predicted height is a bit more than 85 centimeters.

If we have the equation of the line, it is faster and more accurate to substitute 32 for the age in the equation. Our predicted height is

$$\text{height} = 64.93 + (0.635 \times 32) = 85.25 \text{ centimeters}$$

The accuracy of predictions from a regression line depends on how much scatter about the line the data show. In the Kalama example, the data points are all very close to the line, so we are confident that our prediction is accurate. If the data show a linear pattern with considerable spread, we may use a regression line to summarize the pattern but we will put less confidence in predictions based on the line.

EXAMPLE 2.13 | Can we predict the mean height of Kalama residents at age 20 years? Twenty years is 240 months, so we substitute 240 for the age. The prediction is

$$\text{height} = 64.93 + (0.635 \times 240) = 217.33 \text{ centimeters}$$

This is more than 7 feet. Blind calculation has produced an unreasonable result. Our data cover only ages from 18 to 29 months. As people grow older, they gain height more slowly, so our fitted line is not a good model at ages far removed from the data that produced it.

> **Extrapolation**
>
> **Extrapolation** is the use of a regression line for prediction far outside the range of values of the explanatory variable x that you used to obtain the line. Such predictions are often not accurate.

Notice that age 0 (birth) lies well outside the range of the data in Table 2.5. Taking the intercept 64.93 cm (over 25 inches) as an estimate of height at birth is therefore another example of extrapolation. The linear growth pattern does not begin at birth, and this estimate of height at birth is not accurate. In addition, we know that the pattern we discovered in Figure 2.13 does not hold for all children everywhere. We cannot use data from Kalama to predict the height of children in Cairo, much less Los Angeles. Statistical methods such as fitting regression lines cannot go beyond the limitations of the data that feed them. Do not let calculation override common sense.

Least-squares regression

Different people might draw different lines by eye on a scatterplot. This is especially true when the points are widely scattered. We need a way to draw a regression line that doesn't depend on our guess as to where the line should go. No line will pass exactly through all the points, but we want one that is as close as possible. We will use the line to predict y from x, so we want a line that is as close as possible to the points in the *vertical* direction. That's because the prediction errors we make are errors in y, which is the vertical direction in the scatterplot. If we predict 85.25 centimeters for the mean height at age 32 months and the actual mean turns out to be 84 centimeters, our error is

$$\text{error} = \text{observed height} - \text{predicted height}$$
$$= 84 - 85.25 = -1.25 \text{ centimeters}$$

We want a regression line that makes these prediction errors as small as possible. Figure 2.14 illustrates the idea. For clarity, the plot shows only three of the points from Figure 2.13, along with the line, on an expanded scale. The line passes below two of the points and above one of them. The vertical distances of the data points from the line appear as vertical line segments. A "good" regression line makes these distances as small as possible. There are many ways to make "as small as possible" precise. The most common is the *least-squares* idea. The line in Figures 2.13 and 2.14 is in fact the least-squares regression line for predicting mean height from age.*

---

*The method of least squares was first published by the French mathematician Adrien-Marie Legendre (1752–1833) in 1805. It at once became the standard method for combining observations in astronomy and surveying and remains one of the most-used statistical tools.

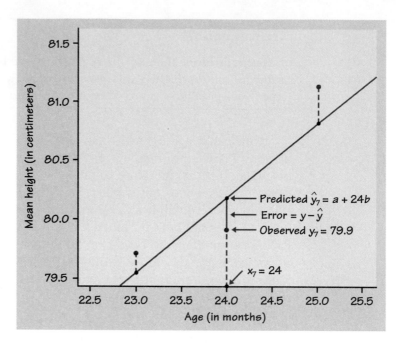

FIGURE 2.14　The least-squares idea: make the errors in predicting y as small as possible by minimizing the sum of their squares.

Least-Squares Regression Line

The **least-squares regression line of y on x** is the line that makes the sum of the squares of the vertical distances of the data points from the line as small as possible.

Here is the least-squares idea expressed as a mathematical problem. We represent n observations on two variables x and y as

$$(x_1, y_1), \ (x_2, y_2), \ \dots, \ (x_n, y_n)$$

If we draw a line $y = a + bx$ through the scatterplot of these observations, the line predicts the value of y corresponding to x_i as $\hat{y}_i = a + bx_i$. We write $\hat{y}$ (read "y hat") in the equation of a regression line to emphasize that the line gives a *predicted* response $\hat{y}$ for any x. The predicted response will usually not be exactly the same as the actually *observed* response y. For example, the seventh observation in Table 2.5 is $x_7 = 24$, $y_7 = 79.9$. Figure 2.14 shows the prediction error for this point, which is

$$
\begin{aligned}
\text{error} &= \text{observed } y - \text{predicted } y \\
&= y_7 - \hat{y}_7 \\
&= y_7 - a - bx_7 \\
&= 79.9 - a - 24b
\end{aligned}
$$

The method of least-squares chooses the line that makes the sum of the squares of these errors as small as possible. To find this line, we must find the values of the intercept a and the slope b that minimize

$$\sum(\text{error})^2 = \sum(y_i - a - bx_i)^2$$

for the given observations x_i and y_i. For the Kalama growth data in Table 2.5, we must find the a and b that minimize

$$(76.1 - a - 18b)^2 + (77.0 - a - 19b)^2 + \cdots + (83.5 - a - 29b)^2$$

The equation of the least-squares line is the solution to this mathematical problem.

You will use software or a calculator with a regression function to find the equation of the least-squares regression line from data on x and y. We will therefore give the equation of the least-squares line in a form that helps our understanding but is not efficient for calculation.

Equation of the Least-Squares Regression Line

We have data on an explanatory variable x and a response variable y for n individuals. The means and standard deviations of the sample data are $\bar{x}$ and s_x for x and $\bar{y}$ and s_y for y, and the correlation between x and y is r. The equation of the least-squares regression line of y on x is

$$\hat{y} = a + bx$$

with **slope**

$$b = r\frac{s_y}{s_x}$$

and **intercept**

$$a = \bar{y} - b\bar{x}$$

EXAMPLE 2.14   Verify from the data in Table 2.5 that the mean and standard deviations of the 12 ages are

$$\bar{x} = 23.5 \text{ months} \quad \text{and} \quad s_x = 3.606 \text{ months}$$

The mean and standard deviation of the 12 heights are

$$\bar{y} = 79.85 \text{ cm} \quad \text{and} \quad s_y = 2.302 \text{ cm}$$

The correlation between height and age is $r = 0.9944$. The least-squares regression line of height y on age x therefore has slope

$$b = r\frac{s_y}{s_x} = 0.9944\frac{2.302}{3.606}$$
$$= 0.6348 \text{ cm per month}$$

and intercept

$$a = \bar{y} - b\bar{x} = 79.85 - (0.6348)(23.5)$$
$$= 64.932 \text{ cm}$$

The equation of the least-squares line is

$$\hat{y} = 64.932 + 0.6348x$$

Interpreting the regression line

The slope $b = 0.6348$ centimeters per month in Example 2.14 is the rate of change in the mean height as age increases. The units "centimeters per month" come from the units of y (centimeters) and x (months). Although the correlation does not change when we change the units of measurement, the equation of the least-squares line does change. The slope in centimeters per month is 2.54 times as large as the slope in inches per month, because there are 2.54 centimeters in an inch.

The expression $b = rs_y/s_x$ for the slope says that along the regression line, *a change of one standard deviation in x corresponds to a change of r standard deviations in y*. When the variables are perfectly correlated ($r = 1$ or $r = -1$), the change in the predicted response $\hat{y}$ is the same (in standard deviation units) as the change in x. Otherwise, because $-1 \le r \le 1$, the change in $\hat{y}$ is less than the change in x. As the correlation grows less strong, the prediction $\hat{y}$ moves less in response to changes in x.

The least-squares regression line always passes through the point $(\overline{x}, \overline{y})$ on the graph of y against x. Check that when you substitute $\overline{x} = 23.5$ into the equation of the regression line in Example 2.14, the result is $\hat{y} = 79.8498$, equal to the mean of y except for roundoff error. So the least-squares regression line of y on x is the line with slope rs_y/s_x that passes through the point $(\overline{x}, \overline{y})$. We can describe regression entirely in terms of the basic descriptive measures $\overline{x}$, s_x, $\overline{y}$, s_y, and r. If both x and y are standardized variables, so that their means are 0 and their standard deviations are 1, then the regression line has slope r and passes through the origin.

Figure 2.15 displays the basic regression output for the Kalama data from a graphing calculator and two statistical software packages. You can find the slope and intercept of the least-squares line, calculated to more decimal places than we need, in all these outputs. The software provides information that we do not yet need—part of the art of using software is to ignore the extra information that is almost always present. We will make use of other parts of the output in Chapter 10.

Correlation and regression

Least-squares regression looks at the distances of the data points from the line only in the y direction. So the two variables x and y play different roles in regression.

EXAMPLE 2.15

Figure 2.16 is a scatterplot of data that played a central role in the discovery that the universe is expanding. They are the distances from Earth of 24 spiral galaxies and the speed at which these galaxies are moving away from us, reported by the astronomer Edwin Hubble in 1929.[9] There is a positive linear relationship, $r = 0.7842$. More distant galaxies are moving away more rapidly. Astronomers believe that there is in fact a perfect linear relationship, and that the scatter is caused by imperfect measurements.

The two lines on the plot are the two least-squares regression lines. The regression line of velocity on distance is solid. The regression line of distance on velocity is dashed. *Although there is only one correlation between velocity and distance, regression of velocity on distance and regression of distance on velocity give different lines.* In the regression setting you must choose which variable is explanatory.

Although the correlation r ignores the distinction between explanatory and response variables, there is a close connection between correlation and regression. We saw that the slope of the least-squares line involves r. Another connection between correlation and regression is even more important. In fact, the numerical value of r as a measure of the strength of a linear relationship is best interpreted by thinking about regression. Here is the fact we need.

TI-83

```
Lin Reg
 y=ax+b
 a=.634965035
 b=64.92832168
 r²=.9887639371
 r=.9943660981
```

Data Desk

```
Dependent variable is: height

R squared = 98.9%          R squared(adjusted) = 98.8%
s = 0.2560 with 12 - 2 = 10 degrees of freedom

Variable Coefficient  s.e. of Coeff  t-ratio   prob
Constant       64.9283        0.5084      128    0.0001
Var1          0.634965        0.0214     29.7    0.0001
```

Minitab

```
The regression equation is
height = 64.9 + 0.635 age

Predictor      Coef      Stdev    t-ratio          p
Constant    64.9283     0.5084     127.71      0.000
height      0.63497    0.02140      29.66      0.000

s = 0.2560       R-sq = 98.9%      R-sq(adj) = 98.8%
```

FIGURE 2.15 The least-squares line for the Kalama data, output from a graphing calculator and two statistical software packages.

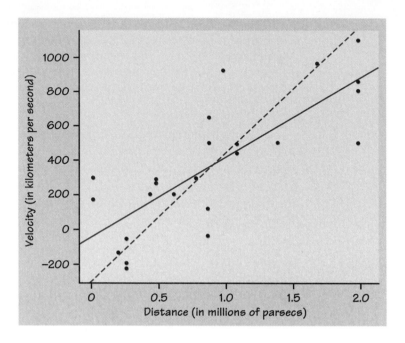

FIGURE 2.16 Hubble's data on the velocity and distance of 24 galaxies. The lines are least-squares regression lines of velocity on distance (solid) and of distance on velocity (dashed).

---

r^2 in Regression

The **square of the correlation, r^2,** is the fraction of the variation in the values of y that is explained by the least-squares regression of y on x.

The correlation between the mean height and age of Kalama children is $r = 0.9944$. Because $r^2 = 0.9888$, the straight-line relationship between height and age explains 98.88% (almost all) of the variation in heights in Table 2.5. When you report a regression, give r^2 as a measure of how successful the regression was in explaining the response. All the software outputs in Figure 2.15 include r^2, either in decimal form or as a percent. When you see a correlation, square it to get a better feel for the strength of the association. Perfect correlation ($r = -1$ or $r = 1$) means the points lie exactly on a line. Then $r^2 = 1$ and all of the variation in one variable is accounted for by the linear relationship with the other variable. If $r = -0.7$ or $r = 0.7$, $r^2 = 0.49$ and about half the variation is accounted for by the linear relationship. In the r^2 scale, correlation ± 0.7 is about halfway between 0 and ± 1.

The use of r^2 to describe the success of regression in explaining the response y is very common. It rests on the fact that there are two sources of variation in the responses y in a regression setting. One reason the Kalama heights vary is that height changes with age—as age x increases from 18 months to 29 months, it pulls height y with it along the regression line. The line explains this part of the variation in heights. The heights do not lie exactly on the line, however, but are scattered above and below it. This is the second source of variation in y, and the regression line tells us nothing about how large it is. We use r^2 to measure variation along the line as a fraction of the total variation in the heights.

For a pictorial grasp of what r^2 tells us, look at Figure 2.17. Both scatterplots resemble the Kalama data, but with many more observations. In both cases, height y increases with age x, and the least-squares regression line is the same as we computed from the Kalama data. The heights in Figure 2.17(a) have the same degree of spread about the line as the Kalama data, so $r = 0.994$ and $r^2 = 0.989$. Heights for age 28 months are higher than for age 20 months, and the spread of heights once we fix age is rather small. Little variation in y remains once we fix x. The figure shows that a prediction of y for $x = 28$ by the "up-and-over" method that takes account of the variation in y for a fixed x pins down height quite closely.

The heights in Figure 2.17(b) have the same overall linear pattern expressed by the regression line, but there is much more scatter about the fitted line. Here, $r = 0.921$ and $r^2 = 0.849$. That is, the link between height and age explains 85% of the variation in heights. The remaining 15% is variation in height when we hold age fixed. Heights for age 28 months are still greater than heights for age 20 months—the line still explains part of the variation in height. But "up-and-over" prediction of height y when $x = 28$ is much less precise because there is much more variation in the heights for individuals with ages near 28 months.

Understanding r^2 *

Here is a more specific interpretation of r^2. The heights y in Table 2.5 range from 76.1 centimeters to 83.5 centimeters. The variance of these heights, a measure of how variable they are, is

$$\text{variance of observed values } y = 5.30091$$

Most of this variability is due to the fact that as age x increases from 18 months to 29 months, it pulls height y along with it. If the only variability in the observed heights were due to the straight-line dependence of height on age, the heights would lie exactly on the regression line. That is, they would be the same as the predicted heights $\hat{y}$. We can compute the predicted heights by substituting each age into the equation of the least-squares line and then describe their variability by their variance. The result is

$$\text{variance of predicted values } \hat{y} = 5.24135$$

There isn't much variance left over—the variance of points exactly on the line is almost as large as the variance of the observed heights. This shows that the

---

*This explanation is optional reading.

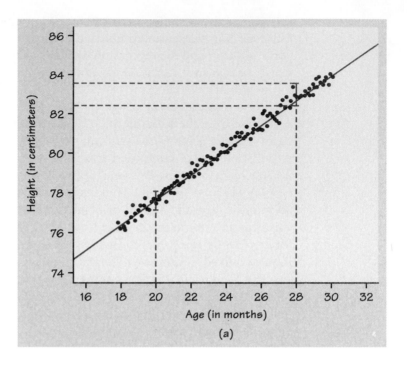

(a)

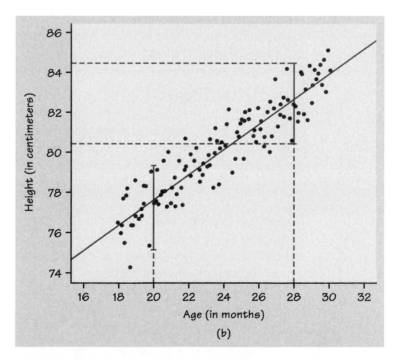

(b)

FIGURE 2.17 Explained versus unexplained variation. In (a), almost all of the variation in height is explained by the linear relationship between height and age. The remaining variation (the spread of heights when $x = 28$ months, for example) is small. In (b), the linear relationship explains a smaller part of the variation in height. The remaining variation (illustrated again for $x = 28$) is larger.

line accounts for most of the variation in y, but what does it have to do with r^2? Here is the fact we need:

$$r^2 = \frac{\text{variance of predicted values } \hat{y}}{\text{variance of observed values } y}$$
$$= \frac{5.24135}{5.30091} = 0.9888$$

This fact is always true. The squared correlation gives us the variance of the predicted responses as a fraction of the variance of the actual responses. The numerator of this fraction is the variance the heights would have if there were no scatter about the least-squares line, so that all the variation in y comes from the straight-line tie between y and x. The denominator is the variance the heights actually do have, including both the tie to x and scatter about the fitted line. This gives an exact meaning for "fraction of variation explained" as an interpretation of r^2.

These connections with correlation are special properties of least-squares regression. They are not true for other methods of fitting a line to data. One reason that least squares is the most common method for fitting a regression line to data is that it has many convenient special properties.

SUMMARY

A **regression line** is straight line that describes how a response variable y changes as an explanatory variable x changes.

The most common method of fitting a line to a scatterplot is least squares. The **least-squares regression line** is the straight line $\hat{y} = a + bx$ that minimizes the sum of the squares of the vertical distances of the observed y-values from the line.

You can use a regression line to **predict** the value of y for any value of x by substituting this x into the equation of the line. **Extrapolation** beyond the range of x values spanned by the data is risky.

The **slope** b of a regression line $\hat{y} = a + bx$ is the rate at which the predicted response $\hat{y}$ changes along the line as the explanatory variable x changes. Specifically, b is the change in $\hat{y}$ when x increases by 1.

The **intercept** a of a regression line $\hat{y} = a + bx$ is the predicted response $\hat{y}$ when the explanatory variable $x = 0$. This prediction is of no statistical use unless x can actually take values near 0.

The least-squares regression line of y on x is the line with slope rs_y/s_x and intercept $a = \overline{y} - b\overline{x}$. This line always passes through the point $(\overline{x}, \overline{y})$.

Correlation and regression are closely connected. The correlation r is the slope of the least-squares regression line when we measure both x and y in standardized units. The square of the correlation r^2 is the fraction of the variance of one variable that is explained by least-squares regression on the other variable.

SECTION 2.3 EXERCISES

2.33   (Review of straight lines) Fred keeps his savings in his mattress. He begins with $500 from his mother and adds $100 each year. His total savings y after x years are given by the equation

$$y = 500 + 100x$$

   (a) Draw a graph of this equation. (*Hint:* Choose two values of x, such as 0 and 10. Find the corresponding values of y from the equation. Plot these two points on graph paper and draw the straight line joining them.)

   (b) After 20 years, how much will Fred have in his mattress?

   (c) If Fred adds $200 instead of $100 each year to his initial $500, what is the equation that describes his savings after x years?

2.34   (Review of straight lines) Sound travels at a speed of 1500 meters per second in seawater. You dive into the sea from your yacht. Give an equation for the distance y at which a shark can hear your splash in terms of the number of seconds x since you hit the water.

2.35   (Review of straight lines) During the period after birth, a male white rat gains 40 grams (g) per week. (This rat is unusually regular in his growth, but 40 g per week is a realistic rate.)

   (a) If the rat weighed 100 g at birth, give an equation for his weight after x weeks. What is the slope of this line?

   (b) Draw a graph of this line between birth and 10 weeks of age.

   (c) Would you be willing to use this line to predict the rat's weight at age 2 years? Do the prediction and think about the reasonableness of the result. (There are 454 g in a pound. To help you assess the result, note that a large cat weighs about 10 pounds.)

2.36   (Review of straight lines) A cellular telephone company offers analog service for $19.99 a month plus $0.85 per minute, or digital service for $24.99 a month plus $0.60 per minute. Give an equation for the amount y of your monthly bill for each service in terms of the number of minutes x that you talk. Which service is more economical if you talk 30 minutes a month? If you talk one hour a month?

2.37   Researchers studying acid rain measured the acidity of precipitation in an isolated wilderness area in Colorado for 150 consecutive weeks in the years 1975 to 1978. The acidity of a solution is measured by pH, with lower pH values indicating that the solution is more acid. The acid rain researchers observed a linear pattern over time. They reported that the least-squares line

$$\text{pH} = 5.43 - (0.0053 \times \text{weeks})$$

fit the data well. (Data from William M. Lewis and Michael C. Grant, "Acid precipitation in the western United States," *Science,* 207 (1980), pp. 176–177.)

(a) Draw a graph of this line. Note that the linear change is decreasing rather than increasing.

(b) According to the fitted line, what was the pH at the beginning of the study (weeks = 1)? At the end (weeks = 150)?

(c) What is the slope of the fitted line? Explain clearly what this slope says about the change in the pH of the precipitation in this wilderness area.

2.38 Concrete road pavement gains strength over time as it cures. Highway builders use regression lines to predict the strength after 28 days (when curing is complete) from measurements made after 7 days. Let x be strength after 7 days (in pounds per square inch) and y the strength after 28 days. One set of data gives this least-squares regression line:

$$\hat{y} = 1389 + 0.96x$$

(a) Draw a graph of this line, with x running from 3000 to 4000 pounds per square inch.

(b) Explain what the slope $b = 0.96$ in this equation says about how concrete gains strength as it cures.

(c) A test of some new pavement after 7 days shows that its strength is 3300 pounds per square inch. Use the equation of the regression line to predict the strength of this pavement after 28 days. Also draw the "up-and-over" lines from $x = 3300$ on your graph, as in Figure 2.13.

2.39 Manatees are large, gentle sea creatures that live along the Florida coast. Many manatees are killed or injured by powerboats. Table 2.6 gives data on powerboat registrations (in thousands) and the number of manatees killed by boats in Florida in the years 1977 to 1990.

(a) Make a scatterplot of these data. Describe the form and direction of the relationship.

(b) Find the correlation. What fraction of the variation in manatee deaths can be explained by the number of boats registered? Does it appear that the number of manatees killed can be predicted accurately from powerboat registrations?

TABLE 2.6   Florida powerboats and manatee deaths, 1977–1990

Year	Boats (thousands)	Manatees killed	Year	Boats (thousands)	Manatees killed
1977	447	13	1984	559	34
1978	460	21	1985	585	33
1979	481	24	1986	614	33
1980	498	16	1987	645	39
1981	513	24	1988	675	43
1982	512	20	1989	711	50
1983	526	15	1990	719	47

(c) Find the least-squares regression line. Predict the number of manatees that will be killed by boats in a year when 716,000 powerboats are registered.

(d) Suppose that in some far future year 2 million powerboats are registered in Florida. Use the regression line to predict manatees killed. Explain why this prediction is very unreliable.

(e) Here are four more years of manatee data, in the same form as in Table 2.6:

1991	716	53	1993	716	35
1992	716	38	1994	735	49

Add these points to your scatterplot. Florida took stronger measures to protect manatees during these years. Do you see any evidence that these measures succeeded?

(f) In part (c) you predicted manatee deaths in a year with 716,000 powerboat registrations. In fact, powerboat registrations remained at 716,000 for the next three years. Compare the mean manatee deaths in these years with your prediction from part (c). How accurate was your prediction?

2.40 Joan is concerned about the amount of energy she uses to heat her home in the Midwest. She keeps a record of the natural gas she consumes each month over one year's heating season. Because the months are not all the same length, she divides each month's consumption by the number of days in the month to get the average number of cubic feet of gas used per day. Demand for heating is strongly influenced by the outside temperature. From local weather records, Joan obtains the average number of heating degree-days per day for each month. (One heating degree-day is accumulated for each degree a day's average temperature falls below 65°F. An average temperature of 20°F, for example, corresponds to 45 degree-days.) Here are Joan's data (provided by Robert Dale, Purdue University):

Month	Oct.	Nov.	Dec.	Jan.	Feb.	Mar.	Apr.	May	June
Degree-days	15.6	26.8	37.8	36.4	35.5	18.6	15.3	7.9	0.0
Gas consumed	520	610	870	850	880	490	450	250	110

(a) Make a scatterplot of these data. There is a strongly linear pattern with no outliers.

(b) Find the equation of the least-squares regression line for predicting gas use from degree-days. Draw this line on your graph. Explain in simple language what the slope of the regression line tells us about how gas use responds to outdoor temperature.

(c) Joan adds insulation in her attic during the summer, hoping to reduce her gas consumption. The next February, there are an average of 40 degree-days per day and her gas consumption is 870 cubic feet per day. Predict from the regression equation how much gas the house

TABLE 2.7 Annual total return on overseas and U.S. stocks

Year	Overseas % return	U.S. % return	Year	Overseas % return	U.S. % return
1971	29.6	14.6	1984	7.4	6.1
1972	36.3	18.9	1985	56.2	31.6
1973	−14.9	−14.8	1986	69.4	18.6
1974	−23.2	−26.4	1987	24.6	5.1
1975	35.4	37.2	1988	28.5	16.8
1976	2.5	23.6	1989	10.6	31.5
1977	18.1	−7.4	1990	−23.0	−3.1
1978	32.6	6.4	1991	12.8	30.4
1979	4.8	18.2	1992	−12.1	7.6
1980	22.6	32.3	1993	32.9	10.1
1981	−2.3	−5.0	1994	6.2	1.3
1982	−1.9	21.5	1995	11.2	37.6
1983	23.7	22.4	1996	6.4	23.0

would have used at 40 degree-days per day last winter before the extra insulation. Did the insulation reduce gas consumption?

2.41 Investors ask about the relationship between returns on investments in the United States and on investments overseas. Table 2.7 gives the total returns on U.S. and overseas common stocks over a 26-year period. (The total return is change in price plus any dividends paid, converted into U.S. dollars. Both returns are averages over many individual stocks.)[10]

(a) Make a scatterplot suitable for predicting overseas returns from U.S. returns.

(b) Find the correlation and r^2. Describe the relationship between U.S. and overseas returns in words, using r and r^2 to make your description more precise.

(c) Find the least-squares regression line of overseas returns on U.S. returns. Add this line to your scatterplot.

(d) In 1993, the return on U.S. stocks was 10.1%. Use the regression line to predict the return on overseas stocks and find the prediction error $y - \hat{y}$. Are you confident that predictions using the regression line will be quite accurate? Why?

2.42 Find the mean and standard deviation of the degree-day and gas consumption data in Exercise 2.40. Find the correlation between the two variables. Use these five numbers to find the equation of the regression line for predicting gas use from degree-days. Verify that your work agrees with your results in Exercise 2.40. Use the same five numbers to find the equation of the regression line for predicting degree-days from gas use. What units does each of these slopes have?

2.43 In Professor Smith's economics course the correlation between the students' total scores before the final examination and their final examination scores is $r = 0.6$. The pre-exam totals for all students in the course have mean 280 and standard deviation 30. The final exam scores have mean 75 and standard deviation 8. Professor Smith has lost Julie's final exam but knows that her total before the exam was 300. He decides to predict her final exam score from her pre-exam total.

(a) What is the slope of the least-squares regression line of final exam scores on pre-exam total scores in this course? What is the intercept?

(b) Use the regression line to predict Julie's final exam score.

(c) Julie doesn't think this method accurately predicts how well she did on the final exam. Calculate r^2 and use the value you get to argue that her actual score could have been much higher or much lower than the predicted value.

2.44   A study of class attendance and grades among first-year students at a state university showed that in general students who attended a higher percent of their classes earned higher grades. Class attendance explained 16% of the variation in grade index among the students. What is the numerical value of the correlation between percent of classes attended and grade index?

2.45   A study of erosion produced the following data on the rate (in liters per second) at which water flows across a soil test bed and the weight (in kilograms) of soil washed away:

Flow rate	0.31	0.85	1.26	2.47	3.75
Eroded soil	0.82	1.95	2.18	3.01	6.07

What percent of the variation in the amount of erosion can be explained by the fact that as the flow rate increases, erosion increases with it in a linear manner?

2.46   The mean height of American women in their early twenties is about 64.5 inches and the standard deviation is about 2.5 inches. The mean height of men the same age is about 68.5 inches, with standard deviation about 2.7 inches. If the correlation between the heights of husbands and wives is about $r = 0.5$, what is the equation of the regression line of the husband's height on the wife's height in young couples? Draw a graph of this regression line. Predict the height of the husband of a woman who is 67 inches tall.

2.47   A student wonders if people of similar heights tend to date each other. She measures herself, her dormitory roommate, and the women in the adjoining rooms; then she measures the next man each woman dates. Here are the data (heights in inches):

Women	66	64	66	65	70	65
Men	72	68	70	68	71	65

(a) Find the equations of two least-squares lines, for the regression of male height on female height and for the regression of female height on male height. How are the slopes of these two lines related? (*Hint:* Look at the expression $b = rs_y/s_x$ for regression slope.)

(b) If both lines were drawn on the same graph, with female height on the x axis, at what point would they intersect?

(c) Consider the regression of male height on female height. How would the slope and intercept of this line change if we measured all heights in centimeters rather than inches?

2.48 Compute the mean and the standard deviation of the metabolic rates and mean body masses in Exercise 2.10 (page 121) and the correlation between these two variables. Use these values to find the slope of the regression line of metabolic rate on lean body mass. Also find the slope of the regression line of lean body mass on metabolic rate. What are the units for each of the two slopes?

2.49 Exercise 2.10 (page 121) gives data on the lean body mass and metabolic rate for 12 women and 7 men. You examined the correlation between these variables in Exercise 2.19 (page 131).

(a) Find the equation of the least-squares regression line for predicting metabolic rate from lean body mass. Explain in simple language what the slope of this line tells us.

(b) Lean body mass x is measured in kilograms and metabolic rate y in calories. What are the numerical values and units of measurement for each of these descriptive measures: $\bar{x}$, s_x, $\bar{y}$, s_y, the correlation r, the slope $b = rs_y/s_x$ of the regression line, and the intercept $a = \bar{y} - b\bar{x}$ of the line?

(c) There are about 2.2 pounds in a kilogram. What would be the numerical values and units for each of the quantities in (b) if we measured lean body mass in pounds? What would be the equation of the least-squares line?

2.50 Table 1.6 (page 33) reports data on 78 seventh-grade students. We want to know how well each of IQ score and self-concept score predicts GPA using least-squares regression. We also want to know which of these explanatory variables predicts GPA better. Give numerical measures that answer these questions, and explain your answers.

2.4 Cautions about Regression and Correlation

Correlation and regression are among the most common statistical tools. They are used in more elaborate form to study relationships among many variables, a situation in which we cannot grasp the essentials by studying a single scatterplot. We need a firm grasp of the use and limitations of these tools, both now and as a foundation for more advanced statistics.

Residuals

A regression line is a mathematical model for the overall pattern of a linear relationship between an explanatory variable and a response variable. Deviations from the overall pattern are also important. In the regression setting, we see deviations by looking at the scatter of the data points about the regression line. The vertical distances from the points to the least-squares regression line are as small as possible, in the sense that they have the smallest possible sum of squares. Because they represent "left-over" variation in the response after fitting the regression line, these distances are called *residuals*.

Residuals

A **residual** is the difference between an observed value of the response variable and the value predicted by the regression line. That is,

$$\text{residual} = \text{observed } y - \text{predicted } y$$
$$= y - \hat{y}$$

EXAMPLE 2.16

According to Table 2.5, the observed mean height of Kalama children at 24 months was 79.9 cm. That is, the observed y is 79.9 when $x = 24$. The least-squares regression line predicts the height to be

$$\hat{y} = 64.93 + (0.635 \times 24) = 80.17 \text{ cm}$$

The residual at 24 months is therefore

$$\text{residual} = \text{observed } y - \text{predicted } y$$
$$= y - \hat{y}$$
$$= 79.9 - 80.17 = -0.27 \text{ cm}$$

The 12 data points used in fitting the least-squares line produce 12 residuals. They are

−0.26	0.01	0.47	−0.06	−0.10	0.17
−0.27	0.30	−0.24	−0.27	0.09	0.16

Most regression software will calculate and store these residuals for you.

Because the residuals show how far the data fall from our regression line, examining the residuals helps assess how well the line describes the data. Although residuals can be calculated from any model fitted to the data, the residuals from the least-squares line have a special property: **the mean of the least-squares residuals is always zero.** Figure 2.18(a) is a scatterplot of the Kalama data with the regression line fitted. Compare this with the *residual plot* in Figure 2.18(b).

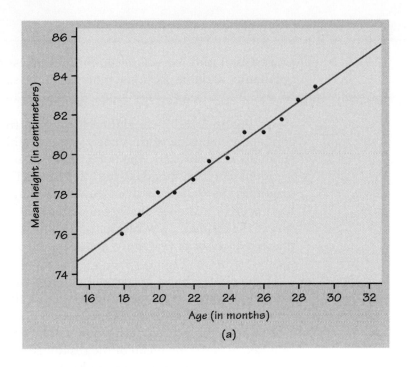

(a)

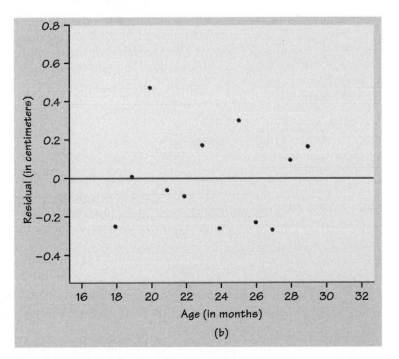

(b)

FIGURE 2.18   (a) The Kalama growth data with the least-squares line. (b) Plot of the residuals from the regression in (a) against the explanatory variable.

> **Residual Plots**
>
> A **residual plot** is a scatterplot of the regression residuals against the explanatory variable. Residual plots help us assess the fit of a regression line.

In Figure 2.18(b), we plot each residual from Example 2.16 against the corresponding value of the explanatory variable x. Because the mean of the residuals is always zero, the horizontal line at zero helps orient us. This line (residual = 0) corresponds to the fitted line in Figure 2.18(a). The residual plot magnifies the deviations from the line to make patterns easier to see. If the regression line catches the overall pattern of the data, there should be *no pattern* in the residuals. The residuals in Figure 2.18(b) have the irregular scatter that is typical of data that do not deviate from the model in any systematic way.

Figure 2.19 shows the overall patterns of some typical residual plots in simplified form. The residuals are plotted in the vertical direction against the corresponding values of the explanatory variable x in the horizontal direction. If all is well, the pattern of this plot will be an unstructured horizontal band centered at zero (the mean of the residuals) and symmetric about zero, as in Figure 2.19(a). A curved pattern like the one in Figure 2.19(b) indicates that the relationship between y and x is curved rather than linear. A straight line is not a good description of such a relationship. A fan-shaped pattern like the one in Figure 2.19(c) shows that the variation of y about the line increases as x increases. Predictions of y will therefore be more precise for smaller values of x, where y shows less spread about the line.

Lurking variables

The basic plot of the residuals against x magnifies patterns that we can see in the original scatterplot of y against x if we add the fitted line to that plot. Other plots of the residuals, however, can reveal new information. They may, for example, suggest the presence of *lurking variables*.

> **Lurking Variable**
>
> A **lurking variable** is a variable that has an important effect on the relationship among the variables in a study but is not included among the variables studied.

Never forget that the relationship between two variables can be strongly influenced by other variables that are lurking in the background. Lurking variables can dramatically change the conclusions of a regression study. Because lurking variables are often unrecognized and unmeasured, detecting their effect is a challenge. Many lurking variables change systematically over time. One useful method for detecting lurking variables is therefore to *plot both*

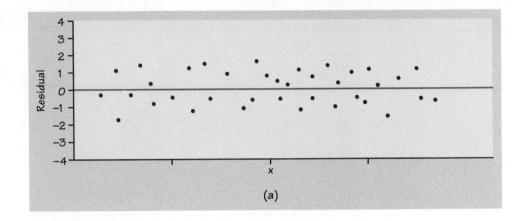

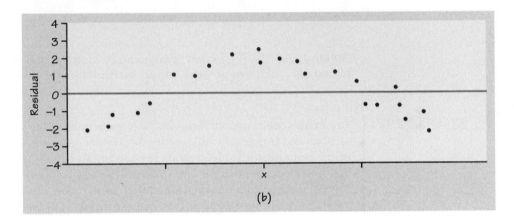

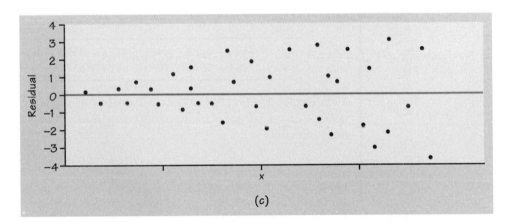

FIGURE 2.19 Simplified patterns in plots of least-squares residuals.

the response variable and the regression residuals against the time order of the observations whenever the time order is available. An understanding of the background of the data then allows you to guess what lurking variables might be present. Here is an example of plotting and interpreting residuals that uncovered a lurking variable.

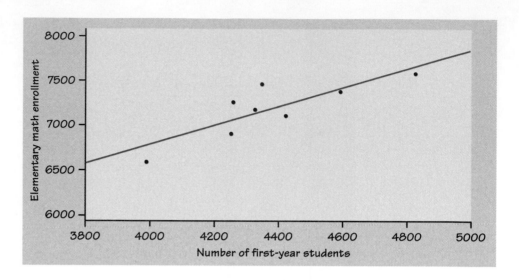

FIGURE 2.20 The regression of elementary mathematics enrollment on number of first-year students at large university, for Example 2.17.

EXAMPLE 2.17 | The mathematics department of a large state university must plan the number of sections and instructors required for its elementary courses. The department hopes that the number of students in these courses can be predicted from the number of first-year students, which is known before the new students actually choose courses. The table below contains data for several years.[11] The explanatory variable x is the number of first-year students. The response variable y is the number of students who enroll in elementary mathematics courses.

Year	1990	1991	1992	1993	1994	1995	1996	1997
x	4595	4827	4427	4258	3995	4330	4265	4351
y	7364	7547	7099	6894	6572	7156	7232	7450

A scatterplot (Figure 2.20) shows a reasonably linear pattern with a cluster of points near the center. We use regression software to obtain the equation of the least-squares regression line:

$$\hat{y} = 2492.69 + 1.0663x$$

The software also tells us that $r^2 = 0.694$. That is, linear dependence on x explains about 70% of the variation in y. The line appears to fit reasonably well.

A plot of the residuals against x (Figure 2.21) magnifies the vertical deviations of the points from the line. We can see that a slightly different line would fit the five lower points very well, so that the three points above the line represent a somewhat different relation between the number of first-year students x and mathematics enrollments y.

A second plot of the residuals clarifies the situation. Figure 2.22 is a plot of the residuals against year. We now see that the five negative residuals are from

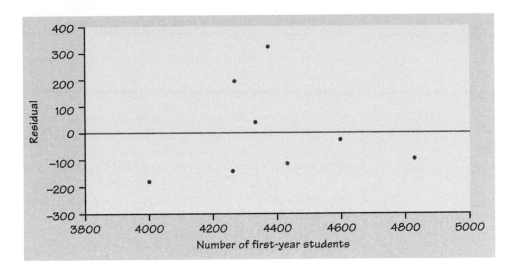

FIGURE 2.21   Residual plot for the regression of mathematics enrollment on number of first-year students, for Example 2.17.

the years 1990 to 1994, and the three positive residuals represent the years 1995 to 1997. This plot suggests that a change took place between 1994 and 1995 that caused a higher proportion of students to take mathematics courses beginning in 1995. In fact, one of the schools in the university changed its program to require that entering students take another mathematics course. This change is the lurking variable that explains the pattern we observed. The mathematics department should not use data from years before 1995 for predicting future enrollment.

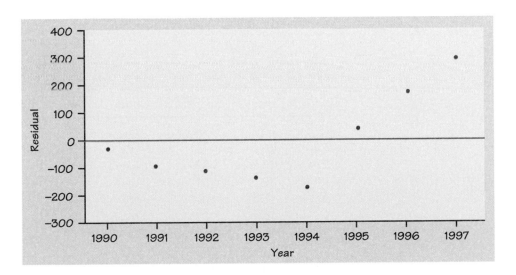

FIGURE 2.22   Plot of the residuals against year for the regression of mathematics enrollment on number of first-year students, for Example 2.17.

Outliers and influential observations

We have seen that the overall pattern of scatterplots and residual plots can be informative. We should also look for striking individual points. Here is an example of data that contain some unusual cases.

EXAMPLE 2.18 | Does the age at which a child begins to talk predict a later score on a test of mental ability? A study of cognitive development in young children recorded the age in months at which each of 21 children spoke their first word and their Gesell Adaptive Score, the result of an ability test taken much later. The data appear in Table 2.8 and in the scatterplot of Figure 2.23.[12] Children 3 and 13, and also Children 16 and 21, have identical values of both variables. We use a different plotting symbol to show that one point stands for two individuals.

The plot shows a moderately linear negative relationship. That is, children who begin to speak later tend to have lower test scores than early talkers. The line on the plot is the least-squares regression line for predicting Gesell score from age at first word. Its equation is

$$\hat{y} = 109.8738 - 1.1270x$$

The data are not strongly linear, but the proportion of variation in Gesell score explained by regression on age at first word, $r^2 = 0.41$, is quite respectable for a study of this kind. We will see that this result is misleading.

Two unusual cases are marked in Figure 2.23. Children 18 and 19 are unusual in different ways. Child 19 lies far from the regression line. Child 18 is close to the line but far out in the x direction. The residual plot in Figure 2.24 confirms that Child 19 has a large residual and that Child 18 does not. Child 19 is an outlier in the y direction, with a Gesell score so high that we should check for a mistake in recording it. In fact, the score is correct.

TABLE 2.8 Age at first word and Gesell score

Child	Age	Score	Child	Age	Score
1	15	95	11	7	113
2	26	71	12	9	96
3	10	83	13	10	83
4	9	91	14	11	84
5	15	102	15	11	102
6	20	87	16	10	100
7	18	93	17	12	105
8	11	100	18	42	57
9	8	104	19	17	121
10	20	94	20	11	86
			21	10	100

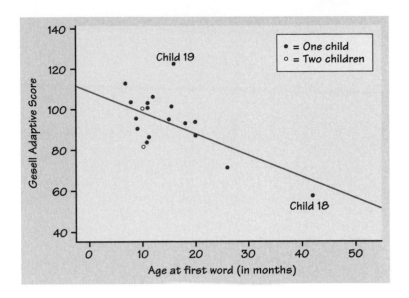

FIGURE 2.23   Regression of Gesell Adaptive Score on age at first word for 21 children, from Table 2.8.

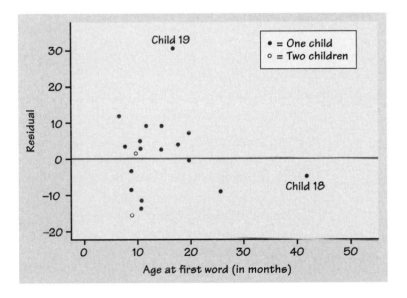

FIGURE 2.24   Residual plot for the regression of Gesell score on age at first word. Child 19 is an outlier and child 18 is an influential case that does not have a large residual.

Child 18 began to speak much later than any of the other children. *Because of its extreme position on the age scale, this point has a strong influence on the position of the regression line.* Figure 2.25 adds a second regression line, calculated after leaving out Child 18. You can see that this one point moves the line quite a bit. Least-squares lines make the sum of squares of the vertical

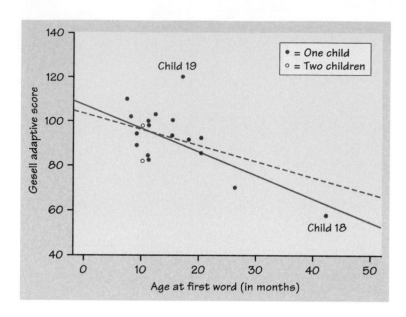

FIGURE 2.25 Least-squares regression of Gesell score on age at first word, with and without an influential observation. The dashed line was calculated leaving out Child 18.

distances to the points as small as possible. A point that is extreme in the x direction with no other points near it pulls the line toward itself. We call such points *influential*.

> ### Outliers and Influential Observations in Regression
>
> An **outlier** is an observation that lies outside the overall pattern of the other observations. Points that are outliers in the y direction of a scatterplot have large regression residuals, but other outliers need not have large residuals.
>
> An observation is **influential** for a statistical calculation if removing it would markedly change the result of the calculation. Points that are outliers in the x direction of a scatterplot are often influential for the least-squares regression line.

We did not need the distinction between outliers and influential observations in Chapter 1. A single large salary that pulls up the mean salary $\bar{x}$ for a group of workers is an outlier because it lies far above the other salaries. It is also influential, because the mean would change if it were removed. In the regression setting, however, not all outliers are influential. The least-squares regression line is most likely to be heavily influenced by observations that are outliers in the x direction. Because influential observations draw the regression line toward themselves, we cannot spot them by looking for large residuals. The basic scatterplot of y against x will alert you to observations that are extreme in x and may therefore be influential. The surest way to verify that a point is influential is to find the regression line both with and without the suspect point, as in Figure 2.25. If the line moves more than a small amount when the point is deleted, the point is influential.

EXAMPLE 2.19

The strong influence of Child 18 makes the original regression of Gesell score on age at first word misleading. The original data have $r^2 = 0.41$. That is, the age at which a child begins to talk explains 41% of the variation on a later test of mental ability. This relationship is strong enough to be interesting to parents. If we leave out Child 18, r^2 drops to only 11%. The apparent strength of the association was largely due to a single influential observation.

What should the child development researcher do? She must decide whether Child 18 is so slow to speak that this individual should not be allowed to influence the analysis. If she excludes Child 18, much of the evidence for a connection between the age at which a child begins to talk and later ability score vanishes. If she keeps Child 18, she needs data on other children who were also slow to begin talking, so that the conclusion no longer depends so heavily on just one child.

Beyond the basics ▶ regression diagnostics

Statisticians have developed a wide variety of special calculations and plots designed to detect difficulties in applying regression. These tools are called *regression diagnostics*.[13] One class of diagnostic tools uses the idea illustrated in Figure 2.25: do the regression without each case in turn, and compare the result without one case with the regression based on all the data. Implementing this idea requires not only a computer but clever software that does the calculations without actually fitting the regression as many times as there are individuals. Here are two examples of these diagnostic tools:

DFFITS

1. Find the predicted response $\hat{y}_i$ for the ith individual from the full data set and then from the data with the ith individual left out. Standardize the difference between these predictions so that they have the standard scale of z-scores. These quantities are called **DFFITS** (an abbreviation for "difference between fitted values"). DFFITS is a direct measure of how much the ith observation moves the fitted line in the y direction.

2. Find the residuals as usual. Because the residuals have mean zero, we can standardize them by dividing by their standard deviation. One very large residual can make the standard deviation large, however. To avoid this, standardize each residual using a standard deviation based on doing the regression with this one individual left out. These are called **studentized residuals**

studentized residuals

. ("Studentized" is an older term that means "standardized" and has remained in use.)

Figure 2.26 shows the ordinary residuals, studentized residuals, and DFFITS for the Gesell data. Large studentized residuals point to outliers, and large DFFITS point to influential observations. As usual, it is easier to examine plots than tables. We recommend making normal quantile plots of these variables.

Obs	Residual	Rstudent	Dffits
1	2.0310	0.1840	0.0413
2	-9.5721	-0.9416	-0.4025
3	-15.6040	-1.5108	-0.3911
4	-8.7309	-0.8143	-0.2243
5	9.0310	0.8329	0.1869
6	-0.3341	-0.0306	-0.0086
7	3.4120	0.3112	0.0772
8	2.5230	0.2297	0.0563
9	3.1421	0.2899	0.0854
10	6.6659	0.6177	0.1728
11	11.0151	1.0508	0.3320
12	-3.7309	-0.3428	-0.0944
13	-15.6040	-1.5108	-0.3911
14	-13.4770	-1.2798	-0.3137
15	4.5230	0.4132	0.1013
16	1.3960	0.1274	0.0330
17	8.6500	0.7983	0.1872
18	-5.5403	-0.8451	-1.1558
19	30.2850	3.6070	0.8537
20	-11.4770	-1.0765	-0.2638
21	1.3960	0.1274	0.0330

FIGURE 2.26 Software output of some regression diagnostics for the Gesell data in Table 2.8.

EXAMPLE 2.20

Figure 2.27 is a normal quantile plot of the studentized residuals. We see a roughly normal distribution with one strong outlier. Child 19 has a residual of 30.285 with a standardized value of 3.607. Because the distribution of residuals for these data is roughly normal, we know that a value 3.6 standard deviations above the mean is extreme.

Figure 2.28 is the normal quantile plot for the DFFITS values. Although DF-FITS are standardized as part of their definition, they tend to be smaller than normal observations. Look for points that stand out in the plot rather than for values that would seem large by the 68–95–99.7 rule. Child 18, with DFFITS = −1.1558, certainly stands out as influential. Child 19 is next (DFFITS = 0.8537), but this outlier is only about half as influential as Child 18 by the DFFITS measure.

Regression diagnostics are helpful even when we regress a response variable on just one explanatory variable, as in this example. DFFITS in particular makes influence specific—it is easier to look at DFFITS than to try to guess which is the second most influential observation in the scatterplot of Figure 2.23. Diagnostic tools become essential in *multiple regression*, which uses several explanatory variables to predict a response, because a scatterplot can no longer fully display the relationships among all the variables. We will have more to say about multiple regression in Chapter 11.

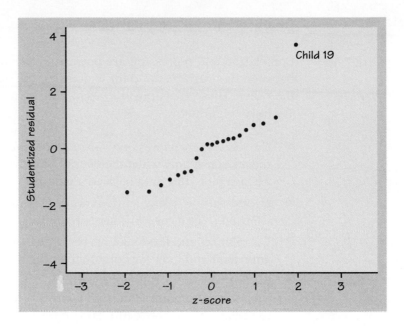

FIGURE 2.27 Normal quantile plot of the studentized residuals for the Gesell data. Use this plot to spot outliers from the pattern represented by the regression line.

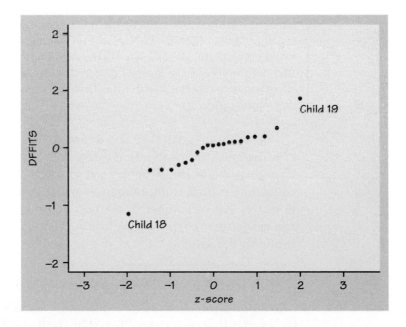

FIGURE 2.28 Normal quantile plot of the DFFITS values for the Gesell data. Use this plot to spot influential observations.

Beware the lurking variable

Correlation and regression are powerful tools for measuring the association between two variables and for expressing the dependence of one variable on the other. These tools must be used with an awareness of their limitations. We have seen that:

- Correlation measures *only linear association,* and fitting a straight line makes sense only when the overall pattern of the relationship is linear. Always plot your data before calculating.

- *Extrapolation* (using a fitted model far outside the range of the data that we used to fit it) can produce predictions that are unreliable.

- Correlation and least-squares regression are *not resistant.* Always plot your data and look for potentially influential points.

- *Lurking variables* can make a correlation or regression misleading. Plot the residuals against time and against other variables that may influence the relationship between x and y.

Here are two more examples that illustrate how lurking variables influence the relationship between an explanatory variable and a response variable.

EXAMPLE 2.21   Measure the number of television sets per person x and the average life expectancy y for the world's nations. There is a high positive correlation—nations with many TV sets have higher life expectancies. Could we lengthen the lives of people in Rwanda by shipping them TV sets? No. Rich nations have more TV sets than poor nations. Rich nations also have longer life expectancies because they offer better nutrition, clean water, and better health care. There is no cause-and-effect tie between TV sets and length of life.

Correlations such as that in Example 2.21 are sometimes called "nonsense correlations." The correlation is real. What is nonsense is the conclusion that changing one of the variables *causes* changes in the other. The question of causation is important enough to merit separate treatment in Section 2.7. For now, just remember that an association between two variables x and y can reflect many types of relationship between x, y, and other variables not explicitly recorded.

> **Association Does Not Imply Causation**
>
> An association between an explanatory variable x and a response variable y, even if it is very strong, is not by itself good evidence that changes in x actually cause changes in y.

Lurking variables sometimes create "nonsense correlations" between x and y, as in Example 2.21. They can also hide a true relationship between x and y, as the following example illustrates.

EXAMPLE 2.22

A study of housing conditions and health in the city of Hull, England, measured a large number of variables for each of the wards into which the city is divided. Two of the variables were an index x of overcrowding and an index y of the lack of indoor toilets. Because x and y are both measures of inadequate housing, we expect a high correlation. Yet the correlation was only $r = 0.08$. How can this be? Investigation disclosed that some poor wards were dominated by public housing. These wards had high values of x but low values of y because public housing always includes indoor toilets. Other poor wards lacked public housing, and in these wards high values of x were accompanied by high values of y. Because the relationship between x and y differed in the two types of wards, analyzing all wards together obscured the nature of the relationship.[14]

Figure 2.29 shows in simplified form how groups formed by a categorical lurking variable as in the housing example can make the correlation r misleading. The groups appear as clusters of points in the scatterplot. There is a strong relationship between x and y within each of the clusters. In fact, $r = 0.85$ and $r = 0.91$ in the two clusters. However, because similar values of x correspond to quite different values of y in the two clusters, x alone is of little value for predicting y. The correlation for all points displayed is therefore low, $r = 0.14$. This example is another reminder to plot the data rather than simply calculate numerical measures such as the correlation.

Beware correlations based on averaged data

Many regression or correlation studies work with averages or other measures that combine information from many individuals. We saw, for example, a very strong linear relationship ($r = 0.994$) between age and the mean height of children in the Egyptian village of Kalama.

Averaging over many children smooths out the variation in height among children the same age, resulting in a very high correlation between mean height and age. Because children of each age vary in height, the correlation

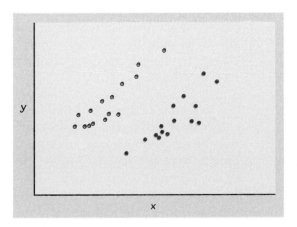

FIGURE 2.29   This scatterplot has low r^2 even though there is a strong correlation within each of the two clusters.

between age and height would be much smaller if we used the data on the heights of all the individual children at each age. **A correlation based on averages over many individuals is usually higher than the correlation between the same variables based on data for individuals.** This fact reminds us again of the importance of noting exactly what variables a statistical study involves.

The restricted-range problem

A regression line is often used to *predict* the response y to a given value x of the explanatory variable. This is clearly valid when the regression reflects a cause-and-effect relationship and r^2 is high enough to give us confidence that changes in x explain most of the variation in y. However, **successful prediction does not require a cause-and-effect relationship.** If both x and y respond to the same underlying unmeasured variables, it may be possible to predict y from x even though x has no direct influence on y.

EXAMPLE 2.23   American law schools use the Law School Admission Test (LSAT) to help predict performance and to aid in admissions decisions.[15] There is no cause-and-effect relationship between the score x that a college senior achieves on the LSAT and the student's grade point average (GPA) y as a first-year law student the following year. Because both x and y reflect the student's level of ability and knowledge, however, it is plausible to predict y from x.

The success of this prediction depends on the strength of the association between LSAT scores and later GPA in law school. The correlation between LSAT scores and the first-year grades in law school of admitted applicants is only about $r = 0.36$. This is discouraging, for such a correlation means that LSAT scores explain only 13% of the variance in law school grades. Nonetheless, LSAT scores are a better predictor than the applicants' undergraduate grades, which have a correlation of only about $r = 0.25$ with later grades in law school.

When no single explanatory variable has a high correlation with a response variable, we can use several explanatory variables together to predict the response. This is called *multiple regression*. What is more, there is a *multiple correlation coefficient r* whose square is the proportion of variation in the response explained by the multiple regression. When both LSAT scores and undergraduate grades are used to predict law school grades, the multiple correlation coefficient is about $r = 0.45$. The two explanatory variables together explain about 20% of the variability in first-year law school grades.

Example 2.23 suggests that LSAT scores and undergraduate grades do a rather poor job of predicting law school GPA. The values of r^2 in this example may be misleadingly low, however, because the data are incomplete in a systematic way. We have GPA information only for students who were admitted to law school. Because many students with low LSAT scores and low undergraduate grades were not admitted, the data refer primarily to students with *restricted range* high scores and high grades. This is the **restricted-range** problem: the data do not contain information on the full range of the explanatory variables. We suspect that if all students who take the LSAT were allowed to try law school, those with low scores and weak undergraduate grades would as a group have lower law school GPAs than the present students. In that case, r^2 would have a higher value.

SUMMARY

You can examine the fit of a regression line by plotting the **residuals,** which are the differences between the observed and predicted values of y. Be on the lookout for points with unusually large residuals and also for nonlinear patterns and uneven variation about the line.

Evidence of the effects of **lurking variables** on the relationship between x and y may be provided by plots of y and the residuals against the time order of the observations.

Also look for **influential observations,** individual points that substantially change the regression line. Influential observations are often outliers in the x direction, but they need not have large residuals.

Correlation and regression must be **interpreted with caution.** Plot the data to be sure the relationship is roughly linear and to detect outliers and influential observations.

Lurking variables that you did not measure may explain the relationship between the variables you did measure. Correlation and regression can be misleading if you ignore important lurking variables.

We cannot conclude that there is a cause-and-effect relationship between two variables just because they are strongly associated. **High correlation does not imply causation.**

A correlation based on averages is usually higher than if we had data for individuals. A correlation based on data with a **restricted range** is often lower than would be the case if we could observe the full range of the variables.

SECTION 2.4 EXERCISES

2.51   A study of nutrition in developing countries collected data from the Egyptian village of Nahya. Here are the mean weights (in kilograms) for 170 infants in Nahya who were weighed each month during their first year of life. (Data from Zeinab E. M. Afifi, "Principal components analysis of growth of Nahya infants: size, velocity and two physique factors," *Human Biology,* 57 (1985), pp. 659–669.)

Age (months)	1	2	3	4	5	6	7	8	9	10	11	12
Weight (kg)	4.3	5.1	5.7	6.3	6.8	7.1	7.2	7.2	7.2	7.2	7.5	7.8

(a) Plot weight against time.

(b) A hasty user of statistics enters the data into software and computes the least-squares line without plotting the data. The result is

```
THE REGRESSION EQUATION IS
WEIGHT = 4.88 + 0.267 AGE
```

Plot this line on your graph. Is it an acceptable summary of the overall pattern of growth? Remember that you can calculate the least-squares line for *any* set of two-variable data. It's up to you to decide if it makes sense to fit a line.

(c) Fortunately for the hasty user, the software also prints out the residuals from the least-squares line. In order of age along the rows, they are

−0.85	−0.31	0.02	0.35	0.58	0.62
0.45	0.18	−0.08	−0.35	−0.32	−0.28

Verify that the residuals have sum 0 (except for roundoff error). Plot the residuals against age and add a horizontal line at 0. Describe carefully the pattern that you see.

2.52 Table 1.8 (page 40) gives the calories and sodium content for each of 17 brands of meat hot dogs. The discussion on page 50 identified "Eat Slim Veal Hot Dogs" as an outlier in the distribution of calories.

(a) Make a scatterplot of sodium content y against calories x. Describe the main features of the relationship. Is "Eat Slim Veal Hot Dogs" an outlier in this plot? Circle its point.

(b) Calculate two least-squares regression lines, one using all of the observations and the other omitting "Eat Slim." Draw both lines on your plot. Does a comparison of the two regression lines show that the "Eat Slim" brand is influential? Explain your answer.

(c) A new brand of meat hot dog (not made with veal) has 150 calories per frank. How many milligrams of sodium do you estimate that one of these hot dogs contains?

2.53 Here are the golf scores of 11 members of a women's golf team in two rounds of tournament play:

Player	1	2	3	4	5	6	7	8	9	10	11
Round 1	89	90	87	95	86	81	105	83	88	91	79
Round 2	94	85	89	89	81	76	89	87	91	88	80

(a) Plot the data with the Round 1 scores on the x axis and the Round 2 scores on the y axis. There is a generally linear pattern except for one potentially influential observation. Circle this observation on your graph.

(b) Here are the equations of two least-squares lines. One of them is calculated from all 11 data points and the other omits the influential observation.

$$\hat{y} = 20.49 + 0.754x$$
$$\hat{y} = 50.01 + 0.410x$$

Draw both lines on your scatterplot. Which line omits the influential observation? How do you know this?

2.54 Gas chromatography is a technique used to detect very small amounts of a substance, for example, a contaminant in drinking water. Laboratories use regression to calibrate such techniques. The data below show the results of five measurements for each of four amounts of the substance being investigated. The explanatory variable x is the amount of substance in the specimen, measured in nanograms (ng), units of 10^{-9} gram. The response variable y is the reading from the gas chromatograph. (Data from D. A. Kurtz (ed.), *Trace Residue Analysis*, American Chemical Society Symposium Series No. 284, 1985, Appendix.)

Amount (ng)	Response				
0.25	6.55	7.98	6.54	6.37	7.96
1.00	29.7	30.0	30.1	29.5	29.1
5.00	211	204	212	213	205
20.00	929	905	922	928	919

(a) Make a scatterplot of these data. The relationship appears to be approximately linear, but the wide variation in the response values makes it hard to see detail in this graph.

(b) Compute the least-squares regression line of y on x, and plot this line on your graph.

(c) Now compute the residuals and make a plot of the residuals against x. It is much easier to see deviations from linearity in the residual plot. Describe carefully the pattern displayed by the residuals.

2.55 Runners are concerned about their form when racing. One measure of form is the stride rate, the number of steps taken per second. As running speed increases, the stride rate should also increase. In a study of 21 of the best American female runners, researchers measured the stride rate for different speeds. The following table gives the speeds (in feet per second) and the mean stride rates for these runners. (Data from R. C. Nelson, C. M. Brooks, and N. L. Pike, "Biomechanical comparison of male and female distance runners," in P. Milvy (ed.), *The Marathon: Physiological, Medical,*

Epidemiological, and Psychological Studies, New York Academy of Sciences, 1977, pp. 793–807.)

Speed	15.86	16.88	17.50	18.62	19.97	21.06	22.11
Stride rate	3.05	3.12	3.17	3.25	3.36	3.46	3.55

(a) Plot the data with speed on the x axis and stride rate on the y axis. Does a straight line adequately describe these data?

(b) Find the equation of the regression line of stride rate on speed. Draw this line on your plot.

(c) For each of the speeds given, obtain the predicted value and the residual. Verify that the residuals add to 0.

(d) Plot the residuals against speed. Describe the pattern. What does the plot indicate about the adequacy of the linear fit? Are there any potentially influential observations? Can you plot the residuals against the time at which the observations were made?

2.56   Research on digestion requires accurate measurements of blood flow through the lining of the stomach. A promising way to make such measurements easily is to inject mildly radioactive microscopic spheres into the blood stream. The spheres lodge in tiny blood vessels at a rate proportional to blood flow; their radioactivity allows blood flow to be measured from outside the body. Medical researchers compared blood flow in the stomachs of dogs, measured by use of microspheres, with simultaneous measurements taken using a catheter inserted into a vein. The data, in milliliters of blood per minute (ml/minute), appear below. (Based on L. H. Archibald, F. G. Moody, and M. Simons, "Measurement of gastric blood flow with radioactive microspheres," *Journal of Applied Physiology,* 38 (1975), pp. 1051–1056.)

Spheres	4.0	4.7	6.3	8.2	12.0	15.9	17.4	18.1	20.2	23.9
Vein	3.3	8.3	4.5	9.3	10.7	16.4	15.4	17.6	21.0	21.7

(a) Make a scatterplot of these data, with the microsphere measurement as the explanatory variable. There is a strongly linear pattern.

(b) Calculate the least-squares regression line of venous flow on microsphere flow. Draw your regression line on the scatterplot.

(c) Predict the venous measurement for microsphere measurements 6, 12, and 18 ml/minute. If the predicted venous flow is within about 10% to 15% of the actual venous flow, the researchers will use the microsphere measurements in future studies. Is this condition satisfied over this range of blood flow?

2.57 Exercise 2.41 (page 151) examined the relationship between returns on U.S. and overseas stocks. Return to the scatterplot and regression line for predicting overseas returns from U.S. returns.

(a) Circle the point that has the largest residual (either positive or negative). What year is this? Redo the regression without this point and add the new regression line to your plot. Was this observation very influential?

(b) Whenever we regress two variables that both change over time, we should plot the residuals against time as a check for time-related lurking variables. Make this plot for the stock returns data. Are there any suspicious patterns in the residuals?

2.58 Table 2.9 gives data on the amount of beef consumed (pounds per person) and average retail price of beef (dollars per pound) in the United States for the years 1970 to 1993. Because all prices were generally rising during this period, the prices given are "real prices" in 1993 dollars. These are dollars with the buying power that a dollar had in 1993.

(a) Economists expect consumption of an item to fall when its real price rises. Make a scatterplot of beef consumption y against beef price x. Do you see a relationship of the type expected?

(b) Find the equation of the least-squares line and draw the line on your plot. What proportion of the variation in beef consumption is explained by regression on beef price?

(c) Although it appears that price helps explain consumption, the scatterplot seems to show some nonlinear patterns. Find the residuals from your regression in (b) and plot them against time. Connect the successive points by line segments to help see the pattern. Are there

TABLE 2.9 Price and consumption of beef, 1970–1993

Year	Price per pound (1993 dollars)	Consumption (lbs. per capita)	Year	Price per pound (1993 dollars)	Consumption (lbs. per capita)
1970	3.721	84.62	1982	3.570	77.03
1971	3.789	83.93	1983	3.396	78.64
1972	4.031	85.27	1984	3.274	78.41
1973	4.543	80.51	1985	3.069	79.19
1974	4.212	85.58	1986	2.989	78.83
1975	4.106	88.15	1987	3.032	73.84
1976	3.698	94.36	1988	3.057	72.65
1977	3.477	91.76	1989	3.096	69.34
1978	3.960	87.29	1990	3.107	67.78
1979	4.423	78.10	1991	3.059	66.79
1980	4.098	76.56	1992	2.931	66.48
1981	3.731	77.26	1993	2.934	65.06

systematic effects of time remaining after we regress consumption on price? (A partial explanation is that beef production responds to price changes only after some time lag.)

2.59   The price of seafood varies with species and time. The following table gives the prices in cents per pound received in 1970 and 1980 by fishermen and vessel owners for several species.

Species	1970 price	1980 price
Cod	13.1	27.3
Flounder	15.3	42.4
Haddock	25.8	38.7
Menhaden	1.8	4.5
Ocean perch	4.9	23.0
Salmon, chinook	55.4	166.3
Salmon, coho	39.3	109.7
Tuna, albacore	26.7	80.1
Clams, soft	47.5	150.7
Clams, blue, hard	6.6	20.3
Lobsters, American	94.7	189.7
Oysters, eastern	61.1	131.3
Sea scallops	135.6	404.2
Shrimp	47.6	149.0

(a) Plot the data with the 1970 price on the x axis and the 1980 price on the y axis.

(b) Describe the overall pattern. Are there any outliers? If so, circle them on your graph. Do these unusual points have large residuals from a fitted line? Are they influential in the sense that removing them would change the fitted line?

(c) Compute the correlation for the entire set of data. What percent of the variation in 1980 prices is explained by the 1970 prices?

(d) Recompute the correlation discarding the cases that you circled in (b). Do these observations have a strong effect on the correlation? Explain why or why not.

(e) Does the correlation provide a good measure of the relationship between the 1970 and 1980 prices for this set of data? Explain your answer.

2.60   Table 2.10 presents four sets of data prepared by the statistician Frank Anscombe to illustrate the dangers of calculating without first plotting the data. (Frank J. Anscombe, "Graphs in statistical analysis," *The American Statistician*, 27 (1973), pp. 17–21.)

(a) Without making scatterplots, find the correlation and the least-squares regression line for all four data sets. What do you notice? Use the regression line to predict y for $x = 10$.

TABLE 2.10 Four data sets for exploring correlation and regression										
Data Set A										
x	10	8	13	9	11	14	6	4	12	7	5
y	8.04	6.95	7.58	8.81	8.33	9.96	7.24	4.26	10.84	4.82	5.68
Data Set B										
x	10	8	13	9	11	14	6	4	12	7	5
y	9.14	8.14	8.74	8.77	9.26	8.10	6.13	3.10	9.13	7.26	4.74
Data Set C										
x	10	8	13	9	11	14	6	4	12	7	5
y	7.46	6.77	12.74	7.11	7.81	8.84	6.08	5.39	8.15	6.42	5.73
Data Set D										
x	8	8	8	8	8	8	8	8	8	8	19
y	6.58	5.76	7.71	8.84	8.47	7.04	5.25	5.56	7.91	6.89	12.50

SOURCE: Frank J. Anscombe, "Graphs in statistical analysis," *The American Statistician*, 27 (1973), pp. 17–21.

(b) Make a scatterplot for each of the data sets and add the regression line to each plot.

(c) In which of the four cases would you be willing to use the regression line to describe the dependence of y on x? Explain your answer in each case.

2.61 In each of the following settings, make a scatterplot and draw the least-squares regression line on the plot. We would be reluctant to use these lines for prediction. State carefully what general caution about regression each example illustrates.

(a) We want to predict a car's fuel consumption from its speed. Data appear in Exercise 2.8 on page 120.

(b) We want to predict average household spending on tobacco in a region of Great Britain from spending on alcohol in the region. Data appear in Exercise 2.24 on page 133.

2.62 If you need medical care, should you go to a hospital that handles many cases like yours? Figure 2.30 presents some data for heart attacks. The figure plots mortality rate (the proportion of patients who died) against the number of heart attack patients treated for a large number of hospitals in a recent year. The line on the plot is the least-squares regression line for predicting mortality from number of patients.

(a) Do the plot and regression generally support the thesis that mortality is lower at hospitals that treat more heart attacks? Is the relationship very strong?

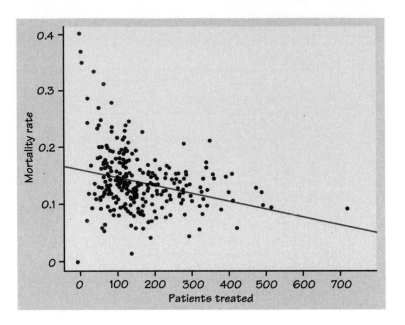

FIGURE 2.30 Mortality of heart attack patients and number of heart attack cases treated for a large group of hospitals, for Exercise 2.62.

(b) In what way is the pattern of the plot nonlinear? (Apply a scatterplot smoother if your software provides one.) Does the nonlinearity strengthen or weaken the conclusion that heart attack patients should avoid hospitals that treat few heart attacks? Why?

2.63 Exercise 2.41 (page 151) examined the relationship between returns on U.S. and overseas stocks. Investors also want to know what typical returns are and how much year-to-year variability (called *volatility* in finance) there is. Regression and correlation don't answer questions about center and spread.

(a) Find the five-number summaries for both U.S. and overseas returns, and make side-by-side boxplots to compare the two distributions.

(b) Were returns generally higher in the United States or overseas during this period? Explain your answer.

(c) Were returns more volatile (more variable) in the United States or overseas during this period? Explain your answer.

2.64 Airborne particles such as dust and smoke are an important part of air pollution. To measure particulate pollution, a vacuum motor draws air through a filter for 24 hours. Weigh the filter at the beginning and end of the period. The weight gained is a measure of the concentration of particles in the air. A study of air pollution made measurements every 6 days with identical instruments in the center of a small city and at a rural location 10 miles southwest of the city. Because the prevailing winds blow from the west, we suspect that the rural readings will be generally lower than the

TABLE 2.11 Particulate levels (grams) in two nearby locations

Day	1	2	3	4	5	6	7	8	9	10	11	12
Rural	NA	67	42	33	46	NA	43	54	NA	NA	NA	NA
City	39	68	42	34	48	82	45	NA	NA	60	57	NA

Day	13	14	15	16	17	18	19	20	21	22	23	24
Rural	38	88	108	57	70	42	43	39	NA	52	48	56
City	39	NA	123	59	71	41	42	38	NA	57	50	58

Day	25	26	27	28	29	30	31	32	33	34	35	36
Rural	44	51	21	74	48	84	51	43	45	41	47	35
City	45	69	23	72	49	86	51	42	46	NA	44	42

city readings, but that the city readings can be predicted from the rural readings. Table 2.11 gives readings taken every 6 days over a 7-month period. The entry NA means that the reading for that date is not available, usually because of equipment failure. (Data provided by Matthew Moore.)

(a) We hope to use the rural particulate level to predict the city level on the same day. Make a graph to examine the relationship. Does the graph suggest that using the least-squares regression line for prediction will give approximately correct results over the range of values appearing in the data? Calculate the least-squares line for predicting city pollution from rural pollution. What percent of the observed variation in city pollution levels does this straight-line relationship account for?

(b) Find the residuals from your least-squares fit. Plot the residuals both against x and against the time order of the observations, and comment on the results.

(c) Which observation appears to be the most influential? Circle this observation on your plot. Is it the observation with the largest residual?

(d) On the fourteenth date in the series, the rural reading was 88 and the city reading was not available. What do you estimate the city reading to be for that date?

(e) Make a normal quantile plot of the residuals. (Make a stemplot or histogram if your software does not make normal quantile plots.) Is the distribution of the residuals nearly symmetric? Does it appear to be approximately normal?

2.65 A multimedia statistics learning system includes a test of skill in using the computer's mouse. The software displays a circle at a random location on the computer screen. The subject tries to click in the circle with the mouse as quickly as possible. A new circle appears as soon as the subject clicks the

TABLE 2.12 Reaction times in a computer game

Time	Distance	Hand	Time	Distance	Hand
115	190.70	right	240	190.70	left
96	138.52	right	190	138.52	left
110	165.08	right	170	165.08	left
100	126.19	right	125	126.19	left
111	163.19	right	315	163.19	left
101	305.66	right	240	305.66	left
111	176.15	right	141	176.15	left
106	162.78	right	210	162.78	left
96	147.87	right	200	147.87	left
96	271.46	right	401	271.46	left
95	40.25	right	320	40.25	left
96	24.76	right	113	24.76	left
96	104.80	right	176	104.80	left
106	136.80	right	211	136.80	left
100	308.60	right	238	308.60	left
113	279.80	right	316	279.80	left
123	125.51	right	176	125.51	left
111	329.80	right	173	329.80	left
95	51.66	right	210	51.66	left
108	201.95	right	170	201.95	left

old one. Table 2.12 gives data for one subject's trials, 20 with each hand. Distance is the distance from the cursor location to the center of the new circle, in units whose actual size depends on the size of the screen. Time is the time required to click in the new circle, in milliseconds.[16]

(a) We suspect that time depends on distance. Make a scatterplot of time against distance, using separate symbols for each hand.

(b) Describe the pattern. How can you tell that the subject is right-handed?

(c) Find the regression line of time on distance separately for each hand. Draw these lines on your plot. Which regression does a better job of predicting time from distance? Give numerical measures that describe the success of the two regressions.

(d) It is possible that the subject got better in later trials due to learning. It is also possible that he got worse due to fatigue. Plot the residuals from each regression against the time order of the trials (down the columns in Table 2.12). Is either of these systematic effects of time visible in the data?

2.66   There is a strong positive correlation between years of education and income for economists employed by business firms. (In particular, economists with doctorates earn more than economists with only a bachelor's degree.) There is also a strong positive correlation between years of education and income for economists employed by colleges and universities. But when all economists are considered, there is a *negative* correlation between education and income. The explanation for this is that business pays high salaries and employs mostly economists with bachelor's degrees, while colleges pay lower salaries and employ mostly economists with doctorates. Sketch a scatterplot with two groups of cases (business and academic) that illustrates how a strong positive correlation within each group and a negative overall correlation can occur together. (*Hint*: Begin by studying Figure 2.29.)

2.67 Exercise 2.55 gives data on the mean stride rate of a group of 21 elite female runners at various running speeds. Find the correlation between speed and stride rate. Would you expect this correlation to increase or decrease if we had data on the individual stride rates of all 21 runners at each speed? Why?

2.68 Are baseball players paid according to their performance? To study this question, a statistician analyzed the salaries of over 260 major league hitters along with such explanatory variables as career batting average, career home runs per time at bat, and years in the major leagues. This is a *multiple regression* with several explanatory variables. More detail on multiple regression appears in Chapter 11, but the fit of the model is assessed just as we have done in this chapter, by calculating and plotting the residuals:

$$\text{residual} = \text{observed } y - \text{predicted } y$$

(This analysis was done by Crystal Richard.)

(a) Figure 2.31(a) is a plot of the residuals against the predicted salary. When points are too close together to plot separately, the plot uses letters of the alphabet to show how many points there are at each position. Describe the pattern that appears on this residual plot. Will the regression model predict high or low salaries more precisely?

(b) After studying the residuals in more detail, the statistician decided to predict the logarithm of salary rather than the salary itself. One reason was that while salaries are not normally distributed (the distribution is skewed to the right), their logarithms are nearly normal. When the response variable is the logarithm of salary, a plot of the residuals against the predicted value is satisfactory—it looks like Figure 2.19(a). Figure 2.30(b) is a plot of the residuals against the number of years the player has been in the major leagues. Describe the pattern that you see. Will the model overestimate or underestimate the salaries of players who are new to the majors? Of players who have been in the major leagues about 8 years? Of players with more than 15 years in the majors?

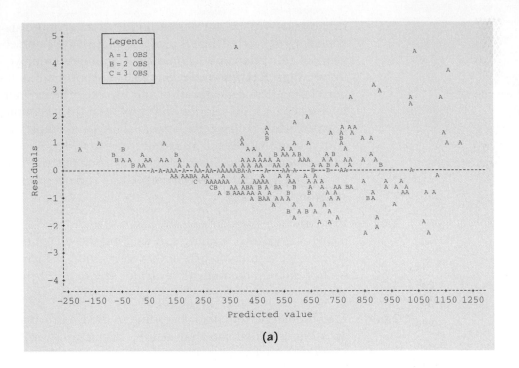

(a)

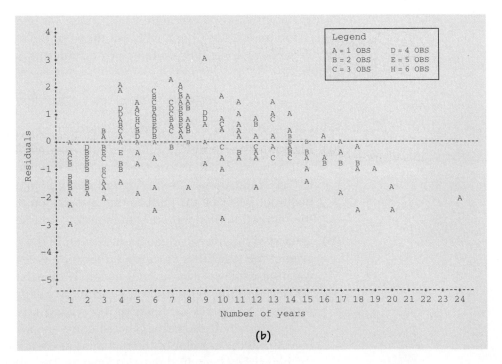

(b)

FIGURE 2.31   Two residual plots for the regression of baseball players' salaries on their performance, for Exercise 2.68.

2.5 An Application: Exponential Growth and World Oil Production*

Petroleum is the most important source of energy for the developed nations and has been the cause of economic dislocation and even war. Table 2.13 and Figure 2.32 show the growth of annual world crude oil production, measured in millions of barrels per year.[17] There is an increasing trend, but the overall pattern is not linear. Oil production has increased much faster than linear growth. The pattern of growth follows a smooth curve until 1973, when a Mideast war touched off a vast price increase and a change in the previous pattern of production. Can we describe the growth of oil production from 1880 to 1973 by a simple mathematical model? We can indeed, and the resulting model for *exponential growth* is second only to straight lines as a useful simple pattern for data. A fitted line does not describe exponential growth, but we will see that a simple transformation of the data enables us to use a regression line and residuals to examine the history of oil production.

The nature of exponential growth

A variable grows linearly over time if it *adds* a fixed increment in each equal time period. Exponential growth occurs when a variable is *multiplied* by a fixed number in each time period. To grasp the effect of multiplicative growth, consider a population of bacteria in which each bacterium splits into two each hour. Beginning with a single bacterium, we have 2 after one hour, 4 at the end

TABLE 2.13 Annual world crude oil production, 1880–1994 (millions of barrels)

Year	Mbbl.	Year	Mbbl.	Year	Mbbl.
1880	30	1945	2,595	1976	20,188
1890	77	1950	3,803	1978	21,922
1900	149	1955	5,626	1980	21,722
1905	215	1960	7,674	1982	19,411
1910	328	1962	8,882	1984	19,837
1915	432	1964	10,310	1986	20,246
1920	689	1966	12,016	1988	21,338
1925	1,069	1968	14,104	1990	22,100
1930	1,412	1970	16,690	1992	22,028
1935	1,655	1972	18,584	1994	22,234
1940	2,150	1974	20,389		

---

*Exponential growth is important in biology, business, and some other areas of application, but the topic can be omitted without loss of continuity.

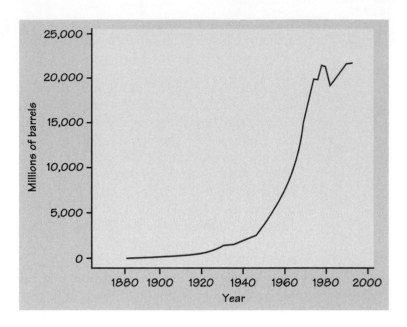

FIGURE 2.32   Annual world crude oil production, 1880 to 1994, from Table 2.13.

of two hours, 8 after three hours, then 16, 32, 64, 128, and so on. These first few numbers are deceiving. After 1 day of doubling each hour, there are 2^{24} (16,777,216) bacteria in the population. That number then doubles the next hour! Try successive multiplications by 2 on your calculator to see for yourself the very rapid increase after a slow start. Figure 2.33 shows the growth of the bacteria population over 24 hours. For the first 15 hours, the population is too small to rise visibly above the zero level on the graph.

It is characteristic of exponential growth that the increase appears slow for a long period, then seems to explode. Both Figures 2.32 and 2.33 display explosive growth after a long period of gradual increase. Our first reaction to the apparently sudden surge is to search for some explanation, something new that distinguishes the high-growth present from the low-growth past. But the nature of the increase has not changed at all. It is our understanding of the long-term consequences of exponential growth that needs correcting. An old story tells how a king of Persia learned this lesson. There was once, the story goes, a courtier who asked the king for a simple reward: a grain of rice on the first square of a chess board, 2 grains on the second square, then 4, 8, 16, and so on through all 64 squares. The king foolishly granted the request, though we assume that he removed the courtier's head rather than hand over his entire kingdom.

Linear versus Exponential Growth

Linear growth increases by a fixed *amount* in each equal time period. **Exponential growth** increases by a fixed *percentage* of the previous total.

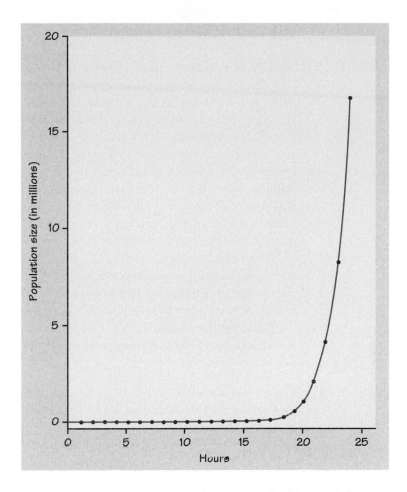

FIGURE 2.33   Growth of a bacteria population that doubles each hour.

Populations of living things—like bacteria—tend to grow exponentially if not restrained by outside limits such as lack of food or space. More pleasantly, money also displays exponential growth when returns to an investment are compounded. Compounding means that last period's income earns income this period.

EXAMPLE 2.24

A dollar invested at an annual rate of 6% turns into $1.06 in a year. The original dollar remains and has earned $0.06 in interest. That is, 6% annual interest means that any amount on deposit for the entire year is multiplied by 1.06. If the $1.06 remains invested for a second year, the new amount is therefore 1.06×1.06, or 1.06^2. That is only $1.12, but this in turn is multiplied by 1.06 during the third year, and so on. After n years, the dollar has become 1.06^n dollars.

If the Native Americans who sold Manhattan Island for $24 in 1626 had deposited the $24 in a savings account at 6% annual interest, they would now have over $40 billion. Our savings accounts don't make us billionaires, because we don't stay around long enough. A century of growth at 6% per year turns $24 into $8143. That's 1.06^{100} times $24. By 1826, two centuries after the sale, the account would hold a bit over $2.7 million. Only after a patient 302 years do we finally reach $1 billion. That's real money, but 302 years is a long time.

The logarithm transformation

We can see multiplication at work in the growth of a biological population and of money deposited at compound interest. It is not surprising that the exponential model fits these cases. But there is no obvious multiplication going on in the case of world oil production. Is this really exponential growth? The shape of the growth curve for oil production in Figure 2.32 does look like the exponential curve in Figure 2.33, but our eyes are not very good at comparing curves of roughly similar shape. We need a better way to check whether growth is exponential. Our eyes are quite good at judging whether or not points lie along a straight line. So we will apply a mathematical transformation that changes exponential growth into linear growth—and patterns of growth that are not exponential into something other than linear.

The necessary transformation is carried out by taking the *logarithm* of the data points. Use a calculator with a `log` button to compute logarithms. Better yet, most statistical software will calculate the logarithms of all the values of a variable in response to a single command. The essential property of the logarithm for our purposes is that it straightens an exponential growth curve. **If a variable grows exponentially, its logarithm grows linearly.**

EXAMPLE 2.25 | Figure 2.33 shows the exponential growth of a population of bacteria that doubles each hour. The logarithms of the number of bacteria should show straight-line growth. Use your calculator to find some of the logarithms. The population counts for the first n hours are 2, 4, 8, ..., 2^n. The logarithms of these counts are

$$\log 2 = 0.301$$
$$\log 4 = 0.602$$
$$\log 8 = 0.903$$
$$\vdots$$
$$\log 2^n = n \log 2 = 0.301n$$

Each hour multiplies the previous count by 2 and adds $\log 2$ to the previous logarithm.

Figure 2.34 graphs the logarithms of the bacteria population counts from Figure 2.33 against time. The promised straight line has indeed appeared. After 15 hours, for example, the population contains 2^{15} (32,768) bacteria. The logarithm of 32,768 is 4.515, and this point appears above the 15-hour mark in Figure 2.34.

Our artificial example of bacterial growth was exactly exponential. Real data will not fit the exponential model perfectly, so applying logarithms will not produce a perfectly straight line. Figure 2.35 is a graph of the logarithms of annual world crude oil production from Table 2.13. The graph is quite close to a straight line from 1880 to 1973. A clear deviation from the line begins in 1973. Closer examination shows a smaller deviation during the 1930s, when growth slowed during the Great Depression. The increase in oil production has indeed been close to exponential over most of the past century, but with important deviations.

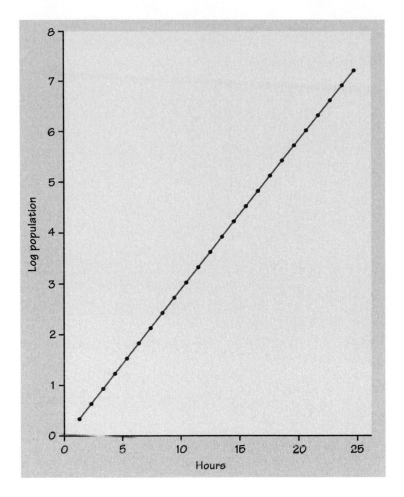

FIGURE 2.34 The logarithms of the bacteria population size from Figure 2.33, for Example 2.25.

The logarithmic transformation enables us to judge the appropriateness of the exponential growth model by reducing the question to judging whether points lie on a straight line. Comparison of Figures 2.32 and 2.35 reveals another advantage: logarithms compress the vertical scale of the plot, making it easier to see deviations from fit in the earlier stages of exponential growth. The slowdown in oil production during the Depression is visible in Figure 2.35 but not in Figure 2.32.

Residuals again

Just as in the case of linear growth, calculating and plotting residuals allow a closer examination of the fit of the exponential growth model to data. Before we can find the residuals, however, we must fit to the data an explicit model described by an equation. Since taking logarithms of data that display exponential growth produces a straight line, we can fit a straight line to the

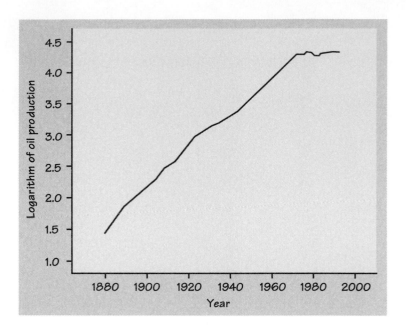

FIGURE 2.35   The logarithms of annual world crude oil production, 1880 to 1994.

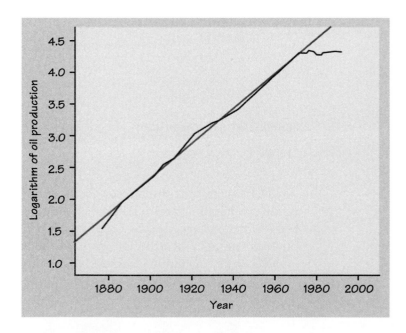

FIGURE 2.36   The least-squares line for world oil production based on data for 1880 to 1972.

logarithms and obtain residuals as the deviations of the actual logarithms from this line. Figure 2.36 shows the result of fitting the least-squares regression line to the logarithms of annual oil production from 1880 to 1972. We did not allow data for years after 1972 to enter into the calculation because we know that the pattern of exponential growth ended in 1973.

EXAMPLE 2.26 | Software finds the regression line to be

$$\text{log production} = -52.686 + (0.02888 \times \text{year})$$

The residuals of the data from the fitted line are computed in the usual way, as

$$\text{residual} = \text{observed value} - \text{predicted value}$$

For the year 1945, the value of the logarithm predicted by the regression line is

$$\text{log production} = -52.686 + (0.02888 \times 1945)$$
$$= 3.4856$$

Because 2595 million barrels of oil were actually produced in 1945, the residual for that year is

$$\text{residual} = \text{observed value} - \text{predicted value}$$
$$= \log 2595 - 3.4856$$
$$= 3.4141 - 3.4856$$
$$= -0.0715$$

Most regression software will compute and print a table of the residuals, and many will also produce residual plots. In Figure 2.37 we have added the residuals for the years 1974 to 1994, which were not used in fitting the line

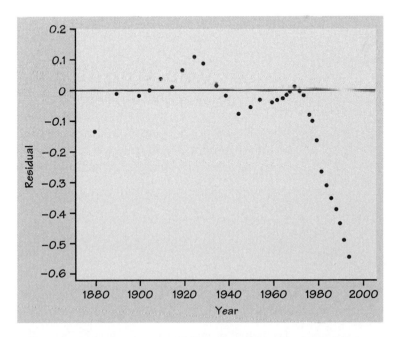

FIGURE 2.37   Residual plot for the least-squares line for world oil production, for Example 2.26.

The deviations from the overall pattern of exponential growth in oil production are easier to see in the residual plot (Figure 2.37) than in the previous Figure 2.36. The pattern of residuals about the horizontal line at 0 shows systematic departures from the overall fit. Because we are plotting logarithms, straight-line patterns of residuals have a specific interpretation. The line of fit (zero residual) represents exponential growth at the average rate observed in the entire period 1880 to 1972. A straight-line rising pattern of residuals shows a period of growth at a faster rate, while a declining pattern shows a slower rate of growth.

The most dramatic departure of the residuals from zero marks the end of exponential growth after 1972. Oil production after 1972 is much lower than extrapolation of the previous trend would have predicted. We can also see other systematic deviations. Oil production was increasing more rapidly than the long-term rate in the years between 1900 and the beginning of the Depression in 1929. Production increased more slowly not only during the Depression but also during World War II. That is, the turning points in the residual plot occur in 1925 (the last point before 1929) and 1945 (the end of World War II). Only after the war did oil production return to an above-average growth rate, which lasted until 1973. The production of oil was of course increasing during the entire period up to 1973—the runs of declining residuals show periods of slower growth, not an actual drop in oil production.

Our tools for data analysis have given us a quite detailed understanding of the history of world oil production. The long-term picture is one of exponential growth from the beginnings of commercial petroleum use to the Arab boycott of 1973. But the long-term rate of growth is an average over several periods of somewhat slower or somewhat faster growth, periods that coincide with major events in the economic history of the twentieth century.

Prediction in the exponential growth model

The use of logarithms to obtain a linear pattern from exponential growth complicates prediction a bit. Because the regression line is fitted to the logarithms of the observations, it also predicts the logarithm. To express the prediction in easily understood terms, we must restate it in terms of the original units of measurement. To do this, use the fact that any number y can be obtained from its common (base 10) logarithm $\log y$ by

$$y = 10^{\log y}$$

EXAMPLE 2.27 The regression line obtained in Example 2.26 predicts that in 1977

$$\log \text{ production} = -52.686 + (0.02888 \times 1977)$$
$$= 4.4098$$

The predicted production in the original units (millions of barrels of oil) is therefore

$$\text{production} = 10^{4.4098} = 25,692$$

The actual production in 1977 was 21,787 million barrels. The prediction is much too high because the exponential growth model fails to fit post-1973 oil production.

SUMMARY

When a variable is multiplied by a fixed amount greater than 1 in each equal time period, **exponential growth** results.

A plot of exponential growth against time becomes linear (a straight line) if we take the **logarithm** before plotting. This fact makes it easy to detect exponential growth.

Deviations from the overall pattern of exponential growth are most easily examined by fitting a line to the logarithms of the data and plotting the **residuals** from this line.

SECTION 2.5 EXERCISES

2.69 (Exact exponential growth) The common intestinal bacterium *E. coli* is one of the fastest-growing bacteria. Under ideal conditions, the number of *E. coli* in a colony doubles about every 15 minutes until restrained by lack of resources. Starting from a single bacterium, how many *E. coli* will there be in 1 hour? In 5 hours?

2.70 (Exact exponential growth) A clever courtier, offered a reward by an ancient king of Persia, asked for a grain of rice on the first square of a chess board, 2 grains on the second square, then 4, 8, 16, and so on.

(a) Make a table of the number of grains on each of the first 10 squares of the board.

(b) Plot the number of grains on each square against the number of the square for squares 1 to 10, and connect the points with a smooth curve. This is an exponential curve.

(c) How many grains of rice should the king deliver for the 64th (and final) square?

(d) Take the logarithm of each of your numbers of grains from (a). Plot these logarithms against the number of squares from 1 to 10. You should get a straight line.

(e) From your graph in (d) find the approximate values of the slope b and the intercept a for the line. Use the equation $y = a + bx$ to predict the logarithm of the amount for the 64th square. Check your result by taking the logarithm of the amount you found in (c).

2.71 (Exact exponential growth) Maria is given a savings bond at birth. The bond is initially worth $500 and earns interest at 7.5% each year. This means that the value is multiplied by 1.075 each year.

(a) Find the value of the bond at the end of 1 year, 2 years, and so on up to 10 years.

(b) Plot the value y against years x. Connect the points with a smooth curve. This is an exponential curve.

(c) Take the logarithm of each of the values y that you found in (a). Plot the logarithm $\log y$ against years x. You should obtain a straight line.

2.72 (Exact exponential growth) Fred and Alice were born the same year, and each began life with $500. Fred added $100 each year, but earned no interest. Alice added nothing, but earned interest at 7.5% annually. After 25 years, Fred and Alice are getting married. Who has more money?

2.73 (Exact exponential growth) The rate of return on U.S. stocks between 1970 and 1995 was 11.34% per year, compounded annually.[18]

(a) If you invested $1000 in an index fund that represents the entire stock market in 1970, how much would your investment be worth at the end of 1995?

(b) The result of (a) does not take into account inflation, the declining buying power of the dollar. The dollar amount needed to have the same buying power that $1000 had in 1970 increased at the rate of 5.62% per year, compounded annually, between 1970 and 1995. How much money in 1995 matched the value of $1000 in 1970?

2.74 Biological populations can grow exponentially if not restrained by predators or lack of food. The gypsy moth outbreaks that occasionally devastate the forests of the Northeast illustrate approximate exponential growth. It is easier to count the number of acres defoliated by the moths than to count the moths themselves. Here are data on an outbreak in Massachusetts. (Data provided by Chuck Schwalbe, U.S. Department of Agriculture.)

Year	Acres
1978	63,042
1979	226,260
1980	907,075
1981	2,826,095

(a) Plot the number y of acres defoliated against the year x. The pattern of growth appears exponential.

(b) Verify that y is being multiplied by about 4 each year by calculating the ratio of acres defoliated each year to the previous year. (Start with 1979 to 1978, when the ratio is $226,260/63,042 = 3.6$.)

(c) Take the logarithm of each number y and plot the logarithms against the year x. The linear pattern confirms that the growth is exponential.

(d) Find the least-squares line fitted to the four points on your graph from (c). Use this line to predict the number of acres defoliated in 1982.

(Predict $\log y$ by substituting $x = 1982$ in the equation. Then use the fact that $y = 10^{\log y}$ to predict y.) The actual number for 1982 was 1,383,265, far less than the prediction. The exponential growth of the gypsy moth population was cut off by a viral disease, and the population quickly collapsed back to a low level.

2.75 We read much about the threat of exploding federal spending on social insurance programs (chiefly social security and Medicare). Here are the amounts spent, in millions of dollars:

Year	1960	1965	1970	1975	1980	1985	1990
Spending	14,307	21,807	45,246	99,715	191,162	313,108	422,257

(a) Examine these data graphically. Does the pattern appear closer to linear growth or to exponential growth?

(b) Plot the logarithms of social insurance spending. Fit the least-squares line and add it to your graph. During which periods was growth faster than the long-term trend? Slower than the trend?

(c) The government has taken some actions to slow growth in spending on Medicare. The 1990 amount lies below the long-term trend line. The latest available figure for federal social insurance spending is 495,710 for 1992. Use your line from (b) to predict 1992 spending. Was the actual amount spent less than the prediction?

2.76 The following table shows the growth of the population of Europe (millions of persons) between 400 B.C. and 1950.

Date	Pop.	Date	Pop.	Date	Pop.
400 B.C.	23	1200	61	1600	90
A.D. 1	37	1250	69	1650	103
200	67	1300	73	1700	115
700	27	1350	51	1750	125
1000	42	1400	45	1800	187
1050	46	1450	60	1850	274
1100	48	1500	69	1900	423
1150	50	1550	78	1950	594

(a) Plot population against time.

(b) The graph shows that the population of Europe dropped at the collapse of the Roman Empire (A.D. 200 to 500) and at the time of the Black

Death (A.D. 1348). Growth has been uninterrupted since 1400. Plot the logarithm of population against time, beginning in 1400.

(c) Was growth exponential between 1400 and 1950? If not, what overall pattern do you see?

2.77 The following table gives the resident population of the United States from 1790 to 1990, in millions of persons:

Date	Pop.	Date	Pop.	Date	Pop.
1790	3.9	1860	31.4	1930	122.8
1800	5.3	1870	39.8	1940	131.7
1810	7.2	1880	50.2	1950	151.3
1820	9.6	1890	62.9	1960	179.3
1830	12.9	1900	76.0	1970	203.3
1840	17.1	1910	92.0	1980	226.5
1850	23.2	1920	105.7	1990	248.7

(a) Plot population against time. The growth of the American population appears roughly exponential.

(b) Plot the logarithms of population against time. The pattern of growth is now clear. An expert says that "the population of the United States increased exponentially from 1790 to about 1880. After 1880 growth was still approximately exponential, but at a slower rate." Explain how this description is obtained from the graph.

(c) Fit the least-squares line to the logarithms. Use your line to predict population in 1997. Actual population at the end of 1997 was 268.9 million. Has population growth since 1990 been faster or slower than the long-term trend?

2.78 The number of motor vehicles (cars, trucks, and buses) registered in the United States has grown as follows (vehicle counts in millions):

Year	1940	1945	1950	1955	1960	1965
Vehicles	32.4	31.0	49.2	62.7	73.9	90.4

Year	1970	1975	1980	1985	1990	1995
Vehicles	108.8	132.9	155.8	171.7	188.8	203.1

(a) Plot the number of vehicles against time. Also plot the logarithm of the number of vehicles against time.

(b) Look at the years 1950 to 1980. Was the growth in motor vehicle registrations more nearly linear or more nearly exponential during this period? (Use a ruler to compare the two graphs.)

(c) Examine the logarithm graph. The year 1940 fits the pattern well, but 1945 and the years after 1980 do not. All years after 1980 have negative residuals. Does the pattern after 1980 suggest exponential growth at a lower rate? Why?

(d) The year 1945 is a single unusual point. Can you suggest an explanation for the negative residual in 1945?

2.79 The least-squares line fitted to the 1950 to 1980 data from the previous exercise is approximately

$$\log y = -30.62 + (0.01657 \times \text{year})$$

Use this line to predict motor vehicle registrations in 1945 and 1995. As your work in the previous exercise shows, these predictions are too high.

2.80 The productivity of American agriculture has grown rapidly due to improved technology (crop varieties, fertilizers, mechanization). Here are data on the output per hour of labor on American farms. The variable is an "index number" that gives productivity as a percent of the 1967 level.

Year	1940	1945	1950	1955	1960	1965	1970	1975	1980	1985	1990
Productivity	21	27	35	47	67	91	113	137	166	217	230

Plot these data and also plot the logarithms of the productivity values against year. Then briefly describe the pattern of growth that your plots suggest.

2.6 Relations in Categorical Data

We have concentrated on relationships in which at least the response variable was quantitative. Now we turn to describing relationships between two or more categorical variables. Some variables—such as gender, race, and occupation—are inherently categorical. Other categorical variables are created by grouping values of a quantitative variable into classes. Published data are often reported in grouped form to save space. To analyze categorical data, we use the *counts* (frequencies) or *percents* (relative frequencies) of individuals that fall into various categories.

TABLE 2.14 Years of school completed, by age, 1995 (thousands of persons)

Education	Age group			Total
	25 to 34	35 to 54	55 and over	
Did not complete high school	5,325	9,152	16,035	30,512
Completed high school	14,061	24,070	18,320	56,451
College 1 to 3 years	11,659	19,926	9,662	41,247
College, 4 or more years	10,342	19,878	8,005	38,225
Total	41,388	73,028	52,022	166,438

EXAMPLE 2.28

two-way table

row and column variables

Table 2.14 presents Census Bureau data on the years of school completed by Americans of different ages. Many people under 25 years of age have not completed their education, so they are left out of the table. Both variables, age and education, are grouped into categories. This is a **two-way table** because it describes two categorical variables. Education is the **row variable** because each horizontal row in the table describes people with one level of education. Age is the **column variable** because each vertical column describes one age group. The entries in the table are the counts of persons in each age-by-education class. Although both age and education in this table are categorical variables, both have a natural order from least to most. The order of the rows and the columns in Table 2.14 reflects the order of the categories.

Marginal distributions

How can we best grasp the information contained in Table 2.14? First, *look at the distribution of each variable separately.* The distribution of a categorical variable says how often each outcome occurred. The "Total" column at the right of the table contains the totals for each of the rows. These row totals give the distribution of education level (the row variable) among all people over 25 years of age: 30,512,000 did not complete high school, 56,451,000 finished high school but did not attend college, and so on. In the same way, the "Total" row at the bottom of the table gives the age distribution. If the row and column totals are missing, the first thing to do in studying a two-way table is to calculate them. The distributions of education alone and age alone are **marginal distribution** called **marginal distributions** because they appear at the right and bottom margins of the two-way table.

If you check the row and column totals in Table 2.14, you will notice a few discrepancies. For example, the sum of the entries in the "25 to 34" column is 41,387. The entry in the "Total" row for that column is 41,388. The explanation **roundoff error** is **roundoff error.** The table entries are in thousands of persons, and each is rounded to the nearest thousand. The Census Bureau obtained the "Total" entry by rounding the exact number of people aged 25 to 34 to the nearest thousand. The result was 41,388,000. Adding the row entries, each of which is already rounded, gives a slightly different result.

Percents are often easier to grasp than counts. We can display the marginal distribution of education level in terms of percents by dividing each row total by the table total and converting to a percent.

EXAMPLE 2.29

The percent of people over 25 years of age who have at least 4 years of college is

$$\frac{\text{total with 4 years of college}}{\text{table total}} = \frac{38,225}{166,438} = 0.230 = 23.0\%$$

Do three more such calculations to obtain the marginal distribution of education level in percents. Here it is:

	Did not complete high school	Completed high school	1 to 3 years of college	≥ 4 years of college
Percent	18.3	33.9	24.8	23.0

The total is 100% because everyone is in one of the four education categories.

Each marginal distribution from a two-way table is a distribution for a single categorical variable. As we saw in Chapter 1 (page 6), we can use a bar graph or a pie chart to display such a distribution. Figure 2.38 is a bar graph of the distribution of years of schooling. We see that people with at least some college education make up about half of the over-25 population.

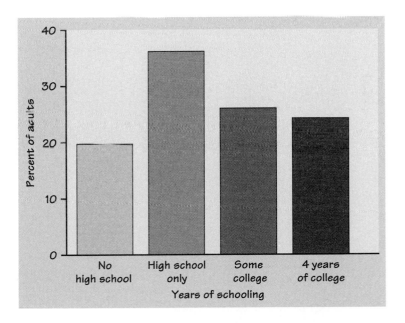

FIGURE 2.38 A bar graph of the distribution of years of schooling completed among people age 25 years and over. This is one of the marginal distributions for Table 2.14.

Describing relationships

Table 2.14 contains much more information than the two marginal distributions of age alone and education alone. The nature of the relationship between age and education cannot be deduced from the separate distributions but requires the full table. **Relationships among categorical variables are described by calculating appropriate percents from the counts given.** We use percents because counts are often hard to compare. For example, 19,878,000 persons aged 35 to 54 have completed college, and only 8,005,000 persons in the 55 and up age group have done so. These counts do not accurately describe the association, however, because there are many more people in the younger age group.

EXAMPLE 2.30

What percent of people aged 25 to 34 have completed 4 years of college? This is the count of those who are 25 to 34 and have 4 years of college as a percent of the age group total:

$$\frac{10,342}{41,388} = 0.250 = 25.0\%$$

"People aged 25 to 34" is the group we want a percent of, so the count for that group is the denominator. In the same way, find the percent of people in each age group who have completed college. The comparison of all three groups is

Age group	25 to 34	35 to 54	55 and over
Percent with 4 years of college	25.0	27.2	15.4

These percentages make it clear that a college education is less common among Americans over age 55 than among younger adults. This is an important aspect of the association between age and education.

Conditional distributions

Example 2.30 does not compare the complete distributions of years of schooling in the three age groups. It compares only the percents who finished college. Let's look at the complete picture.

EXAMPLE 2.31

Information about the 25 to 34 age group occupies the first column in Table 2.14. To find the complete distribution of education in this age group, look only at that column. Compute each count as a percent of the column total, which is 41,388. Here is the distribution:

	Did not complete high school	Completed high school	1 to 3 years of college	≥ 4 years of college
Percent	12.9	34.0	28.2	25.0

These percents should add to 100% because all 25- to 34-year-olds fall in one of the educational categories. (In fact, they add to 100.1% because of roundoff error.) The four percents together are the **conditional distribution** of education, given that a person is 25 to 34 years of age. We use the term "conditional" because the distribution refers only to people who satisfy the condition that they are 25 to 34 years old.

conditional distribution

TABLE OF EDU BY AGE

EDU	AGE			
Frequency Col Pct	25-34	35-54	55 over	Total
NoHS	5325 12.87	9152 12.53	16035 30.82	30512
HSonly	14061 33.97	24070 32.96	18320 35.22	56451
SomeColl	11659 28.17	19926 27.29	9662 18.57	41247
Coll4yrs	10342 24.99	19878 27.22	8005 15.39	38225
Total	41387	73026	52022	166435

FIGURE 2.39   SAS output of the two-way table of education by age with the three conditional distributions of education in each age group. The percents in each column add to 100%.

We now focus in turn on the second column (people aged 35 to 54) and then the third column (people 55 and over) of Table 2.14 in order to find two more conditional distributions. Statistical software will find each entry in a two way table as a percent of its column total in response to a single command. Figure 2.39 displays output from the SAS software.

Each cell in this table contains a count from Table 2.14 along with that count as a percent of the column total. The percents in each column form the conditional distribution of years of schooling for one age group. The percents in each column add to 100% because everyone in the age group is accounted for. All three conditional distributions differ from the marginal distribution in Example 2.29, which describes all people age 25 and over. Comparing the conditional distributions reveals the nature of the association between age and education. The distributions of education in the two younger groups are quite similar, but higher education is less common in the 55 and over group.

Bar graphs can help make the association visible. We could make three side-by-side bar graphs, each resembling Figure 2.38, to present the three conditional distributions. Figure 2.40 shows an alternative form of bar graph. Each set of three bars compares the percents in the three age groups who have reached a specific educational level. We see at once that the "25 to 34" and "35 to 54" bars are similar for all four levels of education, and that the "55 and over" bars show that many more people in this group did not finish high school and that many fewer have any college.

No single graph (such as a scatterplot) portrays the form of the relationship between categorical variables, and no single numerical measure (such as the correlation) summarizes the strength of an association. Bar graphs are flexible enough to be helpful, but you must think about what comparisons you want to display. For numerical measures, we must rely on well-chosen percents or on more advanced statistical methods.[19]

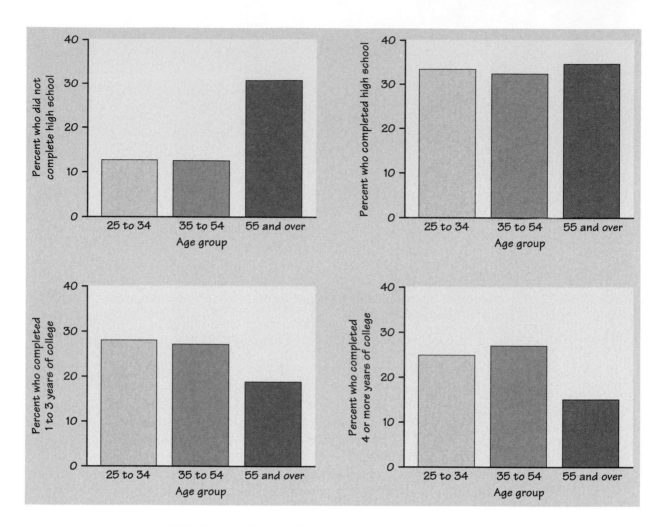

FIGURE 2.40 Bar graphs to compare the education levels of three age groups. Each graph compares the three groups for one of the four education levels.

We have compared the education of different age groups. That is, we have treated age as the explanatory variable and education as the response variable. We might also be interested in the distribution of age among persons having a certain level of education. To do this, look only at one *row* in Table 2.14. Calculate each entry in that row as a percent of the row total, the total for that education group. The result is another conditional distribution, the conditional distribution of age given a certain level of education.

A two-way table contains a great deal of information in compact form. Making that information clear almost always requires finding percents. You must decide which percents you need. If you are studying trends in the training of the American workforce, comparing the distributions of education for different age groups reveals the more extensive education of younger people. If, on the other hand, you are planning a program to improve the skills of people who did not finish high school, the age distribution within this educational group is important information.

Simpson's paradox

As is the case with quantitative variables, the effects of lurking variables can strongly influence relationships between two categorical variables. Here is an example that demonstrates the surprises that can await the unsuspecting consumer of data.

EXAMPLE 2.32

To help consumers make informed decisions about health care, the government releases data about patient outcomes in hospitals. You want to compare Hospital A and Hospital B, which serve your community. Here is a two-way table of data on the survival of patients after surgery in these two hospitals. All patients undergoing surgery in a recent time period are included. "Survived" means that the patient lived at least 6 weeks following surgery.

	Hospital A	Hospital B
Died	63	16
Survived	2037	784
Total	2100	800

The evidence seems clear: Hospital A loses 3% (63/2100) of its surgery patients, and Hospital B loses only 2% (16/800). It seems that you should choose Hospital B if you need surgery.

Not all surgery cases are equally serious, however. Later in the government report you find data on the outcome of surgery broken down by the condition of the patient before the operation. Patients are classified as being in either "poor" or "good" condition. Here are the more detailed data. Check that the entries in the original two-way table are just the sums of the "poor" and "good" entries in this pair of tables.

Good condition			Poor condition		
	Hospital A	Hospital B		Hospital A	HospitalB
Died	6	8	Died	57	8
Survived	594	592	Survived	1443	192
Total	600	600	Total	1500	200

Hospital A beats Hospital B for patients in good condition: only 1% (6/600) died in Hospital A, compared with 1.3% (8/600) in Hospital B. Hospital A wins again for patients in poor condition, losing 3.8% (57/1500) to Hospital B's 4% (8/200). So Hospital A is safer for both patients in good condition and patients in poor condition. If you are facing surgery, you should choose Hospital A.

The patient's condition is a lurking variable when we compare the death rates at the two hospitals. When we ignore the lurking variable, Hospital B seems safer, even though Hospital A does better for both classes of patients. How can A do better in each group, yet do worse overall? Look at the data. Hospital A is a medical center that attracts seriously ill patients from a wide region. It had 1500 patients in poor condition. Hospital B had only 200 such cases. Because patients in poor condition are more likely to die, Hospital A has a higher death rate despite its superior performance for each class of patients. The original two-way table, which did not take account of the condition of the patients, was misleading. Example 2.32 illustrates *Simpson's paradox*.

> **Simpson's Paradox**
>
> An association or comparison that holds for all of several groups can re-verse direction when the data are combined to form a single group. This reversal is called **Simpson's paradox.**

The lurking variables in Simpson's paradox are categorical. That is, they break the individuals into groups, as when surgery patients are classified as "good condition" or "poor condition." Simpson's paradox is an extreme form of the fact that observed associations can be misleading when there are lurking variables. Although the counts in Example 2.32 were made up, the setting is real. The hospital death rates compiled and released by the Federal Health Care Financing Agency do allow for each patient's condition and diagnosis. Hospitals continue to complain that there are differences among patients that can make the government's mortality data misleading.

The perils of aggregation

three-way table

The table of outcome by hospital by patient condition in Example 2.32 is a **three-way table** that reports the frequencies of each combination of levels of three categorical variables. We present a three-way table as several two-way tables, one for each level of the third variable. In Example 2.32, there is a separate table of patient outcome by hospital for each patient condition. We can obtain the original two-way table by adding the corresponding entries in these tables. This is called **aggregating** the data. Aggregation has the effect of ignoring patient condition, which then becomes a lurking variable. Conclu-sions that seem obvious when we look only at aggregated data often become debatable when we examine lurking variables. Here is a real-life example.

aggregation

EXAMPLE 2.33

A study by the National Science Foundation[20] found that the median salary of newly graduated female engineers and scientists was only 73% of the median salary for males. When the new graduates were broken down by field, however, the picture changed. Women's median salaries as a percent of the male median in the 16 fields studied were

94%	96%	98%	95%	85%	85%	84%	100%
103%	100%	107%	93%	104%	93%	106%	100%

How can women do nearly as well as men in every field, and better in some, yet fall far behind men when we look at all young engineers and scientists? Women are concentrated in the life and social sciences, where salaries are lower than in the physical sciences and engineering. When salaries from all fields are lumped together, the women place lower because the fields they favor pay less. Field of science is a lurking variable that influences the observed association between gender and salary.

SUMMARY

A **two-way table** of counts organizes data about two categorical variables. Values of the **row variable** label the rows that run across the table, and values of the **column variable** label the columns that run down the table. Two-way tables are often used to summarize large amounts of data by grouping outcomes into categories.

The **row totals** and **column totals** in a two-way table give the **marginal distributions** of the two variables separately. It is clearer to present these distributions as percents of the table total. Marginal distributions do not give any information about the relationship between the variables.

To find the **conditional distribution** of the row variable for one specific value of the column variable, look only at that one column in the table. Find each entry in the column as a percent of the column total.

There is a conditional distribution of the row variable for each column in the table. Comparing these conditional distributions is one way to describe the association between the row and the column variables. It is particularly useful when the column variable is the explanatory variable. When the row variable is explanatory, find the conditional distribution of the column variable for each row and compare these distributions.

Bar graphs are a flexible means of presenting categorical data. There is no single best way to describe an association between two categorical variables.

We present data on three categorical variables in a **three-way table**, printed as separate two-way tables for each level of the third variable. A comparison between two variables that holds for each level of a third variable can be changed or even reversed when the data are **aggregated** by summing over all levels of the third variable. **Simpson's paradox** refers to the reversal of a comparison by aggregation. It is an example of the potential effect of lurking variables on an observed association.

SECTION 2.6 EXERCISES

Exercises 2.81 to 2.85 are based on Table 2.15. This two-way table reports data from the National Center for Education Statistics on undergraduate students enrolled in U.S. colleges and universities in the fall of 1993. It classifies students by age group and the type of program in which they are enrolled. Students whose age is unknown are not included.

2.81 (a) How many undergraduate students were enrolled in colleges and universities?

(b) What percent of all undergraduate students were 18 to 21 years old in the fall of the academic year?

Age	2-year full-time	2-year part-time	4-year full-time	4-year part-time
Under 18	36	98	75	37
18 to 21	1126	711	3270	223
22 to 34	634	1575	2267	1380
35 and up	221	1092	390	915
Total	2017	3476	6002	2555

TABLE 2.15 Undergraduate college enrollment, fall 1993 (thousands of students)

(c) Find the percent of the undergraduates enrolled in each of the four types of program who were 18 to 21 years old. Make a bar graph to compare these percents.

(d) The 18 to 21 group is the traditional age group for college students. Briefly summarize what you have learned from the data about the extent to which this group predominates in different kinds of college programs.

2.82 (a) An association of two-year colleges asks: "What percent of students enrolled part-time at 2-year colleges are 22 to 34 years old?"

(b) A bank that makes education loans to adults asks: "What percent of all 22- to 34-year-old students are enrolled part-time at 2-year colleges?"

2.83 (a) Find the marginal distribution of age among all undergraduate students, first in counts and then in percents. Make a bar graph of the distribution.

(b) Find the conditional distribution of age (in percents) among students enrolled part-time in 2-year colleges and make a bar graph of this distribution.

(c) Briefly describe the most important differences between the two age distributions.

2.84 Call students aged 35 and up "older students." Compare the presence of older students in the four types of program with numbers, a graph, and a brief summary of your findings.

2.85 Find the conditional distribution of age for each type of program. (This is tedious unless you use software.) Compare the four conditional distributions and write a brief discussion of how the four types of program differ in the ages of the students they attract.

2.86 Here are the row and column totals for a two-way table with two rows and two columns:

a	b	50
c	d	50
60	40	100

Find *two different* sets of counts a, b, c, and d for the body of the table that give these same totals. This shows that the relationship between two variables cannot be obtained from the two individual distributions of the variables.

2.87   A business school conducted a survey of companies in its state. They mailed a questionnaire to 200 small companies, 200 medium-sized companies, and 200 large companies. The rate of nonresponse is important in deciding how reliable survey results are. Here are the data on response to this survey:

	Response	No response	Total
Small	125	75	200
Medium	81	119	200
Large	40	160	200

(a) What was the overall percent of nonresponse?

(b) Describe how nonresponse is related to the size of the business. (Use percents to make your statements precise.)

(c) Draw a bar graph to compare the nonresponse percents for the three size categories.

(d) Using the total number of responses as a base, compute the percent of responses that come from each of small, medium, and large businesses.

(e) The sampling plan was designed to obtain equal numbers of responses from small, medium, and large companies. In preparing an analysis of the survey results, do you think it would be reasonable to proceed as if the responses represented companies of each size equally?

2.88   A study of the career plans of young women and men sent questionnaires to all 722 members of the senior class in the College of Business Administration at the University of Illinois. One question asked which major within the business program the student had chosen. Here are the data from the students who responded. (From F. D. Blau and M. A. Ferber, "Career plans and expectations of young women and men," *Journal of Human Resources*, 26 (1991), pp. 581–607.)

	Female	Male
Accounting	68	56
Administration	91	40
Economics	5	6
Finance	61	59

(a) Describe the differences between the distributions of majors for women and men with percents, with a graph, and in words.

(b) What percent of the students did not respond to the questionnaire? The nonresponse weakens conclusions drawn from these data.

2.89 Cocaine addiction is hard to break. Addicts need cocaine to feel any pleasure, so perhaps giving them an antidepressant drug will help. A 3-year study with 72 chronic cocaine users compared an antidepressant drug called desipramine with lithium and a placebo. (Lithium is a standard drug to treat cocaine addiction. A placebo is a dummy drug, used so that the effect of being in the study but not taking any drug can be seen.) One-third of the subjects, chosen at random, received each drug. Here are the results. (Data from D. M. Barnes, "Breaking the cycle of addiction," *Science*, 241 (1988), pp. 1029–1030.)

	Cocaine relapse?	
	Yes	No
Desipramine	10	14
Lithium	18	6
Placebo	20	4

Compare the effectiveness of the three treatments in preventing relapse. Use percents and draw a bar graph. Write a brief summary of your conclusions.

2.90 Asia has become a major competitor of the United States and Western Europe in education as well as economics. Here are counts of first university degrees in science and engineering in the three regions. (Data from National Science Foundation Science Resources Studies Division, *Data Brief* (1996), no. 12, p. 1.)

Field	United States	Western Europe	Asia
Engineering	61,941	158,931	280,772
Natural science	111,158	140,126	242,879
Social science	182,166	116,353	236,018

Direct comparison of counts of degrees would require us to take into account Asia's much larger population. We can, however, compare the distribution of degrees by field of study in the three regions. Do this using calculations and graphs, and write a brief summary of your findings.

2.91 The following two-way table describes the age and marital status of American women in 1995. The table entries are in thousands of women.

	Marital status				
Age (years)	Never married	Married	Widowed	Divorced	Total
18–24	9,289	3,046	19	260	12,613
25–39	6,948	21,437	206	3,408	32,000
40–64	2,307	26,679	2,219	5,508	36,713
≥ 65	768	7,767	8,636	1,091	18,264
Total	19,312	58,931	11,080	10,266	99,588

(a) Find the sum of the entries in the "Married" column. Why does this sum differ from the "Total" entry for that column?

(b) Give the marginal distribution of marital status for all adult women (use percents). Draw a bar graph to display this distribution.

(c) Compare the conditional distributions of marital status for women aged 18 to 24 and women aged 40 to 64. Briefly describe the most important differences between the two groups of women, and back up your description with percents.

(d) You are planning a magazine aimed at women who have never been married. Find the conditional distribution of age among single women and display it in a bar graph. What age group or groups should your magazine aim to attract?

2.92 Here is a two-way table of suicides committed in 1993, categorized by the sex of the victim and the method used. ("Hanging" also includes suffocation.) Write a brief account of differences in suicide between males and females. Use calculations and a graph to justify your statements.

Method	Male	Female
Firearms	16,381	2,559
Poison	3,569	2,110
Hanging	3,824	803
Other	1,641	623

2.93 Upper Wabash Tech has two professional schools, business and law. Here is a three-way table of applicants to these professional schools, categorized by gender, school, and admission decision. (Although these data are fictitious, similar though less simple situations occur in reality. See P. J. Bickel and J. W. O'Connell, "Is there a sex bias in graduate admissions?" *Science,* 187 (1975), pp. 398–404.)

	Business			Law	
	Admit	Deny		Admit	Deny
Male	480	120	Male	10	90
Female	180	20	Female	100	200

(a) Make a two-way table of gender by admission decision for the combined professional schools by summing entries in the three-way table.

(b) Compute separately the percents of male and female applicants admitted from your two-way table. Upper Wabash Tech's professional schools admit a higher percent of male applicants than of female applicants.

(c) Now compute separately the percents of male and female applicants admitted by the business school and by the law school. Each school admits a higher percent of female applicants.

(d) Explain carefully, as if speaking to a skeptical reporter, how it can happen that Upper Wabash appears to favor males when each school individually favors females.

2.94 Here are the numbers of flights on time and delayed for two airlines at five airports in June 1991. Overall on-time percentages for each airline are often reported in the news. Lurking variables can make such reports misleading. (These data, from reports submitted by airlines to the Department of Transportation, appear in A. Barnett, "How numbers can trick you," *Technology Review,* October 1994, pp. 38–45.)

	Alaska Airlines		America West	
	On time	Delayed	On time	Delayed
Los Angeles	497	62	694	117
Phoenix	221	12	4840	415
San Diego	212	20	383	65
San Francisco	503	102	320	129
Seattle	1841	305	201	61

(a) What percent of all Alaska Airlines flights were delayed? What percent of all America West flights were delayed? These are the numbers usually reported.

(b) Now find the percent of delayed flights for Alaska Airlines at each of the five airports. Do the same for America West.

(c) America West does worse at *every one* of the five airports, yet does better overall. That sounds impossible. Explain carefully, referring to the data, how this can happen. (The weather in Phoenix and Seattle lies behind this example of Simpson's paradox.)

2.95 Recent studies have shown that earlier reports seriously underestimated the health risks associated with being overweight. The error was due to overlooking some important lurking variables. In particular, smoking tends both to reduce weight and to lead to earlier death. Illustrate Simpson's paradox by a simplified version of this situation. That is, make up a three-way table of overweight (yes or no) by early death (yes or no) by smoker (yes or no) such that

- Overweight smokers and overweight nonsmokers both tend to die earlier than those who are not overweight.

- When smokers and nonsmokers are combined into a two-way table of overweight by early death, people who are not overweight tend to die earlier.

2.96   The influence of race on imposition of the death penalty for murder has been much studied and contested in the courts. The following three-way table classifies 326 cases in which the defendant was convicted of murder. The three variables are the defendant's race, the victim's race, and whether the defendant was sentenced to death. (Data from M. Radelet, "Racial characteristics and imposition of the death penalty," *American Sociological Review*, 46 (1981), pp. 918–927.)

	White defendant Death penalty			Black defendant Death penalty	
	Yes	No		Yes	No
White victim	19	132	White victim	11	52
Black victim	0	9	Black victim	6	97

(a)   From these data make a two-way table of defendant's race by death penalty.

(b)   Show that Simpson's paradox holds: a higher percent of white defendants are sentenced to death overall, but for both black and white victims a higher percent of black defendants are sentenced to death.

(c)   Basing your reasoning on the data, explain why the paradox holds in language that a judge could understand.

2.7   The Question of Causation

In many studies of the relationship between two variables, the goal is to establish that changes in the explanatory variable *cause* changes in the response variable. Even when a strong association is present, the conclusion that this association is due to a causal link between the variables is often elusive. What ties between two variables (and others lurking in the background) can explain an observed association? What constitutes good evidence for causation? We begin our consideration of these questions with a set of examples. In each case, there is a clear association between an explanatory variable x and a response variable y. Moreover, the association is positive whenever the direction makes sense.

EXAMPLE 2.34   Examples of observed association between x and y:

1. x = mother's adult height
    y = daughter's adult height

2. x = amount of the artificial sweetener saccharin in a rat's diet
    y = count of tumors in the rat's bladder

3. x = a student's SAT score as a high school senior
    y = the student's first-year college grade point average

4. x = monthly flow of money into stock mutual funds
    y = monthly rate of return for the stock market

5. x = the anesthetic used in surgery
    y = whether the patient survives the surgery

6. x = the number of years of education a worker has
    y = the worker's income

Explaining association: causation

Figure 2.41 shows in outline form how a variety of underlying links between variables can explain association. The dashed line represents an observed association between the variables x and y. Some associations are explained by a direct cause-and-effect link between the variables. The first diagram in Figure 2.41 shows "x causes y" by an arrow running from x to y.

EXAMPLE 2.35   Items 1 and 2 in Example 2.34 are examples of direct causation. Thinking about these examples, however, shows that "causation" is not a simple idea.

1. Height is in part determined by heredity. Daughters inherit half of their genes from their mothers. There is therefore a direct causal link between the height of mothers and the height of daughters. Of course, the causal

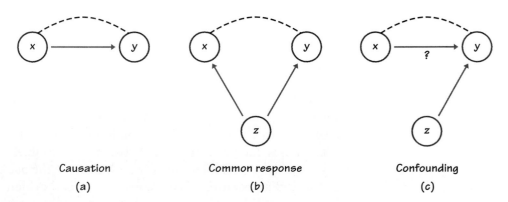

FIGURE 2.41   Some explanations for an observed association. The broken lines show an association. The arrows show a cause-and-effect link. We observe the variables x and y, but z is a lurking variable.

link is not perfect (the correlation r is not 1). Other factors such as nutrition influence the height of children. Even when there is a direct connection (such as heredity), "x causes y" is usually an incomplete explanation for the behavior of y.

2. The best evidence for causation comes from experiments that actually change x while holding all other factors fixed. If y changes, we have good reason to think that x caused the change in y. Experiments show conclusively that large amounts of saccharin in the diet cause bladder tumors in rats. Should we avoid saccharin as a replacement for sugar in food? Rats are not people. Although we can't experiment with people, studies of people who consume different amounts of saccharin show little association between saccharin and bladder tumors.[21] Even well-established causal relations may not generalize to other settings.

Explaining association: common response

common response

"Beware the lurking variable" is good advice when thinking about an association between two variables. The second diagram in Figure 2.41 illustrates **common response.** The observed association between the variables x and y is explained by a lurking variable z. Both x and y change in response to changes in z. This common response creates an association even though there may be no direct causal link between x and y.

EXAMPLE 2.36

The third and fourth items in Example 2.34 illustrate how common response can create an association.

3. Students who are smart and who have learned a lot tend to have both high SAT scores and high college grades. The positive correlation is explained by this common response to students' ability and knowledge.

4. There is a strong positive correlation between how much money individuals add to mutual funds each month and how well the stock market does the same month. Is the new money driving the market up? The correlation may be explained in part by common response to underlying investor sentiment: when optimism reigns, individuals send money to funds and large institutions also invest more. The institutions would drive up prices even if individuals did nothing. What is more, what causation there is may operate in the other direction: when the market is doing well, individuals rush to add money to their mutual funds.[22]

Explaining association: confounding

confounding

Two variables (whether explanatory variables or lurking variables) are **confounded** when their effects on a response variable are mixed together. When many variables interact with each other, confounding of several variables can prevent us from drawing conclusions about causation. The third diagram in Figure 2.41 illustrates confounding. Both the explanatory variable x and the lurking variable z may influence the response variable y. Because x is confounded with z, we cannot distinguish the influence of x from the influence of z. We cannot say how strong the direct effect of x on y is. In fact, it can be hard to say if x influences y at all.

EXAMPLE 2.37 The last two associations in Example 2.34 (Items 5 and 6) are explained in part by confounding.

5. Surgeons choose different anesthetics based in part on the seriousness of the operation and the condition of the patient. An anesthetic that is used mainly for difficult operations on gravely ill patients will have a higher death rate than one used mainly for minor surgery. This confounding of anesthetic x with the type of surgery z makes it hard to see the direct effect of x on patient deaths y.

6. It is likely that more education is a cause of higher income—many highly paid professions require advanced education. However, confounding is also present. People who have high ability and come from prosperous homes are more likely to get many years of education than people who are less able and poorer. Of course, people who start out able and rich are more likely to have high earnings even without much education. We can't say how much of the higher income of well-educated people is actually caused by their education.

Many observed associations are at least partly explained by lurking variables. Both common response and confounding involve the influence of a lurking variable (or variables) z on the response variable y. The distinction between these two types of relationship is less important than the common element, the influence of lurking variables. The most important lesson of these examples is one we have already emphasized: **even a very strong association between two variables is not by itself good evidence that there is a cause-and-effect link between the variables**.

Establishing causation

How can a direct causal link between x and y be established? The best method—indeed, the only fully compelling method—of establishing causation is to conduct a carefully designed experiment in which the effects of possible lurking variables are controlled. Much of Chapter 3 is devoted to the art of designing convincing experiments.

Many of the sharpest disputes in which statistics plays a role involve questions of causation that cannot be settled by experiment. Does taking the drug Bendectin cause birth defects? Does living near power lines cause cancer? Has increased free trade helped to create the growing gap between the incomes of more educated and less educated American workers? All of these questions have become public issues. All concern associations among variables. And all have this in common: they try to pinpoint cause and effect in a setting involving complex relations among many interacting variables. Common response and confounding, along with the number of potential lurking variables, make observed associations misleading. Experiments are not possible for moral or practical reasons. We can't assign some people to live near power lines or compare the same nation with and without free trade agreements.

EXAMPLE 2.38

Are children born to women who take a prescription drug during pregnancy more likely to have birth defects? This question has been asked about many drugs. It is not easy to answer. Consider Bendectin, once commonly prescribed for nausea during pregnancy, then accused of causing birth defects and studied intensively in and out of court.[23]

In the absence of an experiment, we must rely on comparing children whose mothers took Bendectin with similar children whose mothers did not. Who took Bendectin? Mothers of children with birth defects are more likely to remember taking a drug during pregnancy than those with normal children. This recall bias will overstate the rate of birth defects in the Bendectin group. Perhaps we should instead look at prescription records. One study considered a child exposed to Bendectin early in pregnancy if a prescription was filled between 365 and 250 days before birth. There is no way of knowing whether the mothers actually took the drug, or how regularly they took it.

Constructing a group of "similar children" whose mothers did not take Bendectin is even more difficult than deciding who took the drug. What potential lurking variables should we consider? For example, users of Bendectin also took more other drugs than nonusers. Perhaps the alleged effects of Bendectin are due to confounding with some other drug, or to a combination of drugs. Or perhaps birth defects and use of Bendectin and other drugs are common responses to some other medical, economic, or psychological condition. It is hard to say what makes children "similar" to the Bendectin children.

Any study must choose some specific rule for forming the groups to be compared. Any specific rule can be criticized. In the case of Bendectin, more than 20 studies mostly found that the drug appeared safe. Most of the lawsuits charging that Bendectin caused birth defects failed in court. The legal expense nonetheless forced the manufacturer to withdraw the drug from the market. Bendectin probably carries little or no risk, but all we can say for sure is "not enough evidence."

EXAMPLE 2.39

Despite the difficulties, it is sometimes possible to build a very strong case for causation in the absence of experiments. The evidence that smoking causes lung cancer is about as strong as nonexperimental evidence can be.

Doctors had long observed that most lung cancer patients were smokers. Comparison of smokers and "similar" nonsmokers showed a very strong association between smoking and death from lung cancer. Could the association be due to common response? Might there be, for example, a genetic factor that predisposes people both to nicotine addiction and to lung cancer? Smoking and lung cancer would then be positively associated even if smoking had no direct effect on the lungs. Or perhaps confounding is to blame. It might be that smokers live unhealthy lives in other ways (diet, alcohol, lack of exercise) and that some other habit confounded with smoking is a cause of lung cancer. How were these objections overcome?

Let's answer this question in general terms: What are the criteria for establishing causation when we cannot do an experiment?

- *The association is strong.* The association between smoking and lung cancer is very strong. That between Bendectin and birth defects is weak.

- *The association is consistent.* Many studies of different kinds of people in many countries link smoking to lung cancer. That reduces the chance that a lurking variable specific to one group or one study explains the association.

- *Higher doses are associated with stronger responses*. People who smoke more cigarettes per day or who smoke over a longer period get lung cancer more often. People who stop smoking reduce their risk.

- *The alleged cause precedes the effect in time*. Lung cancer develops after years of smoking. The number of men dying of lung cancer rose as smoking became more common, with a lag of about 30 years. Lung cancer kills more men than any other form of cancer. Lung cancer was rare among women until women began to smoke. Lung cancer in women rose along with smoking, again with a lag of about 30 years, and has now passed breast cancer as the leading cause of cancer death among women.

- *The alleged cause is plausible*. Experiments with animals show that tars from cigarette smoke do cause cancer.

Medical authorities do not hesitate to say that smoking causes lung cancer. The U.S. Surgeon General states that cigarette smoking is "the largest avoidable cause of death and disability in the United States."[24] The evidence for causation is overwhelming—but it is not as strong as the evidence provided by well-designed experiments.

SUMMARY

Some observed associations between two variables are due to a **cause-and-effect** relationship between these variables, but others are explained by **lurking variables**.

The effect of lurking variables can operate through **common response** if changes in both the explanatory and response variables are caused by changes in lurking variables. **Confounding** of two variables (either explanatory or lurking variables) means that we cannot distinguish their effects on the response variable.

That an association is due to causation is best established by an **experiment** that changes the explanatory variable while controlling other influences on the response.

In the absence of experimental evidence, be cautious in accepting claims of causation. Good evidence of causation requires a strong association that appears consistently in many studies, a clear explanation for the alleged causal link, and careful examination of possible lurking variables.

SECTION 2.7 EXERCISES

2.97   A study of grade-school children aged 6 to 11 years found a high positive correlation between reading ability y and shoe size x. Explain why common response to a lurking variable z accounts for this correlation. Present your suggestion in a diagram like one of those in Figure 2.41.

2.98 There is a negative correlation between the number of flu cases y reported each week through the year and the amount of ice cream x sold that week. It is unlikely that ice cream prevents flu. What is a more plausible explanation for this correlation? Draw a diagram like one of those in Figure 2.41 to illustrate the relationships among the variables.

2.99 The National Halothane Study was a major investigation of the safety of the anesthetics used in surgery. Records of over 850,000 operations performed in 34 major hospitals showed the following death rates y for four common anesthetics x (L. E. Moses and F. Mosteller, "Safety of anesthetics," in J. M. Tanur et al. (eds.), *Statistics: A Guide to the Unknown*, 3rd ed., Wadsworth, Pacific Grove, Calif., 1989, pp. 15–24):

Anesthetic	A	B	C	D
Death rate	1.7%	1.7%	3.4%	1.9%

Do these data prove that Anesthetic C is causing more deaths than the others?

2.100 A study showed that women who work in the production of computer chips have abnormally high numbers of miscarriages. The union claimed that exposure to chemicals used in production caused the miscarriages. Another possible explanation is that these workers spend most of their work time standing up. Illustrate these relationships in a diagram like one of those in Figure 2.41.

2.101 A study shows that there is a positive correlation between the size of a hospital (measured by its number of beds x) and the median number of days y that patients remain in the hospital. Does this mean that you can shorten a hospital stay by choosing a small hospital? Use a diagram like one of those in Figure 2.41 to explain the association.

2.102 Observational studies suggest that children who watch many hours of television get lower grades in school than those who watch less TV. Explain clearly why these studies do *not* show that watching TV *causes* poor grades. In particular, suggest some other variables that may be confounded with heavy TV viewing and may contribute to poor grades.

2.103 People who use artificial sweeteners in place of sugar tend to be heavier than people who use sugar. Does this mean that artificial sweeteners cause weight gain? Give a more plausible explanation for this association.

2.104 A group of college students believes that herb tea has remarkable powers. To test this belief, they make weekly visits to a local nursing home, visiting with the residents and serving them herb tea. The nursing home staff reports that after several months many of the residents are more cheerful and healthy. A skeptical sociologist commends the students for their good deeds but scoffs at the idea that herb tea helped the residents. It's all confounding, says the sociologist. Identify the explanatory and response variables in this informal study. Then explain what other variables are confounded with the explanatory variable.

2.105   Members of a high school language club believe that study of a foreign language improves a student's command of English. From school records, they obtain the scores on an English achievement test given to all seniors. The mean score of seniors who had studied a foreign language for at least two years is much higher than the mean score of seniors who studied no foreign language. The club's advisor says that these data are not good evidence that language study strengthens English skills. Identify the explanatory and response variables in this study. Then explain what lurking variable prevents the conclusion that language study improves students' English scores.

2.106   It has been suggested that electromagnetic fields of the kind present near power lines can cause leukemia in children. Experiments with children and power lines are not ethical. Careful studies have found no association between exposure to electromagnetic fields and childhood leukemia. (See Gary Taubes, "Magnetic field–cancer link: will it rest in peace?" *Science*, 277 (1997), p. 29.) Describe the kind of information you would seek to investigate the claim that electromagnetic fields are associated with cancer.

2.107   Return to the study of the safety of anesthetics from Exercise 2.99. For ethical reasons, surgeons are unwilling to do an experiment in which similar patients are given different anesthetics. However, you have very detailed records for the 850,000 operations that were investigated in the study cited in Exercise 2.99. What kinds of information would you seek in these records in order to establish that Anesthetic C does or does not cause extra deaths among patients?

CHAPTER 2 EXERCISES

2.108   Studies of disease often ask people about their diet in years past in order to discover links between diet and disease. But how well do people remember their past diet? Can we predict actual past diet as well or better from what subjects eat now as from their memory of past habits? Data on actual past diet are available for 91 people who were part of a study that started in the 1930s and were asked about their diet when they were 18 years old and again when they were 30. Researchers asked them again at age about 55 to describe their eating habits at ages 18 and 30 and also their current diet. The study results appear in J. T. Dwyer et al., "Memory of food intake in the distant past," *American Journal of Epidemiology*, 130 (1989), pp. 1033–1046. The authors say (page 1038):

> The first study aim, to determine how accurately this group of participants remembered past consumption, was addressed by correlations between recalled and historical consumption in each time period. To evaluate the second study aim, that is, whether recalled intake or current intake more accurately predicts historical intake of food groups at age 30 years, we performed regression analysis.

(a)   You must present the results of this paper to a group of people who know no statistics. Tell them in nontechnical language what "correlation" means, why correlation suits the first aim of the study, what "regression" means, and why regression fits the second study aim. Be sure to point out the distinction between correlation and regression.

(b) The study looked at the correlations between actual intake of many foods at age 18 and the intake the subjects now remember for age 18. The median correlation was $r = 0.217$. The authors say, "We conclude that memory of food intake in the distant past is fair to poor." Explain to your audience why $r = 0.217$ points to this conclusion.

(c) The authors used regression to predict the intake of a number of foods at age 30 from current intake of those foods and from what the subjects now remember about their intake at age 30. They conclude that "recalled intake more accurately predicted historical intake at age 30 years than did current diet." As evidence, they present r^2 values for the regressions. Explain to your audience why comparing r^2 is one way to compare how well different explanatory variables predict a response.

2.109 To study the energy savings due to adding solar heating panels to a house, researchers measured the natural gas consumption of the house for more than a year, then installed solar panels and observed the natural gas consumption for almost 2 years. The explanatory variable x is degree-days per day during the several weeks covered by each observation, and the response variable y is gas consumption (in hundreds of cubic feet) per day during the same period. Figure 2.42 plots y against x, with separate symbols for observations taken before and after the installation of the solar panels. The least-squares regression lines were computed separately for the before and after data and are drawn on the plot. The regression lines are

$$\text{Before: } \hat{y} = 1.089 + 0.189x$$
$$\text{After: } \hat{y} = 0.853 + 0.157x$$

(Data provided by Robert Dale, Purdue University.)

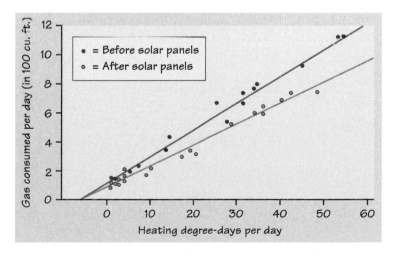

FIGURE 2.42 The regression of residential natural gas consumption on heating degree-days before and after installation of solar heating panels, for Exercise 2.109.

(a) Does the scatterplot suggest that a straight line is an appropriate description of the relationship between degree-days and natural gas consumption? Do any individual observations appear to have large residuals or to be highly influential?

(b) About how much additional natural gas was consumed per day for each additional degree-day before the panels were added? After the panels were added?

(c) The daily average temperature during January in this location is about 30°, which corresponds to 35 degree-days per day. Use the regression lines to predict daily gas usage for a day with 35 degree-days before and after installation of the panels. If the price of natural gas is $0.75 per hundred cubic feet, how much money do the solar panels save in the 31 days of a typical January?

2.110   We have seen that investment theory uses the standard deviation of returns to describe the volatility or risk of an investment. To describe how the risk of a specific security is related to that of the market as a whole, we use least-squares regression. Figure 2.43 plots the monthly percent total return y on Philip Morris common stock against the monthly return x on the Standard & Poor's 500 stock index, which represents the market, for the period between July 1990 and June 1997. The one clear outlier turns out not to be very influential.

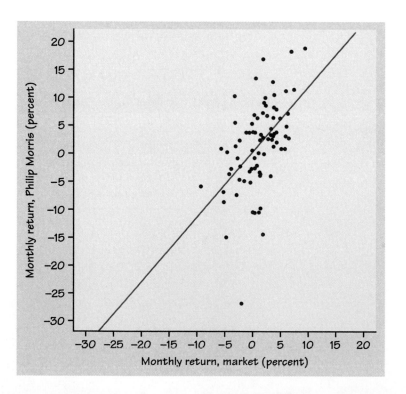

FIGURE 2.43   Monthly returns on an individual stock plotted against the returns for the stock market as a whole, with the least-squares line. See Exercise 2.110.

Here are the basic descriptive measures:

$$\overline{x} = 1.304 \qquad s_x = 3.392 \qquad r = 0.5251$$
$$\overline{y} = 1.878 \qquad s_y = 7.554$$

(a) Find the equation of the least-squares line from this information. What percent of the volatility in Philip Morris stock is explained by the linear relationship with the market as a whole?

(b) Explain carefully what the slope of the line tells us about how Philip Morris stock responds to changes in the market. This slope is called "beta" in investment theory.

(c) Returns on most individual stocks have a positive correlation with returns on the entire market. That is, when the market goes up, an individual stock tends to also go up. Explain why an investor should prefer stocks with beta > 1 when the market is rising and stocks with beta < 1 when the market is falling.

2.111 Lamb's-quarter is a common weed that interferes with the growth of corn. An agriculture researcher planted corn at the same rate in 16 small plots of ground, then weeded the plots by hand to allow a fixed number of lamb's-quarter plants to grow in each meter of corn row. No other weeds were allowed to grow. Here are the yields of corn (bushels per acre) in each of the plots (data provided by Samuel Phillips, Purdue University):

Weeds per meter	Corn yield	Weeds per meter	Corn yield	Weeds per meter	Corn yield	Weeds per meter	Corn yield
0	166.7	1	166.2	3	158.6	9	162.8
0	172.2	1	157.3	3	176.4	9	142.4
0	165.0	1	166.7	3	153.1	9	162.8
0	176.9	1	161.1	3	156.0	9	162.4

(a) What are the explanatory and response variables in this experiment?

(b) Make side-by-side stemplots of the yields, after rounding to the nearest bushel. What do you conclude about the effect of this weed on corn yield?

(c) Make a scatterplot of corn yield against weeds per meter. Find the least-squares regression line and add it to your plot. The advantage of regression over the side-by-side comparison in (b) is that we can use the fitted model to draw conclusions for counts of weeds other than the ones the researcher actually used. What does the slope of the fitted line tell us about the effect of lamb's-quarter on corn yield?

(d) Predict the yield for corn grown under these conditions with 6 lamb's-quarter plants per meter of row.

2.112 The table below gives data on the relationship between running speed (feet per second) and stride rate (steps taken per second) for elite female runners. (See Exercise 2.55 on page 171 for the source of the data.)

Speed	15.86	16.88	17.50	18.62	19.97	21.06	22.11
Stride rate	3.05	3.12	3.17	3.25	3.36	3.46	3.55

Here are the corresponding data from the same source for male runners:

Speed	15.86	16.88	17.50	18.62	19.97	21.06	22.11
Stride rate	2.92	2.98	3.03	3.11	3.22	3.31	3.41

(a) Plot the data for both groups on one graph using different symbols to distinguish between the points for females and those for males.

(b) Suppose now that the data came to you without identification as to gender. Compute the least-squares line from all of the data and plot it on your graph.

(c) Compute the residuals from this line for each observation. Make a plot of the residuals against speed. How does the fact that the data come from two distinct groups show up in the residual plot?

2.113 Wood scientists are interested in replacing solid-wood building material by less expensive products made from wood flakes. They collected the following data to examine the relationship between the length (in inches) and the strength (in pounds per square inch) of beams made from wood flakes. (Data provided by Jim Bateman and Michael Hunt, Purdue University.)

Length	5	6	7	8	9	10	11	12	13	14
Strength	446	371	334	296	249	254	244	246	239	234

(a) Make a scatterplot that shows how the length of a beam affects its strength.

(b) Describe the overall pattern of the plot. Are there any outliers?

(c) Fit a least-squares line to the entire set of data. Graph the line on your scatterplot. Does a straight line adequately describe these data?

(d) The scatterplot suggests that the relation between length and strength can be described by *two* straight lines, one for lengths of 5 to 9 inches and another for lengths of 9 to 14 inches. Fit least-squares lines to these two subsets of the data, and draw the lines on your plot. Do they describe the data adequately? What question would you now ask the wood experts?

2.114 Do mold colonies grow exponentially? In an investigation of the growth of molds, biologists inoculated flasks containing a growth medium with equal amounts of spores of the mold *Aspergillus nidulans*. They measured the size of a colony by analyzing how much remains of a radioactive tracer substance that is consumed by the mold as it grows. Each size measurement requires destroying that colony, so that the data below refer to 30 separate colonies. To smooth the pattern, we take the mean size of the three colonies measured at each time.

Hours	Colony size			Mean
0	1.25	1.60	0.85	1.23
3	1.18	1.05	1.32	1.18
6	0.80	1.01	1.02	0.94
9	1.28	1.46	2.37	1.70
12	2.12	2.09	2.17	2.13
15	4.18	3.94	3.85	3.99
18	9.95	7.42	9.68	9.02
21	16.36	13.66	12.78	14.27
24	25.01	36.82	39.83	33.89
36	138.34	116.84	111.60	122.26

(Similar experiments are described by A. P. J. Trinci, "A kinetic study of the growth of *Aspergillus nidulans* and other fungi," *Journal of General Microbiology*, 57 (1969), pp. 11–24. These data were provided by Thomas Kuczek, Purdue University.)

(a) Graph the mean colony size against time. Then graph the logarithm of the mean colony size against time.

(b) On the basis of data like these, microbiologists divide the growth of mold colonies into three phases that follow each other in time. Exponential growth occurs during only one of these phases. Briefly describe the three phases, making specific reference to the graphs to support your description.

(c) The exponential growth phase for these data lasts from about 6 hours to about 24 hours. Find the least-squares regression line of the logarithms of mean size on hours for only the data between 6 and 24 hours. Use this line to predict the size of a colony 10 hours after inoculation. (The line predicts the logarithm. You must obtain the size from its logarithm.)

(d) Find the correlation between the logarithm of mean size and hours for the data between 6 and 24 hours. Then make a scatterplot of the logarithms of the individual size measurements against hours for this same period and find the correlation. Why do we expect the second r to be smaller? Is it smaller?

2.115 North Carolina State University studied student performance in a course required by its chemical engineering major. One question of interest is the relationship between time spent in extracurricular activities and whether a student earned a C or better in the course. Here are the data for the 119 students who answered a question about extracurricular activities. (From Richard M. Felder et al., "Who gets it and who doesn't: a study of student performance in an introductory chemical engineering course," *1992*

ASEE Annual Conference Proceedings, American Society for Engineering Education, Washington, D.C., 1992, pp. 1516–1519.)

		Course grade	
		A, B, C	D, F
Activity	< 2	11	9
(hours	2 to 12	68	23
per week)	> 12	3	5

(a) Find the proportion of successful students (C or better) in each of the three extracurricular activity groups. Make a bar graph to compare the three proportions of successes. What kind of relationship between extracurricular activities and succeeding in the course do these proportions seem to show?

(b) Does the North Carolina State study convince you that spending more or less time on extracurricular activities *causes* changes in academic success? Explain your answer.

2.116 Firearms are second to motor vehicles as a cause of nondisease deaths in the United States. Here are counts from a study of all firearm-related deaths in Milwaukee, Wisconsin, between 1990 and 1994. (Data from S. W. Hargarten et al., "Characteristics of firearms involved in fatalities," *Journal of the American Medical Association*, 275 (1996), pp. 42–45.) We want to compare the types of firearms used in homicides and in suicides. We suspect that long guns (shotguns and rifles) will more often be used in suicides because many people keep them at home for hunting. Make a careful comparison of homicides and suicides, with a bar graph. What do you find about long guns versus handguns?

	Homicides	Suicides
Handgun	468	124
Shotgun	28	22
Rifle	15	24
Not specified	13	5
Total	524	175

2.117 A study of the salaries of full professors at Upper Wabash Tech shows that the median salary for female professors is considerably less than the median male salary. Further investigation shows that the median salaries for male and female full professors are about the same in every department (English, physics, etc.) of the university. Explain how equal salaries in every department can still result in a higher overall median salary for men.

2.118 Someone says, "There is a strong positive correlation between the number of firefighters at a fire and the amount of damage the fire does. So sending lots of firefighters just causes more damage." Explain why this reasoning is wrong.

2.119   The following table gives the U.S. resident population of voting age and the votes cast for president, both in thousands, for presidential elections between 1960 and 1996:

Year	Population	Votes
1960	109,672	68,838
1964	114,090	70,645
1968	120,285	73,212
1972	140,777	77,719
1976	152,308	81,556
1980	163,945	86,515
1984	173,995	92,653
1988	181,956	91,595
1992	189,524	104,425
1996	196,511	96,456

(a) For each year compute the percent of people who voted. Make a time plot of the percent who voted. Describe the change over time in participation in presidential elections.

(b) Before proposing political explanations for this change, we should examine possible lurking variables. The minimum voting age in presidential elections dropped from 21 to 18 years in 1970. Use this fact to propose a partial explanation for the trend you saw in (a).

2.120   Environmentalists, government officials, and vehicle manufacturers are all interested in studying the exhaust emissions produced by motor vehicles. The major pollutants in vehicle exhaust are hydrocarbons (HC), carbon monoxide (CO), and nitrogen oxides (NOX). Table 2.16 gives data for these three pollutants for a sample of 46 light-duty engines of the same type. In the table EN is an engine identifier. Figure 2.8 (page 118) is a plot of NOX against CO for these engines. Now we will make a more detailed study. (Data from T. J. Lorenzen, "Determining statistical characteristics of a vehicle emissions audit procedure," *Technometrics*, 22 (1980), pp. 483–493.)

(a) Plot HC against CO, HC against NOX, and NOX against CO.

(b) For each plot, describe the nature of the association. In particular, which pollutants are positively associated and which are negatively associated?

(c) Inspect your plots carefully and circle any points that appear to be outliers or influential observations. Next to the circles, indicate the EN identifier. Does any consistent pattern of unusual points appear in all three plots?

(d) Compute correlations for each pair of pollutants. If you found unusual points, compute the correlations both with and without these points.

(e) Find the least-squares lines corresponding to the three plots. Use CO to predict HC and NOX, and NOX to predict HC.

TABLE 2.16 Amounts of three pollutants emitted by light–duty engines (grams per mile)

EN	HC	CO	NOX	EN	HC	CO	NOX
1	0.50	5.01	1.28	2	0.65	14.67	0.72
3	0.46	8.60	1.17	4	0.41	4.42	1.31
5	0.41	4.95	1.16	6	0.39	7.24	1.45
7	0.44	7.51	1.08	8	0.55	12.30	1.22
9	0.72	14.59	0.60	10	0.64	7.98	1.32
11	0.83	11.53	1.32	12	0.38	4.10	1.47
13	0.38	5.21	1.24	14	0.50	12.10	1.44
15	0.60	9.62	0.71	16	0.73	14.97	0.51
17	0.83	15.13	0.49	18	0.57	5.04	1.49
19	0.34	3.95	1.38	20	0.41	3.38	1.33
21	0.37	4.12	1.20	22	1.02	23.53	0.86
23	0.87	19.00	0.78	24	1.10	22.92	0.57
25	0.65	11.20	0.95	26	0.43	3.81	1.79
27	0.48	3.45	2.20	28	0.41	1.85	2.27
29	0.51	4.10	1.78	30	0.41	2.26	1.87
31	0.47	4.74	1.83	32	0.52	4.29	2.94
33	0.56	5.36	1.26	34	0.70	14.83	1.16
35	0.51	5.69	1.73	36	0.52	6.35	1.45
37	0.57	6.02	1.31	38	0.51	5.79	1.51
39	0.36	2.03	1.80	40	0.48	4.62	1.47
41	0.52	6.78	1.15	42	0.61	8.43	1.06
43	0.58	6.02	0.97	44	0.46	3.99	2.01
45	0.47	5.22	1.12	46	0.55	7.47	1.39

(f) Compute the least-squares lines omitting the unusual points as a check on the influence of these points.

(g) Summarize the results of your analysis. In particular, is any of these pollutants a good predictor of another pollutant?

2.121   There are different ways to measure the amount of money spent on education. Average salary paid to teachers and expenditures per pupil are two possible measures. Table 2.17, based on information from the 1996 *Statistical Abstract of the United States*, gives the 1995 values for these variables by state. The states are classified according to region: NE (New England), MA (Middle Atlantic), ENC (East North Central), WNC (West North Central), SA (South Atlantic), ESC (East South Central), WSC (West South Central), MN (Mountain), and PA (Pacific).

(a) Make a stemplot or histogram for teachers' pay. Is the distribution roughly symmetric or clearly skewed? Find the five-number summary. Are there any suspected outliers by the $1.5 \times IQR$ criterion? Which states may be outliers? Do the same for spending per pupil. Are the same states outliers in both distributions?

State	Region	Pay	Spend	State	Region	Pay	Spend
Me.	NE	32.0	6.41	N.H.	NE	29.0	6.13
Vt.	NE	35.4	7.37	Mass.	NE	42.2	6.17
R.I.	NE	40.7	7.36	Conn.	NE	50.0	8.50
N.Y.	MA	47.6	9.45	N.J.	MA	46.1	9.86
Pa.	MA	44.5	7.20	Ohio	ENC	36.8	5.62
Ind.	ENC	36.8	6.00	Ill.	ENC	39.4	5.26
Mich.	ENC	47.4	6.93	Wis.	ENC	37.7	7.00
Minn.	WNC	35.9	5.11	Iowa	WNC	31.5	5.56
Mo.	WNC	31.2	4.97	N.Dak.	WNC	26.3	4.60
S.Dak.	WNC	26.0	4.84	Nebr.	WNC	30.9	5.38
Kans.	WNC	34.7	5.76	Del.	SA	39.1	7.17
Md.	SA	40.7	6.72	D.C.	SA	43.7	8.21
Va.	SA	34.0	5.66	W.Va.	SA	31.9	6.52
N.C.	SA	30.8	4.95	S.C.	SA	30.3	4.93
Ga.	SA	32.6	5.40	Fla.	SA	32.6	5.72
Ky.	ESC	32.3	5.61	Tenn.	ESC	32.5	4.54
Ala.	ESC	31.1	4.46	Miss.	ESC	26.8	4.12
Ark.	WSC	28.9	4.26	La.	WSC	26.5	4.71
Okla.	WSC	28.2	4.38	Tex.	WSC	31.2	5.42
Mont.	MN	28.8	5.83	Idaho	MN	29.8	6.03
Wyo.	MN	31.3	6.07	Colo.	MN	34.6	5.50
N.Mex.	MN	28.5	5.42	Ariz.	MN	32.2	4.25
Utah	MN	29.1	3.67	Nev.	MN	34.8	5.13
Wash.	PA	36.2	5.81	Oreg.	PA	38.6	6.25
Calif.	PA	41.1	4.73	Alaska	PA	48.0	9.93
Hawaii	PA	38.5	6.16				

TABLE 2.17 Mean teachers' pay and per-pupil educational spending by state, 1995 (thousands of dollars)

(b) Make a scatterplot of teachers' pay y against spending x. Describe the pattern of the relationship between pay and spending. Is there a strong association? If so, is it positive or negative? Explain why you might expect to see an association of this kind.

(c) Find the least-squares regression line for predicting teachers' pay from education spending and draw it on your scatterplot. How much on the average does mean teachers' pay increase when spending increases by $1000 per pupil from one state to another? Give a numerical measure of the success of overall spending on education in explaining variations in teachers' pay among states.

(d) On your plot, circle any outlying points found in (a). Label the circled points with the state identifier. Do these points have large residuals? (You need not actually calculate the residuals.) The states you have identified lie close together on the plot. To see if they are influential as a group, find the regression line with all of these states removed from the calculation. Draw this new line on your plot. Was this group of states influential?

2.122 Continue the analysis of teachers' pay and education spending begun in the previous exercise by looking for regional effects. We will compare these three groups:

Coastal	Middle Atlantic, New England, and Pacific
South	South Atlantic, East South Central, and West South Central
Midwest	East North Central and West North Central

Omit the District of Columbia, which is a city rather than a state.

(a) Make side-by-side boxplots for education spending in the three regions. For each region, label any outliers (points identified by the $1.5 \times IQR$ criterion) with the state identifier.

(b) Repeat part (a) for teachers' pay.

(c) Do you see important differences in spending and pay by region? Are the differences consistent for the two variables? That is, are regions that are high in spending also high in pay and vice versa?

(d) Find the residuals from the least-squares lines from the previous exercise (with all states included). Use side-by-side boxplots to display the residuals by region. Draw a line across your plot at zero. Do the residuals appear to differ by region? Describe any patterns that are noteworthy.

2.123 Can we predict college grade point average from SAT scores and high school grades? The CSDATA data set contains information on this issue for a large group of computer science students. We will look only at SAT mathematics scores as a predictor of later college GPA, using the variables SATM and GPA from CSDATA. Make a scatterplot, obtain r and r^2, and draw on your plot the least-squares regression line for predicting GPA from SATM. Then write a brief discussion of the ability of SATM alone to predict GPA. (In Chapter 11 we will see how combining several explanatory variables improves our ability to predict.)

2.124 The WOOD data set comes from a study of the strength of wood products. The researchers measured the modulus of elasticity for each of the 50 strips of wood two times. The results are the variables T1 and T2. Lack of a strong relationship between the two measurements means that the measurement process does not produce consistent results. The correlation is often used to assess the strength of the relationship between repeated measurements. It is called the *reliability* of the measurement process.

(a) Plot the data with T1 (the first measurement) on the x axis and T2 on the y axis. Describe the relationship. Why would you expect it to be linear and positive?

(b) Calculate the correlation. Is this measurement process highly reliable?

(c) We have two ways to calculate the least-squares regression line for predicting T2 from T1. First find the line using your software's regression command. Then find the means and standard deviations for T1 and T2 and use these numbers along with the correlation to find the slope of the least-squares line. Verify that both methods give the same slope.

2.125 We will investigate the effect of adding an unusual observation to the WOOD data set. Add the observation T1 = 1.6, T2 = 1.9 to the data set.

(a) The new point is an outlier in the relationship. To confirm this, make a scatterplot of T2 against T1. Also regress T2 on T1 and plot the residuals against T1. Circle the new point on both graphs.

(b) The new point is *not* an outlier in T1 alone and it is also not an outlier in T2 alone. Verify these statements by constructing stemplots, boxplots, or normal quantile plots for both variables. You have now learned that an observation can be an outlier in two dimensions even though neither x and y lie outside the pattern of their one-dimensional distributions.

(c) How does the new point change the regression equation? Find the equation with and without the new point. Plot both regression lines on your scatterplot. Is this point influential?

(d) How does the new point change the correlation? Report the correlation with and without the new point.

2.126 Once again we will add an unusual observation to the WOOD data set. Add the observation T1 = 2.2, T2 = 1.2 to the original data set. (Discard the extra observation from the previous exercise.)

(a) The new point is somewhat extreme in T1 and T2 but it is not far away from other values of these variables. Verify these statements by constructing stemplots, boxplots, or normal quantile plots for both T1 and T2.

(b) Calculate the mean and standard deviation for T1 and T2 with and without the new point. Does the new point affect these summary measures substantially for either variable?

(c) Plot T2 against T1 with the new data point included. Does the new point appear to be influential? Find the equation for the regression line of T2 on T1 both with and without the new point and draw both lines on your plot. Is the new point in fact influential?

(d) Calculate the correlation with and without the new point. Does the new point change the correlation substantially?

NOTES

1. A sophisticated treatment of improvements and additions to scatterplots is W. S. Cleveland and R. McGill, "The many faces of a scatterplot," *Journal of the American Statistical Association*, 79 (1984), pp. 807–822.

2.   Marilyn A. Houck et al., "Allometric scaling in the earliest fossil bird, *Archaeopteryx lithographica*," *Science*, 247 (1990), pp. 195–198. The authors conclude from a variety of evidence that all specimens represent the same species.

3.   The data are from W. L. Colville and D. P. McGill, "Effect of rate and method of planting on several plant characters and yield of irrigated corn," *Agronomy Journal*, 54 (1962), pp. 235–238.

4.   The motorcycle crash test data are often used to test scatterplot smoothers. They appear in, for example, W. Hardle, *Applied Nonparametric Regression*, Cambridge University Press, New York, 1990. The smoother used to produce Figures 2.5 and 2.6 is called "lowess." A full account appears in section 4.6 of J. M. Chambers, W. S. Cleveland, B. Kleiner, and P. A. Tukey, *Graphical Methods for Data Analysis*, Wadsworth, Belmont, Calif., 1983.

5.   The draft lottery is analyzed in detail by S. E. Fienberg, "Randomization and social affairs: the 1970 draft lottery," *Science*, 171 (1971), pp. 255–261. In 1971, the Department of Defense asked statisticians from the National Bureau of Standards to design a truly random selection procedure. The design of the 1971 lottery is described in another article in the same issue of *Science*.

6.   Based on T. N. Lam, "Estimating fuel consumption from engine size," *Journal of Transportation Engineering*, 111 (1985), pp. 339–357. The data for 10 to 50 km/h are measured; those for 60 and higher are calculated from a model given in the paper and are therefore smoothed.

7.   A careful study of this phenomenon is W. S. Cleveland, P. Diaconis, and R. McGill, "Variables on scatterplots look more highly correlated when the scales are increased," *Science*, 216 (1982), pp. 1138–1141.

8.   Data provided by Linda McCabe and the Agency for International Development.

9.   E. P. Hubble, "A relation between distance and radial velocity among extra-galactic nebulae," *Proceedings of the National Academy of Sciences*, 15 (1929), pp. 168–173.

10.   The U.S. returns are for the Standard & Poor's 500 stock index. The overseas returns are for the Morgan Stanley Europe, Australasia, Far East (EAFE) index.

11.   Data provided by Peter Cook, Department of Mathematics, Purdue University.

12.   These data were originally collected by L. M. Linde of UCLA but were first published by M. R. Mickey, O. J. Dunn, and V. Clark, "Note on the use of stepwise regression in detecting outliers," *Computers and Biomedical Research*, 1 (1967), pp. 105–111. The data have been used by several authors. We found them in N. R. Draper and J. A. John, "Influential observations and outliers in regression," *Technometrics*, 23 (1981), pp. 21–26.

13.   A good but somewhat advanced reference is D. A. Belsley, E. Kuh, and R. E. Welsch, *Regression Diagnostics: Identifying Influential Data and Sources of Collinearity*, Wiley, New York, 1980.

14.   This example is drawn from M. Goldstein, "Preliminary inspection of multivariate data," *The American Statistician*, 36 (1982), pp. 358–362.

15.   The facts about the LSAT, along with much useful information about the alleged cultural bias of such tests, appear in D. Kaye, "Searching for the truth about testing," *The Yale Law Journal*, 90 (1980), pp. 431–457.

16.   P. Velleman, *ActivStats 2.0*, Addison Wesley Interactive, Reading, Mass., 1997.

17.   Data from the Energy Information Administration, recorded in Robert H. Romer, *Energy: An Introduction to Physics*, W. H. Freeman, San Francisco, 1976, for 1880 to 1972, and in the *Statistical Abstract of the United States* for more recent years.

18.   This annually compounded rate is lower than the mean of the annual percent returns for the same period. Cognoscenti will recognize that it is the geometric mean of the annual returns.

19.   You can find a clear and comprehensive discussion of numerical measures of association for categorical data in Chapter 3 of A. M. Liebetrau, *Measures of Association,* Sage Publications, Beverly Hills, Calif., 1983.

20.   National Science Board, *Science and Engineering Indicators, 1991,* U.S. Government Printing Office, Washington, D.C., 1991. The detailed data appear in Appendix Table 3-5, p. 274.

21.   Saccharin appears on the National Institutes of Health's list of suspected carcinogens but remains in wide use. In October 1997, an expert panel recommended by a 4 to 3 vote to keep it on the list, despite recommendations from other scientific groups that it be removed.

22.   A detailed study of this correlation appears in E. M. Remolona, P. Kleinman, and D. Gruenstein, "Market returns and mutual fund flows," *Federal Reserve Bank of New York Economic Policy Review,* 3, no. 2 (1997), pp. 33–52.

23.   See L. Lasagna and S. R. Schulman, "Bendectin and the language of causation," in K. R. Foster, D. E. Bernstein, and P. W. Huber (eds.), *Phantom Risk: Scientific Inference and the Law,* The MIT Press, Cambridge, Mass., 1994, pp. 101–122. See also the first chapter in this book and the review by J. L. Gastwirth, *Journal of the American Statistical Association,* 90 (1995), pp. 388–390.

24.   *The Health Consequences of Smoking: 1983,* U.S. Public Health Service, Washington, D.C., 1983.

Prelude

Statistical designs for producing trustworthy data are perhaps the single most influential contribution of statistics to the advance of knowledge. Random sampling and randomized comparative experiments have in this century revolutionized the practice of many fields of applied science. This chapter presents the basic ideas of statistical designs for choosing a sample that represents some wider population and for laying out an experiment to study the effects of explanatory variables on a response. The deliberate use of chance is central to these designs, so we conclude the chapter with a first look at the consequences of using chance in producing data. Here are some of the examples we will encounter:

- Does regularly taking aspirin reduce the risk of a heart attack?

- Will giving people meters that estimate their electric bills lead them to use less electricity?

- Ann Landers says that 70% of parents would not have children if they could make the decision again. Is this sense or nonsense?

- An opinion poll interviews 2500 adults and finds that 66% say that shopping is time-consuming and frustrating. Why is that 66% almost surely close to the truth about *all* adults?

CHAPTER **3**

Hilary S. Lorenz, *Atomic Graph*, 1994.

Producing Data

Introduction

Numerical data are raw material for the growth of knowledge. Statistics is a tool that helps data produce knowledge rather than confusion. As such, it must be concerned with producing data as well as with interpreting already available data.

Chapters 1 and 2 explored the art of *data analysis*. They showed how to use graphs and numbers to uncover the nature of a set of data. We can distinguish two different attitudes toward data. We may ask, "What do we see in the data?" Or we may ask, "What answer do the data give to this specific question?" This distinction becomes important as we turn to producing data, because trustworthy answers to specific questions often require data produced for exactly that purpose.

exploratory data
analysis

"What do we see?" is the attitude of **exploratory data analysis.** Sometimes we meet data that seem interesting and that deserve careful exploration. We do not know what the data may show, and we may have no specific questions in mind before we begin our examination. We have explored the relationship between SAT scores in the states and the percent of students who take the SAT, hunted for patterns in the results of a draft lottery, and examined the relationship between age and education. Exploratory analysis relies heavily on plotting the data. It seeks patterns that suggest interesting conclusions or questions for further study. However, exploratory analysis alone can rarely provide convincing evidence for its conclusions, because striking patterns in data can arise from many sources.

We may also use data to provide clear answers to specific questions such as "What is the speed of light?" or "How many plants per acre give the highest yield of corn?" or "Does the age at which a child starts to talk predict a later test score?" Data that answer such questions are as much the product of intelligent effort as hybrid tomatoes and compact disc players. This chapter is devoted to developing the skills needed to produce trustworthy data and to judge the quality of data produced by others. The techniques for producing data discussed in this chapter involve no formulas or graphs, but they are among the most important ideas in statistics.

statistical inference

Statistical techniques for producing data open the door to a further advance in data analysis, **formal statistical inference,** which answers specific questions with a known degree of confidence. Statistical inference uses the descriptive tools we have already met, in combination with new kinds of reasoning. It is primarily numerical rather than graphical. The later chapters of this book are devoted to inference. We will see that careful design of data production is the most important prerequisite for trustworthy inference.

3.1   First Steps

The first questions you must answer when you plan to produce data are familiar: "What individuals shall I study?" "What variables shall I measure?"

Answering these questions gives a clear statement of your interests. We will take for granted in this chapter that all the variables we work with are clearly defined and satisfactorily measured.

Statistics begins to play an essential role when the focus turns from measurements on individuals (whether ball bearings, rats, or people) to arrangements for collecting data from many individuals. These arrangements or patterns for producing data are called **designs.** Some questions that a design must address are: How many individuals shall we collect data from? How shall we select the individuals to be studied? If, as in many experiments, several groups of individuals are to receive different treatments, how shall we form the groups? If there is no systematic design for producing data, we may be misled by haphazard or incomplete data or by confounding.

designs

Where to find data: the library and the Internet

You want data on a question of interest to you. What are the social and economic backgrounds of college undergraduates? What causes of death are most common among young people? Or you are writing a paper on homelessness and wonder how many people are homeless. How will you get relevant data?

It is tempting to simply base our conclusions on our own experience, making no use of systematic data. We think (without really thinking) that the students at our college are typical. Or we recall an unusual incident that sticks in our memory exactly because it is unusual. We remember an airplane crash that killed several hundred people. We may ignore the fact that data on all flights show that flying is much safer than driving. We are relying on *anecdotes* rather than data.

Anecdotal Evidence

Anecdotal evidence is based on haphazardly selected individual cases, which often come to our attention because they are striking in some way. These cases need not be representative of any larger group of cases.

A better tactic is to head for the library or the Internet. There we find abundant data, not gathered specifically to answer our questions, but available for our use. The annual *Statistical Abstract of the United States* contains data that shed light on both the population of college students and causes of death broken down by age. You can learn, for example, that 30.4% of college students are employed full-time and that as of 1995 AIDS was the leading cause of death among men aged 25 to 44 but ranked only fourth in killing women in that age group. The *Statistical Abstract* has nothing on the homeless, however.

The Web sites of government statistical offices are prime sources of data. Here is the home page of the U.S. Census Bureau.

More detailed and more recent data are often available on the Internet, though locating them can be time-consuming. Government statistical offices are the primary source for demographic, economic, and social data. The United States does not have a central government statistical office. More than 70 different federal agencies collect data. Fortunately, you can reach most of them through the government's handy FedStats site (http://www.fedstats.gov). If you select the National Center for Education Statistics, you can search the more than 400 tables of *The Digest of Education Statistics* to satisfy your curiosity about undergraduates. Many nations do have a single national statistical office, such as Statistics Canada (http://www.statcan.ca) or Mexico's INEGI (http://www.inegi.gob.mx).

EXAMPLE 3.1   You seek data on homeless people for a paper you are writing. Your Internet harvest will change, probably from week to week, but here is an example of search results. Searching government sites through FedStats discloses that "the Census Bureau does not have a definition and will not provide a total count of 'the homeless.'" At the Department of Housing and Urban Development, you find detailed descriptions of legislation to help the homeless, but again no data. It was much faster to use the keyword search facilities at these agencies than to spend a fruitless afternoon in the library.

One of the best sources of data on public opinion and social trends is the General Social Survey (GSS) operated by the University of Chicago's National Opinion Research Center. A search of the GSS archives (http://www.icpsr.umich.edu/GSS) shows that in 1991 the GSS asked if during the past year a person has "had to temporarily live with others or in a shelter or 'on the street.'" At that time, 986 of those asked said "No," 26 said "Yes," and 5 did not answer. One of the strengths of the GSS is that it asks many questions repeatedly so that we can follow trends. Unfortunately, this question was asked only once.

After a further search, you locate a helpful "Fact Sheet" at the National Coalition for the Homeless (http://nch.ari.net) that explains why homelessness is hard to measure and collects several different estimates of the number of homeless people, with references and comments. These estimates are not all equally trustworthy, but the data there will strengthen your paper.

The library and the Internet are sources of *available data*.

Available Data

Available data are data that were produced in the past for some other purpose but that may help answer a present question.

Available data are the only data used in most student reports. Because producing new data is expensive, we all use available data whenever possible. However, the clearest answers to present questions often require data produced to answer those specific questions. Statistical designs for producing data rely on either *sampling* or *experiments*.

Sampling

sample
population

The General Social Survey interviews 3000 adult residents of the United States every second year. That is, the GSS selects a **sample** of adults to represent the larger **population** of all adults living in the United States. The idea of *sampling* is to study a part in order to gain information about the whole. Data are often produced by sampling a population of people or things. Opinion polls, for example, report the views of the entire country based on interviews with a sample of about 1000 to 1500 people. Government reports on employment and unemployment are produced from a monthly sample of about 60,000 households. The quality of manufactured items is monitored by inspecting small samples each hour or each shift.

census

In all of these examples, the expense of examining every item in the population makes sampling a practical necessity. Timeliness is another reason for preferring a sample to a **census,** which is an attempt to contact every individual in the entire population. We want information on current unemployment and public opinion next week, not next year. Moreover, a carefully conducted sample is often more accurate than a census. Accountants, for example, sample a firm's inventory to verify the accuracy of the records. Attempting to count every last item in the warehouse would be not only expensive but inaccurate. Bored people do not count carefully.

If conclusions based on a sample are to be valid for the entire population, a sound design for selecting the sample is required. Sampling designs are the topic of Section 3.3.

Experiments

A sample survey collects information about a population by selecting and measuring a sample from the population. The goal is a picture of the population, disturbed as little as possible by the act of gathering information. Sample surveys are one kind of *observational study*.

Observation Versus Experiment

An **observational study** observes individuals and measures variables of interest but does not attempt to influence the responses.

An **experiment,** on the other hand, deliberately imposes some treatment on individuals in order to observe their responses.

An observational study, even one based on a statistical sample, is a poor way to gauge the effect of an intervention. To see the response to a change, we must actually impose the change. When our goal is to understand cause and effect, experiments are the only source of fully convincing data.

EXAMPLE 3.2   A majority of adult recipients of welfare are mothers of young children. Observational studies of welfare mothers show that many women are able to increase their earnings and leave the welfare system even though they have preschool children. Some of them take advantage of voluntary job-training programs to improve their skills. Should participation in job-training and job-search programs be required of all able-bodied welfare mothers? Observational studies cannot tell us what the effects of such a policy would be. Even if the mothers studied are a properly chosen sample of all welfare recipients, those who seek out training and find jobs may differ in many ways from those who do not. They are observed to have more education, for example, but they may also differ in values and motivation, things that cannot be observed.

To see if a required jobs program will help mothers escape welfare, such a program must actually be tried. Choose two similar groups of mothers when they apply for welfare. Require one group to participate in a job-training program, but do not offer the program to the other group. Comparing the income and work record of the two groups after several years will show whether requiring training has the desired effect.

In Example 3.2, the effect of voluntary training programs on success in finding work is *confounded* with (mixed up with) the characteristics of mothers who seek out training on their own. Observational studies of the effect of one variable on another often fail because the explanatory variable is confounded with lurking variables. Because experiments allow us to pin down the effects of specific variables of interest to us, they are the preferred method for gaining knowledge in science, medicine, and industry.

Experiments, like samples, require careful design. Because of the unique importance of experiments as a method for producing data, we begin the discussion of statistical designs for data collection in Section 3.2 with the principles underlying the design of experiments.

SUMMARY

Data analysis is sometimes **exploratory** in nature. Exploratory analysis asks what the data tell us about the variables and their relations to each other. The conclusions of an exploratory analysis may not generalize beyond the specific data studied.

Statistical inference produces answers to specific questions, along with a statement of how confident we can be that the answer is correct. The conclusions of statistical inference are usually intended to apply beyond the individuals actually studied. Successful statistical inference usually requires **production of data** intended to answer the specific questions posed.

Anecdotal evidence based on a few individual cases is rarely trustworthy. **Available data** collected for other purposes, such as the data produced by government statistical offices, are often helpful.

We can produce data intended to answer specific questions by sampling or experimentation. **Sampling** selects a part of a population of interest to represent the whole. **Experiments** are distinguished from **observational studies** such as sample surveys by the active imposition of some treatment on the subjects of the experiment.

SECTION 3.1 EXERCISES

3.1   A letter to the editor of *Organic Gardening* magazine (August 1980) said, "Today I noticed about eight stinkbugs on the sunflower stalks. Immediately I checked my okra, for I was sure that they'd be under attack. There wasn't one stinkbug on them. I'd never read that stinkbugs are attracted to sunflowers, but I'll surely interplant them with my okra from now on."

Explain briefly why this anecdote does not provide good evidence that sunflowers attract stinkbugs away from okra. In your explanation, suggest some factors that might account for all of the bugs being on the sunflowers.

3.2   When the discussion turns to the pros and cons of wearing automobile seat belts, Herman always brings up the case of a friend who survived an accident because he was not wearing a seat belt. The friend was thrown out of the car and landed on a grassy bank, suffering only minor injuries, while the car burst into flames and was destroyed. Explain briefly why this anecdote does not provide good evidence that it is safer not to wear seat belts.

3.3   There may be a "gender gap" in political party preference in the United States, with women more likely than men to prefer Democratic candidates. A political scientist selects a large sample of registered voters, both men and women. She asks every voter whether they voted for the Democratic or the Republican candidate in the last congressional election. Is this an observational study or an experiment? Why? What are the explanatory and response variables?

3.4   An educator wants to compare the effectiveness of computer software that teaches reading with that of a standard reading curriculum. She tests the reading ability of each student in a class of fourth graders, then divides them into two groups. One group uses the computer regularly, while the other studies a standard curriculum. At the end of the year, she retests all the students and compares the increase in reading ability in the two groups.

(a)   Is this an observational study or an experiment? Why?

(b)   What are the explanatory and response variables?

3.5   What is the preferred treatment for breast cancer that is detected in its early stages? The most common treatment was once mastectomy (removal of the breast). It is now usual to remove the tumor and nearby lymph nodes, followed by radiation. To study whether these treatments differ in their effectiveness, a medical team examines the records of 25 large hospitals and compares the survival times after surgery of all women who have had either treatment.

(a)   What are the explanatory and response variables?

(b)   Explain carefully why this study is not an experiment.

(c)   Do you think this study will show whether a mastectomy causes longer average survival time? Explain your answer carefully.

3.6   A study of the effect of living in public housing on family stability and other variables in poverty-level households was carried out as follows. The researchers obtained a list of all applicants for public housing during the previous year. Some applicants had been accepted, while others had been turned down by the housing authority. Both groups were interviewed and compared. Was this study an experiment? Why or why not? What are the explanatory and response variables in the study?

3.7   Some people believe that exercise raises the body's metabolic rate for as long as 12 to 24 hours, enabling us to continue to burn off fat after our workout has ended. In a study of this effect, subjects were asked to walk briskly on a treadmill for several hours. Their metabolic rates were measured before, immediately after, and 12 hours after the exercise. The study was criticized because eating raises the metabolic rate, and no record was kept of what the subjects ate after exercising. Was this study an experiment? Why or why not? What are the explanatory and response variables?

3.8 The National Halothane Study was a major investigation of the safety of anesthetics used in surgery. Records of over 850,000 operations performed in 34 major hospitals showed the following death rates for four common anesthetics (L. E. Moses and F. Mosteller, "Safety of anesthetics," in J. M. Tanur et al. (eds.), *Statistics: A Guide to the Unknown*, 3rd ed., Wadsworth, Belmont, Calif., 1989, pp. 15–24):

Anesthetic	A	B	C	D
Death rate	1.7%	1.7%	3.4%	1.9%

There is a clear association between the anesthetic used and the death rate of patients. Anesthetic C appears dangerous.

(a) Explain why we call the National Halothane Study an observational study rather than an experiment, even though it compared the results of using different anesthetics in actual surgery.

(b) When the study looked at other variables that are confounded with a doctor's choice of anesthetic, it found that Anesthetic C was not causing extra deaths. Suggest several variables that are mixed up with what anesthetic a patient receives.

3.2 Design of Experiments

A study is an experiment when we actually do something to people, animals, or objects in order to observe the response. Here is the basic vocabulary of experiments.

Experimental Units, Subjects, Treatment

The individuals on which the experiment is done are the **experimental units.** When the units are human beings, they are called **subjects.** A specific experimental condition applied to the units is called a **treatment.**

---

factors

level of a factor

Because the purpose of an experiment is to reveal the response of one variable to changes in other variables, the distinction between explanatory and response variables is important. The explanatory variables in an experiment are often called **factors.** Many experiments study the joint effects of several factors. In such an experiment, each treatment is formed by combining a specific value (often called a **level**) of each of the factors.

EXAMPLE 3.3   Does regularly taking aspirin help protect people against heart attacks? The Physi-cians' Health Study was a medical experiment that helped answer this question. In fact, the Physicians' Health Study looked at the effects of two drugs: aspirin and beta carotene. The body converts beta carotene into vitamin A, which may help prevent some forms of cancer. The *subjects* were 21,996 male physicians. There were two *factors*, each having two levels: aspirin (yes or no) and beta carotene (yes or no). Combinations of the levels of these factors form the four *treatments* shown in Figure 3.1. One-fourth of the subjects were assigned to each of these treatments.

placebo   On odd-numbered days, the subjects took a white tablet that contained either aspirin or a **placebo,** a dummy pill that looked and tasted like the aspirin but had no active ingredient. On even-numbered days, they took a red capsule containing either beta carotene or a placebo. There were several *response variables*—the study looked for heart attacks, several kinds of cancer, and other medical outcomes. After several years, 239 of the placebo group but only 139 of the aspirin group had suffered heart attacks. This difference is large enough to give good evidence that taking aspirin does reduce heart attacks.[1] It did not appear, however, that beta carotene had any effect.

EXAMPLE 3.4   Does studying a foreign language in high school increase students' verbal ability in English? Julie obtains lists of all seniors in her high school who did and did not study a foreign language. Then she compares their scores on a standard test of English reading and grammar given to all seniors. The average score of the students who studied a foreign language is much higher than the average score of those who did not.

This observational study gives no evidence that studying another language builds skill in English. Students decide for themselves whether or not to elect a foreign language. Those who choose to study a language are mostly students who are already better at English than most students who avoid foreign languages. The difference in average test scores just shows that students who choose to study a language differ (on the average) from those who do not. We can't say whether studying languages *causes* this difference.

Examples 3.3 and 3.4 illustrate the big advantage of experiments over observational studies. **In principle, experiments can give good evidence for causation**. All the doctors in the Physicians' Health Study took a pill every other day, and all got the same schedule of checkups and information. The only difference was the content of the pill. When one group had many fewer heart attacks, we conclude that it was the content of the pill that made the difference. Julie's observational study—a census of all seniors in her high school—does a good job of describing differences between seniors who have studied foreign languages and those who have not. But she can say nothing about cause and effect.

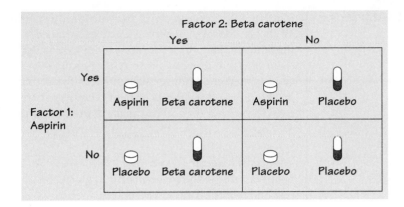

FIGURE 3.1 The treatments in the Physicians' Health Study, for Example 3.3.

Another advantage of experiments is that they allow us to study the specific factors we are interested in, while controlling the effects of lurking variables. The subjects in the Physicians' Health Study were all middle-aged male doctors and all followed the same schedule of medical checkups. These similarities reduce variation among the subjects and make any effects of aspirin or beta carotene easier to see. Experiments also allow us to study the combined effects of several factors. The interaction of several factors can produce effects that could not be predicted from looking at the effects of each factor alone. The Physicians' Health Study tells us that aspirin helps prevent heart attacks, at least in middle-aged men, and that beta carotene taken with the aspirin neither helps nor hinders aspirin's protective powers.

Comparative experiments

Laboratory experiments in science and engineering often have a simple design with only a single treatment, which is applied to all of the experimental units. The design of such an experiment can be outlined as

Treatment $\longrightarrow$ Observe response

For example, we may subject a beam to a load (treatment) and measure its deflection (observation). We rely on the controlled environment of the laboratory to protect us from lurking variables. When experiments are conducted in the field or with living subjects, such simple designs often yield invalid data. That is, we cannot tell whether the response was due to the treatment or to lurking variables. Another medical example will show what can go wrong.

EXAMPLE 3.5 "Gastric freezing" is a clever treatment for ulcers in the upper intestine. The patient swallows a deflated balloon with tubes attached, then a refrigerated liquid is pumped through the balloon for an hour. The idea is that cooling the stomach will reduce its production of acid and so relieve ulcers. An experiment reported in the *Journal of the American Medical Association* showed that gastric freezing did reduce acid production and relieve ulcer pain. The treatment was safe and easy and was widely used for several years. The design of the experiment was

Gastric freezing ⟶ Observe pain relief

placebo effect The gastric freezing experiment was poorly designed. The patients' response may have been due to the **placebo effect.** A placebo is a dummy treatment. Many patients respond favorably to any treatment, even a placebo, presumably because of trust in the doctor and expectations of a cure. This response to a dummy treatment is the placebo effect.

A later experiment divided ulcer patients into two groups. One group was treated by gastric freezing as before. The other group received a placebo treatment in which the liquid in the balloon was at body temperature rather than freezing. The results: 34% of the 82 patients in the treatment group improved, but so did 38% of the 78 patients in the placebo group. This and other properly designed experiments showed that gastric freezing was no better than a placebo, and its use was abandoned.[2]

The first gastric freezing experiment gave misleading results because the effects of the explanatory variable were *confounded* with (mixed up with) the placebo effect. We can defeat confounding by *comparing* two groups of patients, as in the second gastric freezing experiment. The placebo effect and other lurking variables now operate on both groups. The only difference between the groups is the actual effect of gastric freezing. The group **control group** of patients who received a sham treatment is called a **control group,** because it enables us to control the effects of outside variables on the outcome. Control is the first basic principle of statistical design of experiments. Comparison of several treatments in the same environment is the simplest form of control.

Without control, experimental results in medicine and the behavioral sciences can be dominated by such influences as the details of the experimental arrangement, the selection of subjects, and the placebo effect. The result is often *bias*.

> **Bias**
>
> The design of a study is **biased** if it systematically favors certain outcomes.

An uncontrolled study of a new medical therapy, for example, is biased in favor of finding the treatment effective because of the placebo effect. It should not surprise you to learn that uncontrolled studies in medicine give new therapies a much higher success rate than proper comparative experiments. Well-designed experiments, like the Physicians' Health Study and the second gastric freezing study, usually compare several treatments.

Randomization

The design of an experiment first describes the response variable or variables, the factors (explanatory variables), and the layout of the treatments, with comparison as the leading principle. Figure 3.1 illustrates this aspect of the design of the Physicians' Health Study. The second aspect of design is the rule used to assign the experimental units to the treatments. Comparison of the effects of several treatments is valid only when all treatments are applied to similar groups of experimental units. If one corn variety is planted on more fertile ground, or if one cancer drug is given to less seriously ill patients, comparisons among treatments are meaningless. Systematic differences among the groups of experimental units in a comparative experiment cause bias. How can we assign experimental units to treatments in a way that is fair to all of the treatments?

Experimenters often attempt to match groups by elaborate balancing acts. Medical researchers, for example, try to match the patients in a "new drug" experimental group and a "standard drug" control group by age, sex, physical condition, smoker or not, and so on. Matching is helpful but not adequate—there are too many lurking variables that might affect the outcome. The experimenter is unable to measure some of these variables and will not think of others until after the experiment. Some important variables, such as how advanced a cancer patient's disease is, are so subjective that an experimenter might bias the study by, for example, assigning less advanced cancer cases to a promising new treatment in the unconscious hope that it will help them.

The statistician's remedy is to rely on chance to make an assignment that does not depend on any characteristic of the experimental units and that does not rely on the judgment of the experimenter in any way. The use of chance can be combined with matching, but the simplest design creates groups by chance alone. Here is an example.

EXAMPLE 3.6

A food company assesses the nutritional quality of a new "instant breakfast" product by feeding it to newly weaned male white rats and measuring their weight gain over a 28-day period. A control group of rats receives a standard diet for comparison. This nutrition experiment has a single factor (the diet) with two levels. The researchers use 30 rats for the experiment and so must divide them into two groups of 15. To do this in a completely unbiased fashion, they put the cage numbers of the 30 rats in a hat, mix them up, and draw 15. These rats form the experimental group, and the remaining 15 make up the control group. Figure 3.2 outlines the design of this experiment.

randomization

The use of chance to divide experimental units into groups is called **randomization.*** The design in Figure 3.2 combines comparison and

------

*The use of randomization in experimental design was pioneered by R. A. Fisher (1890–1962). Fisher worked with agricultural field experiments and genetics at the Rothamsted Experiment Station in England. He invented statistical design of experiments and formal statistical methods for analyzing experimental data and developed them to an advanced state.

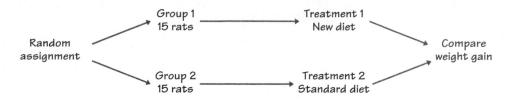

FIGURE 3.2   Outline of a randomized comparative experiment, for Example 3.6.

randomization to arrive at the simplest randomized comparative design. This "flowchart" outline presents all the essentials: randomization, the sizes of the groups and which treatment they receive, and the response variable. There are, as we will see later, statistical reasons for generally using treatment groups about equal in size.

Randomized comparative experiments

The logic behind the randomized comparative design in Figure 3.2 is as follows:

- Randomization produces groups of rats that should be similar in all respects before the treatments are applied.
- Comparative design ensures that influences other than the diets operate equally on both groups.
- Therefore, differences in average weight gain must be due either to the diets or to the play of chance in the random assignment of rats to the two diets.

That "either-or" deserves more thought. We cannot say that *any* difference in the average weight gains of rats fed the two diets must be caused by a difference between the diets. There would be some difference even if both groups received the same diet, because the natural variability among rats means that some grow faster than others. Chance assigns the faster-growing rats to one group or the other, and this creates a chance difference between the groups. We would not trust an experiment with just one rat in each group, for example. The results would depend too much on which group got lucky and received the faster-growing rat. If we assign many rats to each diet, however, the effects of chance will average out and there will be little difference in the average weight gains in the two groups unless the diets themselves cause a difference. "Use enough experimental units to reduce chance variation" is the third big idea of statistical design of experiments.

Principles of Experimental Design

The basic principles of statistical design of experiments are

1. **Control** of the effects of lurking variables on the response, most simply by comparing several treatments.
2. **Randomization,** the use of impersonal chance to assign experimental units to treatments.
3. **Replication** of the experiment on many units to reduce chance variation in the results.

We hope to see a difference in the responses so large that it is unlikely to happen just because of chance variation. We can use the laws of probability, which give a mathematical description of chance behavior, to learn if the treatment effects are larger than we would expect to see if only chance were operating. If they are, we call them *statistically significant*.

Statistical Significance

An observed effect so large that it would rarely occur by chance is called **statistically significant.**

You will often see the phrase "statistically significant" in reports of investigations in many fields of study. It tells you that the investigators found good evidence for the effect they were seeking. The Physicians' Health Study, for example, reported statistically significant evidence that aspirin reduces the number of heart attacks compared with a placebo.

How to randomize

The idea of randomization is to assign subjects to treatments by drawing names from a hat. In practice, experimenters use software to carry out randomization. Most statistical software will choose 15 out of a list of 30 at random, for example. The list can contain the cage numbers of 30 rats or the names of 30 human subjects. The 15 chosen form one group, and the 15 that remain form the second group.

We can randomize without software by using a *table of random digits*. Thinking about random digits helps understand randomization even if you will use software in practice. Table B at the back of the book and on the back endpaper is a table of random digits.

Random Digits

A **table of random digits** is a list of the digits 0, 1, 2, 3, 4, 5, 6, 7, 8, 9 that has the following properties:

1. The digit in any position in the list has the same chance of being any one of 0, 1, 2, 3, 4, 5, 6, 7, 8, 9.
2. The digits in different positions are independent in the sense that the value of one has no influence on the value of any other.

You can think of Table B as the result of asking an assistant (or a computer) to mix the digits 0 to 9 in a hat, draw one, then replace the digit drawn, mix again, draw a second digit, and so on. The assistant's mixing and drawing saves us the work of mixing and drawing when we need to randomize. Table B begins with the digits 19223950340575628713. To make the table easier to read, the digits appear in groups of five and in numbered rows. The groups and rows have no meaning—the table is just a long list of digits having the properties 1 and 2 described above.

Our goal is to use random digits for experimental randomization. We need the following facts about random digits, which are consequences of the basic properties 1 and 2:

3. Any *pair* of random digits has the same chance of being any of the 100 possible pairs: 00, 01, 02, ..., 98, 99.
4. Any *triple* of random digits has the same chance of being any of the 1000 possible triples: 000, 001, 002, ..., 998, 999.
5. ... and so on for groups of four or more random digits.

EXAMPLE 3.7

In the nutrition experiment of Example 3.6, we must divide 30 rats at random into two groups of 15 rats each.

Step 1: Label. Give each rat a numerical label, using as few digits as possible. Two digits are needed to label 30 rats, so we use labels

$$01, 02, 03, \ldots, 29, 30$$

It is also correct to use labels 00 to 29 or some other choice of 30 two-digit labels.

Step 2: Table. Start anywhere in Table B and read two-digit groups. Suppose we begin at line 130, which is

$$69051 \ 64817 \ 87174 \ 09517 \ 84534 \ 06489 \ 87201 \ 97245$$

The first 10 two-digit groups in this line are

$$69 \ 05 \ 16 \ 48 \ 17 \ 87 \ 17 \ 40 \ 95 \ 17$$

Each of these two-digit groups is a label. The labels 00 and 31 to 99 are not used in this example, so we ignore them. The first 15 labels between 01 and 30 that we encounter in the table choose rats for the experimental group. Of the first 10 labels in line 130, we ignore five because they are too high (over 30). The others are 05, 16, 17, 17, and 17. The rats labeled 05, 16, and 17 go into the experimental group. Ignore the second and third 17s because that rat is already in the group. Run your finger across line 130 (and continue to line 131 if needed) until you have chosen 15 rats. They are the rats labeled

$$05, \ 16, \ 17, \ 20, \ 19, \ 04, \ 25, \ 29, \ 18, \ 07, \ 13, \ 02, \ 23, \ 27, \ 21$$

These rats form the experimental group. The remaining 15 are the control group.

As Example 3.7 illustrates, randomization requires two steps: assign labels to the experimental units and then use Table B to select labels at random. Don't try to scramble the labels as you assign them. Table B will do the required randomizing, so assign labels in any convenient manner, such as in alphabetical order for human subjects. You can read digits from Table B in any order—along a row, down a column, and so on—because the table has no order. As an easy standard practice, we recommend reading along rows.

EXAMPLE 3.8   Many utility companies have programs to encourage their customers to conserve energy. An electric company is considering placing electronic meters in households to show what the cost would be if the electricity use at that moment continued for a month. Will meters reduce electricity use? Would cheaper methods work almost as well? The company decides to design an experiment.

One cheaper approach is to give customers a chart and information about monitoring their electricity use. The experiment compares these two approaches (meter, chart) with each other and also with a control group of customers who receive no help in monitoring electricity use. The response variable is total electricity used in a year. The company finds 60 single-family residences in the same city willing to participate, so it assigns 20 residences at random to each of the three treatments. The outline of the design appears in Figure 3.3.

To carry out the random assignment, label the 60 houses 01 to 60. Then enter Table B and read two-digit groups until you have selected 20 houses to receive the meters. Continue in Table B to select 20 more to receive charts. The remaining 20 form the control group. The process is simple but tedious.

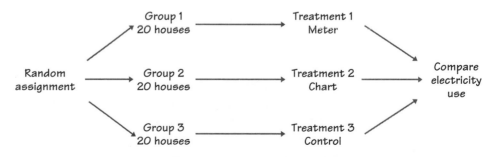

FIGURE 3.3 Outline of a completely randomized design comparing three treatments, for Example 3.8.

completely randomized design

When all experimental units are allocated at random among all treatments, the experimental design is **completely randomized.** The designs in Figures 3.2 and 3.3 are both completely randomized. Completely randomized designs can compare any number of treatments. The treatments can be formed by more than one factor. The Physicians' Health Study had two factors, which combine to form the four treatments shown in Figure 3.1. The study used a completely randomized design that assigned 5499 of the 21,996 subjects to each of the four treatments.

Cautions about experimentation

The logic of a randomized comparative experiment depends on our ability to treat all the experimental units identically in every way except for the actual treatments being compared. Good experiments therefore require careful attention to details. For example, the subjects in both the Physicians' Health Study (Example 3.3) and the second gastric freezing experiment (Example 3.5) all got the same medical attention over the several years the studies continued. Moreover, these studies were **double-blind**—neither the subjects themselves nor the medical personnel who worked with them knew which treatment any subject had received. The double-blind method avoids unconscious bias by, for example, a doctor who doesn't think that "just a placebo" can benefit a patient.

double-blind

lack of realism

The most serious potential weakness of experiments is **lack of realism.** The subjects or treatments or setting of an experiment may not realistically duplicate the conditions we really want to study. Here are some examples.

EXAMPLE 3.9

A study compares two television advertisements by showing TV programs to student subjects. The students know it's "just an experiment." We can't be sure that the results apply to everyday television viewers. Many behavioral science experiments use as subjects students who know they are subjects in an experiment. That's not a realistic setting.

An industrial experiment uses a small-scale pilot production process to find the choices of catalyst concentration and temperature that maximize yield. These may not be the best choices for the operation of a full-scale plant.

Lack of realism can limit our ability to apply the conclusions of an experiment to the settings of greatest interest. Most experimenters want to generalize their conclusions to some setting wider than that of the actual experiment. Statistical analysis of the original experiment cannot tell us how far the results will generalize. Nonetheless, the randomized comparative experiment, because of its ability to give convincing evidence for causation, is one of the most important ideas in statistics.

Matched pairs designs

Completely randomized designs are the simplest statistical designs for experiments. They illustrate clearly the principles of control, randomization, and replication. However, completely randomized designs are often inferior to more elaborate statistical designs. In particular, matching the subjects in various ways can produce more precise results than simple randomization.

EXAMPLE 3.10

Are cereal leaf beetles more strongly attracted by the color yellow or by the color green? Agriculture researchers want to know, because they detect the presence of the pests in farm fields by mounting sticky boards to trap insects that land on them. The board color should attract beetles as strongly as possible. We must design an experiment to compare yellow and green by mounting boards on poles in a large field of oats.

The experimental units are locations within the field far enough apart to represent independent observations. We erect a pole at each location to hold the boards. We might employ a completely randomized design in which we randomly select half the poles to receive a yellow board while the remaining poles receive green. The locations vary widely in the number of beetles present. For example, the alfalfa that borders the oats on one side is a natural host of the beetles, so locations near the alfalfa will have extra beetles. This variation among experimental units can hide the systematic effect of the board color.

matched pairs design

It is more efficient to use a **matched pairs design** in which we mount boards of both colors on each pole. The observations (numbers of beetles trapped) are matched in pairs from the same poles. We compare the number of trapped beetles on a yellow board with the number trapped by the green board on the same pole. Because the boards are mounted one above the other, we select the color of the top board at random. Just toss a coin for each board—if the coin falls heads, the yellow board is mounted above the green board.

Matched pairs designs compare just two treatments. We choose *blocks* of two units that are as closely matched as possible. In Example 3.10, two boards on the same pole form a block. We assign one of the treatments to each unit by tossing a coin or reading odd and even digits from Table B. Alternatively, each block in a matched pairs design may consist of just one subject, who gets both treatments one after the other. Each subject serves as his or her own control. The *order* of the treatments can influence the subject's response, so we randomize the order for each subject, again by a coin toss.

Block designs

The matched pairs design of Example 3.10 uses the principles of comparison of treatments, randomization, and replication on several experimental units. However, the randomization is not complete (all locations randomly assigned to treatment groups) but restricted to assigning the order of the boards at each location. The matched pairs design reduces the effect of variation among locations in the field by comparing the pair of boards at each location. Matched pairs are an example of *block designs*.

Block Design

A **block** is a group of experimental units or subjects that are known before the experiment to be similar in some way that is expected to affect the response to the treatments. In a **block design,** the random assignment of units to treatments is carried out separately within each block.

Block designs can have blocks of any size. A block design combines the idea of creating equivalent treatment groups by matching with the principle of forming treatment groups at random. Blocks are another form of *control*. They control the effects of some outside variables by bringing those variables into the experiment to form the blocks. Here are some typical examples of block designs.

EXAMPLE 3.11   The progress of a type of cancer differs in women and men. A clinical experiment to compare three therapies for this cancer therefore treats sex as a blocking variable. Two separate randomizations are done, one assigning the female subjects to the treatments and the other assigning the male subjects. Figure 3.4 outlines the design of this experiment. Note that there is no randomization involved in making up the blocks. They are groups of subjects who differ in some way (sex in this case) that is apparent before the experiment begins.

EXAMPLE 3.12   The soil type and fertility of farmland differ by location. Because of this, a test of the effect of tillage type (two types) and pesticide application (three application schedules) on soybean yields uses small fields as blocks. Each block is divided into six plots, and the six treatments are randomly assigned to plots separately within each block.

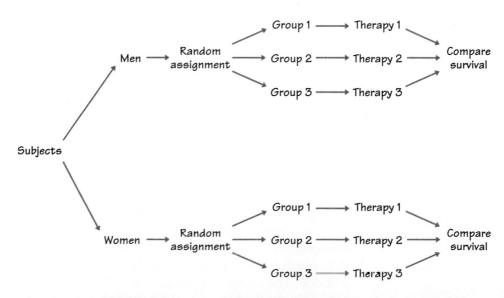

FIGURE 3.4   Outline of a block design. The blocks consist of male and female subjects. The treatments are three therapies for cancer.

EXAMPLE 3.13 | A social policy experiment will assess the effect on family income of several proposed new welfare systems and compare them with the present welfare system. Because the income of a family under any welfare system is strongly related to its present income, the families who agree to participate are divided into blocks of similar income levels. The families in each block are then allocated at random among the welfare systems.

Blocks allow us to draw separate conclusions about each block, for example, about men and women in the cancer study in Example 3.11. Blocking also allows more precise overall conclusions, because the systematic differences between men and women can be removed when we study the overall effects of the three therapies. The idea of blocking is an important additional principle of statistical design of experiments. A wise experimenter will form blocks based on the most important unavoidable sources of variability among the experimental units. Randomization will then average out the effects of the remaining variation and allow an unbiased comparison of the treatments.

SUMMARY

In an experiment, one or more **treatments** are imposed on the **experimental units** or **subjects.** Each treatment is a combination of **levels** of the explanatory variables, which we call **factors.**

The **design** of an experiment refers to the choice of treatments and the manner in which the experimental units or subjects are assigned to the treatments. The basic principles of statistical design of experiments are **control, randomization,** and **replication.**

The simplest form of control is **comparison.** Experiments should compare two or more treatments in order to prevent **confounding** the effect of a treatment with other influences, such as lurking variables.

Randomization uses chance to assign subjects to the treatments. Randomization creates treatment groups that are similar (except for chance variation) before the treatments are applied. Randomization and comparison together prevent **bias,** or systematic favoritism, in experiments.

You can carry out randomization by giving numerical labels to the experimental units and using a **table of random digits** to choose treatment groups.

Replication of the treatments on many units reduces the role of chance variation and makes the experiment more sensitive to differences among the treatments.

Good experiments require attention to detail as well as good statistical design. Many behavioral and medical experiments are **double-blind. Lack of realism** in an experiment can prevent us from generalizing its results.

In addition to comparison, a second form of control is to restrict randomization by forming **blocks** of experimental units that are similar in some way that is important to the response. Randomization is then carried out separately within each block.

Matched pairs are a common form of blocking for comparing just two treatments. In some matched pairs designs, each subject receives both treatments in a random order. In others, the subjects are matched in pairs as closely as possible, and one subject in each pair receives each treatment.

SECTION 3.2 EXERCISES

For each of the experimental situations described in Exercises 3.9 to 3.12, identify the experimental units or subjects, the factors, the treatments, and the response variables.

3.9   Sickle cell disease is an inherited disorder of the red blood cells that in the United States affects mostly blacks. It can cause severe pain and many complications. Can the drug hydroxyurea reduce the severe pain caused by sickle cell disease? A study by the National Institutes of Health gave the drug to 150 sickle cell sufferers and a placebo to another 150. The researchers then counted the episodes of pain reported by each subject.

3.10   A manufacturer of food products uses package liners that are sealed at the top by applying heated jaws after the package is filled. The customer peels the sealed pieces apart to open the package. What effect does the temperature of the jaws have on the force required to peel the liner? To answer this question, the engineers prepare 20 pairs of pieces of package liner. They seal five pairs at each of 250° F, 275° F, 300° F, and 325° F. Then they measure the peel strength of each seal.

3.11   What are the effects of repeated exposure to an advertising message? The answer may depend both on the length of the ad and on how often it is repeated. An experiment investigated this question using undergraduate students as subjects. All subjects viewed a 40-minute television program that included ads for a digital camera. Some subjects saw a 30-second commercial; others, a 90-second version. The same commercial was repeated either 1, 3, or 5 times during the program. After viewing, all of the subjects answered questions about their recall of the ad, their attitude toward the camera, and their intention to purchase it.

3.12   New varieties of corn with altered amino acid patterns may have higher nutritive value than standard corn, which is low in the amino acid lysine. An experiment compares two new varieties, called opaque-2 and floury-2, with normal corn. Corn-soybean meal diets using each type of corn are prepared at three different protein levels: 12%, 16%, and 20%. There are thus nine diets in all. Researchers assign 10 one-day-old male chicks to each diet and record their weight gains after 21 days. The weight gain of the chicks is a measure of the nutritive value of their diet.

3.13 Exercise 3.9 describes a medical study of a new treatment for sickle cell disease.

(a) Outline the design of this experiment.

(b) Use of a placebo is considered ethical if there is no effective standard treatment to give the control group. It might seem humane to give all the subjects hydroxyurea in the hope that it will help them. Explain clearly why this would not provide information about the effectiveness of the drug. (In fact, the experiment was stopped ahead of schedule because the hydroxyurea group had only half as many pain episodes as the control group. Ethical standards required stopping the experiment as soon as significant evidence became available.)

3.14 Some medical researchers suspect that added calcium in the diet reduces blood pressure. You have available 40 men with high blood pressure who are willing to serve as subjects.

(a) Outline an appropriate design for the experiment, taking the placebo effect into account.

(b) The names of the subjects appear below. Do the randomization required by your design, and list the subjects to whom you will give the drug. (If you use Table B, enter the table at line 131.)

Alomar	Denman	Han	Liang	Rosen
Asihiro	Durr	Howard	Maldonado	Solomon
Bennett	Edwards	Hruska	Marsden	Tompkins
Bikalis	Farouk	Imani	Moore	Townsend
Chen	Fratianna	James	O'Brian	Tullock
Clemente	George	Kaplan	Ogle	Underwood
Cranston	Green	Krushchev	Plochman	Willis
Curtis	Guillen	Lawless	Rodriguez	Zhang

3.15 Will providing child care for employees make a company more attractive to women, even those who are unmarried? You are designing an experiment to answer this question. You prepare recruiting material for two fictitious companies, both in similar businesses in the same location. Company A's brochure does not mention child care. There are two versions of Company B's material, identical except that one describes the company's on-site child-care facility. Your subjects are 40 unmarried women who are college seniors seeking employment. Each subject will read recruiting material for both companies and choose the one she would prefer to work for. You will give each version of Company B's brochure to half the women. You suspect that a higher percentage of those who read the description that includes child care will choose Company B.

(a) Outline the design of the experiment. Be sure to identify the response variable.

(b) The names of the subjects appear below. Do the randomization required by your design, and list the subjects who will read the version that mentions child care. (If you use Table B, begin at line 121.)

Abrams	Danielson	Gutierrez	Lippman	Rosen
Adamson	Durr	Howard	Martinez	Sugiwara
Afifi	Edwards	Hwang	McNeill	Thompson
Brown	Fluharty	Iselin	Morse	Travers
Cansico	Garcia	Janle	Ng	Turing
Chen	Gerson	Kaplan	Quinones	Ullmann
Cortez	Green	Kim	Rivera	Williams
Curzakis	Gupta	Lattimore	Roberts	Wong

3.16 Outline the design of the package liner experiment of Exercise 3.10. Label the pairs of liner pieces 01 to 20 and carry out the randomization that your design calls for. (If you use Table B, start at line 120.)

3.17 A horticulturist is comparing two methods (call them A and B) of growing potatoes. Standard potato cuttings will be planted in small plots of ground. The response variables are number of tubers per plant and fresh weight (weight when just harvested) of vegetable growth per plant. There are 20 plots available for the experiment. Sketch the outline of a rectangular field divided into 5 rows of 4 plots each. Then outline the experimental design and do the required randomization. (If you use Table B, start at line 145.) Mark on your sketch which growing method you will use in each plot.

3.18 Does regular exercise reduce the risk of a heart attack? Here are two ways to study this question. Explain clearly why the second design will produce more trustworthy data.

1. A researcher finds 2000 men over 40 who exercise regularly and have not had heart attacks. She matches each with a similar man who does not exercise regularly, and she follows both groups for 3 years.

2. Another researcher finds 4000 men over 40 who have not had heart attacks and are willing to participate in a study. She assigns 2000 of the men to a regular program of supervised exercise. The other 2000 continue their usual habits. The researcher follows both groups for 3 years.

3.19 Example 3.8 (page 245) describes an experiment to learn whether providing households with electronic meters or charts will reduce their electricity consumption. An executive of the electric company objects to including a control group. He says, "It would be simpler to just compare electricity use last year [before the meter or chart was provided] with consumption in the same period this year. If households use less electricity this year, the indicator or chart must be working." Explain clearly why this design is inferior to that in Example 3.8.

3.20 The following situations were not experiments. Can an experiment be done to answer the questions raised? If so, briefly outline its design. If not, explain why an experiment is not feasible.

    (a) The "gender gap" issue of Exercise 3.3 (page 236).

    (b) The comparison of two surgical procedures for breast cancer in Exercise 3.5 (page 236).

3.21   You want to determine the best color for attracting cereal leaf beetles to boards on which they will be trapped. You will compare four colors: blue, green, white, and yellow. The response variable is the count of beetles trapped. You will mount one board on each of 16 poles evenly spaced in a square field, with four poles in each of four rows. Sketch the field with the locations of the 16 poles. Outline the design of a completely randomized experiment to compare the colors. Randomly assign colors to the poles, and mark on your sketch the color assigned to each pole. (If you use Table B, start at line 115.)

3.22   A mathematics education researcher is studying where in high school mathematics texts it is most effective to insert questions. She wants to know whether it is better to present questions as motivation before the text passage or as review after the passage. The result may depend on the type of question asked: simple fact, computation, or word problem.

    (a) This experiment has two factors. What are they? How many treatments do all combinations of levels of these factors form? List the treatments.

    (b) Because it is disruptive to assign high school students at random to the treatment groups, the researcher will assign two classes of the same grade level to each treatment. The response variable is score on a mathematics test taken by all the students in these classes. Outline the design of the experiment. Carry out the random assignment required.

3.23   You have 100 students from a management course available to serve as subjects in the experiment of Exercise 3.11 on effectiveness of television advertising. Outline a completely randomized design for this experiment in detail. (You need not do any randomization.)

3.24   You read a news report of an experiment that claims to show that a meditation technique lowered the anxiety level of subjects. The experimenter interviewed the subjects and assessed their levels of anxiety. The subjects then learned how to meditate and did so regularly for a month. The experimenter reinterviewed them at the end of the month and assessed whether their anxiety levels had decreased or not.

    (a) There was no control group in this experiment. Why is this a blunder? What lurking variables might be confounded with the effect of meditation?

    (b) The experimenter who diagnosed the effect of the treatment knew that the subjects had been meditating. Explain how this knowledge could bias the experimental conclusions.

    (c) Briefly discuss a proper experimental design, with controls and blind diagnosis, to assess the effect of meditation on anxiety level.

3.25   Is the right hand generally stronger than the left in right-handed people? You can crudely measure hand strength by placing a bathroom scale on a shelf with the end protruding, then squeezing the scale between the thumb below and the four fingers above it. The reading of the scale shows the force exerted. Describe the design of a matched pairs experiment to compare the strength of the right and left hands, using 15 right-handed people as subjects. Use a coin to do the required randomization.

3.26   Some investment advisors believe that charts of past trends in the prices of securities can help predict future prices. Most economists disagree. In an experiment to examine the effects of using charts, business students trade (hypothetically) a foreign currency at computer screens. There are 20 student subjects available, named for convenience A, B, C, ..., T. Their goal is to make as much money as possible, and the best performances are rewarded with small prizes. The student traders have the price history of the foreign currency in dollars in their computers. They may or may not also be given software that charts past trends. Describe *two* designs for this experiment, a completely randomized design and a matched pairs design in which each student serves as his or her own control. In both cases, carry out the randomization required by your design.

3.27   Twenty overweight females have agreed to participate in a study of the effectiveness of four reducing regimens, A, B, C, and D. The researcher first calculates how overweight each subject is by comparing the subject's actual weight with her "ideal" weight. The subjects and their excess weights in pounds are

Birnbaum	35	Hernandez	25	Moses	25	Smith	29
Brown	34	Jackson	33	Nevesky	39	Stall	33
Brunk	30	Kendall	28	Obrach	30	Tran	35
Dixon	34	Loren	32	Rodriguez	30	Wilansky	42
Festinger	24	Mann	28	Santiago	27	Williams	22

The response variable is the weight lost after eight weeks of treatment. Because the initial amount overweight will influence the response variable, a block design is appropriate.

(a) Arrange the subjects in order of increasing excess weight. Form five blocks by grouping the four least overweight, then the next four, and so on.

(b) Use Table B to do the required random assignment of subjects to the four reducing regimens separately within each block. Be sure to explain exactly how you used the table.

3.28   Return to the experiment of Exercise 3.21 on the attractiveness of several colors to cereal leaf beetles. The researchers decide to use two oat fields in different locations and to space eight poles equally within each field. Outline a randomized block design using the fields as blocks. Then use

Table B, beginning at line 105, to carry out the random assignment of colors to poles. Report your results by means of a sketch of the two fields with the color at each pole noted.

3.29 It is often suggested that people will spend less on health care if their health insurance requires them to pay some part of the cost of medical treatment themselves. One experiment on this issue asked if the percent of medical costs that are paid by health insurance has an effect either on the amount of medical care that people use or on their health. The treatments were four insurance plans. Each plan paid all medical costs above a ceiling. Below the ceiling, the plans paid 100%, 75%, 50%, or 0% of costs incurred.

(a) Outline the design of a randomized comparative experiment suitable for this study.

(b) Describe briefly the practical and ethical difficulties that might arise in such an experiment.

3.30 Men and women often react differently to advertising. You can use this fact to improve on the completely randomized design for the advertising experiment of Exercises 3.11 and 3.23. Describe the design you recommend. (You need not do any randomization.)

3.31 Table B is a table of random digits. Software can also generate random digits for use in random selection. Which of the following statements are true of a table of random digits, and which are false?

(a) There are exactly four 0s in each row of 40 digits.

(b) Each pair of digits has chance 1/100 of being 00.

(c) The digits 0000 can never appear as a group, because this pattern is not random.

3.32 Let us examine the effects of randomization in designing an experiment. You plan to compare two methods of teaching math skills to seventh-grade students. Your subjects are the 78 students described in Table 1.6 (page 33). You will choose 39 students at random to be taught by Method A. The remaining 39 will be taught using Method B. Because IQ may affect learning, you want the two groups to have about the same mean IQ. Randomization is supposed to make the groups similar.

(a) Use software to choose 39 of the 78 IQ scores at random. (This is the same as choosing 39 of the students to form Group A.) What is the mean IQ of the 39 students chosen? What is the mean IQ of the 39 not chosen, who make up Group B? Do the two groups have similar mean IQs?

(b) You might be unusually lucky (or unlucky) on one random choice. Will randomization *usually* produce two groups with similar mean IQs? To find out, repeat (a) nine more times. Make a table of the mean IQs for Groups A and B on each of your 10 trials. What do you conclude?

3.3   Sampling Design

A political scientist wants to know what percent of the voting-age population consider themselves conservatives. An automaker hires a market research firm to learn what percent of adults aged 18 to 35 recall seeing television advertisements for a new sport utility vehicle. Government economists inquire about average household income. In all these cases, we want to gather information about a large group of individuals. We will not, as in an experiment, impose a treatment in order to observe the response. Time, cost, and inconvenience forbid contacting every individual. In such cases, we gather information about only part of the group in order to draw conclusions about the whole.

Population and Sample

The entire group of individuals that we want information about is called the **population**.

A **sample** is a part of the population that we actually examine in order to gather information.

Notice that "population" is defined in terms of our desire for knowledge. If we wish to draw conclusions about all U.S. college students, that group is our population even if only local students are available for questioning. The sample is the part from which we draw conclusions about the whole. The **design** of a sample refers to the method used to choose the sample from the population. Poor sample designs can produce misleading conclusions, as the following examples illustrate.

sample design

EXAMPLE 3.14   A mill produces large coils of thin steel for use in manufacturing home appliances. The quality engineer wants to submit a sample of 5-centimeter squares to detailed laboratory examination. She asks a technician to cut a sample of 10 such squares. Wanting to provide "good" pieces of steel, the technician carefully avoids the visible defects in the coil material when cutting the sample. The laboratory results are wonderful but the customers complain about the material they are receiving.

EXAMPLE 3.15   The advice columnist Ann Landers once asked her readers, "If you had it to do over again, would you have children?" A few weeks later, her column was headlined "70% OF PARENTS SAY KIDS NOT WORTH IT." Indeed, 70% of the nearly 10,000 parents who wrote in said they would not have children if they could make the choice again.

The results of this sample are worthless as indicators of opinion in the population of all American parents. Because the sample was self-selected, it consisted of people who felt strongly enough to take the trouble to write Ann Landers. Many of them, as their letters showed, were angry at their children. It is not surprising that a statistically designed opinion poll on the same issue a few months later found that 91% of parents *would* have children again. Ann Landers's poorly designed sampling method produced a 70% "No" result when the truth about the population was close to 90% "Yes."

In both Examples 3.14 and 3.15, the sample was selected in a manner that guaranteed that it would not be representative of the entire population. These sampling schemes display *bias*, or systematic error, in favoring some parts of the population over others. Ann Landers's *voluntary response sample* is a particularly common form of biased sample, common because it is inexpensive.

Voluntary Response Sample

A **voluntary response sample** consists of people who choose themselves by responding to a general appeal. Voluntary response samples are biased because people with strong opinions, especially negative opinions, are most likely to respond.

The remedy for bias in choosing a sample is to allow impersonal chance to do the choosing, so that there is neither favoritism by the sampler (as in Example 3.14) nor voluntary response (as in Example 3.15). Random selection of a sample eliminates bias by giving all individuals an equal chance to be chosen, just as randomization eliminates bias in assigning experimental subjects.

Simple random samples

The simplest sampling design amounts to placing names in a hat (the population) and drawing out a handful (the sample). This is simple random sampling.

Simple Random Sample

A **simple random sample (SRS)** of size n consists of n individuals from the population chosen in such a way that every set of n individuals has an equal chance to be the sample actually selected.

Each treatment group in a completely randomized experimental design is an SRS drawn from the available experimental units. We select an SRS by labeling all the individuals in the population and using a table of random digits to select a sample of the desired size, just as in experimental randomization. Notice that an SRS not only gives each individual an equal chance to be chosen (thus avoiding bias in the choice) but gives every possible sample an equal chance to be chosen. There are other random sampling designs that give each individual, but not each sample, an equal chance. One such design, systematic random sampling, is described in Exercise 3.41.

EXAMPLE 3.16 An academic department wishes to choose a three-member advisory committee at random from the members of the department. To choose an SRS of size 3 from the 28 faculty listed below, first label the members of the population as shown.

00	Abbott	07	Goodwin	14	Pillotte	21	Theobald
01	Cicirelli	08	Haglund	15	Raman	22	Vader
02	Cuellar	09	Johnson	16	Riemann	23	Wang
03	Dunsmore	10	Keegan	17	Rodriguez	24	Wieczorek
04	Engle	11	Luo	18	Rowe	25	Williams
05	Fitzpatrick	12	Martinez	19	Sommers	26	Wilson
06	Garcia	13	Nguyen	20	Stone	27	Wong

Now enter Table B, and read two-digit groups until you have chosen three committee members. If you enter at line 140, the committee consists of Martinez (12), Nguyen (13), and Engle (04). Most statistical software systems will select an SRS for you, eliminating the need for Table B.

Stratified samples

The general framework for designs that use chance to choose a sample is a *probability sample*.

Probability Sample

A **probability sample** gives each member of the population a known chance (greater than zero) to be selected.

Some probability sampling designs (such as an SRS) give each member of the population an *equal* chance to be selected. This may not be true in more elaborate sampling designs. In every case, however, the use of chance to select the sample is the essential principle of statistical sampling.

Designs for sampling from large populations spread out over a wide area are usually more complex than an SRS. For example, it is common to sample important groups within the population separately, then combine these samples. This is the idea of a *stratified sample*.

Stratified Random Sample

To select a **stratified random sample,** first divide the population into groups of similar individuals, called **strata.** Then choose a separate SRS in each stratum and combine these SRSs to form the full sample.

Choose the strata based on facts known before the sample is taken. For example, a population of election districts might be divided into urban, suburban, and rural strata. A stratified design can produce more exact information than an SRS of the same size by taking advantage of the fact that individuals in the same stratum are similar to one another. If all individuals in each stratum are identical, for example, just one individual from each stratum is enough to completely describe the population. Strata for sampling are similar to blocks in experiments. We have two names because the idea of grouping similar units before randomizing arose separately in sampling and in experiments.

EXAMPLE 3.17 A radio or television station that broadcasts a piece of music owes a royalty to the composer. ASCAP (the American Society of Composers, Authors, and Publishers) sells licenses that permit broadcast of works by any of its members. ASCAP must pay the proper royalties to the composers whose music was played. Television networks keep program logs of all music played, but local radio and television stations do not. Because there are over a billion ASCAP-licensed performances each year, a detailed accounting is too expensive and cumbersome. Here is a job for sampling.

ASCAP divides its royalties among its members by taping a stratified sample of broadcasts. The sample of local commercial radio stations, for example, consists of 60,000 hours of broadcast time each year. Radio stations are stratified by type of community (metropolitan, rural), geographic location (New England, Pacific, etc.), and the size of the license fee paid to ASCAP, which reflects the size of the audience. In all, there are 432 strata. Tapes are made at random hours for randomly selected members of each stratum. The tapes are reviewed by experts who can recognize almost every piece of music ever written, and the composers are then paid according to their popularity.[3]

Multistage samples

Another common means of restricting random selection is to choose the sample in stages. This is usual practice for national samples of households or people. For example, government data on employment and unemployment are gathered by the Current Population Survey, which conducts interviews in about 60,000 households each month. It is not practical to maintain a list of all U.S. households from which to select an SRS. Moreover, the cost of sending interviewers to the widely scattered households in an SRS would be too high. The Current Population Survey therefore uses a **multistage sampling design.** The final sample consists of clusters of nearby households that an interviewer can easily visit. Most opinion polls and other national samples are also multistage samples, though interviewing in most national samples today is done by telephone rather than in person, eliminating the economic need for clustering. A national multistage sample proceeds somewhat as follows:

Stage 1. Select a sample from the 3000 counties in the United States.

Stage 2. Select a sample of townships within each of the counties chosen. Lists of counties and townships are available, so the first two stages are easy to carry out.

Stage 3. Select a sample of city blocks or other small areas within each chosen township. These samples can be based on local maps.

Stage 4. Take a sample of households within each block.

The sample at any stage of a multistage sampling design may be an SRS. Stratified samples are also common—for example, we might group counties into rural, suburban, and urban strata before sampling. Analysis of data from sampling designs more complex than an SRS takes us beyond basic statistics. But the SRS is the building block of more elaborate designs, and analysis of other designs differs more in complexity of detail than in fundamental concepts.

Cautions about sample surveys

Random selection eliminates bias in the choice of a sample from a list of the population. Sample surveys of large human populations, however, require much more than a good sampling design.[4] To begin, we need an accurate and complete list of the population. Because such a list is rarely available, most samples suffer from some degree of *undercoverage*. A sample survey of households, for example, will miss not only homeless people but prison inmates and students in dormitories. An opinion poll conducted by telephone will miss the 7% to 8% of American households without residential phones. The results of national sample surveys therefore have some bias if the people not covered—who most often are poor people—differ from the rest of the population.

A more serious source of bias in most sample surveys is *nonresponse*, which occurs when a selected individual cannot be contacted or refuses to cooperate. Nonresponse to sample surveys often reaches 30% or more, even with careful planning and several callbacks. The most recent General Social Survey (Example 3.1) had a nonresponse rate of 26%. Because nonresponse is higher in urban areas, most sample surveys substitute other people in the same area to avoid favoring rural areas in the final sample. If the people contacted differ from those who are rarely at home or who refuse to answer questions, some bias remains.

Undercoverage and Nonresponse

Undercoverage occurs when some groups in the population are left out of the process of choosing the sample.

Nonresponse occurs when an individual chosen for the sample can't be contacted or does not cooperate.

EXAMPLE 3.18

Even the U.S. census, backed by the resources of the federal government, suffers from undercoverage and nonresponse. The census begins by mailing forms to every household in the country. The Census Bureau's list of addresses is incomplete, resulting in undercoverage. Despite special efforts to count homeless people (who can't be reached at any address), homelessness causes more undercoverage.

In 1990, about 35% of the households that were mailed census forms did not mail them back. In New York City, 47% did not return the form. That's nonresponse. The Census Bureau sent interviewers to these households. In inner-city areas, the interviewers could not contact about one in five of the nonresponders, even after six tries.

The Census Bureau estimates that the 1990 census missed about 1.8% of the total population due to undercoverage and nonresponse. Because the undercount was greater in the poorer sections of large cities, the Census Bureau estimates that it failed to count 4.4% of blacks and 5.0% of Hispanics.[5]

For the 2000 census, the Bureau would like to replace follow-up of all nonresponders with more intense pursuit of a probability sample of nonresponding households. This approach has the support of statisticians and social scientists but is opposed by politicians who don't mind an undercount concentrated in the poor areas of large cities.

response bias

The behavior of the respondent or of the interviewer causes **response bias** in sample results. Respondents may lie, especially if asked about illegal or unpopular behavior. The sample then underestimates the occurrence of such behavior in the population. An interviewer whose attitude suggests that some answers are more desirable than others will get these answers more often. The race or sex of the interviewer can influence responses to questions about race relations or attitudes toward feminism. Answers to questions that ask respondents to recall past events are often inaccurate because of faulty memory. For example, many people "telescope" events in the past, bringing them forward in memory to more recent time periods. "Have you visited a dentist in the last 6 months?" will often elicit a "Yes" from someone who last visited a dentist 8 months ago.[6] Careful training of interviewers and careful supervision to avoid variation among the interviewers can greatly reduce response bias. Good interviewing technique is another aspect of a well-done sample survey.

wording of questions

The **wording of questions** is the most important influence on the answers given to a sample survey. Confusing or leading questions can introduce strong bias, and even minor changes in wording can change a survey's outcome. Leading questions are common in surveys that are intended to persuade rather than inform. Here are two examples.[7]

EXAMPLE 3.19

When Levi Strauss & Co. asked college students to choose the most popular clothing item from a list, 90% chose Levi's 501 jeans—but they were the only jeans listed.

A survey paid for by makers of disposable diapers found that 84% of the sample opposed banning disposable diapers. Here is the actual question:

> It is estimated that disposable diapers account for less than 2% of the trash in today's landfills. In contrast, beverage containers, third-class mail and yard wastes are estimated to account for about 21% of the trash in landfills. Given this, in your opinion, would it be fair to ban disposable diapers?

This question gives information on only one side of an issue, then asks an opinion. That's a sure way to bias the responses. A different question that described how long disposable diapers take to decay and how many tons they contribute to landfills each year would draw a quite different response.

Never trust the results of a sample survey until you have read the exact questions posed. The sampling design, the amount of nonresponse, and the date of the survey are also important. Good statistical design is a part, but only a part, of a trustworthy survey.

SUMMARY

A sample survey selects a **sample** from the **population** of all individuals about which we desire information. We base conclusions about the population on data about the sample.

The **design** of a sample refers to the method used to select the sample from the population. **Probability sampling designs** use random selection to give each member of the population a known chance (greater than zero) to be selected for the sample.

The basic probability sample is a **simple random sample (SRS).** An SRS gives every possible sample of a given size the same chance to be chosen.

Choose an SRS by labeling the members of the population and using a **table of random digits** to select the sample. Software can automate this process.

To choose a **stratified random sample,** divide the population into **strata,** groups of individuals that are similar in some way that is important to the response. Then choose a separate SRS from each stratum and combine them to form the full sample.

Multistage samples select successively smaller groups within the population in stages, resulting in a sample consisting of clusters of individuals. Each stage may employ an SRS, a stratified sample, or another type of sample.

Failure to use probability sampling often results in **bias,** or systematic errors in the way the sample represents the population. **Voluntary response** samples, in which the respondents choose themselves, are particularly prone to large bias.

In human populations, even probability samples can suffer from bias due to **undercoverage** or **nonresponse,** from **response bias** due to the behavior of the interviewer or the respondent, or from misleading results due to **poorly worded questions.**

SECTION 3.3 EXERCISES

3.33   A sociologist wants to know the opinions of employed adult women about government funding for day care. She obtains a list of the 520 members of a local business and professional women's club and mails a questionnaire to 100 of these women selected at random. Only 48 questionnaires are returned. What is the population in this study? What is the sample from whom information is actually obtained? What is the rate (percentage) of nonresponse?

3.34   Different types of writing can sometimes be distinguished by the lengths of the words used. A student interested in this fact wants to study the lengths of words used by Tom Clancy in his novels. She opens a Clancy novel at random and records the lengths of each of the first 250 words on the page. What is the population in this study? What is the sample? What is the variable measured?

3.35   For each of the following sampling situations, identify the population as exactly as possible. That is, say what kind of individuals the population consists of and say exactly which individuals fall in the population. If the information given is not complete, complete the description of the population in a reasonable way.

(a)   Each week, the Gallup Poll questions a sample of about 1500 adult U.S. residents in order to discover the opinions of Americans on a wide variety of issues.

(b)   Every 10 years, the Census Bureau tries to gather basic information from every household in the United States. But a "long form" requesting much additional information is sent to a sample of about 17% of U.S. households.

(c)   A machinery manufacturer purchases voltage regulators from a supplier. There are reports that variation in the output voltage of the regulators is affecting the performance of the finished products. To assess the quality of the supplier's production, the manufacturer chooses a sample of 5 regulators from the last shipment for careful laboratory analysis.

3.36   A newspaper article about an opinion poll says that "43% of Americans approve of the president's overall job performance." Toward the end of the article, you read: "The poll is based on telephone interviews with 1210 adults from around the United States, excluding Alaska and Hawaii." What variable did this poll measure? What population do you think the newspaper wants information about? What was the sample? Are there any sources of bias in the sampling method used?

3.37   The students listed below are enrolled in a statistics course taught on television. Choose an SRS of 6 students to be interviewed in detail about the quality of the course. (If you use Table B, start at line 139.)

Agarwal	Dewald	Hixson	Puri
Anderson	Fernandez	Klassen	Rodriguez
Baxter	Frank	Liang	Rubin
Bowman	Fuhrmann	Moser	Santiago
Bruvold	Goel	Naber	Shen
Casella	Gupta	Petrucelli	Shyr
Choi	Hicks	Pliego	Sundheim

3.38   You want to choose an SRS of 25 of a city's 440 voting precincts for special voting-fraud surveillance on election day. How will you label the 440 precincts? Choose the SRS, and list the precincts you selected. (If you use Table B, enter at line 117 and select only the first 5 precincts in the sample.)

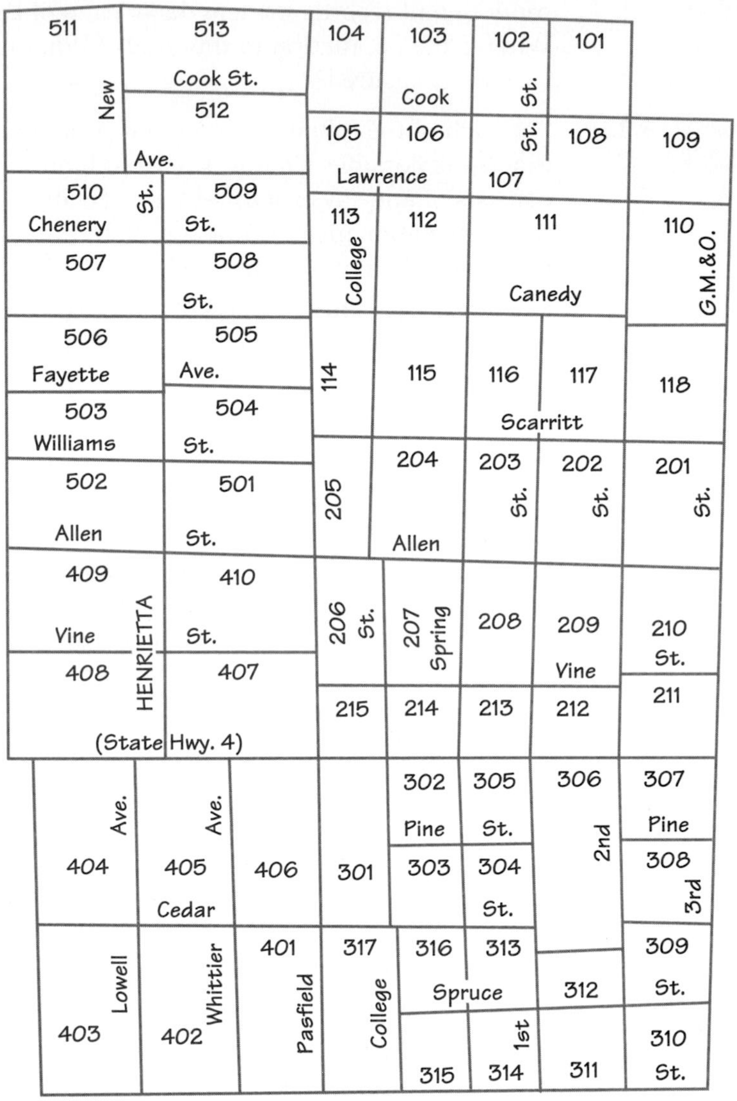

FIGURE 3.5 Map of a census tract, for Exercise 3.39.

3.39 Census tracts are small, homogeneous areas with an average population of about 4000. Figure 3.5 is a map of a census tract in a midwestern city. Each block in the tract is marked with a Census Bureau identifying number. A sample of blocks from a census tract is often the next-to-last stage in a multistage sample. Use Table B at line 125 to choose an SRS of 5 blocks from the tract illustrated.

3.40 In using Table B repeatedly to choose samples or do experimental randomization, you should not always begin at the same place, such as line 101. Why not?

systematic
random sample
3.41 **Systematic random samples** are often used to choose a sample of apartments in a large building or dwelling units in a block at the last stage

of a multistage sample. An example will illustrate the idea of a systematic sample. Suppose that we must choose 4 addresses out of 100. Because 100/4 = 25, we can think of the list as four lists of 25 addresses. Choose 1 of the first 25 at random, using Table B. The sample contains this address and the addresses 25, 50, and 75 places down the list from it. If 13 is chosen, for example, then the systematic random sample consists of the addresses numbered 13, 38, 63, and 88.

(a) Use Table B to choose a systematic random sample of 5 addresses from a list of 200. Enter the table at line 120.

(b) Like an SRS, a systematic sample gives all individuals the same chance to be chosen. Explain why this is true, then explain carefully why a systematic sample is nonetheless *not* an SRS.

3.42 A club has 30 student members and 10 faculty members. The students are

Abel	Fisher	Huber	Moran	Reinmann
Carson	Golomb	Jimenez	Moskowitz	Santos
Chen	Griswold	Jones	Neyman	Shaw
David	Hein	Kiefer	O'Brien	Thompson
Deming	Hernandez	Klotz	Pearl	Utts
Elashoff	Holland	Liu	Potter	Vlasic

and the faculty members are

Andrews	Fernandez	Kim	Moore	Rabinowitz
Besicovitch	Gupta	Lightman	Phillips	Yang

The club can send four students and two faculty members to a convention and decides to choose those who will go by random selection. Use Table B to choose a stratified random sample of 4 students and 2 faculty members.

3.43 A university has 2000 male and 500 female faculty members. The equal opportunity employment officer wants to poll the opinions of a random sample of faculty members. In order to give adequate attention to female faculty opinion, he decides to choose a stratified random sample of 200 males and 200 females. He has alphabetized lists of female and male faculty members. Explain how you would assign labels and use random digits to choose the desired sample. Enter Table B at line 122 and give the labels of the first 5 females and the first 5 males in the sample.

3.44 A university has 2000 male and 500 female faculty members. A stratified random sample of 50 female and 200 male faculty members gives each member of the faculty 1 chance in 10 to be chosen. This sample design gives every individual in the population the same chance to be chosen for the sample. Is it an SRS? Explain your answer.

3.45 The list of individuals from which a sample is actually selected is called the **sampling frame.** Ideally, the frame should list every individual in the

sampling frame

population, but in practice this is often difficult. A frame that leaves out part of the population is a common source of undercoverage.

(a) Suppose that a sample of households in a community is selected at random from the telephone directory. What households are omitted from this frame? What types of people do you think are likely to live in these households? These people will probably be underrepresented in the sample.

(b) It is usual in telephone surveys to use random digit dialing equipment that selects the last four digits of a telephone number at random after being given the exchange (the first three digits). Which of the households that you mentioned in your answer to (a) will be included in the sampling frame by random digit dialing?

3.46   A common form of nonresponse in telephone surveys is "ring-no-answer." That is, a call is made to an active number but no one answers. The Italian National Statistical Institute looked at nonresponse to a government survey of households in Italy during the periods January 1 to Easter and July 1 to August 31. All calls were made between 7 and 10 p.m., but 21.4% gave "ring-no-answer" in one period versus 41.5% "ring-no-answer" in the other period. (Data from Giuliana Coccia, "An overview of non-response in Italian telephone surveys," *Proceedings of the 99th Session of the International Statistical Institute, 1993,* Book 3, pp. 271–272.) Which period do you think had the higher rate of no answers? Why? Explain why a high rate of nonresponse makes sample results less reliable.

3.47   A newspaper advertisement for *USA Today: The Television Show* said:

> Should handgun control be tougher? You call the shots in a special call-in poll tonight. If yes, call 1-900-720-6181. If no, call 1-900-720-6182.
>
> Charge is 50 cents for the first minute.

Explain why this opinion poll is almost certainly biased.

3.48   Should the United Nations continue to have its headquarters in the United States? A television program asked its viewers to call in with their opinions on that question. There were 186,000 callers, 67% of whom said "No." A nationwide random sample of 500 adults found that 72% answered "Yes" to the same question. Explain to someone who knows no statistics why the opinions of only 500 randomly chosen respondents are a better guide to what all Americans think than the opinions of 186,000 callers.

3.49   Here are two wordings for the same question. The first question was asked by presidential candidate Ross Perot, and the second by a Time/CNN poll, both in March 1993.

A. Should laws be passed to eliminate all possibilities of special interests giving huge sums of money to candidates?

B. Should laws be passed to prohibit interest groups from contributing to campaigns, or do groups have a right to contribute to the candidates they support?

One of these questions drew 40% favoring banning contributions; the other drew 80% with this opinion. Which question produced the 40% and which got 80%? Explain why the results were so different. (W. Mitofsky, "Mr. Perot, you're no pollster," *New York Times,* March 27, 1993.)

3.50 Comment on each of the following as a potential sample survey question. Is the question clear? Is it slanted toward a desired response?

(a) "Does your family use food stamps?"

(b) "Which of the following best represents your opinion on gun control?

    1. The government should confiscate our guns.

    2. We have the right to keep and bear arms."

(c) "A national system of health insurance should be favored because it would provide health insurance for everyone and would reduce administrative costs."

(d) "In view of escalating environmental degradation and incipient resource depletion, would you favor economic incentives for recycling of resource-intensive consumer goods?"

3.4 Toward Statistical Inference

The deliberate use of chance in the process of producing data is a central idea of statistical designs for both sampling and experimentation. In both cases, randomization eliminates favoritism, or bias, in assigning subjects to treatments or choosing individuals to be in the sample. However, reducing bias is not the only justification for randomization. We usually produce data in order to draw conclusions about some wider population, a process called **statistical inference** because we infer conclusions about the wider population from data on selected individuals. The reasoning of inference relies on the laws of probability, which describe random behavior. Any use of statistical inference therefore requires that the data arise from some process governed by probability. The most direct way to meet this requirement is to actually use randomization in producing our data.

statistical inference

Inference will occupy the rest of this book, beginning with the necessary probability background in the next chapter. We can see more clearly how statistical inference works and why probability is important by a preliminary study of the most straightforward setting: inference about a population on the basis of a simple random sample drawn from that population. Because statistics deals with numerical variables, we commonly use a number computed from the sample to make inferences about an unknown number that describes the population.

Parameters and Statistics

A **parameter** is a number that describes the **population.** A parameter is a fixed number, but in practice we do not know its value.

A **statistic** is a number that describes a **sample.** The value of a statistic is known when we have taken a sample, but it can change from sample to sample. We often use a statistic to estimate an unknown parameter.

EXAMPLE 3.20

Are attitudes toward shopping changing? Sample surveys show that fewer people enjoy shopping than in the past. A recent survey asked a nationwide random sample of 2500 adults if they agreed or disagreed that "I like buying new clothes, but shopping is often frustrating and time-consuming." Of the respondents, 1650, or 66%, said they agreed.[8] The number 66% is a *statistic.* The population that the poll wants to draw conclusions about is all U.S. residents age 18 and over. The *parameter* of interest is the percent of all adult U.S. residents who would have said "Agree" if asked the same question. We don't know the value of this parameter.

Sampling variability

The sample survey of Example 3.20 produced information on attitudes about shopping. Let's examine it in more detail. We want to estimate the proportion of the population that find clothes shopping frustrating. Call this population proportion p. It is a parameter. The poll found that 1650 out of 2500 randomly selected adults agreed with the statement that shopping is often frustrating. The proportion of the sample who agreed was

$$\hat{p} = \frac{1650}{2500} = 0.66$$

The sample proportion $\hat{p}$ is a statistic. Because an SRS gives all adults the same chance to be sampled, we expect the proportion of frustrated shoppers in the sample to be close to the proportion in the entire population. We therefore use $\hat{p}$ to estimate the unknown parameter p.

How can $\hat{p}$, based on only 2500 of the more than 190 million American adults, be an accurate estimate of p? After all, a second random sample taken at the same time would choose different people and no doubt produce a different value of $\hat{p}$. This basic fact is called **sampling variability:** the value of a statistic varies in repeated random sampling.

sampling variability

To understand why sampling variability is not fatal, we ask, "What would happen if we took many samples?" Here's how to answer that question:

- Take a large number of samples from the same population.
- Calculate the sample proportion $\hat{p}$ for each sample.
- Make a histogram of the values of $\hat{p}$.
- Examine the distribution displayed in the histogram for shape, center, and spread, as well as outliers or other deviations.

simulation

In practice it is too expensive to take many samples from a large population such as all adult U.S. residents. But we can imitate many samples by using random digits. Using random digits from a table or computer software to imitate chance behavior is called **simulation.**

EXAMPLE 3.21

We will simulate drawing simple random samples (SRSs) of size 100 from the population of all adult U.S. residents. Suppose that in fact 60% of the population find clothes shopping time-consuming and frustrating. Then the true value of the parameter we want to estimate is $p = 0.6$. (Of course, we would not sample in practice if we already knew that $p = 0.6$. We are sampling here to learn how sampling behaves.)

We can imitate the population by a table of random digits, with each entry standing for a person. Six of the ten digits (say 0 to 5) stand for people who find shopping frustrating. The remaining four digits, 6 to 9, stand for those who do not. Because all digits in a random number table are equally likely, this assignment produces a population proportion of frustrated shoppers equal to $p = 0.6$. We then imitate an SRS of 100 people from the population by taking 100 consecutive digits from Table B. The statistic $\hat{p}$ is the proportion of 0s to 5s in the sample.

The first 100 entries in Table B contain 63 digits between 0 and 5, so $\hat{p} = 63/100 = 0.63$. A second SRS based on the second 100 entries in Table B gives a different result, $\hat{p} = 0.56$. The two sample results are different, and neither is equal to the true population value $p = 0.6$. That's sampling variability.

Sampling distributions

Simulation is a powerful tool for studying chance. Now that we see how simulation works, it is much faster to abandon Table B and to use a computer programmed to produce random digits. Figure 3.6 is the histogram of the values of $\hat{p}$ from 1000 separate SRSs of size 100 drawn from a population with $p = 0.6$. This histogram shows what would happen if we drew many samples. It displays the *sampling distribution* of $\hat{p}$.

> ### Sampling Distribution
>
> The **sampling distribution** of a statistic is the distribution of values taken by the statistic in all possible samples of the same size from the same population.

Strictly speaking, the sampling distribution is the ideal pattern that would emerge if we looked at all possible samples of size 100 from our population. A distribution obtained from a fixed number of trials, like the 1000 trials in Figure 3.6, is only an approximation to the sampling distribution. One of the uses of probability theory in statistics is to obtain exact sampling distributions without simulation. The interpretation of a sampling distribution is the same, however, whether we obtain it by simulation or by the mathematics of probability.

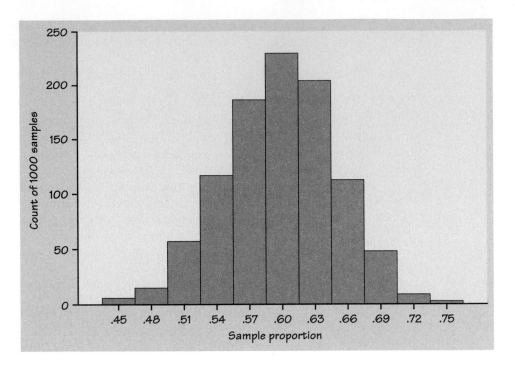

FIGURE 3.6 The sampling distribution of the sample proportion $\hat{p}$ from SRSs of size 100 drawn from a population with population proportion $p = 0.6$. The histogram shows the results of drawing 1000 SRSs.

We can use the tools of data analysis to describe any distribution. Let's apply those tools to Figure 3.6.

■ *Shape:* It looks normal! The normal quantile plot in Figure 3.7 confirms that the distribution of $\hat{p}$ from many samples does have a distribution that is close to normal. There is some granularity, but there are no outliers or other important deviations from the normal pattern.

■ *Center:* The mean of the 1000 $\hat{p}$'s is 0.598 and their median is exactly 0.6. We see that the center of the sampling distribution is very close to the actual value $p = 0.6$ of the population parameter.

■ *Spread:* The standard deviation of the 1000 $\hat{p}$'s is 0.051.

We see from Figure 3.6 that the sample result $\hat{p}$ is sometimes quite far from the true parameter $p = 0.6$. That is why the sample survey of attitudes toward shopping interviewed not 100 but 2500 people. Let's repeat our simulation, this time taking 1000 SRSs of size 2500 from a population with proportion $p = 0.6$ who find shopping frustrating. Figure 3.8 compares a histogram of the 1000 new $\hat{p}$'s with the histogram from Figure 3.6 drawn to the same scale. We see at once that the sample proportions from the larger samples are much more tightly clustered about the parameter value $p = 0.6$ than is the case for the smaller samples. Detailed examination shows that the distribution of the 1000 $\hat{p}$'s from samples of size 2500 is very normal. The

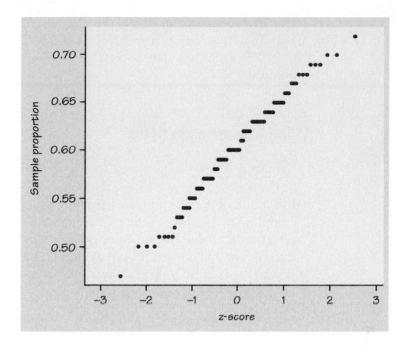

FIGURE 3.7   Normal quantile plot of the sample proportions in Figure 3.6. The distribution is close to normal except for some granularity due to the fact that sample proportions from a sample of size 100 can take only values that are multiples of 0.01. Because a plot of 1000 points is hard to read, this plot presents only every 10th value.

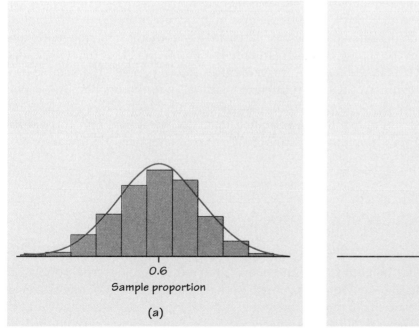

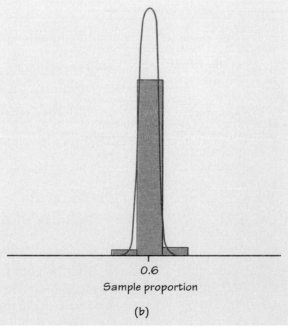

FIGURE 3.8   The sampling distributions for sample proportions $\hat{p}$ for SRSs of two sizes drawn from a population having population proportion $p = 0.6$. (a) Sample size 100. (b) Sample size 2500. Both statistics are unbiased because the mean of their distributions equals the true population value $p = 0.6$. The statistic from the larger sample is less variable.

two normal curves in Figure 3.8 help us compare the two sampling distributions. The mean of the 1000 $\hat{p}$'s is 0.6002, again very close to the value $p = 0.6$ of the population parameter. The standard deviation is 0.0097, or about 0.01.

The bias of a statistic

Sampling distributions allow us to describe *bias* more precisely by speaking of the bias of a statistic rather than bias in a sampling method. Bias concerns the center of the sampling distribution. The center of both of our sets of 1000 samples is very close to the parameter value $p = 0.6$ of the population from which we sampled. In fact, mathematics tells us that the mean of the distribution we would get if we looked at all possible samples, rather than just 1000, is exactly equal to p. The sample proportion $\hat{p}$ from an SRS is an *unbiased estimator* of the parameter p.

> **Unbiased Estimator**
>
> A statistic used to estimate a parameter is **unbiased** if the mean of its sampling distribution is equal to the true value of the parameter being estimated.

An unbiased statistic will sometimes fall above the true value of the parameter and sometimes below if we take many samples. Because its sampling distribution is centered at the true value, however, there is no systematic tendency to overestimate or underestimate the parameter. This makes the idea of lack of bias in the sense of "no favoritism" more precise. The sample proportion $\hat{p}$ from an SRS is an unbiased estimator of the population proportion p. If we draw an SRS from a population in which 60% find shopping frustrating, the mean of the sampling distribution of $\hat{p}$ is 0.6. If we draw an SRS from a population with 50% frustrated shoppers, the mean of $\hat{p}$ is then 0.5.

The variability of a statistic

The statistics whose sampling distributions appear in Figure 3.8 are both unbiased. That is, both distributions are centered at 0.6, the true population parameter. The sample proportion $\hat{p}$ from a random sample of any size is an unbiased estimate of the parameter p. Larger samples have a clear advantage, however. They are much more likely to produce an estimate close to the true value of the parameter because there is much less variability among large samples than among small samples.

EXAMPLE 3.22 | Simulation has shown us that if we repeat our SRS of size 100 many times, the sample proportion $\hat{p}$ who find shopping frustrating will vary according to a sampling distribution that is close to the normal distribution with mean 0.6 and standard deviation 0.051. Apply the 68–95–99.7 rule to this distribution. If we take many samples, 95% of the sample proportions will lie in the range

$$\text{mean} \pm (2 \times \text{standard deviation}) = 0.6 \pm (2 \times 0.051)$$
$$= 0.6 \pm 0.1$$

This is a statement about how much we can trust an SRS of this size. If in fact 60% of the population find clothes shopping frustrating, the estimates from repeated SRSs of size 100 will usually fall between 50% and 70%. That's not very satisfactory.

For samples of size 2500, the sampling distribution of $\hat{p}$ is close to normal with mean 0.6 and standard deviation 0.01. So 95% of these samples will give an estimate within about 0.02 of the mean, that is, between 0.58 and 0.62. An SRS of size 2500 can be trusted to give sample estimates that are very close to the truth about the entire population.

More knowledge of probability will soon enable us to give the standard deviation of the sampling distribution for any size sample. We will then see Example 3.22 as part of a general rule that shows exactly how the variability of sample results decreases for larger samples. One important and surprising aspect of this conclusion is that the spread of the sampling distribution does *not* depend very much on the size of the population.

Variability of a Statistic

The **variability of a statistic** is described by the spread of its sampling distribution. This spread is determined by the sampling design and the sample size n. Larger samples have smaller spreads.

As long as the population is much larger than the sample (say, at least 10 times as large), the spread of the sampling distribution for a sample of fixed size n is approximately the same for any population size.

Why does the size of the population have little influence on the behavior of statistics from random samples? To see why this is plausible, imagine sampling harvested corn by thrusting a scoop into a lot of corn kernels. The scoop doesn't know whether it is surrounded by a bag of corn or by an entire truckload. As long as the corn is well mixed (so that the scoop selects a random sample), the variability of the result depends only on the size of the scoop.

The fact that the variability of sample results is controlled by the size of the sample has important consequences for sampling design. An SRS of size 2500 from the 270 million residents of the United States gives results as precise as an SRS of size 2500 from the 740,000 inhabitants of San Francisco. This is good news for designers of national samples but bad news for those who want accurate information about the citizens of San Francisco. If both use an SRS, both must use the same size sample to obtain equally trustworthy results.

Bias and variability

We can think of the true value of the population parameter as the bull's-eye on a target and of the sample statistic as a bullet fired at the bull's-eye. Both bias and variability describe what happens when we take many shots at the target. *Bias* means that our sight is misaligned and we shoot consistently off the bull's-eye in the same direction. Our sample values do not center about the population value. *High variability* means that repeated shots are widely scattered on the target. Repeated samples do not give similar results but differ widely among themselves. Figure 3.9 shows this target illustration of the two types of error.

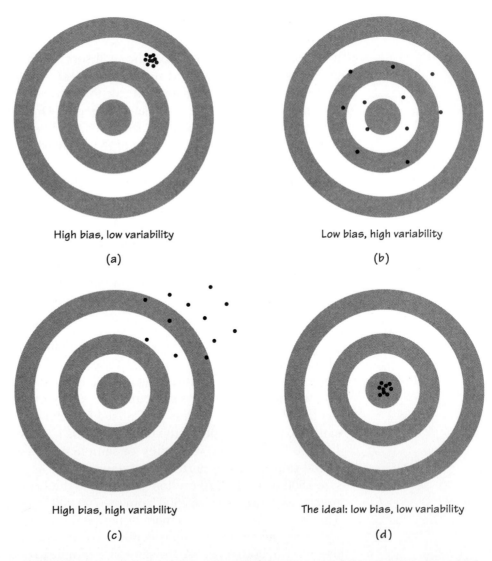

FIGURE 3.9 Bias and variability. (a) High bias, low variability (b) Low bias, high variability (c) High bias, high variability (d) The ideal: low bias, low variability.

Notice that low variability (repeated shots are close together) can accompany high bias (the shots are consistently away from the bull's-eye in one direction). And low bias (the shots center on the bull's-eye) can accompany high variability (repeated shots are widely scattered). A good sampling scheme, like a good shooter, must have both low bias and low variability.

Why randomize?

Why randomize? The act of randomizing guarantees that the results of analyzing our data are subject to the laws of probability. The behavior of statistics is described by a sampling distribution. The form of the distribution is known, and in many cases is approximately normal. Often the center of the distribution lies at the true parameter value, so that the notion that randomization eliminates bias is made more precise. The spread of the distribution describes the variability of the statistic and can be made as small as we wish by choosing a large enough sample. In a randomized experiment, we can reduce variability by choosing larger groups of subjects for each treatment.

These facts are at the heart of formal statistical inference. Later chapters will have much to say in more technical language about sampling distributions and the way statistical conclusions are based on them. What any user of statistics must understand is that all the technical talk has its basis in a simple question: What would happen if the sample or the experiment were repeated many times? The reasoning applies not only to an SRS but also to the complex sampling designs actually used by opinion polls and other national sample surveys. The same conclusions hold as well for randomized experimental designs. The details vary with the design but the basic facts are true whenever randomization is used to produce data.

Remember that proper statistical design is not the only aspect of a good sample or experiment. The sampling distribution shows only how a statistic varies due to the operation of chance in randomization. It reveals nothing about possible bias due to undercoverage or nonresponse in a sample, or to lack of realism in an experiment. The true distance of a statistic from the parameter it is estimating can be much larger than the sampling distribution suggests. What is worse, there is no way to say how large the added error is. The real world is less orderly than statistics textbooks imply.

Beyond the basics ▶ capture–recapture sampling

Sockeye salmon return to reproduce in the river where they were hatched four years earlier. How many salmon survived natural perils and heavy fishing to make it back this year? How many mountain sheep are there in Colorado? Are migratory songbird populations in North America decreasing or holding their own? These questions concern the size of animal populations. Biologists address them with a special kind of repeated sampling, called *capture-recapture sampling*.

EXAMPLE 3.23   You are interested in the number of least flycatchers migrating along a major route in the north-central United States. You set up "mist nets" that capture the birds but do not harm them. The birds caught in the net are fitted with a small aluminum leg band and released. Last year you banded and released 200 least flycatchers. This year you repeat the process. Your net catches 120 least flycatchers, 12 of which have tags from last year's catch.

The proportion of your second sample that have bands should estimate the proportion in the entire population that are banded. So if N is the unknown number of least flycatchers, we should have approximately

$$\text{proportion banded in sample} = \text{proportion banded in population}$$
$$\frac{12}{120} = \frac{200}{N}$$

Solve for N to estimate that the total number of flycatchers migrating while your net was up this year is approximately

$$N = 200 \times \frac{120}{12} = 2000$$

The capture-recapture idea extends the use of a sample proportion to estimate a population proportion. The idea works well if both samples are SRSs from the population and the population remains unchanged between samples. In practice, complications arise because, for example, some of the birds tagged last year died before this year's migration. Variations on capture-recapture samples are widely used in wildlife studies and are now finding other applications. One way to estimate the census undercount in a district is to consider the census as "capturing and marking" the households that respond. Census workers then visit the district, take an SRS of households, and see how many of those counted by the census show up in the sample. Capture-recapture estimates the total count of households in the district. As with estimating wildlife populations, there are many practical pitfalls. Our final word is as before: the real world is less orderly than statistics textbooks imply.

SUMMARY

A number that describes a population is a **parameter.** A number that can be computed from the data is a **statistic.** The purpose of sampling or experimentation is usually to use statistics to make statements about unknown parameters.

A statistic from a probability sample or randomized experiment has a **sampling distribution** that describes how the statistic varies in repeated data production. The sampling distribution answers the question, "What would happen if we repeated the sample or experiment many times?" Formal statistical inference is based on the sampling distributions of statistics.

A statistic as an estimator of a parameter may suffer from **bias** or from high **variability.** Bias means that the center of the sampling distribution is not equal to the true value of the parameter. The variability of the statistic is described by the spread of its sampling distribution.

Properly chosen statistics from randomized data production designs have no bias resulting from the way the sample is selected or the way the experimental units are assigned to treatments. We can reduce the variability of the statistic by increasing the size of the sample or the size of the experimental groups.

SECTION 3.4 EXERCISES

State whether each boldface number in Exercises 3.51 to 3.54 is a parameter or a statistic.

3.51   The Bureau of Labor Statistics announces that last month it interviewed all members of the labor force in a sample of 60,000 households; **6.2%** of the people interviewed were unemployed.

3.52   A carload lot of ball bearings has a mean diameter of **2.503** centimeters (cm). This is within the specifications for acceptance of the lot by the purchaser. The inspector happens to inspect 100 bearings from the lot with a mean diameter of **2.515** cm. This is outside the specified limits, so the lot is mistakenly rejected.

3.53   A telemarketing firm in Los Angeles uses a device that dials residential telephone numbers in that city at random. Of the first 100 numbers dialed, **43** are unlisted. This is not surprising, because **52%** of all Los Angeles residential phones are unlisted.

3.54   A researcher investigating the effects of a toxic compound in food conducts a randomized comparative experiment with young male white rats. A control group is fed a normal diet, while the experimental group is fed a diet with 2500 parts per million of the toxic material. After 8 weeks, the mean weight gain is **335** grams for the control group and **289** grams for the experimental group.

3.55   Figure 3.10 shows histograms of four sampling distributions of statistics intended to estimate the same parameter. Label each distribution relative to the others as high or low bias and as high or low variability.

3.56   A management student is planning to take a survey of student attitudes toward part-time work while attending college. He develops a questionnaire and plans to ask 25 randomly selected students to fill it out. His faculty advisor approves the questionnaire but urges that the sample size be increased to at least 100 students. Why is the larger sample helpful?

3.57   An agency of the federal government plans to take an SRS of residents in each state to estimate the proportion of owners of real estate in each state's population. The population of the states ranges from about 485,000 people in Wyoming to more than 32 million in California.

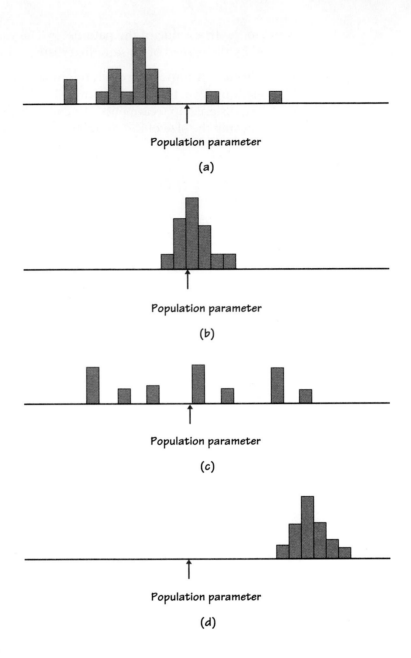

FIGURE 3.10 Which of these sampling distributions displays high or low bias and high or low variability? (Exercise 3.55)

(a) Will the variability of the sample proportion change from state to state if an SRS of size 2000 is taken in each state? Explain your answer.

(b) Will the variability of the sample proportion change from state to state if an SRS of 1/10 of 1% (0.001) of the state's population is taken in each state? Explain your answer.

3.58 A national opinion poll recently estimated that 44% ($\hat{p} = 0.44$) of all American adults agree that parents of school-age children should be given vouchers good for education at any public or private school of their choice. The polling organization used a probability sampling method for which the sample proportion has a normal distribution with standard deviation about 0.015. The poll therefore announced a "margin of error" of 0.03 (two standard deviations) for its result. If a sample were drawn by the same method from the state of New Jersey (population 8 million) instead of from the entire United States (population 270 million), would this margin of error be larger or smaller? Explain your answer.

3.59 Coin tossing can illustrate the idea of a sampling distribution. The population is all outcomes (heads or tails) we would get if we tossed a coin forever. The parameter p is the proportion of heads in this population. We suspect that p is close to 0.5; that is, we think the coin will show about one-half heads in the long run. The sample is the outcomes of 20 tosses, and the statistic $\hat{p}$ is the proportion of heads in these 20 tosses.

(a) Toss a coin 20 times and record the value of $\hat{p}$.

(b) Repeat this sampling process 10 times. Make a stemplot of the 10 values of $\hat{p}$. Is the center of this distribution close to 0.5? (Ten repetitions give only a crude approximation to the sampling distribution. If possible, pool your work with that of other students to obtain several hundred repetitions and make a histogram of the values of $\hat{p}$.)

3.60 Let us illustrate the idea of a sampling distribution in the case of a very small sample from a very small population. The population contains 10 students. Here are their scores on an exam:

Student	0	1	2	3	4	5	6	7	8	9
Score	82	62	80	58	72	73	65	66	74	62

The parameter of interest is the mean score, which is 69.4. The sample is an SRS of $n = 4$ students drawn from this population. The students are labeled 0 to 9 so that a single random digit from Table B chooses one student for the sample.

(a) Use Table B to draw an SRS of size 4 from this population. Write the four scores in your sample and calculate the mean $\bar{x}$ of the sample scores. This statistic is an estimate of the population parameter.

(b) Repeat this process 10 times. Make a histogram of the 10 values of $\bar{x}$. You are constructing the sampling distribution of $\bar{x}$. Is the center of your histogram close to 69.4? (Ten repetitions give only a crude approximation to the sampling distribution. If possible, pool your work with that of other students—using different parts of Table B—to obtain several hundred repetitions and make a histogram of the values of $\bar{x}$. This histogram is a better approximation to the sampling distribution.)

3.61 An entomologist samples a field for egg masses of a harmful insect by placing a yard-square frame at random locations and examining the ground within the frame carefully. She wishes to estimate the proportion of square yards in which egg masses are present. Suppose that in a large field egg masses are present in 20% of all possible yard-square areas—that is, $p = 0.2$ in this population.

(a) Use Table B to simulate the presence or absence of egg masses in each square yard of an SRS of 10 square yards from the field. Be sure to explain clearly which digits you used to represent the presence and the absence of egg masses. What proportion of your 10 sample areas had egg masses?

(b) Repeat (a) with different lines from Table B, until you have simulated the results of 20 SRSs of size 10. What proportion of the square yards in each of your 20 samples had egg masses? Make a stemplot from these 20 values to display the sampling distribution of $\hat{p}$ in this case. What is the mean of this distribution? What is its shape?

3.62   An opinion poll asks, "Are you afraid to go outside at night within a mile of your home because of crime?" Suppose that the proportion of all adult U. S. residents who would say "Yes" to this question is $p = 0.4$.

(a) Use Table B to simulate the result of an SRS of 20 adults. Be sure to explain clearly which digits you used to represent each of "Yes" and "No." What proportion of your 20 responses were "Yes"?

(b) Repeat (a) using different lines in Table B until you have simulated the results of 10 SRSs of size 20 from the same population. Compute the proportion of "Yes" responses in each sample. Find the mean of these 10 proportions. Is it close to p?

3.63   We will illustrate sampling variability and sampling distributions by a small example. Table 1.9 (page 98) gives the positions and weights of the 95 players on a college football team. Call the quarterbacks, running backs, and wide receivers "offensive backs." There are 27 offensive backs on the team. The 95 players are our population.

(a) The proportion of offensive backs in the population is a parameter p. What is the value of p?

(b) Choose an SRS of 15 players from the team. The proportion who are offensive backs is a statistic $\hat{p}$. What is the value of $\hat{p}$?

(c) Now use software to choose 20 SRSs of size 15. For each sample, find the proportion $\hat{p}$ of offensive backs. Make a histogram of the 20 values of $\hat{p}$. This is a rough approximation to the sampling distribution of this statistic.

(d) Find the mean of the 20 $\hat{p}$ values. Is it close to the parameter p? Mark the value of p on the axis of your histogram. How is the lack of bias of $\hat{p}$ apparent from your graph?

3.64   We will illustrate sampling variability and sampling distributions by a small example. The CSDATA data file contains the college grade point averages (GPAs) of all 224 students in a university entering class who planned to major in computer science. This is our population. Use software to do the following.

(a) Describe this population both with a histogram and with numerical summaries. In particular, what is the mean GPA in the population?

(b) Choose an SRS of 20 members from this population. Make a histogram of the GPAs in the sample, and find their mean. Briefly compare the distributions of GPA in the sample and in the population.

(c) Repeat the process of choosing an SRS of size 20 four more times (five in all). Record the five histograms of your sample GPAs. Does it seem reasonable to you from this small trial that an SRS will usually produce a sample that is generally representative of the population?

(d) Now choose 20 more SRSs of size 20 (25 in all). Don't make histograms of these latest samples. Record the mean GPA of each of your 25 samples and make a histogram of the 25 means. This histogram is a rough approximation to the sampling distribution of the mean. One sign of bias would be that the distribution of the sample means was systematically on one side of the true population mean. Mark the population mean GPA on your histogram of the 25 sample means. Is there a clear bias?

(e) Find the mean and standard deviation of your 25 sample means. We expect that the mean will be close to the true mean of the population. Is it? We also expect that the standard deviation of the sampling distribution will be smaller than the standard deviation of the population. Is it?

CHAPTER 3 EXERCISES

3.65 A study of the effect of abortions on the health of subsequent children was conducted as follows. The researchers obtained names of women who had had abortions from medical records in New York City hospitals. They then searched birth records to locate all women in this group who bore a child within five years of the abortion. Then hospital records were examined again for information about the health of the newborn child. Was this an observational study or an experiment? Why?

3.66 In a study of the relationship between physical fitness and personality, middle-aged college faculty who have volunteered for an exercise program are divided into low-fitness and high-fitness groups on the basis of a physical examination. All subjects then take the Cattell Sixteen Personality Factor Questionnaire, and the results for the two groups are compared. Is this study an observational study or an experiment? Explain your answer.

3.67 Do consumers prefer the taste of Pepsi or Coke in a blind test in which neither cola is identified? Describe briefly the design of a matched pairs experiment to investigate this question. How will you use randomization?

3.68 Is the number of days a letter takes to reach another city affected by the time of day it is mailed and whether or not the zip code is used? Describe briefly the design of a two-factor experiment to investigate this question.

Be sure to specify the treatments exactly and to tell how you will handle lurking variables such as the day of the week on which the letter is mailed.

3.69 The previous two exercises illustrate the use of statistically designed experiments to answer questions that arise in everyday life. Select a question of interest to you that an experiment might answer and carefully discuss the design of an appropriate experiment.

3.70 There are several psychological tests available to measure the extent to which Mexican Americans are oriented toward Mexican/Spanish or Anglo/English culture. Two such tests are the Bicultural Inventory (BI) and the Acculturation Rating Scale for Mexican Americans (ARSMA). To study the correlation between the scores on these two tests, researchers will give both tests to a group of 22 Mexican Americans.

  (a) Briefly describe a matched pairs design for this study. In particular, how will you use randomization in your design?

  (b) You have an alphabetized list of the subjects (numbered 1 to 22). Carry out the randomization required by your design and report the result.

3.71 A medical study of heart surgery investigates the effect of drugs called beta-blockers on the pulse rate of the patient during surgery. The pulse rate will be measured at a specific point during the operation. The investigators decide to use as subjects 30 patients facing heart surgery who have consented to take part in the study. You have a list of these patients, numbered 1 to 30 in alphabetical order.

  (a) Outline in graphical form a completely randomized experimental design for this study.

  (b) Carry out the randomization required by your design and report the result. (If you use Table B, enter at line 125.)

3.72 A university's financial aid office wants to know how much it can expect students to earn from summer employment. This information will be used to set the level of financial aid. The population contains 3478 students who have completed at least one year of study but have not yet graduated. A questionnaire will be sent to an SRS of 100 of these students, drawn from an alphabetized list.

  (a) Describe how you will label the students in order to select the sample.

  (b) Use Table B, beginning at line 105, to select the *first five* students in the sample.

3.73 A labor organization wants to study the attitudes of college faculty members toward collective bargaining. These attitudes appear to be different depending on the type of college. The American Association of University Professors classifies colleges as follows:

   Class I: Offer doctorate degrees and award at least 15 per year.

   Class IIA: Award degrees above the bachelor's but are not in Class I.

Class IIB: Award no degrees beyond the bachelor's.

Class III: Two-year colleges.

Discuss the design of a sample of faculty from colleges in your state, with total sample size about 200.

3.74   You want to investigate the attitudes of students at your school about the faculty's commitment to teaching. The student government will pay the costs of contacting about 500 students.

(a)   Specify the exact population for your study; for example, will you include part-time students?

(b)   Describe your sample design. Will you use a stratified sample?

(c)   Briefly discuss the practical difficulties that you anticipate; for example, how will you contact the students in your sample?

3.75   You are participating in the design of a medical experiment to investigate whether a calcium supplement in the diet will reduce the blood pressure of middle-aged men. Preliminary work suggests that calcium may be effective and that the effect may be greater for black men than for white men.

(a)   Outline in graphic form the design of an appropriate experiment.

(b)   Choosing the sizes of the treatment groups requires more statistical expertise. We will learn more about this aspect of design in later chapters. Explain in plain language the advantage of using larger groups of subjects.

3.76   A chemical engineer is designing the production process for a new product. The chemical reaction that produces the product may have higher or lower yield, depending on the temperature and the stirring rate in the vessel in which the reaction takes place. The engineer decides to investigate the effects of combinations of two temperatures (50° C and 60° C) and three stirring rates (60 rpm, 90 rpm, and 120 rpm) on the yield of the process. Two batches of the feedstock will be processed at each combination of temperature and stirring rate.

(a)   How many factors are there in this experiment? How many treatments? Identify each of the treatments. How many experimental units (batches of feedstock) does the experiment require?

(b)   Outline in graphic form the design of an appropriate experiment.

(c)   The randomization in this experiment determines the order in which batches of the feedstock will be processed according to each treatment. Use Table B starting at line 128 to carry out the randomization and state the result.

3.77   A psychologist is interested in the effect of room temperature on the performance of tasks requiring manual dexterity. She chooses temperatures of 70° F and 90° F as treatments. The response variable is the number of correct insertions, during a 30-minute period, in an elaborate peg-and-hole apparatus that requires the use of both hands simultaneously. Each subject is trained on the apparatus and then asked to make as many insertions as possible in 30 minutes of continuous effort.

(a) Outline a completely randomized design to compare dexterity at 70° and 90°. Twenty subjects are available.

(b) Because individuals differ greatly in dexterity, the wide variation in individual scores may hide the systematic effect of temperature unless there are many subjects in each group. Describe in detail the design of a matched pairs experiment in which each subject serves as his or her own control.

3.78 The requirement that human subjects give their informed consent to participate in an experiment can greatly reduce the number of available subjects. In trials of treatments for cancer, for example, patients must agree to be randomly assigned to the standard therapy or to an experimental therapy. The patients who do not wish their treatment to be decided by a coin toss are dropped from the experiment and given the standard therapy. Why is it not correct to keep these patients in the experiment as part of the control group, where they would receive the standard therapy?

3.79 You are on the staff of a member of Congress who is considering a controversial bill that would provide for government-sponsored insurance to cover care in nursing homes. You report that 1128 letters dealing with the issue have been received, of which 871 oppose the legislation. "I'm surprised that most of my constituents oppose the bill. I thought it would be quite popular," says the congresswoman. Are you convinced that a majority of the voters oppose the bill? State briefly how you would explain the statistical issue to the congresswoman.

3.80 A study of the effects of running on personality involved 231 male runners who each ran about 20 miles a week. The runners were given the Cattell Sixteen Personality Factor Questionnaire, a 187-item multiple-choice test often used by psychologists. A news report (*New York Times*, February 15, 1988) stated, "The researchers found statistically significant personality differences between the runners and the 30-year-old male population as a whole." A headline on the article said, "Research has shown that running can alter one's moods."

(a) Explain carefully, to someone who knows no statistics, what "statistically significant" means.

(b) Explain carefully, to someone who knows no statistics, why the headline is misleading.

3.81 To demonstrate how randomization reduces confounding, return to the nutrition experiment described in Example 3.7 (page 244). Suppose that the 30 rats are labeled 01 to 30. Suppose also that, unknown to the experimenter, the 10 rats labeled 01 to 10 have a genetic defect that will cause them to grow more slowly than normal rats. If the experimenter simply puts rats 01 to 15 in the experimental group and rats 16 to 30 in the control group, this lurking variable will bias the experiment against the new food product.

Use Table B to assign 15 rats at random to the experimental group as in Example 3.7. Record how many of the 10 rats with genetic defects are placed in the experimental group and how many are in the control group. Repeat the randomization using different lines in Table B until you have done five random assignments. What is the mean number of genetically defective rats in experimental and control groups in your five repetitions?

3.82 Statistical software allows you to simulate the results of, say, 100 SRSs of size n drawn from a population in which proportion p would say "Yes" to a certain question. We will see in Chapter 5 that "binomial" is the key word to look for in the software menus.

Draw 100 samples of size $n = 50$ from populations with $p = 0.1$, $p = 0.3$, and $p = 0.5$. Make a stemplot of the 100 values of $\hat{p}$ obtained in each simulation. Compare your three stemplots. Do they show about the same variability? How does changing the parameter p affect the sampling distribution? If your software permits, make a normal quantile plot of the $\hat{p}$ values from the population with $p = 0.5$. Is the sampling distribution approximately normal?

3.83 Draw 100 samples of each of the sizes $n = 50$, $n = 200$, and $n = 800$ from a population with $p = 0.6$. Prepare a frequency histogram of the $\hat{p}$ values for each simulation, using the same horizontal and vertical scales so that the three graphs can be compared easily. How does increasing the size of an SRS affect the sampling distribution of $\hat{p}$?

NOTES

1. Steering Committee of the Physicians' Health Study Research Group, "Final report on the aspirin component of the ongoing Physicians' Health Study," *New England Journal of Medicine,* 321 (1989), pp. 129–135.

2. L. L. Miao, "Gastric freezing: an example of the evaluation of medical therapy by randomized clinical trials," in J. P. Bunker, B. A. Barnes, and F. Mosteller (eds.), *Costs, Risks, and Benefits of Surgery,* Oxford University Press, New York, 1977, pp. 198–211.

3. The information in this example is taken from *The Ascap Survey and Your Royalties,* ASCAP, New York, undated.

4. For more detail on the material of this section and complete references, see P. E. Converse and M. W. Traugott, "Assessing the accuracy of polls and surveys," *Science,* 234 (1986), pp. 1094–1098.

5. The estimates of the census undercount come from the Census Bureau's *Report to Congress: The Plan for Census 2000,* July 1997, available at the Census Bureau Web site, http://www.census.gov. The information about nonresponse appears in Eugene P. Eriksen and Teresa K. DeFonso, "Beyond the net undercount: how to measure census error," *Chance,* 6, no. 4 (1993), pp. 38–43 and 14.

6. For more detail on the limits of memory in surveys, see N. M. Bradburn, L. J. Rips, and S. K. Shevell, "Answering autobiographical questions: the impact of memory and inference on surveys," *Science,* 236 (1987), pp. 157–161.

7. The Levi jeans and disposable diaper examples are taken from Cynthia Crossen, "Margin of error: studies galore support products and positions, but are they reliable?" *Wall Street Journal,* November 14, 1991.

8. The survey question and result are reported in Trish Hall, "Shop? Many say 'Only if I must,'" *New York Times,* November 28, 1990.

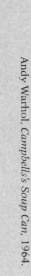

PART 2

Andy Warhol, *Campbell's Soup Can*, 1964.

Probability and Inference

Prelude

Chance is all around us. Sometimes chance results from human design, as in the casino's games of chance and the statistician's random samples. Sometimes nature uses chance, as in choosing the sex of a child. Sometimes the reasons for chance behavior are mysterious, as when the number of deaths each year in a large population is as regular as the number of heads in many tosses of a coin. Probability is the branch of mathematics that describes the pattern of chance outcomes. Probability is the topic of this chapter, but not for its own sake. We will concentrate on the ideas from probability that we need to go more deeply into statistics. After thinking about randomness in Section 4.1, we move quickly from general facts about probability in Section 4.2 to tools for describing the pattern of numerical chance outcomes in Sections 4.3 and 4.4. Section 4.5 offers more of the mathematics of probability for those with an interest in the subject. Later chapters apply basic probability to describe the behavior of statistics from random samples and randomized comparative experiments. Even our brief acquaintance with probability enables us to answer questions like these:

■ How can gambling, which depends on the unpredictable fall of dice and cards, be a profitable business for a casino?

■ Give a test for the AIDS virus to the employees of a small company. What is the chance of at least one positive test if all the people tested are free of the virus?

■ Gregor Mendel found that inheritance is governed by chance. What do the laws of probability say about the color of Mendel's peas or the sex of a couple's children?

■ If you buy a ticket to your state's "Pick 3" lottery game every day for many years, how much will each ticket win on the average?

CHAPTER 4

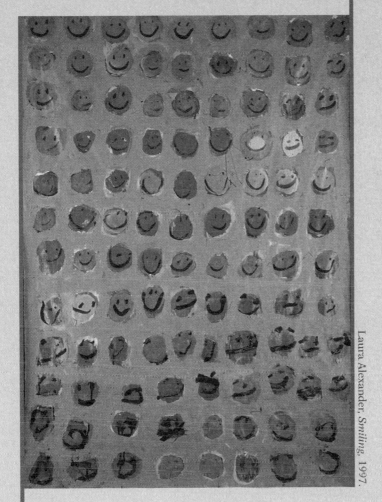

Laura Alexander, *Smiling*, 1997.

Probability: The Study of Randomness

The reasoning of statistical inference rests on asking, "How often would this method give a correct answer if I used it very many times?" When we produce data by random sampling or randomized comparative experiments, the laws of probability answer the question "What would happen if we did this many times?"

4.1   Randomness

Toss a coin, or choose an SRS. The result can't be predicted in advance, because the result will vary when you toss the coin or choose the sample repeatedly. But there is still a regular pattern in the results, a pattern that emerges clearly only after many repetitions. This remarkable fact is the basis for the idea of probability.

EXAMPLE 4.1

When you toss a coin, there are only two possible outcomes, heads or tails. Figure 4.1 shows the results of tossing a coin 1000 times. For each number of tosses from 1 to 1000, we have plotted the proportion of those tosses that gave a head. The first toss was a head, so the proportion of heads starts at 1. The second toss was a tail, reducing the proportion of heads to 0.5 after two tosses. The next three tosses gave a tail followed by two heads, so the proportion of heads after five tosses is 3/5, or 0.6.

The proportion of tosses that produce heads is quite variable at first, but it settles down as we make more and more tosses. Eventually this proportion gets close to 0.5 and stays there. We say that 0.5 is the *probability* of a head. The probability 0.5 appears as a horizontal line on the graph.

The language of probability

"Random" in statistics is not a synonym for "haphazard" but a description of a kind of order that emerges only in the long run. We often encounter the unpredictable side of randomness in our everyday experience, but we rarely see enough repetitions of the same random phenomenon to observe the long-term regularity that probability describes. You can see that regularity emerging in Figure 4.1. In the very long run, the proportion of tosses that give a head is 0.5. This is the intuitive idea of probability. Probability 0.5 means "occurs half the time in a very large number of trials."

We might suspect that a coin has probability 0.5 of coming up heads just because the coin has two sides. As Exercises 4.1 and 4.2 illustrate, such suspicions are not always correct. The idea of probability is empirical. That is, it is based on observation rather than theorizing. Probability describes what happens in very many trials, and we must actually observe many trials to pin down a probability. In the case of tossing a coin, some diligent people have in fact made thousands of tosses.

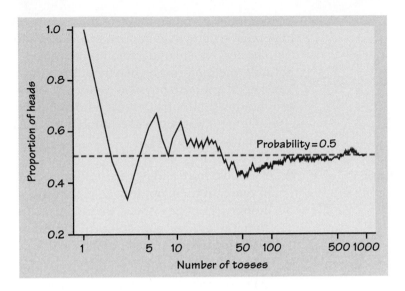

FIGURE 4.1 The proportion of tosses of a coin that give a head changes as we make more tosses. Eventually, however, the proportion approaches 0.5, the probability of a head.

EXAMPLE 4.2 | The French naturalist Count Buffon (1707–1788) tossed a coin 4040 times. Result: 2048 heads, or proportion 2048/4040 = 0.5069 for heads.

Around 1900, the English statistician Karl Pearson heroically tossed a coin 24,000 times. Result: 12,012 heads, a proportion of 0.5005.

While imprisoned by the Germans during World War II, the South African mathematician John Kerrich tossed a coin 10,000 times. Result: 5067 heads, proportion of heads 0.5067.

Randomness and Probability

We call a phenomenon **random** if individual outcomes are uncertain but there is nonetheless a regular distribution of outcomes in a large number of repetitions.

The **probability** of any outcome of a random phenomenon is the proportion of times the outcome would occur in a very long series of repetitions. That is, probability is long-term relative frequency.

Thinking about randomness

That some things are random is an observed fact about the world. The outcome of a coin toss, the time between emissions of particles by a radioactive source, and the sexes of the next litter of lab rats are all random. So is the outcome of a random sample or a randomized experiment. Probability theory is the branch of mathematics that describes random behavior. Of course, we can never observe a probability exactly. We could always continue tossing the coin, for example. Mathematical probability is an idealization based on imagining what would happen in an indefinitely long series of trials.

The best way to understand randomness is to observe random behavior—not only the long-run regularity but the unpredictable results of short runs. You can do this with physical devices, as in Exercises 4.1 to 4.5, but computer simulations (imitations) of random behavior allow faster exploration. Exercises 4.7 to 4.9 suggest some simulations of random behavior. As you explore randomness, remember:

independence

- ▪ You must have a long series of **independent** trials. That is, the outcome of one trial must not influence the outcome of any other. Imagine a crooked gambling house where the operator of a roulette wheel can stop it where she chooses—she can prevent the proportion of "red" from settling down to a fixed number. These trials are not independent.

- ▪ The idea of probability is empirical. Computer simulations start with given probabilities and imitate random behavior, but we can estimate a real-world probability only by actually observing many trials.

- ▪ Nonetheless, computer simulations are very useful because we need long runs of trials. In situations such as coin tossing, the proportion of an outcome often requires several hundred trials to settle down to the probability of that outcome. The kinds of physical random devices suggested in the exercises are too slow for this. Short runs give only rough estimates of a probability.

The uses of probability

Probability theory originated in the study of games of chance. Tossing dice, dealing shuffled cards, and spinning a roulette wheel are examples of deliberate randomization that are similar to random sampling. Although games of chance are ancient, they were not studied by mathematicians until the sixteenth and seventeenth centuries. It is only a mild simplification to say that probability as a branch of mathematics arose when seventeenth-century French gamblers asked the mathematicians Blaise Pascal and Pierre de Fermat for help. Gambling is still with us, in casinos and state lotteries. We will make use of games of chance as simple examples that illustrate the principles of probability.

Careful measurements in astronomy and surveying led to further advances in probability in the eighteenth and nineteenth centuries because the results of repeated measurements are random and can be described by distributions much like those arising from random sampling. Similar distributions appear in data on human life span (mortality tables) and in data on lengths or weights in a population of skulls, leaves, or cockroaches.[1] In the twentieth century, we employ the mathematics of probability to describe the flow of traffic through a highway system, a telephone interchange, or a computer processor; the genetic makeup of individuals or populations; the energy states of subatomic particles; the spread of epidemics or rumors; and the rate of return on risky investments. Although we are interested in probability because of its usefulness in statistics, the mathematics of chance is important in many fields of study.

SUMMARY

A **random phenomenon** has outcomes that we cannot predict but that nonetheless have a regular distribution in very many repetitions.

The **probability** of an event is the proportion of times the event occurs in many repeated trials of a random phenomenon.

SECTION 4.1 EXERCISES

4.1 Hold a penny upright on its edge under your forefinger on a hard surface, then snap it with your other forefinger so that it spins for some time before falling. Based on 50 spins, what is the probability of heads?

4.2 You may feel that it is obvious that the probability of a head in tossing a coin is about 1/2 because the coin has two faces. Such opinions are not always correct. The previous exercise asked you to spin a penny rather than toss it—that changes the probability of a head. Now try another variation. Stand a penny on edge on a hard, flat surface. Pound the surface with your hand so that the penny falls over. What is the probability that it falls with heads upward? Make at least 50 trials to estimate the probability of a head.

4.3 When we toss a penny, experience shows that the probability (long-term proportion) of a head is close to 1/2. Suppose now that we toss the penny repeatedly until we get a head. What is the probability that the first head comes up in an odd number of tosses (1, 3, 5, and so on)? To find out, repeat this experiment 50 times, and keep a record of the number of tosses needed to get a head on each of your 50 trials.

  (a) From your experiment, estimate the probability of a head on the first toss. What value should we expect this probability to have?

  (b) Use your results to estimate the probability that the first head appears on an odd-numbered toss.

4.4 Toss a thumbtack on a hard surface 100 times. How many times did it land with the point up? What is the approximate probability of landing point up?

4.5 Obtain 10 identical thumbtacks and toss all 10 at once. Do this 50 times and record the number that land point up on each trial.

  (a) What is the approximate probability that at least one lands point up?

  (b) What is the approximate probability that more than one lands point up?

4.6 You read in a book on poker that the probability of being dealt three of a kind in a five-card poker hand is 1/50. Explain in simple language what this means.

4.7   In the game of Heads or Tails, Betty and Bob toss a coin four times. Betty wins a dollar from Bob for each head and pays Bob a dollar for each tail—that is, she wins or loses the difference between the number of heads and the number of tails. For example, if there are one head and three tails, Betty loses $2. You can check that Betty's possible outcomes are

$$\{-4, -2, 0, 2, 4\}$$

Assign probabilities to these outcomes by playing the game 20 times and using the proportions of the outcomes as estimates of the probabilities. If possible, combine your trials with those of other students to obtain long-run proportions that are closer to the probabilities.

4.8   A recent opinion poll showed that about 73% of married women agree that their husbands do at least their fair share of household chores. Suppose that this is exactly true. Choosing a married woman at random then has probability 0.73 of getting one who agrees that her husband does his share. We can use software to simulate choosing many women independently. (In most software, the key phrase to look for is "Bernoulli trials." This is the technical term for independent trials with "Yes/No" outcomes. Our outcomes here are "Agree" and "Disagree.")

  (a) Simulate drawing 20 women, then 80 women, then 320 women. What proportion agree in each case? We expect (but because of chance variation we can't be sure) that the proportion will be closer to 0.73 in longer runs of trials.

  (b) Simulate drawing 20 women 10 times and record the percents in each trial who agree. Then simulate drawing 320 women 10 times and again record the 10 percents. Which set of 10 results is less variable? We expect the results of 320 trials to be more predictable (less variable) than the results of 20 trials. That is "long-run regularity" showing itself.

4.9   The basketball player Shaquille O'Neal makes about half of his free throws over an entire season. Use software to simulate 100 free throws shot independently by a player who has probability 0.5 of making each shot. (In most software, the key phrase to look for is "Bernoulli trials." This is the technical term for independent trials with "Yes/No" outcomes. Our outcomes here are "Hit" and "Miss.")

  (a) What percent of the 100 shots did he hit?

  (b) Examine the sequence of hits and misses. How long was the longest run of shots made? Of shots missed? (Sequences of random outcomes often show runs longer than our intuition thinks likely.)

4.10   Continue the exploration begun in Exercise 4.8. Software allows you to simulate many independent "Yes/No" trials more quickly if all you want to save is the count of "Yes" outcomes. The key word "Binomial" simulates n independent Bernoulli trials, each with probability p of a "Yes," and records just the count of "Yes" outcomes.

  (a) Simulate 100 draws of 20 women from the population in Exercise 4.8. Record the number who say "Agree" on each draw. What is the approximate probability that out of 20 women drawn at random at least 14 agree?

(b) Convert the counts who agree into percents of the 20 women in each trial who agree. Make a histogram of these 100 percents. Describe the shape, center, and spread of this distribution.

(c) Now simulate drawing 320 women. Do this 100 times and record the percent who agree on each of the 100 draws. Make a histogram of the percents and describe the shape, center, and spread of the distribution.

(d) In what ways are the distributions in parts (b) and (c) alike? In what ways do they differ? (Because regularity emerges in the long run, we expect the results of drawing 320 women to be less variable than the results of drawing 20 women.)

4.2 Probability Models

Earlier chapters gave mathematical models for linear relationships (in the form of the equation of a line) and for some distributions of data (in the form of normal density curves). Now we must give a mathematical description or model for randomness. To see how to proceed, think first about a very simple random phenomenon, tossing a coin once. When we toss a coin, we cannot know the outcome in advance. What do we know? We are willing to say that the outcome will be either heads or tails. Because the coin appears to be balanced, we believe that each of these outcomes has probability 1/2. This description of coin tossing has two parts:

- A list of possible outcomes
- A probability for each outcome

Such a description is the basis for all probability models. We will begin by describing the outcomes of a random phenomenon and then learn how to assign probabilities to the outcomes.

Sample spaces

A probability model first tells us what outcomes are possible.

> **Sample Space**
>
> The **sample space** S of a random phenomenon is the set of all possible outcomes.

The name "sample space" is natural in random sampling, where each possible outcome is a sample and the sample space contains all possible samples.

To specify S, we must state what constitutes an individual outcome and then state which outcomes can occur. We often have some freedom in defining the sample space, so the choice of S is a matter of convenience as well as correctness. The idea of a sample space, and the freedom we may have in specifying it, are best illustrated by examples.

EXAMPLE 4.3 | Toss a coin. There are only two possible outcomes, and the sample space is

$$S = \{heads, tails\}$$

or more briefly, $S = \{H, T\}$.

EXAMPLE 4.4 | Let your pencil point fall blindly into Table B of random digits and record the value of the digit you land on. The possible outcomes are

$$S = \{0, 1, 2, 3, 4, 5, 6, 7, 8, 9\}$$

EXAMPLE 4.5 | Toss a coin four times and record the results. That's a bit vague. To be exact, record the results of each of the four tosses in order. A typical outcome is then HTTH. Counting shows that there are 16 possible outcomes. The sample space S is the set of all 16 strings of four H's and T's.

Suppose that our only interest is the number of heads in four tosses. Now we can be exact in a simpler fashion. The random phenomenon is to toss a coin four times and count the number of heads. The sample space contains only five outcomes:

$$S = \{0, 1, 2, 3, 4\}$$

This example illustrates the importance of carefully specifying what constitutes an individual outcome.

Although these examples seem remote from the practice of statistics, the connection is surprisingly close. Suppose that in conducting an opinion poll you select four people at random from a large population and ask each if he or she favors reducing federal spending on low-interest student loans. The answers are "Yes" or "No." The possible outcomes—the sample space—are exactly as in Example 4.3 if we replace heads by "Yes" and tails by "No." Similarly, the possible outcomes of an SRS of 1500 people are the same in principle as the possible outcomes of tossing a coin 1500 times. One of the great advantages of mathematics is that the essential features of quite different phenomena can be described by the same mathematical model.

EXAMPLE 4.6 | Many computing systems have a function that will generate a random number between 0 and 1. The sample space is

$$S = \{all\ numbers\ between\ 0\ and\ 1\}$$

This S is a mathematical idealization. Any specific random number generator produces numbers with some limited number of decimal places so that, strictly speaking, not all numbers between 0 and 1 are possible outcomes. The entire interval from 0 to 1 is easier to think about. It also has the advantage of being a suitable sample space for different computers that produce random numbers with different numbers of significant digits.

Intuitive probability

A sample space S lists the possible outcomes of a random phenomenon. To complete a mathematical model for the random phenomenon, we must also give the probabilities with which these outcomes occur.

The true long-term relative frequency of any outcome—say, "exactly 2 heads in four tosses of a coin"—can be found only empirically, and then only approximately. How then can we describe probability mathematically? Rather than immediately attempt to give "correct" probabilities, let's confront the easier task of laying down rules that any assignment of probabilities must satisfy.* We need to assign probabilities not only to single outcomes but also to sets of outcomes.

Event

An **event** is an outcome or a set of outcomes of a random phenomenon. That is, an event is a subset of the sample space.

EXAMPLE 4.7   Take the sample space S for four tosses of a coin to be the 16 possible outcomes in the form HTHH. Then "exactly 2 heads" is an event. Call this event A. The event A expressed as a set of outcomes is

$$A = \{\text{HHTT, HTHT, HTTH, THHT, THTH, TTHH}\}$$

In a probability model, events have probabilities. What properties must any assignment of probabilities to events have? Here are some basic facts about any probability model. These facts follow from the idea of probability as "the long-run proportion of repetitions on which an event occurs."

1. **Any probability is a number between 0 and 1.** Any proportion is a number between 0 and 1, so any probability is also a number between 0 and 1. An event with probability 0 never occurs, and an event with probability 1 occurs on every trial. An event with probability 0.5 occurs in half the trials in the long run.

2. **All possible outcomes together must have probability 1.** Because some outcome must occur on every trial, the sum of the probabilities for all possible outcomes must be exactly 1.

3. **The probability that an event does not occur is 1 minus the probability that the event does occur.** If an event occurs in (say) 70% of all trials, it fails to occur in the other 30%. The probability that an event occurs and the probability that it does not occur always add to 100%, or 1.

---

*Probability is an old branch of mathematics, but it was only in 1933 that the Russian mathematician A. N. Kolmogorov (1903–1987) showed that the entire theory could be derived by starting with a few basic rules.

4. **If two events have no outcomes in common, the probability that one or the other occurs is the sum of their individual probabilities.** If one event occurs in 40% of all trials, a different event occurs in 25% of all trials, and the two can never occur together, then one or the other occurs in 65% of all trials because 40% + 25% = 65%.

Probability rules

Formal probability uses mathematical notation to state Facts 1 to 4 more concisely. We use capital letters near the beginning of the alphabet to denote events. If A is any event, we write its probability as $P(A)$. Here are our probability facts in formal language. As you apply these rules, remember that they are just another form of intuitively true facts about long-run proportions.

Probability Rules

Rule 1. The probability $P(A)$ of any event A satisfies $0 \leq P(A) \leq 1$.

Rule 2. If S is the sample space in a probability model, then $P(S) = 1$.

Rule 3. The **complement** of any event A is the event that A does not occur, written as A^c. The **complement rule** states that

$$P(A^c) = 1 - P(A)$$

Rule 4. Two events A and B are **disjoint** if they have no outcomes in common and so can never occur simultaneously. If A and B are disjoint,

$$P(A \text{ or } B) = P(A) + P(B)$$

This is the **addition rule for disjoint events.**

You may find it helpful to draw a picture to remind yourself of the meaning of complements and disjoint events. A picture like Figure 4.2 that shows the sample space S as a rectangular area and events as areas within S is called

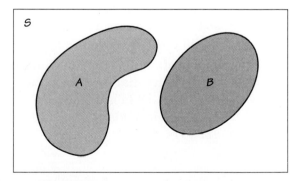

FIGURE 4.2   Venn diagram showing disjoint events A and B.

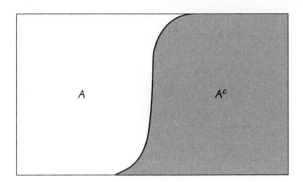

FIGURE 4.3   Venn diagram showing the complement A^c of an event A.

Venn diagram

a **Venn diagram.** The events A and B in Figure 4.2 are disjoint because they do not overlap. As Figure 4.3 shows, the complement A^c contains exactly the outcomes that are not in A.

EXAMPLE 4.8

Draw a woman aged 25 to 29 years old at random and record her marital status. "At random" means that we give every such woman the same chance to be the one we choose. That is, we choose an SRS of size 1. The probability of any marital status is just the proportion of all women aged 25 to 29 who have that status—if we drew many women, this is the proportion we would get. Here is the probability model:

Marital status	Never married	Married	Widowed	Divorced
Probability	0.353	0.574	0.002	0.071

Each probability is between 0 and 1. The probabilities add to 1, because these outcomes together make up the sample space S.

The probability that the woman we draw is not married is, by the complement rule,

$$P(\text{not married}) = 1 - P(\text{married})$$
$$= 1 - 0.574 = 0.426$$

That is, if 57.4% are married, then the remaining 42.6% are not married.

"Never married" and "divorced" are disjoint events, because no woman can be both never married and divorced. So the addition rule says that

$$P(\text{never married or divorced}) = P(\text{never married}) + P(\text{divorced})$$
$$= 0.353 + 0.071 = 0.424$$

That is, 42.4% of women in this age group are either never married or divorced.

Assigning probabilities: finite number of outcomes

The events "never married" and "divorced" in Example 4.8 contain just one outcome each. The addition rule applies to individual outcomes and provides a way to assign probabilities to events with more than one outcome: start with probabilities for individual outcomes and add to get probabilities for events. This idea works well when there are only a finite (fixed and limited) number of outcomes.

Probabilities in a Finite Sample Space

Assign a probability to each individual outcome. These probabilities must be numbers between 0 and 1 and must have sum 1.

The probability of any event is the sum of the probabilities of the outcomes making up the event.

EXAMPLE 4.9

A sociologist studies social mobility in England by recording the social class of a large sample of fathers and their sons. Social class is determined by such factors as education and occupation. The social classes are ordered from Class 1 (lowest) to Class 5 (highest). Here are the probabilities that the son of a lower-class (Class 1) father will end up in each social class:

Son's class	1	2	3	4	5
Probability	0.48	0.38	0.08	0.05	0.01

Consider the events

$$A = \{\text{son remains in Class 1}\}$$
$$B = \{\text{son reaches one of the two highest classes}\}$$

From the table of probabilities,

$$P(A) = 0.48$$
$$P(B) = 0.05 + 0.01 = 0.06$$

What is the probability that the son of a Class 1 father does *not* remain in Class 1? This is

$$P(A^c) = 1 - P(A)$$
$$= 1 - 0.48 = 0.52$$

The events A and B are disjoint, so the probability that the son of a lower-class father either remains in the lower class or reaches one of the two top classes is

$$P(A \text{ or } B) = P(A) + P(B)$$
$$= 0.48 + 0.06 = 0.54$$

Assigning probabilities: equally likely outcomes

Assigning correct probabilities to individual outcomes often requires long observation of the random phenomenon. In some special circumstances, however, we are willing to assume that individual outcomes are equally likely because of some balance in the phenomenon. Ordinary coins have a physical balance that should make heads and tails equally likely, for example, and the table of random digits comes from a deliberate randomization.

EXAMPLE 4.10 The successive digits in Table B (Example 4.4) were produced by a careful randomization that makes each entry equally likely to be any of the 10 candidates. Because the total probability must be 1, the probability of each of the 10 outcomes must be 1/10. That is, the assignment of probabilities to outcomes is

Outcome	0	1	2	3	4	5	6	7	8	9
Probability	0.1	0.1	0.1	0.1	0.1	0.1	0.1	0.1	0.1	0.1

The probability of any event is the sum of the probabilities of the outcomes making up the event. For example, the probability that an odd digit is chosen is

$$P(\text{odd outcome}) = P(1) + P(3) + P(5) + P(7) + P(9) = 0.5$$

Here are two events, with the outcomes that they contain:

$$A = \{\text{odd outcome}\} = \{1, 3, 5, 7, 9\}$$
$$B = \{\text{outcome less than or equal to 3}\} = \{0, 1, 2, 3\}$$

We saw that $P(A) = 0.5$. Because the event B contains 4 outcomes, $P(B) = 0.4$. The event $\{A \text{ or } B\}$ contains 7 outcomes,

$$\{A \text{ or } B\} = \{0, 1, 2, 3, 5, 7, 9\}$$

so it has probability 0.7. This is *not* the sum of $P(A)$ and $P(B)$, because A and B are *not* disjoint events. Outcomes 1 and 3 belong to both A and B.

In Example 4.10 all outcomes have the same probability. Because there are 10 equally likely outcomes, each must have probability 1/10. Because exactly 5 of the 10 equally likely outcomes are odd, the probability of an odd outcome is 5/10, or 0.5. In the special situation where all outcomes are equally likely, we have a simpler rule for assigning probabilities to events.

Equally Likely Outcomes

If a random phenomenon has k possible outcomes, all equally likely, then each individual outcome has probability $1/k$. The probability of any event A is

$$P(A) = \frac{\text{count of outcomes in } A}{\text{count of outcomes in } S}$$
$$= \frac{\text{count of outcomes in } A}{k}$$

Most random phenomena do not have equally likely outcomes, so the general rule for finite sample spaces is more important than the special rule for equally likely outcomes.

Independence and the multiplication rule

Rule 4, the addition rule for disjoint events, describes the probability that *one or the other* of two events A and B will occur in the special situation when A and B cannot occur together because they are disjoint. Our final rule describes the probability that *both* events A and B occur, again only in a special situation. More general rules appear in Section 4.5, but in our study of statistics we will need only the rules that apply to special situations.

Suppose that you toss a balanced coin twice. You are counting heads, so two events of interest are

$$A = \text{first toss is a head}$$
$$B = \text{second toss is a head}$$

The events A and B are not disjoint. They occur together whenever both tosses give heads. We want to compute the probability of the event {A and B} that *both* tosses are heads. The Venn diagram in Figure 4.4 illustrates the event {A and B} as the overlapping area that is common to both A and B.

The coin tossing of Buffon, Pearson, and Kerrich described at the beginning of this chapter makes us willing to assign probability 1/2 to a head when we toss a coin. So

$$P(A) = 0.5$$
$$P(B) = 0.5$$

What is $P(A$ and $B)$? Our common sense says that it is 1/4. The first coin will give a head half the time and then the second will give a head on half of those trials, so both coins will give heads on $1/2 \times 1/2 = 1/4$ of all trials in the long run. This reasoning assumes that the second coin still has probability 1/2 of a head after the first has given a head. This is true—we can verify it by tossing two coins many times and observing the proportion of heads on the second toss after the first toss has produced a head. We say that the events "head on the first toss" and "head on the second toss" are *independent*. Here is our final probability rule.

The Multiplication Rule for Independent Events

Rule 5. Two events A and B are **independent** if knowing that one occurs does not change the probability that the other occurs. If A and B are independent,

$$P(A \text{ and } B) = P(A)P(B)$$

This is the **multiplication rule for independent events.**

Our definition of independence is rather informal. We will make this informal idea precise in Section 4.5. In practice, though, we rarely need a precise definition of independence, because independence is usually *assumed* as part of a probability model when we want to describe random phenomena that seem to be physically unrelated to each other.

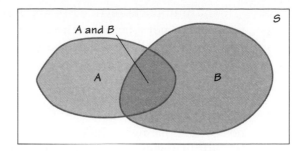

FIGURE 4.4 Venn diagram showing the event {A and B}. This event consists of outcomes common to A and B.

EXAMPLE 4.11

Because a coin has no memory and most coin-tossers cannot influence the fall of the coin, it is safe to assume that successive coin tosses are independent. For a balanced coin this means that after we see the outcome of the first toss, we still assign probability 1/2 to heads on the second toss.

On the other hand, the colors of successive cards dealt from the same deck are not independent. A standard 52-card deck contains 26 red and 26 black cards. For the first card dealt from a shuffled deck, the probability of a red card is $26/52 = 0.50$ (equally likely outcomes). Once we see that the first card is red, we know that there are only 25 reds among the remaining 51 cards. The probability that the second card is red is therefore only $25/51 = 0.49$. Knowing the outcome of the first deal changes the probabilities for the second.

If a doctor measures your blood pressure twice, it is reasonable to assume that the two results are independent because the first result does not influence the instrument that makes the second reading. But if you take an IQ test or other mental test twice in succession, the two test scores are not independent. The learning that occurs on the first attempt influences your second attempt.

When independence is part of a probability model, the multiplication rule applies. Here is an example.

EXAMPLE 4.12

Gregor Mendel used garden peas in some of the experiments that revealed that inheritance operates randomly. The seed color of Mendel's peas can be either green or yellow. Two parent plants are "crossed" (one pollinates the other) to produce seeds. Each parent plant carries two genes for seed color, and each of these genes has probability 1/2 of being passed to a seed. The two genes that the seed receives, one from each parent, determine its color. The parents contribute their genes independently of each other.

Suppose that both parents carry the G and the Y genes. The seed will be green if both parents contribute a G gene; otherwise it will be yellow. If M is the event that the male contributes a G gene and F is the event that the female contributes a G gene, then the probability of a green seed is

$$P(M \text{ and } F) = P(M)P(F)$$
$$= (0.5)(0.5) = 0.25$$

In the long run, 1/4 of all seeds produced by crossing these plants will be green.

The multiplication rule $P(A \text{ and } B) = P(A)P(B)$ holds if A and B are independent but not otherwise. The addition rule $P(A \text{ or } B) = P(A)+P(B)$ holds if A and B are disjoint but not otherwise. Resist the temptation to use these simple formulas when the circumstances that justify them are not present. You must also be certain not to confuse disjointness and independence. If A and B are disjoint, then the fact that A occurs tells us that B cannot occur—look again at Figure 4.2. So disjoint events are not independent. Unlike disjointness or complements, independence cannot be pictured by a Venn diagram, because it involves the probabilities of the events rather than just the outcomes that make up the events.

Applying the probability rules

If two events A and B are independent, then their complements A^c and B^c are also independent and A^c is independent of B. Suppose, for example, that 75% of all registered voters in a suburban district are Republicans. If an opinion poll interviews two voters chosen independently, the probability that the first is a Republican and the second is not a Republican is $(0.75)(0.25) = 0.1875$. The multiplication rule also extends to collections of more than two events, provided that all are independent. Independence of events A, B, and C means that no information about any one or any two can change the probability of the remaining events. The formal definition is a bit messy. Fortunately, independence is usually assumed in setting up a probability model. We can then use the multiplication rule freely, as in this example.

EXAMPLE 4.13 | A transatlantic telephone cable contains repeaters at regular intervals to amplify the signal. If a repeater fails, it must be replaced by fishing the cable to the surface at great expense. Each repeater has probability 0.999 of functioning without failure for 10 years. Repeaters fail independently of each other. (This assumption means that there are no "common causes" such as earthquakes that would affect several repeaters at once.) Denote by A_i the event that the ith repeater operates successfully for 10 years.

The probability that two repeaters both last 10 years is

$$P(A_1 \text{ and } A_2) = P(A_1)P(A_2)$$
$$= 0.999 \times 0.999 = 0.998$$

For a cable with 10 repeaters the probability of no failures in 10 years is

$$P(A_1 \text{ and } A_2 \text{ and } \ldots \text{ and } A_{10}) = P(A_1)P(A_2)\cdots P(A_{10})$$
$$= 0.999 \times 0.999 \times \cdots \times 0.999$$
$$= 0.999^{10} = 0.990$$

Cables with 2 or 10 repeaters are quite reliable. Unfortunately, a transatlantic cable has 300 repeaters. The probability that all 300 work for 10 years is

$$P(A_1 \text{ and } A_2 \text{ and } \ldots \text{ and } A_{300}) = 0.999^{300} = 0.741$$

There is therefore about one chance in four that the cable will have to be fished up for replacement of a repeater sometime during the next 10 years. Repeaters are in fact designed to be much more reliable than 0.999 in 10 years. Some transatlantic cables have served for more than 20 years with no failures.

By combining the rules we have learned, we can compute probabilities for rather complex events. Here is an example.

EXAMPLE 4.14

A diagnostic test for the presence of the AIDS virus has probability 0.005 of producing a false positive. That is, when a person free of the AIDS virus is tested, the test has probability 0.005 of falsely indicating that the virus is present. If the 140 employees of a medical clinic are tested and all 140 are free of AIDS, what is the probability that at least one false positive will occur?

It is reasonable to assume as part of the probability model that the test results for different individuals are independent. The probability that the test is positive for a single person is 0.005, so the probability of a negative result is $1 - 0.005 = 0.995$ by the complement rule. The probability of at least one false positive among the 140 people tested is therefore

$$
\begin{aligned}
P(\text{at least one positive}) &= 1 - P(\text{no positives}) \\
&= 1 - P(\text{140 negatives}) \\
&= 1 - 0.995^{140} \\
&= 1 - 0.496 = 0.504
\end{aligned}
$$

The probability is greater than 1/2 that at least one of the 140 people will test positive for AIDS, even though no one has the virus.

SUMMARY

A **probability model** for a random phenomenon consists of a sample space S and an assignment of probabilities P.

The **sample space** S is the set of all possible outcomes of the random phenomenon. Sets of outcomes are called **events.** P assigns a number $P(A)$ to an event A as its probability.

The **complement** A^c of an event A consists of exactly the outcomes that are not in A. Events A and B are **disjoint** if they have no outcomes in common. Events A and B are **independent** if knowing that one event occurs does not change the probability we would assign to the other event.

Any assignment of probability must obey the rules that state the basic properties of probability:

1. $0 \le P(A) \le 1$ for any event A.
2. $P(S) = 1$.
3. **Complement rule:** For any event A, $P(A^c) = 1 - P(A)$.
4. **Addition rule:** If events A and B are **disjoint,** then $P(A \text{ or } B) = P(A) + P(B)$.
5. **Multiplication rule:** If events A and B are **independent,** then $P(A \text{ and } B) = P(A)P(B)$.

SECTION 4.2 EXERCISES

4.11   Probability is a measure of how likely an event is to occur. Match one of the probabilities that follow with each statement about an event. (The probability is usually a much more exact measure of likelihood than is the verbal statement.)

$$0, 0.01, 0.3, 0.6, 0.99, 1$$

(a) This event is impossible. It can never occur.

(b) This event is certain. It will occur on every trial of the random phenomenon.

(c) This event is very unlikely, but it will occur once in a while in a long sequence of trials.

(d) This event will occur more often than not.

4.12   In each of the following situations, describe a sample space S for the random phenomenon. In some cases, you have some freedom in your choice of S.

(a) A seed is planted in the ground. It either germinates or fails to grow.

(b) A patient with a usually fatal form of cancer is given a new treatment. The response variable is the length of time that the patient lives after treatment.

(c) A student enrolls in a statistics course and at the end of the semester receives a letter grade.

(d) A basketball player shoots two free throws.

(e) A year after knee surgery, a patient is asked to rate the amount of pain in the knee. A seven-point scale is used, with 1 corresponding to no pain and 7 corresponding to extreme discomfort.

4.13   In each of the following situations, describe a sample space S for the random phenomenon. In some cases you have some freedom in specifying S, especially in setting the largest and smallest value in S.

(a) Choose a student in your class at random. Ask how much time that student spent studying during the past 24 hours.

(b) The Physicians' Health Study asked 11,000 physicians to take an aspirin every other day and observed how many of them had a heart attack in a five-year period.

(c) In a test of a new package design, you drop a carton of a dozen eggs from a height of 1 foot and count the number of broken eggs.

(d) Choose a student in your class at random. Ask how much cash that student is carrying.

(e) A nutrition researcher feeds a new diet to a young male white rat. The response variable is the weight (in grams) that the rat gains in 8 weeks.

4.14   Buy a hot dog and record how many calories it has. Give a reasonable sample space S for your possible results. (Table 1.8 on page 40 contains some typical values to guide you.)

4.15 All human blood can be typed as one of O, A, B, or AB, but the distribution of the types varies a bit with race. Here is the distribution of the blood type of a randomly chosen black American:

Blood type	O	A	B	AB
Probability	0.49	0.27	0.20	?

(a) What is the probability of type AB blood? Why?

(b) Maria has type B blood. She can safely receive blood transfusions from people with blood types O and B. What is the probability that a randomly chosen black American can donate blood to Maria?

4.16 If you draw an M&M candy at random from a bag of the candies, the candy you draw will have one of six colors. The probability of drawing each color depends on the proportion of each color among all candies made.

(a) The table below gives the probability of each color for a randomly chosen plain M&M:

Color	Brown	Red	Yellow	Green	Orange	Blue
Probability	0.3	0.2	0.2	0.1	0.1	?

What must be the probability of drawing a blue candy?

(b) The probabilities for peanut M&M's are a bit different. Here they are:

Color	Brown	Red	Yellow	Green	Orange	Blue
Probability	0.2	0.1	0.2	0.1	0.1	?

What is the probability that a peanut M&M chosen at random is blue?

(c) What is the probability that a plain M&M is any of red, yellow, or orange? What is the probability that a peanut M&M has one of these colors?

4.17 Figure 4.5 displays several assignments of probabilities to the six faces of a die. We can learn which assignment is actually *accurate* for a particular die only by rolling the die many times. However, some of the assignments are not *legitimate* assignments of probability. That is, they do not obey the rules. Which are legitimate and which are not? In the case of the illegitimate models, explain what is wrong.

4.18 In each of the following situations, state whether or not the given assignment of probabilities to individual outcomes is legitimate, that is, satisfies the rules of probability. If not, give specific reasons for your answer.

(a) When a coin is spun, $P(H) = 0.55$ and $P(T) = 0.45$.

(b) When two coins are tossed, $P(HH) = 0.4$, $P(HT) = 0.4$, $P(TH) = 0.4$, and $P(TT) = 0.4$.

Outcome	Model 1	Model 2	Model 3	Model 4
⚀	$\frac{1}{3}$	$\frac{1}{6}$	$\frac{1}{7}$	$\frac{1}{3}$
⚁	0	$\frac{1}{6}$	$\frac{1}{7}$	$\frac{1}{3}$
⚂	$\frac{1}{6}$	$\frac{1}{6}$	$\frac{1}{7}$	$-\frac{1}{6}$
⚃	0	$\frac{1}{6}$	$\frac{1}{7}$	$-\frac{1}{6}$
⚄	$\frac{1}{6}$	$\frac{1}{6}$	$\frac{1}{7}$	$\frac{1}{3}$
⚅	$\frac{1}{3}$	$\frac{1}{6}$	$\frac{1}{7}$	$\frac{1}{3}$

FIGURE 4.5   Four assignments of probabilities to the six faces of a die, for Exercise 4.17.

(c) Plain M&M's have not always had the mixture of colors given in Exercise 4.16. In the past there were no red candies and no blue candies. Tan had probability 0.10 and the other four colors had the same probabilities that are given in Exercise 4.16.

4.19   Las Vegas Zeke, when asked to predict the Atlantic Coast Conference basketball champion, follows the modern practice of giving probabilistic predictions. He says, "North Carolina's probability of winning is twice Duke's. North Carolina State and Virginia each have probability 0.1 of winning, but Duke's probability is three times that. Nobody else has a chance." Has Zeke given a legitimate assignment of probabilities to the eight teams in the conference? Explain your answer.

4.20   A sociologist studying social mobility in Denmark finds that the probability that the son of a lower-class father remains in the lower class is 0.46. What is the probability that the son moves to one of the higher classes?

4.21   Government data assign a single cause for each death that occurs in the United States. The data show that the probability is 0.45 that a randomly chosen death was due to cardiovascular (mainly heart) disease, and 0.22 that it was due to cancer. What is the probability that a death was due either to cardiovascular disease or to cancer? What is the probability that the death was due to some other cause?

4.22   Choose an acre of land in Canada at random. The probability is 0.35 that it is forest and 0.03 that it is pasture.

(a) What is the probability that the acre chosen is not forested?

(b) What is the probability that it is either forest or pasture?

(c) What is the probability that a randomly chosen acre in Canada is something other than forest or pasture?

4.23 Select a first-year college student at random and ask what his or her academic rank was in high school. Here are the probabilities, based on proportions from a large sample survey of first-year students:

Rank	Top 20%	Second 20%	Third 20%	Fourth 20%	Lowest 20%
Probability	0.41	0.23	0.29	0.06	0.01

(a) What is the sum of these probabilities? Why do you expect the sum to have this value?

(b) What is the probability that a randomly chosen first-year college student was not in the top 20% of his or her high school class?

(c) What is the probability that a first-year student was in the top 40% in high school?

(d) Now choose two first-year college students at random. What is the probability that both were in the top 20% of their high school classes?

4.24 Choose an American farm at random and measure its size in acres. Here are the probabilities that the farm chosen falls in several acreage categories:

Acres	<10	10–49	50–99	100–179	180–499	500–999	1000–1999	≥2000
Probability	0.09	0.20	0.15	0.16	0.22	0.09	0.05	0.04

Let A be the event that the farm is less than 50 acres in size, and let B be the event that it is 500 acres or more.

(a) Find $P(A)$ and $P(B)$.

(b) Describe A^c in words and find $P(A^c)$ by the complement rule.

(c) Describe {A or B} in words and find its probability by the addition rule.

4.25 Choose an American worker at random and classify his or her occupation into one of the following classes. These classes are used in government employment data.

A   Managerial and professional
B   Technical, sales, administrative support
C   Service occupations
D   Precision production, craft, and repair
E   Operators, fabricators, and laborers
F   Farming, forestry, and fishing

The table below gives the probabilities that a randomly chosen worker falls into each of 12 sex-by-occupation classes.

Class	A	B	C	D	E	F
Male	0.14	0.11	0.06	0.11	0.12	0.03
Female	0.09	0.20	0.08	0.01	0.04	0.01

(a) Verify that this is a legitimate assignment of probabilities to these outcomes.

(b) What is the probability that the worker is female?

(c) What is the probability that the worker is not engaged in farming, forestry, or fishing?

(d) Classes D and E include most mechanical and factory jobs. What is the probability that the worker holds a job in one of these classes?

(e) What is the probability that the worker does not hold a job in Classes D or E?

4.26   A roulette wheel has 38 slots, numbered 0, 00, and 1 to 36. The slots 0 and 00 are colored green, 18 of the others are red, and 18 are black. The dealer spins the wheel and at the same time rolls a small ball along the wheel in the opposite direction. The wheel is carefully balanced so that the ball is equally likely to land in any slot when the wheel slows. Gamblers can bet on various combinations of numbers and colors.

(a) What is the probability that the ball will land in any one slot?

(b) If you bet on "red," you win if the ball lands in a red slot. What is the probability of winning?

(c) The slot numbers are laid out on a board on which gamblers place their bets. One column of numbers on the board contains all multiples of 3, that is, 3, 6, 9, . . . , 36. You place a "column bet" that wins if any of these numbers comes up. What is your probability of winning?

4.27   Abby, Deborah, Julie, Sam, and Roberto work in a firm's public relations office. Their employer must choose two of them to attend a conference in Paris. To avoid unfairness, the choice will be made by drawing two names from a hat. (This is an SRS of size 2.)

(a) Write down all possible choices of two of the five names. This is the sample space.

(b) The random drawing makes all choices equally likely. What is the probability of each choice?

(c) What is the probability that Julie is chosen?

(d) What is the probability that neither of the two men (Sam and Roberto) is chosen?

4.28   A general can plan a campaign to fight one major battle or three small battles. He believes that he has probability 0.6 of winning the large battle and probability 0.8 of winning each of the small battles. Victories or defeats in the small battles are independent. The general must win either the large battle or all three small battles to win the campaign. Which strategy should he choose?

4.29   An automobile manufacturer buys computer chips from a supplier. The supplier sends a shipment containing 5% defective chips. Each chip chosen from this shipment has probability 0.05 of being defective, and each automobile uses 12 chips selected independently. What is the probability that all 12 chips in a car will work properly?

4.30 Government data show that 27% of the civilian labor force have at least 4 years of college and that 16% of the labor force work as laborers or operators of machines or vehicles. Can you conclude that because $(0.27)(0.16) = 0.043$, about 4% of the labor force are college-educated laborers or operators? Explain your answer.

4.31 Choose at random a U.S. resident at least 25 years of age. We are interested in the events

$$A = \{\text{The person chosen completed 4 years of college}\}$$
$$B = \{\text{The person chosen is 55 years old or older}\}$$

Government data recorded in Table 2.14 on page 194 allow us to assign probabilities to these events.

(a) Explain why $P(A) = 0.230$.

(b) Find $P(B)$.

(c) Find the probability that the person chosen is at least 55 years old *and* has 4 years of college education, $P(A \text{ and } B)$. Are the events A and B independent?

4.32 A string of Christmas lights contains 20 lights. The lights are wired in series, so that if any light fails the whole string will go dark. Each light has probability 0.02 of failing during a 3-year period. The lights fail independently of each other. What is the probability that the string of lights will remain bright for 3 years?

4.33 A six-sided die has four green and two red faces and is balanced so that each face is equally likely to come up. The die will be rolled several times. You must choose one of the following three sequences of colors; you will win $25 if the first rolls of the die give the sequence that you have chosen.

> RGRRR
> RGRRRG
> GRRRRR

Which sequence do you choose? Explain your choice. (In a psychological experiment, 63% of 260 students who had not studied probability chose the second sequence. This is evidence that our intuitive understanding of probability is not very accurate. This and similar experiments are reported by A. Tversky and D. Kahneman, "Extensional versus intuitive reasoning: the conjunction fallacy in probability judgment," *Psychological Review*, 90 (1983), pp. 293–315.)

4.34 The gene for albinism in humans is recessive. That is, carriers of this gene have probability 1/2 of passing it to a child, and the child is albino only if both parents pass the albinism gene. Parents pass their genes independently of each other. If both parents carry the albinism gene, what is the probability that their first child is albino? If they have two children (who inherit independently of each other), what is the probability that both are albino? That neither is albino?

4.35   The "random walk" theory of securities prices holds that price movements in disjoint time periods are independent of each other. Suppose that we record only whether the price is up or down each year, and that the probability that our portfolio rises in price in any one year is 0.65. (This probability is approximately correct for a portfolio containing equal dollar amounts of all common stocks listed on the New York Stock Exchange.)

(a) What is the probability that our portfolio goes up for three consecutive years?

(b) If you know that the portfolio has risen in price 2 years in a row, what probability do you assign to the event that it will go down next year?

(c) What is the probability that the portfolio's value moves in the same direction in both of the next 2 years?

4.36   The type of medical care a patient receives may vary with the age of the patient. A large study of women who had a breast lump investigated whether or not each woman received a mammogram and a biopsy when the lump was discovered. Here are some probabilities estimated by the study. The entries in the table are the probabilities that *both* of two events occur; for example, 0.321 is the probability that a patient is under 65 years of age *and* the tests were done. The four probabilities in the table have sum 1 because the table lists all possible outcomes.

	Tests done?	
	Yes	No
Age under 65	0.321	0.124
Age 65 or over	0.365	0.190

(a) What is the probability that a patient in this study is under 65? That a patient is 65 or over?

(b) What is the probability that the tests were done for a patient? That they were not done?

(c) Are the events A = {the patient was 65 or older} and B = {the tests were done} independent? Were the tests omitted on older patients more or less frequently than would be the case if testing were independent of age?

4.3   Random Variables

Sample spaces need not consist of numbers. When we toss four coins, we can record the outcome as a string of heads and tails, such as HTTH. In statistics, however, we are most often interested in numerical outcomes such as the count of heads in the four tosses. It is convenient to use a shorthand notation: Let X be the number of heads. If our outcome is HTTH, then $X = 2$. If the next outcome is TTTH, the value of X changes to $X = 1$. The possible values of X are 0, 1, 2, 3, and 4. Tossing a coin four times will give X one of these possible values. Tossing four more times will give X another and probably different value. We call X a *random variable* because its values vary when the coin tossing is repeated.

Random Variable

A **random variable** is a variable whose value is a numerical outcome of a random phenomenon.

We usually denote random variables by capital letters near the end of the alphabet, such as X or Y. Of course, the random variables of greatest interest to us are outcomes such as the mean $\overline{x}$ of a random sample, for which we will keep the familiar notation.[2] As we progress from general rules of probability toward statistical inference, we will concentrate on random variables. When a random variable X describes a random phenomenon, the sample space S just lists the possible values of the random variable. We usually do not mention S separately. There remains the second part of any probability model, the assignment of probabilities to events. In this section, we will learn two ways of assigning probabilities to the values of a random variable. The two types of probability models that result will dominate our application of probability to statistical inference.

Discrete random variables

We have learned several rules of probability but only one method of assigning probabilities: state the probabilities of the individual outcomes and assign probabilities to events by summing over the outcomes. The outcome probabilities must be between 0 and 1 and have sum 1. When the outcomes are numerical, they are values of a random variable. We will now attach a name to random variables having probability assigned in this way.[3]

Discrete Random Variable

A **discrete random variable** X has a finite number of possible values. The **probability distribution** of X lists the values and their probabilities:

Value of X	x_1	x_2	x_3	$\cdots$	x_k
Probability	p_1	p_2	p_3	$\cdots$	p_k

The probabilities p_i must satisfy two requirements:

1. Every probability p_i is a number between 0 and 1.
2. $p_1 + p_2 + \cdots + p_k = 1$.

Find the probability of any event by adding the probabilities p_i of the particular values x_i that make up the event.

EXAMPLE 4.15

The instructor of a large class gives 15% each of A's and D's, 30% each of B's and C's, and 10% F's. Choose a student at random from this class. To "choose at random" means to give every student the same chance to be chosen. The student's grade on a four-point scale (with A = 4) is a random variable X.

The value of X changes when we repeatedly choose students at random, but it is always one of 0, 1, 2, 3, or 4. Here is the distribution of X:

Grade	0	1	2	3	4
Probability	0.10	0.15	0.30	0.30	0.15

The probability that the student got a B or better is the sum of the probabilities of an A and a B:

$$P(\text{grade is 3 or 4}) = P(X = 3) + P(X = 4)$$
$$= 0.30 + 0.15 = 0.45$$

probability histogram

We can use a **probability histogram** to picture the probability distribution of a discrete random variable. Figure 4.6 displays two probability histograms. Figure 4.6(a) is the distribution of an entry in the table of random digits with probability distributed uniformly over all 10 digits. Part (b) is the distribution of grades in Example 4.15. The horizontal scale shows the possible values of X, and the height of each bar is the probability for the value at its base. A probability histogram is in effect a relative frequency histogram for a very large number of trials.

EXAMPLE 4.16

What is the probability distribution of the discrete random variable X that counts the number of heads in four tosses of a coin? We can derive this distribution if we make two reasonable assumptions:

■ The coin is balanced, so each toss is equally likely to give H or T.
■ The coin has no memory, so tosses are independent.

The outcome of four tosses is a sequence of heads and tails such as HTTH. There are 16 possible outcomes in all. Figure 4.7 lists these outcomes along with the value of X for each outcome. The multiplication rule for independent events tells us that, for example,

$$P(\text{HTTH}) = \frac{1}{2} \times \frac{1}{2} \times \frac{1}{2} \times \frac{1}{2} = \frac{1}{16}$$

Each of the 16 possible outcomes similarly has probability 1/16. That is, these outcomes are equally likely.

The number of heads X has possible values 0, 1, 2, 3, and 4. These values are *not* equally likely. As Figure 4.7 shows, there is only one way that $X = 0$ can occur: namely, when the outcome is TTTT. So $P(X = 0) = 1/16$. But the event $\{X = 2\}$ can occur in six different ways, so that

$$P(X = 2) = \frac{\text{count of ways } X = 2 \text{ can occur}}{16}$$
$$= \frac{6}{16}$$

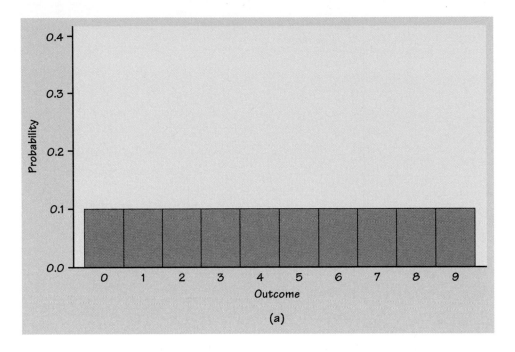

(a)

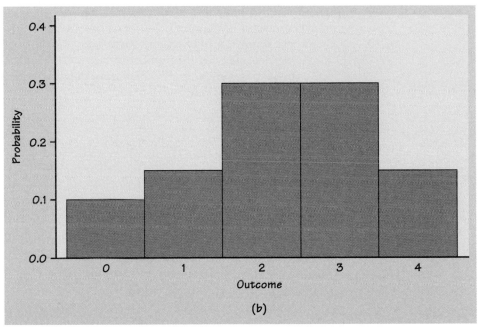

(b)

FIGURE 4.6   Probability histograms for the distributions of discrete random variables: (a) generating a random digit and (b) grades in a large course (Example 4.15).

We can find the probability of each value of X from Figure 4.7 in the same way. Here is the result:

$$P(X = 0) = \frac{1}{16} = 0.0625$$

$$P(X = 1) = \frac{4}{16} = 0.25$$

$$P(X = 2) = \frac{6}{16} = 0.375$$

$$P(X = 3) = \frac{4}{16} = 0.25$$

$$P(X = 4) = \frac{1}{16} = 0.0625$$

These probabilities have sum 1, so this is a legitimate probability distribution. In table form the distribution is

Number of heads	0	1	2	3	4
Probability	0.0625	0.25	0.375	0.25	0.0625

Figure 4.8 is a probability histogram for this distribution. The probability distribution is exactly symmetric. It is an idealization of the relative frequency distribution of the number of heads after many tosses of four coins, which would be nearly symmetric but is unlikely to be exactly symmetric.

Any event involving the number of heads observed can be expressed in terms of X, and its probability can be found from the distribution of X. For example, the probability of tossing at least two heads is

$$P(X \geq 2) = 0.375 + 0.25 + 0.0625 = 0.6875$$

The probability of at least one head is most simply found by use of the complement rule:

$$P(X \geq 1) = 1 - P(X = 0)$$
$$= 1 - 0.0625 = 0.9375$$

Recall that tossing a coin n times is similar to choosing an SRS of size n from a large population and asking a yes-or-no question. We will extend the results of Example 4.16 when we return to sampling distributions in the next chapter.

```
                                    HTTH
                                    HTHT
                     HTTT           THTH           HHHT
                     THTT           HHTT           HHTH
                     TTHT           THHT           HTHH
         TTTT        TTTH           TTHH           THHH           HHHH

         X = 0       X = 1          X = 2          X = 3          X = 4
```

FIGURE 4.7 Possible outcomes in four tosses of a coin. The random variable X is the number of heads.

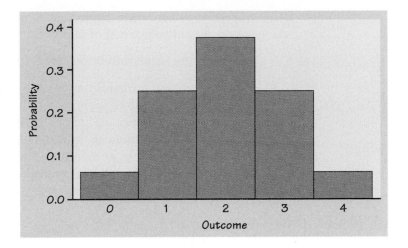

FIGURE 4.8   Probability histogram of the number of heads in four tosses of a coin.

Continuous random variables

When we use the table of random digits to select a digit between 0 and 9, the result is a discrete random variable. The probability model assigns probability 1/10 to each of the 10 possible outcomes, as Figure 4.6(a) shows. Suppose that we want to choose a number at random between 0 and 1, allowing *any* number between 0 and 1 as the outcome. Software random number generators will do this. You can visualize such a random number by thinking of a spinner (Figure 4.9) that turns freely on its axis and slowly comes to a stop. The pointer can

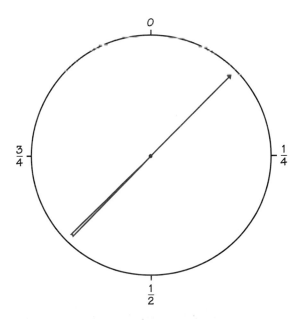

FIGURE 4.9   A spinner that generates a random number between 0 and 1.

come to rest anywhere on a circle that is marked from 0 to 1. The sample space is now an entire interval of numbers:

$$S = \{\text{all numbers } x \text{ such that } 0 \leq x \leq 1\}$$

How can we assign probabilities to such events as $\{0.3 \leq x \leq 0.7\}$? As in the case of selecting a random digit, we would like all possible outcomes to be equally likely. But we cannot assign probabilities to each individual value of x and then sum, because there are infinitely many possible values. Instead we use a new way of assigning probabilities directly to events—as *areas under a density curve*. Any density curve has area exactly 1 underneath it, corresponding to total probability 1.

EXAMPLE 4.17

uniform distribution

The random number generator will spread its output uniformly across the entire interval from 0 to 1 as we allow it to generate a long sequence of numbers. The results of many trials are represented by the density curve of a **uniform distribution** (Figure 4.10). This density curve has height 1 over the interval from 0 to 1. The area under the density curve is 1, and the probability of any event is the area under the density curve and above the event in question.

As Figure 4.10(a) illustrates, the probability that the random number generator produces a number X between 0.3 and 0.7 is

$$P(0.3 \leq X \leq 0.7) = 0.4$$

because the area under the density curve and above the interval from 0.3 to 0.7 is 0.4. The height of the density curve is 1, and the area of a rectangle is the product of height and length, so the probability of any interval of outcomes is just the length of the interval.

Similarly,

$$P(X \leq 0.5) = 0.5$$
$$P(X > 0.8) = 0.2$$
$$P(X \leq 0.5 \text{ or } X > 0.8) = 0.7$$

Notice that the last event consists of two nonoverlapping intervals, so the total area above the event is found by adding two areas, as illustrated by Figure 4.10(b). This assignment of probabilities obeys all of our rules for probability.

Probability as area under a density curve is a second important way of assigning probabilities to events. Figure 4.11 illustrates this idea in general form. We call X in Example 4.17 a *continuous random variable* because its values are not isolated numbers but an entire interval of numbers.

Continuous Random Variable

A **continuous random variable** X takes all values in an interval of numbers. The **probability distribution** of X is described by a density curve. The probability of any event is the area under the density curve and above the values of X that make up the event.

The probability model for a continuous random variable assigns probabilities to intervals of outcomes rather than to individual outcomes. In fact,

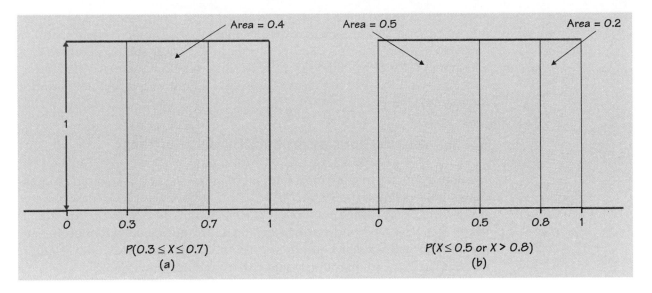

FIGURE 4.10 Assigning probabilities for generating a random number between 0 and 1. The probability of any interval of numbers is the area above the interval and under the curve.

all continuous probability distributions assign probability 0 to every individual outcome. Only intervals of values have positive probability. To see that this is true, consider a specific outcome such as $P(X = 0.8)$ in Example 4.17. The probability of any interval is the same as its length. The point 0.8 has no length so its probability is 0. Although this fact may seem odd at first glance, it does make intuitive as well as mathematical sense. The random number generator produces a number between 0.79 and 0.81 with probability 0.02. An outcome between 0.799 and 0.801 has probability 0.002, and a result

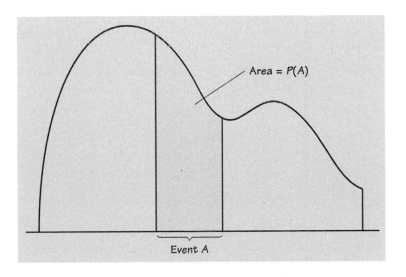

FIGURE 4.11 The probability distribution of a continuous random variable assigns probabilities as areas under a density curve.

between 0.7999 and 0.8001 has probability 0.0002. Continuing to home in on 0.8, we can see why an outcome *exactly* equal to 0.8 should have probability 0. Because there is no probability exactly at $X = 0.8$, the two events $\{X > 0.8\}$ and $\{X \geq 0.8\}$ have the same probability. We can ignore the distinction between $>$ and $\geq$ when finding probabilities for continuous (but not discrete) random variables.

Normal distributions as probability distributions

The density curves that are most familiar to us are the normal curves. (We discussed normal curves in Section 1.3.) Because any density curve describes an assignment of probabilities, *normal distributions are probability distributions*. Recall that $N(\mu, \sigma)$ is our shorthand notation for the normal distribution having mean μ and standard deviation σ. In the language of random variables, if X has the $N(\mu, \sigma)$ distribution, then the standardized variable

$$Z = \frac{X - \mu}{\sigma}$$

is a standard normal random variable having the distribution $N(0, 1)$.

EXAMPLE 4.18

An opinion poll asks an SRS of 1500 American adults what they consider to be the most serious problem facing our schools. Suppose that if we could ask all adults this question, 30% would say "drugs." The proportion $p = 0.3$ is a population parameter. The proportion $\hat{p}$ of the sample who answer "drugs" is a statistic used to estimate p. We will see in the next chapter that $\hat{p}$ is a random variable that has approximately the $N(0.3, 0.0118)$ distribution. The mean 0.3 of this distribution is the same as the population parameter because $\hat{p}$ is an unbiased estimate of p. The standard deviation is controlled mainly by the sample size, which is 1500 in this case.

What is the probability that the poll result differs from the truth about the population by more than two percentage points? Figure 4.12 shows this probability as an area under a normal density curve. By the addition rule for disjoint events, the desired probability is

$$P(\hat{p} < 0.28 \text{ or } \hat{p} > 0.32) = P(\hat{p} < 0.28) + P(\hat{p} > 0.32)$$

You can find the two individual probabilities from software or by standardizing and using Table A.

$$P(\hat{p} < 0.28) = P\left(\frac{\hat{p} - 0.3}{0.0118} < \frac{0.28 - 0.3}{0.0118}\right)$$
$$= P(Z < -1.69) = 0.0455$$
$$P(\hat{p} > 0.32) = P\left(\frac{\hat{p} - 0.3}{0.0118} > \frac{0.32 - 0.3}{0.0118}\right)$$
$$= P(Z > 1.69) = 0.0455$$

Therefore,

$$P(\hat{p} < 0.28 \text{ or } \hat{p} > 0.32) = 0.0455 + 0.0455 = 0.0910$$

The probability that the sample result will miss the truth by more than two percentage points is 0.091. The arrangement of this calculation is familiar from our earlier work with normal distributions. Only the language of probability is new.

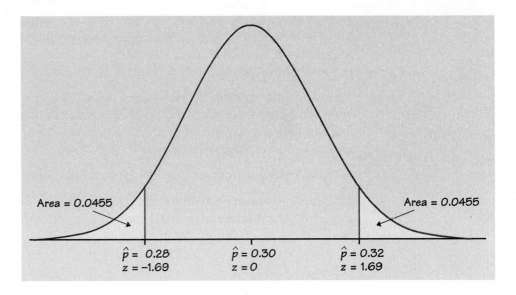

FIGURE 4.12 Probability in Example 4.18 as area under a normal density curve.

We could also do the calculation by first finding the probability of the complement:

$$P(0.28 \leq \hat{p} \leq 0.32) = P\left(\frac{0.28 - 0.3}{0.0118} \leq Z \leq \frac{0.32 - 0.3}{0.0118}\right)$$
$$= P(-1.69 \leq Z \leq 1.69)$$
$$= 0.9545 - 0.0455 = 0.9090$$

Then by the complement rule,

$$P(\hat{p} < 0.28 \text{ or } \hat{p} > 0.32) = 1 - P(0.28 \leq \hat{p} \leq 0.32)$$
$$= 1 - 0.9090 = 0.0910$$

There is often more than one correct way to use the rules of probability to answer a question.

We began this chapter with a general discussion of the idea of probability and the properties of probability models. Two very useful specific types of probability models are distributions of discrete and continuous random variables. In our study of statistics we will employ only these two types of probability models.

SUMMARY

A **random variable** is a variable taking numerical values determined by the outcome of a random phenomenon. The **probability distribution** of a random variable X tells us what the possible values of X are and how probabilities are assigned to those values.

A random variable X and its distribution can be discrete or continuous.

A **discrete random variable** has finitely many possible values. The probability distribution assigns each of these values a probability between 0 and 1 such that the sum of all the probabilities is exactly 1. The probability of any event is the sum of the probabilities of all the values that make up the event.

A **continuous random variable** takes all values in some interval of numbers. A **density curve** describes the probability distribution of a continuous random variable. The probability of any event is the area under the curve above the values that make up the event.

Normal distributions are one type of continuous probability distribution.

You can picture a probability distribution by drawing a **probability histogram** in the discrete case or by graphing the density curve in the continuous case.

SECTION 4.3 EXERCISES

4.37   If a carefully made die is rolled once, it is reasonable to assign probability 1/6 to each of the six faces. What is the probability of rolling a number less than 3?

4.38   A couple plans to have three children. There are 8 possible arrangements of girls and boys. For example, GGB means the first two children are girls and the third child is a boy. All 8 arrangements are (approximately) equally likely.

   (a)   Write down all 8 arrangements of the sexes of three children. What is the probability of any one of these arrangements?

   (b)   Let X be the number of girls the couple has. What is the probability that $X = 2$?

   (c)   Starting from your work in (a), find the distribution of X. That is, what values can X take, and what are the probabilities for each value?

4.39   A study of social mobility in England looked at the social class reached by the sons of lower-class fathers. Social classes are numbered from 1 (low) to 5 (high). Take the random variable X to be the class of a randomly chosen son of a father in Class 1. The study found that the distribution of X is

Son's class	1	2	3	4	5
Probability	0.48	0.38	0.08	0.05	0.01

   (a)   What percent of the sons of lower-class fathers reach the highest class, Class 5?

   (b)   Check that this distribution satisfies the two requirements for a discrete probability distribution.

   (c)   What is $P(X \leq 3)$? (Be careful: the event "$X \leq 3$" includes the value 3.)

(d) What is $P(X < 3)$?

(e) Write the event "a son of a lower-class father reaches one of the two highest classes" in terms of values of X. What is the probability of this event?

4.40 Choose an American household at random and let the random variable X be the number of persons living in the household. If we ignore the few households with more than seven inhabitants, the probability distribution of X is as follows:

Inhabitants	1	2	3	4	5	6	7
Probability	0.25	0.32	0.17	0.15	0.07	0.03	0.01

(a) Verify that this is a legitimate discrete probability distribution and draw a probability histogram to display it.

(b) What is $P(X \geq 5)$?

(c) What is $P(X > 5)$?

(d) What is $P(2 < X \leq 4)$?

(e) What is $P(X \neq 1)$?

(f) Write the event that a randomly chosen household contains more than two persons in terms of the random variable X. What is the probability of this event?

4.41 A study of education followed a large group of fifth-grade children to see how many years of school they eventually completed. Let X be the highest year of school that a randomly chosen fifth grader completes. (Students who go on to college are included in the outcome $X = 12$.) The study found this probability distribution for X:

Years	4	5	6	7	8	9	10	11	12
Probability	0.010	0.007	0.007	0.013	0.032	0.068	0.070	0.041	0.752

(a) What percent of fifth graders eventually finished twelfth grade?

(b) Check that this is a legitimate discrete probability distribution.

(c) Find $P(X \geq 6)$. (Be careful: the event "$X \geq 6$" includes the value 6.)

(d) Find $P(X > 6)$.

(e) What values of X make up the event "the student completed at least one year of high school"? (High school begins with the ninth grade.) What is the probability of this event?

4.42 Some games of chance rely on tossing two dice. Each die has six faces, marked with 1, 2, ..., 6 spots called pips. The dice used in casinos are carefully balanced so that each face is equally likely to come up. When two dice are tossed, each of the 36 possible pairs of faces is equally likely to come up. The outcome of interest to a gambler is the sum of the pips on the two up faces. Call this random variable X.

(a) Write down all 36 possible pairs of faces.

(b) If all pairs have the same probability, what must be the probability of each pair?

(c) Write the value of X next to each pair of faces and use this information with the result of (b) to give the probability distribution of X. Draw a probability histogram to display the distribution.

(d) One bet available in craps wins if a 7 or an 11 comes up on the next roll of two dice. What is the probability of rolling a 7 or an 11 on the next roll?

(e) Several bets in craps lose if a 7 is rolled. If any outcome other than 7 occurs, these bets either win or continue to the next roll. What is the probability that anything other than a 7 is rolled?

4.43   Weary of the low turnout in student elections, a college administration decides to choose an **SRS** of three students to form an advisory board that represents student opinion. Suppose that 40% of all students oppose the use of student fees to fund student interest groups. Then the opinions of the three students on the board are independent and the probability is 0.4 that each opposes the funding of interest groups.

(a) Call the three students A, B, and C. What is the probability that A and B support funding and C opposes it?

(b) List all possible combinations of opinions that can be held by students A, B, and C. (*Hint:* There are eight possibilities.) Then give the probability of each of these outcomes. Note that they are not equally likely.

(c) Let the random variable X be the number of student representatives who oppose the funding of interest groups. Give the probability distribution of X.

(d) Express the event "a majority of the advisory board opposes funding" in terms of X and find its probability.

4.44   Let X be a random number between 0 and 1 produced by the idealized uniform random number generator described in Example 4.17 and Figure 4.10. Find the following probabilities:

(a) $P(0 \le X \le 0.4)$

(b) $P(0.4 \le X \le 1)$

(c) $P(0.3 \le X \le 0.5)$

(d) $P(0.3 < X < 0.5)$

(e) $P(0.226 \le X \le 0.713)$

4.45   Let the random variable X be a random number with the uniform density curve in Figure 4.10. Find the following probabilities:

(a) $P(X \le 0.49)$

(b) $P(X \ge 0.27)$

(c) $P(0.27 < X < 1.27)$

(d) $P(0.1 \le X \le 0.2 \text{ or } 0.8 \le X \le 0.9)$

(e) The probability that X is not in the interval 0.3 to 0.8.

(f) $P(X = 0.5)$

4.46 Many random number generators allow users to specify the range of the random numbers to be produced. Suppose that you specify that the range is to be $0 \leq y \leq 2$. Then the density curve of the outcomes has constant height between 0 and 2, and height 0 elsewhere.

(a) What is the height of the density curve between 0 and 2? Draw a graph of the density curve.

(b) Use your graph from (a) and the fact that probability is area under the curve to find $P(y \leq 1)$.

(c) Find $P(0.5 < y < 1.3)$.

(d) Find $P(y \geq 0.8)$.

4.47 Generate *two* random numbers between 0 and 1 and take Y to be their sum. Then Y is a continuous random variable that can take any value between 0 and 2. The density curve of Y is the triangle shown in Figure 4.13.

(a) Verify by geometry that the area under this curve is 1. $_{.5}$ $A = \frac{1}{2}bh$

(b) What is the probability that Y is less than 1? (Sketch the density curve, shade the area that represents the probability, then find that area. Do this for (c) also.)

(c) What is the probability that Y is less than 0.5? $.25$

4.48 An SRS of 400 American adults is asked, "What do you think is the most serious problem facing our schools?" Suppose that in fact 30% of all adults would answer "drugs" if asked this question. The proportion $\hat{p}$ of the sample who answer "drugs" will vary in repeated sampling. In fact, we can assign probabilities to values of $\hat{p}$ using the normal density curve with mean 0.3 and standard deviation 0.023. Use this density curve to find the probabilities of the following events:

(a) At least half of the sample believes that drugs are the schools' most serious problem.

(b) Less than 25% of the sample believes that drugs are the most serious problem.

(c) The sample proportion is between 0.25 and 0.35.

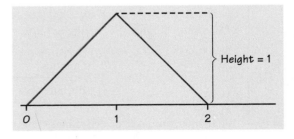

FIGURE 4.13 The density curve for the sum Y of two random numbers, for Exercise 4.47.

4.49   An opinion poll asks an SRS of 1500 adults, "Do you happen to jog?"
        Suppose that the population proportion who jog (a parameter) is $p = 0.15$.
        To estimate p, we use the proportion $\hat{p}$ in the sample who answer "Yes." The
        statistic $\hat{p}$ is a random variable that is approximately normally distributed
        with mean $\mu = 0.15$ and standard deviation $\sigma = 0.0092$. Find the following
        probabilities:

        (a) $P(\hat{p} \geq 0.16)$

        (b) $P(0.14 \leq \hat{p} \leq 0.16)$

4.4   Means and Variances of Random Variables

Probability is the mathematical language that describes the long-run regular
behavior of random phenomena. The probability distribution of a random
variable is an idealized relative frequency distribution. The probability histograms and density curves that picture probability distributions resemble
our earlier pictures of distributions of data. In describing data, we moved
from graphs to numerical measures such as means and standard deviations.
Now we will make the same move to expand our descriptions of the distributions of random variables. We can speak of the mean winnings in a game
of chance or the standard deviation of the randomly varying number of calls
a travel agency receives in an hour. In this section we will learn more about
how to compute these descriptive measures and about the laws they obey.

The mean of a random variable

The mean $\bar{x}$ of a set of observations is their ordinary average. The mean of a
random variable X is also an average of the possible values of X, but with an
essential change to take into account the fact that not all outcomes need be
equally likely. An example will show what we must do.

EXAMPLE 4.19   Thirty-seven states, five Canadian provinces, and most European countries have
government-sponsored lotteries. New Hampshire started the fad in the United States,
in 1964. Here is a simple lottery wager, the Tri-State Pick 3 game that New Hampshire shares with Maine and Vermont. You choose a three-digit number; the state
chooses a three-digit winning number at random and pays you $500 if your number
is chosen. Because there are 1000 three-digit numbers, you have probability 1/1000
of winning. Taking X to be the amount you win, the probability distribution of X is

Outcome	$0	$500
Probability	0.999	0.001

What are your average winnings? The ordinary average of the two possible outcomes $0 and $500 is $250, but that makes no sense as the average winnings because

$500 is much less likely than $0. In the long run you win $500 once in every 1000 tickets and $0 on the remaining 999 of 1000 tickets. Your long-run average winnings from a ticket are

$$\$500\frac{1}{1000} + \$0\frac{999}{1000} = \$0.50$$

or fifty cents. That number is the mean of the random variable X. (Tickets cost $1, so in the long run the state keeps half the money you wager.)

If you play Tri-State Pick 3 several times, we would as usual call the mean of the actual amounts you win $\bar{x}$. The mean in Example 4.19 is a different quantity—it is the long-run average winnings you expect if you play a very large number of times. Just as probabilities are an idealized description of long-run proportions, the mean of a probability distribution describes the long-run average outcome. We can't call this mean $\bar{x}$, so we need a different symbol. The common symbol for the **mean of a probability distribution** is μ, the Greek letter mu. We used μ in Chapter 1 for the mean of a normal distribution, so this is not a new notation. We will often be interested in several random variables, each having a different probability distribution with a different mean. To remind ourselves that we are talking about the mean of X, we often write μ_X rather than simply μ. In Example 4.19, $\mu_X = \$0.50$. Notice that, as often happens, the mean is not a possible value of X. You will often find the mean of a random variable X called the **expected value** of X. This term can be misleading, for we don't necessarily expect one observation on X to be close to its expected value.

The mean of any discrete random variable is found just as in Example 4.19. It is an average of the possible outcomes, but a weighted average in which each outcome is weighted by its probability. Because the probabilities add to 1, we have total weight 1 to distribute among the outcomes. An outcome that occurs half the time has probability one-half and gets one-half the weight in calculating the mean. Here is the general definition.

mean μ

expected value

Mean of a Discrete Random Variable

Suppose that X is a discrete random variable whose distribution is

Value of X	x_1	x_2	x_3	$\cdots$	x_k
Probability	p_1	p_2	p_3	$\cdots$	p_k

To find the **mean** of X, multiply each possible value by its probability, then add all the products:

$$\mu_X = x_1 p_1 + x_2 p_2 + \cdots + x_k p_k$$
$$= \sum x_i p_i$$

EXAMPLE 4.20

The distribution of the count X of heads in four tosses of a balanced coin was found in Example 4.16 to be

Number of heads x_i	0	1	2	3	4
Probability p_i	0.0625	0.25	0.375	0.25	0.0625

The mean of X is therefore

$$\mu_X = (0)(0.0625) + (1)(0.25) + (2)(0.375) + (3)(0.25) + (4)(0.0625)$$
$$= 2$$

This discrete distribution is symmetric, as the probability histogram in Figure 4.14(a) reminds us. The mean therefore falls at the center of symmetry.

EXAMPLE 4.21

What is the mean size of an American household? Here is the distribution of the size of households according to Census Bureau studies:

Inhabitants	1	2	3	4	5	6	7
Proportion of households	0.25	0.32	0.17	0.15	0.07	0.03	0.01

If we imagine selecting a single household at random, the size of the household chosen is a random variable X with probability distribution given by the table. The mean μ_X is the mean household size in the population. This mean is

$$\mu_X = (1)(0.25) + (2)(0.32) + (3)(0.17) + (4)(0.15) + (5)(0.07) + (6)(0.03) + (7)(0.01)$$
$$= 2.6$$

Figure 4.14(b) locates the mean on a probability histogram. (In this example, we have ignored the few households with 8 or more members. The Census Bureau reports that the actual mean size of American households is 2.65 people when these few large households are included.)

What about continuous random variables? The probability distribution of a continuous random variable X is described by a density curve. Chapter 1 showed how to find the mean of the distribution: it is the point at which the area under the density curve would balance if it were made out of solid material. The mean lies at the center of symmetric density curves such as the normal curves. Exact calculation of the mean of a distribution with a skewed density curve requires advanced mathematics.[4] The idea that the mean is the balance point of the distribution applies to discrete random variables as well, but in the discrete case we have a formula that gives us this point.

Statistical estimation and the law of large numbers

We would like to estimate the mean height μ of the population of all American women between the ages of 18 and 24 years. This μ is the mean μ_X of the random variable X obtained by choosing a young woman at random and measuring her height. To estimate μ, we choose an **SRS** of young women and use the sample mean $\bar{x}$ to estimate the unknown population mean μ. In the language of Section 3.4 (page 268), μ is a *parameter* and $\bar{x}$ is a *statistic*.

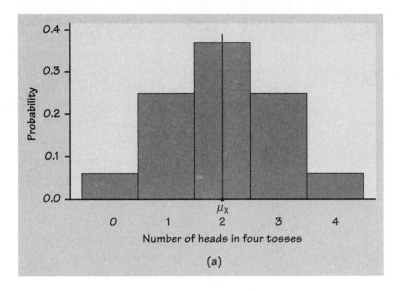

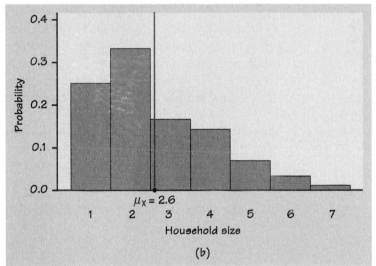

FIGURE 4.14 Locating the mean of a discrete random variable on the probability histogram for (a) the number of heads in four tosses of a coin (Example 4.20) and (b) the number of persons in a household (Example 4.21).

Statistics obtained from probability samples are random variables, because their values would vary in repeated sampling. The sampling distributions of statistics are just the probability distributions of these random variables.

It seems reasonable to use $\bar{x}$ to estimate μ. An SRS should fairly represent the population, so the mean $\bar{x}$ of the sample should be somewhere near the mean μ of the population. Of course, we don't expect $\bar{x}$ to be exactly equal to μ, and we realize that if we choose another SRS, the luck of the draw will probably produce a different $\bar{x}$.

If $\bar{x}$ is rarely exactly right and varies from sample to sample, why is it nonetheless a reasonable estimate of the population mean μ? We gave one answer in Section 3.4: $\bar{x}$ is unbiased and we can control its variability by choosing the sample size. Here is another answer: if we keep on adding observations to our random sample, the statistic $\bar{x}$ is *guaranteed* to get as close as we wish to the parameter μ and then stay that close. We have the comfort of knowing that if we can afford to keep on measuring more young women, eventually we will estimate the mean height of all young women very accurately. This remarkable fact is called the *law of large numbers*. It is remarkable because it holds for *any* population, not just for some special class such as normal distributions.

Law of Large Numbers

Draw independent observations at random from any population with finite mean μ. Decide how accurately you would like to estimate μ. As the number of observations drawn increases, the mean $\bar{x}$ of the observed values eventually approaches the mean μ of the population as closely as you specified and then stays that close.

The behavior of $\bar{x}$ is similar to the idea of probability. In the long run, the proportion of outcomes taking any value gets close to the probability of that value, and the average outcome gets close to the distribution mean. Figure 4.1 (page 291) shows how proportions approach probability in one example. Here is an example of how sample means approach the distribution mean.

EXAMPLE 4.22 | The distribution of the heights of all young women is close to the normal distribution with mean 64.5 inches and standard deviation 2.5 inches. Suppose that $\mu = 64.5$ were exactly true. Figure 4.15 shows the behavior of the mean height $\bar{x}$ of n women chosen at random from a population whose heights follow the $N(64.5, 2.5)$ distribution. The graph plots the values of $\bar{x}$ as we add women to our sample. The first woman drawn had height 64.21 inches, so the line starts there. The second had height 64.35 inches, so for $n = 2$ the mean is

$$\bar{x} = \frac{64.21 + 64.35}{2} = 64.28$$

This is the second point on the line in the graph.

At first, the graph shows that the mean of the sample changes as we take more observations. Eventually, however, the mean of the observations gets close to the population mean $\mu = 64.5$ and settles down at that value. The law of large numbers says that this *always* happens.

The mean μ of a random variable is the average value of the variable in two senses. By its definition, μ is the average of the possible values, weighted by their probability of occurring. The law of large numbers says that μ is also the long-run average of many independent observations on the variable. The law of large numbers can be proved mathematically starting from the basic laws of probability.*

---

*The earliest version of the law of large numbers was proved by the Swiss mathematician Jacob Bernoulli (1654–1705). The Bernoullis were a remarkable mathematical family; five of them contributed to the early study of probability.

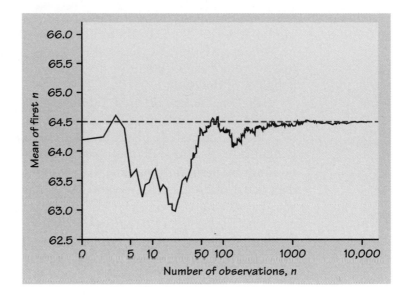

FIGURE 4.15 The law of large numbers in action. As we take more observations, the sample mean $\bar{x}$ always approaches the mean μ of the population.

Thinking about the law of large numbers

The law of large numbers says broadly that the average results of many independent observations are stable and predictable. Casinos are not the only businesses that base forecasts on this fact. A grocery store deciding how many gallons of milk to stock and a fast-food restaurant deciding how many beef patties to prepare can predict demand even though their many customers make independent decisions. The law of large numbers says that these many individual decisions will produce a stable result. It is worth the effort to think a bit more closely about so important a fact.

The "law of small numbers" Both the rules of probability and the law of large numbers describe the regular behavior of chance phenomena *in the long run*. Psychologists have discovered that the popular understanding of randomness is quite different from the true laws of chance.[5] Most people believe in an incorrect "law of small numbers." That is, we expect even short sequences of random events to show the kind of average behavior that in fact appears only in the long run.

Try this experiment: Write down a sequence of heads and tails that you think imitates 10 tosses of a balanced coin. How long was the longest string (called a *run*) of consecutive heads or consecutive tails in your tosses? Most people will write a sequence with no runs of more than two consecutive heads or tails. Longer runs don't seem "random" to us. In fact, the probability of a run of three or more consecutive heads or tails in 10 tosses is greater than 0.8, and the probability of *both* a run of three or more heads and a run of three or more tails is almost 0.2.[6] This and other probability calculations suggest that a short sequence of coin tosses will often not appear random to us. The runs of consecutive heads or consecutive tails that appear in real coin tossing (and that are predicted by the mathematics of probability) seem surprising to us. Because we don't expect to see long runs, we may conclude that the coin tosses are not independent or that some influence is disturbing the random behavior of the coin.

Belief in the law of small numbers influences behavior. If a basketball player makes several consecutive shots, both the fans and his teammates believe that he has a "hot hand" and is more likely to make the next shot. This is doubtful. Careful study suggests that runs of baskets made or missed are no more frequent in basketball than would be expected if each shot were independent of the player's previous shots. Players perform consistently, not in streaks. (Of course, some players make a higher percent of their shots in the long run than others.) Our perception of hot or cold streaks simply shows that we don't perceive random behavior very well.[7]

Gamblers often follow the hot-hand theory, betting that a run will continue. At other times, however, they draw the opposite conclusion when confronted with a run of outcomes. If a coin gives 10 straight heads, some gamblers feel that it must now produce some extra tails to get back to the average of half heads and half tails. Not so. If the next 10,000 tosses give about 50% tails, those 10 straight heads will be swamped by the later thousands of heads and tails. No compensation is needed to get back to the average in the long run. Remember that it is *only* in the long run that the regularity described by probability and the law of large numbers takes over.

Our inability to accurately distinguish random behavior from systematic influences points out once more the need for statistical inference to supplement exploratory analysis of data. Probability calculations can help verify that what we see in the data is more than a random pattern.

How large is a large number?   The law of large numbers says that the actual mean outcome of many trials gets close to the distribution mean μ as more trials are made. It doesn't say how many trials are needed to guarantee a mean outcome close to μ. That depends on the *variability* of the random outcomes. The more variable the outcomes, the more trials are needed to ensure that the mean outcome $\bar{x}$ is close to the distribution mean μ.

The law of large numbers is the foundation of such business enterprises as gambling casinos and insurance companies. Games of chance must be quite variable if they are to hold the interest of gamblers. Even a long evening in a casino has an unpredictable outcome. Gambles with extremely variable outcomes, like state lottos with their very large but very improbable jackpots, require impossibly large numbers of trials to ensure that the average outcome is close to the expected value. Though most forms of gambling are less variable than lotto, the layman's answer to the applicability of the law of large numbers is usually that the house plays often enough to rely on it, but you don't. Much of the psychological allure of gambling is its unpredictability for the player. The business of gambling rests on the fact that the result is not unpredictable for the house. The average winnings of the house on tens of thousands of bets will be very close to the mean of the distribution of winnings. Needless to say, this mean guarantees the house a profit.

The law of large numbers is one of the central facts about probability. It helps us understand the mean μ of a random variable. It explains why gambling casinos and insurance companies make money. It assures us that statistical estimation will be accurate if we can afford enough observations. The basic law of large numbers applies to independent observations that all have the same distribution. Mathematicians have extended the law to many more general settings. Here are two of these.

Is there a winning system for gambling?   Serious gamblers often follow a system of betting in which the amount bet on each play depends on the outcome of previous plays. You might, for example, double your bet on each spin of the roulette wheel until you win—or, of course, until your fortune is exhausted. Such a system tries to take advantage of the fact that you have a memory even though the roulette wheel does not. Can you beat the odds with a system based on the outcomes of past plays? No. Mathematicians have established a stronger version of the law of large numbers that says that if you do not have an infinite fortune to gamble with, your long-run average winnings μ remain the same as long as successive trials of the game (such as spins of the roulette wheel) are independent.

What if observations are not independent?   You are in charge of a process that manufactures video screens for computer monitors. Your equipment measures the tension on the metal mesh that lies behind each screen and is critical to its image quality. You want to estimate the mean tension μ for the process by the average $\bar{x}$ of the measurements. Alas, the tension measurements are not independent. If the tension on one screen is a bit too high, the tension on the next is more likely to also be high. Many real-world processes are like this—the process stays stable in the long run, but observations made close together are likely to be both above or both below the long-run mean. Again the mathematicians come to the rescue: as long as the dependence dies out fast enough as we take measurements farther and farther apart in time, the law of large numbers still holds.

Rules for means

You are studying flaws in the painted finish of refrigerators made by your firm. Dimples and paint sags are two kinds of surface flaw. Not all refrigerators have the same number of dimples: many have none, some have one, some two, and so on. You ask for the average number of imperfections on a refrigerator. The inspectors report finding an average of 0.7 dimples and 1.4 sags per refrigerator. How many total imperfections of both kinds (on the average) are there on a refrigerator? Easy: If the average number of dimples is 0.7 and the average number of sags is 1.4, then counting both gives an average of $0.7 + 1.4 = 2.1$ flaws.

Putting this example into more formal language, the number of dimples on a refrigerator is a random variable X that takes values 0, 1, 2, and so on. X varies as we inspect one refrigerator after another. Only the mean number of dimples $\mu_X = 0.7$ was reported to you. The number of paint sags is a second random variable Y having mean $\mu_Y = 1.4$. (You see how the subscripts keep straight which variable we are talking about.) The total number of both dimples and sags is the sum $X + Y$. That sum is another random variable that varies from refrigerator to refrigerator. Its mean μ_{X+Y} is the average number of dimples and sags together. It is just the sum of the individual means μ_X and μ_Y. That is an important rule for how means of random variables behave.

Another important rule says how the mean of a random variable changes when we multiply every outcome by the same fixed number or add the same fixed number to every outcome. Suppose X is the length in centimeters of a cockroach randomly chosen from those living in a dormitory and that the mean length is $\mu_X = 2.2$ centimeters. If we decide to measure in millimeters, we multiply every value of X by 10 because there are 10 millimeters in a centimeter. The length in millimeters is the new random variable $10X$ formed by multiplying X by 10 to convert centimeters to millimeters. What is the mean μ_{10X} of this new variable? If we multiply every value of X by 10, surely we also multiply the mean (the average value) by 10. That's right: the mean length is $2.2 \times 10 = 22$ millimeters. In more formal language, the mean μ_{10X} of $10X$ is $10\mu_X$. Similarly, if we add the same fixed number to every value of a random variable X, we add that same number to the mean. The first rule in the box below combines the results of multiplying by a fixed number and adding a fixed number.

Rules for Means

Rule 1. If X is a random variable and a and b are fixed numbers, then

$$\mu_{a+bX} = a + b\,\mu_X$$

Rule 2. If X and Y are random variables, then

$$\mu_{X+Y} = \mu_X + \mu_Y$$

Means are a special kind of average. Rules 1 and 2 just say that means behave like averages. Here is an example that applies these rules.

EXAMPLE 4.23

Gain Communications sells aircraft communications units to both the military and civilian markets. Next year's sales depend on market conditions that cannot be predicted exactly. Gain follows the modern practice of using probability estimates of sales. The military division estimates its sales as follows:

Units sold	1000	3000	5000	10,000
Probability	0.1	0.3	0.4	0.2

These are personal probabilities that express the informed opinion of Gain's executives. The corresponding sales estimates for the civilian division are

Units sold	300	500	750
Probability	0.4	0.5	0.1

Take X to be the number of military units sold and Y the number of civilian units. From the probability distributions we compute that

$$\mu_X = (1000)(0.1) + (3000)(0.3) + (5000)(0.4) + (10,000)(0.2)$$
$$= 100 + 900 + 2000 + 2000$$
$$= 5000 \text{ units}$$
$$\mu_Y = (300)(0.4) + (500)(0.5) + (750)(0.1)$$
$$= 120 + 250 + 75$$
$$= 445 \text{ units}$$

Gain makes a profit of $2000 on each military unit sold and $3500 on each civilian unit. Next year's profit from military sales will be $2000X$, $2000 times the number X of units sold. By the first rule for means, the mean military profit is

$$\mu_{2000X} = 2000\mu_X = (2000)(5000) = \$10,000,000$$

Similarly, the civilian profit is $3500Y$ and the mean profit from civilian sales is

$$\mu_{3500Y} = 3500\mu_Y = (3500)(445) = \$1,557,500$$

The total profit is the sum of the military and civilian profit:

$$Z = 2000X + 3500Y$$

The second rule for means says that the mean of this sum of two variables is the sum of the two individual means:

$$\mu_Z = \mu_{2000X} + \mu_{3500Y}$$
$$= \$10,000,000 + \$1,557,500$$
$$= \$11,557,500$$

This mean is the company's best estimate of next year's profit, combining the probability estimates of the two divisions.

Once you have gained some experience in applying the rules for means of random variables, the calculation of mean total profit in Example 4.23 can be done more quickly by applying both rules at once as follows:

$$\mu_Z = \mu_{2000X+3500Y}$$
$$= 2000\mu_X + 3500\mu_Y$$
$$= (2000)(5000) + (3500)(445) = \$11,557,500$$

The variance of a random variable

The mean is a measure of the center of a distribution. Even the most basic numerical description requires in addition a measure of the spread or variability of the distribution. The variance and the standard deviation are the measures of spread that accompany the choice of the mean to measure center. Just as for the mean, we need a distinct symbol to distinguish the variance of a random variable from the variance s^2 of a data set. We write the variance of a random variable X as σ_X^2. Once again the subscript reminds us which variable we have in mind. The definition of the variance σ_X^2 of a random variable is similar to the definition of the sample variance s^2 given in Chapter 1. That is, the variance is an average of the squared deviation $(X - \mu_X)^2$ of the variable X from its mean μ_X. As for the mean, the average we use is a weighted average in which each outcome is weighted by its probability in order to take account of outcomes that are not equally likely. Calculating this weighted average is straightforward for discrete random variables but requires advanced mathematics in the continuous case. Here is the resulting definition.

Variance of a Discrete Random Variable

Suppose that X is a discrete random variable whose distribution is

Value of X	x_1	x_2	x_3	$\cdots$	x_k
Probability	p_1	p_2	p_3	$\cdots$	p_k

and that μ_X is the mean of X. The **variance** of X is

$$\sigma_X^2 = (x_1 - \mu_X)^2 p_1 + (x_2 - \mu_X)^2 p_2 + \cdots + (x_k - \mu_X)^2 p_k$$
$$= \sum (x_i - \mu_X)^2 p_i$$

The **standard deviation** σ_X of X is the square root of the variance.

EXAMPLE 4.24

In Example 4.23 we saw that the number X of communications units sold by the Gain Communications military division has distribution

Units sold	1000	3000	5000	10,000
Probability	0.1	0.3	0.4	0.2

We can find the mean and variance of X by arranging the calculation in the form of a table. Both μ_X and σ_X^2 are sums of columns in this table.

x_i	p_i	$x_i p_i$	$(x_i - \mu_X)^2 p_i$
1,000	0.1	100	$(1,000 - 5,000)^2(0.1) = 1,600,000$
3,000	0.3	900	$(3,000 - 5,000)^2(0.3) = 1,200,000$
5,000	0.4	2,000	$(5,000 - 5,000)^2(0.4) = 0$
10,000	0.2	2,000	$(10,000 - 5,000)^2(0.2) = 5,000,000$
	$\mu_X = 5,000$		$\sigma_X^2 = 7,800,000$

We see that $\sigma_X^2 = 7,800,000$. The standard deviation of X is $\sigma_X = \sqrt{7,800,000} = 2792.8$. The standard deviation is a measure of how variable the number of units sold is. As in the case of distributions for data, the standard deviation of a probability distribution is easiest to understand for normal distributions.

Rules for variances

Rules for variances similar to Rules 1 and 2 for means are often helpful for understanding statistical questions. Although the mean of a sum of random variables is always the sum of their means, this addition rule is not always true for variances. To understand why, take X to be the percent of a family's after-tax income that is spent and Y the percent that is saved. When X increases, Y decreases by the same amount. Though X and Y may vary widely from year to year, their sum $X + Y$ is always 100% and does not vary at all. It is the association between the variables X and Y that prevents their variances from adding. If random variables are independent, this kind of association between their values is ruled out and their variances do add. Two random variables X

independence

and Y are **independent** if any event involving X alone is independent of any event involving Y alone. Probability models often assume independence when the random variables describe outcomes that appear unrelated to each other. You should ask in each instance whether the assumption of independence seems reasonable.

Rules for Variances

Rule 1. If X is a random variable and a and b are fixed numbers, then

$$\sigma_{a+bX}^2 = b^2 \sigma_X^2$$

Rule 2. If X and Y are independent random variables, then

$$\sigma_{X+Y}^2 = \sigma_X^2 + \sigma_Y^2$$
$$\sigma_{X-Y}^2 = \sigma_X^2 + \sigma_Y^2$$

Notice that because a variance is the average of *squared* deviations from the mean, multiplying X by a constant b multiplies σ_X^2 by the *square* of the constant. Adding a constant a to a random variable changes its mean but does not change its variability. The variance of $X + a$ is therefore the same as the variance of X. Because the square of -1 is 1, the addition rule says that the variance of a difference is the *sum* of the variances. The difference $X - Y$ is more variable than either X or Y alone because variations in both X and Y contribute to variation in their difference.

As with data, we prefer the standard deviation to the variance as a measure of variability. The addition rule for variances implies that standard deviations do *not* generally add. Standard deviations are most easily combined by using the rules for variances rather than by giving separate rules for standard deviations. For example, the standard deviations of $2X$ and $-2X$ are both equal to $2\sigma_X$ because this is the square root of the variance $4\sigma_X^2$.

EXAMPLE 4.25

The payoff X of a $1 ticket in the Tri-State Pick 3 game is $500 with probability $1/1000$ and 0 the rest of the time. Here is the combined calculation of mean and variance:

x_i	p_i	$x_i p_i$	$(x_i - \mu_X)^2 p_i$
0	0.999	0	$(0 - 0.5)^2(0.999) =$ 0.24975
500	0.001	0.5	$(500 - 0.5)^2(0.001) = 249.50025$
	$\mu_X = 0.5$		$\sigma_X^2 = 249.75$

The standard deviation is $\sigma_X = \sqrt{249.75} = \$15.80$. It is usual for games of chance to have large standard deviations, because large variability makes gambling exciting.

If you buy a Pick 3 ticket, your winnings are $W = X - 1$ because the dollar you paid for the ticket must be subtracted from the payoff. By the rules for means, the mean amount you win is

$$\mu_W = \mu_X - 1 = -\$0.50$$

That is, you lose an average of 50 cents on a ticket. The rules for variances remind us that the variance and standard deviation of the winnings $W = X - 1$ are the same as those of X. Subtracting a fixed number changes the mean but not the variance.

Suppose now that you buy a $1 ticket on each of two different days. The payoffs X and Y on the two tickets are independent because separate drawings are held each day. Your total payoff $X + Y$ has mean

$$\mu_{X+Y} = \mu_X + \mu_Y = \$0.50 + \$0.50 = \$1.00$$

Because X and Y are independent, the variance of $X + Y$ is

$$\sigma_{X+Y}^2 = \sigma_X^2 + \sigma_Y^2 = 249.75 + 249.75 = 499.5$$

The standard deviation of the total payoff is

$$\sigma_{X+Y} = \sqrt{499.5} = \$22.35$$

This is not the same as the sum of the individual standard deviations, which is $\$15.80 + \$15.80 = \$31.60$. Variances of independent random variables add; standard deviations do not.

If you buy a ticket every day (365 tickets in a year), your mean payoff is the sum of 365 daily payoffs. That's 365 times 50 cents, or $182.50. Of course, it cost $365 to play, so the state's mean take from a daily Pick 3 player is $182.50. Results for individual players will vary, but the law of large numbers assures the state its profit.

EXAMPLE 4.26

A college uses SAT scores as one criterion for admission. Experience has shown that the distribution of SAT scores among its entire population of applicants is such that

$$\text{SAT math score } X : \mu_X = 625 \qquad \sigma_X = 90$$
$$\text{SAT verbal score } Y : \mu_Y = 590 \qquad \sigma_Y = 100$$

What are the mean and standard deviation of the total score $X + Y$ among students applying to this college?

The mean overall SAT score is

$$\mu_{X+Y} = \mu_X + \mu_Y = 625 + 590 = 1215$$

The variance and standard deviation of the total *cannot be computed* from the information given. SAT verbal and math scores are not independent, because students who score high on one exam tend to score high on the other also. Therefore, the addition rule for variances does not apply.

EXAMPLE 4.27

Tom and George play golf at the same club. Tom's score X varies from round to round but has

$$\mu_X = 110 \quad \text{and} \quad \sigma_X = 10$$

George's score Y also varies, with

$$\mu_Y = 100 \quad \text{and} \quad \sigma_Y = 8$$

Tom and George are playing the first round of the club tournament. Because they are not playing together, we will assume that their scores vary independently of each other. The difference between their scores on this first round has mean

$$\mu_{X-Y} = \mu_X - \mu_Y = 110 - 100 = 10$$

The variance of the difference between the scores is

$$\sigma_{X-Y}^2 = \sigma_X^2 + \sigma_Y^2 = 10^2 + 8^2 = 164$$

The standard deviation is found from the variance

$$\sigma_{X-Y} = \sqrt{164} = 12.8$$

The variation in the difference between the scores is greater than the variation in the score of either player because the difference contains two independent sources of variation. It also makes the tournament interesting, because George will not always finish ahead of Tom even though he has the lower mean score. Computations of the variance of sums and differences of statistics play an important role in statistical inference.

SUMMARY

The probability distribution of a random variable X, like a distribution of data, has a **mean μ_X** and a **standard deviation σ_X**.

The **law of large numbers** says that the average of the values of X observed in many trials must approach μ.

The **mean μ** is the balance point of the probability histogram or density curve. If X is discrete with possible values x_i having probabilities p_i, the mean is the average of the values of X, each weighted by its probability:

$$\mu_X = x_1 p_1 + x_2 p_2 + \cdots + x_k p_k$$

The **variance** σ_X^2 is the average squared deviation of the values of the variable from their mean. For a discrete random variable,

$$\sigma_X^2 = (x_1 - \mu)^2 p_1 + (x_2 - \mu)^2 p_2 + \cdots + (x_k - \mu)^2 p_k$$

The **standard deviation** σ_X is the square root of the variance. The standard deviation measures the variability of the distribution about the mean. It is easiest to interpret for normal distributions.

The mean and variance of a continuous random variable can be computed from the density curve, but to do so requires more advanced mathematics.

The means and variances of random variables obey the following rules. If a and b are fixed numbers, then

$$\mu_{a+bX} = a + b\,\mu_X$$
$$\sigma_{a+bX}^2 = b^2 \sigma_X^2$$

If X and Y are any two random variables, then

$$\mu_{X+Y} = \mu_X + \mu_Y$$

and if X and Y are independent, then

$$\sigma_{X+Y}^2 = \sigma_X^2 + \sigma_Y^2$$
$$\sigma_{X-Y}^2 = \sigma_X^2 + \sigma_Y^2$$

SECTION 4.4 EXERCISES

4.50   Keno is a favorite game in casinos, and similar games are popular with the states that operate lotteries. Balls numbered 1 to 80 are tumbled in a machine as the bets are placed, then 20 of the balls are chosen at random. Players select numbers by marking a card. The simplest of the many wagers available is "Mark 1 Number." Your payoff is $3 on a $1 bet if the number you select is one of those chosen. Because 20 of 80 numbers are chosen, your probability of winning is 20/80, or 0.25.

(a) What is the probability distribution (the outcomes and their probabilities) of the payoff X on a single play?

(b) What is the mean payoff μ_X?

(c) In the long run, how much does the casino keep from each dollar bet?

4.51   Example 4.15 gives the distribution of grades (A = 4, B = 3, and so on) in a large course as

Grade	0	1	2	3	4
Probability	0.10	0.15	0.30	0.30	0.15

Find the average (that is, the mean) grade in this course.

4.52   A life insurance company sells a term insurance policy to a 21-year-old male that pays \$100,000 if the insured dies within the next 5 years. The probability that a randomly chosen male will die each year can be found in mortality tables. The company collects a premium of \$250 each year as payment for the insurance. The amount X that the company earns on this policy is \$250 per year, less the \$100,000 that it must pay if the insured dies. Here is the distribution of X. Fill in the missing probability in the table and calculate the mean earnings μ_X.

Age at death	21	22	23	24	25	≥ 26
Payout	−\$99,750	−\$99,500	−\$99,250	−\$99,000	−\$98,750	\$1250
Probability	0.00183	0.00186	0.00189	0.00191	0.00193	

4.53   It would be quite risky for you to insure the life of a 21-year-old friend under the terms of the previous exercise. There is a high probability that your friend would live and you would gain \$1250 in premiums. But if he were to die, you would lose almost \$100,000. Explain carefully why selling insurance is not risky for an insurance company that insures many thousands of 21-year-old men.

4.54   The Tri-State Pick 3 lottery game offers a choice of several bets. You choose a three-digit number. The lottery commission announces the winning three-digit number, chosen at random, at the end of each day. The "box" pays \$83.33 if the number you choose has the same digits as the winning number, in any order. Find the expected payoff for a \$1 bet on the box. (Assume that you chose a number having three distinct digits.)

4.55   For each of the following situations, would you expect the random variables X and Y to be independent? Explain your answers.

   (a) X is the rainfall (in inches) on November 6 of this year and Y is the rainfall at the same location on November 6 of next year.

   (b) X is the amount of rainfall today and Y is the rainfall at the same location tomorrow.

   (c) X is today's rainfall at the Orlando, Florida, airport, and Y is today's rainfall at Disney World just outside Orlando.

4.56   In which of the following games of chance would you be willing to assume independence of X and Y in making a probability model? Explain your answer in each case.

   (a) In blackjack, you are dealt two cards and examine the total points X on the cards (face cards count 10 points). You can choose to be dealt another card and compete based on the total points Y on all three cards.

(b) In craps, the betting is based on successive rolls of two dice. X is the sum of the faces on the first roll, and Y the sum of the faces on the next roll.

4.57 **(a)** A gambler knows that red and black are equally likely to occur on each spin of a roulette wheel. He observes five consecutive reds and bets heavily on red at the next spin. Asked why, he says that "red is hot" and that the run of reds is likely to continue. Explain to the gambler what is wrong with this reasoning.

(b) After hearing you explain why red and black remain equally probable after five reds on the roulette wheel, the gambler moves to a poker game. He is dealt five straight red cards. He remembers what you said and assumes that the next card dealt in the same hand is equally likely to be red or black. Is the gambler right or wrong? Why?

4.58 The baseball player Tony Gwynn gets a hit about 35% of the time over an entire season. After he has failed to hit safely in six straight at-bats, the TV commentator says, "Tony is due for a hit by the law of averages." Is that right? Why? *binomial prob. dist.*

4.59 A time and motion study measures the time required for an assembly line worker to perform a repetitive task. The data show that the time required to bring a part from a bin to its position on an automobile chassis varies from car to car with mean 11 seconds and standard deviation 2 seconds. The time required to attach the part to the chassis varies with mean 20 seconds and standard deviation 4 seconds.

(a) What is the mean time required for the entire operation of positioning and attaching the part?

(b) If the variation in the worker's performance is reduced by better training, the standard deviations will decrease. Will this decrease change the mean you found in (a) if the mean times for the two steps remain as before?

(c) The study finds that the times required for the two steps are independent. A part that takes a long time to position, for example, does not take more or less time to attach than other parts. Would your answer in (a) change if the two variables were dependent?

4.60 Laboratory data show that the time required to complete two chemical reactions in a production process varies. The first reaction has a mean time of 40 minutes and a standard deviation of 2 minutes; the second has a mean time of 25 minutes and a standard deviation of 1 minute. The two reactions are run in sequence during production. There is a fixed period of 5 minutes between them as the product of the first reaction is pumped into the vessel where the second reaction will take place. What is the mean time required for the entire process?

4.61 Find the standard deviation σ_X of the distribution of grades in Exercise 4.51.

4.62   In an experiment on the behavior of young children, each subject is placed in an area with five toys. The response of interest is the number of toys that the child plays with. Past experiments with many subjects have shown that the probability distribution of the number X of toys played with is as follows:

Number of toys x_i	0	1	2	3	4	5
Probability p_i	0.03	0.16	0.30	0.23	0.17	0.11

Calculate the mean μ_X and the standard deviation σ_X.

4.63   In government data, a household consists of all occupants of a dwelling unit, while a family consists of two or more persons who live together and are related by blood or marriage. Here are the distributions of household size and of family size in the United States:

Number of persons	1	2	3	4	5	6	7
Household probability	0.25	0.32	0.17	0.15	0.07	0.03	0.01
Family probability	0	0.42	0.23	0.21	0.09	0.03	0.02

Example 4.21 (page 328) shows the mean and probability histogram for household size. Compare the two distributions using probability histograms, means, and standard deviations. Then write a brief comparison, using your calculations to back up your statements.

4.64   Here is a simple way to create a random variable X that has mean μ and standard deviation σ: X takes only the two values $\mu - \sigma$ and $\mu + \sigma$, each with probability 0.5. Use the definition of the mean and variance for discrete random variables to show that X does have mean μ and standard deviation σ.

4.65   Find the standard deviation of the time required for the two-step assembly operation studied in Exercise 4.59.

4.66   The times for the two reactions in the chemical production process described in Exercise 4.60 are independent. Find the standard deviation of the time required to complete the process.

4.67   Examples 4.23 (page 334) and 4.24 (page 336) concern a probabilistic projection of sales and profits by an electronics firm, Gain Communications.

(a)   Find the variance and standard deviation of the estimated sales X of Gain's civilian unit, using the distribution and mean from Example 4.23.

(b)   Because the military budget and the civilian economy are not closely linked, Gain is willing to assume that its military and civilian sales vary independently. Combine your result from (a) with the results for the military unit from Example 4.24 to obtain the standard deviation of the total sales $X + Y$.

(c) Find the standard deviation of the estimated profit, $Z = 2000X + 3500Y$.

4.68 The academic motivation and study habits of female students as a group are better than those of males. The Survey of Study Habits and Attitudes (SSHA) is a psychological test that measures these factors. The distribution of SSHA scores among the women at a college has mean 120 and standard deviation 28, and the distribution of scores among men students has mean 105 and standard deviation 35. You select a single male student and a single female student at random and give them the SSHA test.

(a) Explain why it is reasonable to assume that the scores of the two students are independent.

(b) What are the mean and standard deviation of the difference (female minus male) between their scores?

(c) From the information given, can you find the probability that the woman chosen scores higher than the man? If so, find this probability. If not, explain why you cannot.

4.69 In a process for manufacturing glassware, glass stems are sealed by heating them in a flame. The temperature of the flame varies a bit. Here is the distribution of the temperature X measured in degrees Celsius:

Temperature	540°	545°	550°	555°	560°
Probability	0.1	0.25	0.3	0.25	0.1

(a) Find the mean temperature μ_X and the standard deviation σ_X.

(b) The target temperature is 550° C. What are the mean and standard deviation of the number of degrees off target $X - 550$?

(c) A manager asks for results in degrees Fahrenheit. The conversion of X into degrees Fahrenheit is given by

$$Y = \frac{9}{5}X + 32$$

What are the mean μ_Y and standard deviation σ_Y of the temperature of the flame in the Fahrenheit scale?

4.70 You have two scales for measuring weights in a chemistry lab. Both scales give answers that vary a bit in repeated weighings of the same item. If the true weight of a compound is 2 grams (g), the first scale produces readings X that have mean 2.000 g and standard deviation 0.002 g. The second scale's readings Y have mean 2.001 g and standard deviation 0.001 g.

(a) What are the mean and standard deviation of the difference $Y - X$ between the readings? (The readings X and Y are independent.)

(b) You measure once with each scale and average the readings. Your result is $Z = (X + Y)/2$. What are μ_Z and σ_Z? Is the average Z more or less variable than the reading Y of the less variable scale?

4.71   The risk of an investment is often measured by the standard deviation of the return on the investment. The more variable the return is (the larger σ is), the riskier the investment. We can measure the great risk of insuring one person's life in Exercise 4.52 by computing the standard deviation of the income X that the insurer will receive. Find σ_X, using the distribution and mean found in Exercise 4.52.

4.72   The risk of insuring one person's life is reduced if we insure many people. Use the result of the previous exercise and rules for means and variances to answer the following questions.

   (a)   Suppose that we insure two 21-year-old males, and that their ages at death are independent. If X and Y are the insurer's income from the two insurance policies, the total income is $T = X + Y$. The mean and variance of each of X and Y are as you computed them in the previous exercise. What are the mean μ_T and the standard deviation σ_T of the insurer's total income T?

   (b)   The insurer's average income on the two policies is

$$ Z = \frac{X + Y}{2} = \frac{1}{2}X + \frac{1}{2}Y $$

   Find the mean and standard deviation of Z. You see that the mean income is the same as for a single policy but the standard deviation is less.

   (c)   If four 21-year-old men are insured, the insurer's average income is

$$ Z = \frac{1}{4}(X_1 + X_2 + X_3 + X_4) $$

   where X_i is the income from insuring one man. The X_i are independent and each has the same distribution as before. Find the mean and standard deviation of Z. Compare your results with the results of (b).

4.73   The psychologist Amos Tversky did many studies of our perception of chance behavior. In its obituary of Tversky (June 6, 1996), the *New York Times* cited the following example.

   (a)   Tversky asked subjects to choose between two public health programs that affect 600 people. One has probability 1/2 of saving all 600 and probability 1/2 that all 600 will die. The other is guaranteed to save exactly 400 of the 600 people. Find the mean number of people saved by the first program.

   (b)   Tversky then offered a different choice. One program has probability 1/2 of saving all 600 and probability 1/2 of losing all 600, while the other will definitely lose exactly 200 lives. What is the difference between this choice and that in (a)?

(c) Given option (a), most subjects choose the second program. Given option (b), most subjects choose the first program. Do the subjects appear to use means in making their decisions? Why do you think their choices differed in the two cases?

4.74 One consequence of the law of large numbers is that once we have a probability distribution for a random variable, we can find its mean by simulating many outcomes and averaging them. The law of large numbers says that if we take enough outcomes, their average value is sure to approach the mean of the distribution.

I have a little bet to offer you. Toss a coin ten times. If there is no run of three or more straight heads or tails in the ten outcomes, I'll pay you $2. If there is a run of three or more, you pay me just $1. Surely you will want to take advantage of me and play this game?

Simulate enough plays of this game (the outcomes are +$2 if you win and −$1 if you lose) to estimate the mean outcome. Is it to your advantage to play?

4.5 General Probability Rules*

Our study of probability has concentrated on random variables and their distributions. Now we return to the laws that govern any assignment of probabilities. The purpose of learning more laws of probability is to be able to give probability models for more complex random phenomena. We have already met and used five rules.

Rules of Probability

Rule 1. $0 \le P(A) \le 1$ for any event A

Rule 2. $P(S) = 1$

Rule 3. **Complement rule:** For any event A,

$$P(A^c) = 1 - P(A)$$

Rule 4. **Addition rule:** If A and B are **disjoint** events, then

$$P(A \text{ or } B) = P(A) + P(B)$$

Rule 5. **Multiplication rule:** If A and B are **independent** events, then

$$P(A \text{ and } B) = P(A)P(B)$$

---

*This section extends the rules of probability discussed in Section 4.2. This material is not needed to understand the statistical methods in later chapters. It can therefore be omitted if desired.

General addition rules

Probability has the property that if A and B are disjoint events, then $P(A \text{ or } B) = P(A) + P(B)$. What if there are more than two events, or if the events are not disjoint? These circumstances are covered by more general addition rules for probability.

Union

The **union** of any collection of events is the event that at least one of the collection occurs.

For two events A and B, the union is the event $\{A \text{ or } B\}$ that A or B or both occur. From the addition rule for two disjoint events we can obtain rules for more general unions. Suppose first that we have several events—say A, B, and C—that are disjoint in pairs. That is, no two can occur simultaneously. The Venn diagram in Figure 4.16 illustrates three disjoint events. The addition rule for two disjoint events extends to the following law:

Addition Rule for Disjoint Events

If events A, B, and C are disjoint in the sense that no two have any outcomes in common, then

$$P(\text{one or more of } A, B, C) = P(A) + P(B) + P(C)$$

This rule extends to any number of disjoint events.

EXAMPLE 4.28

Generate a random number X between 0 and 1. What is the probability that the first digit will be odd? The random number X is a continuous random variable whose density curve has constant height 1 between 0 and 1 and is 0 elsewhere. The event that the first digit of X is odd is the union of five disjoint events. These events are

$$0.10 \leq X < 0.20$$
$$0.30 \leq X < 0.40$$
$$0.50 \leq X < 0.60$$
$$0.70 < X < 0.80$$
$$0.90 \leq X < 1.00$$

Figure 4.17 illustrates the probabilities of these events as areas under the density curve. Each has probability 0.1 equal to its length. The union of the five therefore has probability equal to the sum, or 0.5. As we should expect, a random number is equally likely to begin with an odd or an even digit.

If events A and B are not disjoint, they can occur simultaneously. The probability of their union is then *less* than the sum of their probabilities. As Figure 4.18 suggests, the outcomes common to both are counted twice when we add probabilities, so we must subtract this probability once. Here is the addition rule for the union of any two events, disjoint or not.

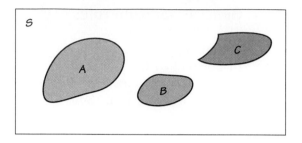

FIGURE 4.16 The addition rule for disjoint events: $P(A \text{ or } B \text{ or } C) = P(A) + P(B) + P(C)$ when events A, B, and C are disjoint.

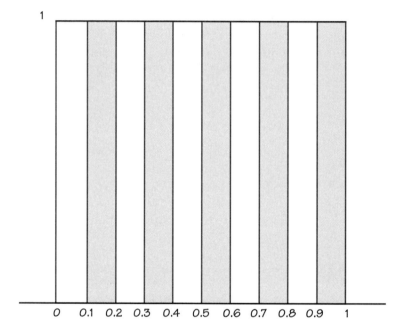

FIGURE 4.17 The probability that the first digit of a random number is odd is the sum of the probabilities of the 5 disjoint events shown.

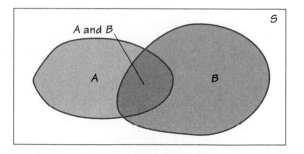

FIGURE 4.18 The general addition rule: $P(A \text{ or } B) = P(A) + P(B) - P(A \text{ and } B)$ for any events A and B.

> ### General Addition Rule for Unions of Two Events
>
> For any two events A and B,
>
> $$P(A \text{ or } B) = P(A) + P(B) - P(A \text{ and } B)$$

If A and B are disjoint, the event $\{A \text{ and } B\}$ that both occur has no outcomes in it. This *empty event* is the complement of the sample space S and must have probability 0. So the general addition rule includes Rule 4, the addition rule for disjoint events.

EXAMPLE 4.29 Deborah and Matthew are anxiously awaiting word on whether they have been made partners of their law firm. Deborah guesses that her probability of making partner is 0.7 and that Matthew's is 0.5. (These are personal probabilities reflecting Deborah's assessment of chance.) This assignment of probabilities does not give us enough information to compute the probability that at least one of the two is promoted. In particular, adding the individual probabilities of promotion gives the impossible result 1.2. If Deborah also guesses that the probability that *both* she and Matthew are made partners is 0.3, then by the addition rule for unions

$$P(\text{at least one is promoted}) = 0.7 + 0.5 - 0.3 = 0.9$$

The probability that *neither* is promoted is then 0.1 by the complement rule.

Venn diagrams are a great help in finding probabilities for unions, because you can just think of adding and subtracting areas. Figure 4.19 shows some events and their probabilities for Example 4.29. What is the probability that Deborah is promoted and Matthew is not? The Venn diagram shows that this is the probability that Deborah is promoted minus the probability that both are promoted, $0.7 - 0.3 = 0.4$. Similarly, the probability that Matthew is promoted and Deborah is not is $0.5 - 0.3 = 0.2$. The four probabilities that

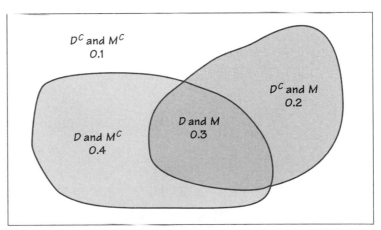

D = Deborah is made partner
M = Matthew is made partner

FIGURE 4.19   Venn diagram and probabilities for Example 4.29.

appear in the figure add to 1 because they refer to four disjoint events whose union is the entire sample space.

Conditional probability

In Section 2.6 we met the idea of a *conditional distribution,* the distribution of a variable given that a condition is satisfied. Now we will introduce the probability language for this idea. Table 4.1 gives data on the age and marital status of adult American women. Choose an adult American woman at random. What is the probability that the woman you choose is married? Table 4.1 shows that there are 99,585,000 adult women in the United States, and that 58,929,000 of them are married. Because "choose at random" gives an equal chance to all women, the probability we want is just the proportion of all women in the population who are married. That is,

$$P(\text{married}) = \frac{58,929}{99,585} = 0.592$$

Suppose we now learn that the woman chosen is between 18 and 24 years old. Does this information change the probability that she is married? We know that young women are less likely to be married, so we are quite sure that the probability for young women is lower than the probability for all women. In fact, there are 12,614,000 women aged 18 to 24, and 3,046,000 of these are married. So the probability that a woman is married *given the information that she is between 18 and 24 years old* is

$$P(\text{married} \mid \text{age 18 to 24}) = \frac{3,046}{12,614} = 0.241$$

conditional probability

This is a **conditional probability.** That is, it gives the probability of one event (the woman chosen is married) under the condition that we know another event (she is between ages 18 and 24). You can read the bar | as "given the information that." It is common sense that if we know a woman is young, we lower our guess of the probability that she is married. If we have the data, we can compute the actual conditional probability. Our task is to turn this common sense into something more general. To do this, we will look at the probabilities of three related events.

TABLE 4.1   Age and marital status of women (thousands of women)

	Age			
	18 to 24	25 to 64	65 and over	Total
Married	3,046	48,116	7,767	58,929
Never married	9,289	9,252	768	19,309
Widowed	19	2,425	8,636	11,080
Divorced	260	8,916	1,091	10,267
Total	12,614	68,709	18,262	99,585

1. 18 to 24 and married.   The counts in Table 4.1 show that if we choose a woman at random, the probability that we draw a married woman between 18 and 24 years old is

$$P(\text{age 18 to 24 and married}) = \frac{3{,}046}{99{,}585} = 0.031$$

This calculation uses a count from the body of the table: there are 3,046,000 women who are both age 18 to 24 and married.

2. 18 to 24.   If we are given no information about marital status, we find probabilities about a woman's age from the "Total" row at the bottom of the table:

18 to 24	25 to 64	65 and over	Total
12,614	68,709	18,262	99,585

The probability that the woman we choose is between 18 and 24 years old is

$$P(\text{age 18 to 24}) = \frac{12{,}614}{99{,}585} = 0.127$$

3. Married, given 18 to 24.   If we are told that the woman is between 18 and 24 years old, we look only at the women in the "18 to 24" column in Table 4.1:

Married	3,046
Never married	9,289
Widowed	19
Divorced	260
Total	12,614

Thus the *conditional* probability that a woman is married, given the information that she is between 18 and 24 years of age, is

$$P(\text{married} \mid \text{age 18 to 24}) = \frac{3{,}046}{12{,}614} = 0.241$$

There is a relationship among these three probabilities. The probability that a woman is both age 18 to 24 *and* married is the product of the probabilities that she is age 18 to 24 and that she is married *given* that she is age 18 to 24. That is,

$$
\begin{aligned}
P(\text{age 18 to 24 } and \text{ married}) &= P(\text{age 18 to 24}) \times P(\text{married} \mid \text{age 18 to 24}) \\
&= \frac{12{,}614}{99{,}585} \times \frac{3{,}046}{12{,}614} \\
&= \frac{3{,}046}{99{,}585} = 0.031 \qquad \text{(as before)}
\end{aligned}
$$

Try to think your way through this in words before looking at the formal notation. We have just discovered the fundamental multiplication rule of probability.

Multiplication Rule

The probability that both of two events A and B happen together can be found by

$$P(A \text{ and } B) = P(A)P(B \mid A)$$

Here $P(B \mid A)$ is the conditional probability that B occurs given the information that A occurs.

In words, this rule says that for both of two events to occur, first one must occur and then, given that the first event has occurred, the second must occur. In our example, the probability that a randomly chosen woman is both age 18 to 24 (event A) and married (event B) is

$$P(A \text{ and } B) = P(A)P(B \mid A)$$
$$= (0.127)(0.241) = 0.031$$

If $P(A)$ and $P(A \text{ and } B)$ are given, we can rearrange the multiplication rule to produce a *definition* of the conditional probability $P(B \mid A)$ in terms of unconditional probabilities.

Definition of Conditional Probability

When $P(A) > 0$, the **conditional probability** of B given A is

$$P(B \mid A) = \frac{P(A \text{ and } B)}{P(A)}$$

Be sure to keep in mind the distinct roles in $P(B \mid A)$ of the event B whose probability we are computing and the event A that represents the information we are given. The conditional probability $P(B \mid A)$ makes no sense if the event A can never occur, so we require that $P(A) > 0$ whenever we talk about $P(B \mid A)$.

EXAMPLE 4.30

What is the conditional probability that a woman is a widow, given that she is at least 65 years old? We see from Table 4.1 that

$$P(\text{at least } 65) = \frac{18{,}262}{99{,}585} = 0.183$$

$$P(\text{widowed } and \text{ at least } 65) = \frac{8{,}636}{99{,}585} = 0.087$$

The conditional probability is therefore

$$P(\text{widowed} \mid \text{at least } 65) = \frac{P(\text{widowed } and \text{ at least } 65)}{P(\text{at least } 65)}$$

$$= \frac{0.087}{0.183} = 0.475$$

Check that this agrees (up to roundoff error) with the result obtained from the "65 and over" column of Table 4.1.

General multiplication rules

The definition of conditional probability reminds us that in principle all probabilities, including conditional probabilities, can be found from the assignment of probabilities to events that describes a random phenomenon. More often, however, conditional probabilities are part of the information given to us in a probability model, and the multiplication rule is used to compute $P(A \text{ and } B)$.

EXAMPLE 4.31

Slim is a professional poker player. At the moment, he wants very much to draw two diamonds in a row. As he sits at the table looking at his hand and at the upturned cards on the table, Slim sees 11 cards. Of these, 4 are diamonds. The full deck contains 13 diamonds among its 52 cards, so 9 of the 41 unseen cards are diamonds. Because the deck was carefully shuffled, each card that Slim draws is equally likely to be any of the cards that he has not seen.

To find Slim's probability of drawing two diamonds, first calculate

$$P(\text{first card diamond}) = \frac{9}{41}$$

$$P(\text{second card diamond} \mid \text{first card diamond}) = \frac{8}{40}$$

Slim finds both probabilities by counting cards. The probability that the first card drawn is a diamond is 9/41 because 9 of the 41 unseen cards are diamonds. If the first card is a diamond, that leaves 8 diamonds among the 40 remaining cards. So the *conditional* probability of another diamond is 8/40. The multiplication rule now says that

$$P(\text{both cards diamonds}) = \frac{9}{41} \times \frac{8}{40} = 0.044$$

Slim will need luck to draw his diamonds.

The union of a collection of events is the event that *any* of them occur. Here is the corresponding term for the event that *all* of them occur.

> **Intersection**
>
> The **intersection** of any collection of events is the event that *all* of the events occur.

The multiplication rule extends to the probability that all of several events occur. The key is to condition each event on the occurrence of *all* of the preceding events. For example, the intersection of three events A, B, and C has probability

$$P(A \text{ and } B \text{ and } C) = P(A)P(B \mid A)P(C \mid A \text{ and } B)$$

EXAMPLE 4.32

Only 5% of male high school basketball, baseball, and football players go on to play at the college level. Of these, only 1.7% enter major league professional sports. About 40% of the athletes who compete in college and then reach the pros have a career of more than 3 years.[8] Define these events:

$$A = \{\text{competes in college}\}$$
$$B = \{\text{competes professionally}\}$$
$$C = \{\text{pro career longer than 3 years}\}$$

What is the probability that a high school athlete competes in college and then goes on to have a pro career of more than 3 years? We know that

$$P(A) = 0.05$$
$$P(B \mid A) = 0.017$$
$$P(C \mid A \text{ and } B) = 0.4$$

The probability we want is therefore

$$P(A \text{ and } B \text{ and } C) = P(A)P(B \mid A)P(C \mid A \text{ and } B)$$
$$= 0.05 \times 0.017 \times 0.40 = 0.00034$$

Only about 3 of every 10,000 high school athletes can expect to compete in college and have a professional career of more than 3 years. High school students would be wise to concentrate on studies rather than on unrealistic hopes of fortune from pro sports.

Tree diagrams

Probability problems often require us to combine several of the basic rules into a more elaborate calculation. Here is an example that illustrates how to solve problems that have several stages.

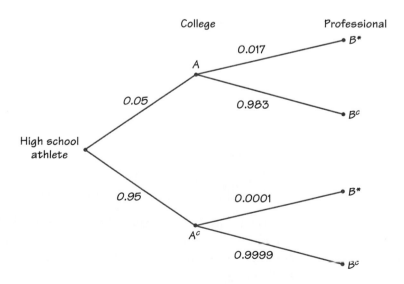

FIGURE 4.20 Tree diagram for Example 4.33. The probability $P(B)$ is the sum of the probabilities of the two marked branches.

EXAMPLE 4.33

tree diagram

What is the probability that a high school athlete will go on to professional sports? In the notation of Example 4.32, this is $P(B)$. To find $P(B)$ from the information in Example 4.32, use the **tree diagram** in Figure 4.20 to organize your thinking.

Each segment in the tree is one stage of the problem. Each complete branch shows a path that an athlete can take. The probability written on each segment is the conditional probability that an athlete follows that segment given that he has reached the point from which it branches. Starting at the left, high school athletes either do or do not compete in college. We know that the probability of competing in college is $P(A) = 0.05$, so the probability of not competing is $P(A^c) = 0.95$. These probabilities mark the leftmost branches in the tree.

Conditional on competing in college, the probability of playing professionally is $P(B \mid A) = 0.017$. So the conditional probability of *not* playing professionally is

$$P(B^c \mid A) = 1 - P(B \mid A) = 1 - 0.017 = 0.983$$

These conditional probabilities mark the paths branching out from A in Figure 4.20.

The lower half of the tree diagram describes athletes who do not compete in college (A^c). It is unusual for these athletes to play professionally, but a few go straight from high school to professional leagues. Suppose that the conditional probability that a high school athlete reaches professional play given that he does not compete in college is $P(B \mid A^c) = 0.0001$. We can now mark the two paths branching from A^c in Figure 4.20.

There are two disjoint paths to B (professional play). By the addition rule, $P(B)$ is the sum of their probabilities. The probability of reaching B through college (top half of the tree) is

$$P(B \text{ and } A) = P(A)P(B \mid A)$$
$$= 0.05 \times 0.017 = 0.00085$$

The probability of reaching B without college is

$$P(B \text{ and } A^c) = P(A^c)P(B \mid A^c)$$
$$= 0.95 \times 0.0001 = 0.000095$$

The final result is

$$P(B) = 0.00085 + 0.000095 = 0.000945$$

About 9 high school athletes out of 10,000 will play professional sports.

Tree diagrams combine the addition and multiplication rules. The multiplication rule says that the probability of reaching the end of any complete branch is the product of the probabilities written on its segments. The probability of any outcome, such as the event B that an athlete reaches professional sports, is then found by adding the probabilities of all branches that are part of that event.

Bayes's rule

There is another kind of probability question that we might ask in the context of studies of athletes. Our earlier calculations look forward toward professional sports as the final stage of an athlete's career. Now let's concentrate on professional athletes and look back at their earlier careers.

EXAMPLE 4.34

What proportion of professional athletes competed in college? In the notation of Examples 4.32 and 4.33 this is the conditional probability $P(A \mid B)$. We start from the definition of conditional probability and then apply the results of Example 4.33:

$$P(A \mid B) = \frac{P(A \text{ and } B)}{P(B)}$$

$$= \frac{0.00085}{0.000945} = 0.8995$$

Almost 90% of professional athletes competed in college.

We know the probabilities $P(A)$ and $P(A^c)$ that a high school athlete does and does not compete in college. We also know the conditional probabilities $P(B \mid A)$ and $P(B \mid A^c)$ that an athlete from each group reaches professional sports. Example 4.33 shows how to use this information to calculate $P(B)$. The method can be summarized in a single expression that adds the probabilities of the two paths to B in the tree diagram:

$$P(B) = P(A)P(B \mid A) + P(A^c)P(B \mid A^c)$$

In Example 4.34 we calculated the "reverse" conditional probability $P(A \mid B)$. The denominator 0.000945 in that example came from the expression just above. Put in this general notation, we have another probability law.

Bayes's Rule

If A and B are any events whose probabilities are not 0 or 1,

$$P(A \mid B) = \frac{P(B \mid A)P(A)}{P(B \mid A)P(A) + P(B \mid A^c)P(A^c)}$$

Bayes's rule is named after Thomas Bayes, who wrestled with arguing from outcomes like B back to antecedents like A in a book published in 1763. It is far better to think your way through problems like Examples 4.33 and 4.34 rather than memorize these formal expressions.

Independence again

The conditional probability $P(B \mid A)$ is generally not equal to the unconditional probability $P(B)$. That is because the occurrence of event A generally gives us some additional information about whether or not event B occurs. If knowing that A occurs gives no additional information about B, then A and B are independent events. The precise definition of independence is expressed in terms of conditional probability.

Independent Events

Two events A and B that both have positive probability are **independent** if
$$P(B \mid A) = P(B)$$

This definition makes precise the informal description of independence given in Section 4.1. We now see that the multiplication rule for independent events, $P(A \text{ and } B) = P(A)P(B)$, is a special case of the general multiplication rule, $P(A \text{ and } B) = P(A)P(B \mid A)$, just as the addition rule for disjoint events is a special case of the general addition rule.

Decision analysis

One kind of decision making in the presence of uncertainty seeks to make the probability of a favorable outcome as large as possible. Here is an example that illustrates how the multiplication and addition rules, organized with the help of a tree diagram, apply to a decision problem.

EXAMPLE 4.35

The kidneys of a patient with end-stage kidney disease will not support life. If the patient is to survive, the available choices are a kidney transplant or regular hemodialysis (use of a kidney machine several times a week). Lynn is faced with this choice. Both treatments are risky. Her doctor gives her the following information for patients in her condition. About 68% of dialysis patients survive for 5 years. Of transplant patients, about 48% survive with the transplanted kidney for 5 years, 43% must undergo regular dialysis because the transplanted kidney fails, and the remaining 9% do not survive the transplant. Of those transplant patients who return to dialysis, about 42% survive 5 years. Which treatment should Lynn choose to give herself the best chance of living for 5 years?[9]

We can organize the information provided by the doctor in a tree diagram (Figure 4.21). Each path through the tree represents a possible outcome of Lynn's case. The probability written beside each branch after the first stage is the conditional probability of the next step given that Lynn has reached this point. For example, 0.68 is the probability that a dialysis patient survives 5 years. The conditional probability that a patient survives 5 years on dialysis given that the patient first had a transplant and then returned to dialysis because the transplant failed is different. It is 0.42, and appears on a different branch of the tree. Study the tree to convince yourself that it organizes all the information available.

The multiplication rule for probabilities says that the probability of reaching the end of a branch in the tree is the product of the probabilities of all the steps along that branch. These probabilities are written at the end of each branch in Figure 4.21. For example, the event that a transplant patient returns to dialysis and then survives 5 years is

$$A = \{D \text{ and } S\}$$

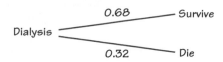

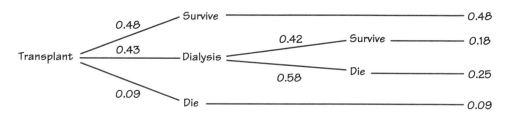

FIGURE 4.21 Tree diagram for the kidney failure decision problem in Example 4.35.

where

$$D = \{\text{transplant patient returns to dialysis}\}$$
$$S = \{\text{patient survives 5 years}\}$$

Therefore

$$P(A) = P(D \text{ and } S)$$
$$= P(D)P(S \mid D)$$
$$= (0.43)(0.42) = 0.18$$

Lynn can now compare her probability of surviving 5 years with regular dialysis with the probability of surviving 5 years with a transplant. For patients who choose to undergo dialysis, the survival probability is 0.68. A transplant patient can survive in either of two ways: with the transplant functioning or, if the transplant has failed, with a return to dialysis. These are *disjoint* events, represented by separate branches of the tree. They have probabilities 0.48 and 0.18, respectively. The addition rule for probability says that the overall probability of surviving 5 years is the sum, 0.48 + 0.18 = 0.66. This is slightly less than the survival probability under dialysis, so dialysis is the better choice. In this example, there are only two decisions to be compared, dialysis or transplant. Other decision problems may have more alternatives.

Where do the conditional probabilities in Example 4.35 come from? They are based in part on data—that is, on studies of many patients with kidney disease. But an individual's chances of survival depend on her age, general health, and other factors. Lynn's doctor considered her individual situation before giving her these particular probabilities. It is characteristic of most decision analysis problems that *personal probabilities* are used to describe the uncertainty of an informed decision maker.

SUMMARY

The **complement** A^c of an event A contains all outcomes that are not in A. The **union** $\{A \text{ or } B\}$ of events A and B contains all outcomes in A, in B, or in both A and B. The **intersection** $\{A \text{ and } B\}$ contains all outcomes that are in both A and B, but not outcomes in A alone or B alone.

The **conditional probability** $P(B \mid A)$ of an event B given an event A is defined by

$$P(B \mid A) = \frac{P(A \text{ and } B)}{P(A)}$$

when $P(A) > 0$ but in practice is most often found from directly available information.

The essential general rules of elementary probability are

Legitimate values: $0 \leq P(A) \leq 1$ for any event A

Total probability 1: $P(S) = 1$

Complement rule: $P(A^c) - 1 - P(A)$

Addition rule: $P(A \text{ or } B) = P(A) + P(B) - P(A \text{ and } B)$

Multiplication rule: $P(A \text{ and } B) = P(A)P(B \mid A)$

If A and B are **disjoint,** then $P(A \text{ and } B) = 0$. The general addition rule for unions then becomes the special addition rule, $P(A \text{ or } B) = P(A) + P(B)$.

A and B are **independent** when $P(B \mid A) = P(B)$. The multiplication rule for intersections then becomes $P(A \text{ and } B) = P(A)P(B)$.

In problems with several stages, draw a **tree diagram** to organize use of the multiplication and addition rules.

SECTION 4.5 EXERCISES

4.75 Call a household prosperous if its income exceeds $75,000. Call the household educated if the householder completed college. Select an American household at random, and let A be the event that the selected household is prosperous and B the event that it is educated. According to the Census Bureau, $P(A) = 0.125$, $P(B) = 0.237$, and the probability that a household is both prosperous and educated is $P(A \text{ and } B) = 0.077$. What is the probability $P(A \text{ or } B)$ that the household selected is either prosperous or educated?

4.76 Consolidated Builders has bid on two large construction projects. The company president believes that the probability of winning the first contract (event A) is 0.6, that the probability of winning the second (event B) is 0.4, and that the probability of winning both jobs (event {*A* and *B*}) is 0.2. What is the probability of the event {*A* or *B*} that Consolidated will win at least one of the jobs?

4.77 Draw a Venn diagram that shows the relation between the events *A* and *B* in Exercise 4.75. Indicate each of the following events on your diagram and use the information in Exercise 4.75 to calculate the probability of each event. Finally, describe in words what each event is.

(a) {*A* and *B*}

(b) {*A* and *B*^{c}}

(c) {*A*^{c} and *B*}

(d) {*A*^{c} and *B*^{c}}

4.78 Draw a Venn diagram that illustrates the relation between events *A* and *B* in Exercise 4.76. Write each of the following events in terms of *A*, *B*, *A*^{c}, and *B*^{c}. Indicate the events on your diagram and use the information in Exercise 4.76 to calculate the probability of each.

(a) Consolidated wins both jobs.

(b) Consolidated wins the first job but not the second.

(c) Consolidated does not win the first job but does win the second.

(d) Consolidated does not win either job.

4.79 Choose an adult American woman at random. Table 4.1 describes the population from which we draw. Use the information in that table to answer the following questions.

(a) What is the probability that the woman chosen is 65 years old or older?

(b) What is the conditional probability that the woman chosen is married, given that she is 65 or over?

(c) How many women are *both* married and in the over-65 age group? What is the probability that the woman we choose is a married woman at least 65 years old?

(d) Verify that the three probabilities you found in (a), (b), and (c) satisfy the multiplication rule.

4.80 Choose an adult American woman at random. Table 4.1 describes the population from which we draw. Use the information in that table to find the following probabilities.

(a) The probability that the woman chosen is a widow.

(b) The conditional probability that the woman chosen is a widow, given that she is at least 65 years old.

(c) The conditional probability that the woman chosen is a widow, given that she is between 25 and 64 years old.

(d) Are the events "widow" and "at least 65 years old" independent? How do you know?

4.81 Choose an adult American woman at random. Table 4.1 describes the population from which we draw.

(a) What is the conditional probability that the woman chosen is 18 to 24 years old, given that she is married?

(b) Earlier, we found that $P(\text{married} \mid \text{age 18 to 24}) = 0.241$. Complete this sentence: 0.241 is the proportion of women who are _____ among those women who are _____.

(c) In (a), you found $P(\text{age 18 to 24} \mid \text{married})$. Write a sentence of the form given in (b) that describes the meaning of this result. The two conditional probabilities give us very different information.

4.82 Here are the counts (in thousands) of earned degrees in the United States in a recent year, classified by level and by the sex of the degree recipient:

	Bachelor's	Master's	Professional	Doctorate	Total
Female	616	194	30	16	856
Male	529	171	44	26	770
Total	1145	365	74	42	1626

(a) If you choose a degree recipient at random, what is the probability that the person you choose is a woman?

(b) What is the conditional probability that you choose a woman, given that the person chosen received a professional degree?

(c) Are the events "choose a woman" and "choose a professional degree recipient" independent? How do you know?

4.83 The previous exercise gives the counts (in thousands) of earned degrees in the United States in a recent year. Use these data to answer the following questions.

(a) What is the probability that a randomly chosen degree recipient is a man?

(b) What is the conditional probability that the person chosen received a bachelor's degree, given that he is a man?

(c) Use the multiplication rule to find the probability of choosing a male bachelor's degree recipient. Check your result by finding this probability directly from the table of counts.

4.84 Here is a two-way table of all suicides committed in a recent year by sex of the victim and method used.

	Male	Female
Firearms	15,802	2,367
Poison	3,262	2,233
Hanging	3,822	856
Other	1,571	571
Total	24,457	6,027

(a) What is the probability that a randomly selected suicide victim is male?

(b) What is the probability that the suicide victim used a firearm?

(c) What is the conditional probability that a suicide used a firearm, given that it was a man? Given that it was a woman?

(d) Describe in simple language (don't use the word "probability") what your results in (c) tell you about the difference between men and women with respect to suicide.

4.85 Common sources of caffeine in the diet are coffee, tea, and cola drinks. Suppose that

55% of adults drink coffee

25% of adults drink tea

45% of adults drink cola

and also that

15% drink both coffee and tea

5% drink all three beverages

25% drink both coffee and cola

5% drink only tea

Draw a Venn diagram marked with this information. Use it along with the addition rules to answer the following questions.

(a) What percent of adults drink only cola?

(b) What percent drink none of these beverages?

4.86 Choose an employed person at random. Let A be the event that the person chosen is a woman, and B the event that the person holds a managerial or professional job. Government data tell us that $P(A) = 0.46$ and the probability of managerial and professional jobs among women is $P(B \mid A) = 0.32$. Find the probability that a randomly chosen employed person is a woman holding a managerial or professional position.

4.87 Functional Robotics Corporation buys electrical controllers from a Japanese supplier. The company's treasurer thinks that there is probability 0.4 that the dollar will fall in value against the Japanese yen in the next month. The treasurer also believes that *if* the dollar falls there is probability 0.8 that the supplier will demand renegotiation of the contract. What probability has the treasurer assigned to the event that the dollar falls and the supplier demands renegotiation?

4.88 In the language of government statistics, the "labor force" includes all civilians over 16 years of age who are working or looking for work. Select a member of the U.S. labor force at random. Let A be the event that the person selected is white, and B the event that he or she is employed. In 1995, 84.6% of the labor force was white. Of the whites in the labor force,

95.1% were employed. Among nonwhite members of the labor force, 91.9% were employed.

(a) Express each of the percents given as a probability involving the events A and B; for example, $P(A) = 0.846$.

(b) Draw a tree diagram for the outcomes of recording first the race (white or nonwhite) of a randomly chosen member of the labor force and then whether or not the person is employed.

(c) Find the probability that the person chosen is an employed white. Also find the probability that an employed nonwhite is chosen. What is the probability $P(B)$ that the person chosen is employed?

4.89 Suppose that in Exercise 4.87 the treasurer also feels that if the dollar does not fall there is probability 0.2 that the Japanese supplier will demand that the contract be renegotiated. What is the probability that the supplier will demand renegotiation?

4.90 Use your results from Exercise 4.88 and the definition of conditional probability to find the probability $P(A \mid B)$ that a randomly selected member of the labor force is white, given that he or she is employed. (Alternatively, you can use Bayes's rule.)

4.91 An examination consists of multiple-choice questions, each having five possible answers. Linda estimates that she has probability 0.75 of knowing the answer to any question that may be asked. If she does not know the answer, she will guess, with conditional probability 1/5 of being correct. What is the probability that Linda gives the correct answer to a question? (Draw a tree diagram to guide the calculation.)

4.92 The voters in a large city are 40% white, 40% black, and 20% Hispanic. (Hispanics may be of any race in official statistics, but in this case we are speaking of political blocks.) A black mayoral candidate anticipates attracting 30% of the white vote, 90% of the black vote, and 50% of the Hispanic vote. Draw a tree diagram with probabilities for the race (white, black, or Hispanic) and vote (for or against the candidate) of a randomly chosen voter. What percent of the overall vote does the candidate expect to get?

4.93 In the setting of Exercise 4.91, find the conditional probability that Linda knows the answer, given that she supplies the correct answer. (You can use the result of Exercise 4.91 and the definition of conditional probability, or you can use Bayes's rule.)

4.94 Choose a point at random in the square with sides $0 \le x \le 1$ and $0 \le y \le 1$. This means that the probability that the point falls in any region within the square is the area of that region. Let X be the x coordinate and Y the y coordinate of the point chosen. Find the conditional probability $P(Y < 1/2 \mid Y > X)$. (*Hint:* Draw a diagram of the square and the events $Y < 1/2$ and $Y > X$.)

4.95   Psychologists use probability models to describe learning in animals. In one experiment, a rat is placed in a dark compartment and a door leading to a light compartment is opened. The rat will not move to the light compartment without reason. A bell is rung and if after 5 seconds the rat has not moved, it gets a shock through the floor of the dark compartment. To avoid the shock, rats soon learn to move when the bell is rung. A simple model for learning says that the rat can only be in one of two states:

> *State A:* The rat will not move from the dark compartment until it receives a shock.

> *State B:* The rat has learned to respond to the bell and moves immediately when the bell rings.

A rat starts in State A and eventually changes to State B. The change from State A to State B is the result of learning. A rat in State A gets a shock every time the bell rings; after it changes to State B, it never gets a shock. Suppose that a rat has probability 0.2 of learning each time it is shocked. Let the random variable X be the number of shocks that this rat receives.[10]

(a) If a rat receives exactly 4 shocks, which state was the rat in at the end of each of Trials 1, 2, 3, and 4?

(b) Use the result of (a) and the multiplication rule to find $P(X = 4)$, the probability that the rat receives exactly 4 shocks.

(c) Based on your work in (a) and (b), give the probability distribution of X. That is, for any positive whole number x, what is $P(X = x)$?

4.96   John has coronary artery disease. He and his doctor must decide between medical management of the disease and coronary bypass surgery. Because John has been quite active, he is concerned about his quality of life as well as the length of life. He wants to make the decision that will maximize the probability of the event A that he survives for 5 years and is able to carry on moderate activity during that time. The doctor makes the following probability estimates for patients of John's age and condition:

- Under medical management, $P(A) = 0.7$.

- There is probability 0.05 that John will not survive bypass surgery, probability 0.10 that he will survive with serious complications, and probability 0.85 that he will survive the surgery without complications.

- If he survives with complications, the conditional probability of the desired outcome A is 0.73. If there are no serious complications, the conditional probability of A is 0.76.

Draw a tree diagram that summarizes this information. Then calculate $P(A)$ assuming that John chooses the surgery. Does surgery or medical management offer him the better chance of achieving his goal? (Based loosely on M. C. Weinstein, J. S. Pliskin, and W. B. Stason, "Coronary artery bypass surgery: decision and policy analysis," in J. P. Bunker, B. A. Barnes, and F. W. Mosteller (eds.), *Costs, Risks and Benefits of Surgery,* Oxford University Press, New York, 1977, pp. 342–371.)

4.97 Zipdrive, Inc., has developed a new disk drive for small computers. The demand for the new product is uncertain but can be described as "high" or "low" in any one year. After 4 years, the product is expected to be obsolete. Management must decide whether to build a plant or to contract with a factory in Hong Kong to manufacture the new drive. Building a plant will be profitable if demand remains high but could lead to a loss if demand drops in future years.

After careful study of the market and of all relevant costs, Zipdrive's planning office provides the following information. Let A be the event that the first year's demand is high, and B be the event that the following 3 years' demand is high. The marketing division's best estimate of the probabilities is

$$P(A) = 0.9$$
$$P(B \mid A) = 0.36$$
$$P(B \mid A^c) = 0$$

The probability that building a plant is more profitable than contracting the production to Hong Kong is 0.95 if demand is high all 4 years, 0.3 if demand is high only in the first year, and 0.1 if demand is low all 4 years.

Draw a tree diagram that organizes this information. The tree will have three stages: first year demand, next 3 years' demand, and whether building or contracting is more profitable. Which decision has the higher probability of being more profitable? (When decision analysis is used for investment decisions like this, firms in fact compare the mean profits rather than the probability of a profit. We ignore this complication.)

CHAPTER 4 EXERCISES

4.98 The original simple form of the Connecticut state lottery (ignoring a few gimmicks) awarded the following prizes for each 100,000 tickets sold. The winners were chosen by drawing tickets at random.

1	$5000 prize
18	$200 prizes
120	$25 prizes
270	$20 prizes

If you hold one ticket in this lottery, what is your probability of winning anything? What is the mean amount of your winnings?

4.99 Rotter Partners is planning a major investment. The amount of profit X is uncertain but a probabilistic estimate gives the following distribution (in millions of dollars):

Profit	1	1.5	2	4	10
Probability	0.1	0.2	0.4	0.2	0.1

(a) Find the mean profit μ_X and the standard deviation of the profit.

(b) Rotter Partners owes its source of capital a fee of $200,000 plus 10% of the profits X. So the firm actually retains

$$Y = 0.9X - 0.2$$

from the investment. Find the mean and standard deviation of Y.

4.100 A grocery store gives its customers cards that may win them a prize when matched with other cards. The back of the card announces the following probabilities of winning various amounts if a customer visits the store 10 times:

Amount	$1000	$200	$50	$10
Probability	1/10,000	1/1000	1/100	1/20

(a) What is the probability of winning nothing?

(b) What is the mean amount won?

(c) What is the standard deviation of the amount won?

4.101 The number of offspring produced by a female Asian stochastic beetle is random, with this pattern: 20% of females die without female offspring, 30% have one female offspring, and 50% have two female offspring. Females of the benign boiler beetle have this reproductive pattern: 40% die without female offspring, 40% have one female offspring, and 20% have two female offspring.

(a) Find the mean number of female offspring for each species of beetle.

(b) Use the law of large numbers to explain why the population should grow if the expected number of female offspring is greater than 1 and die out if this expected value is less than 1.

4.102 A study of the weights of the brains of Swedish men found that the weight X was a random variable with mean 1400 grams and standard deviation 20 grams. Find positive numbers a and b such that $Y = a + bX$ has mean 0 and standard deviation 1.

4.103 You are playing a board game in which the severity of a penalty is determined by rolling three dice and adding the spots on the up faces. The dice are all balanced so that each face is equally likely, and the three dice fall independently.

(a) Give a sample space for the sum X of the spots.

(b) Find $P(X = 5)$.

(c) If X_1, X_2, and X_3 are the number of spots on the up faces of the three dice, then $X = X_1 + X_2 + X_3$. Use this fact to find the mean μ_X and the standard deviation σ_X without finding the distribution of X. (Start with the distribution of each of the X_i.)

4.104 The real return on an investment is its rate of increase corrected for the effects of inflation. You believe that the annual real return X on a portfolio of stocks will vary in the future with mean $\mu_X = 0.11$ and standard deviation $\sigma_X = 0.28$. (That is, you expect stocks to give an average return of 11% in the future, but with large variation from year to year.) You further think that the annual real return Y on Treasury bills will vary with mean $\mu_Y = 0.02$ and standard deviation $\sigma_Y = 0.05$. Even though this is not realistic, assume that returns on stocks and Treasury bills vary independently.

(a) If you put half of your assets into stocks and half into Treasury bills, your overall return will be $Z = 0.5X + 0.5Y$. Calculate μ_Z and σ_Z.

(b) You decide that you are willing to take more risk (greater variation in the return) in exchange for a higher mean return. Choose an allocation of your assets between stocks and Treasury bills that will accomplish this and calculate the mean and standard deviation of the overall return. (If you put a proportion α of your assets into stocks, the total return is $Z = \alpha X + (1 - \alpha)Y$.)

4.105 The most popular game of chance in Roman times was tossing four astragali. An astragalus is a small six-sided bone from the heel of an animal that comes to rest on one of four sides when tossed. (The other two sides are rounded.) The table gives the probabilities of the outcomes for a single astragalus based on modern experiments. The names "broad convex" etc. describe the four sides of the heel bone. The best throw was the "Venus," with all four uppermost sides different. What is the probability of rolling a Venus? (From Florence N. David, *Games, Gods and Gambling*, Charles Griffin, London, 1962, p. 7.)

Side	Broad convex	Broad concave	Narrow flat	Narrow hollow
Probability	0.4	0.4	0.1	0.1

4.106 Ann and Bob are playing the game Two-Finger Morra. Each player shows either one or two fingers and at the same time calls out a guess for the number of fingers the other player will show. If a player guesses correctly and the other player does not, the player wins a number of dollars equal to the total number of fingers shown by both players. If both or neither guesses correctly, no money changes hands. On each play both Ann and Bob choose one of the following options:

Choice	Show	Guess
A	1	1
B	1	2
C	2	1
D	2	2

(a) Give the sample space S by writing all possible choices for both players on a single play of this game.

(b) Let X be Ann's winnings on a play. (If Ann loses \$2, then $X = -2$; when no money changes hands, $X = 0$.) Write the value of the random variable X next to each of the outcomes you listed in (a). This is another choice of sample space.

(c) Now assume that Ann and Bob choose independently of each other. Moreover, they both play so that all four choices listed above are equally likely. Find the probability distribution of X.

(d) If the game is fair, X should have mean zero. Does it? What is the standard deviation of X?

The following exercises require familiarity with the material presented in the optional Section 4.5.

4.107 Here is the distribution of the adjusted gross income X (in thousands of dollars) reported on individual federal income tax returns in 1993:

Income	< 10	10–24	25–49	50–99	≥ 100
Probability	0.29	0.27	0.25	0.14	0.05

(a) What is the probability that a randomly chosen return shows an adjusted gross income of \$50,000 or more?

(b) Given that a return shows an income of at least \$50,000, what is the conditional probability that the income is at least \$100,000?

4.108 You have torn a tendon and are facing surgery to repair it. The orthopedic surgeon explains the risks to you. Infection occurs in 3% of such operations, the repair fails in 14%, and both infection and failure occur together in 1%. What percent of these operations succeed and are free from infection?

4.109 The distribution of blood types among white Americans is approximately as follows: 37% type A, 13% type B, 44% type O, and 6% type AB. Suppose that the blood types of married couples are independent and that both the husband and wife follow this distribution.

(a) An individual with type B blood can safely receive transfusions only from persons with type B or type O blood. What is the probability that the husband of a woman with type B blood is an acceptable blood donor for her?

(b) What is the probability that in a randomly chosen couple the wife has type B blood and the husband has type A?

(c) What is the probability that one of a randomly chosen couple has type A blood and the other has type B?

(d) What is the probability that at least one of a randomly chosen couple has type O blood?

4.110 Exercise 4.25 (page 309) gives the probability distribution of the sex and occupation of a randomly chosen American worker. Use this distribution to answer the following questions:

(a) Given that the worker chosen holds a managerial (Class A) job, what is the conditional probability that the worker is female?

(b) Classes D and E include most mechanical and factory jobs. What is the conditional probability that a worker is female, given that he or she holds a job in one of these classes?

4.111 It is difficult to conduct sample surveys on sensitive issues because many people will not answer questions if the answers might embarrass them. "Randomized response" is an effective way to guarantee anonymity while collecting information on topics such as student cheating or sexual behavior. Here is the idea. To ask a sample of students whether they have plagiarized a term paper while in college, have each student toss a coin in private. If the coin lands "heads" *and* they have not plagiarized, they are to answer "No." Otherwise they are to give "Yes" as their answer. Only the student knows whether the answer reflects the truth or just the coin toss, but the researchers can use a proper random sample with follow-up for nonresponse and other good sampling practices.

Suppose that in fact the probability is 0.3 that a randomly chosen student has plagiarized a paper. Draw a tree diagram in which the first stage is tossing the coin and the second is the truth about plagiarism. The outcome at the end of each branch is the answer given to the randomized-response question. What is the probability of a "No" answer in the randomized-response poll? If the probability of plagiarism were 0.2, what would be the probability of a "No" response on the poll? Now suppose that you get 39% "No" answers in a randomized-response poll of a large sample of students at your college. What do you estimate to be the percent of the population who have plagiarized a paper?

4.112 ELISA tests are used to screen donated blood for the presence of the AIDS virus. The test actually detects antibodies, substances that the body produces when the virus is present. When antibodies are present, ELISA is positive with probability about 0.997 and negative with probability 0.003. When the blood tested is not contaminated with AIDS antibodies, ELISA gives a positive result with probability about 0.015 and a negative result with probability 0.985.[11] (Because ELISA is designed to keep the AIDS virus out of blood supplies, the higher probability 0.015 of a false positive is acceptable in exchange for the low probability 0.003 of failing to detect contaminated blood. These probabilities depend on the expertise of the particular laboratory doing the test.) Suppose that 1% of a large population carries the AIDS antibody in their blood.

   (a) Draw a tree diagram for selecting a person from this population (outcomes: the person does or does not carry the AIDS antibody) and for testing his or her blood (outcomes: positive or negative).

   (b) What is the probability that the ELISA test for AIDS is positive for a randomly chosen person from this population?

   (c) What is the probability that a person has the antibody given that the ELISA test is positive?

(This exercise illustrates a fact that is important when considering proposals for widespread testing for AIDS or illegal drugs: if the condition being tested is uncommon in the population, most positives will be false positives.)

4.113   Toss a balanced coin 10 times. What is the probability of a run of 3 or more consecutive heads? What is the distribution of the length of the longest run of heads? What is the mean length of the longest run of heads? These are quite difficult questions if we must rely on mathematical calculations of probability. Computer simulation of the probability model can provide approximate answers.

   Simulate 50 repetitions of tossing a balanced coin 10 times. (The key word in most software is "Bernoulli.") Now examine your 50 repetitions. Record the length of the longest run of heads in each trial. Combine your results with those of other students so that you have the results of several hundred repetitions.

   (a) Make a table of the (approximate) probability distribution of the length X of the longest run of heads in 10 coin tosses. (The proportion of each outcome in many repetitions is approximately equal to its probability.) Draw a probability histogram of this distribution.

   (b) What is your estimate of the probability of a run of 3 or more heads?

   (c) Find the mean μ_X from your table of probabilities in (a). Then find the average of the 50 values of X in your 50 repetitions. How close to the mean μ_X was the average obtained in 50 repetitions?

NOTES

1.   An informative and entertaining account of the origins of probability theory is Florence N. David, *Games, Gods and Gambling*, Charles Griffin, London, 1962.

2.   We use $\bar{x}$ both for the random variable, which takes different values in repeated sampling, and for the numerical value of the random variable in a particular sample. Similarly, s and $\hat{p}$ stand both for random variables and for specific values. This notation is mathematically imprecise but statistically convenient.

3.   We will consider only the case in which X takes a finite number of possible values. The same ideas, implemented with more advanced mathematics, apply to random variables with an infinite but still countable collection of values.

4. The mean of a continuous random variable X with density function $f(x)$ can be found by integration:

$$\mu_X = \int x f(x) \, dx$$

This integral is a kind of weighted average, analogous to the discrete-case mean

$$\mu_X = \sum x P(X = x)$$

The variance of a continuous random variable X is the average squared deviation of the values of X from their mean, found by the integral

$$\sigma_X^2 = \int (x - \mu)^2 f(x) \, dx$$

5. See A. Tversky and D. Kahneman, "Belief in the law of small numbers," *Psychological Bulletin*, 76 (1971), pp. 105–110, and other writings of these authors for a full account of our misperception of randomness.

6. Probabilities involving runs can be quite difficult to compute. That the probability of a run of three or more heads in 10 independent tosses of a fair coin is $(1/2) + (1/128) = 0.508$ can be found by clever counting, as can the other results given in the text. A general treatment using advanced methods appears in Section XIII.7 of William Feller, *An Introduction to Probability Theory and Its Applications*, vol. 1, 3rd ed., Wiley, New York, 1968.

7. R. Vallone and A. Tversky, "The hot hand in basketball: on the misperception of random sequences," *Cognitive Psychology*, 17 (1985), pp. 295–314. A later series of articles that debate the independence question is A. Tversky and T. Gilovich, "The cold facts about the 'hot hand' in basketball," *Chance*, 2, no. 1 (1989), pp. 16–21; P. D. Larkey, R. A. Smith, and J. B. Kadane, "It's OK to believe in the 'hot hand,'" *Chance*, 2, no. 4 (1989), pp. 22–30; and A. Tversky and T. Gilovich, "The 'hot hand': statistical reality or cognitive illusion?" *Chance*, 2, no. 4 (1989), pp. 31–34.

8. These probabilities come from studies by the sociologist Harry Edwards, reported in the *New York Times*, February 25, 1986.

9. The probabilities in this example are based on the article by Benjamin A. Barnes, "An overview of the treatment of end-stage renal disease and a consideration of some of the consequences," in J. P. Bunker, B. A. Barnes, and F. W. Mosteller (eds.), *Costs, Risks and Benefits of Surgery*, Oxford University Press, New York, 1977, pp. 325–341. See this article for a more realistic analysis.

10. This model is too simple to fit experimental data well. Probability models for learning that give more realistic descriptions of experiments with rats are discussed in R. C. Atkinson, G. H. Bower, and E. J. Crothers, *An Introduction to Mathematical Learning Theory*, Wiley, New York, 1965.

11. These probabilities are estimated from a large national study reported in E. M. Sloand et al., "HIV testing: state of the art," *Journal of the American Medical Association*, 266 (1991), pp. 2861–2866.

Prelude

Exploratory data analysis draws conclusions about the individuals for whom we actually have data. Statistical inference infers from the data conclusions about a wider population or process. Our ability to generalize beyond the individuals we actually measure depends on designs for data production and on the behavior of numerical summaries—statistics— calculated from our data. The reasoning of inference is based on asking, "What would happen if we did this many times?" Answers to this question use the language of probability to describe the long-term behavior of statistics such as sample counts, proportions, and means. The details of this behavior are the topic of this chapter. We will both enlarge our knowledge of probability and prepare for our study of inference. Here are some questions we will soon answer:

- A basketball player who makes 75% of her free throws over the course of a season misses 5 of 12 in a crucial game. The fans think she choked. What is the probability she misses 5 of 12 just by chance?

- A study of the effect of a cholesterol-reducing drug on heart attacks plans to give the drug to 2000 men and a placebo to another 2000. Do we expect to see enough heart attacks among these men to draw conclusions about the drug, or are larger samples needed?

- It is good practice to repeat laboratory measurements several times and report the mean result. What is the advantage of the mean over a single measurement?

CHAPTER 5

Gustavo Ramos Rivera, *Untitled*, 1993.

From Probability to Inference

Introduction

Statistical inference draws conclusions about a population or process on the basis of data. The data are summarized by *statistics* such as means, proportions, and the slopes of least-squares regression lines. When the data are produced by random sampling or randomized experimentation, a statistic is a random variable that obeys the laws of probability theory. The link between probability and data is formed by the *sampling distributions* of statistics. A sampling distribution shows how a statistic would vary in repeated data production. That is, a sampling distribution is a probability distribution that answers the question "What would happen if we did this many times?" In Section 3.4 we looked at several sampling distributions empirically by simulating 1000 random samples. Considered as a probability distribution, a sampling distribution is an idealized mathematical description of the results of an indefinitely large number of samples rather than the results of a particular 1000 samples.

The Distribution of a Statistic

A statistic from a random sample or randomized experiment is a random variable. The probability distribution of the statistic is its **sampling distribution.**

Probability distributions also play a second role in statistical inference. Any quantity that can be measured for each member of a population is described by the distribution of its values for all members of the population. This is the context in which we first met distributions, as models for the overall pattern of data. Imagine choosing one individual at random from the population. The results of repeated choices have a probability distribution that is the same as the distribution for the population.

EXAMPLE 5.1 | The distribution of heights of women between the ages of 18 and 24 is approximately normal with mean 64.5 inches and standard deviation 2.5 inches. Select a woman at random and measure her height. The result is a random variable X. We don't know the height of a randomly chosen woman, but we do know that in repeated sampling X will have the same $N(64.5, 2.5)$ distribution that describes the pattern of heights in the entire population. We call $N(64.5, 2.5)$ the *population distribution*.

Population Distribution

The **population distribution** of a variable is the distribution of its values for all members of the population. The population distribution is also the probability distribution of the variable when we choose one individual from the population at random.

The population of all women between the ages of 18 and 24 actually exists, so that we can in principle draw an SRS from it. Sometimes the population that we want to draw conclusions about does not actually exist in the physical world. For example, we might observe an industrial process producing rods with lengths that vary according to a normal distribution. The population contains all rods that would be produced if the process continued forever in its present state. That population does not actually exist, so we cannot draw an SRS from it. Yet we are often justified in treating the rods as if they were independent observations randomly chosen from this hypothetical population. Once again we describe the population (or the process, if you prefer) by the normal distribution that governs individual observations.

To progress from discussing probability as a topic in itself to probability as a foundation for inference, we must study the sampling distributions of some common statistics. We will meet other sampling distributions later in our study of inference. In each case, the nature of the sampling distribution depends on both the nature of the population distribution and the way we collect the data from the population.

5.1   Sampling Distributions for Counts and Proportions

A sample survey asks 2500 adults whether they agree that "I like buying new clothes, but shopping is often frustrating and time-consuming." The number who say "Yes" is a random variable X. A quality control engineer selects a sample of 10 switches from a supplier's shipment for detailed inspection. A number X fail to meet the specifications. In both of these examples, the random variable X is a **count** of the occurrences of some outcome in a fixed number of observations. There are 2500 observations in the first example and 10 observations in the second. If the number of observations is n, then the **sample proportion** is $\hat{p} = X/n$. For example, if 1650 of the 2500 shoppers say "Yes," the sample proportion is

count

sample proportion

$$\hat{p} = \frac{1650}{2500} = 0.66$$

Sample counts and sample proportions are common statistics. Finding their sampling distributions is the goal of this section.

The binomial distributions for sample counts

The distribution of a count X depends on how the data are produced. Here is a simple but common situation.

> ## The Binomial Setting
>
> 1. There are a fixed number n of observations.
> 2. The n observations are all independent.
> 3. Each observation falls into one of just two categories, which for convenience we call "success" and "failure."
> 4. The probability of a success, call it p, is the same for each observation.

Think of tossing a coin n times as an example of the binomial setting. Each toss gives either heads or tails. The outcomes of successive tosses are independent. If we call heads a success, then p is the probability of a head and remains the same as long as we toss the same coin. The number of heads we count is a random variable X. The distribution of X, and more generally of the count of successes in any binomial setting, is completely determined by the number of observations n and the success probability p.

> ## Binomial Distributions
>
> The distribution of the count X of successes in the binomial setting is called the **binomial distribution** with parameters n and p. The parameter n is the number of observations, and p is the probability of a success on any one observation. The possible values of X are the whole numbers from 0 to n. As an abbreviation, we say that X is $B(n, p)$.

The binomial distributions are an important class of discrete probability distributions. Later in this section we will learn how to assign probabilities to outcomes and how to find the mean and standard deviation of binomial distributions. But the most important aspect of using binomial distributions is the ability to recognize situations to which they apply.

EXAMPLE 5.2

(a) Toss a balanced coin 10 times and count the number X of heads. There are $n = 10$ observations. Successive tosses are independent. If the coin is balanced, the probability of a head is $p = 0.5$ on each toss. The number of heads we observe has the binomial distribution $B(10, 0.5)$.

(b) Deal 10 cards from a shuffled deck and count the number X of red cards. There are 10 observations, and each gives either a red or a black card. But the observations are *not* independent. If the first card is black, the second is more likely to be red because there are now more red cards than black cards remaining in the deck. The count X does *not* have a binomial distribution.

(c) Genetics says that children receive genes from their parents independently. Each child of a particular pair of parents has probability 0.25 of having type O blood. If these parents have 5 children, the number who have type O blood is the count X of successes in 5 independent trials with probability 0.25 of a success on each trial. So X has the $B(5, 0.25)$ distribution.

(d) Engineers define reliability as the probability that an item will perform its function under specific conditions for a specific period of time. If an aircraft engine turbine has probability 0.999 of performing properly for an hour of flight, the number of turbines in a fleet of 350 engines that fly for an hour without failure has the $B(350, 0.999)$ distribution. This binomial distribution is obtained by assuming, as seems reasonable, that the turbines fail independently of each other. A common cause of failure, such as sabotage, would destroy the independence and make the binomial model inappropriate.

Binomial distributions in statistical sampling

The binomial distributions are important in statistics when we wish to make inferences about the proportion p of "successes" in a population. Here is a typical example.

EXAMPLE 5.3

A quality engineer chooses an SRS of 10 switches from a large shipment. Unknown to the engineer, 10% of the switches in the shipment do not conform to the specifications. The engineer counts the number X of nonconforming switches in the sample.

This is not quite a binomial setting. Removing one switch changes the proportion of bad switches in the remaining population, so the state of the second switch chosen is not independent of the first.

If the shipment is large, however, removing a few items has a very small effect on the composition of the remaining population. Successive inspection results are very nearly independent. Suppose that the shipment contains 10,000 switches, of which 1000 are nonconforming. The proportion nonconforming is

$$p = \frac{1,000}{10,000} = 0.1$$

If the first switch chosen is bad, the proportion of bad switches remaining is $999/9999 = 0.0999$. If the first switch is good, the proportion of bad switches left in the shipment is $1000/9999 = 0.10001$. These proportions are so close to 0.1 that for practical purposes we can act as if removing one switch has no effect on the proportion of bad switches remaining. We act as if the count X of switches in the sample that fail inspection has the binomial distribution $B(10, 0.1)$.

In Example 5.3 and similar settings, the population distribution is just the distribution of successes and failures in the population, described by the single parameter p. The sampling distribution of the count of successes in an SRS from the population is nearly binomial if the population is large enough to allow us to ignore the small dependence between observations. The accuracy of this approximation improves as the size of the population increases relative to the size of the sample.

> ### Sampling Distribution of a Count
>
> When the population is much larger than the sample, the count X of successes in an SRS of size n has approximately the $B(n, p)$ distribution if the population proportion of successes is p.
>
> As a rule of thumb, we will use the binomial sampling distribution for counts when the population is at least 10 times as large as the sample.

Finding binomial probabilities: tables

We will later give a formula for the probability that a binomial random variable takes any of its values. In practice, you will rarely have to use this formula for calculations. Some calculators and most statistical software packages calculate binomial probabilities. If you do not have suitable computing facilities, you can still shorten the work of calculating binomial probabilities for some values of n and p by looking up probabilities in Table C in the back of this book. The entries in the table are the probabilities $P(X = k)$ of individual outcomes for a binomial random variable X.

EXAMPLE 5.4

		p
n	k	0.10
10	0	.3487
	1	.3874
	2	.1937
	3	.0574
	4	.0112
	5	.0015
	6	.0001
	7	.0000
	8	.0000
	9	.0000
	10	.0000

A quality engineer selects an SRS of 10 switches from a large shipment for detailed inspection. Unknown to the engineer, 10% of the switches in the shipment fail to meet the specifications. What is the probability that no more than 1 of the 10 switches in the sample fails inspection?

Example 5.3 reminds us that the count X of bad switches in the sample has approximately the $B(10, 0.1)$ distribution. Figure 5.1 is a probability histogram for this distribution. The distribution is strongly skewed. Although X can take any whole-number value from 0 to 10, the probabilities of values larger than 5 are so small that they do not appear in the histogram.

We want to calculate

$$P(X \leq 1) = P(X = 1) + P(X = 0)$$

when X is $B(10, 0.1)$. Your software may do this—look for the key word "binomial." To use Table C for this calculation, look opposite $n = 10$ and under $p = 0.10$. This part of the table appears at the left. The entry opposite each k is $P(X = k)$. We find

$$P(X \leq 1) = P(X = 1) + P(X = 0)$$
$$= 0.3874 + 0.3487 = 0.7361$$

About 74% of all samples will contain no more than 1 bad switch. In fact, 35% of the samples will contain no bad switches. A sample of size 10 cannot be trusted to alert the engineer to the presence of unacceptable items in the shipment.

The excerpt from Table C contains the full $B(10, 0.1)$ distribution. The probabilities are rounded to four decimal places. Outcomes larger than 6 do not have probability exactly 0, but their probabilities are so small that the

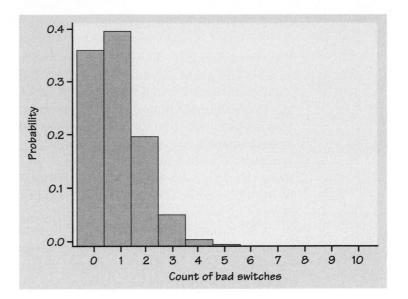

FIGURE 5.1 Probability histogram for the binomial distribution with $n = 10$ and $p = 0.1$.

rounded values are 0.0000. Check that the sum of the probabilities given is 1, as it should be.

The values of p that appear in Table C are all 0.5 or smaller. When the probability of a success is greater than 0.5, restate the problem in terms of the number of failures. The probability of a failure is less than 0.5 when the probability of a success exceeds 0.5. When using the table, always stop to ask whether you must count successes or failures.

EXAMPLE 5.5

Corinne is a basketball player who makes 75% of her free throws over the course of a season. In a key game, Corinne shoots 12 free throws and misses 5 of them. The fans think that she failed because she was nervous. Is it unusual for Corinne to perform this poorly?

To answer this question, assume that free throws are independent with probability 0.75 of a success on each shot. (Studies of long sequences of free throws have found no evidence that they are dependent, so this is a reasonable assumption.) Because the probability of making a free throw is greater than 0.5, we count misses in order to use Table C. The probability of a miss is $1 - 0.75$, or 0.25. The number X of misses in 12 attempts has the $B(12, 0.25)$ distribution.

We want the probability of missing 5 or more. This is

$$P(X \geq 5) = P(X = 5) + P(X = 6) + \cdots + P(X = 12)$$
$$= 0.1032 + 0.0401 + \cdots + 0.0000 = 0.1576$$

Corinne will miss 5 or more out of 12 free throws about 16% of the time, or roughly one of every six games. While below her average level, this performance is well within the range of the usual chance variation in her shooting.

Binomial mean and standard deviation

If a count X is $B(n, p)$, what are the mean μ_X and the standard deviation σ_X? We can guess the mean. If Corinne makes 75% of her free throws, the mean number made in 12 tries should be 75% of 12, or 9. That's μ_X when X is $B(12, 0.75)$. Intuition suggests more generally that the mean of the $B(n, p)$ distribution should be np. Can we show that this is correct and also obtain a short formula for the standard deviation? Because binomial distributions are discrete probability distributions, we could find the mean and variance by using the definitions in Section 4.4. For example, we could obtain the mean by multiplying each outcome by its probability and adding over all outcomes. Here is an easier way.

A binomial random variable X is the count of successes in n independent observations that each have the same probability p of success. Let the random variable S_i indicate whether the ith observation is a success or failure by taking the values $S_i = 1$ if a success occurs and $S_i = 0$ if the outcome is a failure. The S_i are independent because the observations are, and each S_i has the same simple distribution:

Outcome	1	0
Probability	p	$1 - p$

From the definition of the mean of a discrete random variable (page 327), we know that the mean of each S_i is

$$\mu_S = (1)(p) + (0)(1 - p) = p$$

Similarly, the definition of the variance (page 336) shows that $\sigma_S^2 = p(1 - p)$. Because each S_i is 1 for a hit and 0 for a miss, the total number of hits X is just the sum of the S_i's:

$$X = S_1 + S_2 + \cdots + S_n$$

Apply the addition rules for means and variances to this sum. The mean of X is the sum of the means of the S_i's:

$$\mu_X = \mu_{S_1} + \mu_{S_2} + \cdots + \mu_{S_n}$$
$$= n\mu_S = np$$

Similarly, the variance is n times the variance of a single S, so that $\sigma_X^2 = np(1 - p)$. The standard deviation σ_X is the square root of the variance. Here is the result.

Binomial Mean and Standard Deviation

If a count X has the binomial distribution $B(n, p)$, then

$$\mu_X = np$$
$$\sigma_X = \sqrt{np(1 - p)}$$

EXAMPLE 5.6

The Helsinki Heart Study asks whether the anticholesterol drug gemfibrozil will reduce heart attacks. In planning such an experiment, the researchers must be confident that the sample sizes are large enough to enable them to observe enough heart attacks. The Helsinki study plans to give gemfibrozil to 2000 men and a placebo to another 2000. The probability of a heart attack during the five-year period of the study for men this age is about 0.04. What are the mean and standard deviation of the number of heart attacks that will be observed in one group if the treatment does not change this probability?

There are 2000 independent observations, each having probability $p = 0.04$ of a heart attack. The count X of heart attacks is $B(2000, 0.04)$, so that

$$\mu_X = np = (2000)(0.04) = 80$$

$$\sigma_X = \sqrt{np(1-p)} = \sqrt{(2000)(0.04)(0.96)} = 8.76$$

The expected number of heart attacks is large enough to permit conclusions about the effectiveness of the drug.

Sample proportions

What proportion of a large lot of switches fail to meet specifications? What percent of adults favor stronger laws restricting firearms? In statistical sampling we often want to estimate the proportion p of "successes" in a population. Our estimator is the sample proportion of successes:

$$\hat{p} = \frac{\text{count of successes in sample}}{\text{size of sample}}$$

$$= \frac{X}{n}$$

Be sure to distinguish between the *proportion* $\hat{p}$ and the *count* X. The count takes whole-number values between 0 and n, but a proportion is always a number between 0 and 1. In the binomial setting, the count X has a binomial distribution. The proportion $\hat{p}$ does *not* have a binomial distribution. We can, however, do probability calculations about $\hat{p}$ by restating them in terms of the count X and using binomial methods. As an example, we return to the sample survey that we studied by simulation in Example 3.20 (page 268).

EXAMPLE 5.7

A recent survey asked a nationwide random sample of 2500 adults if they agreed or disagreed that "I like buying new clothes, but shopping is often frustrating and time-consuming." Suppose that 60% of all adults would agree if asked this question. What is the probability that the sample proportion who agree is at least 58%?

The count X who agree has the binomial distribution $B(2500, 0.6)$. The sample proportion $\hat{p} = X/2500$ does *not* have a binomial distribution, because it is not a count. We therefore translate any question about the sample proportion $\hat{p}$ into a question about the count X. Because 58% of 2500 is 1450,

$$P(\hat{p} \geq 0.58) = P(X \geq 1450)$$
$$= P(X = 1450) + P(X = 1451) + \cdots + P(X = 2500)$$

This is a rather elaborate calculation. We must add more than 1000 binomial probabilities, and many software binomial probability functions cannot handle an n as large as 2500. Large-scale software tells us that $P(\hat{p} \geq 0.58) = 0.9802$. We need another way to do this calculation.

As a first step, we can obtain the mean and standard deviation of the sample proportion from the mean and standard deviation of the sample count using the rules from Section 4.4 for the mean and variance of a constant times a random variable. Here is the result.

Mean and Standard Deviation of a Sample Proportion

Let $\hat{p}$ be the sample proportion of successes in an SRS of size n drawn from a large population having population proportion p of successes. The mean and standard deviation of $\hat{p}$ are

$$\mu_{\hat{p}} = p$$

$$\sigma_{\hat{p}} = \sqrt{\frac{p(1-p)}{n}}$$

The formula for $\sigma_{\hat{p}}$ is correct in the binomial setting. It is approximately correct for an SRS from a large population. We will use it when the population is at least 10 times as large as the sample.

EXAMPLE 5.8 | The mean and standard deviation of the proportion of the survey respondents in Example 5.7 who find shopping frustrating are

$$\mu_{\hat{p}} = p = 0.6$$

$$\sigma_{\hat{p}} = \sqrt{\frac{p(1-p)}{n}} = \sqrt{\frac{(0.6)(0.4)}{2500}} = 0.0098$$

The fact that the mean of $\hat{p}$ is p states in statistical language that the sample proportion $\hat{p}$ in an SRS is an unbiased estimator of the population proportion p. When a sample is drawn from a new population having a different value of the population proportion p, the sampling distribution of the unbiased estimator $\hat{p}$ changes so that its mean moves to the new value of p. We observed this fact empirically in Section 3.4 and have now verified it from the laws of probability.

The variability of $\hat{p}$ about its mean, as described by the variance or standard deviation, decreases as the sample size increases. So a sample proportion from a large sample will usually lie quite close to the population proportion p. We observed this in the simulation experiment on page 270 in Section 3.4. Now we have discovered exactly how the variability decreases: the standard deviation is $\sqrt{p(1-p)/n}$, or $\sqrt{p(1-p)}/\sqrt{n}$. The $\sqrt{n}$ in the denominator means that the sample size must be multiplied by 4 if we wish to divide the standard deviation in half.

Normal approximation for counts and proportions

We discovered empirically in Section 3.4 that the sampling distribution of a sample proportion $\hat{p}$ is close to normal. Now we have deduced that the distribution of $\hat{p}$ is that of a binomial count divided by the sample size n. This seems at first to be a contradiction. In fact, the exact distribution is indeed based on the binomial, but it is also approximately normal when n is large.

Both the count X and the sample proportion $\hat{p}$ are approximately normal in large samples.

Normal Approximation for Counts and Proportions

Draw an SRS of size n from a large population having population proportion p of successes. Let X be the count of successes in the sample and $\hat{p} = X/n$ the sample proportion of successes. When n is large, the sampling distributions of these statistics are approximately normal:

$$X \text{ is approximately } N\left(np, \sqrt{np(1-p)}\right)$$

$$\hat{p} \text{ is approximately } N\left(p, \sqrt{\frac{p(1-p)}{n}}\right)$$

As a rule of thumb, we will use this approximation for values of n and p that satisfy $np \geq 10$ and $n(1-p) \geq 10$.

Both of these normal approximations are easy to remember because they say that $\hat{p}$ and X are normal with their usual means and standard deviations. The accuracy of the normal approximations improves as the sample size n increases. They are most accurate for any fixed n when p is close to $1/2$, and least accurate when p is near 0 or 1. Whether or not you use the normal approximations should depend on how accurate your calculations need to be. For most statistical purposes great accuracy is not required. Our "rule of thumb" for use of the normal approximations reflects this judgment.

Figure 5.2 summarizes the distribution of a sample proportion in a form that helps you recall the big idea of a sampling distribution. Sampling distributions answer the question "What would happen if we took many samples from the same population?" Keep taking random samples of size n from a population that contains proportion p of successes. Find the sample proportion $\hat{p}$ for each sample. Collect all the $\hat{p}$'s and display their distribution. That's the sampling distribution of $\hat{p}$.

EXAMPLE 5.9   Let's compare the normal approximation for the calculation of Example 5.7 with the exact calculation from software. We want to calculate $P(\hat{p} \geq 0.58)$ when the sample size is $n = 2500$ and the population proportion is $p = 0.6$. Example 5.8 shows that

$$\mu_{\hat{p}} = p = 0.6$$

$$\sigma_{\hat{p}} = \sqrt{\frac{p(1-p)}{n}} = 0.0098$$

Act as if $\hat{p}$ were normal with mean 0.6 and standard deviation 0.0098. The approximate probability, as illustrated in Figure 5.3, is

$$P(\hat{p} \geq 0.58) = P\left(\frac{\hat{p} - 0.6}{0.0098} \geq \frac{0.58 - 0.6}{0.0098}\right)$$

$$\doteq P(Z \geq -2.04) = 0.9793$$

That is, about 98% of all samples have a sample proportion that is at least 0.58. Because the sample was large, this normal approximation is quite accurate. It misses the software value 0.9802 by only 0.0009.

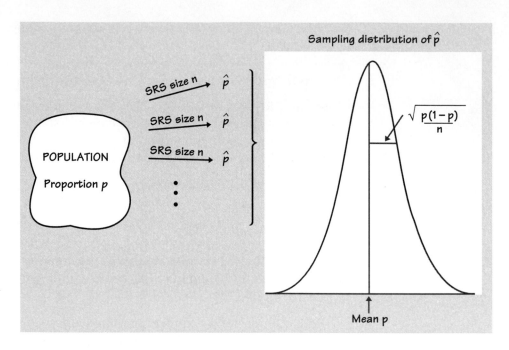

FIGURE 5.2   The sampling distribution of a sample proportion $\hat{p}$ is approximately normal with mean p and standard deviation $\sqrt{p(1-p)/n}$.

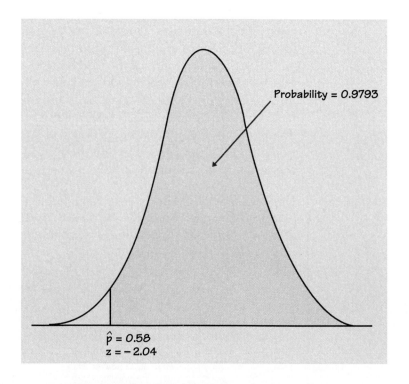

FIGURE 5.3   The normal probability calculation for Example 5.9.

EXAMPLE 5.10 | The quality engineer of Example 5.4 realizes that a sample of 10 switches is inadequate and decides to inspect 100 switches from a large shipment. If in fact 10% of the switches in the shipment fail to meet the specifications, then the count X of bad switches in the sample has approximately the $B(100, 0.1)$ distribution. The distribution of the count of bad switches in a sample of 10 is distinctly nonnormal, as Figure 5.1 showed. When the sample size is increased to 100, however, the shape of the binomial distribution becomes approximately normal.

Software tells us that the actual binomial probability that no more than 9 of the switches in the sample fail inspection is $P(X \leq 9) = 0.4513$. How accurate is the normal approximation?

According to the normal approximation to the binomial distributions, the count X is approximately normal with mean and standard deviation

$$\mu_X = np = (100)(0.1) = 10$$
$$\sigma_X = \sqrt{np(1-p)} = \sqrt{(100)(0.1)(0.9)} = 3$$

Figure 5.4 displays the probability histogram of the binomial distribution with the density curve of the approximating normal distribution superimposed. Both distributions have the same mean and standard deviation, and both the area under the histogram and the area under the curve are 1. (Although the binomial random variable X can take values as large as 100, the probabilities of outcomes larger than 20 are so small that the bars would have no visible height in the figure.)

The normal approximation to the probability of no more than 9 bad switches is the area to the left of $X = 9$ under the normal curve (Figure 5.5). Using Table A,

$$P(X \leq 9) = P\left(\frac{X - 10}{3} \leq \frac{9 - 10}{3}\right)$$
$$\doteq P(Z \leq -0.33) = 0.3707$$

The approximation 0.37 to the binomial probability 0.45 is not very accurate. Notice that $np = (100)(0.1) = 10$, so that this combination of n and p is on the border of the values for which we are willing to use the approximation.

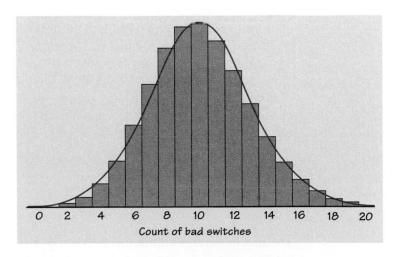

FIGURE 5.4   Probability histogram and normal approximation for the binomial distribution with $n = 100$ and $p = 0.1$, for Example 5.10.

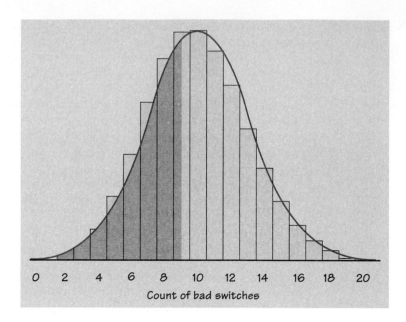

Count of bad switches

FIGURE 5.5   Area under the normal approximation curve for the probability in Example 5.10.

The continuity correction*

Figure 5.5 suggests an idea that greatly improves the accuracy of the normal approximation to binomial probabilities. The discrete binomial distribution in Example 5.10 puts probability exactly on $X = 9$ and $X = 10$ and no probability between these whole numbers. The normal distribution spreads its probability continuously. The bar for $X = 9$ in the probability histogram of Figure 5.5 extends from 8.5 to 9.5, but the normal calculation of $P(X \leq 9)$ includes only the area to the left of the center of this bar. The normal approximation is more accurate if we consider $X = 9$ to extend from 8.5 to 9.5, $X = 10$ to extend from 9.5 to 10.5, and so on.

When we want to include the outcome $X = 9$, we include the entire interval from 8.5 to 9.5 that is the base of the $X = 9$ bar in the histogram. So $P(X \leq 9)$ is calculated as $P(X \leq 9.5)$. On the other hand, $P(X < 9)$ excludes the outcome $X = 9$, so we exclude the entire interval from 8.5 to 9.5 and calculate $P(X \leq 8.5)$ from the normal table. Here is the result of the normal calculation in Example 5.10 improved in this way:

$$P(X \leq 9) = P(X \leq 9.5)$$
$$= P\left(\frac{X - 10}{3} \leq \frac{9.5 - 10}{3}\right)$$
$$\doteq P(Z \leq 0.17) = 0.4325$$

The improved approximation 0.43 is much closer to the binomial probability 0.45. Acting as though a whole number occupies the interval from 0.5 below

---

*This material can be omitted if desired.

continuity correction

to 0.5 above the number is called the **continuity correction** to the normal approximation. If you need accurate values for binomial probabilities, try to use software to do exact calculations. If no software is available, use the continuity correction unless n is very large. Because most statistical purposes do not require extremely accurate probability calculations, we do not emphasize use of the continuity correction.

Binomial formulas*

We can find a formula for the probability that a binomial random variable takes any value by adding probabilities for the different ways of getting exactly that many successes in n observations. Here is the example we will use to show the idea.

EXAMPLE 5.11

Each child born to a particular set of parents has probability 0.25 of having blood type O. If these parents have 5 children, what is the probability that exactly 2 of them have type O blood?

   The count of children with type O blood is a binomial random variable X with $n = 5$ tries and probability $p = 0.25$ of a success on each try. We want $P(X = 2)$.

Because the method doesn't depend on the specific example, we will use "S" for success and "F" for failure for short. In Example 5.11, "S" would stand for type O blood. Do the work in two steps.

Step 1: Find the probability that a specific 2 of the 5 tries give successes, say the first and the third. This is the outcome SFSFF. The multiplication rule for independent events tells us that

$$P(\text{SFSFF}) = P(\text{S})P(\text{F})P(\text{S})P(\text{F})P(\text{F})$$
$$= (0.25)(0.75)(0.25)(0.75)(0.75)$$
$$= (0.25)^2(0.75)^3$$

Step 2: Observe that the probability of *any one* arrangement of 2 S's and 3 F's has this same probability. That's true because we multiply together 0.25 twice and 0.75 three times whenever we have 2 S's and 3 F's. The probability that $X = 2$ is the probability of getting 2 S's and 3 F's in any arrangement whatsoever. Here are all the possible arrangements:

SSFFF	SFSFF	SFFSF	SFFFS	FSSFF
FSFSF	FSFFS	FFSSF	FFSFS	FFFSS

There are 10 of them, all with the same probability. The overall probability of 2 successes is therefore

$$P(X = 2) = 10(0.25)^2(0.75)^3 = 0.2637$$

---

*The formula for binomial probabilities is useful in many settings, but we will not need it in our study of statistical inference. This section can therefore be omitted if desired.

The pattern of this calculation works for any binomial probability. To use it, we need to be able to count the number of arrangements of k successes in n observations without actually listing them. We use the following fact to do the counting.

Binomial Coefficient

The number of ways of arranging k successes among n observations is given by the **binomial coefficient**

$$\binom{n}{k} = \frac{n!}{k!\,(n-k)!}$$

for $k = 0, 1, 2, \ldots, n$.

---

factorial

The formula for binomial coefficients uses the **factorial** notation. The factorial $n!$ for any positive whole number n is

$$n! = n \times (n-1) \times (n-2) \times \cdots \times 3 \times 2 \times 1$$

Also, $0! = 1$. Notice that the larger of the two factorials in the denominator of a binomial coefficient will cancel much of the $n!$ in the numerator. For example, the binomial coefficient we need for Example 5.11 is

$$
\begin{aligned}
\binom{5}{2} &= \frac{5!}{2!\,3!} \\
&= \frac{(5)(4)(3)(2)(1)}{(2)(1) \times (3)(2)(1)} \\
&= \frac{(5)(4)}{(2)(1)} = \frac{20}{2} = 10
\end{aligned}
$$

This agrees with our earlier enumeration.

The notation $\binom{n}{k}$ is *not* related to the fraction $\frac{n}{k}$. A helpful way to remember its meaning is to read it as "binomial coefficient n choose k." Binomial coefficients have many uses in mathematics, but we are interested in them only as an aid to finding binomial probabilities. The binomial coefficient $\binom{n}{k}$ counts the number of ways in which k successes can be distributed among n observations. The binomial probability $P(X = k)$ is this count multiplied by the probability of any specific arrangement of the k successes. Here is the formula we seek.

Binomial Probability

If X has the binomial distribution $B(n, p)$ with n observations and probability p of success on each observation, the possible values of X are 0, 1, 2, $\ldots, n$. If k is any one of these values, the **binomial probability** is

$$P(X = k) = \binom{n}{k} p^k (1-p)^{n-k}$$

Here is an example of the use of the binomial probability formula.

EXAMPLE 5.12

The number X of switches that fail inspection in Example 5.4 has the $B(10, 0.1)$ distribution. The probability that no more than 1 switch fails is

$$P(X \leq 1) = P(X = 1) + P(X = 0)$$

$$= \binom{10}{1}(0.1)^1(0.9)^9 + \binom{10}{0}(0.1)^0(0.9)^{10}$$

$$= \frac{10!}{1!\,9!}(0.1)(0.3874) + \frac{10!}{0!\,10!}(1)(0.3487)$$

$$= (10)(0.1)(0.3874) + (1)(1)(0.3487)$$

$$= 0.3874 + 0.3487 = 0.7361$$

The calculation used the facts that $0! = 1$ and that $a^0 = 1$ for any number $a \neq 0$. The result agrees with that obtained from Table C in Example 5.4.

SUMMARY

A **count** X of successes has the **binomial distribution** $B(n, p)$ in the **binomial setting:** there are n trials, all independent, each resulting in a success or a failure, and each having the same probability p of a success.

Binomial probabilities are most easily found by software. There is an exact formula that is practical for calculations when n is small. Table C contains binomial probabilities for some values of n and p. For large n, you can use the normal approximation.

The binomial distribution $B(n, p)$ is a good approximation to the **sampling distribution of the count of successes** in an SRS of size n from a large population containing proportion p of successes. We will use this approximation when the population is at least 10 times larger than the sample.

The mean and standard deviation of a **binomial count** X and a **sample proportion** of successes $\hat{p} = X/n$ are

$$\mu_X = np, \qquad\qquad \mu_{\hat{p}} = p$$

$$\sigma_X = \sqrt{np(1-p)} \qquad \sigma_{\hat{p}} = \sqrt{\frac{p(1-p)}{n}}$$

The sample proportion $\hat{p}$ is therefore an unbiased estimator of the population proportion p.

The **normal approximation** to the binomial distribution says that if X is a count having the $B(n, p)$ distribution, then when n is large,

$$X \text{ is approximately } N(np, \sqrt{np(1-p)})$$

$$\hat{p} \text{ is approximately } N\left(p, \sqrt{\frac{p(1-p)}{n}}\right)$$

We will use these approximations when $np \geq 10$ and $n(1 - p) \geq 10$. If the **continuity correction** is employed to improve the accuracy, the normal approximations can be used more freely.

The exact **binomial probability formula** is

$$P(X = k) = \binom{n}{k} p^k (1 - p)^{n-k}$$

where the possible values of X are $k = 0, 1, \ldots, n$. The binomial probability formula uses the **binomial coefficient**

$$\binom{n}{k} = \frac{n!}{k!(n - k)!}$$

Here the **factorial** $n!$ is

$$n! = n \times (n - 1) \times (n - 2) \times \cdots \times 3 \times 2 \times 1$$

for positive whole numbers n and $0! = 1$. The binomial coefficient counts the number of ways of distributing k successes among n trials.

SECTION 5.1 EXERCISES

All of the binomial probability calculations required in these exercises can be done by using Table C or the normal approximation. Your instructor may request that you use the binomial probability formula or software. In exercises requiring the normal approximation, you should use the continuity correction only if you studied that topic.

5.1 For each of the following situations, indicate whether a binomial distribution is a reasonable probability model for the random variable X. Give your reasons in each case.

(a) You observe the sex of the next 50 children born at a local hospital; X is the number of girls among them.

(b) A couple decides to continue to have children until their first girl is born; X is the total number of children the couple has.

(c) You want to know what percent of married people believe that mothers of young children should not be employed outside the home. You plan to interview 50 people, and for the sake of convenience you decide to interview both the husband and the wife in 25 married couples. The random variable X is the number among the 50 persons interviewed who think mothers should not be employed.

5.2 In each situation below, is it reasonable to use a binomial distribution for the random variable X? Give reasons for your answer in each case.

(a) An auto manufacturer chooses one car from each hour's production for a detailed quality inspection. One variable recorded is the count X of finish defects (dimples, ripples, etc.) in the car's paint.

(b) The pool of potential jurors for a murder case contains 100 persons chosen at random from the adult residents of a large city. Each person in the pool is asked whether he or she opposes the death penalty; X is the number who say "Yes."

(c) Joe buys a ticket in his state's "Pick 3" lottery game every week; X is the number of times in a year that he wins a prize.

5.3 In each of the following cases, decide whether or not a binomial distribution is an appropriate model, and give your reasons.

(a) Fifty students are taught about binomial distributions by a television program. After completing their study, all students take the same examination. The number of students who pass is counted.

(b) A student studies binomial distributions using computer-assisted instruction. After the initial instruction is completed, the computer presents 10 problems. The student solves each problem and enters the answer; the computer gives additional instruction between problems if the student's answer is wrong. The number of problems that the student solves correctly is counted.

(c) A chemist repeats a solubility test 10 times on the same substance. Each test is conducted at a temperature 10° higher than the previous test. She counts the number of times that the substance dissolves completely.

5.4 Some of the methods in this section are approximations rather than exact probability results. We have given rules of thumb for safe use of these approximations.

(a) You are interested in attitudes toward drinking among the 75 members of a fraternity. You choose 25 members at random to interview. One question is "Have you had five or more drinks at one time during the last week?" Suppose that in fact 20% of the 75 members would say "Yes." Explain why you *cannot* safely use the $B(25, 0.2)$ distribution for the count X in your sample who say "Yes."

(b) The National AIDS Behavioral Surveys found that 0.2% (that's 0.002 as a decimal fraction) of adult heterosexuals had both received a blood transfusion and had a sexual partner from a group at high risk of AIDS. Suppose that this national proportion holds for your region. Explain why you *cannot* safely use the normal approximation for the sample proportion who fall in this group when you interview an SRS of 500 adults.

5.5 You are planning a sample survey of small businesses in your area. You will choose an SRS of businesses listed in the telephone book's Yellow Pages. Experience shows that only about half the businesses you contact will respond.

(a) If you contact 150 businesses, it is reasonable to use the $B(150, 0.5)$ distribution for the number X who respond. Explain why.

(b) What is the expected number (the mean) who will respond?

(c) What is the probability that 70 or fewer will respond? (Use software or the normal approximation.)

(d) How large a sample must you take to increase the mean number of respondents to 100?

5.6 You operate a restaurant. You read that a sample survey by the National Restaurant Association shows that 40% of adults are committed to eating nutritious food when eating away from home. To help plan your menu, you decide to conduct a sample survey in your own area. You will use random digit dialing to contact an SRS of 200 households by telephone.

(a) If the national result holds in your area, it is reasonable to use the $B(200, 0.4)$ distribution to describe the count X of respondents who seek nutritious food when eating out. Explain why.

(b) What is the mean number of nutrition-conscious people in your sample if $p = 0.4$ is true? What is the probability that X lies between 75 and 85? (Use software or the normal approximation.)

(c) You find 100 of your 200 respondents concerned about nutrition. Is this reason to believe that the percent in your area is higher than the national 40%? To answer this question, find the probability that X is 100 or larger if $p = 0.4$ is true. If this probability is very small, that is reason to think that p is actually greater than 0.4.

5.7 Your mail-order company advertises that it ships 90% of its orders within three working days. You select an SRS of 100 of the 5000 orders received in the past week for an audit. The audit reveals that 86 of these orders were shipped on time.

(a) What is the sample proportion of orders shipped on time?

(b) If the company really ships 90% of its orders on time, what is the probability that the proportion in an SRS of 100 orders is as small as the proportion in your sample or smaller? (Use software or the normal approximation.)

(c) A critic says, "Aha! You claim 90%, but in your sample the on-time percentage is lower than that. So the 90% claim is wrong." Explain in simple language why your probability calculation in (b) shows that the result of the sample does not refute the 90% claim.

5.8 The Gallup Poll once found that about 15% of adults jog. Suppose that in fact the proportion of the adult population who jog is $p = 0.15$.

(a) What is the probability that the sample proportion $\hat{p}$ of joggers in an SRS of size $n = 200$ lies between 13% and 17%? (Use software or the normal approximation.)

(b) You just found the probability that an SRS of size 200 gives a sample proportion $\hat{p}$ within ± 2 percentage points of the population proportion $p = 0.15$. Find this probability for SRSs of sizes 800, 1600, and 3200. What general conclusion can you draw from your calculations?

5.9 A national opinion poll found that 44% of all American adults agree that parents should be given vouchers good for education at any public or private school of their choice. Suppose that in fact the population proportion who feel this way is $p = 0.44$.

(a) Many opinion polls have a "margin of error" of about $\pm 3\%$. What is the probability that an SRS of size 300 has a sample proportion $\hat{p}$ that is within $\pm 3\%$ (± 0.03) of the population proportion $p = 0.44$? (Use software or the normal approximation.)

(b) Answer the same question for SRSs of sizes 600 and 1200. What is the effect of increasing the size of the sample?

5.10   A student organization is planning to ask a sample of 50 students if they have noticed AIDS education brochures on campus. The sample percentage who say "Yes" will be reported. The organization's statistical advisor says that the standard deviation of this percentage will be about 7%.

(a) What would the standard deviation be if the sample contained 100 students rather than 50?

(b) How large a sample is required to reduce the standard deviation of the percentage who say "Yes" from 7% to 3.5%? Explain to someone who knows no statistics the advantage of taking a larger sample in a survey of opinion.

5.11   According to government data, 25% of employed women have never been married.

(a) If 10 employed women are selected at random, what is the probability that exactly 2 have never been married? (Use software, Table C, or the binomial probability formula.)

(b) What is the probability that 2 or fewer have never been married?

(c) What is the probability that at least 8 have been married?

5.12   A university that is better known for its basketball program than for its academic strength claims that 80% of its basketball players get degrees. An investigation examines the fate of all 20 players who entered the program over a period of several years that ended 5 years ago. Of these players, 11 graduated and the remaining 9 are no longer in school. If the university's claim is true, the number of players who graduate among the 20 studied should have the $B(20, 0.8)$ distribution.

(a) Find the probability that exactly 11 players graduate under these assumptions. (Use software, Table C, or the binomial probability formula.)

(b) Find the probability that 11 or fewer players graduate. This probability is so small that it casts doubt on the university's claim.

5.13   Children inherit their blood type from their parents, with probabilities that reflect the parents' genetic makeup. Children of Juan and Maria each have probability 1/4 of having blood type A and inherit independently of each other. Juan and Maria plan to have 4 children; let X be the number who have blood type A.

(a) What are n and p in the binomial distribution of X?

(b) Find the probability of each possible value of X, and draw a probability histogram for this distribution. (Use software, Table C, or the binomial probability formula.)

(c) Find the mean number of children with type A blood, and mark the location of the mean on your probability histogram.

5.14 A believer in the "random walk" theory of the behavior of stock prices thinks that an index of stock prices has probability 0.65 of increasing in any year. Moreover, the change in the index in any given year is not influenced by whether it rose or fell in earlier years. Let X be the number of years among the next 6 years in which the index rises.

(a) What are n and p in the binomial distribution of X?

(b) Give the possible values that X can take and the probability of each value. Draw a probability histogram for the distribution of X.

(c) Find the mean of the number X of years in which the stock price index rises, and mark the mean on your probability histogram.

(d) Find the standard deviation of X. What is the probability that X takes a value within one standard deviation of its mean?

5.15 In a test for ESP (extrasensory perception), the experimenter looks at cards that are hidden from the subject. Each card contains either a star, a circle, a wave, or a square. As the experimenter looks at each of 20 cards in turn, the subject names the shape on the card.

(a) If a subject simply guesses the shape on each card, what is the probability of a successful guess on a single card? Because the cards are independent, the count of successes in 20 cards has a binomial distribution.

(b) What is the probability that a subject correctly guesses at least 10 of the 20 shapes?

(c) In many repetitions of this experiment with a subject who is guessing, how many cards will the subject guess correctly on the average? What is the standard deviation of the number of correct guesses?

(d) A standard ESP deck actually contains 25 cards. There are five different shapes, each of which appears on 5 cards. The subject knows that the deck has this makeup. Is a binomial model still appropriate for the count of correct guesses in one pass through this deck? If so, what are n and p? If not, why not?

5.16 A sociology professor asks her class to observe cars having a man and a woman in the front seat and record which of the two is the driver.

(a) Explain why it is reasonable to use the binomial distribution for the number of male drivers in n cars if all observations are made in the same location at the same time of day.

(b) Explain why the binomial model may not apply if half the observations are made outside a church on Sunday morning and half are made on campus after a dance.

(c) The professor requires students to observe 10 cars during business hours in a retail district close to campus. Past observations have shown that the man is driving about 85% of cars in this location. What is the probability that the man is driving 8 or fewer of the 10 cars?

(d) The class has 10 students, who will observe 100 cars in all. What is the probability that the man is driving 80 or fewer of these?

5.17 A study by a federal agency concludes that polygraph (lie detector) tests given to truthful persons have probability about 0.2 of suggesting that the person is deceptive (Office of Technology Assessment, *Scientific Validity of Polygraph Testing: A Research Review and Evaluation*, Government Printing Office, Washington, D.C., 1983).

(a) A firm asks 12 job applicants about thefts from previous employers, using a polygraph to assess their truthfulness. Suppose that all 12 answer truthfully. What is the probability that the polygraph says at least 1 is deceptive?

(b) What is the mean number among 12 truthful persons who will be classified as deceptive? What is the standard deviation of this number?

(c) What is the probability that the number classified as deceptive is less than the mean?

5.18 According to government data, 21% of American children under the age of six live in households with incomes less than the official poverty level. A study of learning in early childhood chooses an SRS of 300 children.

(a) What is the mean number of children in the sample who come from poverty-level households? What is the standard deviation of this number?

(b) Use the normal approximation to calculate the probability that at least 80 of the children in the sample live in poverty. Be sure to check that you can safely use the approximation.

5.19 One way of checking the effect of undercoverage, nonresponse, and other sources of error in a sample survey is to compare the sample with known demographic facts about the population. About 12% of American adults are black. The number X of blacks in a random sample of 1500 adults should therefore vary with the $B(1500, 0.12)$ distribution.

(a) What are the mean and standard deviation of X?

(b) Use the normal approximation to find the probability that the sample will contain 170 or fewer blacks. Be sure to check that you can safely use the approximation.

5.20 A selective college would like to have an entering class of 1200 students. Because not all students who are offered admission accept, the college admits more than 1200 students. Past experience shows that about 70% of the students admitted will accept. The college decides to admit 1500 students. Assuming that students make their decisions independently, the number who accept has the $B(1500, 0.7)$ distribution. If this number is less than 1200, the college will admit students from its waiting list.

(a) What are the mean and the standard deviation of the number X of students who accept?

(b) Use the normal approximation to find the probability that at least 1000 students accept.

(c) The college does not want more than 1200 students. What is the probability that more than 1200 will accept?

(d) If the college decides to increase the number of admission offers to 1700, what is the probability that more than 1200 will accept?

5.21   Here is a simple probability model for multiple-choice tests. Suppose that each student has probability p of correctly answering a question chosen at random from a universe of possible questions. (A strong student has a higher p than a weak student.) The correctness of answers to different questions are independent. Jodi is a good student for whom $p = 0.75$.

(a) Use the normal approximation to find the probability that Jodi scores 70% or lower on a 100-question test.

(b) If the test contains 250 questions, what is the probability that Jodi will score 70% or lower?

(c) How many questions must the test contain in order to reduce the standard deviation of Jodi's proportion of correct answers to half its value for a 100-item test?

(d) Laura is a weaker student for whom $p = 0.6$. Does the answer you gave in (c) for the standard deviation of Jodi's score apply to Laura's standard deviation also?

5.22   When the ESP study of Exercise 5.15 discovers a subject whose performance appears to be better than guessing, the study continues at greater length. The experimenter looks at many cards bearing one of five shapes (star, square, circle, wave, and cross) in an order determined by random numbers. The subject cannot see the experimenter as he looks at each card in turn, in order to avoid any possible nonverbal clues. The answers of a subject who does not have ESP should be independent observations, each with probability 1/5 of success. We record 1000 attempts.

(a) What are the mean and the standard deviation of the count of successes?

(b) What are the mean and standard deviation of the proportion of successes among the 1000 attempts?

(c) What is the probability that a subject without ESP will be successful in at least 24% of 1000 attempts?

(d) The researcher considers evidence of ESP to be a proportion of successes so large that there is only probability 0.01 that a subject could do this well or better by guessing. What proportion of successes must a subject have to meet this standard? (*Hint:* Example 1.27 shows how to do a normal calculation of the type required here.)

5.23 Use the definition of binomial coefficients to show that each of the following facts is true. Then restate each fact in words in terms of the number of ways that k successes can be distributed among n observations.

(a) $\binom{n}{n} = 1$ for any whole number $n \geq 1$.

(b) $\binom{n}{n-1} = n$ for any whole number $n \geq 1$.

(c) $\binom{n}{k} = \binom{n}{n-k}$ for any n and k with $k \leq n$.

5.2 The Sampling Distribution of a Sample Mean

Counts and proportions are discrete random variables that describe categorical data. The statistics most often used to describe measured data, on the other hand, are continuous random variables. The sample mean, percentiles, and standard deviation are examples of statistics based on measured data. Statistical theory describes the sampling distributions of these statistics. In this section we will concentrate on the sample mean. Because sample means are just averages of observations, they are among the most common statistics.

EXAMPLE 5.13

A basic principle of investment is that diversification reduces risk. That is, buying several securities rather than just one reduces the variability of the return on an investment. Figure 5.6 illustrates this principle in the case of common stocks listed on the New York Stock Exchange. Figure 5.6(a) shows the distribution of returns for all 1815 stocks listed on the Exchange for the entire year 1987. This was a year of extreme swings in stock prices, including a record loss of more than 22% on Black Monday, October 19. The mean return for all 1815 stocks was −3.5%, and the distribution shows a very wide spread.

Figure 5.6(b) shows the distribution of returns for all possible portfolios that invested equal amounts in each of 5 stocks in 1987. A portfolio is just a sample of 5 stocks, and its return is the average return for the 5 stocks chosen. The mean return for all portfolios is still −3.5%, but the variation among portfolios is much less than the variation among individual stocks. For example, 11% of all individual stocks had a loss of more than 40%, but only 1% of the portfolios had a loss that large.

The histograms in Figure 5.6 illustrate a fact that we will make precise in this section: **averages are less variable than individual observations.** More detailed examination of the distributions, by normal quantile plots, would point to a second fact: **averages are more normal than individual observations.** These two facts contribute to the popularity of sample means in statistical inference.

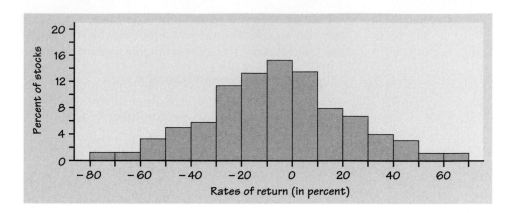

FIGURE 5.6(a)   The distribution of returns for New York Stock Exchange common stock in 1987.

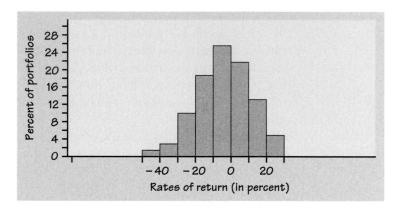

FIGURE 5.6(b)   The distribution of returns for portfolios of 5 stocks in 1987. (Figure 5.6 is taken with permission from John K. Ford, "A method for grading 1987 stock recommendations," *American Association of Individual Investors Journal,* March 1988, pp. 16–17.)

The mean and standard deviation of $\bar{x}$

The sample mean $\bar{x}$ from a sample or an experiment is an estimate of the mean μ of the underlying population, just as a sample proportion $\hat{p}$ is an estimate of a population proportion p. The sampling distribution of $\bar{x}$ is determined by the design used to produce the data, the sample size n, and the population distribution.

Select an SRS of size n from a population, and measure a variable X on each individual in the sample. The data consist of observations on n random variables $X_1, X_2, \ldots, X_n$. A single X_i is a measurement on one individual selected at random from the population and therefore has the distribution of the population. If the population is large relative to the sample, we can consider $X_1, X_2, \ldots, X_n$ to be independent random variables each having the same distribution. This is our probability model for measurements on each individual in an SRS.

The sample mean of an SRS of size n is

$$\bar{x} = \frac{1}{n}(X_1 + X_2 + \cdots + X_n)$$

If the population has mean μ, then μ is the mean of each observation X_i. Therefore, by the addition rule for means of random variables,

$$\mu_{\bar{x}} = \frac{1}{n}(\mu_{X_1} + \mu_{X_2} + \cdots + \mu_{X_n})$$

$$= \frac{1}{n}(\mu + \mu + \cdots + \mu) = \mu$$

That is, *the mean of $\bar{x}$ is the same as the mean of the population.* The sample mean $\bar{x}$ is therefore an unbiased estimator of the unknown population mean μ.

The observations are independent, so the addition rule for variances also applies:

$$\sigma_{\bar{x}}^2 = \left(\frac{1}{n}\right)^2 (\sigma_{X_1}^2 + \sigma_{X_2}^2 + \cdots + \sigma_{X_n}^2)$$

$$= \left(\frac{1}{n}\right)^2 (\sigma^2 + \sigma^2 + \cdots + \sigma^2)$$

$$= \frac{\sigma^2}{n}$$

Just as in the case of a sample proportion $\hat{p}$, the variability of the sampling distribution of a sample mean decreases as the sample size grows. Because the standard deviation of $\bar{x}$ is $\sigma/\sqrt{n}$, it is again true that the standard deviation of the statistic decreases in proportion to the square root of the sample size. Here is a summary of these facts.

Mean and Standard Deviation of a Sample Mean

Let $\bar{x}$ be the mean of an SRS of size n from a population having mean μ and standard deviation σ. The mean and standard deviation of $\bar{x}$ are

$$\mu_{\bar{x}} = \mu$$

$$\sigma_{\bar{x}} = \frac{\sigma}{\sqrt{n}}$$

EXAMPLE 5.14

The height X of a single randomly chosen young woman varies according to the $N(64.5, 2.5)$ distribution. If a medical study asked the height of an SRS of 100 young women, the sampling distribution of the sample mean height $\bar{x}$ would have mean and standard deviation

$$\mu_{\bar{x}} = \mu = 64.5 \text{ inches}$$

$$\sigma_x = \frac{\sigma}{\sqrt{n}} = \frac{2.5}{\sqrt{100}} = 0.25 \text{ inch}$$

The heights of individual women vary widely about the population mean ($\sigma = 2.5$ inches), but the average height of a sample of 100 women has a standard deviation only one-tenth as large ($\sigma_{\bar{x}} = 0.25$ inch).

EXAMPLE 5.15 The fact that a mean of several measurements is less variable than a single measurement is important in science. When Simon Newcomb set out to measure the speed of light, he took repeated measurements of the time required for a beam of light to travel a known distance. Newcomb's 66 measurements appear in Table 1.1 (page 8). We saw that the data contain two outliers that may not be observations from the same population. Disregarding these leaves 64 observations. These are the values of 64 independent random variables, each with a probability distribution that describes the population of all measurements made with Newcomb's apparatus. If Newcomb's procedures were correct, the population mean μ is the true passage time of light. The population variability reflects the random variation in the measurements due to small changes in the environment, the equipment, and the procedure. Suppose that the standard deviation for this population is $\sigma = 5$ nanoseconds.

If Newcomb had taken a single measurement, the standard deviation of the result would be 5, so that a second observation might have a quite different value. Taking 10 measurements and reporting their mean $\bar{x}$ reduces the standard deviation to

$$\sigma_{\bar{x}} = \frac{5}{\sqrt{10}} = 1.58 \text{ nanoseconds}$$

The sample mean of any number of measurements is an unbiased estimate of the population mean. Averaging over more measurements reduces the variability and makes it more likely that $\bar{x}$ is close to μ. The mean $\bar{x}$ of all 64 measurements has standard deviation

$$\sigma_{\bar{x}} = \frac{5}{\sqrt{64}} = 0.625 \text{ nanosecond}$$

Newcomb's observed value of this random variable, $\bar{x} = 27.75$, is a much more reliable estimate of the true passage time than is a single measurement. Increasing the number of measurements from 1 to 64 divides the standard deviation by 8, the square root of 64.

The sampling distribution of $\bar{x}$

We have described the center and spread of the probability distribution of a sample mean $\bar{x}$, but not its shape. The shape of the distribution of $\bar{x}$ depends on the shape of the population distribution. Here is one important case: if the population distribution is normal, then so is the distribution of the sample mean.

Sampling Distribution of a Sample Mean

If a population has the $N(\mu, \sigma)$ distribution, then the sample mean $\bar{x}$ of n independent observations has the $N(\mu, \sigma/\sqrt{n})$ distribution.

EXAMPLE 5.16 Newcomb's population consists of all measurements he might make. A normal quantile plot of his 64 observations (Figure 1.31 on page 82) shows that individual measurements are close to normal. A single measurement has the $N(\mu, 5)$ distribution, where μ is the unknown population mean. By the 95 part of the 68–95–99.7 rule, 95% of these measurements will lie within 2×5, or 10 nanoseconds, of the unknown μ.

The mean $\bar{x}$ of 10 measurements also has a normal distribution, but its standard deviation (Example 5.15) is only 1.58 nanoseconds. So 95% of the time the observed $\bar{x}$ falls within $2 \times 1.58 = 3.16$ nanoseconds of μ. Figure 5.7 compares the distributions of a single measurement and the sample mean of 10 measurements. The picture makes dramatically clear the advantages of averaging over many observations.

The fact that the sample mean of an SRS from a normal population has a normal distribution is a special case of a more general fact: **any linear combination of independent normal random variables is also normally distributed.** That is, if X and Y are independent normal random variables and a and b are any fixed numbers, $aX + bY$ is also normally distributed, and so it is for any number of normal variables. In particular, the sum or difference of independent normal random variables has a normal distribution. The mean and standard deviation of $aX + bY$ are found as usual from the addition rules for means and variances. These facts are often used in statistical calculations.

EXAMPLE 5.17

Tom and George are playing in the club golf tournament. Their scores vary as they play the course repeatedly. Tom's score X has the $N(110, 10)$ distribution, and George's score Y varies from round to round according to the $N(100, 8)$ distribution. If they play independently, what is the probability that Tom will score lower than George and thus do better in the tournament? The difference $X - Y$ between their scores is normally distributed, with mean and variance

$$\mu_{X-Y} = \mu_X - \mu_Y = 110 - 100 = 10$$

$$\sigma^2_{X-Y} = \sigma^2_X + \sigma^2_Y = 10^2 + 8^2 = 164$$

Because $\sqrt{164} = 12.8$, $X - Y$ has the $N(10, 12.8)$ distribution. Figure 5.8 illustrates the probability computation:

$$P(X < Y) = P(X - Y < 0)$$
$$= P\left(\frac{(X - Y) - 10}{12.8} < \frac{0 - 10}{12.8}\right)$$
$$= P(Z < -0.78) = 0.2177$$

Although George's score is 10 strokes lower on the average, Tom will have the lower score in about one of every five matches.

The central limit theorem

The sampling distribution of $\bar{x}$ is normal if the underlying population itself has a normal distribution. What happens when the population distribution is not normal? It turns out that as the sample size increases, the distribution of $\bar{x}$ gets closer to a normal distribution. This is true no matter what shape the population distribution has, as long as the population has a finite standard deviation σ. This famous fact of probability theory is called the *central limit theorem*. It is much more useful than the fact that the distribution of $\bar{x}$ is exactly normal if the population is exactly normal.* For large sample size n, we can regard $\bar{x}$ as having the $N(\mu, \sigma/\sqrt{n})$ distribution.

---

*The first general version of the central limit theorem was established in 1810 by the French mathematician Pierre Simon Laplace (1749–1827).

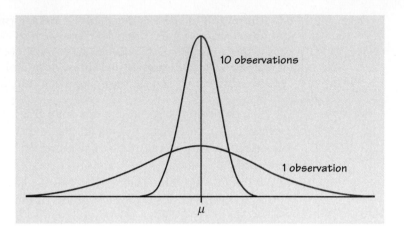

FIGURE 5.7   The sampling distribution of $\bar{x}$ for samples of size 10 compared with the distribution of a single observation.

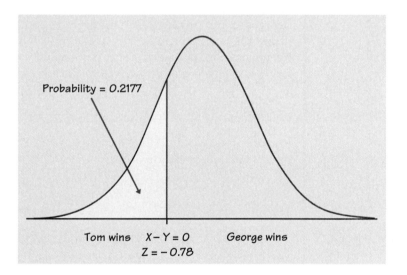

FIGURE 5.8   The normal probability calculation for Example 5.17.

Central Limit Theorem

Draw an SRS of size n from any population with mean μ and finite standard deviation σ. When n is large, the sampling distribution of the sample mean $\bar{x}$ is approximately normal:

$$\bar{x} \text{ is approximately } N\left(\mu, \frac{\sigma}{\sqrt{n}}\right)$$

More generally, the central limit theorem says that the distribution of a sum or average of many small random quantities is close to normal. This is true even if the quantities are not independent (as long as they are not too highly correlated) and even if they have different distributions (as long as no one random quantity is so large that it dominates the others). The central limit theorem suggests why the normal distributions are common models for observed data. Any variable that is a sum of many small influences will have approximately a normal distribution.

How large a sample size n is needed for $\bar{x}$ to be close to normal depends on the population distribution. More observations are required if the shape of the population distribution is far from normal.

EXAMPLE 5.18	Figure 5.9 shows the central limit theorem in action for a very nonnormal population. Figure 5.9(a) displays the density curve of a single observation, that is, of the population. The distribution is strongly right-skewed, and the most probable outcomes are near 0. The mean μ of this distribution is 1, and its standard deviation σ is also 1. This particular continuous distribution is called an **exponential distribution.** Exponential distributions are used as models for the lifetime in service of electronic components and for the time required to serve a customer or repair a machine.

exponential distribution

Figures 5.9(b), (c), and (d) are the density curves of the sample means of 2, 10, and 25 observations from this population. As n increases, the shape becomes more normal. The mean remains at $\mu = 1$, and the standard deviation decreases, taking the value $1/\sqrt{n}$. The density curve for 10 observations is still somewhat skewed to the right but already resembles a normal curve having $\mu = 1$ and $\sigma = 1/\sqrt{10} = 0.32$. The density curve for $n = 25$ is yet more normal. The contrast between the shapes of the population distribution and of the distribution of the mean of 10 or 25 observations is striking.

The central limit theorem allows us to use normal probability calculations to answer questions about sample means from many observations even when the population distribution is not normal.

EXAMPLE 5.19	The time X that a technician requires to perform preventive maintenance on an air-conditioning unit is governed by the exponential distribution whose density curve appears in Figure 5.9(a). The mean time is $\mu = 1$ hour and the standard deviation is $\sigma = 1$ hour. Your company operates 70 of these units. What is the probability that their average maintenance time exceeds 50 minutes?

The central limit theorem says that the sample mean time $\bar{x}$ (in hours) spent working on 70 units has approximately the normal distribution with mean equal to the population mean $\mu = 1$ hour and standard deviation

$$\frac{\sigma}{\sqrt{70}} = \frac{1}{\sqrt{70}} = 0.12 \text{ hour}$$

The distribution of $\bar{x}$ is therefore approximately $N(1, 0.12)$. Figure 5.10 shows this normal curve (*solid*) and also the actual density curve of $\bar{x}$ (*dashed*).

Because 50 minutes is 50/60 of an hour, or 0.83 hour, the probability we want is $P(\bar{x} > 0.83)$. A normal distribution calculation gives this probability as 0.9222. This is the area to the right of 0.83 under the solid normal curve in Figure 5.10. The exactly correct probability is the area under the dashed density curve in the figure. It is 0.9294. The central limit theorem normal approximation is off by only about 0.007.

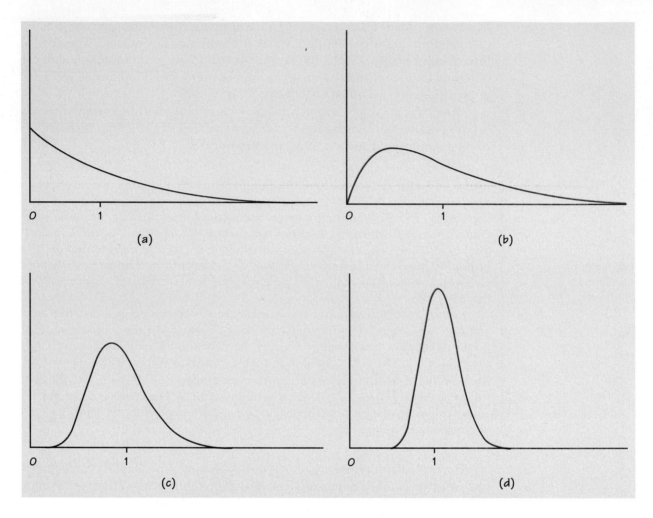

FIGURE 5.9 The central limit theorem in action: the distribution of sample means from a strongly nonnormal population becomes more normal as the sample size increases. (a) The distribution of 1 observation. (b) The distribution of $\overline{x}$ for 2 observations. (c) The distribution of $\overline{x}$ for 10 observations. (d) The distribution of $\overline{x}$ for 25 observations.

The normal approximation for sample proportions and counts is an important example of the central limit theorem. This is true because a sample proportion can be thought of as a sample mean. Recall the idea that we used to find the mean and variance of a $B(n, p)$ random variable X. We wrote the count X as a sum

$$X = S_1 + S_2 + \cdots + S_n$$

of random variables S_i that take the value 1 if a success occurs on the ith trial and the value 0 otherwise. The variables S_i take only the values 0 and 1 and are far from normal. The proportion $\hat{p} = X/n$ is the sample mean of the S_i and, like all sample means, is approximately normal when n is large.

Figure 5.11 summarizes the facts about the sampling distribution of $\overline{x}$ in a way that reminds us of the big idea of a sampling distribution. Keep

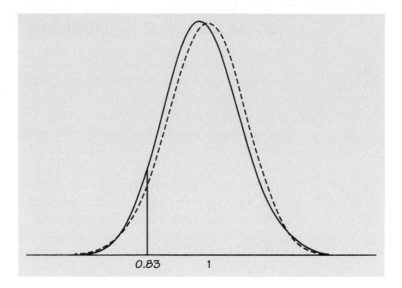

FIGURE 5.10 The exact distribution (*dashed*) and the normal approximation from the central limit theorem (*solid*) for the average time needed to maintain an air conditioner, for Example 5.19.

taking random samples of size n from a population with mean μ. Find the sample mean $\bar{x}$ for each sample. Collect all the $\bar{x}$'s and display their distribution. That's the sampling distribution of $\bar{x}$. Sampling distributions are the key to understanding statistical inference. Keep this figure in mind as you go forward.

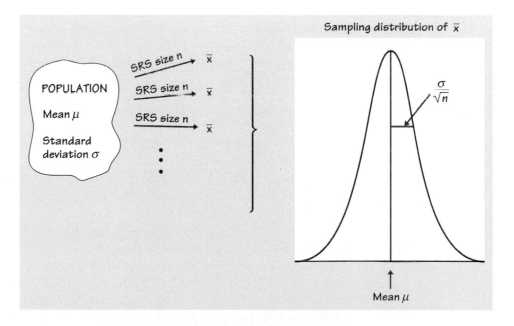

FIGURE 5.11 The sampling distribution of a sample mean $\bar{x}$ has mean μ and standard deviation $\sigma/\sqrt{n}$. The distribution is normal if the population distribution is normal; it is approximately normal for large samples in any case.

Beyond the basics ▶ Weibull distributions

Our discussion of sampling distributions has concentrated on the binomial model for count data and the normal model for quantitative variables. These models are important in statistical practice, but simplicity also contributes to their popularity. The parameters p in the binomial model and μ in the normal model are easy to understand. To estimate them from data we use statistics $\hat{p}$ and $\bar{x}$ that are also easy to understand and that have simple sampling distributions.

There are many other probability distributions that are used to model data in various circumstances. Estimation and sampling distributions are often more complex than for the binomial and normal distributions. Here is an example.

EXAMPLE 5.20

Weibull distributions

The normal distributions are not suitable models for all quantitative variables. The time that a product, such as a computer disk drive, lasts before failing rarely has a normal distribution, for example. Another class of continuous distributions, the **Weibull distributions,** is used to model time to failure. For engineers studying the reliability of products, Weibull distributions are more common than normal distributions.

Figure 5.12 shows the density curves of three members of the Weibull family. Each describes a different type of distribution for the time to failure of a product. Figure 5.12(a) is a model for *infant mortality*. Many of these products fail immediately. If they do not fail at once, then most last a long time. The manufacturer tests these products and only ships the ones that do not fail immediately.

Figure 5.12(b) is a model for *early failure*. These products do not fail immediately, but many fail early in their life after they are in the hands of customers. This is disastrous—the product or the process that makes it must be changed at once.

Figure 5.12(c) is a model for *old-age wearout*. Very few of these products fail until they begin to wear out, and then many fail at about the same age.

A manufacturer certainly wants to know to which of these classes a new product belongs. To find out, engineers operate a random sample of products until they fail. From the failure time data we can estimate the parameter (called the "shape parameter") that distinguishes among the three Weibull distributions in Figure 5.12. The shape parameter has no simple definition like that of a population proportion or mean, and it cannot be estimated by a simple statistic such as $\hat{p}$ or $\bar{x}$.

Two things save the situation. First, statistical theory provides general approaches for finding good estimates of any parameter. These general methods tell us to use $\hat{p}$ and $\bar{x}$ in the binomial and normal settings, and they also tell us how to estimate the Weibull shape parameter. Second, modern software can calculate the estimate from data even though there is no algebraic formula that we can write for the estimate. Statistical practice often relies on both mathematical theory and methods of computation more elaborate than

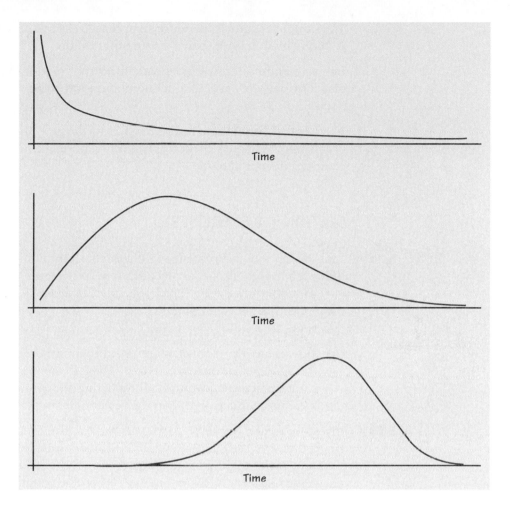

FIGURE 5.12 Density curves for the three members of the Weibull family of distributions, for Example 5.20.

the ones we will meet in this book. Fortunately, big ideas such as sampling distributions carry over to more complicated situations.[1]

SUMMARY

The **sample mean** $\bar{x}$ of an SRS of size n drawn from a large population with mean μ and standard deviation σ has a sampling distribution with mean and standard deviation

$$\mu_{\bar{x}} = \mu$$
$$\sigma_{\bar{x}} = \frac{\sigma}{\sqrt{n}}$$

The sample mean $\bar{x}$ is therefore an unbiased estimator of the population mean μ and is less variable than a single observation.

Linear combinations of independent normal random variables have normal distributions. In particular, if the population has a normal distribution, so does $\bar{x}$.

The **central limit theorem** states that for large n the sampling distribution of $\bar{x}$ is approximately $N(\mu, \sigma/\sqrt{n})$ for any population with mean μ and finite standard deviation σ.

SECTION 5.2 EXERCISES

5.24   Juan makes a measurement in a chemistry laboratory and records the result in his lab report. The standard deviation of students' lab measurements is $\sigma = 10$ milligrams. Juan repeats the measurement 3 times and records the mean $\bar{x}$ of his 3 measurements.

    (a) What is the standard deviation $\sigma_{\bar{x}}$ of Juan's mean result?

    (b) How many times must Juan repeat the measurement to reduce the standard deviation of $\bar{x}$ to 5? Explain to someone who knows no statistics the advantage of reporting the average of several measurements rather than the result of a single measurement.

5.25   Investors remember 1987 as the year stocks lost 22% of their value in a single day. For 1987 as a whole, the mean return of all common stocks on the New York Stock Exchange was $\mu = -3.5\%$. (That is, these stocks lost an average of 3.5% of their value in 1987.) The standard deviation of the returns was about $\sigma = 26\%$. The distribution of annual returns for stocks is roughly normal.

    (a) What percent of stocks lost money? (That is the same as the probability that a stock chosen at random has a return less than 0.)

    (b) Suppose that you held a portfolio of 5 stocks chosen at random from New York Stock Exchange stocks. What are the mean and standard deviation of the returns of randomly chosen portfolios of 5 stocks?

    (c) What percent of such portfolios lost money? Explain the difference between this result and the result of (a).

5.26   The scores of students on the ACT college entrance examination in a recent year had the normal distribution with mean $\mu = 18.6$ and standard deviation $\sigma = 5.9$.

    (a) What is the probability that a single student randomly chosen from all those taking the test scores 21 or higher?

    (b) Now take an SRS of 50 students who took the test. What are the mean and standard deviation of the sample mean score $\bar{x}$ of these 50 students?

(c) What is the probability that the mean score $\bar{x}$ of these students is 21 or higher?

5.27 A laboratory weighs filters from a coal mine to measure the amount of dust in the mine atmosphere. Repeated measurements of the weight of dust on the same filter vary normally with standard deviation $\sigma = 0.08$ milligram (mg) because the weighing is not perfectly precise. The dust on a particular filter actually weighs 123 mg. Repeated weighings will then have the normal distribution with mean 123 mg and standard deviation 0.08 mg.

(a) The laboratory reports the mean of 3 weighings. What is the distribution of this mean?

(b) What is the probability that the laboratory reports a weight of 124 mg or higher for this filter?

5.28 An automatic grinding machine in an auto parts plant prepares axles with a target diameter $\mu = 40.125$ millimeters (mm). The machine has some variability, so the standard deviation of the diameters is $\sigma = 0.002$ mm. A sample of 4 axles is inspected each hour for process control purposes, and records are kept of the sample mean diameter. What will be the mean and standard deviation of the numbers recorded?

5.29 A bottling company uses a filling machine to fill plastic bottles with a popular cola. The bottles are supposed to contain 300 milliliters (ml). In fact, the contents vary according to a normal distribution with mean $\mu = 298$ ml and standard deviation $\sigma = 3$ ml.

(a) What is the probability that an individual bottle contains less than 295 ml?

(b) What is the probability that the mean contents of the bottles in a six-pack is less than 295 ml?

5.30 Judy's doctor is concerned that she may suffer from hypokalemia (low potassium in the blood). There is variation both in the actual potassium level and in the blood test that measures the level. Judy's measured potassium level varies according to the normal distribution with $\mu = 3.8$ and $\sigma = 0.2$. A patient is classified as hypokalemic if the potassium level is below 3.5.

(a) If a single potassium measurement is made, what is the probability that Judy is diagnosed as hypokalemic?

(b) If measurements are made instead on 4 separate days and the mean result is compared with the criterion 3.5, what is the probability that Judy is diagnosed as hypokalemic?

5.31 A roulette wheel has 38 slots, of which 18 are black, 18 are red, and 2 are green. When the wheel is spun, the ball is equally likely to come to rest in any of the slots. Gamblers can place a number of different bets in roulette. One of the simplest wagers chooses red or black. A bet of \$1 on red will pay off an additional dollar if the ball lands in a red slot. Otherwise, the player loses his dollar. When gamblers bet on red or black, the two green slots belong to the house.

(a) A gambler's winnings on a $1 bet are either $1 or −$1. Give the probabilities of these outcomes. Find the mean and standard deviation of the gambler's winnings.

(b) Explain briefly what the law of large numbers tells us about what will happen if the gambler makes a large number of bets on red.

(c) The central limit theorem tells us the approximate distribution of the gambler's mean winnings in 50 bets. What is this distribution? Use the 68–95–99.7 rule to give the range in which the mean winnings will fall 95% of the time. Multiply by 50 to get the middle 95% of the distribution of the gambler's winnings on nights when he places 50 bets.

(d) What is the probability that the gambler will lose money if he makes 50 bets? (This is the probability that the mean is less than 0.)

(e) The casino takes the other side of these bets. If 100,000 bets on red are placed in a week at the casino, what is the distribution of the mean winnings of gamblers on these bets? What range covers the middle 95% of mean winnings in 100,000 bets? Multiply by 100,000 to get the range of gamblers' losses. (Gamblers' losses are the casino's winnings. Part (c) shows that a gambler gets excitement. Now we see that the casino has a business.)

5.32   A company that owns and services a fleet of cars for its sales force has found that the service lifetime of disc brake pads varies from car to car according to a normal distribution with mean $\mu = 55,000$ miles and standard deviation $\sigma = 4500$ miles. The company installs a new brand of brake pads on 8 cars.

(a) If the new brand has the same lifetime distribution as the previous type, what is the distribution of the sample mean lifetime for the 8 cars?

(b) The average life of the pads on these 8 cars turns out to be $\bar{x} = 51,800$ miles. What is the probability that the sample mean lifetime is 51,800 miles or less if the lifetime distribution is unchanged? The company takes this probability as evidence that the average lifetime of the new brand of pads is less than 55,000 miles.

5.33   The number of flaws per square yard in a type of carpet material varies with mean 1.6 flaws per square yard and standard deviation 1.2 flaws per square yard. This population distribution cannot be normal, because a count takes only whole-number values. An inspector studies 200 square yards of the material, records the number of flaws found in each square yard, and calculates $\bar{x}$, the mean number of flaws per square yard inspected.

Use the central limit theorem to find the approximate probability that the mean number of flaws exceeds 2 per square yard.

5.34 The number of accidents per week at a hazardous intersection varies with mean 2.2 and standard deviation 1.4. This distribution takes only whole-number values, so it is certainly not normal.

  (a) Let $\bar{x}$ be the mean number of accidents per week at the intersection during a year (52 weeks). What is the approximate distribution of $\bar{x}$ according to the central limit theorem? 2.2

  (b) What is the approximate probability that $\bar{x}$ is less than 2? $P(x < 2)$?

  (c) What is the approximate probability that there are fewer than 100 accidents at the intersection in a year? (*Hint:* Restate this event in terms of $\bar{x}$.) $P(X < \frac{100}{52}) = P(z < \frac{100/52 - 2.2}{\frac{1.4}{\sqrt{52}}}) = P(zs - 1.4264) = ?$ .0764

5.35 The level of nitrogen oxides (NOX) in the exhaust of a particular car model varies with mean 0.9 grams per mile (g/mi) and standard deviation 0.15 g/mi. A company has 125 cars of this model in its fleet.

  (a) What is the approximate distribution of the mean NOX emission level $\bar{x}$ for these cars?

  (b) What is the level L such that the probability that $\bar{x}$ is greater than L is only 0.01?

5.36 The distribution of annual returns on common stocks is roughly symmetric, but extreme observations are more frequent than in a normal distribution. Because the distribution is not strongly nonnormal, the mean return over even a moderate number of years is close to normal. In the long run, annual real returns on common stocks have varied with mean about 9% and standard deviation about 28%. Andrew plans to retire in 45 years and is considering investing in stocks. What is the probability (assuming that the past pattern of variation continues) that the mean annual return on common stocks over the next 45 years will exceed 15%? What is the probability that the mean return will be less than 5%?

5.37 Children in kindergarten are sometimes given the Ravin Progressive Matrices Test (RPMT) to assess their readiness for learning. Experience at Southwark Elementary School suggests that the RPMT scores for its kindergarten pupils have mean 13.6 and standard deviation 3.1. The distribution is close to normal. Mr. Lavin has 22 children in his kindergarten class this year. He suspects that their RPMT scores will be unusually low because the test was interrupted by a fire drill. To check this suspicion, he wants to find the level L such that there is probability only 0.05 that the mean score of 22 children falls below L when the usual Southwark distribution remains true. What is the value of L?

5.38 The design of an electronic circuit calls for a 100-ohm resistor and a 250-ohm resistor connected in series so that their resistances add. The components used are not perfectly uniform, so that the actual resistances vary independently according to normal distributions. The resistance of 100-ohm resistors has mean 100 ohms and standard deviation 2.5 ohms, while that of 250-ohm resistors has mean 250 ohms and standard deviation 2.8 ohms.

(a) What is the distribution of the total resistance of the two components in series?

(b) What is the probability that the total resistance lies between 345 and 355 ohms?

5.39 An experiment to compare the nutritive value of normal corn and high-lysine corn divides 40 chicks at random into two groups of 20. One group is fed a diet based on normal corn while the other receives high-lysine corn. At the end of the experiment, inference about which diet is superior is based on the difference $\bar{y} - \bar{x}$ between the mean weight gain $\bar{y}$ of the 20 chicks in the high-lysine group and the mean weight gain $\bar{x}$ of the 20 in the normal-corn group. Because of the randomization, the two sample means are independent.

(a) Suppose that $\mu_X = 360$ grams (g) and $\sigma_X = 55$ g in the population of all chicks fed normal corn, and that $\mu_Y = 385$ g and $\sigma_Y = 50$ g in the high-lysine population. What are the mean and standard deviation of $\bar{y} - \bar{x}$?

(b) The weight gains are normally distributed in both populations. What is the distribution of $\bar{x}$? Of $\bar{y}$? What is the distribution of $\bar{y} - \bar{x}$?

(c) What is the probability that the mean weight gain in the high-lysine group exceeds the mean weight gain in the normal-corn group by 25 g or more?

5.40 An experiment on the teaching of reading compares two methods, A and B. The response variable is the Degree of Reading Power (DRP) score. The experimenter uses Method A in a class of 26 students and Method B in a comparable class of 24 students. The classes are assigned to the teaching methods at random. Suppose that in the population of all children of this age the DRP score has the $N(34, 12)$ distribution if Method A is used and the $N(37, 11)$ distribution if Method B is used.

(a) What is the distribution of the mean DRP score $\bar{x}$ for the 26 students in the A group? (Assume that this group can be regarded as an SRS from the population of all children of this age.)

(b) What is the distribution of the mean score $\bar{y}$ for the 24 students in the B group?

(c) Use the results of (a) and (b), keeping in mind that $\bar{x}$ and $\bar{y}$ are independent, to find the distribution of the difference $\bar{y} - \bar{x}$ between the mean scores in the two groups.

(d) What is the probability that the mean score for the B group will be at least 4 points higher than the mean score for the A group?

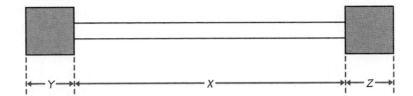

FIGURE 5.13 The dimensions of a mechanical assembly, for Exercise 5.42.

5.41 The two previous exercises illustrate a common setting for statistical inference. This exercise gives the general form of the sampling distribution needed in this setting. We have a sample of n observations from a treatment group and an independent sample of m observations from a control group. Suppose that the response to the treatment has the $N(\mu_X, \sigma_X)$ distribution and that the response of control subjects has the $N(\mu_Y, \sigma_Y)$ distribution. Inference about the difference $\mu_Y - \mu_X$ between the population means is based on the difference $\bar{y} - \bar{x}$ between the sample means in the two groups.

(a) Under the assumptions given, what is the distribution of $\bar{y}$? Of $\bar{x}$?

(b) What is the distribution of $\bar{y} - \bar{x}$?

5.42 A mechanical assembly (Figure 5.13) consists of a shaft with a bearing at each end. The total length of the assembly is the sum $X + Y + Z$ of the shaft length X and the lengths Y and Z of the bearings. These lengths vary from part to part in production, independently of each other and with normal distributions. The shaft length X has mean 11.2 inches and standard deviation 0.002 inch, while each bearing length Y and Z has mean 0.4 inch and standard deviation 0.001 inch.

(a) According to the 68–95–99.7 rule, about 95% of all shafts have lengths in the range $11.2 \pm d_1$ inches. What is the value of d_1? Similarly, about 95% of the bearing lengths fall in the range of $0.4 \pm d_2$. What is the value of d_2?

(b) It is common practice in industry to state the "natural tolerance" of parts in the form used in (a). An engineer who knows no statistics thinks that tolerances add, so that the natural tolerance for the total length of the assembly (shaft and two bearings) is $12 \pm d$ inches, where $d = d_1 + 2d_2$. Find the standard deviation of the total length $X + Y + Z$. Then find the value d such that about 95% of all assemblies have lengths in the range $12 \pm d$. Was the engineer correct?

5.43 Leona and Fred are friendly competitors in high school. Both are about to take the ACT college entrance examination. They agree that if one of them scores 5 or more points better than the other, the loser will buy the winner a pizza. Suppose that in fact Fred and Leona have equal ability, so that each score varies normally with mean 24 and standard deviation 2. (The variation is due to luck in guessing and the accident of the specific questions being familiar to the student.) The two scores are independent. What is the probability that the scores differ by 5 or more points in either direction?

5.44 The study habits portion of the Survey of Study Habits and Attitudes (SSHA) psychological test consists of two sets of questions. One set measures "delay avoidance" and the other measures "work methods." A subject's study habits score is the sum $X + Y$ of the delay avoidance score X and the work methods score Y. The distribution of X in a broad population of first-year college students is close to $N(25, 10)$, while the distribution of Y in the same population is close to $N(25, 9)$.

(a) If a subject's X and Y scores were independent, what would be the distribution of the study habits score $X + Y$?

(b) Using the distribution you found in (a), what percent of the population have a study habits score of 60 or higher?

(c) In fact, the X and Y scores are strongly correlated. In this case, does the mean of $X + Y$ still have the value you found in (a)? Does the standard deviation still have the value you found in (a)?

5.45 A study of working couples measures the income X of the husband and the income Y of the wife in a large number of couples in which both partners are employed. Suppose that you knew the means μ_X and μ_Y and the variances σ_X^2 and σ_Y^2 of both variables in the population.

(a) Is it reasonable to take the mean of the total income $X + Y$ to be $\mu_X + \mu_Y$? Explain your answer.

(b) Is it reasonable to take the variance of the total income to be $\sigma_X^2 + \sigma_Y^2$? Explain your answer.

5.46 Table 1.5 (page 32) gives the survival times of 72 guinea pigs in a medical experiment. The distribution of survival times is strongly skewed to the right. Sampling from this small population can demonstrate how averaging reduces variability and creates a more normal distribution.

(a) Use software to choose 100 SRSs of size 12 from this population. Find the mean survival time $\bar{x}$ for each of the 100 samples.

(b) Find the mean μ of the population. Make a histogram of the 100 $\bar{x}$'s and find their mean. Is the sampling distribution of $\bar{x}$ centered near μ?

(c) Find the standard deviation σ of the population. Then find the standard deviation of the 100 $\bar{x}$'s. In the long run, what do you expect to be the standard deviation of the mean from samples of size $n = 12$? How close to this value did your 100 samples come?

(d) Make normal quantile plots of both the population and the 100 $\bar{x}$'s. Is the sampling distribution of the $\bar{x}$'s markedly closer to normal than the distribution of the population?

SRS = simple random samples

5.3 Control Charts*

There are many situations in which our goal is to hold a variable constant over time. You may monitor your weight or blood pressure and plan to modify your behavior if either changes. Manufacturers watch the results of regular measurements made during production and plan to take action if quality deteriorates. Statistics plays a central role in these situations because of the presence of variation. *All processes have variation.* Your weight fluctuates from day to day, the critical dimension of a machined part varies a bit from item to item. Variation occurs in even the most precisely made product due to small changes in the raw material, the adjustment of the machine, the behavior of the operator, and even the temperature in the plant. Because variation is always present, we can't expect to hold a variable exactly constant over time. The statistical description of stability over time requires that the pattern of variation remain stable, not that there be no variation in the variable measured.

Statistical Control

A variable that continues to be described by the same distribution when observed over time is said to be in statistical control, or simply **in control.**

Control charts are statistical tools that monitor a process and alert us when the process has been disturbed so that it is now **out of control.** This is a signal to find and correct the cause of the disturbance.

Control charts work by distinguishing the natural variation in the process from the additional variation that suggests that the process has changed. A control chart sounds an alarm when it sees too much variation. The most common application of control charts is to monitor the performance of an industrial process. The same methods, however, can be used to check the stability of quantities as varied as the ratings of a television show, the level of ozone in the atmosphere, and the gas mileage of your car. Control charts combine graphical and numerical descriptions of data with use of sampling distributions. They therefore provide a natural bridge between exploratory data analysis and formal statistical inference.†

$\bar{x}$ control charts

The population in the control chart setting is all items that would be produced by the process if it ran on forever in its present state. The items actually

---

*Control charts are important in industry and also illustrate the use of sampling distributions in inference. Nonetheless, this section is not required for an understanding of later material.

†Control charts were invented in the 1920s by Walter Shewhart at the Bell Telephone Laboratories. Shewhart's classic book, *Economic Control of Quality of Manufactured Product* (Van Nostrand, New York, 1931), organized the application of statistics to improving quality.

produced form samples from this population. We generally speak of the process rather than the population. Choose a quantitative variable, such as a diameter or a voltage, that is an important characteristic of an item. The process mean μ is the long-term average value of this variable; μ describes the center of the process. The sample mean $\bar{x}$ of several items estimates μ and helps us judge whether the center of the process has moved away from its proper value. The most common control chart plots the means $\bar{x}$ of small samples taken from the process at regular intervals over time.

EXAMPLE 5.21

A manufacturer of computer monitors must control the tension on the mesh of fine wires that lies behind the surface of the viewing screen. Too much tension will tear the mesh, and too little will allow wrinkles. Tension is measured by an electrical device with output readings in millivolts (mV). The proper tension is 275 mV. Some variation is always present in the production process. When the process is operating properly, the standard deviation of the tension readings is $\sigma = 43$ mV.

The operator measures the tension on a sample of 4 monitors each hour. The mean $\bar{x}$ of each sample estimates the mean tension μ for the process at the time of the sample. Table 5.1 shows the observed $\bar{x}$'s for 20 consecutive hours of production. How can we use these data to keep the process in control?

A plot against time helps us see whether or not the process is stable. Figure 5.14 is a plot of the successive sample means against the order in which the samples were taken. Because the target value for the process mean is $\mu = 275$ mV, we draw a *centerline* at that level across the plot. The means from the later samples fall above this line and are consistently higher than those from earlier samples. This suggests that the process mean μ may have shifted upward, away from its target value of 275 mV. But perhaps the drift in $\bar{x}$ simply reflects the natural variation in the process. We need to back up our graph by calculation.

We expect $\bar{x}$ to have a distribution that is close to normal. Not only are the tension measurements roughly normal, but also the central limit theorem effect implies that sample means will be closer to normal than individual measurements. Because a control chart is a warning device, it is not necessary

TABLE 5.1 $\bar{x}$ from 20 samples of size 4

Sample	$\bar{x}$	Sample	$\bar{x}$
1	269.5	11	264.7
2	297.0	12	307.7
3	269.6	13	310.0
4	283.3	14	343.3
5	304.8	15	328.1
6	280.4	16	342.6
7	233.5	17	338.8
8	257.4	18	340.1
9	317.5	19	374.6
10	327.4	20	336.1

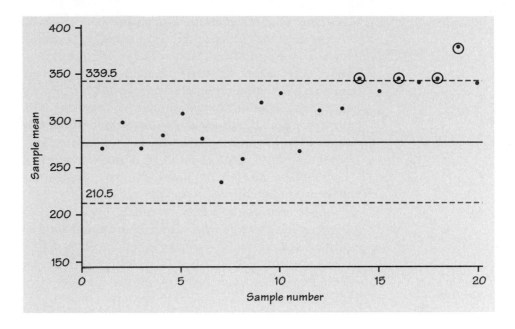

FIGURE 5.14 The $\bar{x}$ control chart for the data of Table 5.1.

that our probability calculations be exactly correct. Approximate normality is good enough. In that same spirit, control charts use the approximate normal probabilities given by the 68–95–99.7 rule rather than more exact calculations using Table A.

If the standard deviation of the individual screens remains at $\sigma = 43$ mV, the standard deviation of $\bar{x}$ from 4 screens is

$$\sigma_{\bar{x}} = \frac{\sigma}{\sqrt{n}} = \frac{43}{\sqrt{4}} = 21.5$$

As long as the mean remains at its target value $\mu = 275$ mV, the 99.7 part of the 68–95–99.7 rule says that almost all values of $\bar{x}$ will lie between

$$\mu - 3\sigma_{\bar{x}} = 275 - (3)(21.5) = 210.5 \text{ mV}$$
$$\mu + 3\sigma_{\bar{x}} = 275 + (3)(21.5) = 339.5 \text{ mV}$$

We therefore draw dashed *control limits* at these two levels on the plot.

$\bar{x}$ Control Chart

To evaluate the control of a process with given standards μ and σ, make an $\bar{x}$ **control chart** as follows:

- Plot the means $\bar{x}$ of regular samples of size n against time.
- Draw a horizontal **centerline** at μ.
- Draw horizontal **control limits** at $\mu \pm 3\sigma/\sqrt{n}$.

Any $\bar{x}$ that does not fall between the control limits is evidence that the process is out of control.

Four points, which are circled in Figure 5.14, lie above the upper control limit of the control chart. It is unlikely (probability less than 0.003) that a particular point would fall outside the control limits if μ and σ remain at their target values. These points are therefore good evidence that the distribution of the production process has changed. It appears that the process mean moved up at about sample number 14. In practice, the operators search for a disturbance in the process as soon as the first out-of-control point is noticed, that is, after sample number 14. Lack of control might be caused by a new operator, a new batch of mesh, or a breakdown in the tensioning apparatus. The out-of-control signal alerts us to the change before a large number of defective monitors are produced.

x chart

An $\bar{x}$ control chart is often called simply an $\bar{x}$ **chart.** Points $\bar{x}$ that vary between the control limits of an $\bar{x}$ chart represent the chance variation that is present in a normally operating process. Points that are out of control suggest that some source of additional variability has disturbed the stable operation of the process. Such a disturbance makes out-of-control points probable rather than unlikely. For example, if the process mean μ in Example 5.21 shifts from 275 mV to 339.5 mV, the process is now centered at the upper control limit. The probability that the next point falls above the upper control limit is now 1/2 rather than about 0.0015.

Statistical process control

The purpose of a control chart is not to ensure good quality by inspecting most of the items produced. Control charts focus on the manufacturing process itself rather than on the products. By checking the process at regular intervals, we can detect disturbances and correct them quickly. This is called

statistical process control

statistical process control. Process control achieves high quality at a lower cost than inspecting all of the products. Small samples of 4 or 5 items are usually adequate for process control.

A process that is in control is stable over time, but stability alone does not guarantee good quality. The natural variation in the process may be so large that many of the products are unsatisfactory. Nonetheless, establishing control brings a number of advantages.

- In order to assess whether the process quality is satisfactory, we must observe the process operating in control free of breakdowns and other disturbances.

- A process in control is predictable. We can predict both the quantity and the quality of items produced.

- When a process is in control we can easily see the effects of attempts to improve the process, which are not hidden by the unpredictable variation that characterizes lack of statistical control.

A process in control is doing as well as it can in its present state. If the process is not capable of producing adequate quality even when undisturbed, we must make some major change in the process, such as installing new machines or retraining the operators.

Using control charts

A control chart gets it name from the statistic that is plotted against time—an $\bar{x}$ chart plots $\bar{x}$, while a $\hat{p}$ chart plots the sample proportion $\hat{p}$. (By an accident of history, a $\hat{p}$ chart is called a p chart in most writing on quality control.) We find the centerline and control limits from the sampling distribution of the statistic. Because you know that $\hat{p}$ from a sample of size n has approximately the $N(p, \sqrt{p(1-p)/n})$ distribution, you should be able to set up a $\hat{p}$ chart to control a population proportion at a target value p. Such $\hat{p}$ charts are most often used to monitor the proportion p of items produced that fail to conform to a set of specifications.

There are many variations on even the $\bar{x}$ chart. In practice we rarely know the process mean μ and the process standard deviation σ. We must then base control limits on estimates of μ and σ from past samples. There are some fine points in using control limits based on past data—for example, we must first check that the process was already in control when the data were gathered. We will content ourselves with the case in which values μ and σ are given.

The probability calculations in a control chart assume that the individual observations are a random sample from the population of interest. If the usual 4 or 5 items in a sample are an SRS from an hour's production, the population is all items produced that hour. It is more common, however, to sample 4 or 5 consecutive items once each hour. In that case the population exists only in our minds. It contains all items that would be produced by the process as it was operating at the time of sampling. The control chart monitors the state of the process once each hour to see if a change has taken place.

The basic signal for lack of control in an $\bar{x}$ chart is a single point beyond the control limits. In practice, however, other signals are used as well. Figure 5.15 illustrates several popular signals. For each signal we can compute the probability that a process in control (that is, with μ and σ at their target values) gives a false signal. The probability of a false signal is about 0.003 for the "one point beyond the limits" signal, by the 99.7 part of the 68–95–99.7 rule.

EXAMPLE 5.22

run signal

One common out-of-control signal is a **run** of 9 consecutive points above or below the centerline (Figure 5.15b). What is the probability of a false signal of this type when the process is in fact in control?

When the process is in control, the centerline is the true process mean μ. Because of the symmetry of the normal distributions, a sample mean $\bar{x}$ is equally likely to fall above or below the line. The means of successive samples are independent because the samples are separated in time. So by the multiplication rule, the probability of a run of 9 points above μ is

$$\left(\frac{1}{2}\right)^9 \doteq 0.002$$

The probability of a run of 9 points below the centerline is the same. The probability that the next 9 points show a run of either kind is therefore about 0.004. The probability that such a run occurs not in the next 9 points but somewhere in a longer sequence of observations would also be useful. Unfortunately, the calculation of such probabilities is quite difficult.

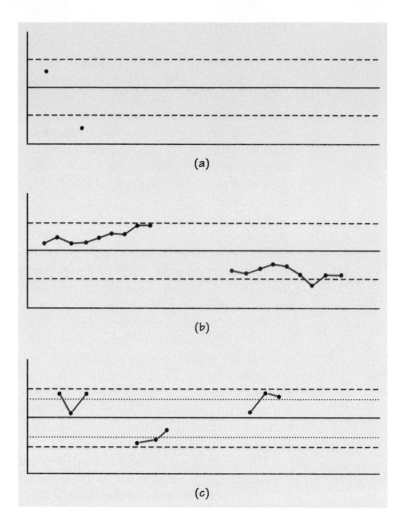

FIGURE 5.15   Out-of-control signals for $\bar{x}$ charts. (a) One point beyond the 3σ level. (b) Run of nine points on one side of the centerline. (c) Two out of three points beyond the 2σ level on the same side of the centerline.

In the $\bar{x}$ chart of Figure 5.14, the run signal does not give an out-of-control signal until sample number 20. The "one point out" signal alerts us at sample 14. In this case the run signal was slow to detect lack of control. When the process mean slowly drifts away from its target value, however, the run signal will often give an out-of-control signal before any individual point falls outside the control limits. That is why it is common practice to use the "one point out" signal and the run signal simultaneously.

EXAMPLE 5.23   Another common out-of-control signal is "2 out of 3 points beyond the 2σ level on the same side of the centerline." That is, at least 2 of the next 3 $\bar{x}$'s fall either below $\mu - 2\sigma/\sqrt{n}$ or above $\mu + 2\sigma/\sqrt{n}$. Figure 5.15(c) illustrates several configurations that give this signal.

    The probability that a process in control produces an $\bar{x}$ larger than $\mu + 2\sigma/\sqrt{n}$ is about 0.025 (that's half of the 0.05 left over from the 95 part of the 68–95–99.7 rule).

The number X of the next 3 samples beyond this level is a count of successes in 3 independent observations; X therefore has the $B(3, 0.025)$ distribution. The binomial formula gives the probability of a false signal as

$$P(X \geq 2) = P(X = 2) + P(X = 3)$$
$$= \binom{3}{2}(0.025)^2(0.975) + \binom{3}{3}(0.025)^3(0.975)^0$$
$$= 0.00183 + 0.00002 \doteq 0.002$$

The probability that the next 3 samples give a false signal on the low side is the same. So the overall probability of a false signal by this criterion is about 0.004.

These examples illustrate the application of probability calculations in inference. They confirm that each of these out-of-control signals is unlikely to occur as long as the process remains in control. So when a signal does occur, we suspect that the process has been disturbed. The probability calculations assumed that μ and σ were given. The results would be a bit different in the more common case in which σ is unknown and must be estimated from past observations. The same out-of-control signals are nonetheless used in both cases. Control charts are an informal tool designed for easy use. We do probability calculations in the simplest case to see roughly how likely false signals are and do not attempt to refine these calculations when conditions are a bit different. We will see in Chapter 7 that more formal inference does adapt its probability calculations to each situation in turn.

SUMMARY

A process that continues over time is **in control** if any variable measured on the process has the same distribution at any time. A process that is in control is operating under stable conditions.

An $\bar{x}$ **control chart** is a graph of sample means plotted against the time order of the samples, with a solid **centerline** at the target value μ of the process mean and dashed **control limits** at $\mu \pm 3\sigma/\sqrt{n}$. An $\bar{x}$ chart helps us decide if a process is in control with mean μ and standard deviation σ.

The probability that the next point lies outside the control limits on an $\bar{x}$ chart is about 0.003 if the process is in control. Such a point is evidence that the process is **out of control**—that is, that the distribution of the process has changed due to some disturbance. A cause for the change in the process should be sought.

There are other common signals for lack of control, such as a **run** of 9 consecutive points on the same side of the centerline. In each case, we can compute the probability that a process in control will give the signal. If this probability is small, we can regard the signal as good evidence of lack of control.

There are other types of control charts, such as $\hat{p}$ **charts.** A control chart is named for the statistic that it plots.

The purpose of **statistical process control** using control charts is to monitor a process so that any changes can be detected and corrected quickly. This is an economical method of maintaining good product quality.

SECTION 5.3 EXERCISES

5.47   A maker of auto air conditioners checks a sample of 4 thermostatic controls from each hour's production. The thermostats are set at 75° F and then placed in a chamber where the temperature is raised gradually. The temperature at which the thermostat turns on the air conditioner is recorded. The standard for the process mean is $\mu = 75°$. Past experience indicates that the response temperature of properly adjusted thermostats varies with $\sigma = 0.5°$. The mean response temperature $\overline{x}$ for each hour's sample is plotted on an $\overline{x}$ control chart. Calculate the centerline and control limits for this chart.

5.48   The width of a slot cut by a milling machine is important to the proper functioning of a hydraulic system for large tractors. The manufacturer checks the control of the milling process by measuring a sample of 5 consecutive items during each hour's production. The mean slot width for each sample is plotted on an $\overline{x}$ control chart. The target width for the slot is $\mu = 0.8750$ inch. When properly adjusted, the milling machine should produce slots with mean width equal to the target value and standard deviation $\sigma = 0.0012$ inch. What centerline and control limits should be drawn on the $\overline{x}$ chart?

5.49   A pharmaceutical manufacturer forms tablets by compressing a granular material that contains the active ingredient and various fillers. The hardness of a sample from each lot of tablets is measured in order to control the compression process. The target values for the hardness are $\mu = 11.5$ and $\sigma = 0.2$. Table 5.2 gives three sets of data, each representing $\overline{x}$ for 20 successive samples of $n = 4$ tablets. One set remains in control at the target value. In a second set, the process mean μ shifts suddenly to a new value. In a third, the process mean drifts gradually.

   (a)   What are the centerline and control limits for an $\overline{x}$ chart for this process?

   (b)   Draw a separate $\overline{x}$ chart for each of the three data sets. Circle any points that are beyond the control limits. Also, check for runs of 9 points above or below the centerline and mark the ninth point of any run as being out of control.

   (c)   Based on your work in (b) and the appearance of the control charts, which set of data comes from a process that is in control? In which case does the process mean shift suddenly and at about which sample do you think that the mean changed? Finally, in which case does the mean drift gradually?

TABLE 5.2	Three sets of $\bar{x}$ from 20 samples of size 4		
Sample	Data Set A	Data Set B	Data Set C
1	11.602	11.627	11.495
2	11.547	11.613	11.475
3	11.312	11.493	11.465
4	11.449	11.602	11.497
5	11.401	11.360	11.573
6	11.608	11.374	11.563
7	11.471	11.592	11.321
8	11.453	11.458	11.533
9	11.446	11.552	11.486
10	11.522	11.463	11.502
11	11.664	11.383	11.534
12	11.823	11.715	11.624
13	11.629	11.485	11.629
14	11.602	11.509	11.575
15	11.756	11.429	11.730
16	11.707	11.477	11.680
17	11.612	11.570	11.729
18	11.628	11.623	11.704
19	11.603	11.472	12.052
20	11.816	11.531	11.905

5.50   The diameter of a bearing deflector in an electric motor is supposed to be 2.205 centimeters (cm). Experience shows that when the manufacturing process is properly adjusted, it produces items with mean 2.2050 cm and standard deviation 0.0010 cm. A sample of 5 consecutive items is measured once each hour. The sample means $\bar{x}$ for the past 12 hours are given below:

Hour	1	2	3	4	5	6
$\bar{x}$	2.2047	2.2047	2.2050	2.2049	2.2053	2.2043

Hour	7	8	9	10	11	12
$\bar{x}$	2.2036	2.2042	2.2038	2.2045	2.2026	2.2040

Make an $\bar{x}$ control chart for the deflector diameter. Use both the "one point out" and the "run of nine" signals to assess the control of the process. At what point should action have been taken to correct the process as the hourly point was added to the chart?

5.51 Ceramic insulators are baked in lots in a large oven. After the baking, 3 insulators are selected at random from each lot and tested for breaking strength. The mean breaking strength for these samples is plotted on a control chart. The specifications call for a mean breaking strength of at least 10 pounds per square inch (psi). Past experience suggests that if the ceramic is properly formed and baked, the standard deviation in the breaking strength is about 1.2 psi. Here are the sample means from the last 15 lots:

Lot	1	2	3	4	5	6	7	8
$\overline{x}$	12.94	11.45	11.78	13.11	12.69	11.77	11.66	12.60

Lot	9	10	11	12	13	14	15
$\overline{x}$	11.23	12.02	10.93	12.38	7.59	13.17	12.14

(a) Find the centerline and control limits for an $\overline{x}$ chart with standards $\mu = 10$ and $\sigma = 1.2$. Make a control chart with these lines drawn on it and plot the $\overline{x}$ points on this chart.

(b) In this case, a process mean breaking strength greater than 10 psi is acceptable, so points out of control in the high direction do not call for remedial action. With this in mind, use both the "one point out" and the "run of nine" signals to assess the control of the process and recommend action.

5.52 The rate of return on a stock varies from month to month. We can use a control chart to see if the pattern of variation is stable over time or whether there are periods during which the stock was unusually volatile by comparison with its own long-run pattern. Here are the mean monthly rates of return (in percent) for Wal-Mart stock for 38 six-month periods, in time order from left to right along the rows. The data begin in January 1973 and end in December 1991.

-11.78	1.68	7.88	-11.01	19.72	1.66	1.13	2.70	-0.60	5.91
2.95	0.05	1.88	6.46	2.25	8.05	4.10	2.08	4.02	11.36
8.13	0.12	1.30	-1.27	6.59	3.01	8.68	-1.52	6.64	-3.20
2.92	0.63	3.49	2.94	5.86	-0.25	6.02	5.87		

We can treat these as the means $\overline{x}$ from 38 samples of 6 consecutive observations each. We will plot them on an $\overline{\overline{x}}$ chart.

(a) For the centerline, use the mean of the $\overline{x}$'s, called $\overline{\overline{x}}$ in quality control. This is the same as the mean of the 228 individual monthly returns.

(b) The standard deviation of the 228 monthly returns includes both long-term and short-term variation and will generally be too large for effective control. Instead, it is common to average the standard deviations of the 38 samples.[2] The result is $\overline{s} = 9.450$. If $\overline{\overline{x}}$ estimates the mean μ and $\overline{s}$ estimates σ, what are the control limits for an $\overline{x}$ chart?

(c) Make the chart. Look for points out of control and for runs of 9. Were there periods during these 19 years when the returns on Wal-Mart stock were out of control?

5.53 A manager who knows no statistics asks you, "What does it mean to say that a process is in control? Is being in control a guarantee that the quality of the product is good?" Answer these questions in plain language that the manager can understand.

5.54 There are other out-of-control signals that are sometimes used with $\bar{x}$ charts. One is "4 out of 5 points beyond the 1σ level" on the same side of the centerline. That is, at least 4 of 5 consecutive points lie above $\mu + \sigma/\sqrt{n}$, or at least 4 of 5 lie below $\mu - \sigma/\sqrt{n}$. Find the probability that the next 5 points give this signal when the process remains in control at the given μ and σ. Use the binomial distribution and approximate probabilities from the 68–95–99.7 rule.

5.55 Another out-of-control signal that is sometimes used is "15 points in a row within the 1σ level." That is, 15 consecutive points fall between $\mu - \sigma/\sqrt{n}$ and $\mu + \sigma/\sqrt{n}$. This signal suggests either that the value of σ used for the chart is too large or that careless measurement is producing results that are suspiciously close to the target. Find the probability that the next 15 points will give this signal when the process remains in control with the given μ and σ.

5.56 The usual American and Japanese practice in making $\bar{x}$ charts is to place the control limits at $\mu \pm 3\sigma/\sqrt{n}$. The probability that a particular $\bar{x}$ falls outside these limits when the process is in control is about 0.003, using the 99.7 part of the 68–95–99.7 rule. European practice, on the other hand, places the control limits at $\mu \pm c\sigma/\sqrt{n}$, where the constant c is chosen to give exactly probability 0.001 of a point $\bar{x}$ falling above $\mu + c\sigma/\sqrt{n}$ when the target μ and σ remain true. (The probability that $\bar{x}$ falls below $\mu - c\sigma/\sqrt{n}$ is also 0.001 because of the symmetry of the normal distributions.) Use Table A to find the value of c.

Sometimes we want to make a control chart for a single measurement x at each time period. Control charts for individual measurements are just $\bar{x}$ charts with the sample size n = 1. We cannot estimate the short-term process standard deviation σ from the individual samples, because a sample of size 1 has no variation. Even with advanced methods[3] of combining the information in several samples, the estimated σ will include some long-term variation and so will be too large. To compensate for this, it is common to use 2σ, rather than 3σ, control limits, that is, control limits $\mu \pm 2\sigma$. The next two exercises deal with control charts for individual measurements.

5.57 Joe has recorded his weight, measured at the gym after a workout, for several years. The mean is 162 pounds and the standard deviation 1.5 pounds, with no signs of lack of control. An injury keeps Joe away from the gym for several months. The data below give his weight, measured once each week for the first 16 weeks after he returns from the injury:

Week	1	2	3	4	5	6	7	8
Weight	168.7	167.6	165.8	167.5	165.3	163.4	163.0	165.5

Week	9	10	11	12	13	14	15	16
Weight	162.6	160.8	162.3	162.7	160.9	161.3	162.1	161.0

The short-term variation in Joe's weight, estimated from these measurements by advanced methods, is about $\sigma = 1.3$ pounds. Joe has a target of $\mu = 162$ pounds for his weight. Make a control chart for his measurements, using control limits $\mu \pm 2\sigma$. Comment on individual points out of control and on runs. Is Joe's weight stable or does it change systematically over this period?

5.58 Professor Moore, who lives a few miles outside a college town, keeps a record of his commuting time. The data (in minutes) appear in Table 5.3. They cover most of the fall semester, although on some dates the professor was out of town or forgot to set his stopwatch. He also noted unusual occurrences on his record sheet: on October 27, a truck backing into a loading dock delayed him, and on December 5, ice on the windshield forced him to stop and clear the glass.

(a) Find $\bar{x}$ and s for the driving times in Table 5.3.

(b) Plot the driving times against the order in which the observations were made. Add a centerline and the control limits $\bar{x} \pm 2s$ to your chart. (The standard deviation s of all observations includes any long-term variation, so these limits are a bit crude.)

TABLE 5.3 Commuting time (minutes), September 10 to December 19

Date	Time	Date	Time	Date	Time
Sept. 10	8.25	Oct. 10	7.75	Nov. 22	7.75
Sept. 11	7.83	Oct. 13	7.92	Nov. 24	7.42
Sept. 12	8.30	Oct. 16	8.00	Nov. 28	6.75
Sept. 15	8.42	Oct. 17	8.08	Nov. 29	7.42
Sept. 17	8.50	Oct. 22	8.42	Dec. 2	8.50
Sept. 22	8.67	Oct. 23	8.75	Dec. 4	8.67
Sept. 23	8.17	Oct. 24	8.08	Dec. 5	10.17
Sept. 24	9.00	Oct. 27	9.75	Dec. 8	8.75
Sept. 25	9.00	Oct. 28	8.33	Dec. 10	8.58
Sept. 29	8.17	Nov. 3	7.83	Dec. 11	8.67
Sept. 30	7.92	Nov. 5	7.92	Dec. 12	9.17
Oct. 3	9.00	Nov. 19	8.58	Dec. 15	9.08
Oct. 8	8.50	Nov. 20	7.83	Dec. 17	8.83
Oct. 9	9.00	Nov. 21	8.42	Dec. 19	8.67

(c) Comment on the control of the process. Can you suggest explanations for individual points that are out of control? Is there any indication of an upward or downward trend in driving time?

5.59 A manufacturer of compact disc players uses statistical process control to monitor the quality of the circuit board that contains most of the player's electronic components. Every circuit board is tested for proper function by a computer-directed test bed after assembly. The plant produces 400 circuit boards per shift, and the proportion $\hat{p}$ of the 400 boards that fail the test is recorded each day. Company standards call for a failure rate of no more than 10%, or $p = 0.1$.

(a) If the process is in control with proportion $p = 0.1$ of defective boards, what are the mean and standard deviation of the sample proportion $\hat{p}$?

(b) According to the normal approximation, what is the distribution of $\hat{p}$ when the process is in control?

(c) Give the centerline (at the mean of $\hat{p}$) and control limits (at three standard deviations of $\hat{p}$ above and below the mean).

(d) The table below lists the proportions defective in 16 consecutive days' production. Make a $\hat{p}$ control chart with centerline and control limits from (c). Plot these values of $\hat{p}$ on the chart. Are any points out of control?

Day	1	2	3	4	5	6	7	8
$\hat{p}$	0.1150	0.1600	0.1300	0.1225	0.1000	0.1225	0.1900	0.1150

Day	9	10	11	12	13	14	15	16
$\hat{p}$	0.1000	0.1600	0.1675	0.1225	0.1375	0.1975	0.1525	0.1675

5.60 In a process control setting, successive samples of n items are inspected. Each item is classified as conforming to specifications or as failing to conform. The proportion $\hat{p}$ of nonconforming items in each sample is recorded and the values of $\hat{p}$ are plotted against the order of the samples. There is a target value p for the population proportion of nonconforming items. Generalize the results of the previous exercise to give formulas in terms of p for the centerline and control limits of a $\hat{p}$ control chart. Use the normal approximation for $\hat{p}$ and follow the model of the $\bar{x}$ chart.

5.61 A producer of agricultural machinery purchases bolts from a supplier in lots of several thousand. The producer submits 80 bolts from each lot to a shear test and accepts the lot if no more than 3 of them fail the test. Past records show that an average of 1.8 bolts per lot have failed the test. This is an opportunity to keep a control chart to monitor the quality level of the supplier over time. Because 1.8/80 = 0.0225, a target proportion of nonconforming bolts is set at $p = 0.0225$ based on past data. The proportion of nonconforming bolts among the 80 tested from each incoming lot will be plotted on a control chart. Give the centerline and control limits to be used on this chart. (Because of the informal nature of process control procedures, the chart is based on the normal approximation to the distribution of the sample proportion $\hat{p}$ even though the approximation is not very accurate for this n and p).

CHAPTER 5 EXERCISES

5.62 Ray is a basketball player who has made about 70% of his free throws over several years. In a tournament game he makes only 2 of 6 free throws. Ray's coach says this was just bad luck. Suppose that Ray's free throws are independent trials with probability 0.7 of a success on each trial. What is the probability that he makes 2 or fewer in 6 attempts? Do you think that his tournament performance is just chance variation?   *M = .7×6*
is expected # same as what he did?

5.63 The distribution of scores for persons over 16 years of age on the Wechsler Adult Intelligence Scale (WAIS) is approximately normal with mean 100 and standard deviation 15. The WAIS is one of the most common "IQ tests" for adults.

    (a) What is the probability that a randomly chosen individual has a WAIS score of 105 or higher?

    (b) What are the mean and standard deviation of the average WAIS score $\bar{x}$ for an SRS of 60 people?

    (c) What is the probability that the average WAIS score of an SRS of 60 people is 105 or higher?

    (d) Would your answers to any of (a), (b), or (c) be affected if the distribution of WAIS scores in the adult population were distinctly nonnormal?

5.64 High school dropouts make up 12.1% of all Americans aged 18 to 24. A vocational school that wants to attract dropouts mails an advertising flyer to 25,000 persons between the ages of 18 and 24.

    (a) If the mailing list can be considered a random sample of the population, what is the mean number of high school dropouts who will receive the flyer? What is the standard deviation of this number?

    (b) What is the probability that at least 3500 dropouts will receive the flyer?

5.65 A political activist is gathering signatures on a petition by going door-to-door asking citizens to sign. She wants 100 signatures. Suppose that the probability of getting a signature at each household is 1/10 and let the random variable X be the number of households visited to collect exactly 100 signatures. Does X have a binomial distribution? If so, give n and p. If not, explain why not.

5.66 According to genetic theory, the blossom color in the second generation of a certain cross of sweet peas should be red or white in a 3:1 ratio. That is, each plant has probability 3/4 of having red blossoms, and the blossom colors of separate plants are independent.

(a) What is the probability that exactly 6 out of 8 of these plants have red blossoms?

(b) What is the mean number of red-blossomed plants when 80 plants of this type are grown from seeds?

(c) What is the probability of obtaining at least 50 red-blossomed plants when 80 plants are grown from seeds?

5.67 The weight of the eggs produced by a certain breed of hen is normally distributed with mean 65 grams (g) and standard deviation 5 g. If cartons of such eggs can be considered to be SRSs of size 12 from the population of all eggs, what is the probability that the weight of a carton falls between 750 g and 825 g?

5.68 According to a market research firm, 52% of all residential telephone numbers in Los Angeles are unlisted. A telemarketing company uses random digit dialing equipment that dials residential numbers at random, regardless of whether they are listed in the telephone directory. The firm calls 500 numbers in Los Angeles.

(a) What is the exact distribution of the number X of unlisted numbers that are called?

(b) Use a suitable approximation to calculate the probability that at least half of the numbers called are unlisted.

5.69 A study of rush hour traffic in San Francisco records the number of people in each car entering a freeway at a suburban interchange. Suppose that this number X has mean 1.5 and standard deviation 0.75 in the population of all cars that enter at this interchange during rush hours.

(a) Does the count X have a binomial distribution? Why or why not?

(b) Could the exact distribution of X be normal? Why or why not?

(c) Traffic engineers estimate that the capacity of the interchange is 700 cars per hour. According to the central limit theorem, what is the approximate distribution of the mean number of persons $\bar{x}$ in 700 randomly selected cars at this interchange?

(d) The count of people in 700 cars is $700\bar{x}$. Use your result from (c) to give an approximate distribution for the count. What is the probability that 700 cars will carry more than 1075 people?

5.70 An opinion poll asks a sample of 500 adults whether they favor giving parents of school-age children vouchers that can be exchanged for education at any public or private school of their choice. Each school would be paid by the government on the basis of how many vouchers it collected. Suppose that in fact 45% of the population favor this idea. What is the probability that at least half of the sample are in favor? (Assume that the sample is an SRS.)

5.71 A machine fastens plastic screw-on caps onto containers of motor oil. If the machine applies more torque than the cap can withstand, the cap will break. Both the torque applied and the strength of the caps vary. The capping machine torque has the normal distribution with mean 7 inch-pounds and standard deviation 0.9 inch-pounds. The cap strength (the torque that would break the cap) has the normal distribution with mean 10 inch-pounds and standard deviation 1.2 inch-pounds.

(a) Explain why it is reasonable to assume that the cap strength and the torque applied by the machine are independent.

(b) What is the probability that a cap will break while being fastened by the capping machine?

5.72 The process that molds the plastic caps referred to in the previous exercise is monitored by testing a sample of 6 caps every 20 minutes. The breaking strength of the caps is measured and an $\bar{x}$ chart is kept. Use the information in the previous exercise to find the centerline and control limits for this $\bar{x}$ chart.

5.73 An important step in the manufacture of integrated circuit chips is etching the lines that will conduct current between components on the chip. The chips contain a line width test pattern that is used for process control measurements. The target width is 3.0 micrometers (μm). Past data show that the line width varies in production according to a normal distribution with mean 2.829 μm and standard deviation 0.1516 μm.

(a) What is the probability that the line width of a randomly chosen chip falls outside the acceptable range 3.0 ± 0.2 μm?

(b) What are control limits for an $\bar{x}$ chart for line width if samples of size 5 are selected at regular intervals during production? (Use the target value 3.0 as your centerline.)

5.74 In an experiment on learning foreign languages, researchers studied the effect of delaying oral practice when beginning language study. The researchers randomly assigned 23 beginning students of Russian to an experimental group and another 23 to a control group. The control group began speaking practice immediately while the experimental group delayed speaking for 4 weeks. At the end of the semester both groups took a standard test of comprehension of spoken Russian. Suppose that in the population of all beginning students, the test scores under the control method vary according to the $N(32, 6)$ distribution. The population distribution when oral practice is delayed is $N(29, 5)$.

(a) What is the sampling distribution of the mean score $\bar{x}$ in the control group in many repetitions of the experiment? What is the sampling distribution of the mean score $\bar{y}$ in the experimental group?

(b) If the experiment were repeated many times, what would be the sampling distribution of the difference $\bar{y} - \bar{x}$ between the mean scores in the two groups?

(c) What is the probability that the experiment will find (misleadingly) that the experimental group has a mean at least as large as that of the control group?

5.75 The amount of nitrogen oxides (NOX) present in the exhaust of a particular type of car varies from car to car according to the normal distribution with mean 1.4 grams per mile (g/mi) and standard deviation 0.3 g/mi. Two cars of this type are tested. One has 1.1 g/mi of NOX, the other 1.9. The test station attendant finds this much variation between two similar cars surprising. If X and Y are independent NOX levels for cars of this type, find the probability

$$P(X - Y \geq 0.8 \text{ or } X - Y \leq -0.8)$$

that the difference is at least as large as the value the attendant observed.

NOTES

1. Statistical methods for dealing with time-to-failure data, including the Weibull model, are presented in Wayne Nelson, *Applied Life Data Analysis*, Wiley, New York, 1982.

2. Although use of $\bar{s}$ is traditional in quality control, this is not an efficient way to estimate the short-term variation σ. Quality control software often implements the estimate of σ from analysis of variance, discussed in Chapter 12.

3. These methods are discussed in all texts on statistical quality control. See, for example, Douglas C. Montgomery, *Introduction to Statistical Quality Control*, 3rd ed., Wiley, New York, 1996.

Prelude

Statistical inference draws conclusions about a population or process based on sample data. Statistical inference also provides a statement, expressed in the language of probability, of how much confidence we can place in our conclusions. Although there are many specific recipes for inference in specific settings, there are only a few general types of statistical inference. This chapter introduces the two most common types: *confidence intervals* and *tests of significance*.

We concentrate in this chapter on the reasoning of inference. We will examine when confidence intervals and significance tests are appropriate, what kinds of conclusions they draw and how, and what their limitations are. Because of the emphasis on understanding the reasoning of inference, we look only at a single simple setting: inference about the mean of a normal population whose standard deviation we know. Later chapters will present recipes for inference in many other situations. The reasoning is more important—first, because it applies to all of the recipes and, second, because a computer can carry out the recipes but not the reasoning. We will deal with questions like these:

- If we have the SAT scores of a random sample of 500 California high school seniors, what can we conclude about the average score of all seniors in the state?

- If you find that the mean contents of the cola bottles in a six-pack is 299 milliliters, is this good evidence that the mean for all bottles is less than the 300 milliliters claimed on the label?

- Psychiatrists compared a group of mentally ill subjects with a group of healthy subjects on 77 different variables describing their childhood and family backgrounds. There were "statistically significant" differences between the groups for 2 of these variables. Is this a sensible scientific procedure?

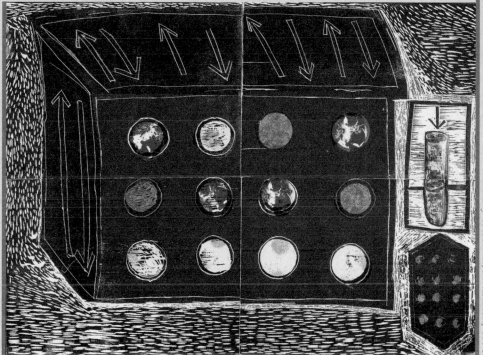

Hilary S. Lorenz, *Incubation*, 1956.

Introduction to Inference

Introduction

The purpose of statistical inference is to draw conclusions from data. We have examined data and arrived at conclusions many times previously. Formal inference adds an emphasis on substantiating our conclusions by probability calculations.

Probability allows us to take chance variation into account and so to correct our judgment by calculation. Was the first Vietnam era draft lottery biased? The correlation between birth date and draft number was $r = -0.226$. Because any two variables will have some chance association in practice, this rather small correlation does not seem convincing. Calculation that a correlation this far from zero has probability less than 0.001 in a truly random lottery gives strong evidence that the lottery was in fact unfair. Our unaided judgment can also err in the opposite direction, seeing a systematic effect when only chance variation is at work. Give a new drug and a placebo to 20 patients each; 12 of those taking the drug show improvement, but only 8 of the placebo patients improve. Is the drug more effective than the placebo? Perhaps, but a difference this large or larger between the results in the two groups would occur about one time in five simply because of chance variation. An effect that could so easily be just chance is not convincing.

In this chapter we introduce the two most prominent types of formal statistical inference. Section 6.1 concerns *confidence intervals* for estimating the value of a population parameter. Section 6.2 presents *tests of significance*, which assess the evidence for a claim. Both types of inference are based on the sampling distributions of statistics. That is, both report probabilities that state *what would happen if we used the inference method many times*. This kind of probability statement is characteristic of standard statistical inference. Users of statistics must understand the nature of the reasoning employed and the meaning of the probability statements that appear, for example, on computer output for statistical procedures.

Because the methods of formal inference are based on sampling distributions, they require a probability model for the data. Trustworthy probability models can arise in many ways, but the model is most secure and inference is most reliable when the data are produced by a properly randomized design. *When you use statistical inference you are acting as if the data come from a random sample or a randomized experiment.* If this is not true, your conclusions may be open to challenge. Do not be overly impressed by the complex details of formal inference. This elaborate machinery cannot remedy basic flaws in producing the data such as voluntary response samples and confounded experiments. Use the common sense developed in your study of the first three chapters of this book, and proceed to detailed formal inference only when you are satisfied that the data deserve such analysis.

The primary purpose of this chapter is to describe the reasoning used in statistical inference. We will discuss only a few specific inference techniques, and these require an unrealistic assumption: that we know the standard deviation σ. Later chapters will present inference methods for use in most of the settings we met in learning to explore data. There are libraries—both of books and of computer software—full of more elaborate statistical techniques. Informed use of any of these methods requires an understanding of the underlying reasoning. A computer will do the arithmetic, but you must still exercise judgment based on understanding.

6.1 Estimating with Confidence

The SAT* tests are widely used measures of readiness for college study. There are two parts, one for verbal reasoning ability (SAT-V) and one for mathematical reasoning ability (SAT-M). In April 1995, the scores were *recentered* so that the mean is approximately 500 in a large "standardization group." This scale is maintained from year to year so that scores have a constant interpretation. In 1996, 1,084,725 college-bound seniors took the SAT. Their mean SAT-V score was 505 with a standard deviation of 110. For the SAT-M the mean was 508 with a standard deviation of 112.

EXAMPLE 6.1

You want to estimate the mean SAT-M score for the more than 250,000 high school seniors in California. You know better than to trust data from the students who choose to take the SAT. Only about 45% of California students take the SAT. These self-selected students are planning to attend college and are not representative of all California seniors. At considerable effort and expense, you give the test to a simple random sample (SRS) of 500 California high school seniors. The mean score for your sample is $\bar{x} = 461$. What can you say about the mean score μ in the population of all 250,000 seniors?

The sample mean $\bar{x}$ is the natural estimator of the unknown population mean μ. We know that $\bar{x}$ is an unbiased estimator of μ. More important, the law of large numbers says that the sample mean must approach the population mean as the size of the sample grows. The value $\bar{x} = 461$ therefore appears to be a reasonable estimate of the mean score μ that all 250,000 students would achieve if they took the test. But how reliable is this estimate? A second sample would surely not give 461 again. Unbiasedness says only that there is no systematic tendency to underestimate or overestimate the truth. Could we plausibly get a sample mean of 530 or 410 on repeated samples? An estimate without an indication of its variability is of little value.

Statistical confidence

Just as unbiasedness of an estimator concerns the center of its sampling distribution, questions about variation are answered by looking at the spread. We know that if the entire population of SAT scores has mean μ and standard deviation σ, then in repeated samples of size 500 the sample mean $\bar{x}$

---

*These tests have been revised and are now called SAT I—Reasoning tests.

follows the $N(\mu, \sigma/\sqrt{500})$ distribution. Let us suppose that we know that the standard deviation σ of SAT-M scores in our California population is $\sigma = 100$. (This is not realistic. We will see in the next chapter how to proceed when σ is not known. For now, we are more interested in statistical reasoning than in details of realistic methods.) In repeated sampling the sample mean $\bar{x}$ follows the normal distribution centered at the unknown population mean μ and having standard deviation

$$\sigma_{\bar{x}} = \frac{100}{\sqrt{500}} \doteq 4.5$$

Now we are in business. Consider this line of thought, which is illustrated by Figure 6.1:

- The 68–95–99.7 rule says that the probability is about 0.95 that $\bar{x}$ will be within 9 points (two standard deviations of $\bar{x}$) of the population mean score μ.
- To say that $\bar{x}$ lies within 9 points of μ is the same as saying that μ is within 9 points of $\bar{x}$.
- So 95% of all samples will capture the true μ in the interval from $\bar{x} - 9$ to $\bar{x} + 9$.

We have simply restated a fact about the sampling distribution of $\bar{x}$. *The language of statistical inference uses this fact about what would happen in the long run to express our confidence in the results of any one sample.* Our sample gave $\bar{x} = 461$. We say that we are *95% confident* that the unknown mean score for all California seniors lies between

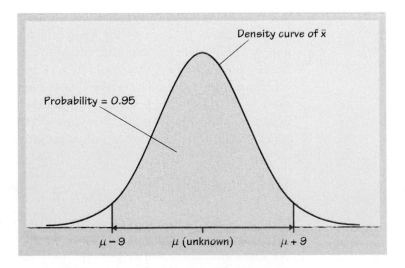

FIGURE 6.1 $\bar{x}$ lies within ± 9 of μ in 95% of all samples, so μ also lies within ± 9 of $\bar{x}$ in those samples.

$$\overline{x} - 9 = 461 - 9 = 452$$

and

$$\overline{x} + 9 = 461 + 9 = 470$$

Be sure you understand the grounds for our confidence. There are only two possibilities:

1. The interval between 452 and 470 contains the true μ.
2. Our SRS was one of the few samples for which $\overline{x}$ is not within 9 points of the true μ. Only 5% of all samples give such inaccurate results.

We cannot know whether our sample is one of the 95% for which the interval $\overline{x} \pm 9$ catches μ or one of the unlucky 5%. The statement that we are 95% confident that the unknown μ lies between 452 and 470 is shorthand for saying, "We arrived at these numbers by a method that gives correct results 95% of the time."

Confidence intervals

The interval of numbers between the values $\overline{x} \pm 9$ is called a *95% confidence interval* for μ. Like most confidence intervals we will meet, this one has the form

estimate $\pm$ margin of error

margin of error

The estimate ($\overline{x}$ in this case) is our guess for the value of the unknown parameter. The **margin of error** ± 9 shows how accurate we believe our guess is, based on the variability of the estimate. The confidence level shows how confident we are that the procedure will catch the true population mean μ.

Figure 6.2 illustrates the behavior of 95% confidence intervals in repeated sampling. The center of each interval is at $\overline{x}$ and therefore varies from sample to sample. The sampling distribution of $\overline{x}$ appears at the top of the figure to show the long-term pattern of this variation. The 95% confidence intervals $\overline{x} \pm 9$ from 25 SRSs appear below. The center x of each interval is marked by a dot. The arrows on either side of the dot span the confidence interval. All except one of the 25 intervals cover the true value of μ. In a very large number of samples, 95% of the confidence intervals would contain μ.*

Statisticians have constructed confidence intervals for many different parameters based on a variety of designs for data collection. We will meet a number of these in later chapters. You need to know two important things about a confidence interval:

---

*Confidence intervals as a systematic method were invented in 1937 by Jerzy Neyman (1894–1981). Neyman was a Pole who moved to England in 1934 and spent the last half of his long life at the University of California at Berkeley. He remained active until his death, almost doubling the list of his publications after his official retirement in 1961.

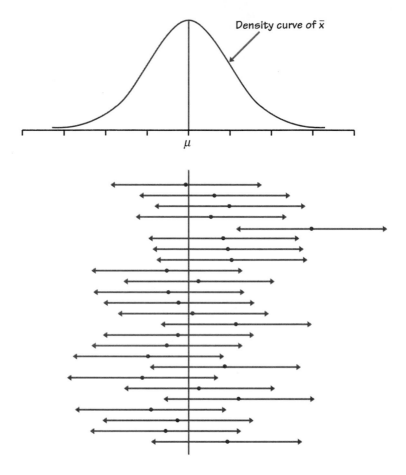

FIGURE 6.2 Twenty-five samples from the same population gave these 95% confidence intervals. In the long run, 95% of all samples give an interval that covers μ.

1. It is an interval of the form (a, b), where a and b are numbers computed from the data.
2. It has a property called a confidence level that gives the probability that the interval covers the parameter.

Users can choose the confidence level, most often 90% or higher because we most often want to be quite sure of our conclusions. We will use C to stand for the confidence level in decimal form. For example, a 95% confidence level corresponds to $C = 0.95$.

Confidence Interval

A level C confidence interval for a parameter is an interval computed from sample data by a method that has probability C of producing an interval containing the true value of the parameter.

Confidence interval for a population mean

We will now construct a level C confidence interval for the mean μ of a population when the data are an SRS of size n. The construction is based on the sampling distribution of the sample mean $\bar{x}$. This distribution is exactly $N(\mu, \sigma/\sqrt{n})$ when the population has the $N(\mu, \sigma)$ distribution. The central limit theorem says that this same sampling distribution is approximately correct for large samples whenever the population mean and standard deviation are μ and σ.

Our construction of a 95% confidence interval for the mean SAT score began by noting that any normal distribution has probability about 0.95 within ± 2 standard deviations of its mean. To construct a level C confidence interval we first catch the central C area under a normal curve. That is, we must find the number z^* such that any normal distribution has probability C within $\pm z^*$ standard deviations of its mean. Because all normal distributions have the same standardized form, we can obtain everything we need from the standard normal curve. Figure 6.3 shows how C and z^* are related. Values of z^* for many choices of C appear in the row labeled z^* in Table D. Here are the most important entries from that part of the table:

z^*	1.645	1.960	2.576	*z values*
C	90%	95%	99%	

Any normal curve has probability C between the point z^* standard deviations below the mean and the point z^* above the mean, as Figure 6.3 reminds us. The sample mean x has the normal distribution with mean μ and standard deviation $\sigma/\sqrt{n}$. So there is probability C that $\bar{x}$ lies between

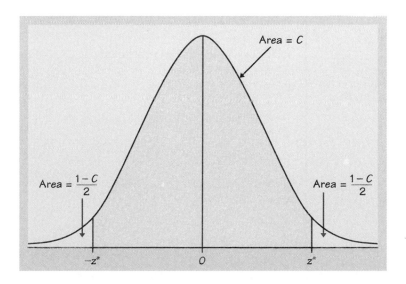

FIGURE 6.3   The area between $-z^*$ and z^* under the standard normal curve is C.

$$\mu - z^* \frac{\sigma}{\sqrt{n}} \quad \text{and} \quad \mu + z^* \frac{\sigma}{\sqrt{n}}$$

This is exactly the same as saying that the unknown population mean μ lies between

$$\overline{x} - z^* \frac{\sigma}{\sqrt{n}} \quad \text{and} \quad \overline{x} + z^* \frac{\sigma}{\sqrt{n}}$$

That is, there is probability C that the interval $\overline{x} \pm z^* \sigma / \sqrt{n}$ contains μ. That is our confidence interval. The estimate of the unknown μ is $\overline{x}$, and the margin of error is $z^* \sigma / \sqrt{n}$.

Confidence Interval for a Population Mean

Choose an SRS of size n from a population having unknown mean μ and known standard deviation σ. A level C confidence interval for μ is

$$\overline{x} \pm z^* \frac{\sigma}{\sqrt{n}}$$

Here z^* is the value on the standard normal curve with area C between $-z^*$ and z^*. This interval is exact when the population distribution is normal and is approximately correct for large n in other cases.

EXAMPLE 6.2 | Tim Kelley has been weighing himself once a week for several years. Last month his four measurements (in pounds) were

$$190.5 \quad 189.0 \quad 195.5 \quad 187.0$$

Give a 90% confidence interval for his mean weight for last month.

We treat the four measurements as an SRS of all possible measurements that Tim could have taken last month. These are estimates of μ, his true mean weight for last month.

Examination of Tim's past data reveals that over relatively short periods of time, his weight measurements are approximately normal with a standard deviation of about 3. For our confidence interval we will use this value as the true standard deviation, that is, $\sigma = 3$.

The mean of Tim's weight readings is

$$\overline{x} = \frac{190.5 + 189.0 + 195.5 + 187.0}{4} = 190.5$$

For 90% confidence, we see from Table D that $z^* = 1.645$. A 90% confidence interval for μ is therefore

$$\overline{x} \pm z^* \frac{\sigma}{\sqrt{n}} = 190.5 \pm 1.645 \frac{3}{\sqrt{4}}$$
$$= 190.5 \pm 2.5$$
$$= (188.0, 193.0)$$

We are 90% confident that Tim's mean weight last month was between 188 and 193 pounds.

In this example we used a value for σ based on past experience. This is a useful choice, particularly when the sample size is very small. Under these circumstances, the sample standard deviation can be less accurate than a value based on past data. When $n = 1$, we have an extreme case where it is not possible to calculate the sample standard deviation. Why not?

EXAMPLE 6.3

Suppose Tim Kelley had weighed himself only once last month, and that his one observation was $x = 190.5$, the same as the mean $\bar{x}$ in Example 6.2. Repeating the calculation with $n = 1$ shows that the 90% confidence interval based on a single measurement is

$$\bar{x} \pm z^* \frac{\sigma}{\sqrt{1}} = 190.5 \pm (1.645)(3)$$
$$= 190.5 \pm 4.9$$
$$= (185.6, 195.4)$$

The mean of four measurements gives a margin of error that is half the size of the margin of error for a sample of size one. Figure 6.4 illustrates the gain from using four observations.

The argument leading to the form of confidence intervals for the population mean μ rested on the fact that the statistic $\bar{x}$ used to estimate μ has a normal distribution. Because many sample estimates have normal distributions (at least approximately), it is useful to notice that the confidence interval has the form

$$\text{estimate} \pm z^* \sigma_{\text{estimate}}$$

The estimate based on the sample is the center of the confidence interval. The margin of error is $z^* \sigma_{\text{estimate}}$. The desired confidence level determines z^* from Table D. The standard deviation of the estimate is found from a knowledge of the sampling distribution in a particular case. When the estimate is $\bar{x}$ from an SRS, the standard deviation of the estimate is $\sigma/\sqrt{n}$.

How confidence intervals behave

The margin of error $z^*\sigma/\sqrt{n}$ for the mean of a normal population illustrates several important properties that are shared by all confidence intervals in common use. The user chooses the confidence level, and the margin of error follows from this choice. High confidence is desirable and so is a small

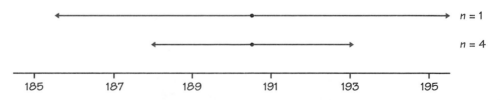

FIGURE 6.4 Confidence intervals for $n = 4$ and $n = 1$, for Examples 6.2 and 6.3.

margin of error. High confidence says that our method almost always gives correct answers. A small margin of error says that we have pinned down the parameter quite precisely.

Suppose that you calculate a margin of error and decide that it is too large. Here are your choices to reduce it:

■ Use a lower level of confidence (smaller C).
■ Increase the sample size (larger n).
■ Reduce σ.

For most problems you would choose a confidence level of 90%, 95%, or 99%. So z^* can be 1.645, 1.960, or 2.576. A look at Figure 6.3 will convince you that z^* will be smaller for lower confidence (smaller C). Table D shows that this is indeed the case. If n and σ are unchanged, a smaller z^* leads to a smaller margin of error. Similarly, increasing the sample size n reduces the margin of error for any fixed confidence level. The square root in the formula implies that we must multiply the number of observations by 4 in order to cut the margin of error in half. The standard deviation σ measures the variation in the population. You can think of the variation among individuals in the population as noise that obscures the average value μ. It is harder to pin down the mean μ of a highly variable population; that is why the margin of error of a confidence interval increases with σ. In practice, we can sometimes reduce σ by carefully controlling the measurement process or by restricting our attention to only part of a large population.

EXAMPLE 6.4

Suppose that Tim Kelley in Example 6.2 wants 99% confidence rather than 90%. Table D tells us that for 99% confidence, $z^* = 2.576$. The margin of error for 99% confidence based on four repeated measurements is

$$z^* \frac{\sigma}{\sqrt{n}} = 2.576 \frac{3}{\sqrt{4}}$$
$$= 3.9$$

The 99% confidence interval is

$$\bar{x} \pm \text{margin of error} = 190.5 \pm 3.9$$
$$= (186.6, 194.4)$$

Requiring 99%, rather than 90%, confidence has increased the margin of error from 2.5 to 3.9. Figure 6.5 compares the two intervals.

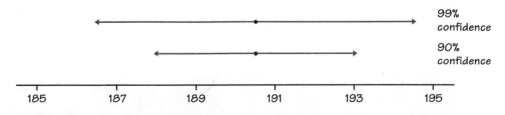

FIGURE 6.5 Confidence intervals for Examples 6.2 and 6.4.

Choosing the sample size

A wise user of statistics never plans data collection without at the same time planning the inference. You can arrange to have both high confidence and a small margin of error. The margin of error of the confidence interval $\bar{x} \pm z^*\sigma/\sqrt{n}$ for a normal mean is

$$z^* \frac{\sigma}{\sqrt{n}}$$

To obtain a desired margin of error m, just set this expression equal to m, substitute the value of z^* for your desired confidence level, and solve for the sample size n. Here is the result.

Sample Size for Desired Margin of Error

The confidence interval for a population mean will have a specified margin of error m when the sample size is

$$n = \left(\frac{z^*\sigma}{m}\right)^2$$

This formula is not the proverbial free lunch. In practice, taking observations costs time and money. The required sample size may be impossibly expensive. Do notice once again that it is the size of the *sample* that determines the margin of error. The size of the *population* (as long as the population is much larger than the sample) does not influence the sample size we need.

EXAMPLE 6.5

Tim Kelley of Example 6.2 has decided that he wants his estimate of his monthly weight accurate to within 2 or 3 pounds with 95% confidence. How many measurements must he take to achieve these margins of error?

For 95% confidence, Table D gives $z^* = 1.960$. For a margin of error of 2 pounds we have

$$n = \left(\frac{z^*\sigma}{m}\right)^2 = \left(\frac{1.96 \times 3}{2}\right)^2 = 8.6$$

Because 8 measurements will give a slightly wider interval than desired and 9 measurements a slightly narrower interval, Tim must take 9 measurements for his estimate to be within 2 pounds of the true value with 95% confidence. (Always round *up* to the next higher whole number when finding n.) For a margin of error of 3 pounds, the formula gives

$$n = \left(\frac{z^*\sigma}{m}\right)^2 = \left(\frac{1.96 \times 3}{3}\right)^2 = 3.8$$

Four observations per month would be sufficient. Tim chooses this option.

Some cautions

We have already seen that small margins of error and high confidence can require large numbers of observations. You should also be keenly aware that *any formula for inference is correct only in specific circumstances.* If the government required statistical procedures to carry warning labels like those on drugs, most inference methods would have long labels indeed. Our handy formula $\bar{x} \pm z^*\sigma/\sqrt{n}$ for estimating a normal mean comes with the following list of warnings for the user:

- The data must be an SRS from the population. We are completely safe if we actually did a randomization and drew an SRS. We are not in great danger if the data can plausibly be thought of as independent observations from a population. That is the case in Examples 6.2 to 6.5, where we have in mind the population that would result if Tim Kelley weighed himself very many times.

- The formula is not correct for probability sampling designs more complex than an SRS. Correct methods for other designs are available. We will not discuss confidence intervals based on multistage or stratified samples. If you plan such samples, be sure that you (or your statistical consultant) know how to carry out the inference you desire.

- There is no correct method for inference from data haphazardly collected with bias of unknown size. Fancy formulas cannot rescue badly produced data.

- Because $\bar{x}$ is not resistant, outliers can have a large effect on the confidence interval. You should search for outliers and try to correct them or justify their removal before computing the interval. If the outliers cannot be removed, ask your statistical consultant about procedures that are not sensitive to outliers.

- If the sample size is small and the population is not normal, the true confidence level will be different from the value C used in computing the interval. Examine your data carefully for skewness and other signs of nonnormality. The interval relies only on the distribution of $\bar{x}$, which even for quite small sample sizes is much closer to normal than that of the individual observations. When $n \geq 15$, the confidence level is not greatly disturbed by nonnormal populations unless extreme outliers or quite strong skewness are present. We will discuss this issue in more detail in the next chapter.

- You must know the standard deviation σ of the population. This unrealistic requirement renders the interval $\bar{x} \pm z^*\sigma/\sqrt{n}$ of little use in statistical practice. We will learn in the next chapter what to do when σ is unknown. If, however, the sample is large, the sample standard deviation s will be close to the unknown σ. The recipe $\bar{x} \pm z^*s/\sqrt{n}$ is then an approximate confidence interval for μ.

The most important caution concerning confidence intervals is a consequence of the first of these warnings. *The margin of error in a confidence interval covers only random sampling errors.* The margin of error is obtained from the sampling distribution and indicates how much error can be expected because of chance variation in randomized data production. Practical difficulties such as undercoverage and nonresponse in a sample survey cause additional errors. These errors can be larger than the random sampling error, particularly when the sample size is large (so that $\sigma/\sqrt{n}$ is small). Remember this unpleasant fact when reading the results of an opinion poll or other sample survey. The practical conduct of the survey influences the trustworthiness of its results in ways that are not included in the announced margin of error.

Every inference procedure that we will meet has its own list of warnings. Because many of the warnings are similar to those above, we will not print the full warning label each time. It is easy to state (from the mathematics of probability) conditions under which a method of inference is exactly correct. These conditions are *never* fully met in practice. For example, no population is exactly normal. Deciding when a statistical procedure should be used in practice often requires judgment assisted by exploratory analysis of the data. Mathematical facts are therefore only a part of statistics. The difference between statistics and mathematics can be stated thus: mathematical theorems are true; statistical methods are often effective when used with skill.

Finally, you should understand what statistical confidence does not say. We are 95% confident that the mean SAT-M score for the California students in Example 6.1 lies between 452 and 470. This says that these numbers were calculated by a method that gives correct results in 95% of all possible samples. It does *not* say that the probability is 95% that the true mean falls between 452 and 470. No randomness remains after we draw a particular sample and get from it a particular interval. The true mean either is or is not between 452 and 470. Probability in its interpretation as long-term relative frequency makes no sense in this situation. The probability calculations of standard statistical inference describe how often the *method* gives correct answers.

Beyond the basics ▸ the bootstrap

Confidence intervals are based on sampling distributions. In this section we have used the fact that the sampling distribution of $\bar{x}$ is $N(\mu, \sigma/\sqrt{n})$ when the data are an SRS from a $N(\mu, \sigma)$ population. If the data are not normal, the central limit theorem tells us that this sampling distribution is still a reasonable approximation as long as the distribution of the data is not strongly skewed and there are no outliers. We used the sampling distribution of $\bar{x}$ to conclude that, for example, 95% of the time, $\bar{x}$ and μ will not differ by more than $1.96\sigma/\sqrt{n}$.

What if the population does not appear to be normal and we have only a small sample? Then we do not know what the sampling distribution of $\bar{x}$ looks like. The **bootstrap** is a new procedure for approximating sampling distributions when theory cannot tell us their shape.[1]

bootstrap

The basic idea is to act as if our sample were the population. We take many samples from it. Each of these is called a **resample.** We calculate the mean, $\overline{x}$, for each resample. We get different results from different resamples because we sample *with replacement*. An individual observation in the original sample can appear more than once in the resample.

For example, if we start with Tim Kelley's weight observations from Example 6.2,

190.5	189.0	195.5	187.0

one resample with replacement might be

189.0	195.5	187.0	189.0

with $\overline{x} = 190.125$. Collect the $\overline{x}$'s from 1000 such resamples. Their distribution will be close to what we would get if we took 1000 samples from the entire population. We treat the distribution of $\overline{x}$'s from our 1000 resamples as if it were the sampling distribution. If we want a 95% confidence interval, for example, we use the middle 95% of this distribution.

The bootstrap is only practical when you can use a computer to take 1000 or more samples quickly. It is an example of how the use of fast and easy computing is changing the way we do statistics.

SUMMARY

The purpose of a **confidence interval** is to estimate an unknown parameter with an indication of how accurate the estimate is and of how confident we are that the result is correct.

Any confidence interval has two parts: an interval computed from the data and a confidence level. The interval often has the form

$$\text{estimate} \pm \text{margin of error}$$

The **confidence level** states the probability that the method will give a correct answer. That is, if you use 95% confidence intervals often, in the long run 95% of your intervals will contain the true parameter value. When you apply the method once, you do not know if your interval gave a correct value (this happens 95% of the time) or not (this happens 5% of the time).

A level C confidence interval for the mean μ of a normal population with known standard deviation σ, based on an SRS of size n, is given by

$$x \pm z^* \frac{\sigma}{\sqrt{n}}$$

Here z^* is obtained from the bottom row in Table D. The probability is C that a standard normal random variable takes a value between $-z^*$ and z^*.

Other things being equal, the margin of error of a confidence interval decreases as

■ the confidence level C decreases,
■ the sample size n increases, and
■ the population standard deviation σ decreases.

The sample size required to obtain a confidence interval of specified margin of error m for a normal mean is

$$n = \left(\frac{z^*\sigma}{m}\right)^2$$

where z^* is the critical point for the desired level of confidence.

A specific confidence interval recipe is correct only under specific conditions. The most important conditions concern the method used to produce the data. Other factors such as the form of the population distribution may also be important.

SECTION 6.1 EXERCISES

6.1 You measure the weights of 24 male runners. You do not actually choose an SRS, but you are willing to assume that these runners are a random sample from the population of male runners in your town or city. Here are their weights in kilograms:

> 67.8, 61.9, 63.0, 53.1, 62.3, 59.7, 55.4, 58.9,
> 60.9, 69.2, 63.7, 68.3, 64.7, 65.6, 56.0, 57.8,
> 66.0, 62.9, 53.6, 65.0, 55.8, 60.4, 69.3, 61.7

Suppose that the standard deviation of the population is known to be $\sigma = 4.5$ kg.

(a) What is $\sigma_{\bar{x}}$, the standard deviation of $\bar{x}$?

(b) Give a 95% confidence interval for μ, the mean of the population from which the sample is drawn. Are you quite sure that the average weight of the population of runners is less than 65 kg?

6.2 Crop researchers plant 15 plots with a new variety of corn. The yields in bushels per acre are

> 138.0, 139.1, 113.0, 132.5, 140.7, 109.7, 118.9, 134.8,
> 109.6, 127.3, 115.6, 130.4, 130.2, 111.7, 105.5

Assume that $\sigma = 10$ bushels per acre.

(a) Find the 90% confidence interval for the mean yield μ for this variety of corn.

(b) Find the 95% confidence interval.

(c) Find the 99% confidence interval.

(d) How do the margins of error in (a), (b), and (c) change as the confidence level increases?

6.3 Suppose that you had measured the weights of the runners in Exercise 6.1 in pounds rather than kilograms. Use your answers to Exercise 6.1 to answer these questions. (One pound is 0.454 kilogram.)

(a) What is the mean weight of these runners?

(b) What is the standard deviation of the mean weight?

(c) Give a 95% confidence interval for the mean weight of the population of runners that these runners represent.

6.4 Find a 99% confidence interval for the mean weight μ of the population of male runners in Exercise 6.1. Is the 99% confidence interval wider or narrower than the 95% interval found in Exercise 6.1? Explain in plain language why this is true.

6.5 Suppose that the crop researchers in Exercise 6.2 obtained the same value of $\overline{x}$ from a sample of 60 plots rather than 15.

(a) Compute the 95% confidence interval for the mean yield μ.

(b) Is the margin of error larger or smaller than the margin of error found for the sample of 15 plots in Exercise 6.2? Explain in plain language why the change occurs.

(c) Will the 90% and 99% intervals for a sample of size 60 be wider or narrower than those for $n = 15$? (You need not actually calculate these intervals.)

6.6 A test for the level of potassium in the blood is not perfectly precise. Moreover, the actual level of potassium in a person's blood varies slightly from day to day. Suppose that repeated measurements for the same person on different days vary normally with $\sigma = 0.2$.

(a) Julie's potassium level is measured once. The result is $x = 3.4$. Give a 90% confidence interval for her mean potassium level.

(b) If three measurements were taken on different days and the mean result is $\overline{x} = 3.4$, what is a 90% confidence interval for Julie's mean blood potassium level?

6.7 A study of the career paths of hotel general managers sent questionnaires to an SRS of 160 hotels belonging to major U.S. hotel chains. There were 114 responses. The average time these 114 general managers had spent with their current company was 11.78 years. Give a 99% confidence interval for the mean number of years general managers of major-chain hotels have spent with their current company. (Take it as known that the standard deviation of time with the company for all general managers is 3.2 years.)

6.8 The Acculturation Rating Scale for Mexican Americans (ARSMA) is a psychological test developed to measure the degree of Mexican/Spanish versus Anglo/English acculturation of Mexican Americans. The distribution of ARSMA scores in a population used to develop the test was approximately normal, with mean 3.0 and standard deviation 0.8. A further study gave ARSMA to 50 first-generation Mexican Americans. The mean of their scores is $\bar{x} = 2.36$. If the standard deviation for the first-generation population is also $\sigma = 0.8$, give a 95% confidence interval for the mean ARSMA score for first-generation Mexican Americans.

6.9 Here are the Degree of Reading Power (DRP) scores for a sample of 44 third-grade students. We first examined these data in Chapter 1, where a normal quantile plot (Figure 1.37 on page 91) showed that the distribution is close to normal except for slightly short tails.

40	26	39	14	42	18	25	43	46	27	19
47	19	26	35	34	15	44	40	38	31	46
52	25	35	35	33	29	34	41	49	28	52
47	35	48	22	33	41	51	27	14	54	45

Suppose that the standard deviation of the population of DRP scores is known to be $\sigma = 11$. Give a 95% confidence interval for the population mean score.

6.10 How large a sample of the hotel managers in Exercise 6.7 would be needed to estimate the mean μ within ± 1 year with 99% confidence?

6.11 Researchers planning a study of the reading ability of third-grade children want to obtain a 95% confidence interval for the population mean score on a reading test, with margin of error no greater than 3 points. They carry out a small pilot study to estimate the variability of test scores. The sample standard deviation is $s = 12$ points in the pilot study, so in preliminary calculations the researchers take the population standard deviation to be $\sigma = 12$.

(a) The study budget will allow as many as 100 students. Calculate the margin of error of the 95% confidence interval for the population mean based on $n = 100$.

(b) There are many other demands on the research budget. If all of these demands were met, there would be funds to measure only 10 children. What is the margin of error of the confidence interval based on $n = 10$ measurements?

(c) Find the smallest value of n that would satisfy the goal of a 95% confidence interval with margin of error 3 or less. Is this sample size within the limits of the budget?

6.12   How large a sample of Julie's potassium levels in Exercise 6.6 would be needed to estimate her mean μ within ±0.06 with 95% confidence?

6.13   In Exercises 6.2 and 6.5, we compared confidence intervals based on corn yields from 15 and 60 small plots of ground. How large a sample is required to estimate the mean yield within ±4 bushels per acre with 90% confidence?

6.14   To assess the accuracy of a laboratory scale, a standard weight known to weigh 10 grams is weighed repeatedly. The scale readings are normally distributed with unknown mean (this mean is 10 grams if the scale has no bias). The standard deviation of the scale readings is known to be 0.0002 gram.

   (a) The weight is weighed five times. The mean result is 10.0023 grams. Give a 98% confidence interval for the mean of repeated measurements of the weight.

   (b) How many measurements must be averaged to get a margin of error of ±0.0001 with 98% confidence?

6.15   The 1990 census "long form" asked the total 1989 income of the householder, the person in whose name the dwelling unit was owned or rented. This census form was sent to a random sample of 17% of the nation's households. Suppose (alas, it is too simple to be true) that the households that returned the long form are an SRS of the population of all households in each district. In Middletown, a city of 40,000 persons, 2621 householders reported their income. The mean of the responses was $\bar{x} = \$23{,}453$, and the standard deviation was $s = \$8721$. The sample standard deviation for so large a sample will be very close to the population standard deviation σ. Use these facts to give an approximate 99% confidence interval for the 1989 mean income of Middletown householders who reported income.

6.16   Refer to the previous problem. Give a 99% confidence interval for the total 1989 income of the households that reported income in Middletown.

6.17   The Gallup Poll asked 1571 adults what they considered to be the most serious problem facing the nation's public schools; 30% said drugs. This sample percent is an estimate of the percent of all adults who think that drugs are the schools' most serious problem. The news article reporting the poll result adds, "The poll has a margin of error—the measure of its statistical accuracy—of three percentage points in either direction; aside from this imprecision inherent in using a sample to represent the whole, such practical factors as the wording of questions can affect how closely a poll reflects the opinion of the public in general."

   The Gallup Poll uses a complex multistage sample design, but the sample percent has approximately a normal distribution. Moreover, it is standard practice to announce the margin of error for a 95% confidence interval unless a different confidence level is stated.

   (a) The announced poll result was 30% $\pm$ 3%. Can we be certain that the true population percent falls in this interval?

(b) Explain to someone who knows no statistics what the announced result 30% ± 3% means.

(c) This confidence interval has the same form we have met earlier:

$$\text{estimate} \pm z^* \sigma_{\text{estimate}}$$

(Actually σ is estimated from the data, but we ignore this for now.) What is the standard deviation σ_{estimate} of the estimated percent?

(d) Does the announced margin of error include errors due to practical problems such as undercoverage and nonresponse?

6.18 When the statistic that estimates an unknown parameter has a normal distribution, a confidence interval for the parameter has the form

$$\text{estimate} \pm z^* \sigma_{\text{estimate}}$$

In a complex sample survey design, the appropriate unbiased estimate of the population mean and the standard deviation of this estimate may require elaborate computations. But when the estimate is known to have a normal distribution and its standard deviation is given, we can calculate a confidence interval for μ from complex sample designs without knowing the formulas that led to the numbers given.

A report based on the Current Population Survey estimates the 1995 median annual earnings of households as $34,076 and also estimates that the standard deviation of this estimate is $200. The Current Population Survey uses an elaborate multistage sampling design to select a sample of about 50,000 households. The sampling distribution of the estimated median income is approximately normal. Give a 95% confidence interval for the 1995 median annual earnings of households.

6.19 As we prepare to take a sample and compute a 95% confidence interval, we know that the probability that the interval we compute will cover the parameter is 0.95. That's the meaning of 95% confidence. If we use several such intervals, however, our confidence that *all* give correct results is less than 95%.

In an agricultural field trial a corn variety is planted in seven separate locations, which may have different mean yields due to differences in soil and climate. At the end of the experiment, seven independent 95% confidence intervals will be calculated, one for the mean yield at each location.

(a) What is the probability that every one of the seven intervals covers the true mean yield at its location? This probability (expressed as a percent) is our overall confidence level for the seven simultaneous statements.

(b) What is the probability that at least six of the seven intervals cover the true mean yields?

6.20 A newspaper headline describing a poll of registered voters taken two weeks before a recent election read "Ringel leads with 52%." The accompanying article describing the poll stated that the margin of error was 2% with 95% confidence.

(a) Explain in plain language to someone who knows no statistics what "95% confidence" means.

(b) The poll shows Ringel leading. But the newspaper article said that the election was too close to call. Explain why.

6.21 A newspaper ad for a manager trainee position contained the statement "Our manager trainees have a first-year earnings average of $20,000 to $24,000." Do you think that the ad is describing a confidence interval? Explain your answer.

6.22 A *New York Times* poll on women's issues interviewed 1025 women and 472 men randomly selected from the United States excluding Alaska and Hawaii. The poll found that 47% of the women said they do not get enough time for themselves.

(a) The poll announced a margin of error of ± 3 percentage points for 95% confidence in conclusions about women. Explain to someone who knows no statistics why we can't just say that 47% of all adult women do not get enough time for themselves.

(b) Then explain clearly what "95% confidence" means.

(c) The margin of error for results concerning men was ± 4 percentage points. Why is this larger than the margin of error for women?

6.23 A student reads that a 95% confidence interval for the mean SAT math score of California high school seniors is 452 to 470. Asked to explain the meaning of this interval, the student says, "95% of California high school seniors have SAT math scores between 452 and 470." Is the student right? Justify your answer.

6.24 A radio talk show invites listeners to enter a dispute about a proposed pay increase for city council members. "What yearly pay do you think council members should get? Call us with your number." In all, 958 people call. The mean pay they suggest is $\bar{x} = \$9740$ per year, and the standard deviation of the responses is $s = \$1125$. For a large sample such as this, s is very close to the unknown population σ. The station calculates the 95% confidence interval for the mean pay μ that all citizens would propose for council members to be $9669 to $9811. Is this result trustworthy? Explain your answer.

6.25 Refer to Exercise 6.1. Suppose that the original set of data you collected contained weights for 25 runners. The extra runner had a weight of 92.3 kg.

(a) Compute the 95% confidence interval for the mean with this observation included.

(b) Would you report the interval you calculated in part (a) of this question or the interval you calculated in Exercise 6.1? Explain the reasons for your choice.

6.2 Tests of Significance

Confidence intervals are one of the two most common types of formal statistical inference. They are appropriate when our goal is to estimate a population parameter. The second common type of inference is directed at a quite different goal: to assess the evidence provided by the data in favor of some claim about the population.

The reasoning of significance tests

A significance test is a formal procedure for comparing observed data with a hypothesis whose truth we want to assess. The hypothesis is a statement about the parameters in a population or model. The results of a test are expressed in terms of a probability that measures how well the data and the hypothesis agree. We use the following example to illustrate these ideas. We start by focusing on the concepts. Don't worry about the details; we will deal with them later.

EXAMPLE 6.6

Tim Kelley of Example 6.2 has a driver's license that gives his weight as 187 pounds. He did not change this information the last time he renewed his license although he suspects that he may have gained a little weight over the past few years. We will examine the truth of the weight on Tim's license using a test of significance. Recall that the four weights recorded this month were

190.5	189.0	195.5	187.0

with sample mean $\bar{x} = 190.5$. As before, we assume that $\sigma = 3$ and that the data are normal.

Are these observations compatible with a true mean weight of $\mu = 187$ pounds? To answer this question, we ask another: what is the probability of observing a sample mean of 190.5 or larger from a population with a true mean of 187? The answer is 0.01. Because this probability is small, we conclude that Tim has gained weight.

What are the key steps in this example? We first started with a question. Is the mean of Tim's four weight measurements last month compatible with a true mean of 187 pounds, or is his suspicion that he has gained weight correct?

Next, we compare the data he collected ($\bar{x} = 190.5$ pounds) with the mean specified in his question ($\mu = 187$ pounds). The answer is given as a probability: if $\mu = 187$, then a sample mean $\bar{x} = 190.5$ or more based on $n = 4$ observations has probability 0.01.

How does the calculated probability of 0.01 lead to the conclusion that Tim's weight has increased? If the true mean is $\mu = 187$ pounds, we would rarely see a sample mean $\bar{x}$ as large as 190.5 pounds. In fact, we would get a result this large only 1 time in 100. So, either

1. we have observed something that is unusual, or

2. our assumption about Tim's true weight ($\mu = 187$ pounds) is not correct—Tim has gained weight.

Each of these possibilities leads to a conclusion. The first accepts the data as compatible with $\mu = 187$ pounds. The second says that Tim's suspicions are true; he has gained weight. We use the calculated probability, 0.01 in this case, to guide our choice. Because this probability is small, we prefer the second conclusion: Tim has gained weight. The chance of observing a weight of $\bar{x} = 190.5$ or higher when the true mean weight is $\mu = 187$ is less than 1 in 100. The following example illustrates a scenario where the data lead to a different conclusion.

EXAMPLE 6.7

Suppose the mean of Tim Kelley's four observations last month was $\bar{x} = 188.5$ pounds. This sample mean is closer to the value $\mu = 187$ pounds given on his driver's license. For a test of significance in this case, we need to calculate the probability that the mean of a sample of size $n = 4$ from a normal population with mean $\mu = 187$ and standard deviation $\sigma = 3$ is greater than or equal to the observed value of $\bar{x} = 188.5$ pounds. The answer is 0.16. Because this probability is not particularly small, we do not conclude that Kelley has gained weight.

The probabilities in Examples 6.6 and 6.7 are measures of the compatibility of the data ($\bar{x} = 188.5$ and $\bar{x} = 190.5$) with the null hypothesis ($\mu = 187$). Figure 6.6 compares the two results graphically. The normal curve centered at 187 is the sampling distribution of $\bar{x}$ when $\mu = 187$. You can see that we are not particularly surprised to observe $\bar{x} = 188.5$, but $\bar{x} = 190.5$ is an unusual observation. We will now consider some of the formal aspects of significance testing and the technical details.

Stating hypotheses

In Example 6.6, we asked whether the data Tim Kelley collected are reasonable if, in fact, his true weight (μ) is 187 pounds. To answer this, we begin

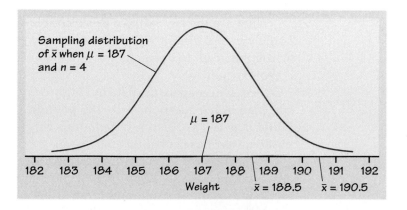

FIGURE 6.6 Comparison of the means in Examples 6.6 and 6.7 relative to the null hypothesized value $\mu = 187$.

by supposing that the statement following the "if" in the previous sentence is true. In other words, we suppose $\mu = 187$. We then ask whether the data provide evidence against the supposition we have made. If so, we have evidence in favor of the effect (Tim has gained weight) we are seeking. The first step in a test of significance is to state a claim that we will try to find evidence *against*.

Null Hypothesis

The statement being tested in a test of significance is called the **null hypothesis.** The test of significance is designed to assess the strength of the evidence against the null hypothesis. Usually the null hypothesis is a statement of "no effect" or "no difference."

We abbreviate "null hypothesis" as H_0. A null hypothesis is a statement about a population, expressed in terms of some parameter or parameters. For example, our null hypothesis for Example 6.6 is

$$H_0: \mu = 187$$

alternative hypothesis

It is convenient also to give a name to the statement we hope or suspect is true instead of H_0. This is called the **alternative hypothesis** and is abbreviated as H_a. In Example 6.6, the alternative hypothesis states that Tim's weight has increased. We write this as

$$H_a: \mu > 187$$

Hypotheses always refer to some population or model, not to a particular outcome. For this reason, we must state H_0 and H_a in terms of population parameters.

Because H_a expresses the effect that we hope to find evidence *for*, we often begin with H_a and then set up H_0 as the statement that the hoped-for effect is not present. Stating H_a is often the more difficult task. It is not always clear, in particular, whether H_a should be **one-sided** or **two-sided.**

one-sided and two-sided alternatives

The alternative $H_a: \mu > 187$ in the weight example is one-sided. Because people tend to gain weight as they age, we are only interested in detecting an increase in μ. On the other hand, it is easy to imagine scenarios where the two-sided alternative is appropriate. Suppose, for example, that μ is mean time required to process orders for the customers of your company. Here you would be interested in knowing whether this mean has increased or decreased. If the historical value is 3.8 days, we would test $H_0: \mu = 3.8$ days versus the alternative $H_a: \mu \neq 3.8$ days.

The alternative hypothesis should express the hopes or suspicions we bring to the data. It is cheating to first look at the data and then frame H_a to fit what the data show. If you do not have a specific direction firmly in mind in advance, use a two-sided alternative. Moreover, some users of statistics argue that we should *always* work with the two-sided alternative.

The choice of the hypotheses in Example 6.6 as

$$H_0: \mu = 187$$
$$H_a: \mu > 187$$

deserves a final comment. We are not concerned with the possibility that Tim may have lost weight. However, we can allow for the possibility that μ is less than 187 pounds by including this case in the null hypothesis. Then we would write

$$H_0: \mu \leq 187$$
$$H_a: \mu > 187$$

This statement is logically satisfying because the hypotheses account for all possible values of μ. However, only the parameter value in H_0 that is closest to H_a influences the form of the test in all common significance-testing situations. We will therefore take H_0 to be the simpler statement that the parameter has a specific value, in this case $H_0: \mu = 187$.

Test statistics

We will learn the form of significance tests in a number of common situations. Here are some principles that apply to most tests and that help in understanding the form of tests:

- The test is based on a statistic that estimates the parameter that appears in the hypotheses. Usually this is the same estimate we would use in a confidence interval for the parameter. When H_0 is true, we expect the estimate to take a value near the parameter value specified by H_0.

- Values of the estimate far from the parameter value specified by H_0 give evidence against H_0. The alternative hypothesis determines which directions count against H_0.

EXAMPLE 6.8 | In Example 6.6, the hypotheses are stated in terms of Kelley's weight as given on his driver's license last month:

$$H_0: \mu = 187$$
$$H_a: \mu > 187$$

The estimate of μ is the sample mean $\bar{x}$. Because H_a is one-sided on the high side, only large values of $\bar{x}$ count as evidence against the null hypothesis.

test statistic   A **test statistic** measures compatibility between the null hypothesis and the data. We use it for the probability calculation that we need for our test of significance. It is a random variable with a distribution that we know.

EXAMPLE 6.9 | In Example 6.6, the null hypothesis is $H_0: \mu = 187$ and $\bar{x} = 190.5$. Recall (page 400) that $\bar{x}$ has the $N(\mu, \sigma/\sqrt{n})$ distribution when the data are normal. The test statistic for this problem is the standardized version of $\bar{x}$:

$$z = \frac{\bar{x} - \mu}{\sigma/\sqrt{n}}$$

This random variable has the standard normal distribution $N(0, 1)$ when we use $\mu = 187$ and the null hypothesis is true. For the data we have, its value is

$$z = \frac{190.5 - 187}{3/\sqrt{4}} = 2.33$$

In this example we were willing to assume that the data are normal based on past experience. However, the central limit theorem (page 402) tells us that z will still be approximately normal when the sample size is large. We will use facts about the normal distribution in what follows. Do you recall the 68–95–99.7 rule and the calculations that we did with the standard normal distribution (Table A)? If not, this would be a good time to refresh your memory by reviewing Section 1.3.

P-values

A test of significance assesses the evidence against the null hypothesis in terms of probability. If the observed outcome is unlikely under the supposition that the null hypothesis is true but is more probable if the alternative hypothesis is true, that outcome is evidence against H_0 in favor of H_a. The less probable the outcome is, the stronger the evidence that H_0 is false. Usually, any *specific* outcome has low probability. We do not expect the mean of Kelley's four weight measurements to be *exactly* 187 pounds, even when $H_0: \mu = 187$ is true. On the other hand, if we observe $\bar{x} = 187$ or something less, we clearly do not have evidence to reject H_0 in favor of $H_a: \mu > 187$.

What about $\bar{x} = 188.5$ or 190 or 191.5 or 202? Clearly, as $\bar{x}$ moves farther away from μ we have stronger evidence that H_0 is not true. Recall that the standard deviation of $\bar{x}$ is $\sigma/\sqrt{n} = 3/\sqrt{4} = 1.5$ pounds. Therefore, if $\bar{x} = 187 + 1.5 = 188.5$, the difference is one standard deviation ($z = 1$); if $x = 187 + 2(1.5) = 190.0$, the difference is two standard deviations ($z = 2$); if $\bar{x} = 187 + 3(1.5) = 191.5$, the difference is three standard deviations ($z = 3$); and if $\bar{x} = 187 + 10(1.5) = 202.0$, the difference is ten standard deviations ($z = 10$). Our conclusions are easy in the first and last cases. From our work with the standard normal distribution, we know that $z = 1$ is not unusual and that $z = 10$ essentially never happens. So if $z = 1$, we do not have evidence to reject H_0; and if $z = 10$, we are on firm ground if we reject H_0. The Supreme Court of the United States has said that "two or three standard deviations" ($z = 2$ or 3) is its criterion for rejecting H_0 (see Exercise 8.66 on page 619), and this is the criterion used in most applications involving the law.

Because not all test statistics are z's (random variables with the $N(0, 1)$ distribution), we translate the value of test statistics into a common language, the language of probability.

A test of significance finds the probability of getting an outcome *as extreme or more extreme than the actually observed outcome.* "Extreme" means "far from what we would expect if H_0 were true." The direction or directions that count as "far from what we would expect" are determined by H_a as well as H_0.

In Example 6.6 we want to know if Kelley's weight has increased, a one-sided H_a. So the evidence against H_0 is measured by the probability

$$P(\bar{x} \geq 190.5)$$

that his sample mean weight would be *as high or higher* than the observed value. This probability is calculated under the assumption that the population mean is $\mu = 187$ pounds.

P-Value

The probability, computed assuming that H_0 is true, that the test statistic would take a value as extreme or more extreme than that actually observed is called the **P-value** of the test. The smaller the P-value, the stronger the evidence against H_0 provided by the data.

The key to calculating the P-value is the sampling distribution of the test statistic. For the problems we consider in this chapter, we only need the standard normal distribution for the test statistic z.

EXAMPLE 6.10   In Example 6.6 the observations are an SRS of size $n = 4$ from a normal population with $\sigma = 3$. The observed average weight is $\bar{x} = 190.5$. In Example 6.9, we found that the test statistic for testing

$$H_0: \mu = 187 \quad \text{versus} \quad H_a: \mu > 187$$

is

$$z = \frac{190.5 - 187}{3/\sqrt{4}} = 2.33$$

If H_0 is true, then z is a single observation from the standard normal, $N(0, 1)$, distribution. Figure 6.7 illustrates this calculation. The P-value is the probability of observing a value of z at least as large as the one that we observed, $z = 2.33$. From Table A, our table of standard normal probabilities, we find

$$P(Z \geq 2.33) = 1 - 0.9901 = 0.01$$

This is the value that was reported in Example 6.6.

Statistical significance

We have finished our discussion of the details of tests of significance. These details are needed to translate $\bar{x}$ and the μ that we specified in our null hypothesis, H_0, into a test statistic z and a P-value. The details are easy and can be programmed into a computer. One important final step is needed. What do we conclude?

We can compare the P-value we calculated with a fixed value that we regard as decisive. This amounts to announcing in advance how much evidence against H_0 we will require to reject H_0. The decisive value of P is called the **significance level.** It is denoted by α, the Greek letter alpha. If we choose $\alpha = 0.05$, we are requiring that the data give evidence against H_0 so strong

significance level

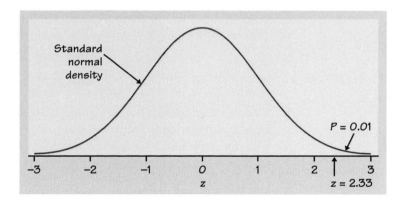

FIGURE 6.7 The P-value for Example 6.10. The P-value here is the probability (when H_0 is true) that $\bar{x}$ takes a value as large or larger than the actually observed value.

that it would happen no more than 5% of the time (1 time in 20) when H_0 is true. If we choose $\alpha = 0.01$, we are insisting on stronger evidence against H_0, evidence so strong that it would appear only 1% of the time (1 time in 100) if H_0 is in fact true. The Supreme Court's criterion ($z = 2$ or 3) translates into $\alpha = 0.0445$ or $\alpha = 0.0027$ for a two-sided test.

Statistical Significance

If the P-value is as small or smaller than α, we say that the data are **statistically significant at level α.**

"Significant" in the statistical sense does not mean "important." The original meaning of the word is "signifying something." In statistics* the term is used to indicate only that the evidence against the null hypothesis reached the standard set by α. Significance at level 0.01 is often expressed by the statement "The results were significant ($P < 0.01$)." Here P stands for the P-value. The P-value is more informative than a statement of significance, because we can then assess significance at any level we choose. For example, a result with $P = 0.03$ is significant at the $\alpha = 0.05$ level but is not significant at the $\alpha = 0.01$ level.

A test of significance is a recipe for assessing the significance of the evidence provided by data against a null hypothesis. The four steps common to all tests of significance are as follows:

1. State the *null hypothesis H_0* and the *alternative hypothesis H_a*. The test is designed to assess the strength of the evidence against H_0; H_a is the statement that we will accept if the evidence enables us to reject H_0.

---

*The reasoning of significance tests has been used sporadically at least since Laplace in the 1820s. A clear exposition by the English statistician Francis Y. Edgeworth (1845–1926) appeared in 1885 in the context of a study of social statistics. Edgeworth introduced the term "significant" as meaning "corresponds to a real difference in fact."

2. Calculate the value of the *test statistic* on which the test will be based. This statistic usually measures how far the data are from H_0.

3. Find the *P-value* for the observed data. This is the probability, calculated assuming that H_0 is true, that the test statistic will weigh against H_0 at least as strongly as it does for these data.

4. State a conclusion. One way to do this is to choose a *significance level* α, how much evidence against H_0 you regard as decisive. If the *P*-value is less than or equal to α, you conclude that the alternative hypothesis is true; if it is greater than α, you conclude that the data do not provide sufficient evidence to reject the null hypothesis. Your conclusion is a sentence that summarizes what you have found by using a test of significance.

We will learn the details of many tests of significance in the following chapters. In most cases, Steps 2 and 3 of our outline are almost automatic once you have practiced a bit. The proper test statistic is determined by the hypotheses and the data collection design. We use computer software or a calculator to find its numerical value and the *P*-value. The computer will not formulate your hypotheses for you, however. Nor will it decide if significance testing is appropriate or help you to interpret the *P*-value that it presents to you. The most difficult and important step is the last one: stating a conclusion. For Tim Kelley's question about his weight we could say, "Tim has gained weight ($z = 2.33$, $P < 0.01$)." A confidence interval such as the one we gave in Example 6.3 would also be useful.

Tests for a population mean

Our discussion has focused on the main ideas of statistical tests, and we have developed the details for one specific procedure. Here is a summary. We have an SRS of size n drawn from a normal population with unknown mean μ. We want to test the hypothesis that μ has a specified value. Call the specified value μ_0. The null hypothesis is

$$H_0\text{: } \mu = \mu_0$$

The test is based on the sample mean $\overline{x}$. Because the standard scale is easy to interpret, we use as our test statistic the *standardized* sample mean z:

$$z = \frac{\overline{x} - \mu_0}{\sigma/\sqrt{n}}$$

The statistic z has the standard normal distribution when H_0 is true. If the alternative is one-sided on the high side

$$H_a\text{: } \mu > \mu_0$$

then the *P*-value is the probability that a standard normal random variable Z takes a value at least as large as the observed z. That is,

$$P = P(Z \geq z)$$

Example 6.10 applied this test. There, $\mu_0 = 187$, the standardized sample mean was $z = 2.33$, and the P-value was $P(Z \geq 2.33) = 0.01$.

Similar reasoning applies when the alternative hypothesis states that the true μ lies below the hypothesized μ_0 (one-sided). When H_a states that μ is simply unequal to μ_0 (two-sided), values of z away from zero in either direction count against the null hypothesis. The P-value is the probability that a standard normal Z is at least as far from zero as the observed z. For example, if the test statistic is $z = 1.7$, the two-sided P-value is the probability that $Z \leq -1.7$ or $Z \geq 1.7$. Because the standard normal distribution is symmetric, we calculate this probability by finding $P(Z \geq 1.7)$ and *doubling* it:

$$P(Z \leq -1.7 \text{ or } Z \geq 1.7) = 2P(Z \geq 1.7)$$
$$= 2(1 - 0.9554) = 0.0892$$

We would make exactly the same calculation if we observed $z = -1.7$. It is the absolute value $|z|$ that matters, not whether z is positive or negative. Here is a statement of the test in general terms.

---

z Test for a Population Mean

To test the hypothesis $H_0: \mu = \mu_0$ based on an SRS of size n from a population with unknown mean μ and known standard deviation σ, compute the test statistic

$$z = \frac{\bar{x} - \mu_0}{\sigma/\sqrt{n}}$$

In terms of a standard normal random variable Z, the P-value for a test of H_0 against

$H_a: \mu > \mu_0$ is $P(Z \geq z)$

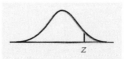

$H_a: \mu < \mu_0$ is $P(Z \leq z)$

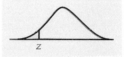

$H_a: \mu \neq \mu_0$ is $2P(Z \geq |z|)$

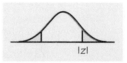

These P-values are exact if the population distribution is normal and are approximately correct for large n in other cases.

EXAMPLE 6.11

Do middle-aged male executives have different average blood pressure than the general population? The National Center for Health Statistics reports that the mean systolic blood pressure for males 35 to 44 years of age is 128 and the standard deviation in this population is 15. The medical director of a company looks at the medical records of 72 company executives in this age group and finds that the mean systolic blood pressure in this sample is $\bar{x} = 126.07$. Is this evidence that executive blood pressures differ from the national average?

The null hypothesis is "no difference" from the national mean $\mu_0 = 128$. The alternative is two-sided, because the medical director did not have a particular direction in mind before examining the data. So the hypotheses about the unknown mean μ of the executive population are

$$H_0: \mu = 128$$
$$H_a: \mu \neq 128$$

As usual in this chapter, we make the unrealistic assumption that the population standard deviation is known, in this case that executives have the same $\sigma = 15$ as the general population of middle-aged males. The z test requires that the 72 executives in the sample are an SRS from the population of all middle-aged male executives in the company. We check this assumption by asking how the data were produced. If medical records are available only for executives with recent medical problems, for example, the data are of little value for our purpose. It turns out that all executives are given a free annual medical exam and that the medical director selected 72 exam results at random.

We can now proceed to compute the test statistic

$$z = \frac{\bar{x} - \mu_0}{\sigma/\sqrt{n}} = \frac{126.07 - 128}{15/\sqrt{72}}$$
$$= -1.09$$

Figure 6.8 illustrates the P-value, which is the probability that a standard normal variable Z takes a value at least 1.09 away from zero. From Table A we find that this probability is

$$P = 2P(Z \geq |-1.09|) = 2P(Z \geq 1.09)$$
$$= 2(1 - 0.8621) = 0.2758$$

That is, more than 27% of the time an SRS of size 72 from the general male population would have a mean blood pressure at least as far from 128 as that of the executive sample. The observed $\bar{x} = 126.07$ is therefore not good evidence that executives differ from other men.

The data in Example 6.11 do *not* establish that the mean blood pressure μ for this company's middle-aged male executives is 128. We sought evidence that μ differed from 128 and failed to find convincing evidence. That is all we can say. No doubt the mean blood pressure of the entire executive population is not exactly equal to 128. A large enough sample would give evidence of the difference, even if it is very small. Tests of significance assess the evidence *against* H_0. If the evidence is strong, we can confidently reject H_0 in favor of the alternative. Failing to find evidence against H_0 means only that the data are consistent with H_0, not that we have clear evidence that H_0 is true.

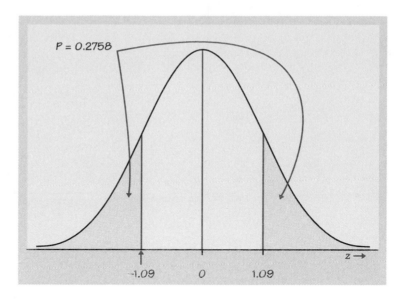

FIGURE 6.8 The P-value for the two-sided test in Example 6.11.

EXAMPLE 6.12

In a discussion of SAT scores, someone comments: "Because only a minority of high school students take the test, the scores overestimate the ability of typical high school seniors. The mean SAT mathematics score is about 475, but I think that if all seniors took the test, the mean score would be no more than 450." You gave the test to an SRS of 500 seniors from California (Example 6.1). These students had a mean score of $x = 461$. Is this good evidence against the claim that the mean for all California seniors is no more than 450? The hypotheses are

$$H_0: \mu = 450$$
$$H_a: \mu > 450$$

As in Example 6.1, we assume that $\sigma = 100$. The z statistic is

$$z = \frac{\bar{x} - \mu_0}{\sigma/\sqrt{n}} = \frac{461 - 450}{100/\sqrt{500}}$$
$$= 2.46$$

Because H_a is one-sided on the high side, large values of z count against H_0. From Table A, we find that the P-value is

$$P = P(Z \geq 2.46) = 1 - 0.9931 = 0.0069$$

Figure 6.9 illustrates this P-value. A mean score as large as that observed would occur fewer than seven times in 1000 samples if the population mean were 450. This is convincing evidence that the mean SAT-M score for all California high school seniors is higher than 450.

Two-sided significance tests and confidence intervals

Recall the confidence intervals that we discussed in the first section of this chapter. The basic idea was that we constructed an interval that would in-

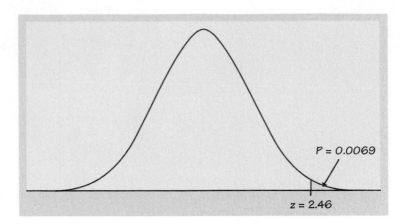

FIGURE 6.9 The P-value for the one-sided test in Example 6.12.

clude the true value of μ with a specified probability C. Suppose we use a 95% confidence interval ($C = 0.95$). Then the values of μ that are not in our interval would seem to be incompatible with the data. This sounds like a significance test with $\alpha = 0.05$ (or 5%) as our standard for drawing a conclusion. The following examples demonstrate that this is correct.

EXAMPLE 6.13 │ The Deely Laboratory analyzes specimens of a pharmaceutical product to determine the concentration of the active ingredient. Such chemical analyses are not perfectly precise. Repeated measurements on the same specimen will give slightly different results. The results of repeated measurements follow a normal distribution quite closely. The analysis procedure has no bias, so that the mean μ of the population of all measurements is the true concentration in the specimen. The standard deviation of this distribution is a property of the analytical procedure and is known to be $\sigma = 0.0068$ grams per liter. The laboratory analyzes each specimen three times and reports the mean result.

Deely's lab has been asked to evaluate the claim that the concentration of the active ingredient in a specimen is 0.86%. The true concentration is the mean μ of the population of repeated analyses. The hypotheses are

$$H_0: \mu = 0.86$$
$$H_a: \mu \neq 0.86$$

The lab chooses the 1% level of significance, $\alpha = 0.01$.

Three analyses of one specimen give concentrations

| 0.8403 | 0.8363 | 0.8447 |

The sample mean of these readings is

$$\bar{x} = \frac{0.8403 + 0.8363 + 0.8447}{3} = 0.8404$$

The test statistic is

$$z = \frac{\bar{x} - \mu_0}{\sigma/\sqrt{n}} = \frac{0.8404 - 0.86}{0.0068/\sqrt{3}} = -4.99 \text{ standard deviations}$$

Because the alternative is two-sided, the P-value is

$$P = 2P(Z \geq |-4.99|) = 2P(Z \geq 4.99)$$

We cannot find this probability in Table A. The largest value of z in that table is 3.49. All that we can say from Table A is that P is less than $2P(Z \geq 3.49) = 2(1-0.9998) = 0.0004$. If we use the bottom row of Table D, we find that the largest value of Z is 3.291, corresponding to a P-value of $1 - 0.999 = 0.001$. Software could be used to give an accurate value of the P-value. However, because the P-value is clearly less than the company's standard of 1%, we reject H_0.

Suppose we compute a 99% confidence interval for the same data.

EXAMPLE 6.14   The 99% confidence interval for μ in Example 6.13 is

$$\bar{x} \pm z^* \frac{\sigma}{\sqrt{n}} = 0.8404 \pm 0.0101$$
$$= (0.8303, 0.8505)$$

The hypothesized value $\mu_0 = 0.86$ in Example 6.13 falls outside the confidence interval we computed in Example 6.14. We are therefore 99% confident that μ is *not* equal to 0.86, so we can reject

$$H_0: \mu = 0.86$$

at the 1% significance level. On the other hand, we cannot reject

$$H_0: \mu = 0.85$$

at the 1% level in favor of the two-sided alternative $H_a: \mu \neq 0.85$, because 0.85 lies inside the 99% confidence interval for μ. Figure 6.10 illustrates both cases.

The calculation in Example 6.13 for a 1% significance test is very similar to the calculation for a 99% confidence interval. In fact, a two-sided test at significance level α can be carried out directly from a confidence interval with confidence level $C = 1 - \alpha$.

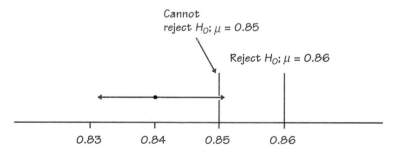

FIGURE 6.10   Values of μ falling outside a 99% confidence interval can be rejected at the 1% significance level; values falling inside the interval cannot be rejected.

> **Confidence Intervals and Two-Sided Tests**
>
> A level α two-sided significance test rejects a hypothesis H_0: $\mu = \mu_0$ exactly when the value μ_0 falls outside a level $1 - \alpha$ confidence interval for μ.

P–values versus fixed α

The observed result in Example 6.13 was $z = -4.99$. The conclusion that this result is significant at the 1% level does not tell the whole story. The observed z is far beyond the z corresponding to 1%, and the evidence against H_0 is far stronger than 1% significance suggests. The P-value

$$2P(Z \geq 4.99) = 0.0000006$$

gives a better sense of how strong the evidence is. *The P-value is the smallest level α at which the data are significant.* Knowing the P-value allows us to assess significance at any level.

EXAMPLE 6.15 | In Example 6.12, we tested the hypotheses

$$H_0: \mu = 450$$
$$H_a: \mu > 450$$

concerning the mean SAT mathematics score μ of California high school seniors. The test had the P-value $P = 0.0069$. This result is significant at the $\alpha = 0.01$ level because $0.0069 \leq 0.01$. It is not significant at the $\alpha = 0.005$ level, because the P-value is larger than 0.005. See Figure 6.11.

A P-value is more informative than a reject-or-not finding at a fixed significance level. But assessing significance at a fixed level α is easier, because no probability calculation is required. You need only look up a number in a table. A value z^* with a specified area to its right under the standard normal curve is called a **critical value** of the standard normal distribution. Because the practice of statistics almost always employs computer software that calculates P-values automatically, the use of tables of critical values is becoming outdated. We include the usual tables of critical values (such as Table D) at the end of the book for learning purposes and to rescue students without good computing facilities. The tables can be used directly to carry out fixed α tests. They also allow us to approximate P-values quickly without a probability calculation. The following example illustrates the use of Table D to find an approximate P-value.

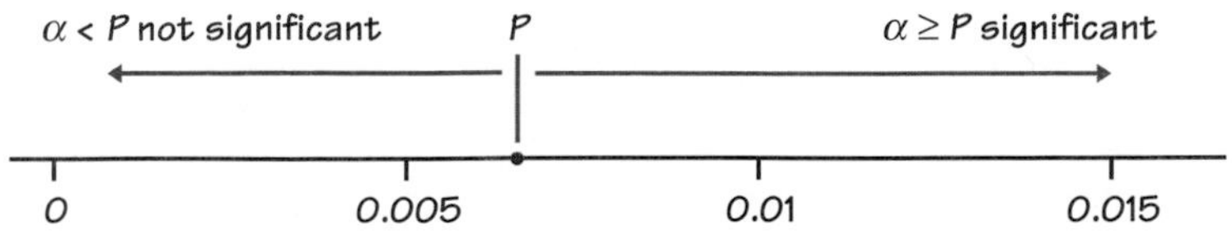

critical value

FIGURE 6.11 An outcome with P-value P is significant at all levels α at or above P and is not significant at smaller levels α.

EXAMPLE 6.16

Bottles of a popular cola drink are supposed to contain 300 milliliters (ml) of cola. There is some variation from bottle to bottle because the filling machinery is not perfectly precise. The distribution of the contents is normal with standard deviation $\sigma = 3$ ml. A student who suspects that the bottler is underfilling measures the contents of six bottles. The results are

| 299.4 | 297.7 | 301.0 | 298.9 | 300.2 | 297.0 |

Is this convincing evidence that the mean contents of cola bottles is less than the advertised 300 ml? The hypotheses are

$$H_0: \mu = 300$$
$$H_a: \mu < 300$$

The sample mean contents of the six bottles measured is $\bar{x} = 299.03$ ml. The z test statistic is therefore

$$z = \frac{\bar{x} - \mu_0}{\sigma/\sqrt{n}} = \frac{299.03 - 300}{3/\sqrt{6}} = -0.792$$

Small values of z count against H_0, because H_a is one-sided on the low side. The P-value is

$$P(Z \leq -0.792)$$

Rather than compute this probability, compare $z = -0.792$ with the critical values in Table D. According to this table, we would

reject H_0 at level $\alpha = 0.25$ if $z \leq -0.674$
reject H_0 at level $\alpha = 0.20$ if $z \leq -0.841$

As Figure 6.12 illustrates, the observed $z = -0.792$ lies between these two critical points. We can reject at $\alpha = 0.25$ but not at $\alpha = 0.20$. That is the same as saying that the P-value lies between 0.25 and 0.20. A sample mean at least as small as that observed will occur in between 20% and 25% of all samples if the population mean is $\mu = 300$ ml. There is no convincing evidence that the mean is below 300, and there is no need to calculate the P-value more exactly.

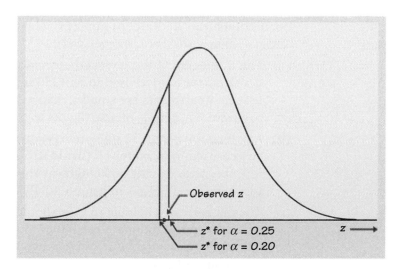

FIGURE 6.12   The P-value of a z statistic can be approximated by noting which levels from Table D it falls between. Here, P lies between 0.20 and 0.25.

SUMMARY

A **test of significance** is intended to assess the evidence provided by data against a **null hypothesis** H_0 in favor of an **alternative hypothesis** H_a.

The hypotheses are stated in terms of population parameters. Usually H_0 is a statement that no effect is present and H_a says that a parameter differs from its null value, in a specific direction (**one-sided alternative**) or in either direction (**two-sided alternative**).

The test is based on a **test statistic**. The **P-value** is the probability, computed assuming that H_0 is true, that the test statistic will take a value at least as extreme as that actually observed. Small P-values indicate strong evidence against H_0. Calculating P-values requires knowledge of the sampling distribution of the test statistic when H_0 is true.

If the P-value is as small or smaller than a specified value α, the data are **statistically significant** at significance level α.

Significance tests for the hypothesis H_0: $\mu = \mu_0$ concerning the unknown mean μ of a population are based on the z **statistic:**

$$z = \frac{\bar{x} - \mu_0}{\sigma/\sqrt{n}}$$

The z test assumes an SRS of size n, known population standard deviation σ, and either a normal population or a large sample. P-values are computed from the normal distribution (Table A). Fixed α tests use the table of **standard normal critical values** (Table D).

SECTION 6.2 EXERCISES

6.26   Each of the following situations requires a significance test about a population mean μ. State the appropriate null hypothesis H_0 and alternative hypothesis H_a in each case.

(a) The mean area of the several thousand apartments in a new development is advertised to be 1250 square feet. A tenant group thinks that the apartments are smaller than advertised. They hire an engineer to measure a sample of apartments to test their suspicion.

(b) Larry's car averages 32 miles per gallon on the highway. He now switches to a new motor oil that is advertised as increasing gas mileage. After driving 3000 highway miles with the new oil, he wants to determine if his gas mileage actually has increased.

(c) The diameter of a spindle in a small motor is supposed to be 5 millimeters. If the spindle is either too small or too large, the motor will not perform properly. The manufacturer measures the diameter in a sample of motors to determine whether the mean diameter has moved away from the target.

6.27 In each of the following situations, a significance test for a population mean μ is called for. State the null hypothesis H_0 and the alternative hypothesis H_a in each case.

(a) Experiments on learning in animals sometimes measure how long it takes a mouse to find its way through a maze. The mean time is 18 seconds for one particular maze. A researcher thinks that a loud noise will cause the mice to complete the maze faster. She measures how long each of 10 mice takes with a noise as stimulus.

(b) The examinations in a large accounting class are scaled after grading so that the mean score is 50. A self-confident teaching assistant thinks that his students have a higher mean score than the class as a whole. His students this semester can be considered a sample from the population of all students he might teach, so he compares their mean score with 50.

(c) A university gives credit in French language courses to students who pass a placement test. The language department wants to know if students who get credit in this way differ in their understanding of spoken French from students who actually take the French courses. Some faculty think the students who test out of the courses are better, but others argue that they are weaker in oral comprehension. Experience has shown that the mean score of students in the courses on a standard listening test is 24. The language department gives the same listening test to a sample of 40 students who passed the credit examination to see if their performance is different.

6.28 In each of the following situations, state an appropriate null hypothesis H_0 and alternative hypothesis H_a. Be sure to identify the parameters that you use to state the hypotheses. (We have not yet learned how to test these hypotheses.)

(a) A sociologist asks a large sample of high school students which academic subject they like best. She suspects that a higher percent of males than of females will name mathematics as their favorite subject.

(b) An education researcher randomly divides sixth-grade students into two groups for physical education class. He teaches both groups basketball skills, using the same methods of instruction in both classes. He encourages Group A with compliments and other positive behavior but acts cool and neutral toward Group B. He hopes to show that positive teacher attitudes result in a higher mean score on a test of basketball skills than do neutral attitudes.

(c) An economist believes that among employed young adults there is a positive correlation between income and the percent of disposable income that is saved. To test this, she gathers income and savings data from a sample of employed persons in her city aged 25 to 34.

6.29 Translate each of the following research questions into appropriate H_0 and H_a.

(a) Census Bureau data show that the mean household income in the area served by a shopping mall is $52,500 per year. A market research firm questions shoppers at the mall to find out whether the mean household income of mall shoppers is higher than that of the general population.

(b) Last year, your company's service technicians took an average of 2.6 hours to respond to trouble calls from business customers who had purchased service contracts. Do this year's data show a different average response time?

6.30   A randomized comparative experiment examined whether a calcium supplement in the diet reduces the blood pressure of healthy men. The subjects received either a calcium supplement or a placebo for 12 weeks. The statistical analysis was quite complex, but one conclusion was that "the calcium group had lower seated systolic blood pressure ($P = 0.008$) compared with the placebo group." Explain this conclusion, especially the P-value, as if you were speaking to a doctor who knows no statistics. (From R. M. Lyle et al.,"Blood pressure and metabolic effects of calcium supplementation in normotensive white and black men," *Journal of the American Medical Association*, 257 (1987), pp. 1772–1776.)

6.31   A social psychologist reports that "ethnocentrism was significantly higher ($P < 0.05$) among church attenders than among nonattenders." Explain what this means in language understandable to someone who knows no statistics. Do not use the word "significance" in your answer.

6.32   A study examined the effect of exercise on how students perform on their final exam in statistics. The P-value was given as 0.87.

(a) State null and alternative hypotheses that could have been used for this study. (Note there is more than one correct answer to this question.)

(b) Do you reject the null hypothesis?

(c) What is your conclusion?

(d) What other facts about the study would you like to know for a proper interpretation of the results?

6.33   The financial aid office of a university asks a sample of students about their employment and earnings. The report says that "for academic year earnings, a significant difference ($P = 0.038$) was found between the sexes, with men earning more on the average. No difference ($P = 0.476$) was found between the earnings of black and white students." Explain both of these conclusions, for the effects of sex and of race on mean earnings, in language understandable to someone who knows no statistics. (From a study by M. R. Schlatter et al., Division of Financial Aid, Purdue University.)

6.34   Statistics can help decide the authorship of literary works. Sonnets by an Elizabethan poet are known to contain an average of $\mu = 6.9$ new words (words not used in the poet's other works). The standard deviation of the number of new words is $\sigma = 2.7$. Now a manuscript with 5 new sonnets has come to light, and scholars are debating whether it is the poet's work. The new sonnets contain an average of $\bar{x} = 11.2$ words not used in the poet's known works. We expect poems by another author to contain more new words, so to see if we have evidence that the new sonnets are not by our poet we test

$$H_0: \mu = 6.9$$
$$H_a: \mu > 6.9$$

Give the z test statistic and its P-value. What do you conclude about the authorship of the new poems?

6.35   The Survey of Study Habits and Attitudes (SSHA) is a psychological test that measures the motivation, attitude toward school, and study habits of students. Scores range from 0 to 200. The mean score for U.S. college students is about 115, and the standard deviation is about 30. A teacher who suspects that older students have better attitudes toward school gives the SSHA to 20 students who are at least 30 years of age. Their mean score is $\bar{x} = 135.2$.

(a)   Assuming that $\sigma = 30$ for the population of older students, carry out a test of

$$H_0: \mu = 115$$
$$H_a: \mu > 115$$

Report the P-value of your test, and state your conclusion clearly.

(b)   Your test in (a) required two important assumptions in addition to the assumption that the value of σ is known. What are they? Which of these assumptions is most important to the validity of your conclusion in (a)?

6.36   The mean yield of corn in the United States is about 120 bushels per acre. A survey of 40 farmers this year gives a sample mean yield of $\bar{x} = 123.8$ bushels per acre. We want to know whether this is good evidence that the national mean this year is not 120 bushels per acre. Assume that the farmers surveyed are an SRS from the population of all commercial corn growers and that the standard deviation of the yield in this population is $\sigma = 10$ bushels per acre. Give the P-value for the test of

$$H_0: \mu - 120$$
$$H_a: \mu \neq 120$$

Are you convinced that the population mean is not 120 bushels per acre? Is your conclusion correct if the distribution of corn yields is somewhat nonnormal? Why?

6.37   In the past, the mean score of the seniors at South High on the American College Testing (ACT) college entrance examination has been 20. This year a special preparation course is offered, and all 53 seniors planning to take the ACT test enroll in the course. The mean of their 53 ACT scores is 22.1. The principal believes that the new course has improved the students' ACT scores.

(a)   Assume that ACT scores vary normally with standard deviation 6. Is the outcome $\bar{x} = 22.1$ good evidence that the population mean score is greater than 20? State H_0 and H_a, compute the test statistic and the P-value, and answer the question by interpreting your result.

(b) The results are in any case inconclusive because of the design of the study. The effects of the new course are confounded with any change from past years, such as other new courses or higher standards. Briefly outline the design of a better study of the effect of the new course on ACT scores.

6.38 The level of calcium in the blood in healthy young adults varies with mean about 9.5 milligrams per deciliter and standard deviation about $\sigma = 0.4$. A clinic in rural Guatemala measures the blood calcium level of 180 healthy pregnant women at their first visit for prenatal care. The mean is $\bar{x} = 9.58$. Is this an indication that the mean calcium level in the population from which these women come differs from 9.5?

(a) State H_0 and H_a.

(b) Carry out the test and give the P-value, assuming that $\sigma = 0.4$ in this population. Report your conclusion.

(c) Give a 95% confidence interval for the mean calcium level μ in this population. We are confident that μ lies quite close to 9.5. This illustrates the fact that a test based on a large sample ($n = 180$ here) will often declare even a small deviation from H_0 to be statistically significant.

6.39 Here are the Degree of Reading Power (DRP) scores for a sample of 44 third-grade students:

40	26	39	14	42	18	25	43	46	27	19
47	19	26	35	34	15	44	40	38	31	46
52	25	35	35	33	29	34	41	49	28	52
47	35	48	22	33	41	51	27	14	54	45

These students can be considered to be an SRS of the third graders in a suburban school district. DRP scores are approximately normal. Suppose that the standard deviation of scores in this school district is known to be $\sigma = 11$. The researcher believes that the mean score μ of all third graders in this district is higher than the national mean, which is 32.

(a) State the appropriate H_0 and H_a to test this suspicion.

(b) Carry out the test. Give the P-value, and then interpret the result in plain language.

6.40 A computer has a random number generator designed to produce random numbers that are uniformly distributed on the interval from 0 to 1. If this is true, the numbers generated come from a population with $\mu = 0.5$ and $\sigma = 0.2887$. A command to generate 100 random numbers gives outcomes with mean $\bar{x} = 0.4365$. Assume that the population σ remains fixed. We want to test

$$H_0: \mu = 0.5$$
$$H_a: \mu \neq 0.5$$

(a) Calculate the value z of the z test statistic.

(b) Is the result significant at the 5% level ($\alpha = 0.05$)?

(c) Is the result significant at the 1% level ($\alpha = 0.01$)?

6.41 To determine whether the mean nicotine content of a brand of cigarettes is greater than the advertised value of 1.4 milligrams, a health advocacy group tests

$$H_0: \mu = 1.4$$
$$H_a: \mu > 1.4$$

The calculated value of the test statistic is $z = 2.42$.

(a) Is the result significant at the 5% level?

(b) Is the result significant at the 1% level?

6.42 There are other z statistics that we have not yet studied. We can use Table D to assess the significance of any z statistic. A study compares the habits of students who are on academic probation with students whose grades are satisfactory. One variable measured is the hours spent watching television last week. The null hypothesis is "no difference" between the means for the two populations. The alternative hypothesis is two-sided. The value of the test statistic is $z = -1.37$.

(a) Is the result significant at the 5% level?

(b) Is the result significant at the 1% level?

6.43 Explain in plain language why a significance test that is significant at the 1% level must always be significant at the 5% level.

6.44 You have performed a two-sided test of significance and obtained a value of $z = 3.3$. Use Table D to find the approximate P-value for this test.

6.45 You have performed a one-sided test of significance and obtained a value of $z = 0.215$. Use Table D to find the approximate P-value for this test.

6.46 You will perform a significance test of

$$H_0: \mu = 0$$

versus

$$H_a: \mu > 0$$

(a) What values of z would lead you to reject H_0 at the 5% level?

(b) If the alternative hypothesis was

$$H_a: \mu \neq 0$$

what values of z would lead you to reject H_0 at the 5% level?

(c) Explain why your answers to parts (a) and (b) are different.

6.47 Between what values from Table D does the P-value for the outcome $z = -1.37$ in Exercise 6.42 lie? (Remember that H_a is two-sided.) Calculate the P-value using Table A, and verify that it lies between the values you found from Table D.

6.48   Radon is a colorless, odorless gas that is naturally released by rocks and soils and may concentrate in tightly closed houses. Because radon is slightly radioactive, there is some concern that it may be a health hazard. Radon detectors are sold to homeowners worried about this risk, but the detectors may be inaccurate. University researchers placed 12 detectors in a chamber where they were exposed to 105 picocuries per liter (pCi/l) of radon over 3 days. Here are the readings given by the detectors. (Data provided by Diana Schellenberg, Purdue University School of Health Sciences.)

91.9	97.8	111.4	122.3	105.4	95.0
103.8	99.6	96.6	119.3	104.8	101.7

Assume (unrealistically) that you know that the standard deviation of readings for all detectors of this type is $\sigma = 9$.

(a) Give a 95% confidence interval for the mean reading μ for this type of detector.

(b) Is there significant evidence at the 5% level that the mean reading differs from the true value 105? State hypotheses and base a test on your confidence interval from (a).

6.49   You measure the weights of 24 male runners. These runners are not a random sample from a population, but you are willing to assume that their weights represent the weights of similar runners. Here are their weights in kilograms:

67.8, 61.9, 63.0, 53.1, 62.3, 59.7, 55.4, 58.9,
60.9, 69.2, 63.7, 68.3, 64.7, 65.6, 56.0, 57.8,
66.0, 62.9, 53.6, 65.0, 55.8, 60.4, 69.3, 61.7

Exercise 6.1 (page 447) asks you to find a 95% confidence interval for the mean weight of the population of all such runners, assuming that the population standard deviation is $\sigma = 4.5$ kg.

(a) Give the confidence interval from that exercise, or calculate the interval if you did not do the exercise.

(b) Based on this confidence interval, does a test of

$$H_0\colon \mu = 61.3 \text{ kg}$$
$$H_a\colon \mu \neq 61.3 \text{ kg}$$

reject H_0 at the 5% significance level?

(c) Would $H_0\colon \mu = 63$ be rejected at the 5% level if tested against a two-sided alternative?

6.50   An understanding of cockroach biology may lead to an effective control strategy for these annoying insects. Researchers studying the absorption of sugar by insects feed cockroaches a diet containing measured amounts of a particular sugar. After 10 hours, the cockroaches are killed and the concentration of the sugar in various body parts is determined by a chemical analysis. The paper that reports the research states that a 95% confidence interval for the mean amount (in milligrams) of the sugar in the hindguts of the cockroaches is 4.2 ± 2.3. (From D. L. Shankland et al., "The effect of 5-thio-D-glucose on insect development and its absorption by insects," *Journal of Insect Physiology,* 14 (1968), pp. 63–72.)

(a) Does this paper give evidence that the mean amount of sugar in the hindguts under these conditions is not equal to 7 mg? State H_0 and H_a and base a test on the confidence interval.

(b) Would the hypothesis that $\mu = 5$ mg be rejected at the 5% level in favor of a two-sided alternative?

6.51   Market pioneers, companies that are among the first to develop a new product or service, tend to have higher market shares than latecomers to the market. What accounts for this advantage? Here is an excerpt from the conclusions of a study of a sample of 1209 manufacturers of industrial goods:

> Can patent protection explain pioneer share advantages? Only 21% of the pioneers claim a significant benefit from either a product patent or a trade secret. Though their average share is two points higher than that of pioneers without this benefit, the increase is not statistically significant ($z = 1.13$). Thus, at least in mature industrial markets, product patents and trade secrets have little connection to pioneer share advantages.

Find the P-value for the given z. Then explain to someone who knows no statistics what "not statistically significant" in the study's conclusion means. Why does the author conclude that patents and trade secrets don't help, even though they contributed 2 percentage points to average market share? (From William T. Robinson, "Sources of market pioneer advantages: the case of industrial goods industries," *Journal of Marketing Research,* 25 (1988), pp. 87–94.)

6.52   An old farmer claims to be able to detect the presence of water with a forked stick. In a test of this claim, he is presented with 5 identical barrels, some containing water and some not. He is right in 4 of the 5 cases.

(a) Suppose the farmer has probability p of being correct. If he is just guessing, $p = 0.5$. State an appropriate H_0 and H_a in terms of p for a test of whether he does better than guessing.

(b) If the farmer is simply guessing, what is the distribution of the number X of correct answers in 5 tries?

(c) The observed outcome is $X = 4$. What is the P-value of the test that takes large values of X to be evidence against H_0?

6.3   Use and Abuse of Tests

Carrying out a test of significance is often quite simple, especially if a fixed significance level α is used or if the P-value is given effortlessly by a computer. Using tests wisely is not so simple. Each test is valid only in certain circumstances, with properly produced data being particularly important. The z test, for example, should bear the same warning label that was attached in Section 6.1 to the corresponding confidence interval. Similar warnings accompany the other tests that we will learn. There are additional caveats that concern tests more than confidence intervals, enough to warrant this separate section. Some hesitation about the unthinking use of significance tests is a sign of statistical maturity.

The reasoning of significance tests has appealed to researchers in many fields, so that tests are widely used to report research results. In this setting H_a is a "research hypothesis" asserting that some effect or difference is present. The null hypothesis H_0 says that there is no effect or no difference. A low P-value represents good evidence that the research hypothesis is true. Here are some comments on the use of significance tests, with emphasis on their use in reporting scientific research.

Choosing a level of significance

The spirit of a test of significance is to give a clear statement of the degree of evidence provided by the sample against the null hypothesis. The P-value does this. But sometimes you will make some decision or take some action if your evidence reaches a certain standard. You can set such a standard by giving a level of significance α. Perhaps you will announce a new scientific finding if your data are significant at the $\alpha = 0.05$ level. Or perhaps you will recommend using a new method of teaching reading if the evidence of its superiority is significant at the $\alpha = 0.01$ level.

Making a decision is different in spirit from testing significance, though the two are often mixed in practice. Choosing a level α in advance makes sense if you must make a decision, but not if you wish only to describe the strength of your evidence. Using tests with fixed α for decision making is discussed at greater length at the end of this chapter.

If you do use a fixed α significance test to make a decision, choose α by asking how much evidence is required to reject H_0. This depends first on how plausible H_0 is. If H_0 represents an assumption that everyone in your field has believed for years, strong evidence (small α) will be needed to reject it. Second, the level of evidence required to reject H_0 depends on the consequences of such a decision. If rejecting H_0 in favor of H_a means making an expensive changeover from one medical therapy or instructional method to another, strong evidence is needed. Both the plausibility of H_0 and H_a and the consequences of any action that rejection may lead to are somewhat subjective. Different persons may feel that different levels of significance are appropriate. It is better to report the P-value, which allows each of us to decide individually if the evidence is sufficiently strong.

Users of statistics have often emphasized certain standard levels of significance, such as 10%, 5%, and 1%. This emphasis reflects the time when tables of critical points rather than computer programs dominated statistical practice. The 5% level ($\alpha = 0.05$) is particularly common. Significance at that level is still a widely accepted criterion for meaningful evidence in research work. *There is no sharp border between "significant" and "insignificant," only increasingly strong evidence as the P-value decreases.* There is no practical distinction between the P-values 0.049 and 0.051. It makes no sense to treat $\alpha = 0.05$ as a universal rule for what is significant.

There is a reason for the common use of $\alpha = 0.05$—the great influence of Sir R. A. Fisher.* Fisher did not originate tests of significance. But because his writings organized statistics, especially as a tool of scientific research, his views on tests were enormously influential. Here is his opinion on choosing a level of significance:

> . . . it is convenient to draw the line at about the level at which we can say: "Either there is something in the treatment, or a coincidence has occurred such as does not occur more than once in twenty trials . . . ."
>
> If one in twenty does not seem high enough odds, we may, if we prefer it, draw the line at one in fifty (the 2 percent point), or one in a hundred (the 1 percent point). Personally, the writer prefers to set a low standard of significance at the 5 percent point, and ignore entirely all results which fail to reach that level. A scientific fact should be regarded as experimentally established only if a properly designed experiment *rarely fails* to give this level of significance.[2]

There you have it. Fisher thought 5% was about right, and who was to disagree with the master? Fisher was of course not an advocate of blind use of significance at the 5% level as a yes-or-no criterion. The last sentence quoted above shows an experienced scientist's feeling for the repeated studies and variable results that mark the advance of knowledge.[3]

What statistical significance doesn't mean

When a null hypothesis ("no effect" or "no difference") can be rejected at the usual levels, $\alpha = 0.05$ or $\alpha = 0.01$, there is good evidence that an effect is present. But that effect may be extremely small. When large samples are available, even tiny deviations from the null hypothesis will be significant. For example, suppose that we are testing the hypothesis of no correlation between two variables. With 1000 observations, an observed correlation of only $r = 0.08$ is significant evidence at the $\alpha = 0.01$ level that the correlation in the population is not zero but positive. The low significance level does not mean there is a strong association, only that there is strong evidence of some association. The true population correlation is probably quite close to the observed sample value, $r = 0.08$. We might well conclude that for practical purposes we can ignore the association between these variables, even though

---

*We have met Fisher as the inventor of randomized experimental designs. He also originated many other statistical techniques, derived the distributions of many common statistics, and introduced such basic terms as "parameter" and "statistic" into statistical writing.

we are confident (at the 1% level) that the correlation is positive. Remember the wise saying: *statistical significance is not the same as practical significance*. On the other hand, if we fail to reject the null hypothesis, it may be because H_0 is true or because our sample size is insufficient to detect the alternative. Exercise 6.55 demonstrates in detail the effect on P of increasing the sample size.

The remedy for attaching too much importance to statistical significance is to pay attention to the actual experimental results as well as to the P-value. Plot your data and examine them carefully. Are there outliers or other deviations from a consistent pattern? A few outlying observations can produce highly significant results if you blindly apply common tests of significance. Outliers can also destroy the significance of otherwise convincing data. The foolish user of statistics who feeds the data to a computer without exploratory analysis will often be embarrassed. Is the effect you are seeking visible in your plots? If not, ask yourself if the effect is large enough to be practically important. It is usually wise to give a confidence interval for the parameter in which you are interested. A confidence interval actually estimates the size of an effect, rather than simply asking if it is too large to reasonably occur by chance alone. Confidence intervals are not used as often as they should be, while tests of significance are perhaps overused.

Don't ignore lack of significance

Researchers typically have in mind the research hypothesis that some effect exists. Following the peculiar logic of tests of significance, they set up as H_0 the null hypothesis that no such effect exists and try their best to get evidence against H_0. A perverse legacy of Fisher's opinion on $\alpha = 0.05$ is that research in some fields has rarely been published unless significance at that level is attained. For example, a survey of four journals of the American Psychological Association showed that of 294 articles using statistical tests, only 8 reported results that did not attain the 5% significance level.[4]

Such a publication policy impedes the spread of knowledge, and not only by declaring that a P-value of 0.051 is "not significant." If a researcher has good reason to suspect that an effect is present and then fails to find significant evidence of it, that may be interesting news—perhaps more interesting than if evidence in favor of the effect at the 5% level had been found. Keeping silent about negative results may condemn other researchers to repeat the attempt to find an effect that isn't there.

Of course, an experiment that fails only causes a stir if it is clear that the experiment would have detected the effect if it were really there. An important aspect of planning a study is to verify that the test you plan to use does have high probability of detecting an effect of the size you hope to find. This probability is the *power* of the test. Power calculations are discussed later in this chapter.

Tests of statistical significance are routinely used to assess the results of research in agriculture, education, engineering, medicine, psychology, and sociology and increasingly in other fields as well. Any tool used routinely is often used unthinkingly. We therefore offer some comments for the thinking researcher on possible abuses of this tool. Thinking consumers of research findings (such as students) should also ponder these comments.

Statistical inference is not valid for all sets of data

We learned long ago that badly designed surveys or experiments often produce invalid results. Formal statistical inference cannot correct basic flaws in the design. There is no doubt a significant difference in English vocabulary scores between high school seniors who have studied a foreign language and those who have not. But because the effect of actually studying a language is confounded with the differences between students who choose language study and those who do not, this statistical significance is hard to interpret. It does indicate that the difference in English scores is greater than would often arise by chance alone. That leaves unsettled the issue of *what* other than chance caused the difference. The most plausible explanation is that students who were already good at English chose to study another language. A randomized comparative experiment would isolate the actual effect of language study and so make significance meaningful.

Tests of significance and confidence intervals are based on the laws of probability. Randomization in sampling or experimentation ensures that these laws apply. But we must often analyze data that do not arise from randomized samples or experiments. To apply statistical inference to such data, we must have confidence in a probability model for the data. The diameters of successive holes bored in auto engine blocks during production, for example, may behave like independent observations on a normal distribution. We can check this probability model by examining the data. If the model appears correct, we can apply the recipes of this chapter to do inference about the process mean diameter μ. Do ask how the data were produced, and don't be too impressed by P-values on a printout until you are confident that the data deserve a formal analysis.

Beware of searching for significance

Statistical significance is a commodity much sought after by researchers. It means (or ought to mean) that you have found an effect that you were looking for. *The reasoning behind statistical significance works well if you decide what effect you are seeking, design an experiment or sample to search for it, and use a test of significance to weigh the evidence you get.* But because a successful search for a new scientific phenomenon often ends with statistical significance, it is all too tempting to make significance itself the object of the search. There are several ways to do this, none of them acceptable in polite scientific society.

One tactic is to make many tests on the same data. Once upon a time three psychiatrists studied a sample of schizophrenic persons and a sample of nonschizophrenic persons. They measured 77 variables for each subject—religion, family background, childhood experiences, and so on.

Their goal was to discover what distinguishes persons who later become schizophrenic. Having measured 77 variables, they made 77 separate tests of the significance of the differences between the two groups of subjects. Pause for a moment of reflection. If you made 77 tests at the 5% level, you would expect a few of them to be significant by chance alone. After all, results significant at the 5% level do occur 5 times in 100 in the long run even when H_0 is true. The psychiatrists found 2 of their 77 tests significant at the 5% level and immediately published this exciting news.[5] Running one test and reaching the $\alpha = 0.05$ level is reasonably good evidence that you have found something; running 77 tests and reaching that level only twice is not.

The case of the 77 tests happened long ago. Such crimes are rarer now—or at least better concealed. The computer has freed us from the labor of doing arithmetic. This is surely a blessing in statistics, where the arithmetic can be long and complicated indeed. Comprehensive statistical software systems are everywhere available, so a few simple commands will set the machine to work performing all manner of complicated tests and operations on your data. The result can be much like the 77 tests of old. We will state it as a law that any large set of data—even several pages of a table of random digits—contains some unusual pattern. Sufficient computer time will discover that pattern, and when you test specifically for the pattern that turned up, the result will be significant. It also will mean exactly nothing.

One lesson here is not to be overawed by the computer. The computer has greatly extended the range of statistical inference, allowing us to handle larger sets of data and to carry out more complex analyses. But it has changed the logic of inference not one bit. Doing 77 tests and finding 2 significant at the $\alpha = 0.05$ level is not evidence of a real discovery. Neither is doing factor analysis followed by discriminant analysis followed by multiple regression and at last discovering a significant pattern in the data. Fancy computer programs are no remedy for bad scientific logic. It is convincing to hypothesize that an effect or pattern will be present, design a study to look for it, and find it at a low significance level. It is not convincing to search for any effect or pattern whatever and find one.

We do not mean that searching data for suggestive patterns is not proper scientific work. It certainly is. Many important discoveries have been made by accident rather than by design. Exploratory analysis of data is an essential part of statistics. We do mean that the usual reasoning of statistical inference does not apply when the search for a pattern is successful. You cannot legitimately test a hypothesis on the same data that first suggested that hypothesis. The remedy is clear. Once you have a hypothesis, design a study to search specifically for the effect you now think is there. If the result of this study is statistically significant, you have real evidence at last.

SUMMARY

P-values are more informative than the reject-or-not result of a fixed level α test. Beware of placing too much weight on traditional values of α, such as $\alpha = 0.05$.

Very small effects can be highly significant (small P), especially when a test is based on a large sample. A statistically significant effect need not be practically important. Plot the data to display the effect you are seeking, and use confidence intervals to estimate the actual value of parameters.

On the other hand, lack of significance does not imply that H_0 is true, especially when the test has low power.

Significance tests are not always valid. Faulty data collection, outliers in the data, and testing a hypothesis on the same data that suggested the hypothesis can invalidate a test. Many tests run at once will probably produce some significant results by chance alone, even if all the null hypotheses are true.

SECTION 6.3 EXERCISES

6.53   Which of the following questions does a test of significance answer?

(a) Is the sample or experiment properly designed?

(b) Is the observed effect due to chance?

(c) Is the observed effect important?

6.54   In a study of the suggestion that taking vitamin C will prevent colds, 400 subjects are assigned at random to one of two groups. The experimental group takes a vitamin C tablet daily, while the control group takes a placebo. At the end of the experiment, the researchers calculate the difference between the percents of subjects in the two groups who were free of colds. This difference is statistically significant ($P = 0.03$) in favor of the vitamin C group. Can we conclude that vitamin C has a strong effect in preventing colds? Explain your answer.

6.55   Every user of statistics should understand the distinction between statistical significance and practical importance. A sufficiently large sample will declare very small effects statistically significant. Let us suppose that SAT mathematics (SAT-M) scores in the absence of coaching vary normally with mean $\mu = 475$ and $\sigma = 100$. Suppose further that coaching may change μ but does not change σ. An increase in the SAT-M score from 475 to 478 is of no importance in seeking admission to college, but this unimportant change can be statistically very significant. To see this, calculate the P-value for the test of

$$H_0: \mu = 475$$
$$H_a: \mu > 475$$

in each of the following situations:

(a) A coaching service coaches 100 students; their SAT-M scores average $\bar{x} = 478$.

(b) By the next year, the service has coached 1000 students; their SAT-M scores average $\bar{x} = 478$.

(c) An advertising campaign brings the number of students coached to 10,000; their average score is still $\bar{x} = 478$.

6.56   Give a 99% confidence interval for the mean SAT-M score μ after coaching in each part of the previous exercise. For large samples, the confidence interval says, "Yes, the mean score is higher after coaching, but only by a small amount."

6.57   As in the previous exercises, suppose that SAT-M scores vary normally with $\sigma = 100$. One hundred students go through a rigorous training program designed to raise their SAT-M scores by improving their mathematics skills. Carry out a test of

$$H_0: \mu = 475$$
$$H_a: \mu > 475$$

in each of the following situations:

(a) The students' average score is $\overline{x} = 491.4$. Is this result significant at the 5% level?

(b) The average score is $\overline{x} = 491.5$. Is this result significant at the 5% level?

The difference between the two outcomes in (a) and (b) is of no importance. Beware attempts to treat $\alpha = 0.05$ as sacred.

6.58   A local television station announces a question for a call-in opinion poll on the six o'clock news and then gives the response on the eleven o'clock news. Today's question concerns a proposed gun-control ordinance. Of the 2372 calls received, 1921 oppose the new law. The station, following standard statistical practice, makes a confidence statement: "81% of the Channel 13 Pulse Poll sample oppose gun control. We can be 95% confident that the proportion of all viewers who oppose the law is within 1.6% of the sample result." Is the station's conclusion justified? Explain your answer.

6.59   A researcher looking for evidence of extrasensory perception (ESP) tests 500 subjects. Four of these subjects do significantly better ($P < 0.01$) than random guessing.

(a) Is it proper to conclude that these four people have ESP? Explain your answer.

(b) What should the researcher now do to test whether any of these four subjects have ESP?

6.60   One way to deal with the problem of misleading P-values when performing more than one significance test is to adjust the criterion you use for statistical significance. The **Bonferroni procedure** does this in a simple way. If you perform 2 tests and want to use the $\alpha = 5\%$ significance level, you would require a P-value of $0.05/2 = 0.025$ to declare either one of the null hypotheses significant. In general, if you perform k tests and want protection at level α, use α/k as your cutoff for statistical significance. You perform 6 tests and calculate P values of 0.476, 0.032, 0.241, 0.008, 0.010, 0.001. Which of these are statistically significant using the Bonferroni procedure with $\alpha = 0.05$?

Bonferroni procedure

6.61 Refer to the previous problem. A researcher has performed 12 tests of significance and wants to apply the Bonferroni procedure with $\alpha = 0.05$. The calculated P-values are 0.041, 0.569, 0.050, 0.416, 0.001, 0.004, 0.256, 0.041 0.888, 0.010, 0.002, 0.433. Which of the null hypotheses are rejected with this procedure?

6.62 The text cites an example in which researchers carried out 77 separate significance tests, of which 2 were significant at the 5% level. Suppose that these tests are independent of each other. (In fact they were not independent, because all involved the same subjects.) If all of the null hypotheses are true, each test has probability 0.05 of being significant at the 5% level.

(a) What is the distribution of the number X of tests that are significant?

(b) Find the probability that 2 or more of the tests are significant.

6.4 Power and Inference as a Decision*

Although we prefer to use P-values rather than the reject-or-not view of the fixed α significance test, the latter view is very important for planning studies and for understanding statistical decision theory. We will discuss these two topics in this section.

Power

In examining the usefulness of a confidence interval, we are concerned with both the level of confidence and the margin of error. The confidence level tells us how reliable the method is in repeated use. The margin of error tells us how sensitive the method is, that is, how closely the interval pins down the parameter being estimated. Fixed level α significance tests are closely related to confidence intervals—in fact, we saw that a two-sided test can be carried out directly from a confidence interval. The significance level, like the confidence level, says how reliable the method is in repeated use. If we use 5% significance tests repeatedly when H_0 is in fact true, we will be wrong (the test will reject H_0) 5% of the time and right (the test will fail to reject H_0) 95% of the time.

High confidence is of little value if the interval is so wide that few values of the parameter are excluded. Similarly, a test with a small level of α is of little value if it almost never rejects H_0 even when the true parameter value is far from the hypothesized value. We must be concerned with the ability of a test to detect that H_0 is false, just as we are concerned with the margin of error of a confidence interval. This ability is measured by the probability that the test will reject H_0 when an alternative is true. The higher this probability is, the more sensitive the test is.

---

*Although the topics in this section are important in planning and interpreting significance tests, they can be omitted without loss of continuity.

> **Power**
>
> The probability that a fixed level α significance test will reject H_0 when a particular alternative value of the parameter is true is called the **power** of the test against that alternative.

EXAMPLE 6.17 Can a 6-month exercise program increase the total body bone mineral content (TBBMC) of young women? A team of researchers is planning a study to examine this question. Based on the results of a previous study, they are willing to assume that $\sigma = 2$ for the percent change in TBBMC over the 6-month period. A change in TBBMC of 1% would be considered important, and the researchers would like to have a reasonable chance of detecting a change this large or larger. Is 25 subjects a large enough sample for this project?

We will answer this question by calculating the power of the significance test that will be used to evaluate the data to be collected. The calculation consists of three steps:

1. State H_0, H_a, the particular alternative we want to detect, and the significance level α.
2. Find the values of $\bar{x}$ that will lead us to reject H_0.
3. Calculate the probability of observing these values of $\bar{x}$ when the alternative is true.

Step 1 The null hypothesis is that the exercise program has no effect on TBBMC. In other words, the mean percent change is zero. The alternative is that exercise is beneficial; that is, the mean change is positive. Formally, we have

$$H_0: \mu = 0$$
$$H_a: \mu > 0$$

The alternative of interest is $\mu = 1\%$. A 5% test of significance will be used.

Step 2 The z test rejects H_0 at the $\alpha = 0.05$ level whenever

$$z = \frac{\bar{x} - \mu_0}{\sigma/\sqrt{n}} = \frac{\bar{x} - 0}{2/\sqrt{25}} \geq 1.645$$

Be sure you understand why we use 1.645. Rewrite this in terms of $\bar{x}$:

$$\bar{x} \geq 1.645 \frac{2}{\sqrt{25}}$$
$$\bar{x} \geq 0.658$$

Because the significance level is $\alpha = 0.05$, this event has probability 0.05 of occurring *when the population mean μ is 0.*

Step 3   The power against the alternative $\mu = 1\%$ increase in TBBMC is the probability that H_0 will be rejected *when in fact* $\mu = 1$. We calculate this probability by standardizing $\bar{x}$, using the value $\mu = 1$, the population standard deviation $\sigma = 2$, and the sample size $n = 25$. The power is

$$P(\bar{x} \geq 0.658 \text{ when } \mu = 1) = P\left(\frac{\bar{x} - \mu}{\sigma/\sqrt{n}} \geq \frac{0.658 - 1}{2/\sqrt{25}}\right)$$
$$= P(Z \geq -0.855) = 0.80$$

Figure 6.13 illustrates the power with the sampling distribution of $\bar{x}$ when $\mu = 1$. This significance test rejects the null hypothesis that exercise has no effect on TBBMC 80% of the time if the true effect of exercise is a 1% increase in TBBMC. If the true effect of exercise is a greater percent increase, the test will have greater power; it will reject with a higher probability.

High power is desirable. Along with 95% confidence intervals and 5% significance tests, 80% power is becoming a standard. Many U.S. government agencies that provide research funds require that the sample size for the funded studies be sufficient to detect important results 80% of the time using a 5% test of significance.

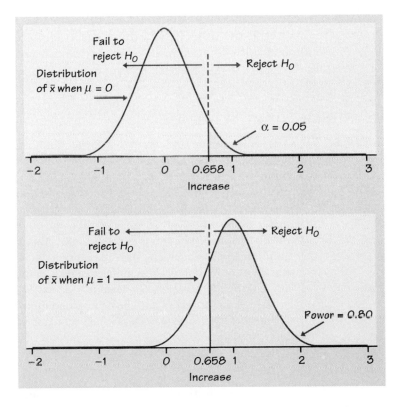

FIGURE 6.13   The sampling distributions of $\bar{x}$ when $\mu = 0$ and when $\mu = 1$ with the α and the power. Power is the probability that the test rejects H_0 when the alternative is true.

Increasing the power

Suppose you have performed a power calculation and found that the power is too small. What can you do to increase it? Here are four ways:

- Increase α. A 5% test of significance will have a greater chance of rejecting the alternative than a 1% test because the strength of evidence required for rejection is less.

- Consider a particular alternative that is farther away from μ_0. Values of μ that are in H_a but lie close to the hypothesized value μ_0 are harder to detect (lower power) than values of μ that are far from μ_0.

- Increase the sample size. More data will provide more information about $\bar{x}$ so we have a better chance of distinguishing values of μ.

- Decrease σ. This has the same effect as increasing the sample size: more information about μ. Improving the measurement process and restricting attention to a subpopulation are two common ways to decrease σ.

Power calculations are important in planning studies. Using a significance test with low power makes it unlikely that you will find a significant effect even if the truth is far from the null hypothesis. A null hypothesis that is in fact false can become widely believed if repeated attempts to find evidence against it fail because of low power. The following example illustrates this point.

EXAMPLE 6.18   The "efficient market hypothesis" for the time series of stock prices says that future stock prices (when adjusted for inflation) show only random variation. No information available now will help us predict stock prices in the future, because the efficient working of the market has already incorporated all available information in the present price. Many studies have tested the claim that one or another kind of information is helpful. In these studies, the efficient market hypothesis is H_0, and the claim that prediction is possible is H_a. Almost all the studies have failed to find good evidence against H_0. As a result, the efficient market theory is quite popular. But an examination of the significance tests employed finds that the power is generally low. Failure to reject H_0 when using tests of low power is not evidence that H_0 is true. As one expert says, "The widespread impression that there is strong evidence for market efficiency may be due just to a lack of appreciation of the low power of many statistical tests."[6]

Here is another example of a power calculation, this time for a two-sided z test.

EXAMPLE 6.19 | Example 6.13 presented a test of

$$H_0\colon \mu = 0.86$$
$$H_a\colon \mu \neq 0.86$$

at the 1% level of significance. What is the power of this test against the specific alternative $\mu = 0.845$?

The test rejects H_0 when $|z| \geq 2.576$. The test statistic is

$$z = \frac{\bar{x} - 0.86}{0.0068/\sqrt{3}}$$

Some arithmetic shows that the test rejects when either of the following is true:

$$z \geq 2.576 \quad \text{(in other words, } \bar{x} \geq 0.870\text{)}$$
$$z \leq -2.576 \quad \text{(in other words, } \bar{x} \leq 0.850\text{)}$$

These are disjoint events, so the power is the sum of their probabilities, *computed assuming that the alternative* $\mu = 0.845$ *is true.* We find that

$$P(\bar{x} \geq 0.87) = P\left(\frac{\bar{x} - \mu}{\sigma/\sqrt{n}} \geq \frac{0.87 - 0.845}{0.0068/\sqrt{3}}\right)$$
$$= P(Z \geq 6.37) \doteq 0$$
$$P(\bar{x} \leq 0.85) = P\left(\frac{\bar{x} - \mu}{\sigma/\sqrt{n}} \leq \frac{0.85 - 0.845}{0.0068/\sqrt{3}}\right)$$
$$= P(Z \leq 1.27) = 0.8980$$

Figure 6.14 illustrates this calculation. Because the power is about 0.9, we are quite confident that the test will reject H_0 when this alternative is true.

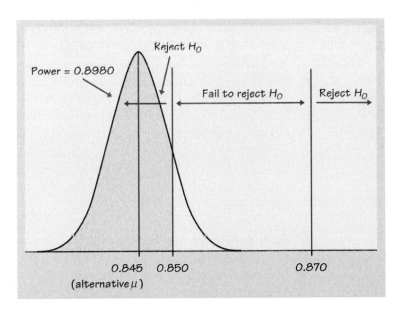

FIGURE 6.14   The power for Example 6.19.

Inference as decision*

We have presented tests of significance as methods for assessing the strength of evidence against the null hypothesis. This assessment is made by the P-value, which is a probability computed under the assumption that H_0 is true. The alternative hypothesis (the statement we seek evidence for) enters the test only to help us see what outcomes count against the null hypothesis. Such is the reasoning of tests of significance as advocated by Fisher and as practiced by many users of statistics.

But signs of another way of thinking were present in the discussion of significance tests with fixed level α. A level of significance α chosen in advance points to the outcome of the test as a *decision*. If the P-value is less than α, we reject H_0 in favor of H_a. Otherwise we fail to reject H_0. The transition from measuring the strength of evidence to making a decision is not a small step. Many statisticians agree with Fisher that making decisions is too grand a goal, especially in scientific inference. A decision is reached only after the evidence of many studies is weighed. Indeed, the goal of research is not "decision" but a gradually evolving understanding. Statistical inference should content itself with confidence intervals and tests of significance. Many users of statistics are content with such methods. It is rare to set up a level α in advance as a rule for making a decision in a scientific problem. More commonly, users think of significance at level 0.05 as a description of good evidence. This is made clearer by giving the P-value.

acceptance sampling

Yet there are circumstances that call for a decision or action as the end result of inference. **Acceptance sampling** is one such circumstance. A producer of bearings and the consumer of the bearings agree that each carload lot shall meet certain quality standards. When a carload arrives, the consumer chooses a sample of bearings to be inspected. On the basis of the sample outcome, the consumer will either accept or reject the carload. Fisher agreed that this is a genuine decision problem. But he insisted that acceptance sampling is completely different from scientific inference. Other eminent statisticians have argued that if "decision" is given a broad meaning, almost all problems of statistical inference can be posed as problems of making decisions in the presence of uncertainty. We will not venture further into the arguments over how we ought to think about inference. We do want to show how a different concept—inference as decision—changes the reasoning used in tests of significance.

Two types of error

Tests of significance concentrate on H_0, the null hypothesis. If a decision is called for, however, there is no reason to single out H_0. There are simply two hypotheses, and we must accept one and reject the other. It is convenient to

---

*The purpose of this section is to clarify the reasoning of significance tests by contrast with a related type of reasoning. It can be omitted without loss of continuity.

call the two hypotheses H_0 and H_a, but H_0 no longer has the special status (the statement we try to find evidence against) that it had in tests of significance. In the acceptance sampling problem, we must decide between

H_0: the lot of bearings meets standards
H_a: the lot does not meet standards

on the basis of a sample of bearings.

We hope that our decision will be correct, but sometimes it will be wrong. There are two types of incorrect decisions. We can accept a bad lot of bearings, or we can reject a good lot. Accepting a bad lot injures the consumer, while rejecting a good lot hurts the producer. To help distinguish these two types of error, we give them specific names.

Type I and Type II Errors

If we reject H_0 (accept H_a) when in fact H_0 is true, this is a **Type I error.**
If we accept H_0 (reject H_a) when in fact H_a is true, this is a **Type II error.**

The possibilities are summed up in Figure 6.15. If H_0 is true, our decision is either correct (if we accept H_0) or is a Type I error. If H_a is true, our decision is either correct or is a Type II error. Only one error is possible at one time. Figure 6.16 applies these ideas to the acceptance sampling example.

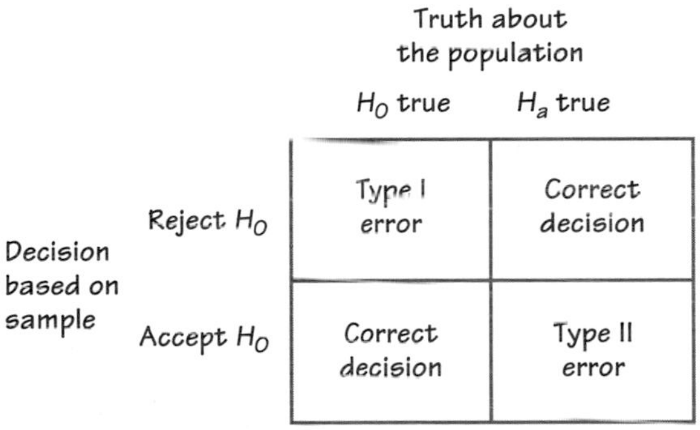

		Truth about the population	
		H_0 true	H_a true
Decision based on sample	Reject H_0	Type I error	Correct decision
	Accept H_0	Correct decision	Type II error

FIGURE 6.15   The two types of error in testing hypotheses.

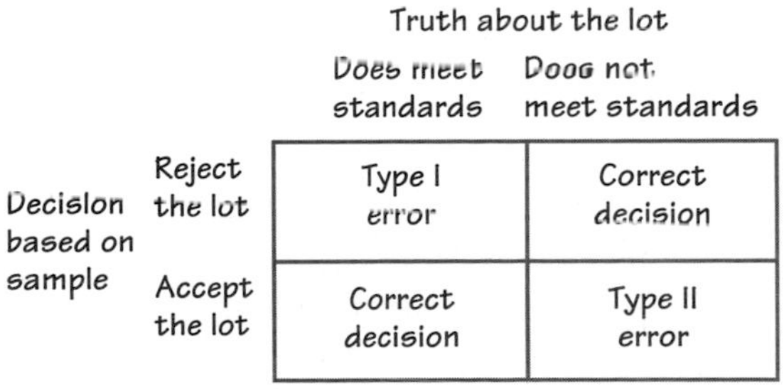

		Truth about the lot	
		Does meet standards	Does not meet standards
Decision based on sample	Reject the lot	Type I error	Correct decision
	Accept the lot	Correct decision	Type II error

FIGURE 6.16   The two types of error in the acceptance sampling setting.

Error probabilities

Any rule for making decisions is assessed in terms of the probabilities of the two types of error. This is in keeping with the idea that statistical inference is based on probability. We cannot (short of inspecting the whole lot) guarantee that good lots of bearings will never be rejected and bad lots never be accepted. But by random sampling and the laws of probability, we can say what the probabilities of both kinds of error are.

Significance tests with fixed level α give a rule for making decisions, because the test either rejects H_0 or fails to reject it. If we adopt the decision-making way of thought, failing to reject H_0 means deciding that H_0 is true. We can then describe the performance of a test by the probabilities of Type I and Type II error.

EXAMPLE 6.20

The mean diameter of a type of bearing is supposed to be 2.000 centimeters (cm). The bearing diameters vary normally with standard deviation $\sigma = 0.010$ cm. When a lot of the bearings arrives, the consumer takes an SRS of five bearings from the lot and measures their diameters. The consumer rejects the bearings if the sample mean diameter is significantly different from 2 at the 5% significance level.

This is a test of the hypotheses

$$H_0: \mu = 2$$
$$H_a: \mu \neq 2$$

To carry out the test, the consumer computes the z statistic:

$$z = \frac{\bar{x} - 2}{0.01/\sqrt{5}}$$

and rejects H_0 if

$$z < -1.96 \quad \text{or} \quad z > 1.96$$

A Type I error is to reject H_0 when in fact $\mu = 2$.

What about Type II errors? Because there are many values of μ in H_a, we will concentrate on one value. The producer and the consumer agree that a lot of bearings with mean 0.015 cm away from the desired mean 2.000 should be rejected. So a particular Type II error is to accept H_0 when in fact $\mu = 2.015$.

Figure 6.17 shows how the two probabilities of error are obtained from the two sampling distributions of $\bar{x}$, for $\mu = 2$ and for $\mu = 2.015$. When $\mu = 2$, H_0 is true and to reject H_0 is a Type I error. When $\mu = 2.015$, accepting H_0 is a Type II error. We will now calculate these error probabilities.

The probability of a Type I error is the probability of rejecting H_0 when it is really true. In Example 6.20, this is the probability that $|z| \geq 1.96$ when $\mu = 2$. But this is exactly the significance level of the test. The critical value 1.96 was chosen to make this probability 0.05, so we do not have to compute it again. The definition of "significant at level 0.05" is that sample outcomes this extreme will occur with probability 0.05 when H_0 is true.

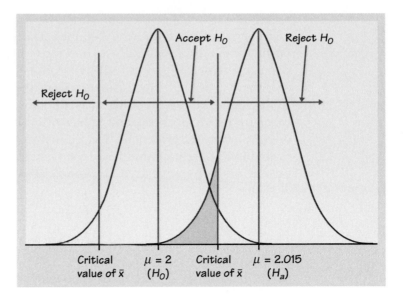

FIGURE 6.17 The two error probabilities for Example 6.20. The probability of a Type I error (*light tan area*) is the probability of rejecting H_0: $\mu = 2$ when in fact $\mu = 2$. The probability of a Type II error (*dark tan area*) is the probability of accepting H_0 when in fact $\mu = 2.015$.

Significance and Type I Error

The significance level α of any fixed level test is the probability of a Type I error. That is, α is the probability that the test will reject the null hypothesis H_0 when H_0 is in fact true.

The probability of a Type II error for the particular alternative $\mu = 2.015$ in Example 6.20 is the probability that the test will fail to reject H_0 when μ has this alternative value. The *power* of the test against the alternative $\mu = 2.015$ is just the probability that the test *does* reject H_0. By following the method of Example 6.19, we can calculate that the power is about 0.92. The probability of a Type II error is therefore $1 - 0.92$, or 0.08.

Power and Type II Error

The power of a fixed level test against a particular alternative is 1 minus the probability of a Type II error for that alternative.

The two types of error and their probabilities give another interpretation of the significance level and power of a test. The distinction between tests of significance and tests as rules for deciding between two hypotheses does not lie in the calculations but in the reasoning that motivates the calculations. In a test of significance we focus on a single hypothesis (H_0) and a single probability (the P-value). The goal is to measure the strength of the sample evidence against H_0. Calculations of power are done to check the sensitivity of the test. If we cannot reject H_0, we conclude only that there is not sufficient evidence against H_0, not that H_0 is actually true. If the same inference problem is thought of as a decision problem, we focus on two hypotheses and give a rule for deciding between them based on the sample evidence. We therefore must focus equally on two probabilities, the probabilities of the two types of error. We must choose one or the other hypothesis and cannot abstain on grounds of insufficient evidence.

The common practice of testing hypotheses

Such a clear distinction between the two ways of thinking is helpful for understanding. In practice, the two approaches often merge. We continued to call one of the hypotheses in a decision problem H_0. The common practice of *testing hypotheses* mixes the reasoning of significance tests and decision rules as follows:

1. State H_0 and H_a just as in a test of significance.
2. Think of the problem as a decision problem, so that the probabilities of Type I and Type II errors are relevant.
3. Because of Step 1, Type I errors are more serious. So choose an α (significance level) and consider only tests with probability of Type I error no greater than α.
4. Among these tests, select one that makes the probability of a Type II error as small as possible (that is, power as large as possible). If this probability is too large, you will have to take a larger sample to reduce the chance of an error.

Testing hypotheses may seem to be a hybrid approach. It was, historically, the effective beginning of decision-oriented ideas in statistics. An impressive mathematical theory of hypothesis testing was developed between 1928 and 1938 by Jerzy Neyman and the English statistician Egon Pearson. The decision-making approach came later (1940s). Because decision theory in its pure form leaves you with two error probabilities and no simple rule on how to balance them, it has been used less often than either tests of significance or tests of hypotheses. Decision ideas have been applied in testing problems mainly by way of the Neyman-Pearson hypothesis-testing theory. That theory asks you first to choose α, and the influence of Fisher often has led users of hypothesis testing comfortably back to $\alpha = 0.05$ or $\alpha = 0.01$. Fisher, who was exceedingly argumentative, violently attacked the Neyman-Pearson decision-oriented ideas, and the argument still continues.

SUMMARY

The **power** of a significance test measures its ability to detect an alternative hypothesis. Power against a specific alternative is calculated as the probability that the test will reject H_0 when that alternative is true. This calculation requires knowledge of the sampling distribution of the test statistic under the alternative hypothesis. Increasing the size of the sample increases the power when the significance level remains fixed.

An alternative to significance testing regards H_0 and H_a as two statements of equal status that we must decide between. This **decision theory** point of view regards statistical inference in general as giving rules for making decisions in the presence of uncertainty.

In the case of testing H_0 versus H_a, decision analysis chooses a decision rule on the basis of the probabilities of two types of error. A **Type I error** occurs if H_0 is rejected when it is in fact true. A **Type II error** occurs if H_0 is accepted when in fact H_a is true.

In a fixed level α significance test, the significance level α is the probability of a Type I error, and the power against a specific alternative is 1 minus the probability of a Type II error for that alternative.

SECTION 6.4 EXERCISES

6.63   Example 6.12 gives a test of a hypothesis about the SAT scores of California high school students based on an SRS of 500 students. The hypotheses are

$$H_0: \mu = 450$$
$$H_a: \mu > 450$$

Assume that the population standard deviation is $\sigma = 100$. The test rejects H_0 at the 1% level of significance when $z \geq 2.326$, where

$$z = \frac{\bar{x} - 450}{100/\sqrt{500}}$$

Is this test sufficiently sensitive to usually detect an increase of 10 points in the population mean SAT score? Answer this question by calculating the power of the test against the alternative $\mu = 460$.

6.64   Example 6.16 discusses a test about the mean contents of cola bottles. The hypotheses are

$$H_0: \mu = 300$$
$$H_a: \mu < 300$$

The sample size is $n = 6$, and the population is assumed to have a normal distribution with $\sigma = 3$. A 5% significance test rejects H_0 if $z \leq -1.645$, where the test statistic z is

$$z = \frac{\bar{x} - 300}{3/\sqrt{6}}$$

Power calculations help us see how large a shortfall in the bottle contents the test can be expected to detect.

(a) Find the power of this test against the alternative $\mu = 299$.

(b) Find the power against the alternative $\mu = 295$.

(c) Is the power against $\mu = 290$ higher or lower than the value you found in (b)? Explain why this result makes sense.

6.65 Increasing the sample size increases the power of a test when the level α is unchanged. Suppose that in the previous exercise a sample of n bottles had been measured. In that exercise, $n = 6$. The 5% significance test still rejects H_0 when $z \le -1.645$, but the z statistic is now

$$z = \frac{\bar{x} - 300}{3/\sqrt{n}}$$

where we substitute the sample size for n.

(a) Find the power of this test against the alternative $\mu = 299$ when $n = 25$.

(b) Find the power against $\mu = 299$ when $n = 100$.

6.66 In Example 6.11, a company medical director failed to find significant evidence that the mean blood pressure of a population of executives differed from the national mean $\mu = 128$. The medical director now wonders if the test used would detect an important difference if one were present. For the SRS of size 72 from a population with standard deviation $\sigma = 15$, the z statistic is

$$z = \frac{\bar{x} - 128}{15/\sqrt{72}}$$

The two-sided test rejects

$$H_0: \mu = 128$$

at the 5% level of significance when $|z| \ge 1.96$.

(a) Find the power of the test against the alternative $\mu = 134$.

(b) Find the power of the test against $\mu = 122$. Can the test be relied on to detect a mean that differs from 128 by 6?

(c) If the alternative were farther from H_0, say $\mu = 136$, would the power be higher or lower than the values calculated in (a) and (b)?

6.67 You have an SRS of size $n = 9$ from a normal distribution with $\sigma = 1$. You wish to test

$$H_0: \mu = 0$$
$$H_a: \mu > 0$$

You decide to reject H_0 if $\bar{x} > 0$ and to accept H_0 otherwise.

(a) Find the probability of a Type I error, that is, the probability that your test rejects H_0 when in fact $\mu = 0$.

(b) Find the probability of a Type II error when $\mu = 0.3$. This is the probability that your test accepts H_0 when in fact $\mu = 0.3$.

(c) Find the probability of a Type II error when $\mu = 1$.

6.68 Use the result of Exercise 6.64 to give the probabilities of Type I and Type II errors for the test discussed there. Take the alternative hypothesis to be $\mu = 295$.

6.69 Use the result of Exercise 6.63 to give the probability of a Type I error and the probability of a Type II error for the test in that exercise when the alternative is $\mu = 460$.

6.70 You must decide which of two discrete distributions a random variable X has. We will call the distributions p_0 and p_1. Here are the probabilities they assign to the values x of X:

x	0	1	2	3	4	5	6
p_0	0.1	0.1	0.1	0.1	0.2	0.1	0.3
p_1	0.2	0.1	0.1	0.2	0.2	0.1	0.1

You have a single observation on X and wish to test

$$H_0: p_0 \text{ is correct}$$
$$H_a: p_1 \text{ is correct}$$

One possible decision procedure is to accept H_0 if $X = 4$ or $X = 6$ and reject H_0 otherwise.

(a) Find the probability of a Type I error, that is, the probability that you reject H_0 when p_0 is the correct distribution.

(b) Find the probability of a Type II error.

6.71 You are designing a computerized medical diagnostic program. The program will scan the results of routine medical tests (pulse rate, blood pressure, urinalysis, etc.) and either clear the patient or refer the case to a doctor. The program will be used as part of a preventive medicine system to screen many thousands of persons who do not have specific medical complaints. The program makes a decision about each patient.

(a) What are the two hypotheses and the two types of error that the program can make? Describe the two types of error in terms of "false positive" and "false negative" test results.

(b) The program can be adjusted to decrease one error probability, at the cost of an increase in the other error probability. Which error probability would you choose to make smaller, and why? (This is a matter of judgment. There is no single correct answer.)

6.72 **(Optional)** The acceptance sampling test in Example 6.20 has probability 0.05 of rejecting a good lot of bearings and probability 0.08 of accepting a bad lot. The consumer of the bearings may imagine that acceptance sampling guarantees that most accepted lots are good. Alas, it is not so. Suppose that 90% of all lots shipped by the producer are bad.

(a) Draw a tree diagram for shipping a lot (the branches are "bad" and "good") and then inspecting it (the branches at this stage are "accept" and "reject").

(b) Write the appropriate probabilities on the branches, and find the probability that a lot shipped is accepted.

(c) Use the definition of conditional probability or Bayes's formula (page 356) to find the probability that a lot is bad, given that the lot is accepted. This is the proportion of bad lots among the lots that the sampling plan accepts.

CHAPTER 6 EXERCISES

6.73 Patients with chronic kidney failure may be treated by dialysis, using a machine that removes toxic wastes from the blood, a function normally performed by the kidneys. Kidney failure and dialysis can cause other changes, such as retention of phosphorus, that must be corrected by changes in diet. A study of the nutrition of dialysis patients measured the level of phosphorus in the blood of several patients on six occasions. Here are the data for one patient (in milligrams of phosphorus per deciliter of blood):

5.6	5.1	4.6	4.8	5.7	6.4

The measurements are separated in time and can be considered an SRS of the patient's blood phosphorus level. Assuming that this level varies normally with $\sigma = 0.9$ mg/dl, give a 90% confidence interval for the mean blood phosphorus level. (Data provided by Joan M. Susic.)

6.74 The normal range of phosphorus in the blood is considered to be 2.6 to 4.8 mg/dl. Is there strong evidence that the patient in the previous exercise has a mean phosphorus level that exceeds 4.8?

6.75 Because sulfur compounds cause "off-odors" in wine, oenologists (wine experts) have determined the odor threshold, the lowest concentration of a compound that the human nose can detect. For example, the odor threshold for dimethyl sulfide (DMS) is given in the oenology literature as 25 micrograms per liter of wine (μg/l). Untrained noses may be less sensitive, however. Here are the DMS odor thresholds for 10 beginning students of oenology:

31	31	43	36	23	34	32	30	20	24

Assume (this is not realistic) that the standard deviation of the odor threshold for untrained noses is known to be $\sigma = 7$ μg/l.

(a) Make a stemplot to verify that the distribution is roughly symmetric with no outliers. (A normal quantile plot confirms that there are no systematic departures from normality.)

(b) Give a 95% confidence interval for the mean DMS odor threshold among all beginning oenology students.

(c) Are you convinced that the mean odor threshold for beginning students is higher than the published threshold, 25 μg/l? Carry out a significance test to justify your answer.

6.76 A government report gives a 90% confidence interval for the 1995 median annual household income as \$34,076 ± \$324. This result was calculated by advanced methods from the Current Population Survey, a multistage random sample of about 50,000 households.

(a) Would a 95% confidence interval be wider or narrower? Explain your answer.

(b) Would the null hypothesis that the 1995 median household income was \$35,000 be rejected at the 10% significance level in favor of the two-sided alternative?

6.77 Refer to the previous problem. Give a 90% confidence interval for the 1995 median *weekly* household income. Use 52.14 as the number of weeks in a year.

6.78 An agronomist examines the cellulose content of a variety of alfalfa hay. Suppose that the cellulose content in the population has standard deviation $\sigma = 8$ milligrams per gram (mg/g). A sample of 15 cuttings has mean cellulose content $\bar{x} = 145$ mg/g.

(a) Give a 90% confidence interval for the mean cellulose content in the population.

(b) A previous study claimed that the mean cellulose content was $\mu = 140$ mg/g, but the agronomist believes that the mean is higher than that figure. State H_0 and H_a and carry out a significance test to see if the new data support this belief.

(c) The statistical procedures used in (a) and (b) are valid when several assumptions are met. What are these assumptions?

6.79 In a study of possible iron deficiency in infants, researchers compared several groups of infants who were following different feeding patterns. One group of 26 infants was being breast-fed. At 6 months of age, these children had a mean hemoglobin level of $\bar{x} = 12.9$ grams per 100 milliliters of blood and a standard deviation of 1.6. Taking the standard deviation to be the population value σ, give a 95% confidence interval for the mean hemoglobin level of breast-fed infants. What assumptions are required for the validity of the method you used to get the confidence interval?

6.80 Consumers can purchase nonprescription medications at food stores, mass merchandise stores such as Kmart and Wal-Mart, or pharmacies. About

45% of consumers make such purchases at pharmacies. What accounts for the popularity of pharmacies, which often charge higher prices?

A study examined consumers' perceptions of overall performance of the three types of stores, using a long questionnaire that asked about such things as "neat and attractive store," "knowledgeable staff," and "assistance in choosing among various types of nonprescription medication." A performance score was based on 27 such questions. The subjects were 201 people chosen at random from the Indianapolis telephone directory. (From a study by Mugdha Gore and Joseph Thomas, Purdue University School of Pharmacy.) Here are the means and standard deviations of the performance scores for the sample:

Store type	$\overline{x}$	s
Food stores	18.67	24.95
Mass merchandisers	32.38	33.37
Pharmacies	48.60	35.62

We do not know the population standard deviation, but a sample standard deviation s from so large a sample is usually close to σ. Use s in place of the unknown σ in this exercise.

(a) What population do you think the authors of the study want to draw conclusions about? What population are you certain they can draw conclusions about?

(b) Give 95% confidence intervals for the mean performance for each type of store in the population.

(c) Based on these confidence intervals, are you convinced that consumers think that pharmacies offer higher performance than the other types of stores? Note that in Chapter 12, we will study a statistical method for comparing means of several groups.

6.81 A study of the pay of corporate chief executive officers (CEOs) examined the increase in cash compensation of the CEOs of 104 companies, adjusted for inflation, in a recent year. The mean increase in real compensation was $\overline{x} = 6.9\%$, and the standard deviation of the increases was $s = 55\%$. Is this good evidence that the mean real compensation μ of all CEOs increased that year? The hypotheses are

$$H_0: \mu = 0 \quad \text{(no increase)}$$
$$H_a: \mu > 0 \quad \text{(an increase)}$$

Because the sample size is large, the sample s is close to the population σ, so take $\sigma = 55\%$.

(a) Sketch the normal curve for the sampling distribution of $\overline{x}$ when H_0 is true. Shade the area that represents the P-value for the observed outcome $\overline{x} = 6.9\%$.

(b) Calculate the P-value.

(c) Is the result significant at the $\alpha = 0.05$ level? Do you think the study gives strong evidence that the mean compensation of all CEOs went up?

6.82 Statisticians prefer large samples. Describe briefly the effect of increasing the size of a sample (or the number of subjects in an experiment) on each of the following:

(a) The width of a level C confidence interval.

(b) The P-value of a test, when H_0 is false and all facts about the population remain unchanged as n increases.

(c) The power of a fixed level α test, when α, the alternative hypothesis, and all facts about the population remain unchanged.

6.83 A roulette wheel has 18 red slots among its 38 slots. You observe many spins and record the number of times that red occurs. Now you want to use these data to test whether the probability of a red has the value that is correct for a fair roulette wheel. State the hypotheses H_0 and H_a that you will test. (We will describe the test for this situation in Chapter 8.)

6.84 When asked to explain the meaning of "statistically significant at the $\alpha = 0.05$ level," a student says, "This means there is only probability 0.05 that the null hypothesis is true." Is this an essentially correct explanation of statistical significance? Explain your answer.

6.85 Another student, when asked why statistical significance appears so often in research reports, says, "Because saying that results are significant tells us that they cannot easily be explained by chance variation alone." Do you think that this statement is essentially correct? Explain your answer.

6.86 A study compares two groups of mothers with young children who were on welfare two years ago. One group attended a voluntary training program offered free of charge at a local vocational school and advertised in the local news media. The other group did not choose to attend the training program. The study finds a significant difference ($P < 0.01$) between the proportions of the mothers in the two groups who are still on welfare. The difference is not only significant but quite large. The report says that with 95% confidence the percent of the nonattending group still on welfare is 21% ± 4% higher than that of the group who attended the program. You are on the staff of a member of Congress who is interested in the plight of welfare mothers and who asks you about the report.

(a) Explain briefly and in nontechnical language what "a significant difference ($P < 0.01$)" means.

(b) Explain clearly and briefly what "95% confidence" means.

(c) Is this study good evidence that requiring job training of all welfare mothers would greatly reduce the percent who remain on welfare for several years?

6.87   Use a computer to generate $n = 5$ observations from a normal distribution with mean 20 and standard deviation 5—$N(20, 5)$. Find the 95% confidence interval for μ. Repeat this process 100 times and then count the number of times that the confidence interval includes the value $\mu = 20$. Explain your results.

6.88   Use a computer to generate $n = 5$ observations from a normal distribution with mean 20 and standard deviation 5—$N(20, 5)$. Test the null hypothesis that $\mu = 20$ using a two-sided significance test. Repeat this process 100 times and then count the number of times that you reject H_0. Explain your results.

6.89   Use the same procedure for generating data as in the previous exercise. Now test the null hypothesis that $\mu = 22.5$. Explain your results.

6.90   Figure 6.2 (page 438) demonstrates the behavior of a confidence interval in repeated sampling by showing the results of 25 samples from the same population. Now you will do a similar demonstration. Suppose that (unknown to the researcher) the mean SAT-M score of all California high school seniors is $\mu = 460$, and that the standard deviation is known to be $\sigma = 100$. The scores vary normally.

   (a) Simulate the drawing of 25 SRSs of size $n = 100$ from this population.

   (b) The 95% confidence interval for the population mean μ has the form $\bar{x} \pm m$. What is the margin of error m? (Remember that we know $\sigma = 100$.)

   (c) Use your software to calculate the 95% confidence interval for μ when $\sigma = 100$ for each of your 25 samples. Verify the computer's calculations by checking the interval given for the first sample against your result in (b). Use the $\bar{x}$ reported by the software.

   (d) How many of the 25 confidence intervals contain the true mean $\mu = 460$? If you repeated the simulation, would you expect exactly the same number of intervals to contain μ? In a very large number of samples, what percent of the confidence intervals would contain μ?

6.91   In the previous exercise you simulated the SAT-M scores of 25 SRSs of 100 California seniors. Now use these samples to demonstrate the behavior of a significance test. We know that the population of all SAT-M scores is normal with standard deviation $\sigma = 100$.

   (a) Use your software to carry out a test of

$$H_0: \mu = 460$$
$$H_a: \mu \neq 460$$

   for each of the 25 samples.

   (b) Verify the computer's calculations by using Table A to find the P-value of the test for the first of your samples. Use the $\bar{x}$ reported by your software.

(c) How many of your 25 tests reject the null hypothesis at the $\alpha = 0.05$ significance level? (That is, how many have P-value 0.05 or smaller?) Because the simulation was done with $\mu = 460$, samples that lead to rejecting H_0 produce the wrong conclusion. In a very large number of samples, what percent would falsely reject the hypothesis?

6.92 Suppose that in fact the mean SAT-M score of California high school seniors is $\mu = 480$. Would the test in the previous exercise usually detect a mean this far from the hypothesized value? This is a question about the power of the test.

(a) Simulate the drawing of 25 SRSs from a normal population with mean $\mu = 480$ and $\sigma = 100$. These represent the results of sampling when in fact the alternative $\mu = 480$ is true.

(b) Repeat on these new data the test of

$$H_0: \mu = 460$$
$$H_a: \mu \neq 460$$

that you did in the previous exercise. How many of the 25 tests have P-values 0.05 or smaller? These tests reject the null hypothesis at the $\alpha = 0.05$ significance level, which is the correct conclusion.

(c) The power of the test against the alternative $\mu = 480$ is the probability that the test will reject $H_0: \mu = 460$ when in fact $\mu = 480$. Calculate this power. In a very large number of samples from a population with mean 480, what percent would reject H_0?

NOTES

1. The standard reference here is Bradley Efron and Robert J. Tibshirani, *An Introduction to the Bootstrap,* Chapman Hall, New York, 1993. A less technical overview is in Bradley Efron and Robert J. Tibshirani, "Statistical data analysis in the computer age," *Science,* 253 (1991), pp. 390–395.

2. R. A. Fisher, "The arrangement of field experiments," *Journal of the Ministry of Agriculture of Great Britain,* 33 (1926), p. 504, quoted in Leonard J. Savage, "On rereading R. A. Fisher," *The Annals of Statistics,* 4 (1976), p. 471.

3. Fisher's work is described in a biography by his daughter: Joan Fisher Box, *R. A. Fisher: The Life of a Scientist,* Wiley, New York, 1978.

4. T. D. Sterling, "Publication decisions and their possible effects on inferences drawn from tests of significance—or vice versa," *Journal of the American Statistical Association,* 54 (1959), pp. 30–34. Related comments appear in J. K. Skipper, A. L. Guenther, and G. Nass, "The sacredness of 0.05: a note concerning the uses of statistical levels of significance in social science," *American Sociologist,* 1 (1967), pp. 16–18.

5. This example is cited by William Feller, "Are life scientists overawed by statistics?" *Scientific Research,* February 3, 1969, p. 26.

6. Robert J. Schiller, "The volatility of stock market prices," *Science,* 235 (1987), pp. 33–36.

Prelude

We began our study of data analysis in Chapter 1 by learning graphical and numerical tools for describing the distribution of a single variable and for comparing several distributions. Our study of the practice of statistical inference begins in the same way, with inference about a single distribution and comparison of two distributions. Comparing more than two distributions requires more elaborate methods, which appear in Chapters 12 and 13.

Two important aspects of any distribution are its center and spread. If the distribution is normal, we describe its center by the mean μ and its spread by the standard deviation σ. In this chapter, we will meet confidence intervals and significance tests for inference about a population mean μ and for comparing the means of two populations. The previous chapter emphasized the reasoning of tests and confidence intervals; now we emphasize statistical practice, so we no longer assume that population standard deviations are known. The t procedures for inference about means are among the most common statistical methods. Inference about the spread of a population, as we will see, poses some difficult practical problems. The methods of this chapter allow us to answer questions like these:

- Twenty high school French teachers spend 4 weeks at a summer institute that emphasizes spoken French. They gain an average of 2.5 points on a 36-point test of understanding of spoken French. Is this good evidence that the institute improved the teachers' comprehension of spoken French?

- Do male and female college students differ in "social insight," their ability to appraise other people? We have the scores of almost 300 students on a test that measures social insight. What conclusions can we draw from the data?

- Preliminary studies suggest that adding calcium to the diet reduces blood pressure, especially among black males. Now researchers perform a randomized comparative experiment to compare calcium with a placebo. Do the data convince us that calcium works?

- A bank wants to increase the amount its credit card users charge on their cards. Will eliminating the annual fee for cardholders who charge large amounts do this, or will cash rebates that are a percent of the amount charged be more effective?

CHAPTER 7

Rebecca Mendoza, *Detail of Koan,* 1997.

Inference
for Distributions

7.1   Inference for the Mean of a Population

Both confidence intervals and tests of significance for the mean μ of a normal population are based on the sample mean $\bar{x}$, which estimates the unknown μ. The sampling distribution of $\bar{x}$ depends on σ. This fact causes no difficulty when σ is known. When σ is unknown, however, we must estimate σ even though we are primarily interested in μ. The sample standard deviation s is used to estimate the population standard deviation σ.

The t distributions

Suppose that we have a simple random sample (SRS) of size n from a normally distributed population with mean μ and standard deviation σ. The sample mean $\bar{x}$ then has the normal distribution with mean μ and standard deviation $\sigma/\sqrt{n}$. When σ is not known, we estimate it with the sample standard deviation s and the standard deviation of $\bar{x}$ by $s/\sqrt{n}$. This quantity is called the *standard error* of the sample mean $\bar{x}$ and we denote it by $\mathrm{SE}_{\bar{x}}$.

Standard Error

When the standard deviation of a statistic is estimated from the data, the result is called the **standard error** of the statistic. The standard error of the sample mean is

$$\mathrm{SE}_{\bar{x}} = \frac{s}{\sqrt{n}}$$

The term "standard error" is sometimes used for the actual standard deviation of a statistic, $\sigma/\sqrt{n}$ in the case of $\bar{x}$. The estimated value $s/\sqrt{n}$ is then called the "estimated standard error." In this book we will use the term "standard error" only when the standard deviation of a statistic is estimated from the data. The term has this meaning in the output of many statistical computer packages and in reports of research in many fields that apply statistical methods.

The standardized sample mean, or one-sample z statistic,

$$z = \frac{\bar{x} - \mu}{\sigma/\sqrt{n}}$$

is the basis of the z procedures for inference about μ when σ is known. This statistic has the standard normal distribution $N(0, 1)$. When we substitute the standard error $s/\sqrt{n}$ for the standard deviation $\sigma/\sqrt{n}$ of $\bar{x}$, the statistic does *not* have a normal distribution. It has a distribution that is new to us, called a t *distribution*.

> ### The *t* Distributions
>
> Suppose that an SRS of size n is drawn from an $N(\mu, \sigma)$ population. Then the **one-sample *t* statistic**
>
> $$t = \frac{\bar{x} - \mu}{s/\sqrt{n}}$$
>
> has the *t* **distribution** with $n - 1$ **degrees of freedom.**

degrees of freedom

There is a different t distribution for each sample size. A particular t distribution is specified by giving the **degrees of freedom.** The degrees of freedom for this t statistic come from the sample standard deviation s in the denominator of t. We saw in Chapter 1 (page 53) that s has $n - 1$ degrees of freedom. There are other t statistics with different degrees of freedom, some of which we will meet later in this chapter.

We use $t(k)$ to stand for the t distribution with k degrees of freedom.* The density curves of the $t(k)$ distributions are similar in shape to the standard normal curve. That is, they are symmetric about 0 and are bell-shaped. The spread of the t distributions is a bit greater than that of the standard normal distribution. This is due to the extra variability caused by substituting the random variable s for the fixed parameter σ. As the degrees of freedom k increase, the $t(k)$ density curve approaches the $N(0, 1)$ curve ever more closely. This reflects the fact that s approaches σ as the sample size increases. Figure 7.1 compares the density curves of the standard normal distribution and the t distribution with 5 degrees of freedom. The similarity in shape is apparent, as is the fact that the t distribution has more probability in the tails and less in the center than does the standard normal distribution.

Table D in the back of the book gives critical values for the t distributions. For convenience, we have labeled the table entries by both the critical value p needed for significance tests, and by the confidence level C (in percent) required for confidence intervals. The standard normal critical values in the bottom row of entries are labeled z^*. As in the case of the normal table (Table A), computer software often makes Table D unnecessary.

The one–sample *t* confidence interval

With the t distributions to help us, we can analyze samples from normal populations with unknown σ by replacing the standard deviation $\sigma/\sqrt{n}$ of $\bar{x}$ by

---

*The t distributions were discovered in 1908 by William S. Gosset. Gosset was a statistician employed by the Guinness brewing company, which required that he not publish his discoveries under his own name. He therefore wrote under the pen name "Student." The t distribution is often called "Student's t" in his honor.

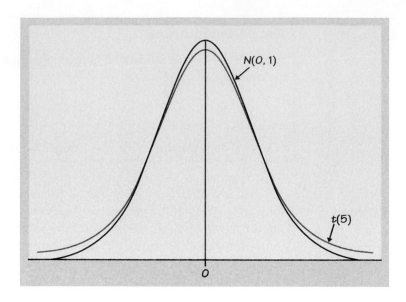

FIGURE 7.1 Density curves for the standard normal and $t(5)$ distributions. Both are symmetric with center 0. The t distributions have more probability in the tails than does the standard normal distribution.

its standard error $s/\sqrt{n}$ in the z procedures of Chapter 6. The z statistic then becomes the one-sample t statistic. We must now employ P-values or critical values from t in place of the corresponding normal values (z). Here are the details for the confidence interval.

The One-Sample t Confidence Interval

Suppose that an SRS of size n is drawn from a population having unknown mean μ. A level C **confidence interval** for μ is

$$\bar{x} \pm t^* \frac{s}{\sqrt{n}}$$

where t^* is the value for the $t(n-1)$ density curve with area C between $-t^*$ and t^*. This interval is exact when the population distribution is normal and is approximately correct for large n in other cases.

The one-sample t confidence interval is similar in both reasoning and computational detail to the z confidence interval of Chapter 6. There, the margin of error for the population mean was $z^*\sigma/\sqrt{n}$. Here, we replace σ by its

margin of error   estimate s and z^* by t^*. So the **margin of error** for the population mean when we use that data to estimate σ is $t^*s/\sqrt{n}$.

EXAMPLE 7.1 In fiscal year 1996, the U.S. Agency for International Development provided 238,300 metric tons of corn soy blend (CSB) for development programs and emergency relief in countries throughout the world. CSB is a highly nutritious, low-cost fortified food that is partially precooked and can be incorporated into different food preparations by the recipients. As part of a study to evaluate appropriate vitamin C levels in this commodity, measurements were taken on samples of CSB produced in a factory.[1]

The following data are the amounts of vitamin C, measured in milligrams per 100 grams (mg/100 g) of blend (dry basis), for a random sample of size 8 from a production run:

26	31	23	22	11	22	14	31

We want to find a 95% confidence interval for μ, the mean vitamin C content of the CSB produced during this run.

The sample mean is $\bar{x} = 22.50$ and the standard deviation is $s = 7.19$ with degrees of freedom $n - 1 = 7$. The standard error is

$$\text{SE}_{\bar{x}} = s / \sqrt{n} = 7.19 / \sqrt{8} = 2.54$$

From Table D we find $t^* = 2.365$. The 95% confidence interval is

$$\begin{aligned}
\bar{x} \pm t^* \frac{s}{\sqrt{n}} &= 22.50 \pm 2.365 \frac{7.19}{\sqrt{8}} \\
&= 22.5 \pm (2.365)(2.54) \\
&= 22.5 \pm 6.0 \\
&= (16.5, 28.5)
\end{aligned}$$

We are 95% confident that the mean vitamin C content of the CSB for this run is between 16.5 and 28.5 mg/100 g.

In this example we have given the actual interval $(16.5, 28.5)$ as our answer. Sometimes, we prefer to report the mean and margin of error: the mean vitamin C content is 22.5 mg/100 g with a margin of error of 6.0 mg/100 g.

The use of the t confidence interval in Example 7.1 rests on assumptions that cannot easily be checked but are reasonable in this case. The samples from the production run were taken at regular time intervals, and experience has shown that this procedure gives data that can be treated as random samples from a normal population. The assumption that the population distribution is normal cannot be effectively checked with only 8 observations. On the other hand, we can see clearly that there are no outliers by making a stemplot:

```
3 | 1 1
2 | 2 2 3 6
1 | 1 4
```

The one-sample t test

Significance tests using the estimated standard error are very similar to the z test that we studied in the last chapter.

The One-Sample t Test

Suppose that an SRS of size n is drawn from a population having unknown mean μ. To test the hypothesis $H_0: \mu = \mu_0$ based on an SRS of size n, compute the one-sample t statistic

$$t = \frac{\bar{x} - \mu_0}{s/\sqrt{n}}$$

In terms of a random variable T having the $t(n-1)$ distribution, the P-value for a test of H_0 against

$H_a: \mu > \mu_0$ is $P(T \geq t)$

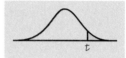

$H_a: \mu < \mu_0$ is $P(T \leq t)$

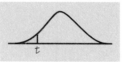

$H_a: \mu \neq \mu_0$ is $2P(T \geq |t|)$

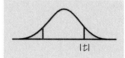

These P-values are exact if the population distribution is normal and are approximately correct for large n in other cases.

EXAMPLE 7.2

The specifications for the CSB described in Example 7.1 state that the mixture should contain 2 pounds of vitamin premix for every 2000 pounds of product. These specifications are designed to produce a mean (μ) vitamin C content in the final product of 40 mg/100 g. We can test a null hypothesis that the mean vitamin C content of the production run in Example 7.1 conforms to these specifications. Specifically, we test

$$H_0: \mu = 40$$
$$H_a: \mu \neq 40$$

Recall that $n = 8$, $\bar{x} = 22.50$, and $s = 7.19$. The t test statistic is

$$t = \frac{\bar{x} - \mu_0}{s/\sqrt{n}} = \frac{22.5 - 40}{7.2/\sqrt{8}}$$
$$= -6.88$$

df = 7

p	0.001	0.0005
t^*	4.785	5.408

Because the degrees of freedom are $n - 1 = 7$, this t statistic has the $t(7)$ distribution. Figure 7.2 shows that the P-value is $2P(T \geq 6.88)$, where T has the $t(7)$ distribution. From Table D we see that $P(T \geq 5.408) = 0.0005$. Therefore, we conclude that the P-value is less than 2×0.0005. Software gives the exact value as $P = 0.0002$. Clearly, these data are incompatible with a process mean of $\mu = 40$ mg/100 g. We reject H_0 and conclude that the vitamin C content for this run is below the specifications.

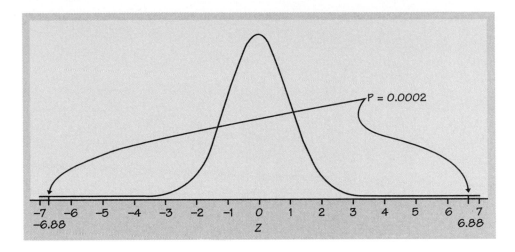

FIGURE 7.2   The P-value for Example 7.2.

In this example we tested the specifications $\mu = 40$ mg/100 g against the two-sided alternative $\mu \neq 40$ mg/100 g because we had no prior suspicion that the production run would produce CSB with too much or too little vitamin C. If we knew that the proper amount of vitamin C was added to the mixture but suspected that some of the vitamin might be lost or destroyed by the production process, we should use a one-sided test. It is *wrong*, however, to examine the data first and then decide to do a one-sided test in the direction indicated by the data. If in doubt, use a two-sided test.

EXAMPLE 7.3   For the vitamin C problem described in the previous example, we want to test whether or not vitamin C is lost or destroyed in the production process. Here we test

$$H_0: \mu = 40$$

versus

$$H_a: \mu < 40$$

The t test statistic does not change: $t = -6.88$. As Figure 7.3 illustrates, however, the P-value is now $P(T \leq -6.88)$, half of the value in the previous exercise. From Table D we can determine that $P < 0.0005$; software gives the exact value as $P = 0.0001$. We conclude that the production process has lost or destroyed some of the vitamin C.

Note that the conclusion drawn in this example relies heavily on the assumption that the proper amount of vitamin C was added to the blend. If this assumption is not correct, it is still true that the CSB produced has a vitamin C content below the specification, but the explanation that this is caused by loss or destruction of the vitamin C in the production process is not necessarily true.

For small data sets, such as the one in Example 7.1, it is easy to perform the computations for confidence intervals and significance tests with an ordinary calculator. For larger data sets, however, we prefer to use software or a statistical calculator.

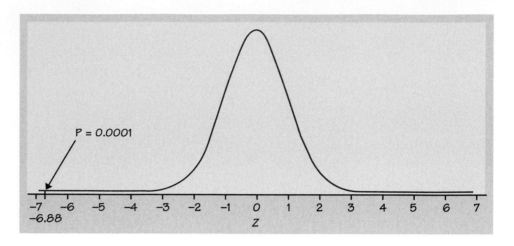

FIGURE 7.3 The *P*-value for Example 7.3.

EXAMPLE 7.4 | The first data that we looked at were Simon Newcomb's measurements of the passage time of light. Table 1.1 (page 8) records his 66 measurements. A normal quantile plot (Figure 7.4) reminds us that when the 2 outliers to the left are omitted, the remaining 64 observations follow a normal distribution quite closely. What result should Newcomb report from these 64 observations?

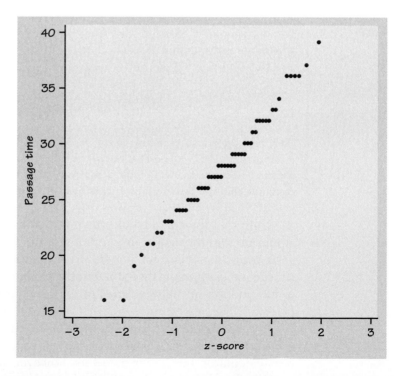

FIGURE 7.4 Normal quantile plot for Newcomb's passage time data with the outliers omitted, for Example 7.4.

Minitab

	N	MEAN	STDEV	SE MEAN	99.0 PERCENT C.I.
C1	64	27.750	5.083	0.635	(26.062, 29.438)

Texas Instruments TI-83

```
TInterval
  (26.062,29.438
  x=27.75
  Sx=5.083430912
  n=64
```

FIGURE 7.5   Minitab and TI-83 output for Example 7.4.

We want to estimate the mean μ of the distribution of measurements from which Newcomb's 64 are a sample. The sample mean $\overline{x} = 27.750$ estimates μ, but to indicate the precision of this estimate we will give a 99% confidence interval. The sample standard deviation is $s = 5.083$ with $n - 1 = 63$ degrees of freedom.

Figure 7.5 displays the outputs from Minitab and the TI-83. They give the results in slightly different forms but it is easy to find what we need. You should have no trouble reading the output from other software systems as well.

The 99% confidence interval for Newcomb's estimated passage time of light is (26.06, 29.44).

We can also use software or a statistical calculator to perform significance tests easily.

EXAMPLE 7.5   The best modern measurements of the speed of light correspond to a passage time of 33.02 in Newcomb's experiment. Is his result significantly different from this modern value? To answer this question, we test

$$H_0: \mu = 33.02$$
$$H_a: \mu \neq 33.02$$

The outputs from Minitab and the TI-83 are given in Figure 7.6. These are also easy to read. Here is one way to report the conclusion: the mean of Newcomb's passage times, $\overline{x} = 27.75$, is lower than the value calculated by modern methods, $\mu = 33.02$ ($t = -8.29$, df $= 63$, $P < 0.0001$).

We have concluded that Newcomb's mean is lower than the hypothesized value even though we have tested the two-sided alternative. At the end of our conclusion statement we give the test statistic, degrees of freedom, and the

Minitab

```
TTEST OF MU = 33.020 VS MU N.E.  33.020

         N      MEAN      STDEV     SE MEAN        T     P VALUE
C1      64     27.750     5.083       0.635     -8.29     0.0000
```

Texas Instruments TI-83

```
T-Test
 μ≠33.02
 t=-82936112889
 p=1.094873E-11
 x̄=27.75
 Sx=5.083430912
 n=64
```

FIGURE 7.6   Minitab and TI-83 output for Example 7.5.

P-value in parentheses. The software outputs do not provide the degrees of freedom; we need to know that it is $n - 1 = 64 - 1 = 63$ in this case. We reported $P < 0.0001$. Minitab gave the value as 0.0000, a rounded value that represents anything less than 0.00005. TI-83 gave $P = 1.094873\mathrm{E}-11$. The $\mathrm{E}-11$ means that we move the decimal point 11 places to the left in the number given—that is, $P = 0.00000000001094873$, or one in one hundred billion. We have chosen to report $P < 0.0001$; most people would accept 1 in ten thousand as an unlikely event. The very small P-value is not the most important result here. The message in this example is that Newcomb's measurements averaged 27.75 while the modern value is 33.02 and that we have sufficient data to conclude that the difference is statistically significant.

Newcomb's measurements can be regarded as independent observations drawn from the population of all measurements he might make. We verified the normality of the measurements with the normal quantile plot in Figure 7.4. Moreover, the sample mean of 64 observations will have a distribution that is nearly normal even if the population is not normal. Only normality of $\bar{x}$ is required by the t procedures. Use of these procedures in this example is very well justified.

The following example reminds us of the relationship between confidence intervals and significance tests.

EXAMPLE 7.6   In Example 7.4 we found a 99% confidence interval for μ, the mean passage time using modern statistical methods. It was (26.06, 29.44). Example 7.5 gave us the results of the hypothesis test that Newcomb was trying to estimate $\mu = 33.02$. Because the confidence interval does not include the hypothesized value, $\mu = 33.02$, we can reject the null hypothesis with $\alpha = 0.01$.

Matched pairs *t* procedures

Newcomb wanted to estimate a constant of nature, so only a single population was involved. We saw in discussing data production that comparative studies are usually preferred to single-sample investigations because of the protection they offer against confounding. For that reason, inference about a parameter of a single distribution is less common than comparative inference. One common comparative design, however, makes use of single-sample procedures. In a *matched pairs* study, subjects are matched in pairs and the outcomes are compared within each matched pair. The experimenter can toss a coin to assign two treatments to the two subjects in each pair. Matched pairs are also common when randomization is not possible. One situation calling for matched pairs is before-and-after observations on the same subjects, as illustrated in the next example.

EXAMPLE 7.7

The National Endowment for the Humanities sponsors summer institutes to improve the skills of high school teachers of foreign languages. One such institute hosted 20 French teachers for 4 weeks. At the beginning of the period, the teachers were given the Modern Language Association's listening test of understanding of spoken French. After 4 weeks of immersion in French in and out of class, the listening test was given again. (The actual French spoken in the two tests was different, so that simply taking the first test should not improve the score on the second test.) Table 7.1 gives the pretest and posttest scores. The maximum possible score on the test is 36.[2]

To analyze these data, we first subtract the pretest score from the posttest score to obtain the improvement for each student. These 20 differences form a single sample. They appear in the "Gain" columns in Table 7.1. The first teacher, for example, improved from 32 to 34, so the gain is $34 - 32 = 2$.

To assess whether the institute significantly improved the teachers' comprehension of spoken French, we test

$$H_0: \mu = 0$$
$$H_a: \mu > 0$$

Here μ is the mean improvement that would be achieved if the entire population of French teachers attended a summer institute. The null hypothesis says that no improvement occurs, and H_a says that posttest scores are higher on the average.

The 20 differences have

$$\bar{x} = 2.5 \quad \text{and} \quad s = 2.893$$

The one-sample t statistic is therefore

$$t = \frac{\bar{x} - 0}{s/\sqrt{n}} = \frac{2.5}{2.893/\sqrt{20}}$$
$$= 3.86$$

df = 19

p	0.001	0.0005
t^*	3.579	3.883

The P-value is found from the $t(19)$ distribution (remember that the degrees of freedom are 1 less than the sample size). Table D shows that 3.86 lies between the upper 0.001 and 0.0005 critical values of the $t(19)$ distribution. The P-value therefore lies between these values. Software gives the value $P = 0.00053$. The improvement in listening scores is very unlikely to be due to chance alone. We have strong evidence that the institute was effective in raising scores. In scholarly publications, the details of routine statistical procedures are omitted; our test would be reported in the form: "The improvement in scores was significant ($t = 3.86$, df = 19, $P = 0.00053$)."

TABLE 7.1 Modern Language Association listening scores for French teachers

Teacher	Pretest	Posttest	Gain	Teacher	Pretest	Posttest	Gain
1	32	34	2	11	30	36	6
2	31	31	0	12	20	26	6
3	29	35	6	13	24	27	3
4	10	16	6	14	24	24	0
5	30	33	3	15	31	32	1
6	33	36	3	16	30	31	1
7	22	24	2	17	15	15	0
8	25	28	3	18	32	34	2
9	32	26	−6	19	23	26	3
10	20	26	6	20	23	26	3

The results of the significance test allow us to conclude that the mean scores have improved. A statistically significant but very small improvement would not be sufficient grounds to say that the summer institute substantially improved the skills of the teachers who participated. A confidence interval tells us how much they improved with a margin of error.

EXAMPLE 7.8

A 90% confidence interval for the mean improvement in the entire population requires the critical value $t^* = 1.729$ from Table D. The confidence interval is

$$\bar{x} \pm t^* \frac{s}{\sqrt{n}} = 2.5 \pm 1.729 \frac{2.893}{\sqrt{20}}$$
$$= 2.5 \pm 1.12$$
$$= (1.38, 3.62)$$

The estimated average improvement is 2.5 points, with margin of error 1.12 for 90% confidence. Though statistically significant, the effect of the institute was rather small.

The following are key points to remember concerning matched pairs:

1. A matched pairs analysis is needed when there are two measurements or observations on each individual and we want to examine the change from the first to the second. Typically, the observations are "before" and "after" measures in some sense.

2. For each individual, subtract the "before" measure from the "after" measure.

3. Analyze the difference using the one-sample confidence interval and significance-testing procedures that we learned in this section.

Use of the t procedures in Examples 7.7 and 7.8 faces several difficulties. First, the teachers are not an SRS from the population of high school French teachers. There is some selection bias in favor of energetic, committed teach-

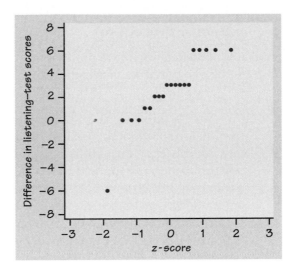

FIGURE 7.7 Normal quantile plot for the change in French listening score, for Example 7.7.

ers who are willing to give up four weeks of their summer vacation. It is therefore not clear to what population the results apply. Second, a look at the data shows that several of the teachers had pretest scores close to the maximum of 36. They could not improve their scores very much even if their mastery of French increased substantially. This is a weakness in the listening test that is the measuring instrument in this study. The differences in scores may not adequately indicate the effectiveness of the institute. This is one reason why the average increase was small.

A final difficulty is that the data show departures from normality. In a matched pairs analysis, we assume that the population of differences has a normal distribution because the t procedures are applied to the differences. In these examples, one teacher actually lost 6 points between the pretest and the posttest. This one subject lowered the sample mean from 2.95 for the other 19 subjects to 2.5 for all 20. A normal quantile plot (Figure 7.7) displays this outlier as well as granularity due to the fact that only whole-number scores are possible. The overall pattern of the plot is otherwise roughly straight. Does this nonnormality forbid use of the t test? The behavior of the t procedures when the population does not have a normal distribution is one of their most important properties.

Robustness of the *t* procedures

The results of one-sample t procedures are exactly correct only when the population is normal. Real populations are never exactly normal. The usefulness of the t procedures in practice therefore depends on how strongly they are affected by nonnormality. Procedures that are not strongly affected are called *robust*.

> **Robust Procedures**
>
> A statistical inference procedure is called **robust** if the probability calculations required are insensitive to violations of the assumptions made.

The assumption that the population is normal rules out outliers, so the presence of outliers shows that this assumption is not valid. The t procedures are not robust against outliers, because $\bar{x}$ and s are not resistant to outliers. If we dropped the single outlier in Example 7.7, the test statistic would change from $t = 3.86$ to $t = 5.98$, and the P-value would be much smaller. In this case, the outlier makes the test result *less* significant and the margin of error of a confidence interval *larger* than they would otherwise be. The results of the t procedures in Example 7.7 are conservative in the sense that the conclusions show a smaller effect than would be the case if the outlier were not present.

Fortunately, the t procedures are quite robust against nonnormality of the population except in the case of outliers or strong skewness. Larger samples improve the accuracy of P-values and critical values from the t distributions when the population is not normal. This is true for two reasons. First, the sampling distribution of the sample mean $\bar{x}$ from a large sample is close to normal (that's the central limit theorem). We need be less concerned about the normality of the individual observations when the sample is large. Second, as the size n of a sample grows, the sample standard deviation s approaches the population standard deviation σ. This fact is closely related to the law of large numbers. In large samples, s will be an accurate estimate of σ whether or not the population has a normal distribution.

A normal quantile plot, stemplot, or boxplot to check for skewness and outliers is an important preliminary to the use of t procedures for small samples. For most purposes, the one-sample t procedures can be safely used when $n \geq 15$ unless an outlier or clearly marked skewness is present. Except in the case of small samples, the assumption that the data are an SRS from the population of interest is more crucial than the assumption that the population distribution is normal. Here are practical guidelines for inference on a single mean:[3]

- *Sample size less than 15:* Use t procedures if the data are close to normal. If the data are clearly nonnormal or if outliers are present, do not use t.

- *Sample size at least 15:* The t procedures can be used except in the presence of outliers or strong skewness.

- *Large samples:* The t procedures can be used even for clearly skewed distributions when the sample is large, roughly $n \geq 40$.

Consider, for example, some of the data we studied in Chapter 1. Newcomb's data on the passage time of light in Figure 1.30 (page 82) contain two outliers, which make the use of t procedures risky even though the sample size is $n = 66$. When the outliers are removed (Figure 1.31 on page 82), the data are quite normal, and the t procedures are well justified. We would not employ t

for inference about the meat hot dog calorie data on page 40, a small sample ($n = 17$) with two clusters and an outlier. On the other hand, the grocery spending data in Figure 1.16 (page 49), although clearly skewed, have no outliers and a moderate sample size ($n = 50$). We would apply the t procedures in this case.

The power of the t test*

The power of a statistical test measures its ability to detect deviations from the null hypothesis. In practice we carry out the test in the hope of showing that the null hypothesis is false, so high power is important. The power of the one-sample t test against a specific alternative value of the population mean μ is the probability that the test will reject the null hypothesis when the alternative value of the mean is true. To calculate the power, we assume a fixed level of significance, often $\alpha = 0.05$.

Calculation of the exact power of the t test takes into account the estimation of σ by s and is a bit complex. But an approximate calculation that acts as if σ were known is almost always adequate for planning a study. This calculation is very much like that for the z test, presented in Section 6.4: write the event that the test rejects H_0 in terms of $\overline{x}$, and then find the probability of this event when the population mean has the alternative value.

EXAMPLE 7.9 It is the winter before the summer language institute of Example 7.7. The director of the institute, thinking ahead to the report he must write, hopes that the planned 20 students will enable him to be quite certain of detecting an average improvement of 2 points in the mean listening score. Is this realistic?

We wish to compute the power of the t test for

$$H_0: \mu = 0$$
$$H_a: \mu > 0$$

against the alternative $\mu = 2$ when $n = 20$. We must have a rough guess of the size of σ in order to compute the power. In planning a large study, a pilot study is often run for this and other purposes. In this case, listening-score improvements in past summer language institutes have had sample standard deviations of about 3. We therefore take both $\sigma = 3$ and $s = 3$ in our approximate calculation.

The t test with 20 observations rejects H_0 at the 5% significance level if the t statistic

$$t = \frac{\overline{x} - 0}{s/\sqrt{20}}$$

exceeds the upper 5% point of $t(19)$, which is 1.729. Taking $s = 3$, the event that the test rejects H_0 is therefore

$$t = \frac{\overline{x}}{3/\sqrt{20}} \geq 1.729$$
$$\overline{x} \geq 1.729\frac{3}{\sqrt{20}}$$
$$\overline{x} \geq 1.160$$

---

*This section can be omitted without loss of continuity.

The power is the probability that $\bar{x} \geq 1.160$ when $\mu = 2$. Taking $\sigma = 3$, this probability is found by standardizing $\bar{x}$:

$$P(\bar{x} \geq 1.160 \text{ when } \mu = 2) = P\left(\frac{\bar{x} - 2}{3/\sqrt{20}} \geq \frac{1.160 - 2}{3/\sqrt{20}}\right)$$
$$= P(Z \geq -1.252)$$
$$= 1 - 0.1056 = 0.8944$$

A true difference of 2 points in the population mean scores will produce significance at the 5% level in 89% of all possible samples. The director can be reasonably confident of detecting a difference this large.

Inference for nonnormal populations*

We have not discussed how to do inference about the mean of a clearly nonnormal distribution based on a small sample. If you face this problem, you should consult an expert. Three general strategies are available.

- In some cases a distribution other than a normal distribution will describe the data well. There are many nonnormal models for data, and inference procedures for these models are available.

- Because skewness is the chief barrier to the use of t procedures on data without outliers, you can attempt to transform skewed data so that the distribution is symmetric and as close to normal as possible. Confidence levels and P-values from the t procedures applied to the transformed data will be quite accurate for even moderate sample sizes.

distribution–free

- The third strategy is to use a **distribution-free** inference procedure. Such procedures do not assume that the population distribution has any specific form, such as normal. Distribution-free procedures are often called **nonparametric procedures.**

nonparametric procedures

Each of these strategies can be effective, but each quickly carries us beyond the basic practice of statistics. We emphasize procedures based on normal distributions because they are the most common in practice, because their robustness makes them widely useful, and (most important) because we are first of all concerned with understanding the principles of inference. We will therefore not discuss procedures for nonnormal continuous distributions and will present only one of the many distribution-free procedures that do not require that the population have any specific type of distribution. We will be content with illustrating by example the use of a transformation and of a simple distribution-free procedure.

Transforming data   When the distribution of a variable is skewed, it often happens that a simple transformation results in a variable whose distribution is symmetric and even close to normal. The most common transformation is

---

*This section can be omitted without loss of continuity.

log transformation

the **logarithm** or **log.** The logarithm tends to pull in the right tail of a distribution. For example, the data 2, 3, 4, 20 show an outlier in the right tail. Their logarithms 0.30, 0.48, 0.60, 1.30 are much less skewed. Taking logarithms is a possible remedy for right-skewness. Instead of analyzing values of the original variable X, we first compute their logarithms and analyze the values of $\log X$. Here is an example of this approach.

EXAMPLE 7.10

Table 2.16 (page 222) presents data on the amounts (in grams per mile) of three pollutants in the exhaust of 46 vehicles of the same type, measured under standard conditions prescribed by the Environmental Protection Agency. We will concentrate on emissions of carbon monoxide (CO). We would like to give a confidence interval for the mean emissions μ for this vehicle type.[4]

A normal quantile plot of the CO data from Table 2.16 (Figure 7.8) shows that the distribution is skewed to the right. Because there are no extreme outliers, the sample mean of 46 observations will nonetheless have an approximately normal sampling distribution. The t procedures could be used for approximate inference. For more exact inference, we will seek to transform the data so that the distribution is more nearly normal. Figure 7.9 is a normal quantile plot of the logarithms of the CO measurements. The transformed data are very close to normal, so t procedures will give quite exact results.

The application of the t procedures to the transformed data is straightforward. Call the original CO values from Table 2.16 values of the variable X. The transformed data are values of $X_{\text{new}} = \log X$. In most software packages, it is an easy task to transform data in this way and then analyze the new variable.

EXAMPLE 7.11

Analysis of the logs of the CO values in Minitab produces the following output:

N	MEAN	STDEV	SE MEAN	95.0 PERCENT C.I.
46	0.8198	0.2654	0.0391	(0.7409, 0.8986)

For comparison, the 95% t confidence interval for the original mean μ is found from the original data as follows:

N	MEAN	STDEV	SE MEAN	95.0 PERCENT C.I.
46	7.960	5.261	0.776	(6.398, 9.523)

The advantage of analyzing transformed data is that use of procedures based on the normal distributions is better justified and the results are more exact. The disadvantage is that a confidence interval for the mean μ in the original scale (in our example, emissions measured in grams per mile) cannot be recovered from the confidence interval for the mean of the logs.

The sign test

Perhaps the most straightforward way to cope with nonnormal data is to use a *distribution-free* procedure. As the name indicates, these procedures do not

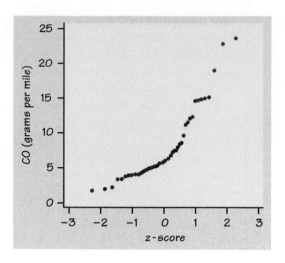

FIGURE 7.8   Normal quantile plot for CO emissions, for Example 7.10. The distribution is skewed to the right.

require the population distribution to have any specific form, such as normal. Distribution-free significance tests are quite simple and are available in most statistical software systems. Distribution-free tests have two drawbacks. First, they are generally less powerful than tests designed for use with a specific distribution, such as the *t* test. Second, we must often modify the statement of the hypotheses in order to use a distribution-free test. A distribution-free test concerning the center of a distribution, for example, is usually stated in terms of the median rather than the mean. This is sensible when the distribution may be skewed. But the distribution-free test does not ask the same question (Has the mean changed?) that the *t* test does. The simplest distribution-free test, and one of the most useful, is the **sign test.** The following example illustrates this test.

sign test

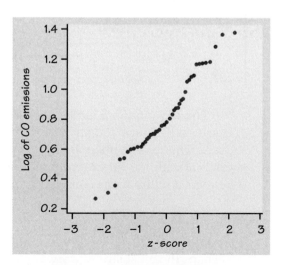

FIGURE 7.9   Normal quantile plot for the logarithms of the CO emissions, for Example 7.10. This distribution is close to normal.

EXAMPLE 7.12

Return to the data of Example 7.7 (page 513) showing the improvement in French listening scores after attending a summer institute. In that example we used the one-sample t test on these data, despite granularity and an outlier that make the P-value only roughly correct. The sign test is based on the following simple observation: of the 17 teachers whose scores changed, 16 improved and only 1 did more poorly. This is evidence that the institute improved French listening skills.

To perform a significance test based on the count of teachers whose scores improved, let p be the probability that a randomly chosen teacher would improve if she attended the institute. The null hypothesis of "no effect" says that the posttest is just a repeat of the pretest with no change in ability, so a teacher is equally likely to do better on either test. We therefore want to test

$$H_0: p = 1/2$$
$$H_a: p > 1/2$$

The 17 teachers whose scores changed are 17 independent trials, so the number who improve has the binomial distribution $B(17, 1/2)$ if H_0 is true. The P-value for the observed count 16 is therefore $P(X \geq 16)$, where X has the $B(17, 1/2)$ distribution. You can compute this probability with software or from the binomial probability formula (page 388):

$$
\begin{aligned}
P(X \geq 16) &= P(X = 16) + P(X = 17) \\
&= \binom{17}{16}\left(\frac{1}{2}\right)^{16}\left(\frac{1}{2}\right)^{1} + \binom{17}{17}\left(\frac{1}{2}\right)^{17}\left(\frac{1}{2}\right)^{0} \\
&= (17)\left(\frac{1}{2}\right)^{17} + \left(\frac{1}{2}\right)^{17} \\
&= 0.00014
\end{aligned}
$$

As in Example 7.7, there is very strong evidence that participation in the institute has improved performance on the listening test.

There are several varieties of sign test, all based on counts and the binomial distribution. The sign test for matched pairs (Example 7.12) is the most useful. The null hypothesis of "no effect" is then always $H_0: p = 1/2$. The alternative can be one-sided in either direction or two-sided, depending on the type of change we are looking for. The test gets its name from the fact that we look only at the signs of the differences, not their actual values.

The Sign Test for Matched Pairs

Ignore pairs with difference 0; the number of trials n is the count of the remaining pairs. The test statistic is the count X of pairs with a positive difference. P-values for X are based on the binomial $B(n, 1/2)$ distribution.

The matched pairs t test in Example 7.7 tested the hypothesis that the mean of the distribution of differences (score after the institute minus score before) is 0. The sign test in Example 7.12 is in fact testing the hypothesis that the *median* of the differences is 0. If p is the probability that a difference is positive, then $p = 1/2$ when the median is 0. This is true because the median of the distribution is the point with probability 1/2 lying to its right. As Figure 7.10 illustrates, $p > 1/2$ when the median is greater than 0, again

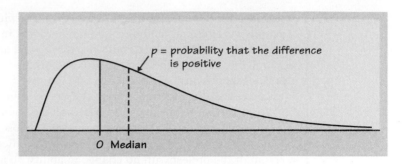

FIGURE 7.10   Why the sign test tests the median distance: when the median is greater than 0, the probability p of a positive difference is greater than 1/2, and vice versa.

because the probability to the right of the median is always 1/2. The sign test of $H_0: p = 1/2$ against $H_a: p > 1/2$ is a test of

$$H_0: \text{population median} = 0$$
$$H_a: \text{population median} > 0$$

The sign test in Example 7.12 makes no use of the actual scores—it just counts how many teachers improved. The teachers whose scores did not change were ignored altogether. Because the sign test uses so little of the available information, it is much less powerful than the t test when the population is close to normal. There are other distribution-free tests that are more powerful than the sign test.[5]

SUMMARY

Significance tests and confidence intervals for the mean μ of a normal population are based on the sample mean $\bar{x}$ of an SRS. Because of the central limit theorem, the resulting procedures are approximately correct for other population distributions when the sample is large.

The standardized sample mean, or **one-sample z statistic,**

$$z = \frac{\bar{x} - \mu}{\sigma/\sqrt{n}}$$

has the $N(0, 1)$ distribution. If the standard deviation $\sigma/\sqrt{n}$ of $\bar{x}$ is replaced by the **standard error** $s/\sqrt{n}$, the **one-sample t statistic**

$$t = \frac{\bar{x} - \mu}{s/\sqrt{n}}$$

has the t distribution with $n - 1$ degrees of freedom.

There is a t distribution for every positive **degrees of freedom** k. All are symmetric distributions similar in shape to normal distributions. The $t(k)$ distribution approaches the $N(0, 1)$ distribution as k increases.

A level C **confidence interval for the mean** μ of a normal population is

$$\bar{x} \pm t^* \frac{s}{\sqrt{n}}$$

where t^* is the value for the $t(n-1)$ density curve with area C between $-t^*$ and t^*. The quantity

$$t^* \frac{s}{\sqrt{n}}$$

is the **margin of error.**

Significance tests for $H_0\colon \mu = \mu_0$ are based on the t statistic. P-values or fixed significance levels are computed from the $t(n-1)$ distribution.

These one-sample procedures are used to analyze **matched pairs** data by first taking the differences within the matched pairs to produce a single sample.

The t procedures are relatively **robust** against nonnormal populations, especially for larger sample sizes. The t procedures are useful for nonnormal data when $n \geq 15$ unless the data show outliers or strong skewness.

The power of the t test is calculated like that of the z test, using an approximate value for both σ and s.

Small samples from skewed populations can sometimes be analyzed by first applying a **transformation** (such as the logarithm) to obtain an approximately normally distributed variable. The t procedures then apply to the transformed data.

The **sign test** is a **distribution-free test** because it uses probability calculations that are correct for a wide range of population distributions.

The sign test for "no treatment effect" in matched pairs counts the number of positive differences. The P-value is computed from the $B(n, 1/2)$ distribution, where n is the number of non-0 differences. The sign test is less powerful than the t test in cases where use of the t test is justified.

SECTION 7.1 EXERCISES

7.1 What critical value t^* from Table D should be used for a confidence interval for the mean of the population in each of the following situations?

(a) A 90% confidence interval based on $n = 12$ observations.

(b) A 95% confidence interval from an SRS of 30 observations.

(c) An 80% confidence interval from a sample of size 18.

7.2 Use software to find the critical values t^* that you would use for each of the following confidence intervals for the mean.

(a) A 99% confidence interval based on $n = 55$ observations.

(b) A 90% confidence interval from an SRS of 35 observations.

(c) An 95% confidence interval from a sample of size 90.

7.3 The one-sample t statistic for testing

$$H_0: \mu = 0$$
$$H_a: \mu > 0$$

from a sample of $n = 15$ observations has the value $t = 1.97$.

(a) What are the degrees of freedom for this statistic?

(b) Give the two critical values t^* from Table D that bracket t.

(c) What are the right-tail probabilities p for these two entries?

(d) Between what two values does the P-value of the test fall?

(e) Is the value $t = 1.97$ significant at the 5% level? Is it significant at the 1% level?

(f) If you have software available, find the exact P-value.

7.4 The one-sample t statistic from a sample of $n = 30$ observations for the two-sided test of

$$H_0: \mu = 64$$
$$H_a: \mu \neq 64$$

has the value $t = 1.12$.

(a) What are the degrees of freedom for t?

(b) Locate the two critical values t^* from Table D that bracket t. What are the right-tail probabilities p for these two values?

(c) How would you report the P-value for this test?

(d) Is the value $t = 1.12$ statistically significant at the 10% level? At the 5% level?

(e) If you have software available, find the exact P-value.

7.5 The one-sample t statistic for a test of

$$H_0: \mu = 20$$
$$H_a: \mu < 20$$

based on $n = 12$ observations has the value $t = -2.45$.

(a) What are the degrees of freedom for this statistic?

(b) How would you report the P-value based on Table D?

(c) If you have software available, find the exact P-value.

7.6 The scores of four roommates on the Law School Admission Test are

628	593	455	503

Find the mean, the standard deviation, and the standard error of the mean. Is it appropriate to calculate a confidence interval for these data? Explain why or why not.

7.7 Here are estimates of the daily intakes of calcium (in milligrams) for 38 women between the ages of 18 and 24 years who participated in a study of women's bone health:

808	882	1062	970	909	802	374	416	784	997
651	716	438	1420	1425	948	1050	976	572	403
626	774	1253	549	1325	446	465	1269	671	696
1156	684	1933	748	1203	2433	1255	1100		

(a) Display the data using a stemplot and make a normal quantile plot. Describe the distribution of calcium intakes for these women.

(b) Calculate the mean, the standard deviation, and the standard error.

(c) Find a 95% confidence interval for the mean.

7.8 Refer to the previous exercise. Eliminate the two largest values and answer parts (a), (b), and (c).

7.9 Refer to Exercise 7.7. Suppose that the recommended daily allowance (RDA) of calcium for women in this age range is 1200 milligrams (this value is changed from time to time on the basis of the statistical analysis of new data). We want to express the results in terms of percent of the RDA.

(a) Divide each intake by the RDA, multiply by 100, and compute the 95% confidence interval from the transformed data.

(b) Verify that you can obtain the same result by similarly transforming the interval you calculated in Exercise 7.7.

7.10 Refer to Exercises 7.7 and 7.9. You want to compare the average calcium intake of these women with the RDA using a significance test.

(a) State appropriate null and alternative hypotheses.

(b) Give the test statistic, the degrees of freedom, and the P-value.

(c) State your conclusion.

7.11 A manufacturer of small appliances employs a market research firm to estimate retail sales of its products by gathering information from a sample of retail stores. An SRS of 50 stores in the Midwest sales region were asked to give the number of the manufacturer's electric can openers that were sold last month. Here are the data:

19	19	16	19	25	26	24	63	22	16
13	26	34	10	48	16	20	14	13	24
34	14	25	16	26	25	25	26	11	79
17	25	18	15	13	35	17	15	21	12
19	20	32	19	24	19	17	41	24	27

(a) Display the data with a stemplot and a normal quantile plot. Describe the distribution.

(b) Give a 95% confidence interval for the mean number of can openers sold by all stores in the region.

(c) The distribution of sales is strongly right-skewed because there are many smaller stores and a few very large stores. The use of t in part (a) is reasonably safe despite this violation of the normality assumption. Why?

7.12   Refer to the previous exercise. Each electric can opener sold generates a profit of $2.15 for the manufacturer.

(a) What is the mean profit per store in the Midwest sales region?

(b) Transform the confidence interval you found in the previous exercise into an interval for the mean profit for stores in the Midwest region.

7.13   Refer to the previous two exercises. There are 3275 stores that sell can openers manufactured by this company.

(a) Estimate the total profit for sales last month in the Midwest region.

(b) Give a 95% confidence interval for the total profit for sales last month in the Midwest region.

7.14   The U.S. Department of Agriculture (USDA) uses many types of surveys to obtain important economic estimates. In one pilot study they used 22 observations to estimate the price received by farmers for corn sold in October. They reported the mean price as $2.08 per bushel with a standard error of $0.176 per bushel. Give a 95% confidence interval for the mean price received by farmers for corn sold in October. (Based on R. G. Hood, "Results of the prices received by farmers for grain—quality assurance project," USDA Report SRB-95-07, Washington, D.C., 1995.)

7.15   The cost of health care is the subject of many studies that use statistical methods. One such study estimated that the average length of service for home health care among people over the age of 65 who use this type of service is 96.0 days with a standard error of 5.1 days. Assuming that the degrees of freedom are large, calculate a 90% confidence for the true mean length of service. (Based on A. N. Dey, "Characteristics of elderly home health care users," National Center for Health Statistics, Washington, D.C., 1996.)

7.16   Many organizations are doing surveys to determine the satisfaction of their customers. Attitudes toward various aspects of campus life were the subject of one such study conducted at Purdue University. Each question was rated on a 1 to 5 scale, with 5 being the highest rating. The average response of 1406 first-year students to "Feeling welcomed at Purdue" was 3.9 with a standard deviation of 0.98. Assuming that the respondents are an SRS, give a 99% confidence interval for the mean of all first-year students.

7.17   Do piano lessons improve the spatial-temporal reasoning of preschool children? Neurobiological arguments suggest that this may be true. A study designed to test this hypothesis measured the spatial-temporal reasoning of 34 preschool children before and after six months of piano lessons. (The study also included children who took computer lessons and a control group; but we are not concerned with those here.) The changes in the reasoning scores are

| 2 | 5 | 7 | −2 | 2 | 7 | 4 | 1 | 0 | 7 | 3 | 4 | 3 | 4 | 9 | 4 | 5 |
| 2 | 9 | 6 | 0 | 3 | 6 | −1 | 3 | 4 | 6 | 7 | −2 | 7 | −3 | 3 | 4 | 4 |

(Data from F. H. Rauscher et al., "Music training causes long-term enhancement of preschool children's spatial-temporal reasoning," *Neurological Research,* 19 (1997), pp. 2–8.)

(a) Display the data and summarize the distribution.

(b) Find the mean, the standard deviation, and the standard error of the mean.

(c) Give a 95% confidence interval for the mean improvement in reasoning scores.

7.18 Refer to the previous exercise. Test the null hypothesis that there is no improvement versus the alternative suggested by the neurobiological arguments. State the hypotheses, and give the test statistic with degrees of freedom and the P-value. What do you conclude? From your answer to part (c) of the previous exercise, what can be concluded from this significance test?

7.19 The researchers studying vitamin C in CSB in Example 7.1 were also interested in a similar commodity called wheat soy blend (WSB). Both of these commodities are mixed with other ingredients and cooked. Loss of vitamin C as a result of this process was another concern of the researchers. One preparation used in Haiti called gruel (or "bouillie" in Creole) can be made from WSB, salt, sugar, milk, banana, and other optional items to improve the taste. Samples of gruel prepared in Haitian households were collected. The vitamin C content (in milligrams per 100 grams of blend, dry basis) was measured before and after cooking. Here are the results:

Sample	1	2	3	4	5
Before	73	79	86	88	78
After	20	27	29	36	17

Set up appropriate hypotheses and carry out a significance test for these data. (It is not possible for cooking to increase the amount of vitamin C.)

7.20 Refer to the previous exercise. The fact that vitamin C is destroyed by cooking is neither new nor surprising; so the significance test you performed in the previous exercise simply confirms that this fact is evident in the small sample of data collected. The real question here concerns how much of the vitamin C is lost.

(a) Give a 95% confidence interval for the amount of vitamin C lost by preparing and cooking gruel in Haiti.

(b) The units for your interval are mg/100 g of blend. To meaningfully interpret the amount lost, it is necessary to know something about the amount that we started with. For example, a loss of 50 mg/100 g would probably not be of much concern if we started with

5000 mg/100 g. The specifications call for the blend to contain
98 mg/100 g (dry basis). The difference between this specification
and the "before" values above is due to sample variation in the
manufacturing process and the handling of the product from the time
it was manufactured until it was used to prepare gruel in these Haitian
homes. Express the "after" data as percent of specification and give a
95% confidence interval for the mean percent.

7.21   A bank wonders whether omitting the annual credit card fee for customers
who charge at least $2400 in a year would increase the amount charged on
its credit card. The bank makes this offer to an SRS of 250 of its existing
credit card customers. It then compares how much these customers charge
this year with the amount that they charged last year. The mean increase is
$342, and the standard deviation is $108.

  (a) Is there significant evidence at the 1% level that the mean amount
  charged increases under the no-fee offer? State H_0 and H_a and carry
  out a t test.

  (b) Give a 95% confidence interval for the mean amount of the increase.

  (c) The distribution of the amount charged is skewed to the right, but
  outliers are prevented by the credit limit that the bank enforces on each
  card. Use of the t procedures is justified in this case even though the
  population distribution is not normal. Explain why.

  (d) A critic points out that the customers would probably have charged
  more this year than last even without the new offer because the
  economy is more prosperous and interest rates are lower. Briefly
  describe the design of an experiment to study the effect of the no-fee
  offer that would avoid this criticism.

7.22   In an experiment on the metabolism of insects, American cockroaches were
fed measured amounts of a sugar solution after being deprived of food for
a week and of water for 3 days. After 2, 5, and 10 hours, the researchers
dissected some of the cockroaches and measured the amount of sugar in
various tissues. Five cockroaches fed the sugar D-glucose and dissected
after 10 hours had the following amounts (in micrograms) of D-glucose in
their hindguts:

| 55.95 | 68.24 | 52.73 | 21.50 | 23.78 |

Find a 95% confidence interval for the mean amount of D-glucose in
cockroach hindguts under these conditions. (Based on D. L. Shankland et
al., "The effect of 5-thio-D-glucose on insect development and its absorption
by insects," *Journal of Insect Physiology*, 14 (1968), pp. 63–72.)

7.23   The level of various substances in the blood of kidney dialysis patients
is of concern because kidney failure and dialysis can lead to nutritional
problems. A researcher performed blood tests on several dialysis patients
on six consecutive clinic visits. One variable measured was the level of
phosphate in the blood. Phosphate levels for an individual tend to vary

normally over time. The data on one patient, in milligrams of phosphate per deciliter (mg/dl) of blood, are given below. (Data from Joan M. Susic, "Dietary phosphorus intakes, urinary and peritoneal phosphate excretion and clearance in continuous ambulatory peritoneal dialysis patients," M.S. thesis, Purdue University, 1985.)

5.6	5.1	4.6	4.8	5.7	6.4

(a) Calculate the sample mean $\bar{x}$ and its standard error.

(b) Use the t procedures to give a 90% confidence interval for this patient's mean phosphate level.

7.24 Poisoning by the pesticide DDT causes tremors and convulsions. In a study of DDT poisoning, researchers fed several rats a measured amount of DDT. They then measured electrical characteristics of the rats' nervous systems that might explain how DDT poisoning causes tremors. One important variable was the "absolutely refractory period," the time required for a nerve to recover after a stimulus. This period varies normally. Measurements on four rats gave the data below (in milliseconds). (Data from D. L. Shankland, "Involvement of spinal cord and peripheral nerves in DDT-poisoning syndrome in albino rats," *Toxicology and Applied Pharmacology*, 6 (1964), pp. 97–213.)

1.6	1.7	1.8	1.9

(a) Find the mean refractory period $\bar{x}$ and the standard error of the mean.

(b) Give a 90% confidence interval for the mean "absolutely refractory period" for all rats of this strain when subjected to the same treatment.

7.25 The normal range of values for blood phosphate levels is 2.6 to 4.8 mg/dl. The sample mean for the patient in Exercise 7.23 falls above this range. Is this good evidence that the patient's mean level in fact falls above 4.8? State H_0 and H_a and use the data in Exercise 7.23 to carry out a t test. Between which levels from Table D does the P-value lie? Are you convinced that the patient's phosphate level is higher than normal?

7.26 Suppose that the mean "absolutely refractory period" for unpoisoned rats is known to be 1.3 milliseconds. DDT poisoning should slow nerve recovery and so increase this period. Do the data in Exercise 7.24 give good evidence for this supposition? State H_0 and H_a and do a t test. Between what levels from Table D does the P-value lie? What do you conclude from the test?

7.27 In a randomized comparative experiment on the effect of dietary calcium on blood pressure, 54 healthy white males were divided at random into two groups. One group received calcium; the other, a placebo. At the beginning of the study, the researchers measured many variables on the subjects. The paper reporting the study gives $\bar{x} = 114.9$ and $s = 9.3$ for the seated systolic blood pressure of the 27 members of the placebo group.

(a) Give a 95% confidence interval for the mean blood pressure of the population from which the subjects were recruited.

(b) What assumptions about the population and the study design are required by the procedure you used in (a)? Which of these assumptions are important for the validity of the procedure in this case?

7.28 The Acculturation Rating Scale for Mexican Americans (ARSMA) measures the extent to which Mexican Americans have adopted Anglo/English culture. During the development of ARSMA, the test was given to a group of 17 Mexicans. Their scores, from a possible range of 1.00 to 5.00, had $\bar{x} = 1.67$ and $s = 0.25$. Because low scores should indicate a Mexican cultural orientation, these results helped to establish the validity of the test. (Based on I. Cuellar, L. C. Harris, and R. Jasso, "An acculturation scale for Mexican American normal and clinical populations," *Hispanic Journal of Behavioral Sciences*, 2 (1980), pp. 199–217.)

(a) Give a 95% confidence interval for the mean ARSMA score of Mexicans.

(b) What assumptions does your confidence interval require? Which of these assumptions is most important in this case?

7.29 How accurate are radon detectors of a type sold to homeowners? To answer this question, university researchers placed 12 detectors in a chamber that exposed them to 105 picocuries per liter (pCi/l) of radon. The detector readings were as follows (data provided by Diana Schellenberg, Purdue University School of Health Sciences):

91.9	97.8	111.4	122.3	105.4	95.0
103.8	99.6	96.6	119.3	104.8	101.7

(a) Make a stemplot of the data. The distribution is somewhat skewed to the right, but not strongly enough to forbid use of the t procedures.

(b) Is there convincing evidence that the mean reading of all detectors of this type differs from the true value 105? Carry out a test in detail and write a brief conclusion.

7.30 Gas chromatography is a sensitive technique used by chemists to measure small amounts of compounds. The response of a gas chromatograph is calibrated by repeatedly testing specimens containing a known amount of the compound to be measured. A calibration study for a specimen containing 1 nanogram (ng) (that's 10^{-9} gram) of a compound gave the following response readings:

21.6	20.0	25.0	21.9

The response is known from experience to vary according to a normal distribution unless an outlier indicates an error in the analysis. Estimate the mean response to 1 ng of this substance, and give the margin of error for your choice of confidence level. Then explain to a chemist who knows no statistics what your margin of error means. (Data from the appendix of D. A. Kurtz (ed.), *Trace Residue Analysis*, American Chemical Society Symposium Series, no. 284, 1985.)

7.31 The embryos of brine shrimp can enter a dormant phase in which metabolic activity drops to a low level. Researchers studying this dormant phase measured the level of several compounds important to normal metabolism. The results were reported in a table, with the note, "Values are means ± SEM for three independent samples." The table entry for the compound ATP was 0.84 ± 0.01. Biologists reading the article are presumed to be able to decipher this. (From S. C. Hand and E. Gnaiger, "Anaerobic dormancy quantified in *Artemia* embryos," *Science*, 239 (1988), pp. 1425–1427.)

(a) What does the abbreviation "SEM" stand for?

(b) The researchers made three measurements of ATP, which had $\bar{x} = 0.84$. What was the sample standard deviation s for these measurements?

(c) Give a 90% confidence interval for the mean ATP level in dormant brine shrimp embryos.

7.32 The researchers studying vitamin C in CSB in Example 7.1 were also interested in a similar commodity called wheat soy blend (WSB). A major concern was the possibility that some of the vitamin C content would be destroyed as a result of storage and shipment of the commodity to its final destination. The researchers specially marked a collection of bags at the factory and took a sample from each of these to determine the vitamin C content. Five months later in Haiti they found the specially marked bags and took samples. The data consist of two vitamin C measures for each bag, one at the time of production in the factory and the other five months later in Haiti. The units are mg/100 g as in Example 7.1. Here are the data:

Factory	Haiti	Factory	Haiti	Factory	Haiti
44	40	45	38	39	43
50	37	32	40	52	38
48	39	47	35	45	38
44	35	40	38	37	38
42	35	38	34	38	41
47	41	41	35	44	40
49	37	43	37	43	35
50	37	40	34	39	38
39	34	37	40	44	36

(a) Set up hypotheses to examine the question of interest to these researchers.

(b) Perform the significance test and summarize your results.

(c) Find 95% confidence intervals for the mean at the factory, the mean five months later in Haiti, and for the change.

7.33   The design of controls and instruments has a large effect on how easily people can use them. A student project investigated this effect by asking 25 right-handed students to turn a knob (with their right hands) that moved an indicator by screw action. There were two identical instruments, one with a right-hand thread (the knob turns clockwise) and the other with a left-hand thread (the knob must be turned counterclockwise). The table below gives the times required (in seconds) to move the indicator a fixed distance. (Data provided by Timothy Sturm.)

Subject	Right thread	Left thread	Subject	Right thread	Left thread
1	113	137	14	107	87
2	105	105	15	118	166
3	130	133	16	103	146
4	101	108	17	111	123
5	138	115	18	104	135
6	118	170	19	111	112
7	87	103	20	89	93
8	116	145	21	78	76
9	75	78	22	100	116
10	96	107	23	89	78
11	122	84	24	85	101
12	103	148	25	88	123
13	116	147			

(a)   Each of the 25 students used both instruments. Discuss briefly how the experiment should be arranged and how randomization should be used.

(b)   The project hoped to show that right-handed people find right-hand threads easier to use. State the appropriate H_0 and H_a about the mean time required to complete the task.

(c)   Carry out a test of your hypotheses. Give the P-value and report your conclusions.

7.34   Give a 90% confidence interval for the mean time advantage of right-hand over left-hand threads in the setting of the previous exercise. Do you think that the time saved would be of practical importance if the task were performed many times—for example, by an assembly line worker? To help answer this question, find the mean time for right-hand threads as a percent of the mean time for left-hand threads.

7.35   The table below gives the pretest and posttest scores on the MLA listening test in Spanish for 20 high school Spanish teachers who attended an intensive summer course in Spanish. The setting is identical to the one described in Example 7.7. (Data provided by Joseph A. Wipf, Department of Foreign Languages and Literatures, Purdue University.)

Subject	Pretest	Posttest	Subject	Pretest	Posttest
1	30	29	11	30	32
2	28	30	12	29	28
3	31	32	13	31	34
4	26	30	14	29	32
5	20	16	15	34	32
6	30	25	16	20	27
7	34	31	17	26	28
8	15	18	18	25	29
9	28	33	19	31	32
10	20	25	20	29	32

(a) We hope to show that attending the institute improves listening skills. State an appropriate H_0 and H_a. Be sure to identify the parameters appearing in the hypotheses.

(b) Make a graphical check for outliers or strong skewness in the data that you will use in your statistical test, and report your conclusions on the validity of the test.

(c) Carry out a test. Can you reject H_0 at the 5% significance level? At the 1% significance level?

(d) Give a 90% confidence interval for the mean increase in listening score due to attending the summer institute.

7.36 The ARSMA test (Exercise 7.28) was compared with a similar test, the Bicultural Inventory (BI), by administering both tests to 22 Mexican Americans. Both tests have the same range of scores (1.00 to 5.00) and are scaled to have similar means for the groups used to develop them. There was a high correlation between the two scores, giving evidence that both are measuring the same characteristics. The researchers wanted to know whether the population mean scores for the two tests were the same. The differences in scores (ARSMA − BI) for the 22 subjects had $\bar{x} = 0.2519$ and $s = 0.2767$.

(a) Describe briefly how the administration of the two tests to the subjects should be conducted, including randomization.

(b) Carry out a significance test for the hypothesis that the two tests have the same population mean. Give the P-value and state your conclusion.

(c) Give a 95% confidence interval for the difference between the two population mean scores.

7.37 An agricultural field trial compares the yield of two varieties of tomatoes for commercial use. The researchers divide in half each of 10 small plots of land in different locations and plant each tomato variety on one half of each plot. After harvest, they compare the yields in pounds per plant at each location. The 10 differences (Variety A − Variety B) give the following statistics: $\bar{x} = 0.34$ and $s = 0.83$. Is there convincing evidence that Variety A has the higher mean yield? State H_0 and H_a, and give a P-value to answer this question.

7.38 The following situations all require inference about a mean or means. Identify each as (1) a single sample, (2) matched pairs, or (3) two independent samples. The procedures of this section apply to cases (1) and (2). We will learn procedures for (3) in the next section.

(a) An education researcher wants to learn whether inserting questions before or after introducing a new concept in an elementary school mathematics text is more effective. He prepares two text segments that teach the concept, one with motivating questions before and the other with review questions after. Each text segment is used to teach a different group of children, and their scores on a test over the material are compared.

(b) Another researcher approaches the same problem differently. She prepares text segments on two unrelated topics. Each segment comes in two versions, one with questions before and the other with questions after. Each of a group of children is taught both topics, one topic (chosen at random) with questions before and the other with questions after. Each child's test scores on the two topics are compared to see which topic he or she learned better.

(c) To evaluate a new analytical method, a chemist obtains a reference specimen of known concentration from the National Institute of Standards and Technology. She then makes 20 measurements of the concentration of this specimen with the new method and checks for bias by comparing the mean result with the known concentration.

(d) Another chemist is evaluating the same new method. He has no reference specimen, but a familiar analytic method is available. He wants to know if the new and old methods agree. He takes a specimen of unknown concentration and measures the concentration 10 times with the new method and 10 times with the old method.

7.39 Table 1.3 (page 31) gives the percent of residents 65 years of age and over in each of the 50 states. It does not make sense to use the t procedures (or any other statistical procedures) to give a 95% confidence interval for the mean percent of over-65 residents in the population of the American states. Explain why not.

The following exercises concern the optional material in the sections on the power of the t test and on nonnormal populations.

7.40 The bank in Exercise 7.21 tested a new idea on a sample of 250 customers. Suppose that the bank wanted to be quite certain of detecting a mean increase of $\mu = \$100$ in the amount charged, at the $\alpha = 0.01$ significance level. Perhaps a sample of only $n = 50$ customers would accomplish this. Find the approximate power of the test with $n = 50$ against the alternative $\mu = \$100$ as follows:

(a) What is the t critical value for the one-sided test with $\alpha = 0.01$ and $n = 50$?

(b) Write the criterion for rejecting $H_0: \mu = 0$ in terms of the t statistic. Then take $s = 108$ (an estimate based on the data in Exercise 7.21) and state the rejection criterion in terms of $\bar{x}$.

(c) Assume that $\mu = 100$ (the given alternative) and that $\sigma = 108$ (an estimate from the data in Exercise 7.21). The approximate power is the probability of the event you found in (b), calculated under these assumptions. Find the power. Would you recommend that the bank do a test on 50 customers, or should more customers be included?

7.41   The tomato experts who carried out the field trial described in Exercise 7.37 suspect that the relative lack of significance there is due to low power. They would like to be able to detect a mean difference in yields of 0.5 pound per plant at the 0.05 significance level. Based on the previous study, use 0.83 as an estimate of both the population σ and the value of s in future samples.

(a) What is the power of the test from Exercise 7.37 with $n = 10$ against the alternative $\mu = 0.5$?

(b) If the sample size is increased to $n = 25$ plots of land, what will be the power against the same alternative?

7.42   Exercise 7.36 reports a small study comparing ARSMA and BI, two tests of the acculturation of Mexican Americans. Would this study usually detect a difference in mean scores of 0.2? To answer this question, calculate the approximate power of the test (with $n = 22$ subjects and $\alpha = 0.05$) of

$$H_0: \mu = 0$$
$$H_a: \mu \neq 0$$

against the alternative $\mu = 0.2$. Note that this is a two-sided test.

(a) From Table D, what is the critical value for $\alpha = 0.05$?

(b) Write the criterion for rejecting H_0 at the $\alpha = 0.05$ level. Then take $s = 0.3$, the approximate value observed in Exercise 7.36, and restate the rejection criterion in terms of $\bar{x}$.

(c) Find the probability of this event when $\mu = 0.2$ (the alternative given) and $\sigma = 0.3$ (estimated from the data in Exercise 7.36) by a normal probability calculation. This is the approximate power.

7.43   Apply the sign test to the data in Exercise 7.33 to assess whether the subjects can complete a task with right-hand thread significantly faster than with left-hand thread.

(a) State the hypotheses two ways, in terms of a population median and in terms of the probability of completing the task faster with a right-hand thread.

(b) Carry out the sign test. Find the approximate P-value using the normal approximation to the binomial distributions, and report your conclusion.

7.44 Use the sign test to assess whether the summer institute of Exercise 7.35 improves Spanish listening skills. State the hypotheses, give the P-value using the binomial table, and report your conclusion.

7.45 The paper reporting the results on ARSMA used in Exercise 7.36 does not give the raw data or any discussion of normality. You would like to replace the t procedure used in Exercise 7.36 by a sign test. Can you do this from the available information? Carry out the sign test and state your conclusion, or explain why you are unable to carry out the test.

7.46 Table 1.5 (page 32) gives the survival times of 72 guinea pigs after they were injected with tubercle bacilli in a medical experiment. Figure 1.38 (page 92) is a normal quantile plot of these data. The distribution is strongly skewed to the right. (Data from T. Bjerkedal, "Acquisition of resistance in guinea pigs infected with different doses of virulent tubercle bacilli," *American Journal of Hygiene,* 72 (1960), pp. 130–148.)

(a) Give a 95% confidence interval for the mean survival time by applying the t procedures to these data.

(b) Transform the data by taking the logarithm of each value. Display the transformed data by either a histogram or a normal quantile plot. The distribution of the logarithms remains somewhat right-skewed but is much closer to symmetry than the original distribution. Probability values from the t distribution will be more accurate for the transformed data.

(c) Give a 95% confidence interval for the mean of the log survival time by applying the t procedures to the transformed data.

7.47 A manufacturer of electric motors tests insulation at a high temperature (250° C) and records the number of hours until the insulation fails. The data for 5 specimens are

300	324	372	372	444

(Data from Wayne Nelson, *Applied Life Data Analysis,* Wiley, New York, 1982, p. 471.) The small sample size makes judgment from the data difficult, but engineering experience suggests that the logarithm of the failure time will have a normal distribution. Take the logarithms of the 5 observations, and use t procedures to give a 90% confidence interval for the mean of the log failure time for insulation of this type.

7.2 Comparing Two Means

A medical researcher is interested in the effect on blood pressure of added calcium in our diet. She conducts a randomized comparative experiment in which one group of subjects receives a calcium supplement and a control group gets a placebo. A psychologist develops a test that measures social insight. He compares the social insight of male college students with that of female college students by giving the test to a large group of students of each sex. A bank wants to know which of two incentive plans will most increase the use of its credit cards. It offers each incentive to a random sample of credit card customers and compares the amount charged during the following six months. Two-sample problems such as these are among the most common situations encountered in statistical practice.

Two-Sample Problems

- The goal of inference is to compare the responses in two groups.
- Each group is considered to be a sample from a distinct population.
- The responses in each group are independent of those in the other group.

A two-sample problem can arise from a randomized comparative experiment that randomly divides the subjects into two groups and exposes each group to a different treatment. Comparing random samples separately selected from two populations is also a two-sample problem. Unlike the matched pairs designs studied earlier, there is no matching of the units in the two samples, and the two samples may be of different sizes. Inference procedures for two-sample data differ from those for matched pairs.

We can present two-sample data graphically by a back-to-back stemplot (for small samples) or by side-by-side boxplots (for larger samples). Now we will apply the ideas of formal inference in this setting. When both population distributions are symmetric, and especially when they are at least approximately normal, a comparison of the mean responses in the two populations is most often the goal of inference.

We have two independent samples, from two distinct populations (such as subjects given a treatment and those given a placebo). The same variable is measured for both samples. We will call the variable x_1 in the first population and x_2 in the second because the variable may have different distributions in the two populations. Here is the notation that we will use to describe the two populations:

Population	Variable	Mean	Standard deviation
1	x_1	μ_1	σ_1
2	x_2	μ_2	σ_2

We want to compare the two population means, either by giving a confidence interval for $\mu_1 - \mu_2$ or by testing the hypothesis of no difference, H_0: $\mu_1 = \mu_2$.

Inference is based on two independent SRSs, one from each population. Here is the notation that describes the samples:

Population	Sample size	Sample mean	Sample standard deviation
1	n_1	$\bar{x}_1$	s_1
2	n_2	$\bar{x}_2$	s_2

Throughout this section, the subscripts 1 and 2 show the population to which a parameter or a sample statistic refers.

The two-sample z statistic

The natural estimator of the difference $\mu_1 - \mu_2$ is the difference between the sample means, $\bar{x}_1 - \bar{x}_2$. If we are to base inference on this statistic, we must know its sampling distribution. Our knowledge of probability is equal to the task. First, the mean of the difference $\bar{x}_1 - \bar{x}_2$ is the difference of the means $\mu_1 - \mu_2$. This follows from the addition rule for means (page 334) and the fact that the mean of any $\bar{x}$ is the same as the mean of the population. Because the samples are independent, their sample means $\bar{x}_1$ and $\bar{x}_2$ are independent random variables. The addition rule for variances (page 337) says that the variance of the difference $\bar{x}_1 - \bar{x}_2$ is the sum of their variances, which is

$$\frac{\sigma_1^2}{n_1} + \frac{\sigma_2^2}{n_2}$$

We now know the mean and variance of the distribution of $\bar{x}_1 - \bar{x}_2$ in terms of the parameters of the two populations. If the two population distributions are both normal, then the distribution of $\bar{x}_1 - \bar{x}_2$ is also normal. This is true because each sample mean alone is normally distributed and because a difference of independent normal random variables is also normal.

EXAMPLE 7.13 | A fourth-grade class has 10 girls and 7 boys. The children's heights are recorded on their tenth birthdays. What is the chance that the girls are taller than the boys? Of course, it is very unlikely that all of the girls are taller than all of the boys. We translate the question into the following: what is the probability that the mean height of the girls is greater than the mean height of the boys?

Based on information from the National Center for Health Statistics, we assume that the heights (in inches) of 10-year-old girls are $N(54.5, 2.7)$ and the heights of 10-year-old boys are $N(54.1, 2.4)$. The heights of the students in our class are assumed to be random samples from these populations. The two distributions are shown in Figure 7.11(a).

The difference $\bar{x}_1 - \bar{x}_2$ between the female and male mean heights varies in different random samples. The sampling distribution has mean

$$\mu_1 - \mu_2 = 54.5 - 54.1 = 0.4 \text{ inch}$$

and variance

$$\frac{\sigma_1^2}{n_1} + \frac{\sigma_2^2}{n_2} = \frac{2.7^2}{10} + \frac{2.4^2}{7} = 1.55$$

The standard deviation of the difference in sample means is therefore $\sqrt{1.55} = 1.25$ inches.

If the heights vary normally, the difference in sample means is also normally distributed. The distribution of the difference in heights is shown in Figure 7.11(b). We standardize $\bar{x}_1 - \bar{x}_2$ by subtracting its mean (0.4) and dividing by its standard deviation (1.25). Therefore, the probability that the girls are taller than the boys is

$$P(\bar{x}_1 - \bar{x}_2 > 0) = P\left(\frac{(\bar{x}_1 - \bar{x}_2) - 0.4}{1.25} > \frac{0 - 0.4}{1.25}\right) = P(Z > -0.32) = 0.63$$

Even though the population mean height of 10-year-old girls is greater than the population mean height of 10-year-old boys, the probability that the sample mean of the girls is greater than the sample mean of the boys in our class is only 63%. Large samples are needed to see the effects of small differences.

As Example 7.13 reminds us, any normal random variable has the $N(0, 1)$ distribution when standardized. We have arrived at a new z statistic.

Two-Sample z Statistic

Suppose that $\bar{x}_1$ is the mean of an SRS of size n_1 drawn from an $N(\mu_1, \sigma_1)$ population and that $\bar{x}_2$ is the mean of an independent SRS of size n_2 drawn from an $N(\mu_2, \sigma_2)$ population. Then the **two-sample z statistic**

$$z = \frac{(\bar{x}_1 - \bar{x}_2) - (\mu_1 - \mu_2)}{\sqrt{\dfrac{\sigma_1^2}{n_1} + \dfrac{\sigma_2^2}{n_2}}}$$

has the standard normal $N(0, 1)$ sampling distribution.

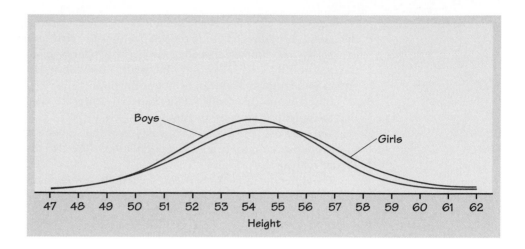

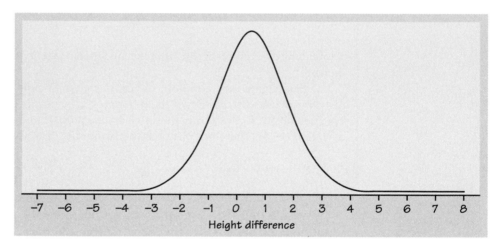

FIGURE 7.11 Distributions for Example 7.13. (a) Distributions of heights of 10-year-old boys and girls. (b) Distribution of the difference between means of 10 girls and 7 boys.

In the unlikely event that both population standard deviations are known, the two-sample z statistic is the basis for inference about $\mu_1 - \mu_2$. Exact z procedures are seldom used, however, because σ_1 and σ_2 are rarely known. In Chapter 6, we discussed the one-sample z procedures in order to introduce the ideas of inference. Now we pass immediately to the more useful t procedures.

The two-sample t procedures

Suppose now that the population standard deviations σ_1 and σ_2 are not known. We estimate them by the sample standard deviations s_1 and s_2 from our two samples. Following the pattern of the one-sample case, we substitute the standard errors $s_i / \sqrt{n_i}$ for the standard deviations $\sigma_i / \sqrt{n_i}$ used in the two-sample z statistic. The result is the *two-sample t statistic*:

$$t = \frac{(\bar{x}_1 - \bar{x}_2) - (\mu_1 - \mu_2)}{\sqrt{\dfrac{s_1^2}{n_1} + \dfrac{s_2^2}{n_2}}}$$

Unfortunately, this statistic does *not* have a t distribution. A t distribution replaces a $N(0, 1)$ distribution only when a single standard deviation (σ) in a z statistic is replaced by a standard error (s). In this case, we replaced two standard deviations (σ_1 and σ_2) by their estimates (s_1 and s_2), which does not produce a statistic having a t distribution.

approximations for the degrees of freedom

Nonetheless, we can approximate the distribution of the two-sample t statistic by using the $t(k)$ distribution with an **approximation for the degrees of freedom k.** We use these approximations to find approximate values of t^* for confidence intervals and to find approximate P-values for significance tests. Here are two approximations:

1. Use a value of k that is calculated from the data. In general it will not be a whole number.

2. Use k equal to the smaller of $n_1 - 1$ and $n_2 - 1$.

Most statistical software uses the first option to approximate the $t(k)$ distribution for two-sample problems unless the user requests another method. Use of this approximation without software is a bit complicated; we will give the details later in this section. If you are not using software, the second approximation is preferred. This approximation is appealing because it is conservative.[6] Margins of error for confidence intervals are a bit larger than they need to be, so the true confidence level is larger than C. True P-values are a bit smaller than we calculate, so we understate the evidence against the null hypothesis. In practice, the choice of approximation almost never makes a difference in our conclusion.

The two-sample t significance test

The Two-Sample t Significance Test

Suppose that an SRS of size n_1 is drawn from a normal population with unknown mean μ_1 and that an independent SRS of size n_2 is drawn from another normal population with unknown mean μ_2. To test the hypothesis H_0: $\mu_1 = \mu_2$, compute the **two-sample t statistic**

$$t = \frac{\bar{x}_1 - \bar{x}_2}{\sqrt{\dfrac{s_1^2}{n_1} + \dfrac{s_2^2}{n_2}}}$$

and use P-values or critical values for the $t(k)$ distribution, where the degrees of freedom k are either approximated by software or are the smaller of $n_1 - 1$ and $n_2 - 1$.

EXAMPLE 7.14

An educator believes that new directed reading activities in the classroom will help elementary school pupils improve some aspects of their reading ability. She arranges for a third-grade class of 21 students to take part in these activities for an eight-week period. A control classroom of 23 third graders follows the same curriculum without the activities. At the end of the eight weeks, all students are given a Degree of Reading Power (DRP) test, which measures the aspects of reading ability that the treatment is designed to improve. The data appear in Table 7.2.[7]

First examine the data.

```
    Control  |   | Treatment
        9 7 0 | 1 |
        8 6 0 | 2 | 4
        7 7 3 | 3 | 3
  8 6 3 2 2 2 1 | 4 | 3 3 3 4 6 9 9
      5 5 4 3 | 5 | 2 3 4 6 7 7 8 9
        2 0 | 6 | 1 2 7
            | 7 | 1
          5 | 8 |
```

A back-to-back stemplot suggests that there is a mild outlier in the control group but no deviation from normality serious enough to forbid use of t procedures. Separate normal quantile plots for both groups (Figure 7.12) confirm that both are approximately normal. The scores of the treatment group appear to be somewhat higher than those of the control group. The summary statistics are

Group	n	$\bar{x}$	s
Treatment	21	51.48	11.01
Control	23	41.52	17.15

Because we hope to show that the treatment (Group 1) is better than the control (Group 2), the hypotheses are

$$H_0: \mu_1 = \mu_2$$
$$H_a: \mu_1 > \mu_2$$

The two-sample t test statistic is

$$t = \frac{\bar{x}_1 - \bar{x}_2}{\sqrt{\dfrac{s_1^2}{n_1} + \dfrac{s_2^2}{n_2}}}$$

$$= \frac{51.48 - 41.52}{\sqrt{\dfrac{11.01^2}{21} + \dfrac{17.15^2}{23}}} = 2.31$$

The P-value for the one-sided test is $P(T \geq 2.31)$. Software gives the approximate P-value as 0.0132 and uses 37.9 as the degrees of freedom. For the second approximation, the degrees of freedom k are equal to the smaller of

$$n_1 - 1 = 21 - 1 = 20 \quad \text{and} \quad n_2 - 1 = 23 - 1 = 22$$

Comparing 2.31 with the entries in Table D for 20 degrees of freedom, we see that P lies between 0.02 and 0.01. The data strongly suggest that directed reading activity improves the DRP score ($t = 2.31$, df $= 20$, $P < 0.02$).

df = 20

p	0.02	0.01
t^*	2.197	2.528

TABLE 7.2   DRP scores for third graders

Treatment group				Control group			
24	61	59	46	42	33	46	37
43	44	52	43	43	41	10	42
58	67	62	57	55	19	17	55
71	49	54		26	54	60	28
43	53	57		62	20	53	48
49	56	33		37	85	42	

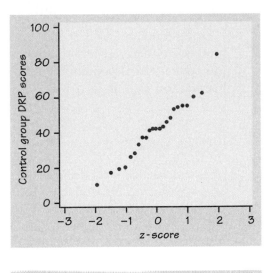

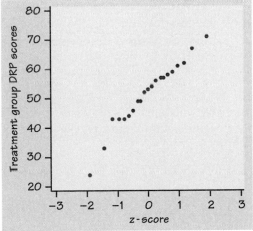

FIGURE 7.12   Normal quantile plots for the DRP scores in Table 7.2.

If your software gives only P-values for the two-sided alternative, $2P(T \geq |t|)$, you need to divide the reported value by 2 after checking that the means differ in the direction specified by the alternative hypothesis.

The two-sample t confidence interval

The same ideas that we used for the two-sample t significance tests also apply to give us *two-sample t confidence intervals*. We can use either software or the conservative approach with Table D to approximate the value of t^*.

The Two-Sample t Confidence Interval

Suppose that an SRS of size n_1 is drawn from a normal population with unknown mean μ_1 and that an independent SRS of size n_2 is drawn from another normal population with unknown mean μ_2. The **confidence interval for $\mu_1 - \mu_2$** given by

$$(\bar{x}_1 - \bar{x}_2) \pm t^* \sqrt{\frac{s_1^2}{n_1} + \frac{s_2^2}{n_2}}$$

has confidence level at least C no matter what the population standard deviations may be. Here, t^* is the value for the $t(k)$ density curve with area C between $-t*$ and $t*$. The value of the degrees of freedom k is either approximated by software or we use the smaller of $n_1 - 1$ and $n_2 - 1$.

To complete analysis of the DRP scores we examined in Example 7.14, we need to describe the size of the treatment effect. We do this with a confidence interval for the difference between the treatment group and the control group means.

EXAMPLE 7.15

We will find a 95% confidence interval for the mean improvement in the entire population of third graders. The interval is

$$(\bar{x}_1 - \bar{x}_2) \pm t^* \sqrt{\frac{s_1^2}{n_1} + \frac{s_2^2}{n_2}} = (51.48 - 41.52) \pm t^* \sqrt{\frac{11.01^2}{21} + \frac{17.15^2}{23}} = 9.96 \pm 4.31 t^*$$

Using software, the degrees of freedom are 37.9 and $t^* = 2.025$. This approximation gives

$$9.96 \pm 4.31 \times 2.025 = 9.96 \pm 8.72 = (1.2, 18.7)$$

The conservative approach uses the $t(20)$ distribution. Table D gives $t^* = 2.086$. With this approximation we have

$$9.96 \pm 4.31 \times 2.086 = 9.96 \pm 8.99$$
$$= (1.0, 18.9)$$

We can see that the conservative approach does, in fact, give a larger interval than the more accurate approximation used by software. However, the difference is rather small.

We estimate the mean improvement to be about 10 points, but with a margin of error of almost 9 points. Although we have good evidence of some improvement, the data do not allow a very precise estimate of the size of the average improvement.

The design of the study in Example 7.14 is not ideal. Random assignment of students was not possible in a school environment, so existing third-grade classes were used. The effect of the reading programs is therefore confounded with any other differences between the two classes. The classes were chosen to be as similar as possible—for example, in the social and economic status of the students. Extensive pretesting showed that the two classes were on the average quite similar in reading ability at the beginning of the experiment. To avoid the effect of two different teachers, the researcher herself taught reading in both classes during the eight-week period of the experiment. We can therefore be somewhat confident that the two-sample test is detecting the effect of the treatment and not some other difference between the classes. This example is typical of many situations in which an experiment is carried out but randomization is not possible.

Robustness of the two–sample procedures

The two-sample t procedures are more robust than the one-sample t methods. When the sizes of the two samples are equal and the distributions of the two populations being compared have similar shapes, probability values from the t table are quite accurate for a broad range of distributions when the sample sizes are as small as $n_1 = n_2 = 5$.[8] When the two population distributions have different shapes, larger samples are needed. The guidelines given on page 516 for the use of one-sample t procedures can be adapted to two-sample procedures by replacing "sample size" with the "sum of the sample sizes" $n_1 + n_2$. These guidelines are rather conservative, especially when the two samples are of equal size. In planning a two-sample study, you should usually choose equal sample sizes. The two-sample t procedures are most robust against nonnormality in this case, and the conservative probability values are most accurate.

Here is an example with large sample sizes. Even if the distributions are not normal, we are confident that the sample means will be approximately normal. The two-sample t test is very robust in this case.

EXAMPLE 7.16 | The Chapin Social Insight Test is a psychological test designed to measure how accurately the subject appraises other people. The possible scores on the test range from 0 to 41. During the development of the Chapin test, it was given to several different groups of people. Here are the results for male and female college students majoring in the liberal arts:[9]

Group	Sex	n	$\bar{x}$	s
1	Male	133	25.34	5.05
2	Female	162	24.94	5.44

Do these data support the contention that female and male students differ in average social insight? Because no specific direction for the male/female difference was hypothesized before looking at the data, we choose a two-sided alternative. The hypotheses are

$$H_0: \mu_1 = \mu_2$$
$$H_a: \mu_1 \neq \mu_2$$

The two-sample t statistic is

$$t = \frac{\bar{x}_1 - \bar{x}_2}{\sqrt{\frac{s_1^2}{n_1} + \frac{s_2^2}{n_2}}}$$

$$= \frac{25.34 - 24.94}{\sqrt{\frac{(5.05)^2}{133} + \frac{(5.44)^2}{162}}}$$

$$= 0.654$$

df = 100

p	0.25	0.20
t^*	0.677	0.845

Using the conservative approach, we find the P-value by comparing 0.654 to critical values for the $t(132)$ distribution and then doubling p because the alternative is two-sided. Table D (for 100 degrees of freedom) shows that 0.654 does not reach the 0.25 critical value, which is the largest entry for p in this table. The P-value is therefore greater than 0.50. The data give no evidence of a gender difference in mean social insight score ($t = 0.654$, df = 132, $P > 0.50$). Software uses the $t(289)$ distribution and gives $P = 0.51$.

The researcher in Example 7.16 did not do an experiment but compared samples from two populations. The large samples imply that the assumption that the populations have normal distributions is of little importance. The sample means will be nearly normal in any case. The major question concerns the population to which the conclusions apply. The student subjects are certainly not an SRS of all liberal arts majors. If they are volunteers from a single college, the sample results may not extend to a wider population.

Inference for small samples

Small samples require special care. We do not have enough observations for boxplots or normal quantile plots, and only the most extreme outliers stand out. Because we have so few observations, we are very reluctant to discard any of them. Furthermore, the power of our significance test tends to be low, and the margin of error of our confidence interval tends to be large. Despite all of these difficulties, we can often draw important conclusions from studies with small sample sizes. If the size of an effect is large, it will be evident even if the n's are small.

EXAMPLE 7.17 | The pesticide DDT causes tremors and convulsions if it is ingested by humans or other mammals. Researchers seek to understand how the convulsions are caused. In a randomized comparative experiment, 6 white rats poisoned with DDT were compared with a control group of 6 unpoisoned rats. Electrical measurements of nerve activity are the main clue to the nature of DDT poisoning. When a nerve is stimulated, its electrical response shows a sharp spike followed by a much smaller second spike. Researchers found that the second spike is larger in rats fed DDT than in normal rats. This observation helps biologists understand how DDT causes tremors.[10]

The researchers measured the amplitude of the second spike as a percentage of the first spike when a nerve in the rat's leg was stimulated. For the poisoned rats the results were

12.207	16.869	25.050	22.429	8.456	20.589

The control group data were

11.074	9.686	12.064	9.351	8.182	6.642

Normal quantile plots (Figure 7.13) show no evidence of outliers or strong skewness. Both populations are reasonably normal, as far as can be judged from six observations. The difference in means is quite large, but in such small samples the sample mean is highly variable. A significance test can help confirm that we are seeing a real effect. Because the researchers did not conjecture in advance that the size of the second spike would increase in rats fed DDT, we test

$$H_0: \mu_1 = \mu_2$$
$$H_a: \mu_1 \neq \mu_2$$

Here is the output from the SAS statistical software system for these data:

```
                          TTEST PROCEDURE

Variable: SPIKE

GROUP       N           Mean            Std Dev           Std Error
---------------------------------------------------------------------
DDT         6       17.60000000        6.34014839         2.58835474
CONTROL     6        9.49983333        1.95005932         0.79610839


Variances        T        DF       Prob>|T|
--------------------------------------------------
Unequal       2.9912      5.9        0.0247
Equal         2.9912     10.0        0.0135
```

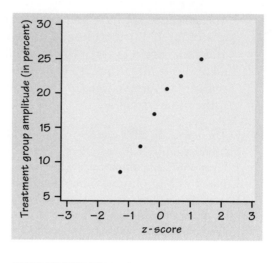

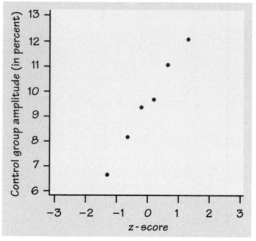

FIGURE 7.13 Normal quantile plots for the amplitude of the second spike as a percent of the first spike in Example 7.17.

SAS reports the results of two t procedures: the general two-sample procedure ("Unequal" variances) and a special procedure that assumes the two population variances to be equal. For now, we are interested in the first of these procedures. We will discuss the second in the section on pooled two-sample t procedures. The two-sample t statistic has the value $t = 2.9912$, the approximate degrees of freedom are df = 5.9, and the P-value from the $t(5.9)$ distribution is 0.0247. There is good evidence that the mean size of the secondary spike is larger in rats fed DDT.

Would the conservative test based on 5 degrees of freedom (both $n_1 - 1$ and $n_2 - 1$ are 5) have given a different result in this example? The conservative P-value is $2P(T \geq 2.9912)$ where T has the $t(5)$ distribution. Table D shows that 2.9912 lies between the values for $p = 0.02$ and 0.01 for the $t(5)$ distribution, so P for the two-sided test lies between 0.02 and 0.04. For practical purposes this is the same result as that given by the software.

Software approximation for the degrees of freedom*

We noted earlier that the two-sample t statistic does not have an exact t distribution. Moreover, the exact distribution changes as the unknown population standard deviations σ_1 and σ_2 change. However, the distribution can be approximated by a t distribution with degrees of freedom given by

$$df = \frac{\left(\dfrac{s_1^2}{n_1} + \dfrac{s_2^2}{n_2}\right)^2}{\dfrac{1}{n_1 - 1}\left(\dfrac{s_1^2}{n_1}\right)^2 + \dfrac{1}{n_2 - 1}\left(\dfrac{s_2^2}{n_2}\right)^2}$$

This is the approximation used by most statistical software. It is quite accurate when both sample sizes n_1 and n_2 are 5 or larger.

EXAMPLE 7.18 | For the DRP study of Example 7.14, the following table summarizes the data:

Group	n	$\overline{x}$	s
1	21	51.48	11.01
2	23	41.52	17.15

For greatest accuracy, we will use critical points from the t distribution with degrees of freedom df given by the equation above.

$$df = \frac{\left(\dfrac{11.01^2}{21} + \dfrac{17.15^2}{23}\right)^2}{\dfrac{1}{20}\left(\dfrac{11.01^2}{21}\right)^2 + \dfrac{1}{22}\left(\dfrac{17.15^2}{23}\right)^2}$$

$$= \frac{344.486}{9.099} = 37.86$$

This is the value that we reported in Examples 7.14 and 7.15 where we gave the results produced by software.

The number df given by the above approximation is always at least as large as the smaller of $n_1 - 1$ and $n_2 - 1$. On the other hand, df is never larger than the sum $n_1 + n_2 - 2$ of the two individual degrees of freedom. The number of degrees of freedom df is generally not a whole number. There is a t distribution with any positive degrees of freedom, even though Table D contains entries only for whole-number degrees of freedom. When df is small and is not a whole number, interpolation between entries in Table D may be needed

---

*This material can be omitted unless you are using statistical software and wish to understand what the software does.

to obtain an accurate critical value or P-value. Because of this and the need to calculate df, we do not recommend regular use of this approximation if a computer is not doing the arithmetic. With a computer, however, the more accurate procedures are painless.

The pooled two-sample t procedures*

There is one situation in which a t statistic for comparing two means has exactly a t distribution. Suppose that the two normal population distributions have the *same* standard deviation. In this case we need only substitute a single standard error in a z statistic, and the resulting t statistic has a t distribution. We will develop the z statistic first, as usual, and from it the t statistic.

Call the common—and still unknown—standard deviation of both populations σ. Both sample variances s_1^2 and s_2^2 estimate σ^2. The best way to combine these two estimates is to average them with weights equal to their degrees of freedom. This gives more weight to the information from the larger sample, which is reasonable. The resulting estimator of σ^2 is

$$s_p^2 = \frac{(n_1 - 1)s_1^2 + (n_2 - 1)s_2^2}{n_1 + n_2 - 2}$$

pooled estimator of σ^2

This is called the **pooled estimator of σ^2** because it combines the information in both samples.

When both populations have variance σ^2, the addition rule for variances says that $\bar{x}_1 - \bar{x}_2$ has variance equal to the *sum* of the individual variances, which is

$$\frac{\sigma^2}{n_1} + \frac{\sigma^2}{n_2} = \sigma^2 \left(\frac{1}{n_1} + \frac{1}{n_2} \right)$$

The standardized difference of means in this equal-variance case is therefore

$$z = \frac{(\bar{x}_1 - \bar{x}_2) - (\mu_1 - \mu_2)}{\sigma \sqrt{\dfrac{1}{n_1} + \dfrac{1}{n_2}}}$$

This is a special two-sample z statistic for the case in which the populations have the same σ. Replacing the unknown σ by the estimate s_p gives a t statistic. The degrees of freedom are $n_1 + n_2 - 2$, the sum of the degrees of freedom of the two sample variances. This statistic is the basis of the pooled two-sample t inference procedures.

---

*This section can be omitted if desired, but it should be read if you plan to read Chapters 12 and 13.

The Pooled Two-Sample t Procedures

Suppose that an SRS of size n_1 is drawn from a normal population with unknown mean μ_1 and that an independent SRS of size n_2 is drawn from another normal population with unknown mean μ_2. Suppose also that the two populations have the same standard deviation. A level C confidence interval for $\mu_1 - \mu_2$ is

$$(\bar{x}_1 - \bar{x}_2) \pm t^* s_p \sqrt{\frac{1}{n_1} + \frac{1}{n_2}}$$

Here t^* is the value for the $t(n_1 + n_2 - 2)$ density curve with area C between $-t^*$ and t^*.

To test the hypothesis H_0: $\mu_1 = \mu_2$, compute the pooled two-sample t statistic

$$t = \frac{\bar{x}_1 - \bar{x}_2}{s_p \sqrt{\dfrac{1}{n_1} + \dfrac{1}{n_2}}}$$

In terms of a random variable T having the $t(n_1 + n_2 - 2)$ distribution, the P-value for a test of H_0 against

$$H_a: \mu_1 > \mu_2 \quad \text{is} \quad P(T \geq t)$$
$$H_a: \mu_1 < \mu_2 \quad \text{is} \quad P(T \leq t)$$
$$H_a: \mu_1 \neq \mu_2 \quad \text{is} \quad 2P(T \geq |t|)$$

EXAMPLE 7.19

Does increasing the amount of calcium in our diet reduce blood pressure? Examination of a large sample of people revealed a relationship between calcium intake and blood pressure, but such observational studies do not establish causation. Animal experiments then showed that calcium supplements do reduce blood pressure in rats, justifying an experiment with human subjects. A randomized comparative experiment gave one group of 10 black men a calcium supplement for 12 weeks. The control group of 11 black men received a placebo that appeared identical. (In fact, a block design with black and white men as the blocks was used. We will look only at the results for blacks, because the earlier survey suggested that calcium is more effective for blacks.) The experiment was double-blind. Table 7.3 gives the seated systolic (heart contracted) blood pressure for all subjects at the beginning and end of the 12-week period, in millimeters (mm) of mercury. Because the researchers were interested in decreasing blood pressure, Table 7.3 also shows the decrease for each subject. An increase appears as a negative entry.[11]

As usual, we first examine the data. To compare the effects of the two treatments, take the response variable to be the amount of the decrease in blood pressure. Inspection of the data reveals that there are no outliers. Normal quantile plots (Figure 7.14) give a more detailed picture. The calcium group has a somewhat short left tail, but there are no departures from normality that will prevent use of t procedures. To examine the question of the researchers who collected these data, we perform a significance test.

TABLE 7.3   Seated systolic blood pressure

Calcium group			Placebo group		
Begin	End	Decrease	Begin	End	Decrease
107	100	7	123	124	−1
110	114	−4	109	97	12
123	105	18	112	113	−1
129	112	17	102	105	−3
112	115	−3	98	95	3
111	116	−5	114	119	−5
107	106	1	119	114	5
112	102	10	112	114	−2
136	125	11	110	121	−11
102	104	−2	117	118	−1
			130	133	−3

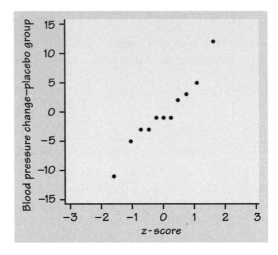

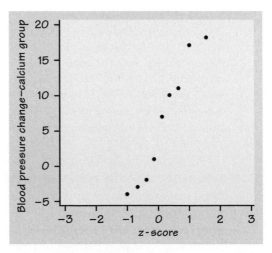

FIGURE 7.14   Normal quantile plots for the decrease in blood pressure from Table 7.3.

EXAMPLE 7.20

Take Group 1 to be the calcium group and Group 2 to be the placebo group. The evidence that calcium lowers blood pressure more than a placebo is assessed by testing

$$H_0: \mu_1 = \mu_2$$
$$H_a: \mu_1 > \mu_2$$

Here are the summary statistics for the decrease in blood pressure:

Group	Treatment	n	$\overline{x}$	s
1	Calcium	10	5.000	8.743
2	Placebo	11	−0.273	5.901

The calcium group shows a drop in blood pressure, and the placebo group has a small increase. The sample standard deviations do not rule out equal population standard deviations. A difference this large will often arise by chance in samples this small. We are willing to assume equal population standard deviations. The pooled sample variance is

$$s_p^2 = \frac{(n_1 - 1)s_1^2 + (n_2 - 1)s_2^2}{n_1 + n_2 - 2}$$
$$= \frac{(9)(8.743)^2 + (10)(5.901)^2}{10 + 11 - 2} = 54.536$$

so that

$$s_p = \sqrt{54.536} = 7.385$$

The pooled two-sample t statistic is

$$t = \frac{\overline{x}_1 - \overline{x}_2}{s_p\sqrt{\dfrac{1}{n_1} + \dfrac{1}{n_2}}} = \frac{5.000 - (-0.273)}{7.385\sqrt{\dfrac{1}{10} + \dfrac{1}{11}}}$$
$$= \frac{5.273}{3.227} = 1.634$$

df = 19

p	0.10	0.05
t^*	1.328	1.729

The P-value is $P(T \geq 1.634)$, where T has the $t(19)$ distribution. From Table D we can see that P falls between the $\alpha = 0.10$ and $\alpha = 0.05$ levels. Statistical software gives the exact value $P = 0.059$. The experiment found evidence that calcium reduces blood pressure, but the evidence falls a bit short of the traditional 5% and 1% levels.

Sample size strongly influences the P-value of a test. An effect that fails to be significant at a specified level α in a small sample can be significant in a larger sample. In the light of the rather small samples in Example 7.20, the evidence for some effect of calcium on blood pressure is rather good. The published account of the study combined these results for blacks with the results for whites and adjusted for pretest differences among the subjects. Using this more detailed analysis, the researchers were able to report the P-value $P = 0.008$.

Of course, a P-value is almost never the last part of a statistical analysis. To make a judgment regarding the size of the effect of calcium on blood pressure, we need a confidence interval.

EXAMPLE 7.21

We estimate that the effect of calcium supplementation is the difference between the sample means of the calcium and the placebo groups, $\bar{x}_1 - \bar{x}_2 = 5.273$ mm. A 90% confidence interval for $\mu_1 - \mu_2$ uses the critical value $t^* = 1.729$ from the $t(19)$ distribution. The interval is

$$(\bar{x}_1 - \bar{x}_2) \pm t^* s_p \sqrt{\frac{1}{n_1} + \frac{1}{n_2}} = [5.000 - (-0.273)] \pm (1.729)(7.385)\sqrt{\frac{1}{10} + \frac{1}{11}}$$

$$= 5.273 \pm 5.579$$

We are 90% confident that the difference in means is in the interval $(-0.306, 10.852)$. The calcium treatment reduced blood pressure by about 5.3 mm more than a placebo on the average, but the margin of error for this estimate is 5.6 mm.

The pooled two-sample t procedures are anchored in statistical theory and so have long been the standard version of the two-sample t in textbooks. But they require the assumption that the two unknown population standard deviations are equal. As we shall see in Section 7.3, this assumption is hard to verify. The pooled t procedures are therefore a bit risky. They are reasonably robust against both nonnormality and unequal standard deviations when the sample sizes are nearly the same. When the samples are quite different in size, the pooled t procedures become sensitive to unequal standard deviations and should be used with caution unless the samples are large. Unequal standard deviations are quite common. In particular, it is common for the spread of data to increase when the center moves up, as happened in Example 7.17. Statistical software usually calculates both the pooled and the unpooled t statistics, as in Example 7.17. The very unequal sample standard deviations in that example suggest that we should not use the pooled procedures.

SUMMARY

Significance tests and confidence intervals for the difference of the means μ_1 and μ_2 of two normal populations are based on the difference $\bar{x}_1 - \bar{x}_2$ of the sample means from two independent SRSs. Because of the central limit theorem, the resulting procedures are approximately correct for other population distributions when the sample sizes are large.

When independent SRSs of sizes n_1 and n_2 are drawn from two normal populations with parameters μ_1, σ_1 and μ_2, σ_2 the **two-sample z statistic**

$$z = \frac{(\bar{x}_1 - \bar{x}_2) - (\mu_1 - \mu_2)}{\sqrt{\dfrac{\sigma_1^2}{n_1} + \dfrac{\sigma_2^2}{n_2}}}$$

has the $N(0, 1)$ distribution.

The **two-sample t statistic**

$$t = \frac{(\bar{x}_1 - \bar{x}_2) - (\mu_1 - \mu_2)}{\sqrt{\dfrac{s_1^2}{n_1} + \dfrac{s_2^2}{n_2}}}$$

does *not* have a t distribution. However, good approximations are available.

Conservative inference procedures for comparing μ_1 and μ_2 are obtained from the two-sample t statistic by using the $t(k)$ distribution with degrees of freedom k equal to the smaller of $n_1 - 1$ and $n_2 - 1$.

More accurate probability values can be obtained by estimating the degrees of freedom from the data. This is the usual procedure for statistical software.

An approximate C level **confidence interval** for $\mu_1 - \mu_2$ is given by

$$(\bar{x}_1 - \bar{x}_2) \pm t^* \sqrt{\frac{s_1^2}{n_1} + \frac{s_2^2}{n_2}}$$

Here, t^* is the value for the $t(k)$ density curve with area C between $-t^*$ and t^*, where k is computed from the data by software or is the smaller of $n_1 - 1$ and $n_2 - 1$. The quantity

$$t^* \sqrt{\frac{s_1^2}{n_1} + \frac{s_2^2}{n_2}}$$

is the **margin of error.**

Significance tests for H_0: $\mu_1 = \mu_2$ use the **two-sample t statistic**

$$t = \frac{\bar{x}_1 - \bar{x}_2}{\sqrt{\dfrac{s_1^2}{n_1} + \dfrac{s_2^2}{n_2}}}$$

The P-value is approximated using the $t(k)$ where k is estimated from the data using software or is the smaller of $n_1 - 1$ and $n_2 - 1$.

The guidelines for practical use of two-sample t procedures are similar to those for one-sample t procedures. Equal sample sizes are recommended.

If we can assume that the two populations have equal variances, **pooled two-sample t procedures** can be used. These are based on the **pooled estimator**

$$s_p^2 = \frac{(n_1 - 1)s_1^2 + (n_2 - 1)s_2^2}{n_1 + n_2 - 2}$$

of the unknown common variance and the $t(n_1 + n_2 - 2)$ distribution.

SECTION 7.2 EXERCISES

In exercises that call for two-sample t procedures, you may use either of the two approximations for the degrees of freedom that we have discussed: the value given by your software or the smaller of $n_1 - 1$ and $n_2 - 1$. Be sure to state clearly which approximation you have used.

7.48   Does bread lose its vitamins when stored? Small loaves of bread were prepared with flour that was fortified with a fixed amount of vitamins. After baking, the vitamin C content of two loaves was measured. Another two loaves were baked at the same time, stored for three days, and then the vitamin C content was measured. The units are milligrams per hundred grams of flour (mg/100 g). Here are the data (provided by Helen Park; see Helen Park et al., "Fortifying bread with each of three antioxidants," *Cereal Chemistry*, 74 (1997), pp. 202–206):

Immediately after baking:	47.62, 49.79
Three days after baking:	21.25, 22.34

(a) When bread is stored, does it lose vitamin C? To answer this question, perform a two-sample t test for these data. Be sure to state your hypotheses, test statistic with degrees of freedom, and the P-value.

(b) Give a 90% confidence interval for the amount of vitamin C lost.

7.49   Refer to the previous exercise. The amount of vitamin E (in mg/100 g of flour) in the same loaves was also measured. Here are the data:

Immediately after baking:	94.6, 96.0
Three days after baking:	97.4, 94.3

(a) When bread is stored, does it lose vitamin E? To answer this question, perform a two-sample t test for these data. Be sure to state your hypotheses, test statistic with degrees of freedom, and the P-value.

(b) Give a 90% confidence interval for the amount of vitamin E lost.

7.50   Refer to the previous two exercises. Some people claim that significance tests with very small samples never lead to rejection of the null hypothesis. Discuss this claim using the results of these two exercises.

7.51   Do piano lessons improve the spatial-temporal reasoning of preschool children? We examined this question in Exercises 7.17 and 7.18 (pages 526–527) by analyzing the change in spatial-temporal reasoning of 34 preschool children after six months of piano lessons. Here we examine the same question by comparing the changes of those students with the changes of 44 children in a control group. Here are the data for the children who took piano lessons:

2	5	7	−2	2	7		4	1	0	7	3		4	3		4	9	4	5
2	9	6		0	3	6	−1	3	4	6	7	−2	7	−3		3	4	4	

The control group scores are:

1	−1	0	1	−4	0	0	1	0	−1	0	1	1	−3	−2	4	−1
2	4	2	2	2	−3	−3	0	2	0	−1	3	−1	5	−1	7	0
4	0	2	1	−6	0	2	−1	0	−2							

(Data from F. H. Rauscher et al., "Music training causes long-term enhancement of preschool children's spatial-temporal reasoning," *Neurological Research,* 19 (1997), pp. 2–8.)

(a) Display the data and summarize the distributions.

(b) Make a table with the sample size, the mean, the standard deviation, and the standard error of the mean for each of the two groups.

(c) Translate the question of interest into hypotheses, test them, and summarize your conclusions.

7.52 Refer to the previous exercise. Give a 95% confidence interval that describes the comparison between the children who took piano lessons and the controls.

7.53 Refer to Exercises 7.17 and 7.18 (pages 526–527) and the previous two exercises. We have used four ways to address the question of interest. Discuss the relative merits of each approach.

7.54 In what ways are companies that fail different from those that continue to do business? A study compared various characteristics of 68 healthy and 33 failed firms. One of the variables was the ratio of current assets to current liabilities. Roughly speaking, this is the amount that the firm is worth divided by what it owes. The data are given in Table 7.4. (Based on C. Papoulias and P. Theodossiou, "Analysis and modeling of recent business failures in Greece," *Managerial and Decision Economics,* 13 (1992), pp. 163–169.)

(a) Display the data so that the two distributions can be compared. Describe the shapes of the distributions and any important characteristics.

(b) We expect that failed firms will have a lower ratio. Describe and test appropriate hypotheses for these data. What do you conclude?

(c) It is not possible to do a randomized experiment for this kind of question. Explain why.

7.55 Physical fitness is related to personality characteristics. In one study of this relationship, middle-aged college faculty who had volunteered for a fitness program were divided into low-fitness and high-fitness groups

TABLE 7.4 Ratio of current assets to current liabilities								
Healthy firms						Failed firms		
1.50	0.10	1.76	1.14	1.84	2.21	0.82	0.89	1.31
2.08	1.43	0.68	3.15	1.24	2.03	0.05	0.83	0.90
2.23	2.50	2.02	1.44	1.39	1.64	1.68	0.99	0.62
0.89	0.23	1.20	2.16	1.80	1.87	0.91	0.52	1.45
1.91	1.67	1.87	1.21	2.05	1.06	1.16	1.32	1.17
0.93	2.17	2.61	3.05	1.52	1.93	0.42	0.48	0.93
1.95	2.61	1.11	0.95	0.96	2.25	0.88	1.10	0.23
2.73	1.56	2.73	0.90	2.12	1.42	1.11	0.19	0.13
1.62	1.76	2.22	2.80	1.85	0.96	2.03	0.51	1.12
1.71	1.02	2.50	1.55	1.69	1.64	0.92	0.26	1.15
1.03	1.80	0.67	2.44	2.30	2.21	0.13	0.88	0.09
1.96	1.81							

based on a physical examination. The subjects then took the Cattell Sixteen Personality Factor Questionnaire. Here are the data for the "ego strength" personality factor. (Based on A. H. Ismail and R. J. Young, "The effect of chronic exercise on the personality of middle-aged men," *Journal of Human Ergology*, 2 (1973), pp. 47–57.)

Low fitness			High fitness		
4.99	5.53	3.12	6.68	5.93	5.71
4.24	4.12	3.77	6.42	7.08	6.20
4.74	5.10	5.09	7.32	6.37	6.04
4.93	4.47	5.40	6.38	6.53	6.51
4.16	5.30		6.16	6.68	

(a) Is the difference in mean ego strength significant at the 5% level? At the 1% level? Be sure to state H_0 and H_a.

(b) You should be hesitant to generalize these results to the population of all middle-aged men. Explain why.

7.56 In a study of cereal leaf beetle damage on oats, researchers measured the number of beetle larvae per stem in small plots of oats after randomly applying one of two treatments: no pesticide or Malathion at the rate of 0.25 pound per acre. Here are the data:

Control:	2	4	3	4	2	3	3	5	3	2	6	3	4
Treatment:	0	1	1	2	1	2	1	1	2	1	1	1	

(Based on M. C. Wilson et al., "Impact of cereal leaf beetle larvae on yields of oats," *Journal of Economic Entomology*, 62 (1969), pp. 699–702.) Is there significant evidence at the 1% level that the mean number of larvae per stem is reduced by Malathion? Be sure to state H_0 and H_a.

7.57   Does cocaine use by pregnant women cause their babies to have low birth weight? To study this question, birth weights of babies of women who tested positive for cocaine/crack during a drug-screening test were compared with the birth weights for women who either tested negative or were not tested, a group we call "other." Here are the summary statistics. The birth weights are measured in grams. (Data from a study conducted at the Medical University of South Carolina in 1989.)

Group	n	$\bar{x}$	s
Positive test	134	2733	599
Other	5974	3118	672

(a) Formulate appropriate hypotheses and carry out the test of significance for these data.

(b) Give a 95% confidence interval for the mean difference in birth weights.

(c) Discuss the limitations of the study design. What do you believe can be concluded from this study?

7.58   The U.S. Department of Agriculture (USDA) uses many types of surveys to obtain important economic estimates. In one pilot study they estimated wheat prices in July and in September using independent samples. Here is a brief summary from the report. (Based on R. G. Hood, "Results of the prices received by farmers for grain—quality assurance project," USDA Report SRB-95-07, Washington, D.C., 1995.)

Month	n	$\bar{x}$	$s/\sqrt{n}$
July	90	$2.95	$0.023
September	45	$3.61	$0.029

(a) Note that the report gave standard errors. Find the standard deviation for each of the samples.

(b) Use a significance test to examine whether or not the price of wheat was the same in July and September. Be sure to give details and carefully state your conclusion.

7.59   Refer to the previous exercise. Give a 95% confidence interval for the increase in price between July and September.

7.60   A market research firm supplies manufacturers with estimates of the retail sales of their products from samples of retail stores. Marketing managers are prone to look at the estimate and ignore sampling error. Suppose that an SRS of 75 stores this month shows mean sales of 52 units of a small appliance, with standard deviation 13 units. During the same month last year, an SRS of 53 stores gave mean sales of 49 units, with standard deviation 11 units. An increase from 49 to 52 is a rise of 6%. The marketing manager is happy, because sales are up 6%.

(a) Use the two-sample t procedure to give a 95% confidence interval for the difference in mean number of units sold at all retail stores.

(b) Explain in language that the manager can understand why he cannot be certain that sales rose by 6%, and that in fact sales may even have dropped.

7.61 A bank compares two proposals to increase the amount that its credit card customers charge on their cards. (The bank earns a percentage of the amount charged, paid by the stores that accept the card.) Proposal A offers to eliminate the annual fee for customers who charge $2400 or more during the year. Proposal B offers a small percent of the total amount charged as a cash rebate at the end of the year. The bank offers each proposal to an SRS of 150 of its existing credit card customers. At the end of the year, the total amount charged by each customer is recorded. Here are the summary statistics:

Group	n	$\overline{x}$	s
A	150	$1987	$392
B	150	$2056	$413

(a) Do the data show a significant difference between the mean amounts charged by customers offered the two plans? Give the null and alternative hypotheses, and calculate the two-sample t statistic. Obtain the P-value (either approximately from Table D or more accurately from software). State your practical conclusions.

(b) The distributions of amounts charged are skewed to the right, but outliers are prevented by the limits that the bank imposes on credit balances. Do you think that skewness threatens the validity of the test that you used in (a)? Explain your answer.

7.62 A study of iron deficiency among infants compared samples of infants following different feeding regimens. One group contained breast-fed infants, while the children in another group were fed a standard baby formula without any iron supplements. Here are summary results on blood hemoglobin levels at 12 months of age. (From M. F. Picciano and R. H. Deering, "The influence of feeding regimens on iron status during infancy," *American Journal of Clinical Nutrition*, 33 (1980), pp. 746–753.)

Group	n	$\overline{x}$	s
Breast-fed	23	13.3	1.7
Formula	19	12.4	1.8

(a) Is there significant evidence that the mean hemoglobin level is higher among breast-fed babies? State H_0 and H_a and carry out a t test. Give the P-value. What is your conclusion?

(b) Give a 95% confidence interval for the mean difference in hemoglobin level between the two populations of infants.

(c) State the assumptions that your procedures in (a) and (b) require in order to be valid.

7.63 In a study of heart surgery, one issue was the effect of drugs called beta-blockers on the pulse rate of patients during surgery. The available subjects were divided at random into two groups of 30 patients each. One group received a beta-blocker; the other, a placebo. The pulse rate of each patient at a critical point during the operation was recorded. The treatment group had mean 65.2 and standard deviation 7.8. For the control group, the mean was 70.3 and the standard deviation 8.3.

(a) Do beta-blockers reduce the pulse rate? State the hypotheses and do a t test. Is the result significant at the 5% level? At the 1% level?

(b) Give a 99% confidence interval for the difference in mean pulse rates.

7.64 What aspects of rowing technique distinguish between novice and skilled competitive rowers? Researchers compared two groups of female competitive rowers: a group of skilled rowers and a group of novices. The researchers measured many mechanical aspects of rowing style as the subjects rowed on a Stanford Rowing Ergometer. One important variable is the angular velocity of the knee (roughly, the rate at which the knee joint opens as the legs push the body back on the sliding seat). This variable was measured when the oar was at right angles to the machine. (Based on W. N. Nelson and C. J. Widule, "Kinematic analysis and efficiency estimate of intercollegiate female rowers," unpublished manuscript, 1983.) The data show no outliers or strong skewness. Here is the SAS computer output:

```
                        TTEST PROCEDURE

Variable: KNEE

GROUP      N              Mean              Std Dev           Std Error
-------------------------------------------------------------------------
SKILLED    10          4.18283335         0.47905935         0.15149187
NOVICE      8          3.01000000         0.95894830         0.33903942


Variances         T         DF      Prob>|T|
-----------------------------------------------------
Unequal       3.1583       9.8       0.0104
Equal         3.3918      16.0       0.0037
```

(a) The researchers believed that the knee velocity would be higher for skilled rowers. State H_0 and H_a.

(b) Give the value of the two-sample t statistic and its P-value (note that SAS provides two-sided P-values). What do you conclude?

(c) Give a 90% confidence interval for the mean difference between the knee velocities of skilled and novice female rowers.

7.65 The novice and skilled rowers in the previous exercise were also compared with respect to several physical variables. Here is the SAS computer output for weight in kilograms:

```
                          TTEST PROCEDURE

Variable: WEIGHT

GROUP        N               Mean          Std Dev         Std Error
------------------------------------------------------------------------
SKILLED     10          70.3700000      6.10034898        1.92909973
NOVICE       8          68.4500000      9.03999930        3.19612240


Variances          T        DF      Prob>|T|
-------------------------------------------------
Unequal         0.5143     11.8      0.6165
Equal           0.5376     16.0      0.5982
```

Is there significant evidence of a difference in the mean weights of skilled and novice rowers? State H_0 and H_a, report the two-sample t statistic and its P-value, and state your conclusion.

7.66 The Johns Hopkins Regional Talent Searches give the SAT (intended for high school juniors and seniors) to 13-year-olds. In all, 19,883 males and 19,937 females took the tests between 1980 and 1982. The mean scores of males and females on the verbal test are nearly equal, but there is a clear difference between the sexes on the mathematics test. The reason for this difference is not understood. Here are the data (from a news article in *Science*, 224 (1983), pp. 1029–1031):

Group	$\bar{x}$	s
Males	416	87
Females	386	74

Give a 99% confidence interval for the difference between the mean score for males and the mean score for females in the population that Johns Hopkins searches.

7.67 The Survey of Study Habits and Attitudes (SSHA) is a psychological test designed to measure the motivation, study habits, and attitudes toward learning of college students. These factors, along with ability, are important in explaining success in school. Scores on the SSHA range from 0 to 200. A selective private college gives the SSHA to an SRS of both male and female first-year students. The data for the women are as follows:

154	109	137	115	152	140	154	178	101
103	126	126	137	165	165	129	200	148

Here are the scores of the men:

108	140	114	91	180	115	126	92	169	146
109	132	75	88	113	151	70	115	187	104

(a) Examine each sample graphically, with special attention to outliers and skewness. Is use of a t procedure acceptable for these data?

(b) Most studies have found that the mean SSHA score for men is lower than the mean score in a comparable group of women. Test this supposition here. That is, state hypotheses, carry out the test and obtain a P-value, and give your conclusions.

(c) Give a 90% confidence interval for the mean difference between the SSHA scores of male and female first-year students at this college.

7.68   Plant scientists have developed varieties of corn that have increased amounts of the essential amino acid lysine. In a test of the protein quality of this corn, an experimental group of 20 one-day-old male chicks was fed a ration containing the new corn. A control group of another 20 chicks received a ration that was identical except that it contained normal corn. Here are the weight gains (in grams) after 21 days. (Based on G. L. Cromwell et al., "A comparison of the nutritive value of *opaque-2, floury-2* and normal corn for the chick," *Poultry Science,* 47 (1968), pp. 840–847.)

Control				Experimental			
380	321	366	356	361	447	401	375
283	349	402	462	434	403	393	426
356	410	329	399	406	318	467	407
350	384	316	272	427	420	477	392
345	455	360	431	430	339	410	326

(a) Present the data graphically. Are there outliers or strong skewness that might prevent the use of t procedures?

(b) State the hypotheses for a statistical test of the claim that chicks fed high-lysine corn gain weight faster. Carry out the test. Is the result significant at the 10% level? At the 5% level? At the 1% level?

(c) Give a 95% confidence interval for the mean extra weight gain in chicks fed high-lysine corn.

7.69   Table 7.3 (page 552) gives data on the blood pressure before and after treatment for two groups of black males. One group took a calcium supplement, and the other group received a placebo. Example 7.20 compares the decrease in blood pressure in the two groups using pooled two-sample t procedures.

(a) Repeat the significance test using a two-sample t test that does not require equal population standard deviations. Compare your P-value with the result $P = 0.059$ for the pooled t test.

(b) Give a 90% confidence interval for the difference in means, again using a procedure that does not require equal standard deviations. How does the margin of error of your interval compare with that in Example 7.21?

7.70 Table 1.8 (page 40) gives data on the calories in a sample of brands of each of three kinds of hot dogs. The boxplot in Figure 1.16 shows that poultry hot dogs have fewer calories than either beef or meat hot dogs.

(a) Give a 95% confidence interval for the difference in mean calorie content between beef and poultry hot dogs.

(b) Based on your confidence interval, can the hypothesis that the population means are equal be rejected at the 5% significance level? Explain your answer.

(c) What assumptions does your statistical procedure in (a) require? Which of these assumptions are justified or not important in this case? Are any of the assumptions doubtful in this case?

7.71 Researchers studying the learning of speech often compare measurements made on the recorded speech of adults and children. One variable of interest is called the voice onset time (VOT). Here are the results for 6-year-old children and adults asked to pronounce the word "bees." The VOT is measured in milliseconds and can be either positive or negative. (From M. A. Zlatin and R. A. Koenigsknecht, "Development of the voicing contrast: a comparison of voice onset time in stop perception and production," *Journal of Speech and Hearing Research*, 19 (1976), pp. 93–111.)

Group	n	$\bar{x}$	s
Children	10	-3.67	33.89
Adults	20	-23.17	50.74

(a) What is the standard error of the sample mean VOT for the 20 adult subjects? What is the standard error of the difference $\bar{x}_1 - \bar{x}_2$ between the mean VOT for children and adults?

(b) The researchers were investigating whether VOT distinguishes adults from children. State H_0 and H_a and carry out a two-sample t test. Give a P-value and report your conclusions.

(c) Give a 95% confidence interval for the difference in mean VOTs when pronouncing the word "bees." Explain why you knew from your result in (b) that this interval would contain 0 (no difference).

7.72 The researchers in the study discussed in the previous exercise looked at VOTs for adults and children pronouncing several different words. Explain why they should not perform a separate two-sample t test for each word and conclude that the words with a significant difference (say, $P < 0.05$)

distinguish children from adults. (The researchers did not make this mistake.)

7.73 College financial aid offices expect students to use summer earnings to help pay for college. But how large are these earnings? One college studied this question by asking a sample of students how much they earned. Omitting students who were not employed, 1296 responses were received. Here are the data in summary form (data for 1982, provided by Marvin Schlatter, Division of Financial Aid, Purdue University):

Group	n	$\bar{x}$	s
Males	675	$1884.52	$1368.37
Females	621	$1360.39	$1037.46

(a) Use the two-sample t procedures to give a 90% confidence interval for the difference between the mean summer earnings of male and female students.

(b) The distribution of earnings is strongly skewed to the right. Nevertheless, use of t procedures is justified. Why?

(c) Once the sample size was decided, the sample was chosen by taking every k th name from an alphabetical list of undergraduates. Is it reasonable to consider the samples as SRSs chosen from the male and female undergraduate populations?

(d) What other information about the study would you request before accepting the results as describing all undergraduates?

The following exercises concern optional material on the pooled two-sample t *procedures and on the power of tests.*

7.74 Example 7.17 (page 526) reports the analysis of some biological data. Starting from the computer's results for $\bar{x}_i$ and s_i, verify the values given for the test statistic $t = 2.99$ and the degrees of freedom df $= 5.9$.

7.75 Example 7.16 (page 526) analyzed the Chapin Social Insight Test scores of male and female college students. The sample standard deviations s_i for the two samples are almost identical. This suggests that the two populations have similar standard deviations σ_i. The use of the pooled two-sample t procedures is justified in this case. Repeat the analysis in Example 7.16 using the pooled procedures. Compare your results with those obtained in the example.

7.76 The pooled two-sample t procedures can be used to analyze the hemoglobin data in Exercise 7.62 because the very similar sample standard deviations suggest that the assumption of equal population standard deviations is justified. Repeat parts (a) and (b) of that exercise using the pooled t procedures. Compare your results with those you obtained with the two-sample t procedures in Exercise 7.62.

7.77   The data on weights of skilled and novice rowers in Exercise 7.65 can be analyzed by the pooled t procedures, which assume equal population variances. Report the value of the t statistic, its degrees of freedom, and its P-value, and then state your conclusion. (The pooled procedures should not be used for the comparison of knee velocities in Exercise 7.64, because the sample standard deviations in the two groups are different enough to cast doubt on the assumption of a common population standard deviation.)

7.78   Repeat the comparison of mean VOTs for children and adults in Exercise 7.71 using a pooled t procedure. (In practice, we would not pool in this case, because the data suggest some difference in the population standard deviations.)

(a)   Carry out the significance test, and give a P-value.

(b)   Give a 95% confidence interval for the difference in population means.

(c)   How similar are your results to those you obtained in Exercise 7.71 from the two-sample t procedures?

7.3   Optional Topics in Comparing Distributions*

In this section we discuss four topics that are related to the material that we have already covered in this chapter. If we can do inference for means, it is natural to ask if we can do something similar for spread. The answer is yes, but there are many cautions. We also discuss robustness and show how to find the power for the two-sample t test. If you plan to design studies, you should become familiar with this last topic.

Inference for population spread

The two most basic descriptive features of a distribution are its center and spread. In a normal population, these aspects are measured by the mean and the standard deviation. We have described procedures for inference about population means for normal populations and found that these procedures are often useful for nonnormal populations as well. It is natural to turn next to inference about the standard deviations of normal populations. Our recommendation here is short and clear: Don't do it without expert advice.

There are indeed inference procedures appropriate for the standard deviations of normal populations. We will describe the most common such procedure: the F test for comparing the spread of two normal populations. Unlike the t procedures for means, the F test and other procedures for standard deviations are extremely sensitive to nonnormal distributions. This lack of robustness does not improve in large samples. It is difficult in practice to tell whether a significant F-value is evidence of unequal population spreads or simply evidence that the populations are not normal.

The deeper difficulty that underlies the very poor robustness of normal population procedures for inference about spread already appeared in our

*This section can be omitted without loss of continuity.

work on describing data. The standard deviation is a natural measure of spread for normal distributions but not for distributions in general. In fact, because skewed distributions have unequally spread tails, no single numerical measure is adequate to describe the spread of a skewed distribution. Thus, the standard deviation is not always a useful parameter, and even when it is, the results of inference about it are not trustworthy. Consequently, we do not recommend use of inference about population standard deviations in basic statistical practice.[12]

It was once common to test equality of standard deviations as a preliminary to performing the pooled two-sample t test for equality of two population means. It is better practice to check the distributions graphically, with special attention to skewness and outliers, and to use the software-based two-sample t that does not require equal standard deviations.

Chapters 12 and 13 discuss analysis of variance procedures for comparing several means. These procedures are extensions of the pooled t test and require that the populations have a common standard deviation. Fortunately, like the t test, analysis of variance comparisons of means are quite robust. (Analysis of variance uses F statistics, but these are not the same as the F statistic for comparing two population standard deviations.) Formal tests for the assumption of equal standard deviations are not very robust and will often give misleading results. In the words of one distinguished statistician, "To make a preliminary test on variances is rather like putting to sea in a rowing boat to find out whether conditions are sufficiently calm for an ocean liner to leave port!"[13]

The *F* test for equality of spread

Because of the limited usefulness of procedures for inference about the standard deviations of normal distributions, we will present only one such procedure. Suppose that we have independent SRSs from two normal populations, a sample of size n_1 from $N(\mu_1, \sigma_1)$ and a sample of size n_2 from $N(\mu_2, \sigma_2)$. The population means and standard deviations are all unknown. The hypothesis of equal spread,

$$H_0: \sigma_1 = \sigma_2$$
$$H_a: \sigma_1 \neq \sigma_2$$

is tested by a simple statistic: the ratio of sample variances.

The *F* Statistic and *F* Distributions

When s_1^2 and s_2^2 are sample variances from independent SRSs of sizes n_1 and n_2 drawn from normal populations, the F statistic

$$F = \frac{s_1^2}{s_2^2}$$

has the F distribution with $n_1 - 1$ and $n_2 - 1$ degrees of freedom when $H_0: \sigma_1 = \sigma_2$ is true.

F distributions

The **F distributions** are a family of distributions with two parameters: the degrees of freedom of the sample variances in the numerator and denominator of the *F* statistic.* The numerator degrees of freedom are always mentioned first. Interchanging the degrees of freedom changes the distribution, so the order is important. Our brief notation will be $F(j, k)$ for the *F* distribution with j degrees of freedom in the numerator and k in the denominator. The *F* distributions are not symmetric but are right-skewed. The density curve in Figure 7.15 illustrates the shape. Because sample variances cannot be negative, the *F* statistic takes only positive values and the *F* distribution has no probability below 0. The peak of the *F* density curve is near 1; values far from 1 in either direction provide evidence against the hypothesis of equal standard deviations.

Tables of *F* critical values are awkward, because a separate table is needed for every pair of degrees of freedom j and k. Table E in the back of the book gives upper p critical values of the *F* distributions for $p = 0.10, 0.05, 0.025, 0.01$, and 0.001. For example, these critical values for the $F(9, 10)$ distribution shown in Figure 7.15 are

p	0.10	0.05	0.025	0.01	0.001
F^*	2.35	3.02	3.78	4.94	8.96

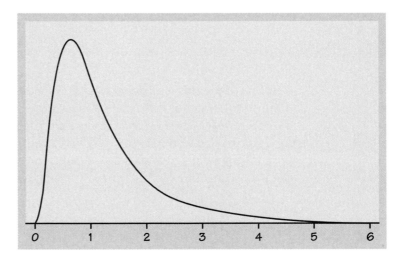

FIGURE 7.15 The density curve for the $F(9, 10)$ distribution. The *F* distributions are skewed to the right.

---

*The *F* distributions are another of R. A. Fisher's contributions to statistics and are called *F* in his honor. Fisher introduced *F* statistics for comparing several means. We will meet these useful statistics in later chapters.

The skewness of the F distributions causes additional complications. In the symmetric normal and t distributions, the point with probability 0.05 below it is just the negative of the point with probability 0.05 above it. This is not true for F distributions. We therefore require either tables of both the upper and lower tails or means of eliminating the need for lower tail critical values. Statistical software that eliminates the need for tables is plainly very convenient. If you do not use statistical software, arrange the F test as follows:

1. Take the test statistic to be

$$F = \frac{\text{larger } s^2}{\text{smaller } s^2}$$

This amounts to naming the populations so that s_1^2 is the larger of the observed sample variances. The resulting F is always 1 or greater.

2. Compare the value of F with critical values from Table E. Then *double* the significance levels from the table to obtain the significance level for the two-sided F test.

The idea is that we calculate the probability in the upper tail and double to obtain the probability of all ratios on either side of 1 that are at least as improbable as that observed. Remember that the order of the degrees of freedom is important in using Table E.

EXAMPLE 7.22

Example 7.19 recounts a medical experiment comparing the effects of calcium and a placebo on the blood pressure of black men. The analysis employed the pooled two-sample t procedures. Because these procedures require equal population standard deviations, it is tempting to first test

$$H_0: \sigma_1 = \sigma_2$$
$$H_a: \sigma_1 \neq \sigma_2$$

The larger of the two sample standard deviations is $s = 8.743$ from 10 observations. The other is $s = 5.901$ from 11 observations. The two-sided test statistic is therefore

$$F = \frac{\text{larger } s^2}{\text{smaller } s^2} = \frac{8.743^2}{5.901^2} = 2.195$$

We compare the calculated value $F = 2.20$ with critical points for the $F(9, 10)$ distribution. Table E shows that 2.20 is *less* than the 0.10 critical value of the $F(9, 10)$ distribution, which is $F^* = 2.35$. Doubling 0.10, we know that the observed F falls short of the 0.20 critical value. The results are not significant at the 20% level (or any lower level). Statistical software shows that the exact upper-tail probability is 0.118, and hence $P = 0.236$. *If* the populations were normal, the observed standard deviations would give little reason to suspect unequal population standard deviations. Because one of the populations shows some nonnormality, we cannot be fully confident of this conclusion.

Robustness of normal inference procedures

We have claimed that

- The t procedures for inference about means are quite robust against nonnormal population distributions. These procedures are particularly robust when the population distributions are symmetric and (for the two-sample case) when the two sample sizes are equal.

- The F test and other procedures for inference about variances are so lacking in robustness as to be of little use in practice.

Simulations with a large variety of nonnormal distributions support these claims. Note 3 at the end of this chapter gives references. One set of simulations was carried out with samples of size 25 and used significance tests with fixed level $\alpha = 0.05$. The three types of tests studied were the one-sample and pooled two-sample t tests and the F test for comparing two variances.

The robustness of the one-sample and two-sample t procedures is remarkable. The true significance level remains between about 4% and 6% for a large range of populations. The t test and the corresponding confidence intervals are among the most reliable tools that statisticians use. Remember, however, that outliers can greatly disturb the t procedures. Also, two-sample procedures are less robust when the sample sizes are not similar.

The lack of robustness of the tests for variances is equally remarkable. The true significance levels depart rapidly from the target 5% as the population distribution departs from normality. The two-sided F test carried out with 5% critical values can have a true level of less than 1% or greater than 11% even in symmetric populations with no outliers. Results such as these are the basis for our recommendation that these procedures not be used.

The power of the two-sample t test

The two-sample t test is one of the most used statistical procedures. Unfortunately, because of inadequate planning, users frequently fail to find evidence for the effects that they believe to be true. Power calculations should be part of the planning of any statistical study. Information from a pilot study or previous research is needed.

In Section 7.1 (optional material), we learned how to find an approximation for the power of the one-sample t test. The basic concepts for the two-sample case are the same. Here, we give the exact method, which involves a **noncentral t distribution** new distribution, the **noncentral t distribution.** To perform the calculations, we simply need software to calculate probabilities for this distribution.

We first present the method for the pooled two-sample t test, where the parameters are μ_1, μ_2, and the common standard deviation σ. Modifications to get approximate results when we do not pool are then described.

To find the power for the pooled two-sample t test, use the following steps. We consider only the case where the null hypothesis is $\mu_1 - \mu_2 = 0$.

1. Specify
   (a) an alternative that you consider important to detect; that is, a value for $\mu_1 - \mu_2$;
   (b) the sample sizes, n_1 and n_2;
   (c) the Type I error for a fixed significance level, α;
   (d) a guess at the standard deviation, σ.

2. Find the degrees of freedom df $= n_1 + n_2 - 2$ and the value of t^* that will lead to rejection of H_0.

noncentrality parameter

3. Calculate the **noncentrality parameter**

$$\delta = \frac{|\mu_1 - \mu_2|}{\sigma\sqrt{\dfrac{1}{n_1} + \dfrac{1}{n_2}}}$$

4. Find the power as the probability that a noncentral t random variable with degrees of freedom df and noncentrality parameter δ will be less than t^*. In SAS this is 1-PROBT(tstar,df,delta). If you do not have software that can perform this calculation, you can approximate the power as the probability that a standard normal random variable is greater than $t^* - \delta$, that is, $P(z > t^* - \delta)$, and use Table A.

Note that the denominator in the noncentrality parameter,

$$\sigma\sqrt{\frac{1}{n_1} + \frac{1}{n_2}}$$

is our guess at the standard error for the difference in the sample means. Therefore, if we wanted to assess a possible study in terms of the margin of error for the estimated difference, we would examine t^* times this quantity.

If we do not assume that the standard deviations are equal, we need to guess both standard deviations and then combine these for our guess at the standard error:

$$\sqrt{\frac{\sigma_1^2}{n_1} + \frac{\sigma_2^2}{n_2}}$$

This guess is then used in the denominator of the noncentrality parameter. For the degrees of freedom, the conservative approximation is appropriate.

EXAMPLE 7.23 | In Example 7.20 we examined the effect of calcium on blood pressure by comparing the means of a treatment group and a placebo group using a two-sample t test. The P-value was 0.059, failing to achieve the usual standard of 0.05 for statistical significance. Suppose that we wanted to plan a new study that would provide convincing evidence, say at the 0.01 level, with high probability. Let's examine a study design with 45 subjects in each group ($n_1 = n_2 = 45$). Based on our previous results, we choose $\mu_1 - \mu_2 = 5$ as an alternative that we would like to be able to detect with $\alpha = 0.01$. For σ we use 7.4, our pooled estimate from Example 7.20. The degrees of freedom are $n_1 + n_2 - 2 = 88$ and $t^* = 2.37$ for the significance test. The noncentrality parameter is

$$\delta = \frac{5}{7.4\sqrt{\dfrac{1}{45} + \dfrac{1}{45}}} = \frac{5}{1.56} = 3.21$$

Software gives the power as 0.7965, or 80%. The normal approximation gives 0.7983, a very accurate result. With this choice of sample sizes we would expect the margin of error for a 95% confidence interval ($t^* = 1.99$) for the difference in means to be

$$t^* \times 7.4\sqrt{\frac{1}{45} + \frac{1}{45}} = 1.99 \times 1.56 = 3.1$$

With software it is very easy to examine the effects of variations on a study design. In the above example, we might want to examine the power for $\alpha = 0.05$ and the effects of reducing the sample sizes.

SUMMARY

Inference procedures for comparing the standard deviations of two normal populations are based on the **F statistic,** which is the ratio of sample variances

$$F = \frac{s_1^2}{s_2^2}$$

If an SRS of size n_1 is drawn from the x_1 population and an independent SRS of size n_2 is drawn from the x_2 population, the F statistic has the **F distribution** $F(n_1 - 1, n_2 - 1)$ if the two population standard deviations σ_1 and σ_2 are in fact equal.

The **F test for equality of standard deviations** tests $H_0: \sigma_1 = \sigma_2$ versus $H_a: \sigma_1 \neq \sigma_2$ using the statistic

$$F = \frac{\text{larger } s^2}{\text{smaller } s^2}$$

and doubles the upper tail probability to obtain the P-value.

The t procedures are quite **robust** when the distributions are not normal. The F tests and other procedures for inference about the spread of one or more normal distributions are so strongly affected by nonnormality that we do not recommend them for regular use.

The **power** of the pooled two-sample t test is found by first computing the critical value for the significance test, the degrees of freedom, and the **non-centrality parameter** for the alternative of interest. These are used to find the power from the **noncentral t distribution.** A normal approximation works quite well. Calculating margins of error for various study designs and assumptions is an alternative procedure for evaluating designs.

SECTION 7.3 EXERCISES

In all exercises calling for use of the F test, assume that both population distributions are very close to normal. The actual data are not always sufficiently normal to justify use of the F test.

7.79   The F statistic $F = s_1^2/s_2^2$ is calculated from samples of size $n_1 = 10$ and $n_2 = 21$. (Remember that n_1 is the numerator sample size.)

   (a)   What is the upper 5% critical value for this F?

   (b)   In a test of equality of standard deviations against the two-sided alternative, this statistic has the value $F = 2.45$. Is this value significant at the 10% level? Is it significant at the 5% level?

7.80   The F statistic for equality of standard deviations based on samples of sizes $n_1 = 21$ and $n_2 = 26$ takes the value $F = 2.88$.

   (a)   Is this significant evidence of unequal population standard deviations at the 5% level?

   (b)   Use Table E to give an upper and a lower bound for the P-value.

7.81   Refer to Exercise 7.51 (page 556), where we examined the effect of piano lessons on spatial-temporal reasoning. Do the data provide evidence that would cause us to suspect that the standard deviation of the children who took piano lessons is different from that of the controls? Set up the hypotheses, perform the significance test, and summarize the results.

7.82   A USDA survey used to estimate wheat prices in July and September is described in Exercise 7.58 (page 559). Using the standard deviations you calculated there, perform the test for equality of standard deviations and summarize your conclusion.

7.83   We studied the loss of vitamin C when bread is stored in Exercise 7.48 (page 556). Recall that two loaves were measured immediately after baking and another two loaves were measured after three days of storage. These are very small sample sizes.

(a) Use Table E to find the value that the ratio of variances would have to exceed for us to reject the null hypothesis (at the 5% level) that the standard deviations are equal. What does this suggest about the power of the test?

(b) Perform the test and state your conclusion.

7.84 Exercise 7.64 (page 561) records the results of comparing a measure of rowing style for skilled and novice female competitive rowers. Is there significant evidence of inequality between the standard deviations of the two populations?

(a) State H_0 and H_a.

(b) Calculate the F statistic. Between which two levels does the P-value lie?

7.85 Answer the same questions for the weights of the two groups, recorded in Exercise 7.65 (page 562).

7.86 The observed inequality between the sample standard deviations of male and female SAT mathematics scores in Exercise 7.66 (page 562) is clearly significant. You can say this without doing any calculations. Find F and look in Table E. Then explain why the significance of F could be seen without arithmetic.

7.87 Return to the SSHA data in Exercise 7.67 (page 562). SSHA scores are generally less variable among women than among men. We want to know whether this is true for this college.

(a) State H_0 and H_a. Note that H_a is one-sided in this case.

(b) Because Table E contains only upper critical values for F, a one-sided test requires that in calculating F the numerator s^2 belongs to the group that H_a claims to have the larger σ. Calculate this F.

(c) Compare F to the entries in Table E (no doubling of p) to obtain the P-value. Be sure the degrees of freedom are in the proper order. What do you conclude about the variation in SSHA scores?

7.88 The data for VOTs of children and adults in Exercise 7.71 (page 564) show quite different sample standard deviations. How statistically significant is the observed inequality?

7.89 In Exercise 7.57 (page 559) data on cocaine use and birth weight are summarized. The study has been criticized because of several design problems. Suppose that you are designing a new study. Based on the results in Exercise 7.57, you think that the true difference in mean birth weights may be about 300 grams (g); a difference this large is clinically important. For planning purposes assume that you will have 100 women in each group and that the common standard deviation is 650 g, a guess that is between the two standard deviations in Exercise 7.57. If you use a pooled two-sample t test with a Type I error of 0.05, what is the power of the test for this design?

7.90   Refer to the previous exercise. Repeat the power calculation, for 25, 50, 75, and 125 women in each group. Summarize these results. Include the results for 100 women per group that you found in the previous exercise.

7.91   Refer to the previous two exercises. For each of the sample sizes considered, what is your guess at the margin of error for the 95% confidence interval for the difference in mean weights? Display these results with a graph or a sketch.

7.92   Suppose that you wanted to compare intramural basketball players and intramural soccer players on the "ego strength" personality factor described in Exercise 7.55 (page 557). With the data from that exercise, you will use $\sigma = 0.7$ for planning purposes. The pooled two-sample t test with $\alpha = 0.01$ will be used to make the comparison. Based on Exercise 7.55, you judge a difference of 0.5 points to be of interest.

   (a) Find the power for the design with 20 players from each sport.

   (b) The power in part (a) is not acceptable. Redo the calculations for 30 players in each group and $\alpha = 0.05$.

CHAPTER 7 EXERCISES

7.93   Some research suggests that women perform better than men in school while men score higher on standardized tests. Table 1.6 (page 33) presents data on a measure of school performance, grade point average (GPA), and a standardized test, IQ, for 78 seventh-grade students. Do these data lend further support to the previously found gender differences? Give graphical displays of the data and describe the distributions. Use significance tests and confidence intervals to examine this question and prepare a short report summarizing your findings.

7.94   Table 1.9 (page 98) gives positions and weight (in pounds) for players on a major college football team. Assume that these players can be viewed as a random sample of players from similar teams. Use a significance test and a confidence interval to compare the mean weights of offensive line (OL) and defensive line (DL) players. Give a summary of your results with a graphical display of the data.

7.95   Do houses that have four bedrooms cost more than houses with three? The data in Table 2.3 (page 123) can be used to address this question for a neighborhood in Boulder, Colorado. Delete the houses with 5 bedrooms and then describe the selling-price distributions of those with 3 and 4 bedrooms. Are there any outliers? Examine the question of interest using a significance test and a confidence interval. Run your analysis with and without any houses you think might be outliers.

7.96   Data on the numbers of manatees killed by boats each year are shown in Table 2.6 (page 149). After a long period of increasing numbers of deaths, the pattern flattens somewhat. In fact, the total for 1990 is 47, less than the total of 50 for 1989. Perhaps the trend has now reversed. We would like to do a significance test to compare these two counts. Theoretical considerations suggest that the standard errors ($\sigma/\sqrt{n}$) for these types of counts can be approximated by the square root of the count. So, for example, the 1990 count, 47, has a standard error that is approximately $\sqrt{47}$. Use this approximation to perform an approximate two-sample z test for the difference between the 1989 and 1990 deaths. Find an approximate 95% confidence interval for the difference. What do you conclude?

7.97 On the morning of March 5, 1996, a train with 14 tankers of propane derailed near the center of the small Wisconsin town of Weyauwega. Six of the tankers were ruptured and burning when the 1700 residents were ordered to evacuate the town. Researchers study disasters like this so that effective relief efforts can be designed for future disasters. About half of the households with pets did not evacuate all of their pets. A study conducted after the derailment focused on problems associated with retrieval of the pets after the evacuation and characteristics of the pet owners. One of the scales measured "commitment to adult animals," and the people who evacuated all or some of their pets were compared with those who did not evacuate any of their pets. Higher scores indicate that the pet owner is more likely to take actions that benefit the pet. Here are the data summaries (data provided by Professor Sebastian Heath, School of Veterinary Medicine, Purdue University):

Group	n	$\overline{x}$	s
Evacuated all or some pets	116	7.95	3.62
Did not evacuate any pets	125	6.26	3.56

Analyze the data and prepare a short report describing the results.

7.98 In a study of the effectiveness of weight-loss programs, 47 subjects who were at least 20% overweight took part in a group support program for 10 weeks. Private weighings determined each subject's weight at the beginning of the program and 6 months after the program's end. The matched pairs t test was used to assess the significance of the average weight loss. The paper reporting the study said, "The subjects lost a significant amount of weight over time, $t(46) = 4.68, p < 0.01$." It is common to report the results of statistical tests in this abbreviated style. (Based loosely on D. R. Black et al., "Minimal interventions for weight control: a cost-effective alternative," *Addictive Behaviors*, 9 (1984), pp. 279–285.)

(a) Why was the matched pairs statistic appropriate?

(b) Explain to someone who knows no statistics but is interested in weight-loss programs what the practical conclusion is.

(c) The paper follows the tradition of reporting significance only at fixed levels such as $\alpha = 0.01$. In fact, the results are more significant than "$p < 0.01$" suggests. What can you say about the P-value of the t test?

7.99 Nitrites are often added to meat products as preservatives. In a study of the effect of these chemicals on bacteria, the rate of uptake of a radiolabeled amino acid was measured for a number of cultures of bacteria, some growing in a medium to which nitrites had been added. Here are the summary statistics from this study:

Group	n	$\bar{x}$	s
Nitrite	30	7880	1115
Control	30	8112	1250

Carry out a test of the research hypothesis that nitrites decrease amino acid uptake, and report your results.

7.100 The one-hole test is used to test the manipulative skill of job applicants. This test requires subjects to grasp a pin, move it to a hole, insert it, and return for another pin. The score on the test is the number of pins inserted in a fixed time interval. In one study, male college students were compared with experienced female industrial workers. Here are the data for the first minute of the test. (Based on G. Salvendy, "Selection of industrial operators: the one-hole test," *International Journal of Production Research*, 13 (1973), pp. 303–321.)

Group	n	$\bar{x}$	s
Students	750	35.12	4.31
Workers	412	37.32	3.83

(a) It was expected that the experienced workers would outperform the students, at least during the first minute, before learning occurs. State the hypotheses for a statistical test of this expectation and perform the test. Give a *P*-value and state your conclusions.

(b) The distribution of scores is slightly skewed to the left. Explain why the procedure you used in (a) is nonetheless acceptable.

(c) One purpose of the study was to develop performance norms for job applicants. Based on the data above, what is the range that covers the middle 95% of experienced workers? (Be careful! This is not the same as a 95% confidence interval for the mean score of experienced workers.)

(d) The five-number summary of the distribution of scores among the workers is

23	33.5	37	40.5	46

for the first minute and

32	39	44	49	59

for the fifteenth minute of the test. Display these facts graphically, and describe briefly the differences between the distributions of scores in the first and fifteenth minute.

7.101   The composition of the earth's atmosphere may have changed over time. One attempt to discover the nature of the atmosphere long ago studies the gas trapped in bubbles inside ancient amber. Amber is tree resin that has hardened and been trapped in rocks. The gas in bubbles within amber should be a sample of the atmosphere at the time the amber was formed. Measurements on specimens of amber from the late Cretaceous era (75 to 95 million years ago) give these percents of nitrogen:

63.4	65.0	64.4	63.3	54.8	64.5	60.8	49.1	51.0

These values are quite different from the present 78.1% of nitrogen in the atmosphere. Assume (this is not yet agreed on by experts) that these observations are an SRS from the late Cretaceous atmosphere. (Data from R. A. Berner and G. P. Landis, "Gas bubbles in fossil amber as possible indicators of the major gas composition of ancient air," *Science*, 239 (1988), pp. 1406–1409.)

(a) Graph the data, and comment on skewness and outliers.

(b) The t procedures will be only approximate in this case. Give a 95% t confidence interval for the mean percent of nitrogen in ancient air.

7.102   Do various occupational groups differ in their diets? A British study of this question compared 98 drivers and 83 conductors of London double-decker buses. The conductors' jobs require more physical activity. The article reporting the study gives the data as "Mean daily consumption (± se)." Some of the study results appear below. (From J. W. Marr and J. A. Heady, "Within- and between-person variation in dietary surveys: number of days needed to classify individuals," *Human Nutrition: Applied Nutrition*, 40A (1986), pp. 347–364.)

	Drivers	Conductors
Total calories	2821 ± 44	2844 ± 48
Alcohol (grams)	0.24 ± 0.06	0.39 ± 0.11

(a) What does "se" stand for? Give $\bar{x}$ and s for each of the four sets of measurements.

(b) Is there significant evidence at the 5% level that conductors consume more calories per day than do drivers? Use the two-sample t method to give a P-value, and then assess significance.

(c) How significant is the observed difference in mean alcohol consumption? Use two-sample t methods to obtain the P-value.

(d) Give a 90% confidence interval for the mean daily alcohol consumption of London double-decker bus conductors.

(e) Give an 80% confidence interval for the difference in mean daily alcohol consumption between drivers and conductors.

7.103 Use of the pooled two-sample t test is justified in part (b) of the previous exercise. Explain why. Find the P-value for the pooled t statistic, and compare with your result in the previous exercise.

7.104 The report cited in Exercise 7.102 says that the distribution of alcohol consumption among the individuals studied is "grossly skew."

(a) Do you think that this skewness prevents the use of the two-sample t test for equality of means? Explain your answer.

(b) Do you think that the skewness of the distributions prevents the use of the F test for equality of standard deviations? Explain your answer.

7.105 Exercise 1.123 (page 99) gives the populations of all 58 counties in the state of California. Is it proper to apply the one-sample t method to these data to give a 95% confidence interval for the mean population of a California county? Explain your answer.

7.106 The amount of lead in a certain type of soil, when released by a standard extraction method, averages 86 parts per million (ppm). A new extraction method is tried on 40 specimens of the soil, yielding a mean of 83 ppm lead and a standard deviation of 10 ppm.

(a) Is there significant evidence at the 1% level that the new method frees less lead from the soil?

(b) A critic argues that because of variations in the soil, the effectiveness of the new method is confounded with characteristics of the particular soil specimens used. Briefly describe a better data production design that avoids this criticism.

7.107 High levels of cholesterol in the blood are not healthy in either humans or dogs. Because a diet rich in saturated fats raises the cholesterol level, it is plausible that dogs owned as pets have higher cholesterol levels than dogs owned by a veterinary research clinic. "Normal" levels of cholesterol based on the clinic's dogs would then be misleading. A clinic compared healthy dogs it owned with healthy pets brought to the clinic to be neutered. The summary statistics for blood cholesterol levels (milligrams per deciliter of blood) appear below. (From V. D. Bass, W. E. Hoffmann, and J. L. Dorner, "Normal canine lipid profiles and effects of experimentally induced pancreatitis and hepatic necrosis on lipids," *American Journal of Veterinary Research*, 37 (1976), pp. 1355–1357.)

Group	n	$\bar{x}$	s
Pets	26	193	68
Clinic	23	174	44

(a) Is there strong evidence that pets have a higher mean cholesterol level than clinic dogs? State the H_0 and H_a and carry out an appropriate test. Give the P-value and state your conclusion.

(b) Give a 95% confidence interval for the difference in mean cholesterol levels between pets and clinic dogs.

(c) Give a 95% confidence interval for the mean cholesterol level in pets.

(d) What assumptions must be satisfied to justify the procedures you used in (a), (b), and (c)? Assuming that the cholesterol measurements have no outliers and are not strongly skewed, what is the chief threat to the validity of the results of this study?

7.108 Exercise 2.64 and Table 2.11 (page 176) give data on the concentration of airborne particulate matter in a rural area upwind from a small city and in the center of the city. In that exercise, the focus was on using the rural readings to predict the city readings for the same day. Now we want to compare the mean level of particulates in the city and in the rural area. We suspect that pollution is higher in the city and hope to find evidence for this suspicion.

(a) State H_0 and H_a.

(b) Which type of t procedure is appropriate: one-sample, matched pairs, or two-sample?

(c) Make a graph to check for outliers or strong skewness that might prevent the use of t procedures. Your graph should reflect the type of procedure that you will use.

(d) Carry out the appropriate t test. Give the P-value and report your conclusion.

(e) Give a 90% confidence interval for the mean amount by which the city particulate level exceeds the rural level.

matchpairs

7.109 The sign test allows us to assess whether city particulate levels are higher than nearby rural levels on the same day without the use of normal distributions. Carry out a sign test for the data used in the previous exercise. State H_0 and H_a and give the P-value and your conclusion.

7.110 Exercise 1.24 (page 30) gives 29 measurements of the density of the earth made in 1798 by Henry Cavendish. Display the data graphically to check for skewness and outliers. Then give an estimate for the density of the earth from Cavendish's data and a margin of error for your estimate.

7.111 Elite distance runners are thinner than the rest of us. Here are data on skinfold thickness, which indirectly measures body fat, for 20 elite runners and 95 ordinary men in the same age group. The data are in millimeters and are given in the form "mean (standard deviation)." (From M. L. Pollock et al., "Body composition of elite class distance runners," in P. Milvey (ed.), *The Marathon: Physiological, Medical, Epidemiological, and Psychological Studies*, New York Academy of Sciences, New York, 1977, p. 366.)

	Runners	Others
Abdomen	7.1 (1.0)	20.6 (9.0)
Thigh	6.1 (1.8)	17.4 (6.6)

Use confidence intervals to describe the difference between runners and typical young men.

7.112 Table 2.16 (page 222) gives the levels of three pollutants in the exhaust of 46 randomly selected vehicles of the same type. You will investigate emissions of nitrogen oxides (NOX).

(a) Make a stemplot and, if your software allows, a normal quantile plot of the NOX levels. Do the plots suggest that the distribution of NOX emissions is approximately normal? Can you safely employ t procedures to analyze these data?

(b) Give a 99% confidence interval for the mean NOX level in vehicles of this type.

(c) Your supervisor hopes the average NOX level is less than 1 gram per mile. You will have to tell him that it's not so. Carry out a significance test to assess the strength of the evidence that the mean NOX level is greater than 1 and then write a short report to your supervisor based on your work in (b) and (c). (Your supervisor has never heard of P-values, so you must use plain language.)

7.113 Is there a difference between the average SAT scores of males and females? The CSDATA data set gives the math (SATM) and verbal (SATV) scores for a group of 224 computer science majors. The variable SEX indicates whether each individual is male or female.

(a) Compare the two distributions graphically, and then use the two-sample t test to compare the average SATM scores of males and females. Is it appropriate to use the pooled t test for this comparison? Write a brief summary of your results and conclusions that refers to both versions of the t test and to the F test for equality of standard deviations. Also give a 99% confidence interval for the difference in the means.

(b) Answer part (a) for the SAT verbal score.

(c) The students in the CSDATA data set were all computer science majors who began college during a particular year. To what extent do you think that your results would generalize to (*i*) computer science students entering in different years, (*ii*) computer science majors at other colleges and universities, and (*iii*) college students in general?

7.114 The WOOD data set gives first (T1) and second (T2) measurements of the modulus of elasticity for 50 strips of wood. We would like to see if the process of taking the first measurement changes the strips so that the second measurement tends to be higher or lower than the first.

(a) Compute the differences $D = T1 - T2$ and describe the distribution of D using graphical and numerical summaries.

(b) Give a 95% confidence interval for the mean of D.

(c) Perform the test of the null hypothesis that the mean of D is 0. Summarize your results and draw a conclusion.

7.115　In the READING data set the response variable Post3 is to be compared for three methods of teaching reading. The Basal method is the standard, or control, method, and the two new methods are DRTA and Strat. We can use the methods of this chapter to compare Basal with DRTA and Basal with Strat. Note that to make comparisons among three treatments it is more appropriate to use the procedures that we will learn in Chapter 12.

(a)　Is the mean reading score with the DRTA method higher than that for the Basal method? Perform an analysis to answer this question and summarize your results.

(b)　Answer part (a) for the Strat method in place of DRTA.

7.116　Example 7.13 (page 539) tells us that the mean height of 10-year-old girls is $N(54.5, 2.7)$ and for boys it is $N(54.1, 2.4)$. The null hypothesis that the mean heights of 10-year-old boys and girls are equal is clearly false. The difference in mean heights is $54.5 - 54.1 = 0.4$ inch. Small differences such as this can require large sample sizes to detect. To simplify our calculations, let's assume that the standard deviations are the same, say $\sigma = 2.5$, and that we will measure the heights of an equal number of girls and boys. How many would we need to measure to have an 80% chance of detecting the (true) alternative hypothesis?

7.117　In Exercise 7.92 (page 575) you found the power for a study designed to compare the "ego strengths" of intramural basketball players and soccer players. Here you will design a study to compare intramural basketball players and volleyball players. Assume the same value of $\sigma = 0.07$ and use $\alpha = 0.05$. You are planning to have 20 players in each group. Calculate the power of the pooled two-sample t test that compares the mean ego strengths of players who participate in these two sports for different values of the true difference. Include values that have a very small chance of being detected and some that are virtually certain to be seen in your sample. Plot the power versus the true difference and write a short summary of what you have found.

7.118　As the degrees of freedom increase, the t distributions get closer and closer to the z $(N(0, 1))$ distribution. One way to see this is to look at how the value of t^* for a 95% confidence interval changes with the degrees of freedom. Make a plot with degrees of freedom from 2 to 100 on the x axis and t^* on the y axis. Draw a horizontal line on the plot corresponding to the value of $z^* = 1.96$. Summarize the main features of the plot.

7.119　The margin of error for a confidence interval depends on the confidence level, the standard deviation, and the sample size. Fix the confidence level at 95% and the standard deviation at 1 to examine the effect of the sample size. Find the margin of error for sample sizes of 5 to 100 by 5's—that is, let $n = 5, 10, 15, \ldots, 100$. Plot the margins of error versus the sample size and summarize the relationship.

NOTES

1. These data are from "Results report on the vitamin C pilot program," prepared by SUSTAIN (Sharing United States Technology to Aid in the Improvement of Nutrition) for the U.S. Agency for International Development. The report was used by the Committee on International Nutrition of the National Academy of Sciences/Institute of Medicine (NAS/IOM) to make recommendations on whether or not the vitamin C content of food commodities used in U.S. food aid programs should be increased. The program was directed by Peter Ranum and Françoise Chomé.

2. Data provided by Joseph A. Wipf, Department of Foreign Languages and Literatures, Purdue University.

3. These recommendations are based on extensive computer work. See, for example, Harry O. Posten, "The robustness of the one-sample t-test over the Pearson system," *Journal of Statistical Computation and Simulation,* 9 (1979), pp. 133–149, and E. S. Pearson and N. W. Please, "Relation between the shape of population distribution and the robustness of four simple test statistics," *Biometrika,* 62 (1975), pp. 223–241.

4. The data and the use of the log transformation come from Thomas J. Lorenzen, "Determining statistical characteristics of a vehicle emissions audit procedure," *Technometrics,* 22 (1980), pp. 483–493.

5. You can find a practical discussion of distribution-free inference in Myles Hollander and Douglas A. Wolfe, *Nonparametric Statistical Methods,* Wiley, New York, 1973.

6. Detailed information about the conservative t procedures can be found in Paul Leaverton and John J. Birch, "Small sample power curves for the two sample location problem," *Technometrics,* 11 (1969), pp. 299–307; in Henry Scheffé, "Practical solutions of the Behrens-Fisher problem," *Journal of the American Statistical Association,* 65 (1970), pp. 1501–1508, and in D. J. Best and J. C. W. Rayner, "Welch's approximate solution for the Behrens-Fisher problem," *Technometrics,* 29 (1987), pp. 205–210.

7. This example is adapted from Maribeth C. Schmitt, "The effects of an elaborated directed reading activity on the metacomprehension skills of third graders," Ph.D. dissertation, Purdue University, 1987.

8. See the extensive simulation studies in Harry O. Posten, "The robustness of the two-sample t test over the Pearson system," *Journal of Statistical Computation and Simulation,* 6 (1978), pp. 295–311.

9. From H. G. Gough, *The Chapin Social Insight Test,* Consulting Psychologists Press, Palo Alto, Calif., 1968.

10. This example is loosely based on D. L. Shankland, "Involvement of spinal cord and peripheral nerves in DDT-poisoning syndrome in albino rats," *Toxicology and Applied Pharmacology,* 6 (1964), pp. 197–213.

11. This study is reported in Roseann M. Lyle et al., "Blood pressure and metabolic effects of calcium supplementation in normotensive white and black men," *Journal of the American Medical Association,* 257 (1987), pp. 1772–1776. The individual measurements in Table 7.3 were provided by Dr. Lyle.

12. The problem of comparing spreads is difficult even with advanced methods. Common distribution-free procedures do not offer a satisfactory alternative to the F test, because they are sensitive to unequal shapes when comparing two distributions. A good introduction to the available methods is W. J. Conover, M. E. Johnson, and M. M. Johnson, "A comparative study of tests for homogeneity of variances, with applications to outer continental shelf bidding data," *Technometrics,* 23 (1981), pp. 351–361. Modern resampling procedures often work well. See Dennis D. Boos and Colin Brownie, "Bootstrap methods for testing homogeneity of variances," *Technometrics,* 31 (1989), pp. 69–82.

13. G. E. P. Box, "Non-normality and tests on variances," *Biometrika,* 40 (1953), pp. 318–335. The quote appears on page 333.

Prelude

Some statistical studies concern variables measured in a scale of equal units, such as dollars or grams. Other studies examine categorical variables, such as the race or occupation of a person, the make of a car, or the type of complaint received from a customer. When we record categorical variables, our data consist of counts or of percents obtained from counts. This chapter and the next present confidence intervals and significance tests for use with count data.

The parameters we want to do inference about are population proportions. Just as in the case of inference about population means, we may be concerned with a single population or with comparing two populations. Inference about proportions in these one-sample and two-sample settings is very similar to inference about means, which we met in Chapter 7. The methods of this chapter answer questions such as these:

■ What is the prevalence of frequent binge drinking among U.S. college students? A survey designed to answer this question sampled 17,096 students on 140 campuses. Are there gender differences in this behavior?

■ Does regularly taking aspirin help prevent heart attacks? A double-blind randomized comparative experiment assigned 11,037 male doctors to take aspirin and another 11,034 to take a placebo. After 5 years, 104 of the aspirin group and 189 of the control group had died of heart attacks. Is this difference large enough to convince us that aspirin works?

Bill Weege.

Inference for Proportions

Many statistical studies produce counts rather than measurements. An opinion poll asks a sample of adults whether they approve of the president's conduct of his office; the data are the counts of "Yes," "No," and "Don't know." An experiment compares the effectiveness of four treatments to prevent the common cold; the data are the number of subjects given each treatment who catch a cold during the next month. A researcher classifies each of a sample of students according to his or her major field of study and political preference; the data are the counts of students in each cell of a two-way table. This chapter presents procedures for statistical inference in these settings.

We begin in Section 8.1 with inference about a single population proportion. The statistical model for a count is then the binomial distribution, which we studied in Section 5.1. Section 8.2 concerns methods for comparing two proportions. Binomial distributions again play an important role.

8.1   Inference for a Single Proportion

We want to estimate the proportion p of some characteristic, such as approving the president's conduct of his office, among the members of a large population. We select a simple random sample (SRS) of size n from the population and record the count X of "successes" (such as "Yes" answers to a question about the president). We will use "success" as shorthand for whatever characteristic we are interested in. The sample proportion of successes $\hat{p} = X/n$ estimates the unknown population proportion p. If the population is much larger than the sample (say, at least 10 times as large), the individual responses are nearly independent and the count X has approximately the binomial distribution $B(n, p)$. In statistical terms, we are concerned with inference about the probability p of a success in the binomial setting.

If the sample size n is small, we must base tests and confidence intervals for p on the binomial distributions. These are awkward to work with because of the discreteness of the binomial distributions.[1] But we know that when the sample is large, both the count X and the sample proportion $\hat{p}$ are approximately normal. We will consider only inference procedures based on the normal approximation. These procedures are similar to those for inference about the mean of a normal distribution.

Confidence interval for a single proportion

The unknown population proportion p is estimated by the sample proportion $\hat{p} = X/n$. We know (Chapter 5, page 383) that if the sample size n is sufficiently large, $\hat{p}$ has approximately the normal distribution, with mean $\mu_{\hat{p}} = p$ and standard deviation $\sigma_{\hat{p}} = \sqrt{p(1-p)/n}$.

To find a confidence interval for p, we must estimate the standard deviation of $\hat{p}$ from the data. To do this, replace p by $\hat{p}$ in the expression for $\sigma_{\hat{p}}$. Recall that we used the term *standard error* in Chapter 7 for the standard deviation of a statistic when it is estimated from data.

Standard Error of a Sample Proportion

The standard error of $\hat{p}$ is

$$SE_{\hat{p}} = \sqrt{\frac{\hat{p}(1-\hat{p})}{n}}$$

The z procedure for a confidence interval involves two approximations: the normal approximation to the binomial distribution and the approximation of p by $\hat{p}$ in the standard deviation.

Large-Sample Confidence Interval for a Population Proportion

Draw an SRS of size n from a large population with unknown proportion p of successes. An **approximate level C confidence interval** for p is

$$\hat{p} \pm z^* SE_{\hat{p}}$$

where z^* is the value for the standard normal density curve with area C between $-z^*$ and z^*.

margin of error   The margin of error is

$$m = z^* SE_{\hat{p}}$$

The large-sample confidence interval is quite accurate and we recommend it when we have at least 10 successes and 10 failures. If this rule of thumb is not met, or if the sample is more than 10% of the population, accurate inference requires more elaborate procedures.

EXAMPLE 8.1   Alcohol abuse has been described by college presidents as the number one problem on campus and it is an important cause of death in young adults. How common is it? A survey of 17,096 students in U.S. four-year colleges collected information on drinking behavior and alcohol-related problems.[2] The researchers defined "frequent binge drinking" as having five or more drinks in a row three or more times in the past two weeks. According to this definition, 3314 students were classified as frequent binge drinkers. The proportion of drinkers is

$$\hat{p} = \frac{3314}{17,096} = 0.194$$

To find a 95% confidence interval, first compute the standard error:

$$SE_{\hat{p}} = \sqrt{\frac{\hat{p}(1-\hat{p})}{n}}$$

$$= \sqrt{\frac{(0.194)(1-0.194)}{17,096}}$$

$$= 0.00302$$

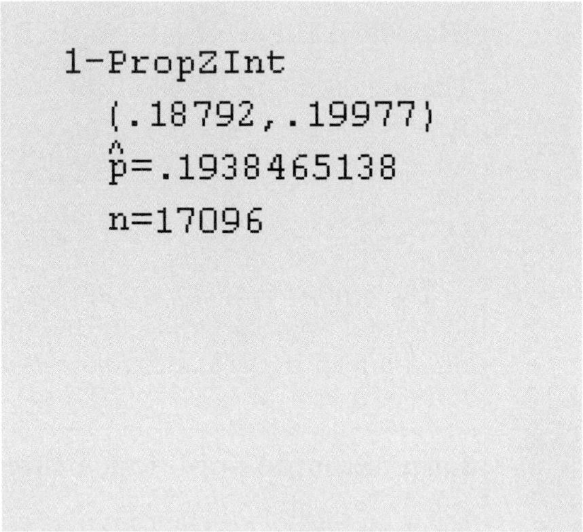

```
1-PropZInt
  (.18792,.19977)
  p=.1938465138
  n=17096
```

FIGURE 8.1 TI-83 output for Example 8.1.

From Table D we find the value of z^* to be 1.960. So the confidence interval is

$$\hat{p} \pm z^*\text{SE}_{\hat{p}} = 0.194 \pm (1.960)(0.00302)$$
$$= 0.194 \pm 0.006$$
$$= (0.188, 0.200)$$

Figure 8.1 gives the output for this example from the TI-83. We estimate with 95% confidence that between 18.8% and 20.0% of college students are frequent binge drinkers. Here is an alternative summary. We estimate that 19.4% of college students are frequent binge drinkers. For 95% confidence, the margin of error is 0.6%.

Remember that the margin of error in this confidence interval includes only random sampling error. There are other sources of error that are not taken into account. This survey used a design where the number of students sampled was proportional to the size of the college they attended; in this way we can treat the data as if we had an SRS. However, as is the case with many such surveys, we are forced to assume that the respondents provided accurate information. If the students did not answer the questions honestly, the results may be biased. Furthermore, we also have the typical problem of nonresponse. The response rate for this survey was 69%, an excellent rate for surveys of this type. Do the students who did not respond have different drinking habits than those who did? If so, this is another source of bias.

Significance test for a single proportion

Recall that the sample proportion $\hat{p} = X/n$ is approximately normal, with mean $\mu_{\hat{p}} = p$ and standard deviation $\sigma_{\hat{p}} = \sqrt{p(1-p)/n}$. For confidence intervals, we substitute $\hat{p}$ for p in the last expression to obtain the standard error. When performing a significance test, however, the null hypothesis specifies a

value for p, and we assume that this is the true value when calculating the
P-value. Therefore, when we test H_0: $p = p_0$, we substitute this p_0 into the
expression for $\sigma_{\hat{p}}$ and then standardize $\hat{p}$. Here are the details.

Large-Sample Significance Test for a Population Proportion

Draw an SRS of size n from a large population with unknown proportion
p of successes. To test the hypothesis H_0: $p = p_0$, compute the **z statistic**

$$z = \frac{\hat{p} - p_0}{\sqrt{\dfrac{p_0(1 - p_0)}{n}}}$$

In terms of a standard normal random variable Z, the approximate P-value
for a test of H_0 against

H_a: $p > p_0$ is $P(Z \geq z)$

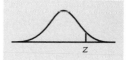

H_a: $p < p_0$ is $P(Z \leq z)$

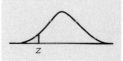

H_a: $p \neq p_0$ is $2P(Z \geq |z|)$

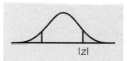

We recommend the large-sample z significance test as long as the expected number of successes (np_0) and the expected number of failures
($n(1 - p_0)$) are both greater than 10. If this rule of thumb is not met, or if
the population is less than 10 times as large as the sample, other procedures
should be used. Here is an example.

EXAMPLE 8.2 | The French naturalist Count Buffon once tossed a coin 4040 times and obtained 2048
heads. This is a binomial experiment with $n = 4040$. The sample proportion is

$$\hat{p} = \frac{2048}{4040} = 0.5069$$

If Buffon's coin was balanced, then the probability of obtaining heads on any toss is
0.5. To assess whether the data provide evidence that the coin was not balanced, we
test

$$H_0: p = 0.5$$
$$H_a: p \neq 0.5$$

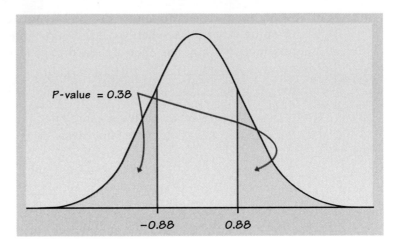

FIGURE 8.2   The P-value for Example 8.2.

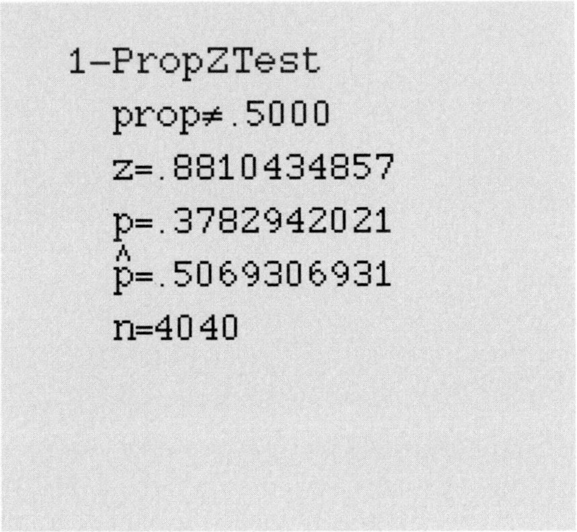

$$
\begin{aligned}
&\texttt{1-PropZTest} \\
&\quad \texttt{prop}\neq\texttt{.5000} \\
&\quad \texttt{z=.8810434857} \\
&\quad \texttt{p=.3782942021} \\
&\quad \hat{\texttt{p}}\texttt{=.5069306931} \\
&\quad \texttt{n=4040}
\end{aligned}
$$

FIGURE 8.3   TI-83 output for Example 8.2.

The test statistic is

$$
z = \frac{\hat{p} - 0.5}{\sqrt{\dfrac{(0.5)(0.5)}{4040}}} = \frac{0.5069 - 0.5}{\sqrt{\dfrac{(0.5)(0.5)}{4040}}} = 0.88
$$

Figure 8.2 illustrates the calculation of the P-value. From Table A we find $P(Z \leq 0.88) = 0.8106$. Therefore, the probability in each tail is $1 - 0.8106 = 0.1894$, and the P-value is $P = 2 \times 0.1894 = 0.38$. Figure 8.3 gives output from the TI-83 for this problem. Because these calculations are so simple, most software packages do not provide them. The data are compatible with the balanced-coin hypothesis ($\hat{p} = 0.5069$, $z = 0.88$, $P = 0.38$).

In this example we have arbitrarily chosen to associate the outcome "heads" with success and "tails" with failure. Suppose we reversed the choice.

If we observed 2048 heads, then the other $4040 - 2048 = 1992$ tosses were tails. Let's repeat the significance test with "tails" as the success outcome.

EXAMPLE 8.3

A coin was tossed $n = 4040$ times and we observed $X = 1992$ tails. We want to test the null hypothesis that the coin is fair—that is, that the probability of tails is 0.5. So p is the probability that the coin comes up tails and we test

$$H_0: p = 0.5$$
$$H_a: p \neq 0.5$$

The test statistic is

$$z = \frac{\hat{p} - 0.5}{\sqrt{\dfrac{(0.5)(0.5)}{4040}}} = \frac{0.4931 - 0.5}{\sqrt{\dfrac{(0.5)(0.5)}{4040}}} = -0.88$$

Using Table A, we find that $P = 0.38$.

When we interchanged heads and tails (or success and failure), we simply changed the sign of the test statistic z. The P-value remained the same. These facts are true in general. Our conclusion does not depend on an arbitrary choice of success and failure.

Unless a coin is perfectly balanced, its true probability of heads will not be *exactly* 0.5. The significance test in Example 8.2 demonstrates only that Buffon's coin is statistically indistinguishable from a fair coin on the basis of 4040 tosses.

Confidence intervals provide additional information

To see what other values of p are compatible with the sample results, we will calculate a confidence interval.

EXAMPLE 8.4

For Buffon's coin-tossing experiment, the standard error for the proportion of heads is

$$\begin{aligned} \text{SE}_{\hat{p}} &= \sqrt{\frac{\hat{p}(1 - \hat{p})}{n}} \\ &= \sqrt{\frac{(0.5069)(1 - 0.5069)}{4040}} \\ &= 0.00787 \end{aligned}$$

We will calculate a 99% confidence interval. From Table D we find $z^* = 2.576$, so the margin of error is

$$z^* \text{SE}_{\hat{p}} = (2.576)(0.00787) = 0.0203$$

So the 99% confidence interval for the probability of heads is

$$\begin{aligned} \hat{p} \pm z^* \text{SE}_{\hat{p}} &= 0.5069 \pm 0.0203 \\ &= (0.4866, \ 0.5272) \end{aligned}$$

We estimate that Buffon's coin comes up heads 50.7% of the time. The margin of error for 99% confidence is 2%. The 99% confidence interval is 48.7% to 52.7%.

The confidence interval of Example 8.4 is much more informative than the significance test of Example 8.2. We have determined the values of p that are consistent with the observed results. This is the range of p_0's that would not be rejected by a test at the $\alpha = 0.01$ level of significance. We would not be surprised if the true probability of heads for Buffon's coin were something like 0.51.

We do not often use significance tests for a single proportion because it is uncommon to have a situation where there is a precise p_0 that we want to test. For physical experiments such as coin tossing or drawing cards from a well-shuffled deck, probability arguments lead to an ideal p_0, as in Example 8.2. Even here, however, it can be argued that no real coin has a probability of heads *exactly* equal to 0.5. Data from past large samples can sometimes provide a p_0 for the null hypothesis of a significance test. In some types of epidemiology research, for example, "historical controls" from past studies serve as the benchmark for evaluating new treatments. Medical researchers argue about the validity of these approaches, because the past never quite resembles the present. In general, we prefer comparative studies whenever possible.

Choosing a sample size

In Chapter 6, we showed how to choose the sample size n to obtain a confidence interval with specified margin of error m for a normal mean. Because we are using a normal approximation for inference about a population proportion, sample size selection proceeds in much the same way.

Recall that the margin of error for the large-sample confidence interval for a population proportion is

$$m = z^* \text{SE}_{\hat{p}}$$

The critical value z^* depends on the confidence level C, and the standard error depends on the value of p and the sample size n:

$$\text{SE}_{\hat{p}} = \sqrt{\frac{\hat{p}(1 - \hat{p})}{n}}$$

Because the value of $\hat{p}$ is not known before the data are gathered, we must guess a value to use in the calculations. Let p^* be the guessed value. The margin of error of the confidence interval is largest when $\hat{p} = 0.5$. So a conservative approach to the problem is to use $p^* = 0.5$ as the guessed value. This ensures that our final confidence interval will have a margin of error less than or equal to the specified value. If m is the desired margin of error, set

$$m = z^* \sqrt{\frac{p^*(1 - p^*)}{n}}$$

and solve this equation for n to find the sample size required. Here is the result.

Sample Size for Desired Margin of Error

The level C confidence interval for a proportion p will have a margin of error approximately equal to a specified value m when the sample size is

$$n = \left(\frac{z^*}{m}\right)^2 p^*(1 - p^*)$$

where p^* is a guessed value for the true proportion.

The approximate margin of error will be less than or equal to m if p^* is chosen to be 0.5. This gives

$$n = \left(\frac{z^*}{2m}\right)^2$$

The value of n obtained by this method is not particularly sensitive to the choice of p^* when p^* is fairly close to 0.5. However, if the value of p is smaller than about 0.3 or larger than about 0.7, use of $p^* = 0.5$ may select a sample size that is much larger than needed.

EXAMPLE 8.5

You are doing a sample survey to estimate the proportion of voters who plan to vote for candidate Garcia in the coming election. Let p be the true proportion of voters who intend to vote for her. Because you judge that the value of p is not very different from 0.5, you use $p^* = 0.5$ for planning the sample size.

You want to estimate p with 95% confidence and a margin of error less than or equal to 3%, or 0.03. The sample size required is

$$n = \left(\frac{z^*}{2m}\right)^2 = \left[\frac{1.96}{(2)(0.03)}\right]^2 = 1067.1$$

which we round up to $n = 1068$. (Rounding down would give a margin of error slightly greater than 0.03.)

Similarly, for a 2.5% margin of error we have (after rounding up)

$$n = \left[\frac{1.96}{(2)(0.025)}\right]^2 = 1537$$

and for a 2% margin of error the required sample size is

$$n = \left[\frac{1.96}{(2)(0.02)}\right]^2 = 2401$$

News reports frequently describe the results of surveys with sample sizes between 1000 and 1500 and a margin of error of about 3%. These surveys generally use sampling procedures more complicated than simple random sampling, so the calculation of confidence intervals is more involved than what we have studied in this section. The calculations in Example 8.5 nonetheless show in principle how such surveys are planned.

In practice, many factors influence the choice of a sample size. The following example illustrates one set of factors.

EXAMPLE 8.6 | An association of Christmas tree growers in Indiana sponsored a sample survey of Indiana households to help improve the marketing of Christmas trees.[3] The researchers decided to use a telephone survey and estimated that each telephone interview would take about 2 minutes. Nine trained students in agribusiness marketing were to make the phone calls between 1:00 P.M. and 8:00 P.M. on a Sunday. After discussing problems related to people not being at home or being unwilling to answer the questions, the survey team proposed a sample size of 500. Several of the questions asked demographic information about the household; the key questions of interest had responses of "Yes" or "No." For example, "Did you have a Christmas tree this year?"

The primary purpose of the survey was to estimate various sample proportions for Indiana households. Would the proposed sample size of $n = 500$ be adequate to provide the sponsors of the research with the information they required? To address this question, we calculate the margins of error of 95% confidence intervals for various values of $\hat{p}$.

EXAMPLE 8.7 | In the Christmas tree market survey, the margin of error of a 95% confidence interval for any value of $\hat{p}$ and $n = 500$ is

$$m = z^*\mathrm{SE}_{\hat{p}}$$

$$= 1.96\sqrt{\frac{\hat{p}(1-\hat{p})}{500}}$$

$$= 0.088\sqrt{\hat{p}(1-\hat{p})}$$

The results for various values of $\hat{p}$ are

$\hat{p}$	m	$\hat{p}$	m
0.05	0.019	0.60	0.043
0.10	0.026	0.70	0.040
0.20	0.035	0.80	0.035
0.30	0.040	0.90	0.026
0.40	0.043	0.95	0.019
0.50	0.044		

The survey team judged these margins of error to be acceptable, and they used a sample size of 500 in their survey.

The table in Example 8.7 illustrates several points. First, the margins of error for $\hat{p} = 0.05$ and $\hat{p} = 0.95$ are the same. The margins of error will always be the same for $\hat{p}$ and $1 - \hat{p}$. This is a direct consequence of the form of the confidence interval. Second, the margin of error varies only between 0.040 and 0.044 as $\hat{p}$ varies from 0.3 to 0.7, and the margin of error is greatest when $\hat{p} = 0.5$, as we claimed earlier. It is true in general that the margin of error will vary relatively little for values of $\hat{p}$ between 0.3 and 0.7. Therefore, when planning a study, it is not necessary to have a very precise guess for p. If

$p^* = 0.5$ is used and the observed $\hat{p}$ is between 0.3 and 0.7, the actual interval will be a little shorter than needed but the difference will be small.

Finally, note that the margins of error m for very small and very large values of $\hat{p}$ are not included in the table. Suppose, for example, that we expect p to be about 0.01. Then for $n = 500$, we have $np = 5$. Recall that for the normal approximation to be reasonably accurate, we require $np \geq 10$ and $n(1 - p) \geq 10$. We would not be justified in using the normal approximation to the binomial to calculate the confidence interval when p is as small as 0.01. More advanced methods are needed in this case.

SUMMARY

Inference about a population proportion p from an SRS of size n is based on the **sample proportion** $\hat{p} = X/n$. When n is large, $\hat{p}$ has approximately the normal distribution with mean p and standard deviation $\sqrt{p(1-p)/n}$.

An **approximate level C confidence interval** for p is

$$\hat{p} \pm z^* \text{SE}_{\hat{p}}$$

where z^* is the value for the standard normal density curve with area C between $-z^*$ and z^*, and

$$\text{SE}_{\hat{p}} = \sqrt{\frac{\hat{p}(1 - \hat{p})}{n}}$$

is the standard error.

Tests of $H_0: p = p_0$ are based on the **z statistic**

$$z = \frac{\hat{p} - p_0}{\sqrt{\dfrac{p_0(1 - p_0)}{n}}}$$

with P-values calculated from the $N(0, 1)$ distribution.

The **sample size** required to obtain a confidence interval of approximate margin of error m for a proportion is

$$n = \left(\frac{z^*}{m}\right)^2 p^*(1 - p^*)$$

where p^* is a guessed value for the proportion, and z^* is the standard normal critical value for the desired level of confidence. To ensure that the margin of error of the interval is less than or equal to m no matter what p may be, use

$$n = \left(\frac{z^*}{2m}\right)^2$$

SECTION 8.1 EXERCISES

8.1   In each of the following cases state whether or not the normal
approximation to the binomial should be used for a confidence interval for
the population proportion p.

(a) $n = 30$ and we observe $\hat{p} = 0.9$.

(b) $n = 25$ and we observe $\hat{p} = 0.5$.

(c) $n = 100$ and we observe $\hat{p} = 0.04$.

(d) $n = 600$ and we observe $\hat{p} = 0.6$.

8.2   When trying to hire managers and executives, companies sometimes
verify the academic credentials described by the applicants. One company
that performs these checks summarized their findings for a six-month
period. Of the 84 applicants whose credentials were checked, 15 lied
about having a degree. (Data provided by Jude M. Werra & Associates,
Brookfield, Wis.)

(a) Find the proportion of applicants who lied about having a degree and
the standard error.

(b) Consider these data to be a random sample of credentials from a large
collection of similar applicants. Give a 90% confidence interval for the
true proportion of applicants who lie about having a degree.

8.3   Refer to the previous exercise. Suppose that 9 applicants lied about their
major. Can we conclude that a total of $24 = 15 + 9$ applicants lied about
having a degree or about their major? Explain your answer.

8.4   In the United States approximately 900 people die in bicycle accidents each
year. One study examined the records of 1711 bicyclists aged 15 or older
who were fatally injured in bicycle accidents between 1987 and 1991 and
were tested for alcohol. Of these, 542 tested positive for alcohol (blood
alcohol concentration of 0.01% or higher). (Data from Guohua Li and
Susan P. Baker, "Alcohol in fatally injured bicyclists," *Accident Analysis
and Prevention*, 26 (1994), pp. 543–548.)

(a) Summarize the data with appropriate descriptive statistics.

(b) To do statistical inference for these data, we think in terms of a model
where p is a parameter that represents the probability that a tested
bicycle rider is positive for alcohol. Find a 95% confidence interval
for p.

(c) Can you conclude from your statistical analysis of this study that
alcohol causes fatal bicycle accidents?

8.5   Refer to the previous exercise. In this study 386 bicyclists had blood alcohol
levels above 0.10%, a level defining legally drunk in many states. Give a 95%
confidence interval for the proportion who were legally drunk according to
this criterion.

8.6 One question in the Christmas tree market survey described in Examples 8.6 and 8.7 (page 594) was "Did you have a Christmas tree this year?" Of the 500 respondents, 421 answered "Yes."

(a) Find the sample proportion and its standard error.

(b) Give a 95% confidence interval for the proportion of Indiana households who had a Christmas tree last year.

8.7 As part of a quality improvement program, your mail-order company is studying the process of filling customer orders. According to company standards, an order is shipped on time if it is sent within 3 working days of the time it is received. You select an SRS of 100 of the 5000 orders received in the past month for an audit. The audit reveals that 86 of these orders were shipped on time. Find a 95% confidence interval for the true proportion of the month's orders that were shipped on time.

8.8 Large trees growing near power lines can cause power failures during storms when their branches fall on the lines. Power companies spend a great deal of time and money trimming and removing trees to prevent this problem. Researchers are developing hormone and chemical treatments that will stunt or slow tree growth. If the treatment is too severe, however, the tree will die. In one series of laboratory experiments on 216 sycamore trees, 41 trees died. Give a 99% confidence interval for the proportion of sycamore trees that would be expected to die from this particular treatment.

8.9 In recent years over 70% of first-year college students responding to a national survey have identified "being well-off financially" as an important personal goal. A state university finds that 132 of an SRS of 200 of its first-year students say that this goal is important. Give a 95% confidence interval for the proportion of all first-year students at the university who would identify being well-off as an important personal goal.

8.10 In each of the following cases state whether or not the normal approximation to the binomial should be used for a significance test on the population proportion p.

(a) $n = 10$ and $H_0: p = 0.4$.

(b) $n = 100$ and $H_0: p = 0.6$.

(c) $n = 1000$ and $H_0: p = 0.996$.

(d) $n = 500$ and $H_0: p = 0.3$.

8.11 Refer to the Christmas tree market survey described in Examples 8.6 and 8.7 (page 594). Of the 500 responding households, 38% were from rural areas (including small towns), and the other 62% were from urban areas (including suburbs). According to the census, 36% of Indiana households are in rural areas, and the remaining 64% are in urban areas. Let p be the proportion of rural respondents. Set up hypotheses about p_0 and perform a test of significance to examine how well the sample represents the state in regard to rural versus urban residence. Summarize your results.

8.12   Refer to the previous exercise. There we arbitrarily chose to state the hypotheses in terms of the proportion of rural respondents. We could as easily have used the proportion of *urban* respondents.

   (a) Write hypotheses to examine how well the sample represents the state in regard to rural versus urban residence in terms of the proportion of urban residents.

   (b) Perform the test of significance and summarize the results.

   (c) Compare your results with the results of the previous exercise. Summarize and generalize your conclusion.

8.13   The Gallup Poll asked a sample of 1785 U.S. adults, "Did you, yourself, happen to attend church or synagogue in the last 7 days?" Of the respondents, 750 said "Yes." Suppose (it is not, in fact, true) that Gallup's sample was an SRS.

   (a) Give a 99% confidence interval for the proportion of all U.S. adults who attended church or synagogue during the week preceding the poll.

   (b) Do the results provide good evidence that less than half of the population attended church or synagogue?

   (c) How large a sample would be required to obtain a margin of error of ±0.01 in a 99% confidence interval for the proportion who attend church or synagogue? (Use Gallup's result as the guessed value of p.)

8.14   A national opinion poll found that 44% of all American adults agree that parents should be given vouchers good for education at any public or private school of their choice. The result was based on a small sample. How large an SRS is required to obtain a margin of error of ±0.03 (that is, $\pm3\%$) in a 95% confidence interval? (Use the previous poll's result to obtain the guessed value p^*.)

8.15   An entomologist samples a field for egg masses of a harmful insect by placing a yard-square frame at random locations and carefully examining the ground within the frame. An SRS of 75 locations selected from a county's pastureland found egg masses in 13 locations. Give a 95% confidence interval for the proportion of all possible locations that are infested.

8.16   Of the 500 respondents in the Christmas tree market survey, 44% had no children at home and 56% had at least one child at home. The corresponding figures for the most recent census are 48% with no children and 52% with at least one child. Test the null hypothesis that the telephone survey technique has a probability of selecting a household with no children that is equal to the value obtained by the census. Give the z statistic and the P-value. What do you conclude?

8.17   The South African mathematician John Kerrich tossed a coin 10,000 times and obtained 5067 heads.

(a) Is this significant evidence at the 5% level that the probability that Kerrich's coin comes up heads is not 0.5?

(b) Use a 95% confidence interval to find the range of probabilities of heads that would not be rejected at the 5% level.

8.18 A matched pairs experiment compares the taste of instant versus fresh-brewed coffee. Each subject tastes two unmarked cups of coffee, one of each type, in random order and states which he or she prefers. Of the 50 subjects who participate in the study, 19 prefer the instant coffee. Let p be the probability that a randomly chosen subject prefers freshly brewed coffee to instant coffee. (In practical terms, p is the proportion of the population who prefer fresh-brewed coffee.)

(a) Test the claim that a majority of people prefer the taste of fresh-brewed coffee. Report the z statistic and its P-value. Is your result significant at the 5% level? What is your practical conclusion?

(b) Find a 90% confidence interval for p.

8.19 Leroy, a starting player for a major college basketball team, made only 38.4% of his free throws last season. During the summer he worked on developing a softer shot in the hope of improving his free-throw accuracy. In the first eight games of this season Leroy made 25 free throws in 40 attempts. Let p be his probability of making each free throw he shoots this season.

(a) State the null hypothesis H_0 that Leroy's free-throw probability has remained the same as last year and the alternative H_a that his work in the summer resulted in a higher probability of success.

(b) Calculate the z statistic for testing H_0 versus H_a.

(c) Do you accept or reject H_0 for $\alpha = 0.05$? Find the P-value.

(d) Give a 90% confidence interval for Leroy's free-throw success probability for the new season. Are you convinced that he is now a better free-throw shooter than last season?

(e) What assumptions are needed for the validity of the test and confidence interval calculations that you performed?

8.20 You want to estimate the proportion of students at your college or university who are employed for 10 or more hours per week while classes are in session. You plan to present your results by a 95% confidence interval. Using the guessed value $p^* = 0.35$, find the sample size required if the interval is to have an approximate margin of error of $m = 0.05$.

8.21 Land's Beginning is a company that sells its merchandise through the mail. It is considering buying a list of addresses from a magazine. The magazine claims that at least 20% of its subscribers have high incomes (they define this to be household income in excess of $100,000). Land's Beginning would like to estimate the proportion of high-income people on the list. Checking income is very difficult and expensive but another company offers this service. Land's Beginning will pay to find incomes for an SRS of people on the magazine's list. They would like the margin of error of the 95% confidence interval for the proportion to be 0.05 or less. Use the guessed value $p^* = 0.2$ to find the required sample size.

8.22   Refer to the previous exercise. For each of the following variations on the design specifications, state whether the required sample size will be higher, lower, or the same as that found above.

(a)   Use a 99% confidence interval.

(b)   Change the allowable margin of error to 0.01.

(c)   Use a planning value of $p^* = 0.15$.

(d)   Use a different company to do the income checks.

8.23   A student organization wants to start a nightclub for students under the age of 21. To assess support for this proposal, they will select an SRS of students and ask each respondent if he or she would patronize this type of establishment. They expect that about 70% of the student body would responds favorably. What sample size is required to obtain a 90% confidence interval with an approximate margin of error of 0.04? Suppose that 50% of the sample responds favorably. Calculate the margin of error of the 90% confidence interval.

8.24   An automobile manufacturer would like to know what proportion of its customers is dissatisfied with the service received from their local dealer. The customer relations department will survey a random sample of customers and compute a 99% confidence interval for the proportion that is dissatisfied. From past studies, they believe that this proportion will be about 0.2. Find the sample size needed if the margin of error of the confidence interval is to be about 0.015. Suppose 10% of the sample say that they are dissatisfied. What is the margin of error of the 99% confidence interval?

8.25   You have been asked to survey students at a large college to determine the proportion who favor an increase in student fees to support an expansion of the student newspaper. Each student will be asked whether he or she is in favor of the proposed increase. Using records provided by the registrar you can select a random sample of students from the college. After careful consideration of your resources, you decide that it is reasonable to conduct a study with a sample of 100 students.

(a)   For this sample size, construct a table of the margins of error for 95% confidence intervals when $\hat{p}$ takes the values 0.1, 0.2, 0.3, 0.4, 0.5, 0.6, 0.7, 0.8, and 0.9.

(b)   For a sample of size 100, would you use a confidence interval based on a normal approximation if $\hat{p} = 0.04$? Why or why not?

8.26   A former editor of the student newspaper agrees to underwrite the study in the previous exercise because she believes the results will demonstrate that most students support an increase in fees. She is willing to provide funds for a sample of size 500. Answer all of the questions posed in the previous exercise. Then write a short summary for your benefactor of why the increased sample size will provide better results.

8.2   Comparing Two Proportions

Because comparative studies are so common, we often want to compare the proportions of two groups (such as men and women) that have some characteristic. We call the two groups being compared Population 1 and Population 2, and the two population proportions of "successes" p_1 and p_2. The data consist of two independent SRSs, of size n_1 from Population 1 and size n_2 from Population 2. The proportion of successes in each sample estimates the corresponding population proportion. Here is the notation we will use in this section:

Population	Population proportion	Sample size	Count of successes	Sample proportion
1	p_1	n_1	X_1	$\hat{p}_1 = \dfrac{X_1}{n_1}$
2	p_2	n_2	X_2	$\hat{p}_2 = \dfrac{X_2}{n_2}$

To compare the two populations, we use the difference between the two sample proportions:

$$D = \hat{p}_1 - \hat{p}_2$$

When both sample sizes are sufficiently large, the sampling distribution of the difference D is approximately normal.

Inference procedures for comparing proportions are z procedures based on the normal approximation and on standardizing the difference D. The first step is to obtain the mean and standard deviation of D. By the addition rule for means, the mean of D is the difference of the means:

$$\mu_D = \mu_{\hat{p}_1} - \mu_{\hat{p}_2} = p_1 - p_2$$

That is, the difference $D = \hat{p}_1 - \hat{p}_2$ between the sample proportions is an unbiased estimator of the population difference $p_1 - p_2$. Similarly, the addition rule for variances tells us that the variance of D is the *sum* of the variances,

$$\sigma_D^2 = \sigma_{\hat{p}_1^2} + \sigma_{\hat{p}_2^2}$$

$$= \frac{p_1(1 - p_1)}{n_1} + \frac{p_2(1 - p_2)}{n_2}$$

Therefore, when n_1 and n_2 are large, D is approximately normal with mean $\mu_D = p_1 - p_2$ and standard deviation

$$\sigma_D = \sqrt{\frac{p_1(1 - p_1)}{n_1} + \frac{p_2(1 - p_2)}{n_2}}$$

Confidence intervals

For both confidence intervals and hypothesis tests, we need to substitute a value for the unknown standard deviation σ_D. First, consider a confidence interval for $p_1 - p_2$. Substitute the sample estimates $\hat{p}_1$ and $\hat{p}_2$ for p_1 and p_2 in the expression for σ_D to obtain

$$\text{SE}_D = \sqrt{\frac{\hat{p}_1(1 - \hat{p}_1)}{n_1} + \frac{\hat{p}_2(1 - \hat{p}_2)}{n_2}}$$

This is the standard error of D that we will use in our confidence interval calculations.

Confidence Intervals for Comparing Two Proportions

Draw an SRS of size n_1 from a large population having proportion p_1 of successes and an independent SRS of size n_2 from another population having proportion p_2 of successes. When n_1 and n_2 are large, an **approximate level C confidence interval** for $p_1 - p_2$ is

$$(\hat{p}_1 - \hat{p}_2) \pm z^* \text{SE}_D$$

where

$$\text{SE}_D = \sqrt{\frac{\hat{p}_1(1 - \hat{p}_1)}{n_1} + \frac{\hat{p}_2(1 - \hat{p}_2)}{n_2}}$$

and z^* is the value for the standard normal density curve with area C between $-z^*$ and z^*.

This interval is approximately correct when the sample sizes n_1 and n_2 are large. As a general rule, we will use this method when the number of successes and failures in each sample is at least 5.

In Example 8.1 (page 587) we estimated the proportion of college students who engage in frequent binge drinking. Are there characteristics of these students that relate to this behavior? For example, how similar is this behavior in men and women? This kind of follow-up question is natural in many studies such as this one. We will first use a confidence interval to examine it.

EXAMPLE 8.8 In the binge-drinking study, data were also summarized by gender. Here is a summary:

Population	n	X	$\hat{p} = X/n$
1 (men)	7,180	1,630	0.227
2 (women)	9,916	1,684	0.170
Total	17,096	3,314	0.194

In this table the $\hat{p}$ column gives the sample proportions of frequent binge drinkers. The last line gives the totals that we studied in Example 8.1. We will find a 95% confidence interval for the difference between the proportion of men and women who are frequent binge drinkers. Output from the TI-83 is given in Figure 8.4. To perform the computations using our formulas, we first calculate the standard error of the observed difference:

$$\text{SE}_D = \sqrt{\frac{\hat{p}_1(1 - \hat{p}_1)}{n_1} + \frac{\hat{p}_2(1 - \hat{p}_2)}{n_2}}$$

$$= \sqrt{\frac{(0.227)(0.773)}{7180} + \frac{(0.170)(0.830)}{9916}}$$

$$= 0.00622$$

The 95% confidence interval is

$$(\hat{p}_1 - \hat{p}_2) \pm z^* \text{SE}_D = (0.227 - 0.170) \pm (1.960)(0.00622)$$

$$= 0.057 \pm 0.012$$

$$= (0.045, 0.069)$$

With 95% confidence we can say that the difference in the proportions is between 0.045 and 0.069. Alternatively, we can report that the men are 5.7% more likely to be frequent binge drinkers with a 95% margin of error of 1.2%.

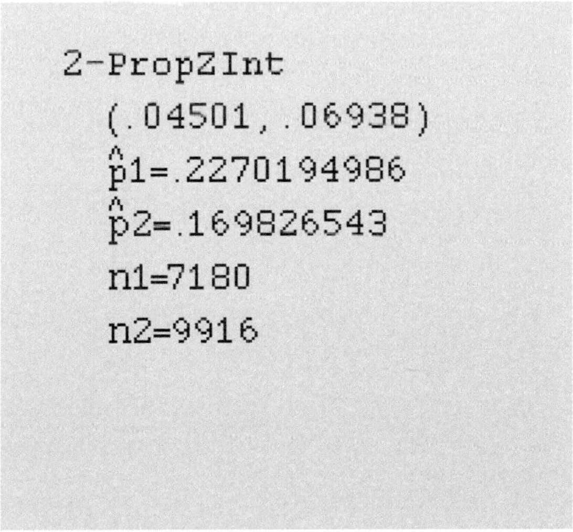

```
2-PropZInt
  (.04501,.06938)
  p̂1=.2270194986
  p̂2=.169826543
  n1=7180
  n2=9916
```

FIGURE 8.4 TI-83 output for Example 8.8.

In this example men and women were not sampled separately. The sample sizes are in fact random and reflect the gender distributions of the colleges that were randomly chosen. Two-sample significance tests and confidence intervals are still approximately correct in this situation. The authors of the report note that women are somewhat overrepresented partly because 6 of the 140 colleges in the study were all-women institutions.

In the example above we chose men to be the first population. Had we chosen women to be the first population, the estimate of the difference would be negative (-0.057). Because it is easier to discuss positive numbers, we generally choose the first population to be the one with the higher proportion.

Significance tests

Although we prefer to compare two proportions by giving a confidence interval for the difference between the two population proportions, it is sometimes useful to test the null hypothesis that the two population proportions are the same.

We standardize $D = \hat{p}_1 - \hat{p}_2$ by subtracting its mean $p_1 - p_2$ and then dividing by its standard deviation

$$\sigma_D = \sqrt{\frac{p_1(1 - p_1)}{n_1} + \frac{p_2(1 - p_2)}{n_2}}$$

If n_1 and n_2 are large, the standardized difference is approximately $N(0, 1)$. For the large-sample confidence interval we used sample estimates in place of the unknown population values in the expression for σ_D. Although this approach would lead to a valid significance test, we instead adopt the more common practice of replacing the unknown σ_D with an estimate that takes into account that our null hypothesis states that $p_1 = p_2$. If these two proportions are equal, then we can view all of the data as coming from a single population. Let p denote the common value of p_1 and p_2; then the standard deviation of $D = \hat{p}_1 - \hat{p}_2$ is

$$\sigma_D = \sqrt{\frac{p(1 - p)}{n_1} + \frac{p(1 - p)}{n_2}}$$

$$= \sqrt{p(1 - p)\left(\frac{1}{n_1} + \frac{1}{n_2}\right)}$$

We estimate the common value of p by the overall proportion of successes in the two samples:

$$\hat{p} = \frac{\text{number of successes in both samples}}{\text{number of observations in both samples}} = \frac{X_1 + X_2}{n_1 + n_2}$$

pooled estimate of p   This estimate of p is called the **pooled estimate** because it combines, or pools, the information from both samples.

To estimate σ_D under the null hypothesis, we substitute $\hat{p}$ for p in the expression for σ_D. The result is a standard error for D that assumes the truth of H_0: $p_1 = p_2$:

$$SE_{Dp} = \sqrt{\hat{p}(1 - \hat{p})\left(\frac{1}{n_1} + \frac{1}{n_2}\right)}$$

The subscript on SE_{Dp} reminds us that we pooled data from the two samples to construct the estimate.

Significance Tests for Comparing Two Proportions

To test the hypothesis

$$H_0: p_1 = p_2$$

compute the z statistic

$$z = \frac{\hat{p}_1 - \hat{p}_2}{SE_{Dp}}$$

where the pooled standard error is

$$SE_{Dp} = \sqrt{\hat{p}(1 - \hat{p})\left(\frac{1}{n_1} + \frac{1}{n_2}\right)}$$

and

$$\hat{p} = \frac{X_1 + X_2}{n_1 + n_2}$$

In terms of a standard normal random variable Z, the P-value for a test of H_0 against

$H_a: p_1 > p_2$ is $P(Z \geq z)$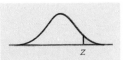

$H_a: p_1 < p_2$ is $P(Z \leq z)$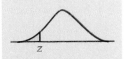

$H_a: p_1 \neq p_2$ is $2P(Z \geq |z|)$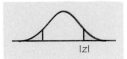

This z test is based on the normal approximation to the binomial distribution. As a general rule, we will use it when the number of successes and the number of failures in each of the samples is at least 5.

EXAMPLE 8.9

Are men and women college students equally likely to be frequent binge drinkers? We examine the survey data in Example 8.8 (page 603) to answer this question. Here is the data summary:

Population	n	X	$\hat{p} = X/n$
1 (men)	7,180	1,630	0.227
2 (women)	9,916	1,684	0.170
Total	17,096	3,314	0.194

The sample proportions are certainly quite different, but we will perform a significance test to see if the difference is large enough to lead us to believe that the population proportions are not equal. Formally, we test the hypotheses

$$H_0: p_1 = p_2$$
$$H_a: p_1 \neq p_2$$

The pooled estimate of the common value of p is

$$\hat{p} = \frac{1630 + 1684}{7180 + 9916} = \frac{3314}{17,096} = 0.194$$

Note that this was the estimate on the bottom line of the data summary above. It is the one we used in Example 8.1 (page 603) to estimate the proportion of all college students who are frequent binge drinkers.

The test statistic is calculated as follows:

$$SE_{Dp} = \sqrt{(0.194)(0.806)\left(\frac{1}{7180} + \frac{1}{9916}\right)} = 0.006126$$

$$z = \frac{\hat{p}_1 - \hat{p}_2}{SE_{Dp}} = \frac{0.227 - 0.170}{0.006126}$$

$$= 9.34$$

The P-value is $2P(Z \geq 9.34)$. The largest value of z in Table A is 3.49, so from this table we can conclude $P < 2 \times 0.0002 = 0.0004$. Most software reports this result as simply 0 or a very small number. Output from the TI-83 is given in Figure 8.5. It reports the P-value as 1.011212E-20. The E-20 means that we should move the decimal point 20 places to the left; this is certainly a very small number. The exact value is not particularly important; it is clear that we should reject the null hypothesis. We report: among college students in the study, 22.7% of the men and 17.0% of the women were frequent binge drinkers; the difference is statistically significant ($z = 9.34, P < 0.0001$).

We could have argued that we expect the proportion to be higher for men than for women in this example. This would justify using the one-sided alternative $H_a: p_1 > p_2$. The P-value would be half of the value obtained for the two-sided test. Because the z statistic is so large, this distinction is of no practical importance.

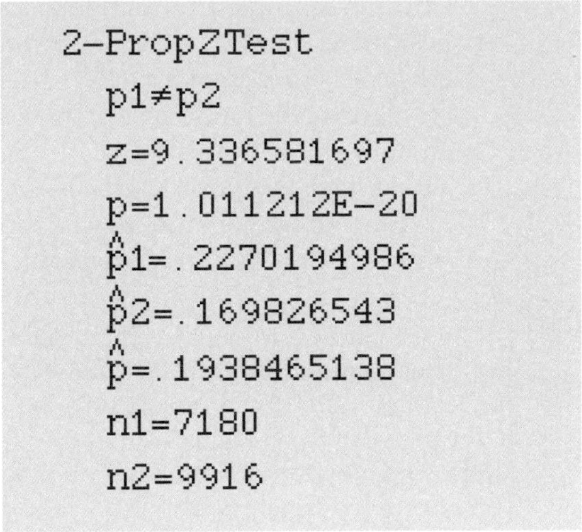

```
2-PropZTest
  p1≠p2
  z=9.336581697
  p=1.011212E-20
  p̂1=.2270194986
  p̂2=.169826543
  p̂=.1938465138
  n1=7180
  n2=9916
```

FIGURE 8.5   TI-83 output for Example 8.9.

Beyond the basics ▶ relative risk

relative risk

We summarized the comparison of the frequent binge drinking proportion for men and women by reporting a confidence interval for the difference in Example 8.8. Another way to summarize the comparison is to use the ratio of the two proportions, which is called the **relative risk** (RR) in many applications. Note that a relative risk of 1 means that the two proportions, $\hat{p}_1$ and $\hat{p}_2$, are equal. The procedure for calculating confidence intervals for relative risk is based on the same kind of principles that we have studied, but the details are somewhat more complicated. Fortunately, we can leave the details to software and concentrate on interpretation and communication of the results.

EXAMPLE 8.10

In Example 8.8 (page 603) we summarized the data on the proportions of men and women who are frequent binge drinkers with the following table:

Population	n	X	$\hat{p} = X/n$
1 (men)	7,180	1,630	0.227
2 (women)	9,916	1,684	0.170
Total	17,096	3,314	0.194

The relative risk is

$$\text{RR} = \frac{\hat{p}_1}{\hat{p}_2} = \frac{0.227}{0.170} = 1.34$$

Software (for example, PROC FREQ with the MEASURES option in SAS) gives the 95% confidence interval as

$$(1.26, 1.42)$$

Men are 1.34 times as likely as women to be frequent binge drinkers; the 95% confidence interval is (1.26, 1.42).

In this example the confidence interval appears to be symmetric about the estimate: that is, 1.34 is in the middle of 1.26 and 1.42. If we reported the results with more accuracy (RR = 1.337, 95% confidence interval = 1.258 to 1.421), we would see that the interval is *not* symmetric, and this is true in general.

Relative risk is particularly useful when the proportions that we are studying are small. In epidemiology and many areas of medical statistics it is routinely used. Here is another example.

EXAMPLE 8.11

There is much evidence that high blood pressure is associated with increased risk of death from cardiovascular disease. A major study of this association examined 3338 men with high blood pressure and 2676 men with low blood pressure. During the period of the study, 21 men in the low-blood-pressure and 55 in the high-blood-pressure group died from cardiovascular disease.[4] We choose the group with the higher proportion to be Population 1 so that the RR will be greater than 1. For men with low blood pressure, the proportion who died is 0.0078, while the proportion for those with high blood pressure is 0.0165. The difference is statistically significant ($z = 2.98, P = 0.0014$) and the 95% confidence interval for the difference in proportions is (0.0032, 0.0141). Let's see if relative risk helps us to communicate the results of this study more effectively. The relative risk is

$$\text{RR} = \frac{\hat{p}_1}{\hat{p}_2} = \frac{0.0165}{0.0078} = 2.1$$

Software gives the 95% confidence interval

$$(1.3, 3.5)$$

The risk of dying from cardiovascular disease is doubled for men with high blood pressure compared with men who have low blood pressure (RR = 2.1, 95% confidence interval = 1.3 to 3.5).

SUMMARY

Comparison of two population proportions from independent SRSs of sizes n_1 and n_2 is based on the **difference of sample proportions** $D = \hat{p}_1 - \hat{p}_2$. The approximate **level C confidence interval** for $p_1 - p_2$ is

$$(\hat{p}_1 - \hat{p}_2) \pm z^* \text{SE}_D$$

where

$$\text{SE}_D = \sqrt{\frac{\hat{p}_1(1 - \hat{p}_1)}{n_1} + \frac{\hat{p}_2(1 - \hat{p}_2)}{n_2}}$$

is the standard error of the difference.

Significance tests of H_0: $p_1 = p_2$ use the **z statistic**

$$z = \frac{\hat{p}_1 - \hat{p}_2}{\text{SE}_{Dp}}$$

with P-values from the $N(0, 1)$ distribution. In this statistic,

$$SE_{Dp} = \sqrt{\hat{p}(1-\hat{p})\left(\frac{1}{n_1} + \frac{1}{n_2}\right)}$$

and $\hat{p}$ is the **pooled estimate** of the common value of p_1 and p_2:

$$\hat{p} = \frac{X_1 + X_2}{n_1 + n_2}$$

Relative risk is the ratio of two sample proportions:

$$RR = \frac{\hat{p}_1}{\hat{p}_2}$$

Confidence intervals for relative risk are often used to summarize the comparison of two proportions.

SECTION 8.2 EXERCISES

8.27   To what extent do syntax textbooks, which analyze the structure of sentences, illustrate gender bias? A study of this question sampled sentences from ten texts. One part of the study examined the use of the words "girl," "boy," "man," and "woman." We will call the first two words juvenile and the last two adult. Is the proportion of female references that are juvenile (girl) equal to the proportion of male references that are juvenile (boy)? Here are data from one of the texts. (From Monica Macaulay and Colleen Brice, "Don't touch my projectile: gender bias and stereotyping in syntactic examples," *Language*, 73 no. 4 (1997), pp. 798–825. Note, the first part of the title is a direct quote from one of the texts.)

Gender	n	X (juvenile)
Female	60	48
Male	132	52

(a) Find the proportion of juvenile references for females and its standard error. Do the same for the males.

(b) Give a 95% confidence interval for the difference and briefly summarize what the data show.

8.28   In Exercise 8.4 (page 596) we examined the percent of fatally injured bicyclists tested for alcohol who tested positive. Here we examine the same data with respect to gender.

Gender	n	X (tested positive)
Female	191	27
Male	1520	515

(a) Summarize the data by giving the proportions and a 90% confidence interval for the difference.

(b) The standard error SE_D is calculated from the standard errors for the two samples, $\sqrt{\hat{p}_1(1 - \hat{p}_1)/n_1}$ and $\sqrt{\hat{p}_2(1 - \hat{p}_2)/n_2}$. Which of these contributes the largest amount to the standard error of the difference? Explain why.

8.29 Is lying about credentials by job applicants changing? In Exercise 8.2 (page 596) we looked at the proportion of applicants who lied about having a degree in a six-month period. To see if there is a change over time, we can compare that period with the following six months. Here are the data:

Period	n	X (lied)
1	84	15
2	106	21

Use a 90% confidence interval to address the question of interest.

8.30 Exercise 8.27 addresses a question about gender bias with a confidence interval. Set up the problem as a significance test. Carry out the test and summarize the results.

8.31 The proportions of female and male fatally injured bicyclists were compared with a confidence interval in Exercise 8.28. Examine the same data with a test of significance.

8.32 Data on the proportion of applicants who lied about having a degree in two consecutive six-month periods are given in Exercise 8.29. Formulate appropriate null and alternative hypotheses that can be addressed with these data, carry out the significance test and summarize the results.

8.33 In the Christmas tree survey introduced in Example 8.6 (page 597), respondents who had a tree during the holiday season were asked whether the tree was natural or artificial. Respondents were also asked if they lived in an urban area or in a rural area. Of the 421 households displaying a Christmas tree, 160 lived in rural areas and 261 were urban residents. The tree growers want to know if there is a difference in preference for natural trees versus artificial trees between urban and rural households. Here are the data:

Population	n	X (natural)
1 (rural)	160	64
2 (urban)	261	89

(a) Give the null and alternative hypotheses that are appropriate for this problem assuming that we have no prior information suggesting that one population would have a higher preference than the other.

(b) Test the null hypothesis. Give the test statistic, the P-value, and summarize the results.

(c) Give a 90% confidence interval for the difference in proportions.

8.34   In the 1996 regular baseball season, the World Series Champion New York Yankees played 80 games at home and 82 games away. They won 49 of their home games and 43 of the games played away. We can consider these games as samples from potentially large populations of games played at home and away. How much advantage does the Yankee home field provide?

(a) Find the proportion of wins for the home games. Do the same for the away games.

(b) Find the standard error needed to compute a confidence interval for the difference in the proportions.

(c) Compute a 90% confidence interval for the difference between the probability that the Yankees win at home and the probability that they win when on the road. Are you convinced that the 1996 Yankees were more likely to win at home?

8.35   The state agriculture department asked random samples of Indiana farmers in each county whether they favored a mandatory corn checkoff program to pay for corn product marketing and research. In Tippecanoe County, 263 farmers were in favor of the program and 252 were not. In neighboring Benton County, 260 were in favor and 377 were not.

(a) Find the proportions of farmers in favor of the program in each of the two counties.

(b) Find the standard error needed to compute a confidence interval for the difference in the proportions.

(c) Compute a 99% confidence interval for the difference between the proportions of farmers favoring the program in Tippecanoe County and in Benton County. Do you think opinions differed in the two counties?

8.36   Return to the New York Yankees baseball data in Exercise 8.34.

(a) Combining all of the games played, what proportion did the Yankees win?

(b) Find the standard error needed for testing that the probability of winning is the same at home and away.

(c) Most people think that it is easier to win at home than away. Formulate null and alternative hypotheses to examine this idea.

(d) Compute the z statistic and its P-value. What conclusion do you draw?

8.37   Return to the survey of farmers described in Exercise 8.35.

(a) Combine the two samples and find the overall proportion of farmers who favor the corn checkoff program.

(b) Find the standard error needed for testing that the population proportions of farmers favoring the program are the same in the two counties.

(c) Formulate null and alternative hypotheses for comparing the two counties.

(d) Compute the z statistic and its P-value. What conclusion do you draw?

8.38 A major court case on the health effects of drinking contaminated water took place in the town of Woburn, Massachusetts. A town well in Woburn was contaminated by industrial chemicals. During the period that residents drank water from this well, there were 16 birth defects among 414 births. In years when the contaminated well was shut off and water was supplied from other wells, there were 3 birth defects among 228 births. The plaintiffs suing the firm responsible for the contamination claimed that these data show that the rate of birth defects was higher when the contaminated well was in use. How statistically significant is the evidence? What assumptions does your analysis require? Do these assumptions seem reasonable in this case?

8.39 A study of chromosome abnormalities and criminality examined data on 4124 Danish males born in Copenhagen. Each man was classified as having a criminal record or not, using the penal registers maintained in the offices of the local police chiefs. Each was also classified as having the normal male XY chromosome pair or one of the abnormalities XYY or XXY. Of the 4096 men with normal chromosomes, 381 had criminal records, while 8 of the 28 men with chromosome abnormalities had criminal records. Some experts believe that chromosome abnormalities are associated with increased criminality. Do these data lend support to this belief? Report your analysis and draw a conclusion. (Data from H. A. Witkin et al., "Criminality in XYY and XXY men," *Science*, 193 (1976), pp. 547–555.)

8.40 The 1958 Detroit Area Study was an important sociological investigation of the influence of religion on everyday life. It is described in Gerhard Lenski, *The Religious Factor*, Doubleday, New York, 1961. The sample "was basically a simple random sample of the population of the metropolitan area." Of the 656 respondents, 267 were white Protestants and 230 were white Catholics. One question asked whether the government was doing enough in areas such as housing, unemployment, and education; 161 of the Protestants and 136 of the Catholics said "No." Is there evidence that white Protestants and white Catholics differed on this issue?

8.41 The respondents in the Detroit Area Study (see the previous exercise) were also asked whether they believed that the right of free speech included the right to make speeches in favor of communism. Of the white Protestants, 104 said "Yes," while 75 of the white Catholics said "Yes." Give a 95% confidence interval for the amount by which the proportion of Protestants who agreed that communist speeches are protected exceeds the proportion of Catholics who held this opinion.

8.42   A university financial aid office polled an SRS of undergraduate students to study their summer employment. Not all students were employed the previous summer. Here are the results for men and women:

	Men	Women
Employed	718	593
Not employed	79	139
Total	797	732

(a) Is there evidence that the proportion of male students employed during the summer differs from the proportion of female students who were employed? State H_0 and H_a, compute the test statistic, and give its P-value.

(b) Give a 99% confidence interval for the difference between the proportions of male and female students who were employed during the summer. Does the difference seem practically important to you?

8.43   The power takeoff driveline on farm tractors is a potentially serious hazard to farmers. A shield covers the driveline on new tractors, but for a variety of reasons, the shield is often missing on older tractors. Two types of shield are the bolt-on and the flip-up. A study initiated by the National Safety Council took a sample of older tractors to examine the proportions of shields removed. The study found that 35 shields had been removed from the 83 tractors having bolt-on shields and that 15 had been removed from the 136 tractors with flip-up shields. (Data from W. E. Sell and W. E. Field, "Evaluation of PTO master shield usage on John Deere tractors," paper presented at the American Society of Agricultural Engineers 1984 Summer Meeting.)

(a) Test the null hypothesis that there is no difference between the proportions of the two types of shields removed. Give the z statistic and the P-value. State your conclusion in words.

(b) Give a 90% confidence interval for the difference in the proportions of removed shields for the bolt-on and the flip-up types. Based on the data, what recommendation would you make about the type of shield to be used on new tractors?

8.44   A clinical trial examined the effectiveness of aspirin in the treatment of cerebral ischemia (stroke). Patients were randomized into treatment and control groups. The study was double-blind in the sense that neither the patients nor the physicians who evaluated the patients knew which patients received aspirin and which the placebo tablet. After 6 months of treatment, the attending physicians evaluated each patient's progress as either favorable or unfavorable. Of the 78 patients in the aspirin group, 63 had favorable outcomes; 43 of the 77 control patients had favorable outcomes. (From William S. Fields et al., "Controlled trial of aspirin in cerebral ischemia," *Stroke,* 8 (1977), pp. 301–315.)

(a) Compute the sample proportions of patients having favorable outcomes in the two groups.

(b) Give a 95% confidence interval for the difference between the favorable proportions in the treatment and control groups.

(c) The physicians conducting the study had concluded from previous research that aspirin was likely to increase the chance of a favorable outcome. Carry out a significance test to confirm this conclusion. State hypotheses, find the P-value, and write a summary of your results.

8.45 The pesticide diazinon is in common use to treat infestations of the German cockroach, *Blattella germanica*. A study investigated the persistence of this pesticide on various types of surfaces. Researchers applied a 0.5% emulsion of diazinon to glass and plasterboard. After 14 days, they placed 18 cockroaches on each surface and recorded the number that died within 48 hours. On glass, 9 cockroaches died, while on plasterboard, 13 died. (Based on Elray M. Roper and Charles G. Wright, "German cockroach (Orthoptera: Blatellidae) mortality on various surfaces following application of diazinon," *Journal of Economic Entomology*, 78 (1985), pp. 733–737.)

(a) Calculate the mortality rates (sample proportion that died) for the two surfaces.

(b) Find a 90% confidence interval for the difference in the two population proportions.

(c) Chemical analysis of the residues of diazinon suggests that it may persist longer on plasterboard than on glass because it binds to the paper covering on the plasterboard. The researchers therefore expected the mortality rate to be greater on plasterboard than on glass. Conduct a significance test to assess the evidence that this is true.

8.46 Refer to the study of undergraduate student summer employment described in Exercise 8.42. Similar results from a smaller number of students may not have the same statistical significance. Specifically, suppose that 72 of 80 men surveyed were employed and 59 of 73 women surveyed were employed. The sample proportions are essentially the same as in the earlier exercise.

(a) Compute the z statistic for these data and report the P-value. What do you conclude?

(b) Compare the results of this significance test with your results in Exercise 8.42. What do you observe about the effect of the sample size on the results of these significance tests?

8.47 Suppose that the experiment of Exercise 8.45 placed more cockroaches on each surface and observed similar mortality rates. Specifically, suppose that 36 cockroaches were placed on each surface and that 26 died on the plasterboard, while 18 died on the glass.

(a) Compute the z statistic for these data and report its P-value. What do you conclude?

(b) Compare the results of this significance test with those you gave in Exercise 8.45. What do you observe about the effect of the sample size on the results of these significance tests?

CHAPTER 8 EXERCISES

8.48   Refer to the study of gender bias and stereotyping described in Exercise 8.27 (page 609). Here are the counts of "girl," "woman," "boy," and "man" for all of the grammar texts studied. The one we analyzed in Exercise 8.27 was number 6.

	Text number									
	1	2	3	4	5	6	7	8	9	10
Girl	2	5	25	11	2	48	38	5	48	13
Woman	3	2	31	65	1	12	2	13	24	5
Boy	7	18	14	19	12	52	70	6	128	32
Man	27	45	51	138	31	80	2	27	48	95

For each text perform the significance test to compare the proportions of juvenile references for females and males. Summarize the results of the significance tests for the ten texts studied. The researchers who conducted the study note that the authors of the last three texts are women, while the other seven were written by men. Do you see any pattern that suggests that the gender of the author is associated with the results?

8.49   Refer to the previous exercise. Let us now combine the categories "girl" with "woman" and "boy" with "man." For each text calculate the proportion of male references and test the hypothesis that male and female references are equally likely (that is, the proportion of male references is equal to 0.5). Summarize the results of your ten tests. Is there a pattern that suggests a relation with the gender of the author?

8.50   Many new products introduced into the market are targeted toward children. The choice behavior of children with regard to new products is of particular interest to companies that design marketing strategies for these products. As part of one study, children in different age groups were compared on their ability to sort new products into the correct product category (milk or juice). Here are some of the data (based on Deborah Roedder John and Ramnath Lakshmi-Ratan, "Age differences in children's choice behavior: the impact of available alternatives," *Journal of Marketing Research*, 29 (1992), pp. 216–226):

Age group	n	Number who sorted correctly
4- to 5-year-olds	50	10
6- to 7-year-olds	53	28

Test the null hypothesis that the two age groups are equally skilled at sorting. Justify your choice of an alternative hypothesis. Also, give a 90%

confidence interval for the difference. Summarize your results with a short paragraph.

8.51   A television news program conducts a call-in poll about a proposed city ban on handgun ownership. Of the 2372 calls, 1921 oppose the ban. The station, following recommended practice, makes a confidence statement: "81% of the Channel 13 Pulse Poll sample opposed the ban. We can be 95% confident that the true proportion of citizens opposing a handgun ban is within 1.6% of the sample result." Is this conclusion justified?

8.52   Eleven percent of the products produced by an industrial process over the past several months fail to conform to the specifications. The company modifies the process in an attempt to reduce the rate of nonconformities. In a trial run, the modified process produces 16 nonconforming items out of a total of 300 produced. Do these results demonstrate that the modification is effective? Support your conclusion with a clear statement of your assumptions and the results of your statistical calculations.

8.53   In the setting of the previous exercise, give a 95% confidence interval for the proportion of nonconforming items for the modified process. Then, taking $p_0 = 0.11$ to be the old proportion and p the proportion for the modified process, give a 95% confidence interval for $p - p_0$.

8.54   In a random sample of 950 students from a large public university, it was found that 444 of the students changed majors sometime during their college years.

   (a) Give a 99% confidence interval for the proportion of students at this university who change majors.

   (b) Express your results from (a) in terms of the *percent* of students who change majors.

   (c) University officials concerned with counseling students are interested in the number of students who change majors rather than the proportion. The university has 35,000 undergraduate students. Convert the confidence interval you found in (a) to a confidence interval for the *number* of students who change majors during their college years.

8.55   Many colleges that once enrolled only male or only female students have become coeducational. Some administrators and alumni were concerned that the academic standards of the institutions would decrease with the change. One formerly all-male college undertook a study of the first class to contain women. The class consisted of 851 students, 214 of whom were women. An examination of first-semester grades revealed that 15 of the top 30 students were female.

   (a) What is the proportion of women in the class? Call this value p_0.

   (b) Assume that the number of females in the top 30 is approximately a binomial random variable with $n = 30$ and unknown probability p of success. In this case success corresponds to the student being female. What is the value of $\hat{p}$?

(c) Are women more likely to be top students than their proportion in the class would suggest? State hypotheses that ask this question, carry out a significance test, and report your conclusion.

8.56 In a study on blood pressure and diet, a random sample of Seventh Day Adventists were interviewed at a national meeting. Because many people who belong to this denomination are vegetarians, they are a very useful group for studying the effects of a meatless diet. Blacks in the population as a whole have a higher average blood pressure than whites. A study of this type should therefore take race into account in the analysis. The 312 people in the sample were categorized by race and whether or not they were vegetarians. The data are given in the following table (data provided by Chris Melby and David Goldflies, Department of Physical Education, Health, and Recreation Studies, Purdue University):

	Black	White
Vegetarian	42	135
Not vegetarian	47	88

Are the proportions of vegetarians the same among all black and white Seventh Day Adventists who attended this meeting? Analyze the data, paying particular attention to this question. Summarize your analysis and conclusions. What can you infer about the proportions of vegetarians among black and white Seventh Day Adventists in general? What about blacks and whites in general?

8.57 In Example 8.11 we discussed a study that examined the association between high blood pressure and increased risk of death from cardiovascular disease. There were 2676 men with low blood pressure and 3338 men with high blood pressure. In the low-blood-pressure group, 21 men died from cardiovascular disease; in the high blood pressure group, 55 died.

(a) Verify the calculations in Example 8.11 by computing the 95% confidence interval for the difference in proportions.

(b) Do the study data confirm that death rates are higher among men with high blood pressure? State hypotheses, carry out a significance test, and give your conclusions.

8.58 An article in the *New York Times* of January 30, 1988, described the results of an experiment on the effects of aspirin on cardiovascular disease. The subjects were 5139 male British medical doctors. The doctors were randomly assigned to two groups. One group of 3429 doctors took one aspirin daily, and the other group did not take aspirin. After 6 years, there were 148 deaths from heart attack or stroke in the first group and 79 in the second group. A similar experiment using male American medical doctors as subjects was reported in the *New York Times* on January 27, 1988. These doctors were also randomly assigned to one of two groups. The 11,037 doctors in the first group took one aspirin every other day, and the 11,034 doctors in the second group took no aspirin. After nearly 5 years there were 104 deaths from heart attacks in the first group and 189 in the second. Analyze the data from these two studies and summarize the results. How do the conclusions of the two studies differ, and why?

8.59   Gastric freezing was once a recommended treatment for ulcers in the upper intestine. A randomized comparative experiment found that 28 of the 82 patients who were subjected to gastric freezing improved, while 30 of the 78 patients in the control group improved. The hypothesis of "no difference" between the two groups can be tested in two ways: using a z statistic or using the chi-square statistic.

   (a)   State the appropriate hypothesis and a two-sided alternative and carry out a z test. What is the P-value?

   (b)   What do you conclude about the effectiveness of gastric freezing as a treatment for ulcers? (See Example 3.5 on page 240 for a discussion of gastric freezing.)

8.60   In this exercise we examine the effect of the sample size on the significance test for comparing two proportions. In each case suppose that $\hat{p}_1 = 0.6$, $\hat{p}_2 = 0.4$, and take n to be the common value of n_1 and n_2. Use the z statistic to test $H_0: p_1 = p_2$ versus the alternative $H_a: p_1 \neq p_2$. Compute the statistic and the associated P-value for the following values of n: 15, 25, 50, 75, 100, and 500. Summarize the results in a table. Explain what you observe about the effect of the sample size on statistical significance when the sample proportions $\hat{p}_1$ and $\hat{p}_2$ are unchanged.

8.61   In the first section of this chapter, we studied the effect of the sample size on the margin of error of the confidence interval for a single proportion. In this exercise we perform some calculations to observe this effect for the two-sample problem. As in the exercise above, suppose that $\hat{p}_1 = 0.6$, $\hat{p}_2 = 0.4$, and n represents the common value of n_1 and n_2. Compute the 95% confidence intervals for the difference in the two proportions for $n = $ 15, 25, 50, 75, 100, and 500. For each interval calculate the margin of error. Summarize and explain your results.

8.62   For a single proportion the margin of error of a confidence interval is largest for any given sample size n and confidence level C when $\hat{p} = 0.5$. This led us to use $p^* = 0.5$ for planning purposes. The same kind of result is true for the two-sample problem. The margin of error of the confidence interval for the difference between two proportions is largest when $\hat{p}_1 = \hat{p}_2 = 0.5$. Use these conservative values in the following calculations, and assume that the sample sizes n_1 and n_2 have the common value n. Calculate the margins of error of the 99% confidence intervals for the difference in two proportions for the following choices of n: 10, 30, 50, 100, 200, and 500. Present the results in a table or with a graph. Summarize your conclusions.

8.63   As the previous problem noted, using the guessed value 0.5 for both $\hat{p}_1$ and $\hat{p}_2$ gives a conservative margin of error in confidence intervals for the difference between two population proportions. You are planning a survey and will calculate a 95% confidence interval for the difference in two proportions when the data are collected. You would like the margin of error of the interval to be less than or equal to 0.05. You will use the same sample size n for both populations.

(a) How large a value of n is needed?

(b) Give a general formula for n in terms of the desired margin of error m and the critical value z^*.

8.64 You are planning a survey in which a 90% confidence interval for the difference between two proportions will present the results. You will use the conservative guessed value 0.5 for $\hat{p}_1$ and $\hat{p}_2$ in your planning. You would like the margin of error of the confidence interval to be less than or equal to 0.1. It is very difficult to sample from the first population, so that it will be impossible for you to obtain more than 20 observations from this population. Taking $n_1 = 20$, can you find a value of n_2 that will guarantee the desired margin of error? If so, report the value; if not, explain why not.

8.65 You are asked to evaluate a proposal for an experiment on the effects of aspirin on cardiovascular disease similar to the experiments described in Exercise 8.59. The researchers will randomly assign subjects to a treatment group or to a control group. The proposed sample sizes are 200 for each group. Write a short evaluation of this proposal, using any relevant information from Exercise 8.59.

8.66 *Castaneda v. Partida* is an important court case in which statistical methods were used as part of a legal argument. When reviewing this case, the Supreme Court used the phrase "two or three standard deviations" as a criterion for statistical significance. This Supreme Court review has served as the basis for many subsequent applications of statistical methods in legal settings. (The two or three standard deviations referred to by the Court are values of the z statistic and correspond to P-values of approximately 0.05 and 0.0026.) In *Castaneda* the plaintiffs alleged that the method for selecting juries in a county in Texas was biased against Mexican Americans. For the period of time at issue, there were 181,535 persons eligible for jury duty, of whom 143,611 were Mexican Americans. Of the 870 people selected for jury duty, 339 were Mexican Americans.

(a) What proportion of eligible voters were Mexican Americans? Let this value be p_0.

(b) Let p be the probability that a randomly selected juror is a Mexican American. The null hypothesis to be tested is $H_0: p = p_0$. Find the value of $\hat{p}$ for this problem, compute the z statistic, and find the P-value. What do you conclude? (A finding of statistical significance in this circumstance does not constitute a proof of discrimination. It can be used, however, to establish a prima facie case. The burden of proof then shifts to the defense.)

(c) We can reformulate this exercise as a two-sample problem. Here we wish to compare the proportion of Mexican Americans among those selected as jurors with the proportion of Mexican Americans among those not selected as jurors. Let p_1 be the probability that a randomly selected juror is a Mexican American, and let p_2 be the probability that a randomly selected nonjuror is a Mexican American. Find the z statistic and its P-value. How do your answers compare with your results in (a)? (For a further discussion of this case, see D. H. Kaye and M. Aickin (eds.), *Statistical Methods in Discrimination Litigation*, Marcel Dekker, New York, 1986.)

NOTES

1.   Details of exact binomial procedures can be found in Chapter 2 of Myles Hollander and Douglas Wolfe, *Nonparametric Statistical Methods,* Wiley, New York, 1973.

2.   Results of this survey are reported in Henry Wechsler et al., "Health and behavioral consequences of binge drinking in college," *Journal of the American Medical Association,* 272 (1994), pp. 1672–1677.

3.   This example is adapted from a survey directed by Professor Joseph N. Uhl of the Department of Agricultural Economics, Purdue University. The survey was sponsored by the Indiana Christmas Tree Growers Association and was conducted in April 1987.

4.   The study and data are described in J. Stamler, "The mass treatment of hypertensive disease: defining the problem," *Mild Hypertension: To Treat or Not to Treat,* New York Academy of Sciences, 1978, pp. 333–358.

PART III

Andy Warhol, *Campbell's Soup Can*, 1964.

Topics in Inference

Prelude

We continue our study of methods for analyzing categorical data in this chapter. Inference about proportions in one-sample and two-sample settings was the focus of Chapter 8. We now study how to compare two or more populations when the response variable has two or more possible values and how to test whether two categorical variables are independent. A single statistical test handles both of these cases. The corresponding procedures for quantitative variables are more elaborate; they will appear in Chapters 12 and 13. The methods of this chapter answer questions such as these:

- Why do college students participate in intramural sports? Are the reasons women participate different from the reasons men participate?

- Does smoking behavior vary according to the socioeconomic status (SES) of adults? To find out, researchers classify several hundred men according to their SES (high, medium, or low) and also according to their smoking behavior (current smoker, former smoker, never smoked). Is there a significant relationship between SES and smoking, and if so, how can we describe it?

Charles Arnoldi, *Untitled (SF)*, 1994.

Inference
for Two-Way Tables

9.1 Inference for Two-Way Tables

When we compared two proportions in Chapter 8, we started by summarizing the raw data by giving the number of observations in each group (n) and how many of these were classified as "successes" (X).

EXAMPLE 9.1

Here is a summary from Example 8.8 (page 603), where we compared frequent binge drinking of men and women by examining the proportions of each gender who engage in this activity.

Population	n	X
1 (men)	7,180	1,630
2 (women)	9,916	1,684
Total	17,096	3,314

The two-way table

two-way table

In this chapter we start with a different summary of the data. Rather than recording just the count of binge drinkers, we record counts for all outcomes in a **two-way table.** We discussed two-way tables in Section 2.6.

EXAMPLE 9.2

Here is the two-way table classifying students by gender and whether or not they are binge drinkers:

Frequent binge drinker	Gender		Total
	Men	Women	
Yes	1,630	1,684	3,314
No	5,550	8,232	13,782
Total	7,180	9,916	17,096

For this problem we could view gender as an explanatory variable and frequent binge drinking as a response variable. This is why we put gender in the columns (like the x axis in a regression) and frequent binge drinking in the rows (like the y axis in a regression). Be sure that you understand how this table is obtained from the one above. Most errors in the use of categorical data methods come from a misunderstanding of how these tables are constructed.

We call this particular two-way table a 2×2 table because there are two rows (yes and no for frequent binge drinking) and two columns (men and women). The advantage of the two-way table is that you can present data for variables having more than two categories by simply increasing the number of rows or columns. For example, we could give a finer classification for the rows: nondrinker, non–binge drinker, infrequent binge drinker, and frequent binge drinker. We studied two-way tables in Section 2.6 (page 194). Now would be a good time to briefly review that material. In this section we study statistical inference for count data that are classified according to two variables. The question of interest is whether there is a relation between the row variable and the column variable. For example, is there a relation between gender and frequent binge drinking? In Example 8.8 we found that there was a relation: men are more likely to be frequent binge drinkers than women. Now we will extend our methods to handle larger tables. Here is an example of a 4×2 table (two categorical variables, one with four categories and one with two categories).

EXAMPLE 9.3

Do men and women participate in sports for the same reasons? One goal for sports participants is social comparison—the desire to win or to do better than other people. Another is mastery—the desire to improve one's skills or to try one's best. A study on why students participate in sports collected data from 67 male and 67 female undergraduates at a large university.[1] Each student was classified into one of four categories based on his or her responses to a questionnaire about sports goals. The four categories were high social comparison–high mastery (HSC-HM), high social comparison–low mastery (HSC-LM), low social comparison–high mastery (LSC-HM), and low social comparison–low mastery (LSC-LM). One purpose of the study was to compare the goals of male and female students. Here are the data displayed in a two-way table:

Observed counts for sports goals

Goal	Female	Male	Total
HSC-HM	14	31	45
HSC-LM	7	18	25
LSC-HM	21	5	26
LSC-LM	25	13	38
Total	67	67	134

The entries in this table are the observed, or sample, counts. For example, there are 14 females in the high social comparison–high mastery group. Note that the marginal totals are given with the table. These are not part of the raw data but are calculated by summing over the rows or columns. The column totals are the numbers of observations sampled in the two populations. The grand total, 134, can be obtained by summing the row totals or the column totals. It is the total number of observations in the study.

The rows and columns of a two-way table represent values of two categorical variables. These are goal and sex in Example 9.3. Each combination of values for these two variables defines a **cell.** A two-way table with r rows and c columns contains $r \times c$ cells. The 4×2 table in Example 9.3 has 8 cells. The objective in this example is to compare men and women. That is, the column variable describes which population an observation comes from. The row variable is a categorical response variable, type of sports goal. It is not always the case that one direction of the table identifies populations to be

cell

compared. Two-way tables can display observations on any two categorical variables.

Analysis of two-way tables is best done using statistical software to carry out the considerable arithmetic required. We will use Example 9.3 and output from a typical statistical computing package to describe inference for two-way tables. In Section 9.2 we will show how to do the work with a calculator if software is not available.

Describing relations in two-way tables

joint distribution
conditional distributions

To describe relations between the variables, we compute and compare percents. The count in each cell can be viewed as a percent of the grand total, the row total, or the column total. In the first case, we are describing the **joint distribution** of the two variables; in the other two cases, we are examining the **conditional distributions.** You must decide which percents are most appropriate. Software usually prints out all three, but not all are of interest in a specific problem.

EXAMPLE 9.4

Figure 9.1 shows the output of the procedure FREQ in the SAS statistical software package for the data of Example 9.3. The two-way table appears in the output in expanded form. Each cell contains five entries, which are labeled in the upper left-hand corner of the output. FREQUENCY is simply the cell count. The row and column totals appear in the margins, just as in Example 9.3. The last three entries in each cell give the cell count as a percent of the overall total, of the row total, and of the column total. These are labeled PERCENT, ROW PCT, and COL PCT. For the HSC-HM females, the three percents are 10.45%, 31.11%, and 20.90%. You can verify these percents by dividing the cell count 14 by the total count 134, by the row total 45, and by the column total 67, in that order.

In this example, we are interested in the effect of sex on the distribution of sports goals. To compare the sexes we examine the column percents.

Column percents for sports goals

	Sex	
Goal	Female	Male
HSC-HM	21	46
HSC-LM	10	27
LSC-HM	31	7
LSC-LM	37	19
Total	100	100

We rounded the percents from the output to get a clearer summary of the data. The total row reminds us that these groups account for 100% of both men and women. (In fact, the sums differ slightly from 100% because of roundoff error.) The bar graphs in Figure 9.2 compare the male and female percents. The data reveal something interesting: it appears that females and males have different goals when they participate in recreational sports.

```
                    TABLE OF GOAL BY SEX

         GOAL         SEX

         FREQUENCY |
         EXPECTED  |
          PERCENT  |
         ROW PCT   |
         COL PCT   | FEMALE  |  MALE   |  TOTAL
         ----------+---------+---------+--------
         HSC-HM    |     14  |     31  |    45
                   |   22.5  |   22.5  |
                   |  10.45  |  23.13  |  33.58
                   |  31.11  |  68.89  |
                   |  20.90  |  46.27  |
         ----------+---------+---------+--------
         HSC-LM    |      7  |     18  |    25
                   |   12.5  |   12.5  |
                   |   5.22  |  13.43  |  18.66
                   |  28.00  |  72.00  |
                   |  10.45  |  26.87  |
         ----------+---------+---------+--------
         LSC-HM    |     21  |      5  |    26
                   |   13.0  |   13.0  |
                   |  15.67  |   3.73  |  19.40
                   |  80.77  |  19.23  |
                   |  31.34  |   7.46  |
         ----------+---------+---------+--------
         LSC-LM    |     25  |     13  |    38
                   |   19.0  |   19.0  |
                   |  18.66  |   9.70  |  28.36
                   |  65.79  |  34.21  |
                   |  37.31  |  19.40  |
         ----------+---------+---------+--------
         TOTAL           67        67       134
                      50.00     50.00    100.00


            STATISTICS FOR TABLE OF GOAL BY SEX

    STATISTIC              DF      VALUE      PROB
    ----------------------------------------------
    CHI-SQUARE              3      24.898     0.000
    SAMPLE SIZE=134
```

FIGURE 9.1 Computer output for the sports goals study, for Example 9.3.

The differences between the distributions of male and female sports goals in the sample appear to be large. A statistical test will tell us whether or not these differences can be plausibly attributed to chance. Specifically, if the two population distributions were the same, how likely is it that a sample would show differences as large or larger than those displayed in Figure 9.2?

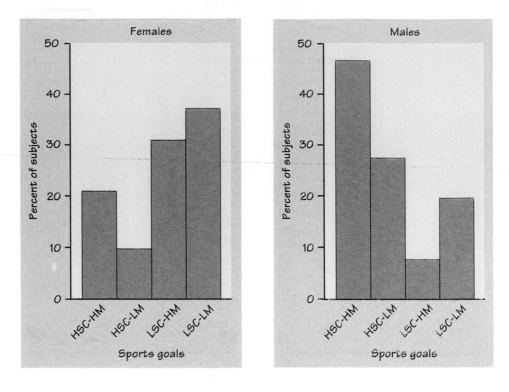

FIGURE 9.2   Comparison of sports goals distributions for females and males, for Example 9.3.

The hypothesis: no association

The null hypothesis H_0 of interest in a two-way table is: there is *no association* between the row variable and the column variable. In the sports goals example, this null hypothesis says that sex and sports goals are not related. The alternative hypothesis H_a is that there is an association between these two variables. The alternative H_a does not specify any particular direction for the association, but it includes possibilities such as "men rate social comparison higher as a goal than do women." We cannot describe H_a as either one-sided or two-sided, because it includes all of the many kinds of association that are possible.

In our example, the hypothesis H_0 that there is no association between sex and sports goals is equivalent to the statement that the distributions of sports goals in the male and female populations are the same. For $r \times c$ tables like that in Example 9.3, where the columns correspond to independent samples from distinct populations, there are c distributions for the row variable, one for each population. The null hypothesis then says that the c distributions of the row variable are identical. The alternative hypothesis is that the distributions are not all the same.

Expected cell counts

expected cell counts

To test the null hypothesis in $r \times c$ tables, we compare the observed cell counts with **expected cell counts** calculated under the assumption that the null hypothesis is true. A numerical summary of the comparison will be our test statistic.

EXAMPLE 9.5

The expected counts for the sports goals example appear in the computer output shown in Figure 9.1. They are labeled EXPECTED and are the second entry in each cell. For example, the expected number of high social comparison–high mastery females is 22.5.

How is this expected count obtained? Look at the percents in the right margin of the table in Figure 9.1. We see that 33.58% of all respondents (female and male together) are in the HSC-HM group. If the null hypothesis of no sex difference in sports goals is true, we expect this overall percentage to apply to both men and women. So we expect 33.58% of the 67 females in the study to be in this group. The expected count is therefore 33.58% of 67, which is 22.5. The other expected counts are calculated similarly. Because the number of males is the same as the number of females in this study, the expected counts for males are the same as the expected counts for females.

Let's examine the calculation of the expected cell counts a bit more carefully. To compute the expected count for HSC-HM females we multiplied the proportion of HSC-HM people in the data (45/134) by the number of females (67). From Figure 9.1 we see that the numbers 45 and 67 are the row and column totals for the cell of interest; 134 is n, the total number of observations for the table. The expected cell count is therefore the product of the row and column totals divided by the table total:

$$\text{expected cell count} = \frac{\text{row total} \times \text{column total}}{n}$$

The chi-square test

To test the H_0 that there is no association between the row and column classifications, we use a statistic that compares the entire set of observed counts with the set of expected counts. First, take the difference between each observed count and its corresponding expected count, and square these values so that they are all 0 or positive. A large difference means less if it comes from a cell we think will have a large count, so divide each squared difference by the expected count, a kind of standardization. Finally, sum over all cells. The result is called the *chi-square statistic* X^2.*

---

*The chi-square statistic was invented by the English statistician Karl Pearson (1857–1936) in 1900, for purposes slightly different from ours. It is the oldest inference procedure still used in its original form. With the work of Pearson and his contemporaries at the beginning of this century, statistics first emerges as a separate discipline.

Chi-Square Statistic

The **chi-square statistic** is a measure of how much the observed cell counts in a two-way table diverge from the expected cell counts. The recipe for the statistic is

$$X^2 = \sum \frac{(\text{observed count} - \text{expected count})^2}{\text{expected count}}$$

where "observed" represents an observed sample count, "expected" represents the expected count for the same cell, and the sum is over all $r \times c$ cells in the table.

If the expected counts and the observed counts are very different, a large value of X^2 will result. So large values of X^2 provide evidence against the null hypothesis. To obtain a P-value for the test, we need the sampling distribution of X^2 under the assumption that H_0 (no association between the row and column variables) is true. We once again use an approximation, related to the normal approximation for binomial distributions. The result is a new distribution, the **chi-square distribution,** which we denote by χ^2 (χ is the Greek letter chi).

chi-square
distribution χ^2

Like the t distributions, the χ^2 distributions form a family described by a single parameter, the degrees of freedom. We use $\chi^2(\text{df})$ to indicate a particular member of this family. Figure 9.3 displays the density curves of the $\chi^2(2)$ and $\chi^2(4)$ distributions. As the figure suggests, χ^2 distributions take only positive values and are skewed to the right. Table F in the back of the book gives upper critical values for the χ^2 distributions.

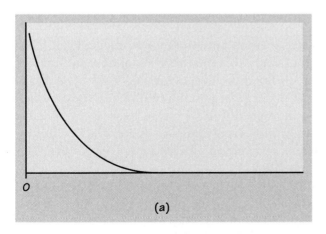

(a)

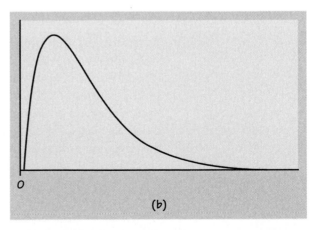

(b)

FIGURE 9.3   (a) The $\chi^2(2)$ density curve. (b) The $\chi^2(4)$ density curve.

> ## Chi-Square Test for Two-Way Tables
>
> The null hypothesis H_0 is that there is no association between the row and column variables in a two-way table. The alternative is that these variables are related.
>
> If H_0 is true, the chi-square statistic X^2 has approximately a χ^2 distribution with $(r - 1)(c - 1)$ degrees of freedom.
>
> The P-value for the chi-square test is
>
>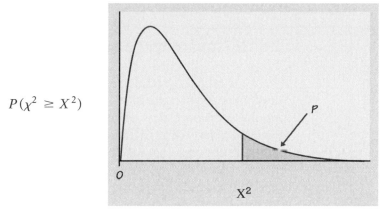
>
> where χ^2 is a random variable having the $\chi^2(\mathrm{df})$ distribution with df $= (r - 1)(c - 1)$.

The chi-square test always uses the upper tail of the χ^2 distribution, because any deviation from the null hypothesis makes the statistic larger. The approximation of the distribution of X^2 by χ^2 becomes more accurate as the cell counts increase. Moreover, it is more accurate for tables larger than 2×2 tables. For tables larger than 2×2, we will use this approximation whenever the average of the expected counts is 5 or more and the smallest expected count is 1 or more. For 2×2 tables, we require that all four expected cell counts be 5 or more.[2]

EXAMPLE 9.6

The results of the chi-square significance test for the sports goals example appear in the lower part of the computer output in Figure 9.1. Because all of the expected cell counts are moderately large, the χ^2 distribution provides accurate P-values. We see that $X^2 = 24.898$, df $= 3$, and the P-value is given as 0.000 under the heading PROB. Because the P-value is rounded to the nearest 0.001, we know that it is less than 0.0005. As a check we verify that the df are correct for a 4×2 table:

$$(r - 1)(c - 1) = (4 - 1)(2 - 1) = 3$$

The chi-square test confirms that the data contain clear evidence against the null hypothesis that female and male students have the same distributions of sports goals. Under H_0, the chance of obtaining a value of X^2 greater than or equal to the calculated value of 24.898 is very small—less than 0.0005.

The test indicates that the male and female distributions are not the same, but it does not say how they differ. Always combine the test with a description that shows what kind of relationship is present. From the percents in Figure 9.1 and the graphs in Figure 9.2, we see that the percent of males in each of the HSC goal classes is more than twice the percent of females. The HSC-HM group contains 46.27% of the males but only 20.90% of the females, and the HSC-LM group contains 26.87% of the males and 10.45% of the females. The pattern is reversed for the LSC goal classes. We conclude that males are more likely to be motivated by social comparison goals and females are more likely to be motivated by mastery goals.

The chi-square test and the z test

The 2×2 table is special in another way. A comparison of the proportions of "successes" in two populations leads to a 2×2 table. We can compare two population proportions either by the chi-square test or by the two-sample z test from Section 8.2. In fact, *these tests always give exactly the same result,* because the chi-square statistic is equal to the square of the z statistic, and $\chi^2(1)$ critical values are equal to the squares of the corresponding $N(0, 1)$ critical values. See, for example, Exercise 9.26. The advantage of the z test is that we can test either one-sided or two-sided alternatives. The chi-square test always tests the two-sided alternative. Of course, the chi-square test can compare more than two populations, whereas the z test compares only two.

Beyond the basics ▶ meta-analysis

meta-analysis

Policymakers wanting to make decisions based on research are sometimes faced with the problem of summarizing the results of many studies. These studies may show effects of different magnitudes, some highly significant and some not significant. What *overall conclusion* can we draw? **Meta-analysis** is a collection of statistical techniques designed to combine information from different but similar studies. The individual studies must be examined with care to ensure that their design and data quality are adequate. The basic idea is to compute a measure of the effect of interest for each study. These are then combined, usually by taking some sort of weighted average, to produce a summary measure for all of the studies. Of course, a confidence interval for the summary is included in the results. Here is an example.

EXAMPLE 9.7 | Vitamin A is often given to young children in developing countries to prevent night blindness. It was observed that children receiving vitamin A appear to have reduced death rates. To investigate the possible relationship between vitamin A supplementation and death, a large field trial with over 25,000 children was undertaken in the Aceh province of Indonesia. About half of the children were given large doses of vitamin A, and the other half were controls. In 1986, the researchers reported a 34% reduction in mortality (deaths) for the treated children who were one to six years old compared with the controls. Several additional studies were then undertaken. Most of the results confirmed the association: treatment of young children in developing countries with vitamin A reduces the death rate; but the size of the effect varied quite a bit.

How can we use the results of these studies to guide policy decisions? To address this question, a meta-analysis was performed on data available from eight studies.[3] Although the designs varied, each study provided a two-way table of counts. Here is the table for Aceh. A total of $n = 25{,}200$ children were enrolled in the study. Approximately half received vitamin A supplements. One year after the start of the study, the number of children who died was determined.

	Vitamin A	Control
Dead	101	130
Alive	12,890	12,079
Total	12,991	12,209

relative risk   The summary measure chosen was the **relative risk:** the ratio of the proportion of children who died in the vitamin A group divided by the proportion of children who died in the control group. For Aceh, the proportion who died in the vitamin A group was

$$\frac{101}{12{,}991} = 0.00777$$

or 7.7 per thousand; for the control group, the proportion who died was

$$\frac{130}{12{,}209} = 0.01065$$

or 10.6 per thousand. The relative risk is therefore

$$\frac{0.00777}{0.01065} = 0.73$$

Note that a value less than 1 occurs whenever the vitamin A group has the lower mortality rate.

The relative risks for the eight studies were

0.73	0.50	0.94	0.71	0.70	1.04	0.74	0.80

These were combined to produce a relative risk estimate of 0.77 with a 95% confidence interval of (0.68, 0.88). That is, vitamin A supplementation reduced the mortality rate to 77% of its value in an untreated group. In other words, there is a 23% reduction in the mortality rate. The confidence interval does not include 1, so the null hypothesis of no effect (a relative risk of 1) can be clearly rejected. The researchers examined many variations of this meta-analysis, such as using different weights and leaving out one study at a time. These variations had little effect on the final estimate.

After these findings were published, large-scale programs to distribute high-potency vitamin A supplements were started. These programs have saved hundreds of thousands of lives since the meta-analysis was conducted and the arguments and uncertainties were resolved.

SUMMARY

The **null hypothesis** for $r \times c$ tables of count data is that there is no relationship between the row variable and the column variable.

Expected cell counts under the null hypothesis are computed using the formula

$$\text{expected count} = \frac{\text{row total} \times \text{column total}}{n}$$

The null hypothesis is tested by the **chi-square statistic,** which compares the observed counts with the expected counts:

$$X^2 = \sum \frac{(\text{observed} - \text{expected})^2}{\text{expected}}$$

Under the null hypothesis, X^2 has approximately the χ^2 distribution with $(r-1)(c-1)$ degrees of freedom. The P-value for the test is

$$P(\chi^2 \geq X^2)$$

where χ^2 is a random variable having the $\chi^2(\text{df})$ distribution with df $= (r-1)(c-1)$.

The chi-square approximation is adequate for practical use when the average expected cell count is 5 or greater and all individual expected counts are 1 or greater, except in the case of 2×2 tables. All four expected counts in a 2×2 table should be 5 or greater.

The section we just completed assumed that you have access to software or a statistical calculator. If you do not, you now need to study the material on computations in the following optional section. All exercises appear at the end of the chapter.

9.2 Formulas and Models for Two-Way Tables*

Computations

The calculations required to analyze a two-way table are straightforward but tedious. In practice, we recommend using software, but it is possible to do the work with a calculator, and some insight can be gained by examining the details. Here is an outline of the steps required.

---

*The analysis of two-way tables is based on computations that are a bit messy and on statistical models that require a fair amount of notation to describe. This section gives the details. By studying this material you will deepen your understanding of the methods described in this chapter, but this section is optional.

Computations for Two-Way Tables

1. Calculate descriptive statistics that convey the important information in the table. Usually these will be column or row percents.
2. Find the expected counts and use these to compute the X^2 statistic.
3. Use chi-square critical values from Table F to find the approximate P-value.
4. Draw a conclusion about the association between the row and column variables.

The following example illustrates these steps. The two-way table in this example does not compare several populations. Instead, it arises by classifying observations on a single population in two ways.

EXAMPLE 9.8

In a study of heart disease in male federal employees, researchers classified 356 volunteer subjects according to their socioeconomic status (SES) and their smoking habits.[4] There were three categories of SES: high, middle, and low. Individuals were asked whether they were current smokers, former smokers, or had never smoked, producing three categories for smoking habits as well. Here is the two way table that summarizes the data:

Observed counts for smoking and SES

Smoking	High	Middle	Low	Total
Current	51	22	43	116
Former	92	21	28	141
Never	68	9	22	99
Total	211	52	93	356

This is a 3×3 table, to which we have added the marginal totals obtained by summing across rows and columns. For example, the first-row total is $51 + 22 + 43 = 116$. The grand total, the number of subjects in the study, can be computed by summing the row totals, $116 + 141 + 99 = 356$, or the column totals, $211 + 52 + 93 = 356$. It is good statistical practice to do both as a check on your arithmetic.

Computing conditional distributions

We start our analysis by computing descriptive statistics that summarize the observed relation between SES and smoking. The researchers suspected that SES helps explain smoking, so in this situation SES is the explanatory variable and smoking is the response variable. In general, the clearest description of the relationship is provided by comparing the conditional distributions of the response variable for each value of the explanatory variable. That is, compare the column percents that give the conditional distribution of smoking for each SES category.

EXAMPLE 9.9

We must calculate the column percents. For the high-SES group, there are 51 current smokers out of a total of 211 people. The column proportion for this cell is

$$\frac{51}{211} = 0.242$$

That is, 24.2% of the high-SES group are current smokers. Similarly, 92 of the 211 people in this group are former smokers. The column proportion is

$$\frac{92}{211} = 0.436$$

or 43.6%. In all, we must calculate nine percents. Here are the results:

Column percents for smoking and SES				
	SES			
Smoking	High	Middle	Low	All
Current	24.2	42.3	46.2	32.6
Former	43.6	40.4	30.1	39.6
Never	32.2	17.3	23.7	27.8
Total	100.0	100.0	100.0	100.0

In addition to the distributions of smoking behavior in each of the three SES categories, the table gives the overall breakdown of smoking among all 356 subjects. These percents appear in the rightmost column, labeled "All." They make up the marginal distribution of the row variable "smoking."

The sum of the percents in each column should be 100, except for possible small roundoff errors. It is good practice to calculate each percent separately and then sum each column as a check. In this way we can find arithmetic errors that would not be uncovered if, for example, we calculated the column percent for the "Never" row by subtracting the sum of the percents for "Current" and "Former" from 100.

Figure 9.4 compares the distributions of smoking behavior in the three SES groups. The percent of current smokers decreases as SES increases from low to middle to high; in particular, relatively few high-SES subjects smoke. The percent of former smokers increases as SES increases, suggesting that higher-SES smokers were more likely to quit. The percent of people who never smoked is highest in the high-SES group, but the middle-SES group has a somewhat lower percentage than the low-SES group. Overall, the column percents suggest that there is a negative association between smoking and SES: higher-SES people tend to smoke less.

The chi-square test assesses whether this observed association is statistically significant. That is, is the SES-smoking relationship in the sample suffi-

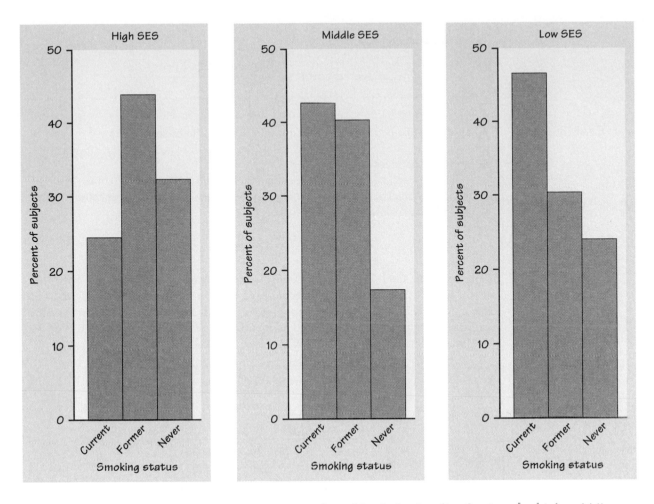

FIGURE 9.4 Comparison of smoking-behavior distributions for high, middle, and low SES, for Example 9.9.

ciently strong for us to conclude that it is due to a relationship between these two variables in the underlying population and not merely to chance? Note that the test only asks whether there is evidence of some relationship. To explore the direction or nature of the relationship we must examine the column or row percents. Note also that in using the chi-square test we are acting as if the subjects were a simple random sample from the population of interest. If the volunteers are a biased sample—for example, if smokers are reluctant to volunteer for a study of employee health—then conclusions about the entire population of employees are not justified.

Computing expected cell counts

The null hypothesis is that there is no relationship between SES and smoking in the population. The alternative is that these two variables are related. Here is the formula for the expected cell counts under the hypothesis of "no relationship."

> **Expected Cell Counts**
>
> $$\text{expected count} = \frac{\text{row total} \times \text{column total}}{n}$$

EXAMPLE 9.10

What is the expected count in the upper left cell in the table of Example 9.8, corresponding to high-SES current smokers, under the null hypothesis that smoking and SES are independent?

The row total, the count of current smokers, is 116. The column total, the count of high-SES subjects, is 211. The total sample size is $n = 356$. The expected number of high-SES current smokers is therefore

$$\frac{(116)(211)}{356} = 68.75$$

Note that $116/356$ is the proportion of subjects who are current smokers. The number of high-SES subjects is 211. Because we *expect* $116/356$ (32.6%) of these to be current smokers, the expected count is $(116/356) \times 211$. This is what our formula specifies. You can find the expected counts for the other cells in the same way. We summarize these calculations in a table of expected counts:

	Expected counts for smoking and SES			
		SES		
Smoking	High	Middle	Low	Total
Current	68.75	16.94	30.30	115.99
Former	83.57	20.60	36.83	141.00
Never	58.68	14.46	25.86	99.00
Total	211.00	52.00	92.99	355.99

We can check our work by adding the expected counts to obtain the row and column totals, as in the table. These should be the same as those in the table of observed counts, except for small roundoff errors, such as 115.99 rather than 116 for the first-row total.

Computing the chi–square statistic

The expected counts are all large, so we proceed with the chi-square test. We compare the table of observed counts with the table of expected counts using the X^2 statistic (page 630).[5] We must calculate the term for each cell, then sum over all nine cells. For the high-SES current smokers, the observed count is 51 and the expected count is 68.75. The contribution to the X^2 statistic for this cell is therefore

$$\frac{(51 - 68.75)^2}{68.75} = 4.583$$

Similarly, the calculation for the middle-SES current smokers is

$$\frac{(22 - 16.94)^2}{16.94} = 1.511$$

The X^2 statistic is the sum of nine such terms:

$$X^2 = \sum \frac{(\text{observed} - \text{expected})^2}{\text{expected}}$$

$$= \frac{(51 - 68.75)^2}{68.75} + \frac{(22 - 16.94)^2}{16.94} + \frac{(43 - 30.30)^2}{30.30}$$

$$+ \frac{(92 - 83.57)^2}{83.57} + \frac{(21 - 20.60)^2}{20.60} + \frac{(28 - 36.83)^2}{36.83}$$

$$+ \frac{(68 - 58.68)^2}{58.68} + \frac{(9 - 14.46)^2}{14.46} + \frac{(22 - 25.86)^2}{25.86}$$

$$= 4.583 + 1.511 + 5.323 + 0.850 + 0.008 + 2.117 + 1.480 + 2.062 + 0.576$$

$$= 18.51$$

Because there are $r = 3$ smoking categories and $c = 3$ SES groups, the degrees of freedom for this statistic are

$$(r - 1)(c - 1) = (3 - 1)(3 - 1) = 4$$

df = 4		
p	0.001	0.0005
χ^2	18.47	20.00

Under the null hypothesis that smoking and SES are independent, the test statistic X^2 has a $\chi^2(4)$ distribution. To obtain the P-value, refer to the row in Table F corresponding to 4 df. The calculated value $X^2 = 18.51$ lies between upper critical points corresponding to probabilities 0.001 and 0.0005. The P-value is therefore between 0.001 and 0.0005. Because the expected cell counts are all large, the P-value from Table F will be quite accurate. There is strong evidence ($X^2 = 18.51$, df $= 4$, $P < 0.001$) of an association between smoking and SES in the population of federal employees. The size and nature of this association are described by the table of percents examined in Example 9.9 and the display of these percents in Figure 9.4. The chi-square test assesses only the significance of the association, so the descriptive percents are essential to an understanding of the data.

Models for two-way tables

The chi-square test for the presence of a relationship between the two directions in a two-way table is valid for data produced from several different study designs. The precise statement of the null hypothesis "no relationship" in terms of population parameters is different for different designs. We now describe two of these settings in detail. The essential requirement is that *each experimental unit or subject is counted only once in the data table*. In the sports goals example, each student is either male or female and is classified as having only one sports goal. In the SES-smoking example, each federal employee fits into one SES-smoking cell. If, for example, the same individual is observed at several different times and so may appear in several cells of the table, more advanced statistical procedures are needed.

Comparing several populations: the first model   The first model for a two-way table is illustrated by the comparison of two proportions discussed in Section 8.2 and Example 9.2 and also by the comparison of the sports goals of men and women in Example 9.3. We have *separate and independent random samples* from each of c populations. Each individual falls into one of r categories, and we wish to compare the distributions of these categories in the populations. In Section 8.2, we compared just two populations ($c = 2$) and we categorized each observation as a success or a failure ($r = 2$). The data could have been displayed there in a two-way 2×2 table. Because the counts of failures are known once we know the counts of successes, we simply compared the two proportions of successes and did not need the two-way table format. The $r \times c$ table allows us to compare more than two populations or to use more than two categories, or both.

EXAMPLE 9.11   In the sports goals example, we compare two populations (men and women) and assign each subject to one of four categories. We have a 4×2 table. The parameters of the model are the population proportions for each of the four categories, one set of proportions for each population. We have samples from two populations of subjects, men and women.

More generally, if we take an independent SRS from each of c populations and classify the outcomes into one of r categories, we have an $r \times c$ table of population proportions. There are c different sets of proportions to be compared. There are c groups of subjects and a single categorical variable with r possible values is measured for each individual.

EXAMPLE 9.12   The null hypothesis for the sports goals example is that there is no relationship between sex and goals. That is, the sports goals distributions for females and males are the same. Therefore, the probability that a randomly selected female is in the high social comparison–high mastery group is equal to the corresponding probability for a randomly selected male. The same is true for each of the other three goals. The null hypothesis H_0, expressed in terms of the population parameters, says that the two probability distributions are equal.

Model for Comparing Several Populations Using Two-Way Tables

Select independent SRSs from each of c populations, of sizes $n_1, n_2, \ldots, n_c$. Classify each individual in a sample according to a categorical response variable with r possible values. There are c different probability distributions, one for each population.

The null hypothesis is that the distributions of the response variable are the same in all c populations. The alternative hypothesis says that these c distributions are not all the same.

Testing independence: the second model   A second model for which our analysis of $r \times c$ tables is valid is illustrated by the SES-smoking study, Example 9.8. There, a *single* sample from a *single* population was classified according to two categorical variables.

EXAMPLE 9.13   In this study, researchers recorded the values of two categorical variables for each of the 356 federal employees that they studied. These variables were SES and smoking. Since each variable has three possible values, there are nine possible outcomes. We summarize the data in a 3×3 table of counts.

In the table of observed cell counts, only the table total is fixed before the data are observed. This total (356) is the number of subjects in the study. The row and column sums are random and would no doubt be different if we took another sample.

More generally, take an SRS from a single population and record the values of two categorical variables, one with r possible values and the other with c possible values. The data are summarized by recording the numbers of individuals for each possible combination of outcomes for the two random variables. This gives an $r \times c$ table of counts. Each of these $r \times c$ possible outcomes

joint distribution   has its own probability. The probabilities give the **joint distribution** of the two categorical variables.

Each of the two categorical random variables has a distribution. We call

marginal distributions   these distributions **marginal distributions** because they are the sums of the population proportions in the rows and columns.

The null hypothesis "no relationship" now states that the row and column variables are independent. The multiplication rule for independent events (page 302) tells us that the joint probabilities are the products of the marginal probabilities.

EXAMPLE 9.14   The joint probability distribution gives a probability for each of the nine cells in our 3×3 table of smoking by SES. The marginal distribution for smoking gives probabilities for each of the three possible smoking categories; the marginal distribution for SES gives probabilities for each of the three possible SES categories.

Independence between smoking and SES implies that the joint distribution can be obtained by multiplying the appropriate terms from the two marginal distributions. For example, the probability that an individual is a high-SES current smoker is the probability that the individual is high-SES times the probability that the individual is a current smoker. The hypothesis that smoking and SES are independent says that the multiplication rule applies to *all* outcomes.

Model for Examining Independence in Two-Way Tables

Select an SRS of size n from a population. Measure two categorical variables for each individual.

The null hypothesis is that the row and column variables are independent. The alternative hypothesis is that the row and column variables are dependent.

The formula that we gave for expected counts (page 638) can also be justified using the independence model. We illustrate the idea with the following example.

EXAMPLE 9.15 | The expected number of high-SES current smokers is simply the sample size n times the probability that an individual is a high-SES current smoker

$$n \times P(\text{high-SES current smoker})$$

In Example 9.14 we saw that the independence model implies that this probability is equal to the probability that the individual is high-SES times the probability that the individual is a current smoker. So,

$$P(\text{high-SES current smoker}) = P(\text{current smoker}) \times P(\text{high-SES})$$

and under the independence model, the expected number of high-SES current smokers is

$$n \times P(\text{current smoker}) \times P(\text{high-SES})$$

Using the estimated marginal distributions from the table of counts, we find that the estimated count is

$$356 \times \frac{116}{356} \times \frac{211}{356} = \frac{116 \times 211}{356} = 68.75$$

or

$$\text{expected count} = \frac{\text{row total} \times \text{column total}}{n}$$

Concluding remarks

You can distinguish between the two models by examining the design of the study. In the independence model, there is a single sample. The column totals and row totals are random variables. The total sample size n is set by the researcher; the column and row sums are only known after the data are collected. For the comparison-of-populations model, on the other hand, there is a sample from each of two or more populations. The column sums are the sample sizes selected at the design phase of the research. The null hypothesis in both models says that there is no relationship between the column variable and the row variable. The precise statement of the hypothesis differs, depending on the sampling design. Fortunately, *the test of the hypothesis of "no relationship" is the same for both models;* it is the chi-square test. There are yet other statistical models for two-way tables that justify the chi-square test of the null hypothesis "no relation," made precise in ways suitable for these models. Statistical methods related to the chi-square test also allow the analysis of three-way and higher-way tables of count data. You can find a discussion of these topics in advanced texts on categorical data.[6]

SUMMARY

For two-way tables we first compute percents or proportions that describe the relationship of interest. Then, we compute expected counts, the X^2 statistic, and the P-value.

Two different models for generating $r \times c$ tables lead to the chi-square test. In the first model, independent SRSs are drawn from each of c populations, and each observation is classified according to a categorical variable with r possible values. The null hypothesis is that the distributions of the row categorical variable are the same for all c populations. In the second model, a single SRS is drawn from a population, and observations are classified according to two categorical variables having r and c possible values. In this model, H_0 states that the row and column variables are independent.

CHAPTER 9 EXERCISES

9.1 To be competitive in global markets, many U.S. corporations are undertaking major reorganizations. Often these involve "downsizing" or a "reduction in force" (RIF), where substantial numbers of employees are terminated. Federal and various state laws require that employees be treated equally regardless of their age. In particular, employees over the age of 40 years are in a "protected" class, and many allegations of discrimination focus on comparing employees over 40 with their younger coworkers. Here are the data for a recent RIF:

	Over 40	
Terminated	No	Yes
Yes	7	41
No	504	765

(a) Make a table that includes the following information for each group (over 40 or not): total number of employees, the proportion of employees who were terminated, and the standard error for the proportion.

(b) Perform the chi-square test for this two-way table. Give the test statistic, the degrees of freedom, the P-value, and your conclusion.

9.2 Refer to the previous exercise. A major issue that arises in this kind of case concerns the extent to which the employees are similar. This often involves examining other variables. For example, suppose that the employees over 40 did not do as good a job as the younger workers. Let's examine the last performance appraisal to explore this idea. The possible values are as follows: partially meets expectations, fully meets expectations, usually exceeds expectations, and continually exceeds expectations. Because there were very few employees who partially exceeded expectations, we combine the first two categories. Here are the data:

	Over 40	
Performance appraisal	No	Yes
Partially or fully meets expectations	82	230
Usually exceeds expectations	353	496
Continually exceeds expectations	61	32

Analyze the data. Do the older employees appear to have lower performance evaluations?

9.3 Euthanasia of healthy but unwanted pets by animal shelters is believed to be the leading cause of death for cats and dogs. A study designed to find factors associated with bringing a cat to an animal shelter compared data on cats that were brought to the Humane Society of Saint Joseph County in Mishawaka, Indiana, with controls from the same county that were not. One of the factors examined was the source of the cat; the categories were private owner or breeder, pet store, and other (includes born in home, stray, and obtained from a shelter). This kind of study is called a **case–control study** by epidemiologists. Here are the data (from Gary J. Patronek et al., "Risk factors for relinquishment of cats to an animal shelter," *Journal of the American Veterinary Medical Association*, 209 (1996), pp. 582–588):

case–control study

	Source		
Group	Private	Pet store	Other
Cases	124	16	76
Controls	219	24	203

(a) Should you use row or column percents to describe these data? Give reasons for your answer and summarize the pattern that you see in these percents.

(b) Use a significance test to examine the association between the two variables. Summarize the results.

9.4 Refer to the previous exercise. The same researchers did a similar study for dogs. The data are given in the following table (from Gary J. Patronek et al., "Risk factors for relinquishment of dogs to an animal shelter," *Journal of the American Veterinary Medical Association*, 209 (1996), pp. 572–581):

		Source	
Group	Private	Pet store	Other
Cases	188	7	90
Controls	518	68	142

Analyze the data and write a short report explaining your conclusions.

9.5 In the studies described in Exercises 9.3 and 9.4, the *control group data* were obtained by a random digit dial telephone survey and can be considered to be an SRS of households with a cat or a dog in this area. Compare the sources of cats and dogs in this area. Write a short report on your analysis; include appropriate descriptive statistics, the results of a significance test, and your conclusion.

9.6 Can you increase the response rate for a mail survey by contacting the respondents before they receive the survey? A study designed to address this question compared three groups of subjects. The first group received a preliminary letter about the survey, the second group was phoned, and the third received no preliminary contact. A positive response was defined as returning the survey within two weeks. Here are the counts (from James E. Stafford, "Influence of preliminary contact on mail returns," *Journal of Marketing Research,* 3 (1966), pp. 410–411):

	Intervention		
Response	Letter	Phone call	None
Yes	171	146	118
No	220	68	455
Total	391	214	573

(a) For each intervention find the proportion of positive responses.

(b) Translate the question of interest into appropriate null and alternative hypotheses for this problem.

(c) Give the test statistic, degrees of freedom, and the P-value for the significance test. What do you conclude?

9.7 The effect of a prenotification letter on the response rate of a survey was examined in a study of physicians. The response rate of those who received the letters was compared with that of a control group who did not. Here are the data (from Patricia H. Shiono and Mark A. Klebanoff, "The effect of two mailing strategies on the response to a survey of physicians," *American Journal of Epidemiology,* 134 (1991), pp. 539–542):

Response	Letter	No letter
Yes	2570	2645
No	2448	2384
Total	5018	5029

(a) Give the percent of positive responses for those who received letters and those who did not.

(b) Analyze the two-way table. State the hypotheses, give the test statistic, degrees of freedom, P-value, and your conclusion.

9.8   The survey in Exercise 9.6 was conducted in 1966 on subjects who were college students in Houston, Texas, and asked about clothing preferences. The data in Exercise 9.7 were collected in 1989 and asked questions about pregnancy among resident physicians. You are planning a study to be conducted next month to assess the needs in your community for a new Internet access provider. Discuss whether and how you would use the results of these two surveys to design your study.

9.9   Psychological and social factors can influence the survival of patients with serious diseases. One study examined the relationship between survival of patients with coronary heart disease (CHD) and pet ownership. Each of 92 patients was classified as having a pet or not and by whether they survived for one year. Here are the data (from Erika Friedmann et al., "Animal companions and one-year survival of patients after discharge from a coronary care unit," *Public Health Reports*, 96 (1980), pp. 307–312):

	Pet ownership	
Patient status	No	Yes
Alive	28	50
Dead	11	3

(a) Was this study an experiment? Why or why not?

(b) The researchers thought that having a pet might improve survival, so pet ownership is the explanatory variable. Compute appropriate percentages to describe the data and state your preliminary findings.

(c) State in words the null hypothesis for this problem. What is the alternative hypothesis?

(d) Find the X^2 statistic, its degrees of freedom, and the P-value.

(e) What do you conclude? Do the data give convincing evidence that owning a pet is an effective treatment for increasing the survival of CHD patients?

9.10 Investors use many "indicators" in their attempts to predict the behavior of the stock market. One of these is the "January indicator." Some investors believe that if the market is up in January, then it will be up for the rest of the year. On the other hand, if it is down in January, then it will be down for the rest of the year. The following table gives data for the Standard & Poor's 500 stock index for the 75 years from 1916 to 1990:

Rest of year	January Up	Down
Up	35	13
Down	13	14

A chi-square analysis is valid for this problem if we assume that the yearly data are independent observations of a process that generates either an "up" or a "down" both in January and for the rest of the year.

(a) Calculate the column percents for this table. Explain briefly what they express.

(b) Do the same for the row percents.

(c) State appropriate null and alternative hypotheses for this problem. Use words rather than symbols.

(d) Find the table of expected counts under the null hypothesis. In which cells do the expected counts exceed the observed counts? In what cells are they less than the observed counts? Explain why the pattern suggests that the January indicator is valid.

(e) Give the value of the X^2 statistic, its degrees of freedom, and the P-value. What do you conclude?

(f) Write a short discussion of the evidence for the January indicator, referring to your analysis for substantiation.

9.11 The baseball player Reggie Jackson had a reputation for hitting better in the World Series than during the regular season. In his 21 year career, Jackson was at bat 9864 times in regular-season play and had 2584 hits. During World Series games, he was at bat 98 times and had 35 hits. We can view Jackson's regular-season at bats as a random sample from a population of potential at bats (he might have batted many more times if the season were longer, for example), and his World Series at bats as a sample from a second population.

(a) Display the data in a 2 × 2 table of counts with "Regular season" and "World Series" as the column headings, and fill in the marginal sums.

(b) Calculate appropriate percents to compare Jackson's regular-season and World Series performances. Did he hit better in World Series games?

(c) Is there a significant difference between Jackson's regular-season and World series performances? State hypotheses (in words), and then calculate the X^2 statistic, its degrees of freedom, and its P-value. What is your conclusion?

9.12   In January 1975, the Committee on Drugs of the American Academy of Pediatrics recommended that tetracycline drugs not be given to children under the age of 8. A two-year study conducted in Tennessee investigated the extent to which physicians had prescribed these drugs between 1973 and 1975. The study categorized family practice physicians according to whether the county of their practice was urban, intermediate, or rural. The researchers examined how many doctors in each of these categories prescribed tetracycline to at least one patient under the age of 8. Here is the table of observed counts (data from Wayne A. Ray et al., "Prescribing of tetracycline to children less than 8 years old," *Journal of the American Medical Association*, 237 (1977), pp. 2069–2074):

	County type		
	Urban	Intermediate	Rural
Tetracycline	65	90	172
No tetracycline	149	136	158

(a) Find the row and column sums and put them in the margins of the table.

(b) For each type of county find the percent of physicians who prescribed tetracycline and the percent of those who did not. Do the same for the combined sample. Display the percents in a table and describe briefly what they show.

(c) Write null and alternative hypotheses to assess whether county type and prescription practices are unrelated.

(d) Carry out a significance test, give a full report of the results, and interpret them in plain language.

9.13   If the performance of a stock fund is due to the skill of the manager, then we would expect a fund that does well this year to perform well next year also. This is called persistence of fund performance. One study classified funds as losers or winners depending on whether their rate of return was less than or greater than the median of all funds. To examine the question of interest we form a two-way table that classifies each fund as a loser or winner in each of two successive years. Here are the data for one such table (from Burton G. Malkeil, "Returns from investing in equity mutual funds, 1971 to 1991," *Journal of Finance*, 50 (1995), pp. 549–572):

	Next year	
This year	Winner	Loser
Winner	85	35
Loser	37	83

Is there evidence in favor of persistence of fund performance in this table? Support your conclusion with a complete analysis of the data.

9.14 Refer to the previous exercise. Rerun the analysis using the method for comparing two proportions of Chapter 8. Verify that the X^2 statistic is the square of the z and that the P-values for both analyses are the same.

9.15 In the previous exercise, it is natural to use "this year" to define the two samples. If we drew separate random samples of winners and losers this year and we recorded the outcome next year, we would call this a **prospective study** (forward looking). On the other hand, if we drew separate random samples of winners and losers "next year" and looked back historically to determine if they were winners or losers in the previous year, we would have a *retrospective study* (backward looking). Verify that you get the same value of z (and therefore the same P-value) using these two different approaches.

prospective study

retrospective study

9.16 If we find evidence in favor of an effect in one set of circumstances, it is natural to want to conclude that it holds in many others. Unfortunately, this reasoning can sometimes lead us to incorrect conclusions. For example, here is another table from the study described in the previous exercise:

	Next year	
This year	Winner	Loser
Winner	96	148
Loser	145	99

Analyze these data in the same way. What do you conclude? This set of data is for 1987 to 1988; in the previous exercise the years were 1977 to 1978. Many things change with time, as we learned in Chapter 2.

9.17 Alcohol and nicotine consumption during pregnancy may harm children. Because drinking and smoking behaviors may be related, it is important to understand the nature of this relationship when assessing the possible effects on children. One study classified 452 mothers according to their alcohol intake prior to pregnancy recognition and their nicotine intake during pregnancy. The data are summarized in the following table (from Ann P. Streissguth et al., "Intrauterine alcohol and nicotine exposure: attention and reaction time in 4-year-old children," *Developmental Psychology*, 20 (1984), pp. 533–541):

Alcohol (ounces/day)	Nicotine (milligrams/day)		
	None	1–15	16 or more
None	105	7	11
0.01–0.10	58	5	13
0.11–0.99	84	37	42
1.00 or more	57	16	17

Carry out a complete analysis of the association between alcohol and nicotine consumption. That is, describe the nature and strength of this association and assess its statistical significance. Include charts or figures to display the association.

9.18   Nutrition and illness are related in a complex way. If the diet is inadequate, the ability to resist infections can be impaired and illness results. On the other hand, some illnesses cause lack of appetite, so that poor nutrition can be the result of illness. In a study of morbidity and nutritional status in 1165 preschool children living in poor conditions in Delhi, India, data were obtained on nutrition and illness. Nutrition was described by a standard method as normal or as one of four levels of inadequate: I, II, III, and IV. For the purpose of analysis, the two most severely undernourished groups, III and IV, were combined. One part of the study examined four categories of illness during the past year: upper respiratory infection (URI), diarrhea, URI and diarrhea, and none. The following table gives the data. (Data from Vimlesh Seth et al., "Profile of morbidity and nutritional status and their effect on the growth potentials in preschool children in Delhi, India," *Tropical Pediatrics and Environmental Health*, 25 (1979), pp. 23–29.)

Illness	Nutritional status			
	Normal	I	II	III and IV
URI	95	143	144	70
Diarrhea	53	94	101	48
URI and diarrhea	27	60	76	27
None	113	48	44	22
Total	288	345	365	167

Carry out a complete analysis of the association between nutritional status and type of illness. That is, describe the association numerically, assess its significance, and write a brief summary of your findings that refers to your analysis for substantiation.

9.19   There is much evidence that high blood pressure is associated with increased risk of death from cardiovascular disease. A major study of this association examined 2676 men with low blood pressure and 3338 men with high blood pressure. During the period of the study, 21 men in the low-blood-pressure and 55 in the high-blood-pressure group died from cardiovascular disease. (Data taken from J. Stamler, "The mass treatment of hypertensive disease: defining the problem," *Mild Hypertension: To Treat or Not to Treat*, New York Academy of Sciences, 1978, pp. 333–358.)

(a) What is the explanatory variable? Describe the association in these data numerically and in words.

(b) Do the study data confirm that death rates are higher among men with high blood pressure? State hypotheses, carry out a significance test, and give your conclusions.

(c) Present the data in a two-way table. Is the chi-square test appropriate for the hypotheses you stated in (b)?

(d) Give a 95% confidence interval for the difference between the death rates for the low- and high-blood-pressure groups.

9.20 It is traditional practice in Egypt to withhold food from children with diarrhea. Because it is known that feeding children with this illness reduces mortality, medical authorities undertook a nationwide program designed to promote feeding sick children. To evaluate the impact of the program, surveys were taken before and after the program was implemented. In the first survey, 457 of 1003 surveyed mothers followed the practice of feeding children with diarrhea. For the second survey, 437 of 620 surveyed followed this practice. (Data taken from O. M. Galal et al., "Feeding the child with diarrhea: a strategy for testing a health education message within the primary health care system in Egypt," *Socio-economic Planning Sciences,* 21 (1987), pp. 139–147.)

(a) Assume that the data come from two independent samples. Test the hypothesis that the program was effective, that is, that the practice of feeding children with diarrhea increased between the time of the first study and the time of the second. State H_0 and H_a, give the test statistic and its P-value, and summarize your conclusion.

(b) Present the data in a two-way table. Can the X^2 statistic test your hypotheses?

(c) Describe the results using a 95% confidence interval for the difference in proportions.

9.21 Aluminum is suspected as a factor in the development of Alzheimer's disease. In one study, researchers compared a group of Alzheimer's patients with a carefully selected control group of people who did not have Alzheimer's but were similar in other ways. (Selection of a matching control group is a difficult task. In epidemiological studies such as this, however, experiments are not possible.) The focus of the study was on the use of antacids that contain aluminum. Each subject was classified according to the use of these antacids. The two-way table below gives the data. (Data from Amy Borenstein Graves et al., "The association between aluminum-containing products and Alzheimer's disease," *Journal of Clinical Epidemiology,* 43 (1990), pp. 35–44.)

	Aluminum-containing antacid use			
	None	Low	Medium	High
Alzheimer's patients	112	3	5	8
Control group	114	9	3	2

Analyze the data and summarize your results. Does the use of aluminum-containing antacids appear to be associated with Alzheimer's disease?

9.22 Are there gender differences in the progress of students in doctoral programs? A major university classified all students entering Ph.D. programs in a given year by their status 6 years later. The categories used were as follows: completed the degree, still enrolled, and dropped out. Here are the data:

Status	Men	Women
Completed	423	98
Still enrolled	134	33
Dropped out	238	98

Assume that these data can be viewed as a random sample giving us information on student progress. Describe the data using whatever percents are appropriate. State and test a null hypothesis and alternative that address the question of gender differences. Summarize your conclusions. What factors not given might be relevant to this study?

9.23   PTC is a compound that has a strong bitter taste for some people and is tasteless for others. The ability to taste this compound is an inherited trait. Many studies have assessed the proportions of people in different populations who can taste PTC. The following table gives results for samples from several countries. (Data from A. E. Mourant et al., *The Distribution of Human Blood Groups and Other Polymorphisms,* Oxford University Press, London, 1976.)

	Ireland	Portugal	Norway	Italy
Tasters	558	345	185	402
Nontasters	225	109	81	134

Complete the table and describe the data. Do they provide evidence that the proportion of PTC tasters varies among the four countries? Give a complete summary of your analysis.

9.24   There are four major blood types in humans: O, A, B, and AB. In a study conducted using blood specimens from the Blood Bank of Hawaii, individuals were classified according to blood type and ethnic group. The ethnic groups were Hawaiian, Hawaiian-white, Hawaiian-Chinese, and white. Assume that the blood bank specimens are random samples from the Hawaiian populations of these ethnic groups. (Data are from the book cited in the previous exercise.)

Blood type	Hawaiians	Hawaiian-white	Hawaiian-Chinese	White
O	1,903	4,469	2,206	53,759
A	2,490	4,671	2,368	50,008
B	178	606	568	16,252
AB	99	236	243	5,001

Summarize the data numerically and with a chart or figure. Is there evidence to conclude that blood type and ethnic group are related? Explain how you arrived at your conclusion.

9.25   An article in the *New York Times* of January 30, 1988, described the results of an experiment on the effects of aspirin on cardiovascular disease. The subjects were 5139 male British medical doctors. The doctors were randomly assigned to two groups. One group of 3429 doctors took one aspirin daily, and the other group did not take aspirin. After 6 years, there were 148 deaths from heart attack or stroke in the first group and 79 in the second group. The Physicians' Health Study was a similar experiment using male American medical doctors as subjects. These doctors were also randomly assigned to one of two groups. The 11,037 doctors in the first group took one aspirin every other day, and the 11,034 doctors in the second group took no aspirin. After nearly 5 years there were 104 deaths from heart attacks in the first group and 189 in the second. Analyze the data from these two studies and summarize the results. How do the conclusions of the two studies differ, and why?

9.26   Gastric freezing was once a recommended treatment for ulcers in the upper intestine. A randomized comparative experiment found that 28 of the 82 patients who were subjected to gastric freezing improved, while 30 of the 78 patients in the control group improved. The hypothesis of "no difference" between the two groups can be tested in two ways: using a z statistic or using the X^2 statistic.

(a)   State the appropriate hypothesis and a two-sided alternative, and carry out a z test. What is the P-value?

(b)   Present the data in a 2×2 table. State the appropriate hypotheses and carry out the chi-square test. What is the P-value? Verify that the X^2 statistic is the square of the z statistic.

(c)   What do you conclude about the effectiveness of gastric freezing as a treatment for ulcers? (See Example 3.5 on page 240 for a discussion of gastric freezing.)

9.27   Example 2.32 (page 199) presents artificial data that illustrate Simpson's paradox. The data concern the survival rates of surgery patients at two hospitals.

(a)   Apply the chi-square test to the data for all patients combined and summarize the results.

(b)   Run separate chi-square analyses for the patients in good condition and for those in poor condition. Summarize these results.

(c)   Are the effects that illustrate Simpson's paradox in this example statistically significant?

9.28   You are asked to evaluate a proposal for an experiment on the effects of aspirin on cardiovascular disease similar to the experiments described in Exercise 9.25. The researchers will randomly assign subjects to a treatment group or to a control group. The proposed sample sizes are 200 for each group. Write a short evaluation of this proposal, using any relevant information from Exercise 9.25.

9.29   The sports goals study described in Example 9.3 (page 625) actually involved *three* categorical variables: sex, social comparison, and mastery. In that example, we combined the social comparison and mastery categories to obtain a two-way table. There are statistical methods for analyzing three-way tables, but these are beyond the scope of this text. We can, however, look at the several two-way tables formed by rearranging the data.

   The analyses and discussion of this example presented earlier suggest that the major sex difference is in the social comparison variable. To investigate this idea, rearrange the data into a 2 × 2 table that classifies the students by sex and social comparison. Analyze this table and summarize the results. Do the same using sex and mastery as the classification variables. Considering these analyses and the analysis presented in Example 9.6 (page 631), what conclusions do you draw?

9.30   A recent study of 865 college students found that 42.5% had student loans. The students were randomly selected from the approximately 30,000 undergraduates enrolled in a large public university. The overall purpose of the study was to examine the effects of student loan burdens on the choice of a career. A student with a large debt may be more likely to choose a field where starting salaries are high so that the loan can more easily be repaid. The following table classifies the students by field of study and whether or not they have a loan. (Data provided by Susan Prohofsky, from her Ph.D. dissertation, "Selection of undergraduate major: the influence of expected costs and expected benefits," Purdue University, 1991.)

Field of study	Student loan	
	Yes	No
Agriculture	32	35
Child development and family studies	37	50
Engineering	98	137
Liberal arts and education	89	124
Management	24	51
Science	31	29
Technology	57	71

Carry out a complete analysis of the association between having a loan and field of study, including a description of the association and an assessment of its statistical significance.

9.31   In the study described in the previous exercise, students were asked to respond to some questions regarding their interests and attitudes. Some of these questions form a scale called PEOPLE that measures altruism, or an interest in the welfare of others. Each student was classified as low, medium, or high on this scale. Is there an association between PEOPLE score and field of study? Here are the data:

	PEOPLE score		
Field of study	Low	Medium	High
Agriculture	5	27	35
Child development and family studies	1	32	54
Engineering	12	129	94
Liberal arts and education	7	77	129
Management	3	44	28
Science	7	29	24
Technology	2	62	64

Analyze the data and summarize your results. Are there some fields of study that have very large or very small proportions of students in the high-PEOPLE category?

9.32 An article in the *New York Times* of April 24, 1991, discussed data from the Centers for Disease Control that showed an increase in cases of measles in the United States. Of particular concern are complications from measles that can lead to death. The article noted that young children, who do not have fully developed immune systems, face an increased risk of death from complications of measles. Here are data on the 23,067 cases of measles reported in 1990. For each age group, the probability of death from measles is a parameter of interest. A comparison of the estimates of these parameters across age groups will provide information about the relationship between age and survival of an attack of measles.

	Survival	
Age group	Dead	Survived
Under 1 year	17	3806
1–4 years	37	7113
5–9 years	3	2208
10–14 years	3	1888
15–19 years	8	2715
20–24 years	6	2209
25–29 years	9	1492
30 years and over	14	1636

Summarize the death rates by age group. Prepare a plot to illustrate the pattern. Test the hypothesis that survival and age are related, report the results, and summarize your conclusion. From the data given, is it possible to study the association between catching measles and age? Explain why or why not.

9.33   In healthy individuals the concentration of various substances in the blood remains within relatively narrow bounds. One such substance is potassium. A person is said to be hypokalemic if the potassium level is too low (less than 3.5 milliequivalents per liter) and hyperkalemic if the level is too high (above 5.5 meq/l). Hypokalemia is associated with a variety of symptoms such as excessive tiredness, while hyperkalemia is generally an indication of a serious problem. Patients being treated with diuretics (pharmaceuticals that help the body to eliminate water) sometimes have abnormal potassium concentrations. In a large study of patients on chronic diuretic therapy, several risk factors were studied to see if they were associated with abnormal potassium levels. Of the 5180 patients studied, 1094 were hypokalemic, 4689 had normal potassium levels, and 27 were hyperkalemic. The following table gives the percentages of patients having each of four risk factors in the three potassium groups. (Data taken from William M. Tierney, Clement J. McDonald, and George P. McCabe, "Serum potassium testing in diuretic-treated outpatients," *Medical Decision Making,* 5 (1985), pp. 91–104.)

	Potassium group		
	Hypokalemic	Normal	Hyperkalemic
n	1094	4689	27
Hypertension	88.3%	78.1%	40.7%
Heart failure	16.5%	24.7%	55.6%
Diabetes	20.6%	25.5%	29.6%
Sex (% female)	72.5%	68.0%	48.1%

For example, 88.3% of the 1094 hypokalemic patients had hypertension, and 78.1% of the 4689 normal patients had hypertension. For each of the four risk factors, use the percents and n's given to compute the counts for the 2×3 table needed to study the association between the factor and potassium. Then analyze each table using the methods presented in this chapter. Note that there are very few patients in the hyperkalemic group. Therefore, reanalyze the data dropping this category from the tables. Write a short summary explaining what you have found.

9.34   The proportion of women entering many professions has undergone considerable change in recent years. A study of students enrolled in pharmacy programs describes the changes in this field. A random sample of 700 students in their third or higher year of study at colleges of pharmacy was taken in each of nine years. The following table gives the numbers of women in each of these samples. (Data are based on *Seventh Report to the President and Congress on the Status of Health Personnel in the United States,* Public Health Service, Washington, D.C., 1990.)

Year	1970	1972	1974	1976	1978	1980	1982	1984	1986
Women	164	195	226	283	302	342	369	385	412

Use the chi-square test to assess the change in the percentage of women pharmacy students over time, and summarize your results. (You will need to calculate the number of male students for each year using the fact that the sample size each year is 700.) Plot the percentage of women versus year. Describe the plot. Is it roughly linear? Find the least-squares line that summarizes the relation between time and the percentage of women pharmacy students.

9.35 Refer to the previous exercise. The actual percents of women pharmacy students for the years 1987 to 1996 are given in the following table. (Data provided by Dr. Susan Meyer of the American Association of Colleges of Pharmacy.)

Year	1987	1988	1989	1990	1991	1992	1993	1994	1995	1996
Women	60.0%	60.6%	61.6%	62.4%	63.0%	63.4%	63.2%	63.3%	63.4%	63.8%

Plot these percents versus year and summarize the pattern. Using your analysis of the data in this and the previous exercise, write a report summarizing the changes that have occurred in the percent of women pharmacy students from 1970 to 1996. Include an estimate of the percent for the year 2000 with an explanation of why you chose this estimate.

9.36 Refer to Exercise 8.66 (page 619) concerning *Castaneda v. Partida,* the case where the Supreme Court review used the phrase "two or three standard deviations" as a criterion for statistical significance. Recall that there were 181,535 persons eligible for jury duty, of whom 143,611 were Mexican Americans. Of the 870 people selected for jury duty, 339 were Mexican Americans. We are interested in finding out if there is an association between being a Mexican American and being selected as a juror. Formulate this problem using a two-way table of counts. Construct the 2×2 table using the variables Mexican American or not and juror or not. Find the X^2 statistic and its P-value. Square the z statistic that you obtained in Exercise 8.67 and verify that the result is equal to the X^2 statistic.

9.37 Refer to Exercise 9.3 (page 644) concerning the case-control study of factors associated with bringing a cat to an animal shelter and the similar study for dogs described in Exercise 9.4. The last category for the source of the pet was given as "other" and includes born in home, stray, and obtained from a shelter. The following two-way table lists these categories separately for cats:

		Source			
Group	Private	Pet store	Home	Stray	Shelter
Cases	124	16	20	38	18
Controls	219	24	38	116	49

Here is the same breakdown for dogs:

		Source			
Group	Private	Pet store	Home	Stray	Shelter
Cases	188	7	11	23	56
Controls	518	68	20	55	67

Analyze these 2×5 tables and compare the results that you obtained for the 2×3 tables in Exercises 9.3 and 9.4. With a large number of cells, the chi-square test sometimes does not have very much power.

9.38   Refer to Exercises 9.1, 9.3, 9.6, and 9.12. For each, state whether you are comparing two populations (the first model for two-way tables) or testing independence between two categorical variables (the second model).

9.39   Refer to Exercise 9.21 (page 651), where we examined the relationship between use of aluminum-containing antacid and Alzheimer's disease. In that exercise the P-value was 0.068, failing to achieve the traditional standard for statistical significance (0.05). Suppose that we did a similar study with more data. In particular, let's double each of the counts in the original table. Perform the analysis on these counts and summarize the effect of increasing the sample size.

Notes

1.   This study is reported in Joan L. Duda, "The relationship between goal perspectives, persistence and behavioral intensity among male and female recreational sport participants," *Leisure Sciences*, 10 (1988), pp. 95–106.

2.   When the expected cell counts are small, it is best to use a test based on the exact distribution rather than the chi-square approximation, particularly for 2×2 tables. Many statistical software systems offer an "exact" test as well as the chi-square test for 2×2 tables.

3.   The full report of the study appeared in George H. Beaton et al., "Effectiveness of vitamin A supplementation in the control of young child morbidity and mortality in developing countries," United Nations ACC/SCN State-of-the-Art Series, Nutrition Policy Discussion Paper no. 13, 1993.

4. These data were taken from Ray H. Rosenman et al., "A 4-year prospective study of the relationship of different habitual vocational physical activity to risk and incidence of ischemic heart disease in volunteer male federal employees," in P. Milvey (ed.), *The Marathon: Physiological, Medical, Epidemiological, and Psychological Studies,* New York Academy of Sciences, 301 (1977), pp. 627–641.

5. An alternative formula that can be used for hand or calculator computations is

$$X^2 = \sum \frac{(\text{observed})^2}{\text{expected}} - n$$

6. See, for example, Alan Agresti, *Categorical Data Analysis,* Wiley, New York, 1990.

Prelude

Throughout this book we have followed several strategies for working with data. Two of these strategies are (1) descriptive analysis and exploration of data precedes formal inference and (2) examination of each variable separately precedes study of relations among quantitative variables. In this chapter we reach the second part of both strategies in the study of formal inference for relations among quantitative variables.

We will describe methods for inference when there is a single response variable and one quantitative explanatory variable. The descriptive tools we learned in Chapter 2—scatterplots, least-squares regression, and correlation—are essential preliminaries to inference and also provide a foundation for confidence intervals and significance tests. Here are some examples of questions we can answer from data using the methods in this chapter:

- A scatterplot shows a straight-line relationship between how much natural gas a household consumes and how cold the outside temperature is. How accurately can we predict gas consumption from temperature?

- What is the relationship between the selling price of a home and the number of square feet that it contains?

660

CHAPTER 10

Clinton Hill, *Untitled*, 1991.

Inference for Regression

We first met the sample mean $\bar{x}$ in Chapter 1 as a measure of the center of a collection of observations. Later we learned that when the data are a random sample from a population, the sample mean is an estimate of the population mean μ. In Chapters 6 and 7, we used $\bar{x}$ as the basis for confidence intervals and significance tests for inference about μ.

Now we will follow the same program for the problem of fitting straight lines to data. In Chapter 2 we met the least-squares regression line $\hat{y} = a + bx$ as a description of a straight-line relationship between a response variable y and an explanatory variable x. At that point we did not distinguish between sample and population. Now we will think of the least-squares line computed from a sample as an estimate of a *true* regression line for the population, just as $\bar{x}$ is an estimate of the true population mean μ. To emphasize the change in thinking, we introduce new notation. Following the common practice of using Greek letters for population parameters, we will write the population line as $\beta_0 + \beta_1 x$. The least-squares line fitted to sample data is now $b_0 + b_1 x$. This notation reminds us that the intercept b_0 of the fitted line estimates the intercept β_0 of the population line, and the slope b_1 estimates the slope β_1.

We are concerned with inference in the setting of simple linear regression, the familiar case of a single explanatory variable x. By extending the ideas in Chapters 6 and 7, we can give confidence intervals and significance tests for inference about the slope β_1 and the intercept β_0. Because regression lines are often used for prediction, we also consider inference about either the mean response or an individual future observation on y for a given value of the explanatory variable x. Finally, we discuss statistical inference about the correlation between two variables x and y that need not have an explanatory-response relationship.

10.1   Simple Linear Regression

Statistical model for linear regression

Simple linear regression studies the relationship between a response variable y and a single explanatory variable x. We expect that different values of x will produce different mean responses. We encountered a similar but simpler situation in Chapter 7 when we discussed methods for comparing two population means. Figure 10.1 illustrates the statistical model for a comparison of the blood pressure in two groups of experimental subjects, one group taking a calcium supplement and the other a placebo. In each population the observed change in blood pressure varies according to a normal distribution. The mean change will generally be different for the two populations. In the figure, μ_1 and μ_2 represent the two population means. The two normal curves in Figure 10.1 have the same spread, indicating that the population standard deviations are assumed to be equal. We can think of the treatment (placebo or calcium) as the explanatory variable in this example.

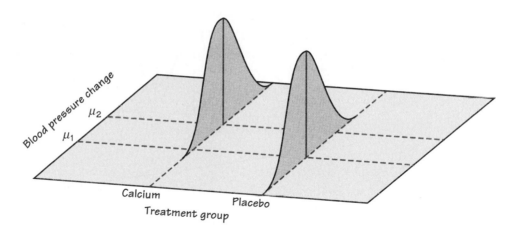

FIGURE 10.1   The statistical model for comparing responses to two treatments; the mean response varies with the treatment.

In linear regression the explanatory variable x can have many different values. Imagine, for example, giving different amounts x of calcium to different groups of subjects. We can think of the values of x as defining different **subpopulations,** one for each possible value of x. Each subpopulation consists of all individuals in the population having the same value of x. If you give $x = 1500$ mg to everyone, then these are the subjects "who actually receive 1500 mg." In this case, the subpopulation would be the same as the sample.

subpopulations

The statistical model for simple linear regression assumes that for each value of x the observed values of the response variable y are normally distributed about a mean that depends on x. We use μ_y to represent these means. Rather than just two means μ_1 and μ_2, we are interested in how the many means μ_y change as x changes. In general the means μ_y can change according to any sort of pattern as x changes. In **simple linear regression** we assume that they all lie on a line when plotted against x. The equation of the line is

simple linear regression

$$\mu_y = \beta_0 + \beta_1 x$$

population regression line

with intercept β_0 and slope β_1. This is the **population regression line;** it describes how the mean response changes with x. Actually observed y's will vary about these means. The model assumes that this variation, measured by the standard deviation σ, is the same for all values of x. Figure 10.2 displays this statistical model. The line is the population regression line, and the three normal curves show how the response y will vary for three different values of the explanatory variable x.

Data for simple linear regression

The data for a linear regression are observed values of y and x. The model takes each x to be a fixed known quantity. In practice, x may not be exactly

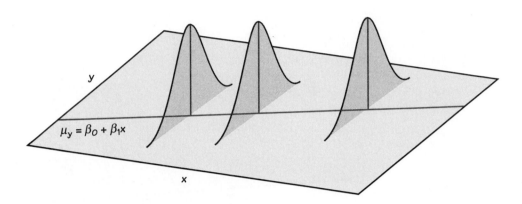

FIGURE 10.2 The statistical model for linear regression; the mean response is a straight-line function of the explanatory variable.

known. If the error in measuring x is large, more advanced inference methods are needed. The response y to a given x is a random variable that will take different values if we have several observations with the same x-value. The model describes the mean and standard deviation of the random variable y. These unknown parameters must be estimated from the data.

We will use the following example to explain the fundamentals of simple linear regression. Because regression calculations in practice are always done by statistical software, we will rely on computer output for the arithmetic in this example. In the next section, we give an example that illustrates how to do the work with a calculator if software is unavailable.

EXAMPLE 10.1

"The fat content of the human body has physiological and medical importance. It may influence morbidity and mortality, it may alter the effectiveness of drugs and anesthetics, and it may affect the ability to withstand exposure to cold and starvation."[1]

In practice, fat content is found by measuring body density, the weight per unit volume of the body. High fat content corresponds to low body density. Body density is hard to measure directly—the standard method requires that subjects be weighed underwater. For this reason, scientists have sought variables that are easier to measure and that can be used to predict body density. Research suggests that *skinfold thickness* can accurately predict body density. To measure skinfold thickness, pinch a fold of skin between calipers at four body locations to determine the thickness, and add the four thicknesses. There is a linear relationship between body density and the logarithm of the skinfold thickness measure. The explanatory variable x is the log of the sum of the skinfold measures, and the response variable y is body density.

In the statistical model for predicting body density from skinfold thickness, subpopulations are defined by the x variable. All individuals with the same skinfold thickness x are in the same subpopulation. Their body densities are assumed to be normally distributed about a mean density that has a straight-line relation to skinfold thickness. The mean density for people with log skinfold thickness x is

$$\mu_y = \beta_0 + \beta_1 x$$

This population regression line gives the mean body density for all values of x. We cannot observe this line, because the observed responses y vary about their means. The statistical model for linear regression consists of the population regression line and a description of the variation of y about the line. Figure 10.2 displays the line and the variation in a picture, and the following equation expresses the idea in an equation:

$$\text{DATA} = \text{FIT} + \text{RESIDUAL}$$

The FIT part of the model consists of the subpopulation means, given by the expression $\beta_0 + \beta_1 x$. The RESIDUAL part represents deviations of the data from the line of population means. We assume that these deviations are normally distributed with standard deviation σ. We use ϵ (the Greek letter epsilon) to stand for the RESIDUAL part of the statistical model. A response y is the sum of its mean and a chance deviation ϵ from the mean. The deviations ϵ represent "noise," that is, variation in y due to other causes that prevent the observed (x, y)-values from forming a perfectly straight line on the scatterplot.

Simple Linear Regression Model

Given n observations on the explanatory variable x and the response variable y,

$$(x_1, y_1), \quad (x_2, y_2), \quad \ldots, \quad (x_n, y_n)$$

the **statistical model for simple linear regression** states that the observed response y_i when the explanatory variable takes the value x_i is

$$y_i = \beta_0 + \beta_1 x_i + \epsilon_i$$

Here $\beta_0 + \beta_1 x_i$ is the mean response when $x = x_i$. The deviations ϵ_i are assumed to be independent and normally distributed with mean 0 and standard deviation σ.

The parameters of the model are β_0, β_1, and σ.

Because the means μ_y lie on the line $\mu_y = \beta_0 + \beta_1 x$, they are all determined by β_0 and β_1. Once we have estimates of β_0 and β_1, the linear relationship determines the estimates of μ_y for all values of x. Linear regression allows us to do inference not only for subpopulations for which we have data but also for those corresponding to x's not present in the data. We will learn how to do inference about

- the slope β_1 and the intercept β_0 of the population regression line,
- the mean response μ_y for a given value of x, and
- an individual future response y for a given value of x.

Estimating the regression parameters

The method of least squares presented in Chapter 2 fits a line to summarize a relationship between the observed values of an explanatory variable and a response variable. Now we want to use the least-squares line as a basis for inference about a population from which our observations are a sample. We can do this only when the statistical model just presented holds. In that setting, the slope b_1 and intercept b_0 of the least-squares line

$$\hat{y} = b_0 + b_1 x$$

estimate the slope β_1 and the intercept β_0 of the population regression line.

Using the formulas from Chapter 2 (page 141) and our new notation, the slope of the least-squares line is

$$b_1 = r \frac{s_y}{s_x}$$

and the intercept is

$$b_0 = \bar{y} - b_1 \bar{x}$$

Here, r is the correlation between y and x, s_y is the standard deviation of y, and s_x is the standard deviation of x. Some algebra based on the rules for means of random variables (Section 4.4) shows that b_0 and b_1 are unbiased estimators of β_0 and β_1. Furthermore, b_0 and b_1 are normally distributed with means β_0 and β_1 and standard deviations that can be estimated from the data. Normality of these sampling distributions is a consequence of the assumption that the ϵ_i are distributed normally. A general form of the central limit theorem tells us that the distributions of b_0 and b_1 will still be approximately normal even if the ϵ_i are not. On the other hand, outliers and influential observations can invalidate the results of inference for regression.

The predicted value of y for a given value x^* of x is the point on the least-squares line $\hat{y} = b_0 + b_1 x^*$. This is an unbiased estimator of the mean response μ_y when $x = x^*$. The **residuals** are

residuals

$$
\begin{aligned}
e_i &= \text{observed response} - \text{predicted response} \\
&= y_i - \hat{y}_i \\
&= y_i - b_0 - b_1 x_i
\end{aligned}
$$

The residuals e_i correspond to the model deviations ϵ_i. The e_i sum to 0, and the ϵ_i come from a population with mean 0.

The remaining parameter to be estimated is σ, which measures the variation of y about the population regression line. Because this parameter is the standard deviation of the model deviations, it should come as no surprise that we use the residuals to estimate it. As usual, we work first with the variance and take the square root to obtain the standard deviation. For simple linear regression the estimate of σ^2 is the average squared residual

$$s^2 = \frac{\sum e_i^2}{n-2} = \frac{\sum (y_i - \hat{y}_i)^2}{n-2}$$

degrees of freedom

We average by dividing the sum by $n - 2$ in order to make s^2 an unbiased estimate of σ^2. The sample variance of n observations uses the divisor $n - 1$ for this same reason. The quantity $n - 2$ is called the **degrees of freedom** for s^2. The estimate of σ is given by

$$s = \sqrt{s^2}$$

In practice, we use software to do the unpleasant arithmetic needed to obtain b_1, b_0, and s. Here are the results for the example of body density and skinfold thickness.

EXAMPLE 10.2 We use the SAS statistical software package to calculate the regression of body density on log of skinfolds for Example 10.1. In entering the data, we chose the names LSKIN and DEN for the explanatory and response variables. It is good practice to use names, rather than just x and y, to remind yourself which data the output describes. The skinfolds were measured in millimeters (mm) at the biceps, triceps, subscapular and suprailiac areas. LSKIN is the log of the sum of these four measures. Density is measured in thousands of kilograms per cubic meter (10^3 kg/m^3). Because the relationship between skinfold and body density varies slightly with age and sex, we look only at data from a sample of 92 males aged 20 to 29. In the discussion that follows, we round off the numbers from the computer output. Software often reports many more digits than are meaningful or useful.

First, we examine the variables and their relationship. Figure 10.3 is a scatterplot of the data. The relationship appears to be linear, and no unusual observations are evident.

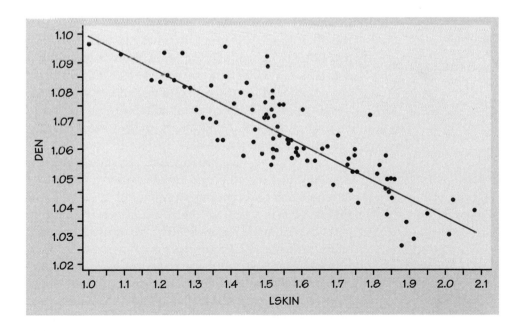

FIGURE 10.3   Scatterplot of DEN versus LSKIN with the least-squares regression line.

```
     Root MSE         0.00854      R-square         0.7204
     Dep Mean         1.06403      Adj R-sq         0.7173
     C.V.             0.80252


Parameter Estimates


                        Parameter      Standard      T for H0:
Variable    DF          Estimate          Error    Parameter=0      Prob > |T|

INTERCEP    1           1.162999     0.00655964        177.296          0.0001
LSKIN       1          -0.063119     0.00414491        -15.228          0.0001
```

FIGURE 10.4   Regression output for the body density example.

We then proceed to fit the least-squares line. Figure 10.4 gives part of the output of the SAS regression procedure for this example. The results of the least-squares fit appear at the bottom of the output. The values of b_0 and b_1 are given in the column labeled Parameter Estimate. The column labeled Variable tells us that the entry in the INTERCEP row is b_0 and the entry for LSKIN is b_1. We see that $b_0 = 1.16$ and $b_1 = -0.0631$, so the least-squares line is

$$\hat{y} = 1.16 - 0.0631x$$

That line appears on the scatterplot of Figure 10.3. Moving to the upper section of the output, the estimated standard deviation s is labeled Root MSE and is 0.00854. We now have estimates of all the parameters β_0, β_1, and σ.

The output contains much other information that we ignore for now. Computer outputs often give more information than we want or need. On the other hand, it is very frustrating to find that a software package does not print out the particular statistics that we want for an analysis. The experienced user of statistical software learns to ignore the parts of the output that are not needed for the problem at hand.

Regression outputs from different software all contain the same basic quantities. Figure 10.5 gives the output for the body density example for Excel, Minitab, and the TI-83.

Now that we have fitted a line, we should examine the residuals. Plot the residuals both against the case number (especially if this reflects the order in which the observations were taken) and against the explanatory variable. Figure 10.6 is a plot of the residuals versus case number, and Figure 10.7 plots the residuals versus LSKIN. No unusual values or patterns are evident in either plot. Finally, Figure 10.8 is a normal quantile plot of the residuals. The plot looks fairly straight, so the assumption of normally distributed deviations appears to be reasonable. That is important for the inference that will follow. We need not make these plots by hand; the software will produce all of them on demand.

For the 92 males aged 20 to 29 in the study, the mean body density was 1.064, and the standard deviation was 0.016. The regression output shows

Excel

	A	B	C	D	E	F	G
1	SUMMARY OUTPUT						
2							
3	*Regression Statistics*						
4	Multiple R	0.848766779					
5	R Square	0.720405045					
6	Adjusted R Square	0.717298435					
7	Standard Error	0.008539025					
8	Observations	92					
9							
10		*Coefficients*	*Standard Error*	*t Stat*	*P-value*	*Lower 95%*	*Upper 95%*
11	Intercept	1.162999075	0.006559641	177.2962	2.67E-116	1.1499672	1.1760309
12	X Variable 1	-0.063118994	0.00414491	-15.2281	1.22E-26	-0.0713536	-0.0548844
13							

Sheet1 / den /

Ready   Sum=0   NUM

Minitab

```
The regression equation is
DEN = 1.16 - 0.0631 LSKIN

Predictor        Coef        StDev            T            P
Constant      1.16300     0.00656       177.30        0.000
LSKIN        -0.063119    0.004145      -15.23        0.000

S = 0.008539    R-Sq = 72.0%        R-Sq(adj) = 71.7%
```

TI-83

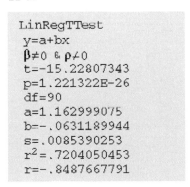

```
LinRegTTest
y=a+bx
β≠0 & ρ≠0
t=-15.22807343
p=1.221322E-26
df=90
a=1.162999075
b=-.0631189944
s=.0085390253
r²=.7204050453
r=-.8487667791
```

FIGURE 10.5 Regression output from Excel, Minitab, and the TI-83 for the body density example.

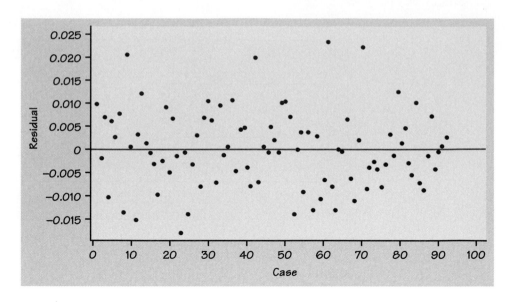

FIGURE 10.6   Plot of residuals versus case number for the body density example.

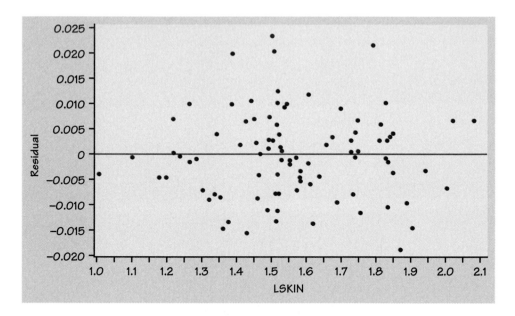

FIGURE 10.7   Plot of residuals versus the explanatory variable for the body density example.

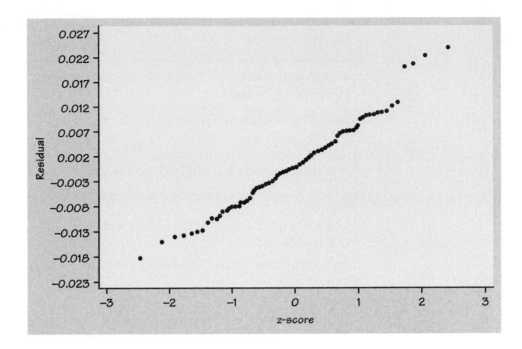

FIGURE 10.8 Normal quantile plot of the residuals for the body density example.

that the standard deviation of body density about the regression line is 0.009, only about half as large. This decrease in the standard deviation is due to the ability of skinfold thickness to explain body density.

Confidence intervals and significance tests

Chapter 7 presented confidence intervals and significance tests for means and differences in means. In each case, inference rested on the standard errors of estimates and on t distributions. Inference for the intercept and slope in a linear regression is similar in principle, although the recipes are more complicated. All of the confidence intervals, for example, have the form

$$\text{estimate} \pm t^* \text{SE}_{\text{estimate}}$$

where t^* is a critical point of a t distribution.

Confidence intervals and tests for the slope and intercept are based on the normal sampling distributions of the estimates b_1 and b_0. Standardizing these estimates gives standard normal z statistics. The standard deviations of these estimates are multiples of σ, the model parameter that describes the variability about the true regression line. Because we do not know σ, we estimate it by s, the variability of the data about the least-squares line. When we do this, we get t distributions with degrees of freedom $n - 2$, the degrees of freedom of s. We give formulas for the standard errors SE_{b_1} and SE_{b_0} in Section 10.2.

For now we will concentrate on the basic ideas and let the computer do the computations.

Confidence Intervals and Significance Tests for Regression Slope and Intercept

A **level C confidence interval for the intercept** β_0 is

$$b_0 \pm t^* \text{SE}_{b_0}$$

A **level C confidence interval for the slope** β_1 is

$$b_1 \pm t^* \text{SE}_{b_1}$$

In these expressions t^* is the value for the $t(n-2)$ density curve with area C between $-t^*$ and t^*.

To test the hypothesis H_0: $\beta_1 = 0$, compute the **test statistic**

$$t = \frac{b_1}{\text{SE}_{b_1}}$$

The **degrees of freedom** are $n-2$. In terms of a random variable T having the $t(n-2)$ distribution, the P-value for a test of H_0 against

H_a: $\beta_1 > 0$ is $P(T \geq t)$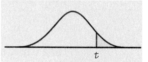

H_a: $\beta_1 < 0$ is $P(T \leq t)$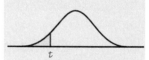

H_a: $\beta_1 \neq 0$ is $2P(T \geq |t|)$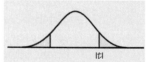

There are similar formulas for significance tests about the intercept β_0, using SE_{b_0} and the $t(n-2)$ distribution. Although computer outputs often include a test of H_0: $\beta_0 = 0$, this information usually has little practical value. From the equation for the population regression line, $\mu_y = \beta_0 + \beta_1 x$, we see that β_0 is the mean response corresponding to $x = 0$. In many practical situations, this subpopulation does not exist or is not interesting.

On the other hand, the test of H_0: $\beta_1 = 0$ is quite useful. When we substitute $\beta_1 = 0$ in the model, the x term drops out and we are left with

$$\mu_y = \beta_0$$

This model says that the mean of y docs not vary with x. All of the y's come from a single population with mean β_0, which we would estimate by $\bar{y}$. The hypothesis $H_0: \beta_1 = 0$ therefore says that there is no straight-line relationship between y and x and that linear regression of y on x is of no value for predicting y.

EXAMPLE 10.3

The computer output in Figure 10.4 for the body density problem contains the information needed for inference about the regression slope and intercept. The column labeled Standard Error gives the standard errors of the estimates. The value of SE_{b_1} appears to the right of the estimated slope $b_1 = -0.0631$; it is 0.0042. (As usual, we have rounded the values from the output.)

The t statistic and P-value for the test of $H_0: \beta_1 = 0$ against the two-sided alternative $H_a: \beta_1 \neq 0$ appear in the columns labeled T for H0: Parameter = 0 and Prob > |T|. We can verify the t calculation from the formula for the standardized estimate:

$$t = \frac{b_1}{SE_{b_1}} = \frac{-0.063119}{0.00414491} = -15.23$$

The P-value is 0.0001. There is strong evidence against the null hypothesis. We conclude that linear regression on the logarithm of the skinfold measures is useful for predicting body density.

Higher body fat reduces body density and increases skinfold thickness, so we expect a negative association. We might therefore choose the one-sided alternative $H_a: \beta_1 < 0$. This choice does not affect our conclusion for this example, but the P-value for the one-sided alternative is half of the value for the two-sided alternative.

A confidence interval for β_1 requires a critical value t^* from the $t(n-2) = t(90)$ distribution. In Table D there are entries for 80 and 100 degrees of freedom. The values for these rows are very similar. To be conservative, we will use the larger critical value, for 80 degrees of freedom. For a 95% confidence interval we use the value in the 95% column, which is $t^* = 1.99$.

To compute the 95% confidence interval for β_1 we combine the estimate of the slope with the margin of error:

$$b_1 \pm t^* SE_{b_1} = -0.063119 \pm (1.99)(0.00414491)$$
$$= -0.0631 \pm 0.0082$$

The interval is $(-0.07, -0.05)$. We estimate that an increase of 1 in the logarithm of skinfold thickness is associated with a decrease of between 0.05 and 0.07 times 10^3 kg/m^3 in body density.

Note that the intercept in this example is not of practical interest. It estimates the mean body density when the logarithm of the sum of the four skinfold thicknesses (that's x) is 0, a value that cannot occur. The values of LSKIN in the data range from 1.00 to 2.08. For this reason, we do not compute a confidence interval for β_0.

Confidence intervals for mean response

For any specific value of x, say x^*, the mean of the response y in this subpopulation is given by

$$\mu_y = \beta_0 + \beta_1 x^*$$

To estimate this mean from the sample, we substitute the estimates b_0 and b_1 for β_0 and β_1:

$$\hat{\mu}_y = b_0 + b_1 x^*$$

A confidence interval for μ_y adds to this estimate a margin of error based on the standard error $\text{SE}_{\hat{\mu}}$. (The formula for the standard error is given in Section 10.2.)

Confidence Interval for a Mean Response

A **level C confidence interval for the mean response** μ_y when x takes the value x^* is

$$\hat{\mu}_y \pm t^* \text{SE}_{\hat{\mu}}$$

where t^* is value for the $t(n-2)$ density curve with area C between $-t^*$ and t^*.

Many computer programs calculate confidence intervals for the mean response corresponding to each of the x-values in the data. Some can calculate an interval for any value x^* of the explanatory variable.[2]

EXAMPLE 10.4

Figure 10.9 gives more of the output generated by the SAS regression procedure for the data on body density and skinfold measures. We show the output for only the first 24 of the 92 cases.

The observed data values y_i appear in the column labeled Dep Var DEN, the predicted values $\hat{\mu}_y$ are given in the column labeled Predict Value, and the standard errors $\text{SE}_{\hat{\mu}}$ of the predicted values appear in the column labeled Std Err Predict. The lower and upper limits for the 95% confidence intervals are in the columns labeled Lower95% Mean and Upper95% Mean.

Consider the first line of the output. The value of LSKIN for this individual is 1.27. The predicted mean DEN for all individuals with LSKIN = 1.27 is

$$\hat{\mu}_{\text{DEN}(1.27)} = 1.0828 \times 10^3 \text{ kg/m}^3$$

with a standard error of

$$\text{SE}_{\hat{\mu}} = 0.002$$

The 95% confidence interval is (1.0798, 1.0859). We conclude with 95% confidence that the mean body density for men aged 20 to 29 whose logarithm of skinfold measures is 1.27 lies between 1.0798 and 1.0859.

We can gain more insight by plotting the upper and lower confidence limits for a range of values of x on a graph with the data and the least-squares line. The 95% confidence limits appear as curved black lines in Figure 10.10. For any x^*, the confidence interval for the mean response extends from the lower black line to the upper black line. The intervals are narrowest for values of x^* near the mean of the observed x's and increase as x^* moves away from $\bar{x}$.

Obs	LSKIN	Dep Var DEN	Predict Value	Std Err Predict	Lower95% Mean	Upper95% Mean
1	1.27	1.0930	1.0828	0.002	1.0798	1.0859
2	1.56	1.0630	1.0645	0.001	1.0628	1.0663
3	1.45	1.0780	1.0715	0.001	1.0695	1.0735
4	1.52	1.0560	1.0671	0.001	1.0652	1.0689
5	1.51	1.0730	1.0677	0.001	1.0659	1.0695
6	1.51	1.0710	1.0677	0.001	1.0659	1.0695
7	1.50	1.0760	1.0683	0.001	1.0665	1.0702
8	1.62	1.0470	1.0607	0.001	1.0589	1.0626
9	1.50	1.0890	1.0683	0.001	1.0665	1.0702
10	1.75	1.0530	1.0525	0.001	1.0502	1.0549
11	1.43	1.0570	1.0727	0.001	1.0706	1.0748
12	1.81	1.0510	1.0488	0.001	1.0461	1.0514
13	1.60	1.0740	1.0620	0.001	1.0602	1.0638
14	1.49	1.0700	1.0690	0.001	1.0671	1.0708
15	1.29	1.0810	1.0816	0.001	1.0787	1.0845
16	1.52	1.0640	1.0671	0.001	1.0652	1.0689
17	1.83	1.0370	1.0475	0.001	1.0447	1.0503
18	1.58	1.0600	1.0633	0.001	1.0615	1.0650
19	1.70	1.0650	1.0557	0.001	1.0536	1.0578
20	1.59	1.0580	1.0626	0.001	1.0609	1.0644
21	2.02	1.0420	1.0355	0.002	1.0314	1.0396
22	1.84	1.0450	1.0469	0.001	1.0440	1.0497
23	1.87	1.0260	1.0450	0.002	1.0419	1.0480
24	1.83	1.0460	1.0475	0.001	1.0447	1.0503

FIGURE 10.9 Computer output giving confidence intervals for subpopulation means for the body density example.

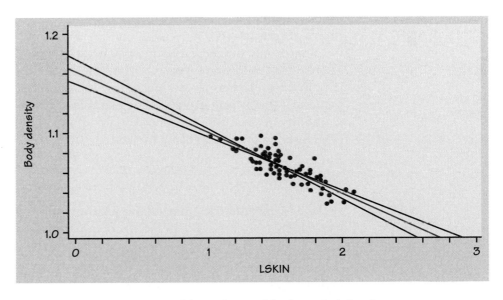

FIGURE 10.10 The 95% confidence limits (black curves) for the mean response for the body density example. The range of LSKIN is extended to show that the intervals grow wider as x^* moves away from the mean of the x observations.

To show this effect clearly, we have extended the x axis in the plot far beyond the range of the actual data. In practice, you would avoid extrapolation for such extreme values.

Prediction intervals

In the last example, we predicted the mean body density for all young men with LSKIN equal to 1.27. Suppose that John has LSKIN = 1.27. We can also predict John's individual body density. The predicted value is the same as if we were predicting the mean of the entire subpopulation, but the margin of error is larger because it is harder to predict one individual value than to predict the mean.

The predicted response y for an individual case with a specific value x^* of the explanatory variable x is

$$\hat{y} = b_0 + b_1 x^*$$

This is the same as the expression for $\hat{\mu}_y$. That is, the fitted line is used both to estimate the mean response when $x = x^*$ and to predict a single future response. We use the two notations $\hat{\mu}_y$ and $\hat{y}$ to remind ourselves of these two distinct uses.

A useful prediction should include a margin of error to indicate its accuracy. The interval used to predict a future observation is called a **prediction interval.** Although the response y that is being predicted is a random variable, the interpretation of a prediction interval is similar to that for a confidence interval. Draw a sample of n observations (x_i, y_i) and then one additional observation (x^*, y). Do this many times, and each time calculate the 95% prediction interval for y based on the sample. Then 95% of the prediction intervals will contain the y-value of the additional observation. In other words, the probability that this method produces an interval that contains the value of a future observation is 0.95.

The form of the prediction interval is very similar to that of the confidence interval for the mean response. The difference is that the standard error $\mathrm{SE}_{\hat{y}}$ used in the prediction interval includes both the variability due to the fact that the least-squares line is not exactly equal to the true regression line *and* the variability of the future response variable y around the subpopulation mean. (The formula for $\mathrm{SE}_{\hat{y}}$ appears in Section 10.2.)

prediction interval (margin note)

Prediction Interval for a Future Observation

A **level C prediction interval for a future observation** on the response variable y from the subpopulation corresponding to x^* is

$$\hat{y} \pm t^* \mathrm{SE}_{\hat{y}}$$

where t^* is the value for the $t(n-2)$ density curve with area C between $-t^*$ and t^*.

Obs	LSKIN	Dep Var DEN	Predict Value	Std Err Predict	Lower95% Predict	Upper95% Predict
1	1.27	1.0930	1.0828	0.002	1.0656	1.1001
2	1.56	1.0630	1.0645	0.001	1.0475	1.0816
3	1.45	1.0780	1.0715	0.001	1.0544	1.0886
4	1.52	1.0560	1.0671	0.001	1.0500	1.0841
5	1.51	1.0730	1.0677	0.001	1.0506	1.0848
6	1.51	1.0710	1.0677	0.001	1.0506	1.0848
7	1.50	1.0760	1.0683	0.001	1.0513	1.0854
8	1.62	1.0470	1.0607	0.001	1.0437	1.0778
9	1.50	1.0890	1.0683	0.001	1.0513	1.0854
10	1.75	1.0530	1.0525	0.001	1.0354	1.0697
11	1.43	1.0570	1.0727	0.001	1.0556	1.0898
12	1.81	1.0510	1.0488	0.001	1.0316	1.0659
13	1.60	1.0740	1.0620	0.001	1.0450	1.0791
14	1.49	1.0700	1.0690	0.001	1.0519	1.0860
15	1.29	1.0810	1.0816	0.001	1.0644	1.0988
16	1.52	1.0640	1.0671	0.001	1.0500	1.0841
17	1.83	1.0370	1.0475	0.001	1.0303	1.0647
18	1.58	1.0600	1.0633	0.001	1.0462	1.0803
19	1.70	1.0650	1.0557	0.001	1.0386	1.0728
20	1.59	1.0580	1.0626	0.001	1.0456	1.0797
21	2.02	1.0420	1.0355	0.002	1.0180	1.0530
22	1.84	1.0450	1.0469	0.001	1.0297	1.0641
23	1.87	1.0260	1.0450	0.002	1.0277	1.0622
24	1.83	1.0460	1.0475	0.001	1.0303	1.0647

FIGURE 10.11 Computer output giving prediction intervals for the body density example.

EXAMPLE 10.5

Figure 10.11 gives prediction output from the SAS regression procedure for the first 24 of the 92 cases in the body density versus skinfold data set. The format is similar to Figure 10.9, but the two rightmost columns now contain the upper and lower limits of the 95% prediction interval for the body density of a single young man with LSKIN equal to the value in the first column. These limits are wider than the corresponding 95% confidence limits for the mean body density that appear in Figure 10.9. (Note that this output does not give $SE_{\hat{y}}$, but only $SE_{\hat{\mu}}$.)

Consider Observation 1, which has LSKIN = 1.27. The 95% prediction interval is 1.0656 to 1.1001. For a male aged 20 to 29 with LSKIN = 1.27, we predict with 95% confidence that his body density will be between 1.0656 and 1.1001 times 10^3 kg/m^3.

Figure 10.12 shows the upper and lower prediction limits for a range of values of x, along with the data and the least-squares line. The 95% prediction limits are indicated by the black curves. Compare this figure with Figure 10.10, which shows the 95% confidence limits drawn to the same scale. The upper and lower limits of the prediction intervals are farther from the

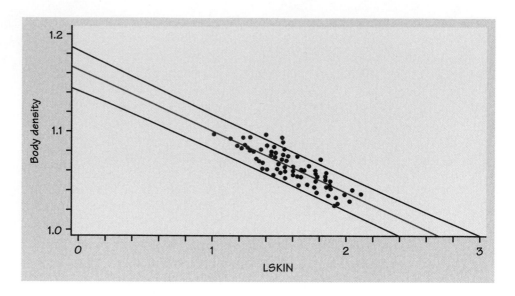

FIGURE 10.12   The 95% prediction limits (black curves) for individual responses for the body density example. Compare with Figure 10.10.

least-squares line than are the confidence limits. The figures remind us that the interval for a single future observation must be larger than an interval for the mean of its subpopulation.

Beyond the basics ▶ nonlinear regression

The regression model that we have studied assumes that the relationship between the response variable and the explanatory variable can be summarized with a straight line. When the relationship is not linear, we can sometimes make it linear by a transformation. In other circumstances, we use models that allow for various types of curved relationships. These models are called **nonlinear** models.

nonlinear models

Technical details are much more complicated for nonlinear models. In general we cannot write down simple formulas for the parameter estimates; we use a computer to solve systems of equations to find the estimates. However, the basic principles are those that we have already learned. For example,

$$\text{DATA} = \text{FIT} + \text{RESIDUAL}$$

still applies. The FIT is a nonlinear (curved) function and the residuals are assumed to be an SRS from the $N(0, \sigma)$ distribution. The nonlinear function contains parameters that must be estimated from the data. Approximate standard errors for these estimates are part of the standard output provided by software. Here is an example.

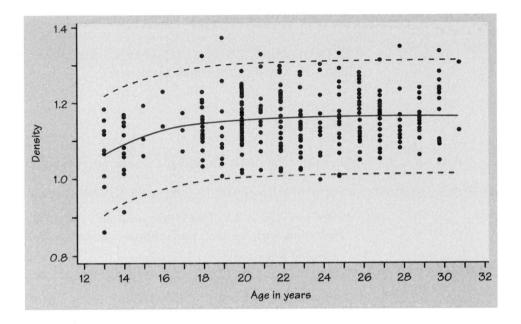

FIGURE 10.13   Plot of total body bone mineral density versus age.

EXAMPLE 10.6

As we age, our bones become weaker and are more likely to break. Osteoporosis (or weak bones) is the major cause of bone fractures in older women. Some researchers have studied this problem by looking at how and when bone mass is accumulated by young women. Understanding the relationship between age and bone mass is an important part of this approach to the problem.

Figure 10.13 displays data for a measure of bone strength, called "total body bone mineral density" (TBBMD), and age for a sample of 256 young women.[3] TBBMD is measured in grams per square centimeter (g/cm^2), and age is recorded in years. The solid curve is the nonlinear fit, and the dashed curves are 95% prediction limits. The fitted nonlinear equation is

$$\hat{y} = 1.162\frac{e^{-1.62+0.28x}}{1+e^{-1.62+0.28x}}$$

In this equation, $\hat{y}$ is the predicted value of TBBMD, the response variable; and x is age, the explanatory variable. A straight line would not do a very good job of summarizing the relationship between TBBMD and age. At first, TBBMD increases with age, but then it levels off as age increases. The value of the function where it is level is called "peak bone mass"; it is a parameter in the nonlinear model. The estimate is 1.162 and the standard error is 0.008. Software gives the 95% confidence interval as (1.146, 1.178).

The long-range goals of the researchers who conducted this study include developing intervention programs (exercise and increasing calcium intake have been shown to be effective) for young women that will increase their TBBMD. What age groups should be the target of these interventions? The fitted nonlinear model can be used to obtain estimates of the age (with a standard error) at which any given percent of the peak bone mass is attained.

We estimate that the age at which the population reaches 95% of peak bone mass is 16.2 years (SE = 1.1 years). For 99% of peak bone mass, the age is 22.1 years (SE = 2.5 years). Intervention programs should be directed toward high school and college-aged women.

SUMMARY

The statistical model for **simple linear regression** is

$$y_i = \beta_0 + \beta_1 x_i + \epsilon_i$$

where $i = 1, 2, \ldots, n$. The ϵ_i are assumed to be independent and normally distributed with mean 0 and standard deviation σ. The **parameters** of the model are β_0, β_1, and σ.

The intercept and slope β_0 and β_1 are estimated by the intercept and slope of the **least-squares regression line**, b_0 and b_1. The parameter σ is estimated by

$$s = \sqrt{\frac{\sum e_i^2}{n - 2}}$$

where the e_i are the **residuals**

$$e_i = y_i - \hat{y}_i$$

A **level C confidence interval for β_1** is

$$b_1 \pm t^* \mathrm{SE}_{b_1}$$

where t^* is the value for the $t(n - 2)$ density curve with area C between $-t^*$ and t^*.

The **test of the hypothesis** $H_0 : \beta_1 = 0$ is based on the **t statistic**

$$t = \frac{b_1}{\mathrm{SE}_{b_1}}$$

and the $t(n - 2)$ distribution. There are similar formulas for confidence intervals and tests for β_0, but these are meaningful only in special cases.

The **estimated mean response** for the subpopulation corresponding to the value x^* of the explanatory variable is

$$\hat{\mu}_y = b_0 + b_1 x^*$$

A **level C confidence interval for the mean response** is

$$\hat{\mu}_y \pm t^* \mathrm{SE}_{\hat{\mu}}$$

where t^* is the value for the $t(n - 2)$ density curve with area C between $-t^*$ and t^*.

The **estimated value of the response variable** y for a future observation from the subpopulation corresponding to the value x^* of the explanatory variable is

$$\hat{y} = b_0 + b_1 x^*$$

A **level C prediction interval** for the estimated response is

$$\hat{y} \pm t^* \text{SE}_{\hat{y}}$$

where t^* is the value for the $t(n-2)$ density curve with area C between $-t^*$ and t^*.

The following section contains material that you should study if you plan to read Chapter 11 on multiple regression. In addition, the section we just completed assumes that you have access to software or a statistical calculator. If you do not, you now need to study the material on computations in the following optional section. The exercises are given at the end of the chapter.

10.2   More Detail about Simple Linear Regression*

In this section we study three optional topics. The first is analysis of variance for regression. If you plan to study the next chapter on multiple regression, you should study this material. The second topic concerns computations for regression inference. Here we present and illustrate the use of formulas for the inference procedures that we have just studied. Finally, we discuss inference for correlation.

Analysis of variance for regression

analysis of variance

The usual computer output for regression includes additional calculations called **analysis of variance.** Analysis of variance, often abbreviated ANOVA, is essential for multiple regression (Chapter 11) and for comparing several means (Chapters 12 and 13). Analysis of variance summarizes information about the sources of variation in the data. It is based on the

$$\text{DATA} = \text{FIT} + \text{RESIDUAL}$$

framework.

The total variation in the response y is expressed by the deviations $y_i - \bar{y}$. If these deviations were all 0, all observations would be equal and there would be no variation in the response. There are two reasons why the individual observations y_i are not all equal to their mean $\bar{y}$. First, the responses y_i correspond to different values of the explanatory variable x and will differ because of that. The fitted value $\hat{y}_i$ estimates the mean response for the specific x_i. The

---

*This material is optional.

differences $\hat{y}_i - \bar{y}$ reflect the variation in mean response due to differences in the x_i. This variation is accounted for by the regression line, because the $\hat{y}$'s lie exactly on the line. Second, individual observations will vary about their mean because of variation within the subpopulation of responses to a fixed x_i. This variation is represented by the residuals $y_i - \hat{y}_i$ that record the scatter of the actual observations about the fitted line. The overall deviation of any y observation from the mean of the y's is the sum of these two deviations:

$$(y_i - \bar{y}) = (\hat{y}_i - \bar{y}) + (y_i - \hat{y}_i)$$

For deviations, this equation expresses the idea that DATA = FIT + RESIDUAL.

We have several times measured variation by an average of squared deviations. If we square each of the three deviations above and then sum over all n observations, it is an algebraic fact that the sums of squares add:

$$\sum (y_i - \bar{y})^2 = \sum (\hat{y}_i - \bar{y})^2 + \sum (y_i - \hat{y}_i)^2$$

We rewrite this equation as

$$\text{SST} = \text{SSM} + \text{SSE}$$

where

$$\text{SST} = \sum (y_i - \bar{y})^2$$
$$\text{SSM} = \sum (\hat{y}_i - \bar{y})^2$$
$$\text{SSE} = \sum (y_i - \hat{y}_i)^2$$

sum of squares

The SS in each abbreviation stands for **sum of squares,** and the T, M, and E stand for total, model, and error, respectively. ("Error" here stands for deviations from the line, which might better be called "residual" or "unexplained variation.") The total variation, as expressed by SST, is the sum of the variation due to the straight-line model (SSM) and the variation due to deviations from this model (SSE). This partition of the variation in the data between two sources is the heart of analysis of variance.

If H_0: $\beta_1 = 0$ were true, there would be no subpopulations and all of the y's should be viewed as coming from a single population with mean μ_y. The variation of the y's would then be described by the sample variance

$$s_y^2 = \frac{\sum (y_i - \bar{y})^2}{n - 1}$$

degrees of freedom

The numerator in this expression is SST. The denominator is the total **degrees of freedom,** or simply DFT.

Just as the total sum of squares SST is the sum of SSM and SSE, the total degrees of freedom DFT is the sum of DFM and DFE, the degrees of freedom for the model and for the error:

$$\text{DFT} = \text{DFM} + \text{DFE}$$

mean square

The model has one explanatory variable x, so the degrees of freedom for this source are DFM $= 1$. Because DFT $= n-1$, this leaves DFE $= n-2$ as the degrees of freedom for error. For each source, the ratio of the sum of squares to the degrees of freedom is called the **mean square,** or simply MS. The general formula for a mean square is

$$\text{MS} = \frac{\text{sum of squares}}{\text{degrees of freedom}}$$

Each mean square is an average squared deviation. MST is just s_y^2, the sample variance that we would calculate if all of the data came from a single population. MSE is familiar to us:

$$\text{MSE} = s^2 = \frac{\sum (y_i - \hat{y}_i)^2}{n-2}$$

It is our estimate of σ^2, the variance about the population regression line.

Sums of Squares, Degrees of Freedom, and Mean Squares

Sums of squares represent variation present in the responses. They are calculated by summing squared deviations. **Analysis of variance** partitions the total variation between two sources.

The sums of squares are related by the formula

$$\text{SST} = \text{SSM} + \text{SSE}$$

That is, the total variation is partitioned into two parts, one due to the model and one due to deviations from the model.

Degrees of freedom are associated with each sum of squares. They are related in the same way:

$$\text{DFT} = \text{DFM} + \text{DFE}$$

To calculate **mean squares,** use the formula

$$\text{MS} = \frac{\text{sum of squares}}{\text{degrees of freedom}}$$

---

interpretation of r^2

In Section 2.3 we noted that r^2 is the fraction of variation in the values of y that is explained by the least-squares regression of y on x. The sums of squares make this interpretation precise. Recall that SST $=$ SSM $+$ SSE. It is an algebraic fact that

$$r^2 = \frac{\text{SSM}}{\text{SST}} = \frac{\sum (\hat{y}_i - \bar{y})^2}{\sum (y_i - \bar{y})^2}$$

Because SST is the total variation in y and SSM is the variation due to the regression of y on x, this equation is the precise statement of the fact that r^2 is the fraction of variation in y explained by x.

The ANOVA F test

The null hypothesis H_0: $\beta_1 = 0$ that y is not linearly related to x can be tested by comparing MSM with MSE. The ANOVA test statistic is an **F statistic,**

$$F = \frac{\text{MSM}}{\text{MSE}}$$

When H_0 is true, this statistic has an F distribution with 1 degree of freedom in the numerator and $n - 2$ degrees of freedom in the denominator. These degrees of freedom are those of MSM and MSE. The F distributions were introduced in Section 7.3. Figure 7.15 (page 568) shows a typical F density curve. Just as there are many t statistics, there are many F statistics. The ANOVA F statistic is not the same as the F statistic of Section 7.3.

When $\beta_1 \neq 0$, MSM tends to be large relative to MSE. So large values of F are evidence against H_0 in favor of the two-sided alternative.

Analysis of Variance F Test

In the simple linear regression model, the hypotheses

$$H_0: \beta_1 = 0$$
$$H_a: \beta_1 \neq 0$$

are tested by the **F statistic**

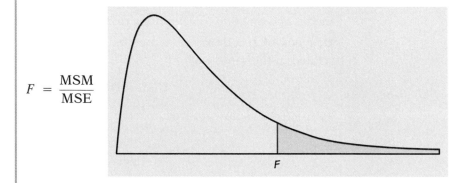

$$F = \frac{\text{MSM}}{\text{MSE}}$$

The P-value is the probability that a random variable having the $F(1, n - 2)$ distribution is greater than or equal to the calculated value of the F statistic.

The F statistic tests the same null hypothesis as one of the t statistics that we encountered earlier in this chapter, so it is not surprising that the two are related. It is an algebraic fact that $t^2 = F$ in this case. For linear regression with one explanatory variable, we prefer the t form of the test because it more easily allows us to test one-sided alternatives and is closely related to the confidence interval for β_1.

ANOVA table

The ANOVA calculations are displayed in an *analysis of variance table,* often abbreviated **ANOVA table.** Here is the format of the table for simple linear regression:

Source	Degrees of freedom	Sum of squares	Mean square	F
Model	1	$\sum(\hat{y}_i - \bar{y})^2$	SSM/DFM	MSM/MSE
Error	$n-2$	$\sum(y_i - \hat{y}_i)^2$	SSE/DFE	
Total	$n-1$	$\sum(y_i - \bar{y})^2$	SST/DFT	

EXAMPLE 10.7

A company with a good reputation can attract highly qualified applicants for jobs, can charge premium prices for its products, and can attract investors. Is reputation, a subjective assessment, related to objective measures of corporate performance? A study designed to answer this and other questions examined the records of 154 Fortune 500 firms.[4]

Reputation was measured by a *Fortune* magazine survey, and performance was defined as profitability, the return on invested capital. We use a regression to predict the response variable profitability (PROFIT) from the explanatory variable reputation (REPUTAT). Figure 10.14 gives the SAS output, including the ANOVA table. The output matches the ANOVA table format just given and has an additional column for the *P*-value. Note that the label C Total is used in the output rather than Total.

```
Dependent Variable: PROFIT

                           Analysis of Variance

                        Sum of            Mean
Source          DF      Squares          Square       F Value      Prob>F

Model            1      0.18957         0.18957        36.492       0.0001
Error          152      0.78963         0.00519
C Total        153      0.97920


     Root MSE        0.07208       R-Square        0.1936
     Dep Mean        0.10000       Adj R-sq        0.1883
     C.V.           72.07575

                         Parameter Estimates

                    Parameter       Standard       T for H0:
     Variable  DF    Estimate          Error     Parameter=0     Prob > |T|

     INTERCEP   1   -0.147573     0.04139259          -3.565         0.0005
     REPUTAT    1    0.039111     0.00647442           6.041         0.0001
```

FIGURE 10.14   Regression output with ANOVA table for Example 10.7.

The calculated value of the F statistic is 36.492 and the P-value is 0.0001. There is strong evidence against the null hypothesis that there is no relationship between reputation and profitability. Note that the t statistic for REPUTAT is 6.041. The square of this t is 36.493, which agrees very well with the F statistic given on the output. Whenever numbers are rounded off in calculations such as this, there will be roundoff error.

Note that the slope b_1 is positive. We conclude that companies that have better reputations also have higher profitability ($F = 36.49$, df $= 1,152$, $P = 0.0001$). We see that $r^2 = 0.1936$; so 19% of the variability in profitability is explained by reputation.

Calculations for regression inference

We recommend using statistical software for regression calculations. With time and care, however, the work is feasible with a calculator. We will use the following example to illustrate how to perform inference for regression analysis using a calculator.

EXAMPLE 10.8 | Like many other businesses, technological advances and new methods have produced dramatic results in agriculture. Here are data on the average yield in bushels per acre of U.S. corn for selected years:[5]

Year	1966	1976	1986	1996
Yield	73.1	88.0	119.4	127.1

Here, year is the explanatory variable x and yield is the response variable y. The data and the least-squares line are plotted in Figure 10.15. The strong straight-line pattern suggests that we can use linear regression to model the relationship between yield and time.

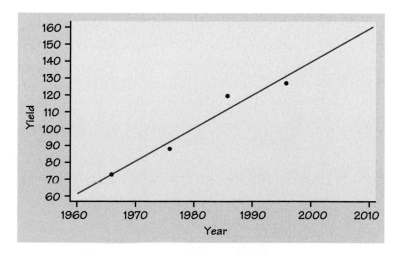

FIGURE 10.15   Data and regression line for Example 10.8.

We begin our regression calculations by fitting the least-squares line. Fitting the line gives estimates b_1 and b_0 of the model parameters β_1 and β_0. Next we examine the residuals from the fitted line and obtain an estimate s of the remaining parameter σ. These calculations are preliminary to inference. Finally, we use s to obtain the standard errors needed for the various interval estimates and significance tests. Roundoff errors that accumulate during these calculations can ruin the final results. Be sure to carry many significant digits and check your work carefully.

Preliminary calculations

After examining the scatterplot (Figure 10.15) to verify that the data show a straight-line pattern with no unusual points, we begin our calculations.

EXAMPLE 10.9

We start by making a table with the mean and standard deviation for each of the variables, the correlation, and the sample size. These calculations should be familiar from Chapters 1 and 2. Here is the summary:

Variable	Mean	Standard deviation	Correlation	Sample size
Year	$\bar{x} = 1981$	$s_x - 12.91$	$r - 0.97584$	$n = 4$
Yield	$\bar{y} = 101.9$	$s_y = 25.5861$		

These quantities are the building blocks for our calculations.

We will need one additional quantity for the calculations to follow. It is the expression $\sum(x_i - \bar{x})^2$. We obtain this quantity as an intermediate step when we calculate s_x. You could also find it using the fact that $\sum(x_i - \bar{x})^2 = (n-1)s_x^2$. You should verify that the value for our example is

$$\sum(x_i - \bar{x})^2 = 500$$

Our first task is to find the least-squares line. This is easy with the building blocks that we have assembled.

EXAMPLE 10.10

Although a calculator that has a correlation function will also give us b_0 and b_1, we can check the basic facts. The slope of the least-squares line is

$$b_1 = r\frac{s_y}{s_x}$$
$$= 0.97584\frac{25.5861}{12.91}$$
$$= 1.934$$

The intercept is

$$b_0 = \bar{y} - b_1\bar{x}$$
$$= 101.9 - (1.934)(1981)$$
$$= -3729.354$$

The equation of the least-squares regression line is therefore

$$\hat{y} = -3729 + 1.9x$$

This is the line shown in Figure 10.15.

We now have estimates of the first two parameters, β_0 and β_1, of our linear regression model. We now find the estimate of the third parameter σ, the standard deviation s about the fitted line. To do this we need to find the predicted values and then the residuals.

The predicted value for 1966 is

$$\begin{aligned}
\hat{y}_1 &= b_0 + b_1 x_1 \\
&= -3729.354 + (1.934)(1966) \\
&= 72.89
\end{aligned}$$

and the residual is

$$\begin{aligned}
e_1 &= y_1 - \hat{y}_1 \\
&= 73.1 - 72.89 \\
&= 0.21
\end{aligned}$$

The residuals for the other years are calculated in the same way. You should verify that they are -4.23, 7.83, and -3.81. Also check that the four residuals sum to zero.

EXAMPLE 10.11   The estimate of σ^2 is s^2, the sum of the squares of the residuals divided by $n - 2$. The estimated standard deviation about the line is the square root of this quantity.

$$\begin{aligned}
s^2 &= \frac{\sum e_i^2}{n - 2} \\
&= \frac{(0.21)^2 + (-4.23)^2 + (7.83)^2 + (-3.81)^2}{2} \\
&= 46.8810
\end{aligned}$$

So the estimate of the standard deviation about the line is

$$s = \sqrt{46.8810} = 6.8470$$

Now that we have estimates of the three parameters of our model, we can proceed to the more detailed calculations needed for regression inference.

Inference for slope and intercept   Confidence intervals and significance tests for the slope β_1 and intercept β_0 of the population regression line make use of the estimates b_1 and b_0 and their standard errors.

Some algebra using the rules for variances (Section 4.4) establishes that the standard deviation of b_1 is

$$\sigma_{b_1} = \frac{\sigma}{\sqrt{\sum (x_i - \bar{x})^2}}$$

Similarly, the standard deviation of b_0 is

$$\sigma_{b_0} = \sigma\sqrt{\frac{1}{n} + \frac{\bar{x}^2}{\sum(x_i - \bar{x})^2}}$$

To estimate these standard deviations, we need only replace σ by its estimate s.

Standard Errors for Estimated Regression Coefficients

The standard error of the slope b_1 of the least-squares regression line is

$$SE_{b_1} = \frac{s}{\sqrt{\sum(x_i - \bar{x})^2}}$$

The standard error of the intercept b_0 is

$$SE_{b_0} = s\sqrt{\frac{1}{n} + \frac{\bar{x}^2}{\sum(x_i - \bar{x})^2}}$$

The plot of the regression line with the data in Figure 10.15 *appears* to show a very strong relationship, but our sample size is very small. We assess the situation with a significance test for the slope.

EXAMPLE 10.12

First we need the standard error of the estimated slope:

$$SE_{b_1} = \frac{s}{\sqrt{\sum(x_i - \bar{x})^2}}$$
$$= \frac{6.8470}{\sqrt{500}} = 0.3062$$

To test

$$H_0: \beta_1 = 0$$
$$H_a: \beta_1 \neq 0$$

calculate the t statistic:

$$t = \frac{b_1}{SE_{b_1}}$$
$$= \frac{1.934}{0.3062} = 6.316$$

Using Table D with $n - 2 = 2$ degrees of freedom, we conclude $P < 0.04$. (The exact value obtained from software is 0.0242.) We conclude that over the period 1966 to 1996 there has been an approximately linear increase in the yield of U.S. corn ($t = 6.32$, df $= 2$, $P < 0.04$).

The significance test tells us that the data provide sufficient information to conclude that there has been an increase. We use the estimate b_1 and its confidence interval to describe what the increase is and how well we have estimated it.

EXAMPLE 10.13 For the corn yield problem, let's find a 95% confidence interval for the slope β_1. The degrees of freedom are $n - 2 = 2$, so t^* from Table D is 4.303. To find the 95% confidence interval for β_1 we compute

$$b_1 \pm t^* \text{SE}_{b_1} = 1.934 \pm (4.303)(0.3062)$$
$$= 1.934 \pm 1.318$$

The interval is (0.62, 3.25). Over the period 1966 to 1996, U.S. corn yields per acre have increased at the rate of about 2 bushels per year. The margin of error for 95% confidence is 1.3 bushels.

Note the effect of the small sample size on the critical value t^*. With one additional observation, it would decrease to 3.182.

In this example, the intercept β_0 does not have a meaningful interpretation. Because β_0 is the mean of y when $x = 0$, β_0 represents corn yield in the year 0. For problems where inference for β_0 is appropriate, the calculations are performed in the same way as those for β_1. Note that there is a different formula for the standard error.

Confidence intervals for the mean response and prediction intervals for a future observation   When we substitute a particular value x^* of the explanatory variable into the regression equation and obtain a value of $\hat{y}$, we can view the result in two ways. We have estimated the mean response μ_y, and we have also predicted a future value of the response y. The margins of error for these two uses are quite different. Prediction intervals for an individual response are wider than confidence intervals for estimating a mean response. We now proceed with the details of these calculations. Once again, standard errors are the essential quantities. And once again, these standard errors are multiples of s, our basic measure of the variability of the responses about the fitted line.

Standard Errors for $\hat{\mu}$ and $\hat{y}$

The standard error of $\hat{\mu}$ is

$$\text{SE}_{\hat{\mu}} = s \sqrt{\frac{1}{n} + \frac{(x^* - \bar{x})^2}{\sum (x_i - \bar{x})^2}}$$

The standard error for predicting an individual response $\hat{y}$ is[6]

$$\text{SE}_{\hat{y}} = s \sqrt{1 + \frac{1}{n} + \frac{(x^* - \bar{x})^2}{\sum (x_i - \bar{x})^2}}$$

Note that the only difference between the recipes for these two standard errors is the extra 1 under the square root sign in the standard error for prediction. This standard error is larger due to the additional variation of individual responses about the mean response. It produces prediction intervals that are wider than the confidence intervals for the mean response.

For problems such as the corn yield example, we often want to estimate values of y in the future. This is a type of forecasting. We use a prediction interval for this purpose.

EXAMPLE 10.14

Assuming that the linear pattern of increasing corn yield continues into the future, let's estimate the yield in the year 2006. We also provide a prediction interval that quantifies the uncertainty in this estimate. For the year 2006, we predict that the corn yield will be

$$
\begin{aligned}
\hat{y} &= b_0 + b_1 x^* \\
&= -3729.354 + (1.934)(2006) = 150.25
\end{aligned}
$$

The standard error is

$$
\begin{aligned}
\text{SE}_{\hat{y}} &= s\sqrt{1 + \frac{1}{n} + \frac{(x^* - \bar{x})^2}{\sum(x_i - \bar{x})^2}} \\
&= 6.8470\sqrt{1 + \frac{1}{4} + \frac{(2006 - 1981)^2}{500}} \\
&= 6.8470\sqrt{1 + \frac{1}{4} + \frac{625}{500}} = 10.83
\end{aligned}
$$

To find the 95% prediction interval we compute

$$
\begin{aligned}
\hat{y} \pm t^* \text{SE}_{\hat{y}} &= 150.25 \pm (4.303)(10.83) \\
&= 150.25 \pm 46.60
\end{aligned}
$$

The interval is $(103.65, 196.85)$. We predict that the U.S. corn yield in the year 2006 will be 150 bushels per acre. The margin of error is 47 bushels per acre for 95% confidence.

Calculations for confidence intervals for the mean response are similar. The only difference is the use of the formula for $\text{SE}_{\hat{\mu}}$ in place of $\text{SE}_{\hat{y}}$.

Inference for correlation

The correlation coefficient is a measure of the strength and direction of the linear association between two variables. Correlation does not require an explanatory-response relationship between the variables. We can consider the sample correlation r as an estimate of the correlation in the population and base inference about the population correlation on r.

The correlation between the variables x and y when they are measured for every member of a population is the **population correlation.** As usual, we use Greek letters to represent population parameters. In this case ρ (the Greek letter rho) is the population correlation. When $\rho = 0$, there is no linear association in the population. In the important case where the two variables x and y are both normally distributed, the condition $\rho = 0$ is equivalent to the statement that x and y are independent. That is, there is no association of any kind between x and y. (Technically, the condition required is that x

population correlation ρ

jointly normal variables

and y be **jointly normal.** This means that the distribution of x is normal and also that the conditional distribution of y given any fixed value of x is normal.) We therefore may wish to test the null hypothesis that a population correlation is 0.

Test for a Zero Population Correlation

To test the hypothesis H_0: $\rho = 0$, compute the t statistic:

$$t = \frac{r \sqrt{n-2}}{\sqrt{1-r^2}}$$

where n is the sample size and r is the sample correlation.

In terms of a random variable T having the $t(n-2)$ distribution, the P-value for a test of H_0 against

H_a: $\rho > 0$ is $P(T \geq t)$

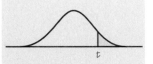

H_a: $\rho < 0$ is $P(T \leq t)$

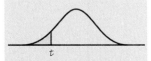

H_a: $\rho \neq 0$ is $2P(T \geq |t|)$

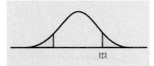

Most computer packages have routines for calculating correlations and some will provide the significance test for the null hypothesis that ρ is zero.

EXAMPLE 10.15

For the body density and skinfold measures problem, the SAS output appears in Figure 10.16. The sample correlation between body density and log of skinfolds is $r = -0.84877$ and is called a Pearson correlation coefficient in this output. The P-value for a two-sided test of H_0: $\rho = 0$ is given below the correlation; it is 0.0001. To test the one-sided alternative that the population correlation is negative, we divide the P-value in the output by 2, after checking that the sample coefficient is in fact negative. We conclude that there is a negative correlation between body density and log of skinfolds.

If your software does not give the significance test, the computations are not very difficult to do with a calculator.

```
Pearson Correlation Coefficients / Prob > |R| under H0: Rho=0 / N = 92


                          LSKIN                      DEN


    LSKIN              1.00000                  -0.84877
                          0.0                      0.0001


    DEN               -0.84877                   1.00000
                       0.0001                      0.0
```

FIGURE 10.16   Correlation output for Example 10.15.

EXAMPLE 10.16   Refer to Example 10.7 (page 685), where we examined the relationship between the reputation of a company and its profitability for a sample of $n = 154$ companies. The correlation between these two variables is 0.44. The t statistic is

$$t = \frac{r\sqrt{n-2}}{\sqrt{1-r^2}}$$

$$= \frac{0.44\sqrt{154-2}}{\sqrt{1-(0.44)^2}} = 6.04$$

From Table D using $n - 2 = 152$ degrees of freedom we have $P < 0.0001$. The data provide clear evidence that reputation and profitability are positively related.

There is a close connection between the significance test for a correlation and the test for the slope in a linear regression. Recall that

$$b_1 = r\frac{s_y}{s_x}$$

From this fact we see that if the slope is 0, so is the correlation, and vice versa. It should come as no surprise to learn that the procedures for testing $H_0: \beta_1 = 0$ and $H_0: \rho = 0$ are also closely related. In fact, the t statistics for testing these hypotheses are numerically equal. That is,

$$\frac{b_1}{s_{b_1}} = \frac{r\sqrt{n-2}}{\sqrt{1-r^2}}$$

Check that this holds in both of our examples.

In our examples, the conclusion that there is a significant correlation between the two variables would not come as a surprise to anyone familiar with the meaning of these variables. The significance test simply tells us whether or not there is evidence in the data to conclude that the population correlation is different from 0. The actual size of the correlation is of considerably more interest. We would therefore like to give a confidence interval for the population correlation. Unfortunately, most software packages do not perform this calculation. Because hand calculation of the confidence interval is very tedious, we do not give the method here.[7]

SUMMARY

The **ANOVA table** for a linear regression gives the degrees of freedom, sum of squares, and mean squares for the model, error, and total sources of variation. The **ANOVA F statistic** is the ratio MSM/MSE. Under H_0: $\beta_1 = 0$, this statistic has an $F(1, n-2)$ distribution and is used to test H_0 versus the two-sided alternative.

The **square of the sample correlation** can be expressed as

$$r^2 = \frac{\text{SSM}}{\text{SST}}$$

and is interpreted as the proportion of the variability in the response variable y that is explained by the explanatory variable x in the linear regression.

The **standard errors** for b_0 and b_1 are

$$\text{SE}_{b_0} = s\sqrt{\frac{1}{n} + \frac{\bar{x}^2}{\sum(x_i - \bar{x})^2}}$$

$$\text{SE}_{b_1} = \frac{s}{\sqrt{\sum(x_i - \bar{x})^2}}$$

The **standard error** that we use for a confidence interval for the estimated mean response for the subpopulation corresponding to the value x^* of the explanatory variable is

$$\text{SE}_{\hat{\mu}} = s\sqrt{\frac{1}{n} + \frac{(x^* - \bar{x})^2}{\sum(x_i - \bar{x})^2}}$$

The **standard error** that we use for a prediction interval for a future observation from the subpopulation corresponding to the value x^* of the explanatory variable is

$$\text{SE}_{\hat{y}} = s\sqrt{1 + \frac{1}{n} + \frac{(x^* - \bar{x})^2}{\sum(x_i - \bar{x})^2}}$$

When the variables y and x are jointly normal, the sample correlation is an estimate of the **population correlation ρ**. The test of H_0: $\rho = 0$ is based on the **t statistic**

$$t = \frac{r\sqrt{n-2}}{\sqrt{1-r^2}}$$

which has a $t(n-2)$ distribution under H_0. This test statistic is numerically identical to the t statistic used to test H_0: $\beta_1 = 0$.

CHAPTER 10 EXERCISES

10.1   We assume that our wages will increase as we gain experience and become more valuable to our employers. Wages also increase because of inflation. By examining a sample of employees at a given point in time, we can look at part of the picture. How does length of service (LOS) relate to wages? Table 10.1 gives data on the LOS in months and wages for 60 women who work in Indiana banks. Wages are yearly total income divided by the number of weeks worked. We have multiplied wages by a constant for reasons of confidentiality. (The data were provided by Professor Shelly MacDermid, Department of Child Development and Family Studies, Purdue University, from a study reported in S. M. MacDermid, "Is small beautiful? Work-family tension, work conditions, and organizational size," *Family Relations*, 44 (1994), pp. 159–167.)

   (a) Plot wages versus LOS. Describe the relationship. There is one woman with relatively high wages for her length of service. Circle this point and do not use it in the rest of this exercise.

   (b) Find the least-squares line. Summarize the significance test for the slope. What do you conclude?

TABLE 10.1   Bank wages and length of service

Wages	LOS	Size	Wages	LOS	Size	Wages	LOS	Size
48.3355	94	Large	64.1026	24	Large	41.2088	97	Small
49.0279	48	Small	54.9451	222	Small	67.9096	228	Small
40.8817	102	Small	43.8095	58	Large	43.0942	27	Large
36.5854	20	Small	43.3455	41	Small	40.7000	48	Small
46.7596	60	Large	61.9893	153	Large	40.5748	7	Large
59.5238	78	Small	40.0183	16	Small	39.6825	74	Small
39.1304	45	Large	50.7143	43	Small	50.1742	204	Large
39.2465	39	Large	48.8400	96	Large	54.9451	24	Large
40.2037	20	Large	34.3407	98	Large	32.3822	13	Small
38.1563	65	Small	80.5861	150	Large	51.7130	30	Large
50.0905	76	Large	33.7163	124	Small	55.8379	95	Large
46.9043	48	Small	60.3792	60	Large	54.9451	104	Large
43.1894	61	Small	48.8400	7	Large	70.2786	34	Large
60.5637	30	Large	38.5579	22	Small	57.2344	184	Small
97.6801	70	Large	39.2760	57	Large	54.1126	156	Small
48.5795	108	Large	47.6564	78	Large	39.8687	25	Large
67.1551	61	Large	44.6864	36	Large	27.4725	43	Small
38.7847	10	Small	45.7875	83	Small	67.9584	36	Large
51.8926	68	Large	65.6288	66	Large	44.9317	60	Small
51.8326	54	Large	33.5775	47	Small	51.5612	102	Large

(c) State carefully what the slope tells you about the relationship between wages and length of service.

(d) Give a 95% confidence interval for the slope.

10.2 Refer to the previous exercise. Analyze the data with the outlier included. How does this change the estimates of the parameters β_0, β_1, and σ? What effect does the outlier have on the results of the significance test for the slope?

10.3 The selling price and the number of square feet for homes sold in a neighborhood in Boulder, Colorado, during the first part of 1995 appear in Table 2.3 and are discussed in Exercise 2.13 (page 123).

(a) Plot the selling price versus the number of square feet. Describe the pattern.

(b) Perform a linear regression analysis. Give the least-squares line and the results of the significance test for the slope.

10.4 Refer to the previous problem. There are five homes with somewhat high selling prices relative to the other houses in this set of data. Rerun the regression analysis without these five observations. Summarize the results and explain the effect of these observations on the least-squares line and the significance test results by comparing these results with those of the previous exercise.

10.5 In Example 10.8 we examined the yield in bushels per acre of corn for the years 1966, 1976, 1986, and 1996. Data for all years between 1957 and 1996 appear in Table 10.2.

(a) Plot the yield versus year. Describe the relationship. Are there any outliers or unusual years?

(b) Perform the regression analysis and summarize the results. How rapidly has yield increased over time?

TABLE 10.2 U.S. corn yield

Year	Yield	Year	Yield	Year	Yield	Year	Yield
1957	48.3	1967	80.1	1977	90.8	1987	119.8
1958	52.8	1968	79.5	1978	101.0	1988	84.6
1959	53.1	1969	85.9	1979	109.5	1989	116.3
1960	54.7	1970	72.4	1980	91.0	1990	118.5
1961	62.4	1971	88.1	1981	108.9	1991	108.6
1962	64.7	1972	97.0	1982	113.2	1992	131.5
1963	67.9	1973	91.3	1983	81.1	1993	100.7
1964	62.9	1974	71.9	1984	106.7	1994	138.6
1965	74.1	1975	86.4	1985	118.0	1995	113.5
1966	73.1	1976	88.0	1986	119.4	1996	127.1

10.6 Utility companies need to estimate the amount of energy that will be used by their customers. The consumption of natural gas required for heating homes depends on the outdoor temperature. When the weather is cold, more gas will be consumed. A study of one home recorded the average daily gas consumption y (in hundreds of cubic feet) for each month during one heating season. The explanatory variable x is the average number of heating degree-days per day during the month. One heating degree-day is accumulated for each degree a day's average temperature falls below 65° F. An average temperature of 50°, for example, corresponds to 15 degree-days. The data for October through June are given in the following table. (Data provided by Professor Robert Dale of Purdue University.)

	Oct.	Nov.	Dec.	Jan.	Feb.	Mar.	Apr.	May	June
x	15.6	26.8	37.8	36.4	35.5	18.6	15.3	7.9	0.0
y	5.2	6.1	8.7	8.5	8.8	4.9	4.5	2.5	1.1

   (a) Find the equation of the least-squares line.

   (b) Test the null hypothesis that the slope is zero and describe your conclusion.

   (c) Give a 95% confidence interval for the slope.

   (d) The parameter β_0 corresponds to natural gas consumption for cooking, hot water, and other uses when there is no demand for heating. Give a 95% confidence interval for this parameter.

10.7 Manatees are large sea creatures that live in the shallow water along the coast of Florida. Many manatees are injured or killed each year by powerboats. Table 2.6 (page 149) gives data on manatees killed and powerboat registrations (in thousands of boats) in Florida for the period 1977 to 1990.

   (a) Make a scatterplot of boats registered and manatees killed. Is there a strong straight-line pattern?

   (b) Find the equation of the least-squares regression line. Draw this line on your scatterplot.

   (c) Is there strong evidence that the mean number of manatees killed increases as the number of powerboats increases? State this question as null and alternative hypotheses about the slope of the population regression line, obtain the t statistic, and give your conclusion.

   (d) Predict the number of manatees that will be killed if there are 716,000 powerboats registered. In 1991, 1992, and 1993, the number of powerboats remained at 716,000. The numbers of manatees killed were 53, 38, and 35. Compare your prediction with these data. Does the comparison suggest that measures taken to protect the manatees in these years were effective?

10.8 The Leaning Tower of Pisa is an architectural wonder. Engineers concerned about the tower's stability have done extensive studies of its increasing tilt. Measurements of the lean of the tower over time provide much useful information. The following table gives measurements for the years 1975 to 1987. The variable "lean" represents the difference between where a point on the tower would be if the tower were straight and where it actually is. The data are coded as tenths of a millimeter in excess of 2.9 meters, so that the 1975 lean, which was 2.9642 meters, appears in the table as 642. Only the last two digits of the year were entered into the computer. (Data from G. Geri and B. Palla, "Considerazioni sulle più recenti osservazioni ottiche alla Torre Pendente di Pisa," *Estratto dal Bollettino della Società Italiana di Topografia e Fotogrammetria*, 2 (1988), pp. 121–135. Professor Julia Mortera of the University of Rome provided valuable assistance with the translation.)

Year	75	76	77	78	79	80	81	82	83	84	85	86	87
Lean	642	644	656	667	673	688	696	698	713	717	725	742	757

(a) Plot the data. Does the trend in lean over time appear to be linear?

(b) What is the equation of the least-squares line? What percentage of the variation in lean is explained by this line? $R^2 \quad \frac{SSE}{SST}$

(c) Give a 95% confidence interval for the average rate of change (tenths of a millimeter per year) of the lean.

10.9   Refer to the previous exercise.

(a) In 1918 the lean was 2.9071 meters. (The coded value is 71.) Using the least-squares equation for the years 1975 to 1987, calculate a predicted value for the lean in 1918. (Note that you must use the coded value 18 for year.)

(b) Although the least-squares line gives an excellent fit to the data for 1975 to 1987, this pattern did not extend back to 1918. Write a short statement explaining why this conclusion follows from the information available. Use numerical and graphical summaries to support your explanation.

10.10   Refer once more to Exercise 10.8.   *prediction*

(a) The engineers working on the Leaning Tower of Pisa are most interested in how much the tower will lean if no corrective action is taken. Use the least-squares equation to predict the tower's lean in the year 1997.

(b) To give a margin of error for the lean in 1997, would you use a confidence interval for a mean response or a prediction interval? Explain your choice.

10.11   Exercise 10.6 (page 697) demonstrates that there is a strong linear relationship between household consumption of natural gas and outdoor temperature, measured by heating degree-days. The slope and intercept depend on the particular house and on the habits of the household living there. Data for two heating seasons (18 months) for another household produce the least-squares line $\hat{y} = 2.405 + 0.26896x$ for predicting average daily gas consumption y from average degree-days per day x. The standard error of the slope is $SE_{b_1} - 0.00815$.

(a) Explain briefly what the slope β_1 of the population regression line represents. Then give a 95% confidence interval for β_1.

(b) This interval is based on twice as many observations as the one calculated in Exercise 10.6 for a different household, and the two standard errors are of similar size. How would you expect the margins of error of the two intervals to be related? Check your answer by comparing the two margins of error.

10.12 The standard error of the intercept in the regression of gas consumption on degree-days for the household in the previous exercise is $SE_{b_0} = 0.20351$.

(a) Explain briefly what the intercept represents in this setting. Find a 95% confidence interval for the intercept.

(b) Compare the width of your interval with the one calculated for a different household in Exercise 10.6. Explain why it is shorter.

10.13 Exercise 10.6 gives information about the regression of natural gas consumption on degree-days for a particular household.

(a) What is the t statistic for testing H_0: $\beta_1 = 0$?

(b) For the alternative H_a: $\beta_1 > 0$, what critical value would you use for a test at the $\alpha = 0.05$ significance level? Do you reject H_0 at this level?

(c) How would you report the P-value for this test?

10.14 Can a pretest on mathematics skills predict success in a statistics course? The 55 students in an introductory statistics class took a pretest at the beginning of the semester. The least-squares regression line for predicting the score y on the final exam from the pretest score x was $\hat{y} = 10.5 + 0.82x$. The standard error of b_1 was 0.38. Test the null hypothesis that there is no linear relationship between the pretest score and the score on the final exam against the two-sided alternative.

10.15 Exercise 2.56 (page 172) gives the following data from a study of two methods for measuring the blood flow in the stomachs of dogs:

Spheres	4.0	4.7	6.3	8.2	12.0	15.9	17.4	18.1	20.2	23.9
Vein	3.3	8.3	4.5	9.3	10.7	16.4	15.4	17.6	21.0	21.7

"Spheres" is an experimental method that the researchers hope will predict "Vein," the standard but difficult method. Examination of the data gives no reason to doubt the validity of the simple linear regression model. The estimated regression line is $\hat{y} = 1.031 + 0.902x$, where y is the response variable Vein and x is the explanatory variable Spheres. The estimate of σ is $s = 1.757$.

(a) Find $\bar{x}$ and $\sum(x_i - \bar{x})^2$ from the data.

(b) We expect x and y to be positively associated. State hypotheses in terms of the slope of the population regression line that express this expectation, and carry out a significance test. What conclusion do you draw?

(c) Find a 99% confidence interval for the slope.

(d) Suppose that we observe a value of Spheres equal to 15.0 for one dog. Give a 90% interval for predicting the variable Vein for that dog.

10.16 Table 2.11 (page 177) gives data on air pollution measurements in two locations, rural and city. Exercise 2.64 gives the background information for the data. Readings for both locations are available for 26 cases. Examination of these data suggests that the simple linear regression model fits. The regression equation for predicting the city reading from the rural one is $\hat{y} = -2.580 + 1.0935x$, and the estimated standard deviation about the line is $s = 4.4792$.

(a) Calculate $\bar{x}$ and $\sum(x_i - \bar{x})^2$ for the 26 cases.

(b) State appropriate null and alternative hypotheses for assessing whether or not there is a linear relationship between the city and rural readings. Give the test statistic and report the P-value for testing your null hypothesis. Summarize your conclusion.

(c) The rural reading is 43 for a period during which the city equipment is out of service. Give a 95% interval for the missing city reading.

10.17 Ohm's law $I = V/R$ states that the current I in a metal wire is proportional to the voltage V applied to its ends and is inversely proportional to the resistance R in the wire. Students in a physics lab performed experiments to study Ohm's law. They varied the voltage and measured the current at each voltage with an ammeter. The goal was to determine the resistance R of the wire. We can rewrite Ohm's law in the form of a linear regression as $I = \beta_0 + \beta_1 V$, where $\beta_0 = 0$ and $\beta_1 = 1/R$. Because voltage is set by the experimenter, we think of V as the explanatory variable. The current I is the response. Here are the data for one experiment (data provided by Sara McCabe):

V	0.50	1.00	1.50	1.80	2.00
I	0.52	1.19	1.62	2.00	2.40

(a) Plot the data. Are there any outliers or unusual points?

(b) Find the least-squares fit to the data, and estimate $1/R$ for this wire. Then give a 95% confidence interval for $1/R$.

(c) If b_1 estimates $1/R$, then $1/b_1$ estimates R. Estimate the resistance R. Similarly, if L and U represent the lower and upper confidence limits for $1/R$, then the corresponding limits for R are given by $1/U$ and $1/L$, as long as L and U are positive. Use this fact and your answer to (b) to find a 95% confidence interval for R.

(d) Ohm's law states that β_0 in the model is 0. Calculate the test statistic for this hypothesis and give an approximate P-value.

10.18 Most statistical software systems have an option for doing regressions in which the intercept is set in advance at 0. If you have access to such software, reanalyze the Ohm's law data given in the previous exercise with this option and report the estimate of R. The output should also include an estimated standard error for $1/R$. Use this to calculate the 95% confidence interval for R. Note: With this option the degrees of freedom for t^* will be 1 greater than for the model with the intercept.

10.19 The human body takes in more oxygen when exercising than when it is at rest. To deliver the oxygen to the muscles, the heart must beat faster. Heart rate is easy to measure, but measuring oxygen uptake requires elaborate equipment. If oxygen uptake (VO₂) can be accurately predicted from heart rate (HR), the predicted values can replace actually measured values for various research purposes. Unfortunately, not all human bodies are the same, so no single prediction equation works for all people. Researchers can, however, measure both HR and VO₂ for one person under varying sets of exercise conditions and calculate a regression equation for predicting that person's oxygen uptake from heart rate. They can then use predicted oxygen uptakes in place of measured uptakes for this individual in later experiments. Here are data for one individual (data provided by Paul Waldsmith from experiments conducted in Don Corrigan's laboratory at Purdue University):

HR	94	96	95	95	94	95	94	104	104	106
VO2	0.473	0.753	0.929	0.939	0.832	0.983	1.049	1.178	1.176	1.292

HR	108	110	113	113	118	115	121	127	131
VO2	1.403	1.499	1.529	1.599	1.749	1.746	1.897	2.040	2.231

(a) Plot the data. Are there any outliers or unusual points?

(b) Compute the least-squares regression line for predicting oxygen uptake from heart rate for this individual.

(c) Test the null hypothesis that the slope of the regression line is 0. Explain in words the meaning of your conclusion from this test.

(d) Calculate a 95% interval for the oxygen uptake of this individual on a future occasion when his heart rate is 95. Repeat the calculation for heart rate 110.

(e) From what you have learned in (a), (b), (c), and (d) of this exercise, do you think that the researchers should use predicted VO₂ in place of measured VO₂ for this individual under similar experimental conditions? Explain your answer.

10.20 Premature infants are often kept in intensive-care nurseries after they are born. It is common practice to measure their blood pressure frequently. The oscillometric method of measuring blood pressure is noninvasive and easy to use. The traditional procedure, called the direct intra-arterial method, is believed to be more accurate but is invasive and more difficult to perform. Several studies have reported high correlations between measurements made by the two methods, ranging from $r = 0.49$ to $r = 0.98$. These correlations are statistically significant. One study that investigated the relation between the two methods reported the regression equation $\hat{y} = 15 + 0.83x$. Here x represents the easy method and y represents the difficult one. (From John A. Wareham et al., "Prediction of arterial blood pressure on the premature neonate using the oscillometric method," *American Journal of Diseases of Children,* 141 (1987), pp. 1108–1110.) The standard error of the slope is 0.065 and the sample size is 81. Calculate the t statistic for testing $H_0\colon \beta_1 = 0$. Specify an appropriate alternative hypothesis for this problem, and give an approximate P-value for the test. Then explain your conclusion in words a physician can understand. (The authors of the study calculated and plotted prediction intervals. They found the widths to be unacceptably large and concluded that statistical significance does not imply that results are clinically useful.)

10.21 Soil aeration and soil water evaporation involve the exchange of gases between the soil and the atmosphere. Experimenters have investigated the effect of the airflow above the soil on this process. One such experiment varied the speed of the air x and measured the rate of evaporation y. The fitted regression equation based on 18 observations was $\hat{y} = 5.0 + 0.00665x$. The standard error of the slope was reported to be 0.00182.

(a) It is reasonable to suppose that greater airflow will cause more evaporation. State hypotheses to test this belief and calculate the test statistic. Find an approximate P-value for the significance test and report your conclusion.

(b) Construct a 95% confidence interval for the additional evaporation experienced when airflow increases by 1 unit.

10.22 In Exercise 10.5 we examined the relationship between time and the yield of corn in the United States. Table 10.3 gives similar data for the yield of soybeans (in bushels per acre). Give a complete analysis of these data. Include a plot of the data, significance test results, examination of the residuals, and your conclusions.

10.23 Refer to Example 10.8 (page 686) and Exercise 10.5 on the changes in the U.S. corn yield over time.

(a) Give a 95% prediction interval for the yield in the year 2006.

(b) Compare this prediction interval with the one given in Example 10.14 (page 691).

(c) Find the margin of error for this prediction interval. Compare it with the margin of error for the prediction interval found in Example 10.14. Why are they different?

TABLE 10.3 U.S. soybean yield

Year	Yield	Year	Yield	Year	Yield	Year	Yield
1957	23.2	1967	24.5	1977	30.6	1987	33.9
1958	24.2	1968	26.7	1978	29.4	1988	27.0
1959	23.5	1969	27.4	1979	32.1	1989	32.3
1960	23.5	1970	26.7	1980	26.5	1990	34.1
1961	25.1	1971	27.5	1981	30.1	1991	34.2
1962	24.2	1972	27.8	1982	31.5	1992	37.6
1963	24.4	1973	27.8	1983	26.2	1993	32.6
1964	22.8	1974	23.7	1984	28.1	1994	41.4
1965	24.5	1975	28.9	1985	34.1	1995	35.3
1966	25.4	1976	26.1	1986	33.3	1996	37.6

10.24 The corn yields of Table 10.2 (page 696) and the soybean yields of Table 10.3 both vary over time for similar reasons, including improved technology and weather conditions. Let's examine the relationship between the two yields.

(a) Plot the two yields with corn on the x axis and soybeans on the y axis. Describe the relationship.

(b) Find the correlation. How well does it summarize the relation?

(c) Use corn yield to predict soybean yield. Give the equation and the results of the significance test for the slope. This test also tests the null hypothesis that the two yields are uncorrelated.

(d) Obtain the residuals from the model in part (c) and plot them versus time. Describe the pattern.

10.25 The corn yield data of Exercise 10.5 show a larger amount of scatter about the least-squares line for the later years, when the yields are higher. This may be an indication that the standard deviation σ of our model is not a constant but is increasing with time. Take logs of the yields and rerun the analyses. Prepare a short report comparing the two analyses. Include plots, a comparison of the significance test results, and the percent of variation explained by each model.

10.26 Return to the data on current versus voltage given in the Ohm's law experiment of Exercise 10.17.

(a) Compute all values for the ANOVA table.

(b) State the null hypothesis tested by the ANOVA F statistic, and explain in plain language what this hypothesis says.

(c) What is the distribution of this F statistic when H_0 is true? Find an approximate P-value for the test of H_0.

10.27 Return to the oxygen uptake and heart rate data given in Exercise 10.19.

(a) Construct the ANOVA table.

(b) What null hypothesis is tested by the ANOVA F statistic? What does this hypothesis say in practical terms?

(c) Give the degrees of freedom for the F statistic and an approximate P-value for the test of H_0.

(d) Verify that the square of the t statistic that you calculated in Exercise 10.19 is equal to the F statistic in your ANOVA table. (Any difference found is due to roundoff error.)

(e) What proportion of the variation in oxygen uptake is explained by heart rate for this set of data?

10.28   A study conducted in the Egyptian village of Kalama examined the relationship between the birth weights of 40 infants and various socioeconomic variables. (The study is reported in M. El-Kholy, F. Shaheen, and W. Mahmoud, "Relationship between socioeconomic status and birth weight, a field study in a rural community in Egypt," *Journal of the Egyptian Public Health Association,* 61 (1986), pp. 349–358.)

(a) The correlation between monthly income and birth weight was $r = 0.39$. Calculate the t statistic for testing the null hypothesis that the correlation is 0 in the entire population of infants.

(b) The researchers expected that higher birth weights would be associated with higher incomes. Express this expectation as an alternative hypothesis for the population correlation.

(c) Determine a P-value for H_0 versus the alternative that you specified in (b). What conclusion does your test suggest?

10.29   Chinese students from public schools in Hong Kong were the subjects of a study designed to investigate the relationship between various measures of parental behavior and other variables. The sample size was 713. The data were obtained from questionnaires filled in by the students. One of the variables examined was parental control, an indication of the amount of control that the parents exercised over the behavior of the students. Another was the self-esteem of the students. (This study is reported in S. Lau and P. C. Cheung, "Relations between Chinese adolescents' perception of parental control and organization and their perception of parental warmth," *Developmental Psychology,* 23 (1987), pp. 726–729.)

(a) The correlation between parental control and self-esteem was $r = -0.19$. Calculate the t statistic for testing the null hypothesis that the population correlation is 0.

(b) Find an approximate P-value for testing H_0 versus the two-sided alternative and report your conclusion.

10.30   The WOOD data set described in the Data Appendix contains two measurements of the strength of each of 50 strips of yellow poplar wood. In this problem we look at this experiment as one in which the process of measuring strength is to be evaluated. Specifically, we ask how well a repeat measurement T2 can be predicted from the first measurement T1 for strips of wood of this type.

(a) Plot T2 versus T1 and describe the pattern you see.

(b) Carry out a linear regression with T1 as the explanatory variable and T2 as the response variable. Write the fitted equation and give s, the estimate of the standard deviation σ of the model.

(c) What is the correlation between T1 and T2? This quantity is sometimes called the **reliability** of the strength-measuring procedure. What proportion of the variability in T2 is explained by T1?

reliability

(d) Give the t statistic for testing the null hypothesis that the slope is 0. State an appropriate alternative hypothesis for this problem and state the P-value for the significance test. Report your conclusions from parts (b), (c), and (d) in plain language.

(e) Verify that the square of the t statistic for the slope hypothesis is equal to the F statistic given in the ANOVA table.

10.31 Refer to the previous exercise. Rerun the regression for the 25 observations with odd-numbered strips; that is, $S = 1, 3, \ldots, 49$. We want to compare the results of this run with those for the full data set from the previous exercise. In particular, we want to see which quantities remain approximately the same and which change.

(a) Construct a table giving b_0, b_1, s, and s_{b_1} for the two runs. Which values are approximately the same and which have changed?

(b) Summarize the differences between the two ANOVA tables.

(c) Compare the two correlations.

(d) Summarize your results in plain language. In particular, did your results in the previous exercise depend strongly on the quite large sample size?

(e) Suppose that we had available another 50 wood strips from the same population measured under the same conditions. If all 100 observations were included in a single analysis, approximately what values would you expect to see for b_0, b_1, s, and r?

10.32 This exercise uses the Gesell Adaptive Score data given in Table 2.8 (page 160). Run a linear regression to predict the Gesell score from age at first word. Use all 21 cases for this analysis and assume that the simple linear regression model holds.

(a) Write the fitted regression equation.

(b) What is the value of the t statistic for testing the null hypothesis that the slope is 0? Give an approximate P-value for the test of H_0 versus the alternative that the slope is negative.

(c) Give a 95% confidence interval for the slope.

(d) What percentage of the variation in Gesell scores is explained by the age at first word?

(e) What is s, the estimate of the model standard deviation σ?

(f) In Chapter 2 we noted that Cases 18 and 19 in the Gesell data set were somewhat unusual. The simple linear regression model is therefore suspect when all cases are included. Rerun your analysis with these two cases excluded. Compare your results with those obtained in parts (a) through (e). In particular, to what extent did the conclusions from the first regression run depend on these two cases?

10.33 The data on gas consumption and degree-days from Exercise 10.6 (page 697) are as follows:

	Oct.	Nov.	Dec.	Jan.	Feb.	Mar.	Apr.	May	June
x	15.6	26.8	37.8	36.4	35.5	18.6	15.3	7.9	0.0
y	5.2	6.1	8.7	8.5	8.8	4.9	4.5	2.5	1.1

Suppose that the gas consumption for December was incorrectly recorded as 87 instead of 8.7.

(a) Calculate the least-squares regression line for the incorrect set of data.

(b) Find the standard error of b_1.

(c) Compute the test statistic for H_0: $\beta_1 = 0$ and find the P-value. How do the results compare with those for the correct set of data?

10.34 We studied the *Archaeopteryx* data in Example 2.4 (page 110) and Exercise 2.18 (page 131). The data consist of measurements on the lengths in centimeters of the femur (a leg bone) and the humerus (a bone in the upper arm) for the five specimens that preserve both bones:

Femur	38	56	59	64	74
Humerus	41	63	70	72	84

(a) Plot the data and describe the pattern. Is it reasonable to summarize this kind of relationship with a correlation?

(b) Find the correlation and perform the significance test. Summarize the results and report your conclusion.

10.35 Here are the golf scores of 12 members of a college women's golf team in two rounds of tournament play. (A golf score is the number of strokes required to complete the course, so that low scores are better.)

Player	1	2	3	4	5	6	7	8	9	10	11	12
Round 1	89	90	87	95	86	81	102	105	83	88	91	79
Round 2	94	85	89	89	81	76	107	89	87	91	88	80

(a) Plot the data and describe the relationship between the two scores.

(b) Find the correlation between the two scores and test the null hypothesis that the population correlation is 0. Summarize your results.

(c) The plot shows one outlier. Recompute the correlation and redo the significance test without this observation. Write a short summary explaining the effect of the outlier on the correlation and significance test in (b).

10.36 Concern about the quality of life for chronically ill patients is becoming as important as treating their physical symptoms. The SF-16, a questionnaire for measuring the health quality of life, was given to 50 patients with chronic obstructive lung disease. A correlation of 0.68 was reported between the component of the questionnaire called general health perceptions (GHP) and a measure of lung function called forced vital capacity (FVC), expressed as a percent of normal. The mean and standard deviation of GHP are 43.5 and 20.3, and for FVC the values are 80.9 and 17.2. (From Donald A. Mahler and John I. Mackowiak, "Evaluation of the short-form 36-item questionnaire to measure health-related quality of life in patients with COLD," *Chest,* 107 (1995), pp. 1585–1589).

(a) Find the equation of the least-squares line for predicting GHP from FVC.

(b) Give the results of the significance test for the null hypothesis that the slope is 0. (*Hint:* What is the relation between this test and the test for a zero correlation?)

10.37 A study reported a correlation $r = 0.5$ based on a sample size of $n = 20$; another reported the same correlation based on a sample size of $n = 10$. For each, perform the test of the null hypothesis that $\rho = 0$. Describe the results and explain why the conclusions are different.

10.38 Refer to the bank wages data given in Table 10.1 and described in Exercise 10.1 (page 695). The data also include a variable "Size," which classifies the bank as large or small. Obtain the residuals from the regression used to predict wages from LOS, and plot them versus LOS using different symbols for the large and small banks. Include on your plot a horizontal line at 0 (the mean of the residuals). Describe the important features of this plot. Explain what they indicate about wages in this set of data.

10.39 The SAT and the ACT are the two major standardized tests that colleges use to evaluate candidates. Most students take just one of these tests. However, some students take both. Table 10.4 gives the scores of 60 students who did this. How can we relate the two tests?

(a) Plot the data with SAT on the x axis and ACT on the y axis. Describe the overall pattern and any unusual observations.

(b) Find the least-squares regression line and draw it on your plot. Give the results of the significance test for the slope.

(c) What is the correlation between the two tests?

10.40 Refer to the previous exercise. Find the predicted value of ACT for each observation in the data set.

TABLE 10.4 SAT and ACT scores

SAT	ACT	SAT	ACT	SAT	ACT	SAT	ACT
1000	24	870	21	1090	25	800	21
1010	24	880	21	860	19	1040	24
920	17	850	22	740	16	840	17
840	19	780	22	500	10	1060	25
830	19	830	20	780	12	870	21
1440	32	1190	30	1120	27	1120	25
490	7	800	16	590	12	800	18
1050	23	830	16	990	24	960	27
870	18	890	23	700	16	880	21
970	21	880	24	930	22	1020	24
920	22	980	27	860	23	790	14
810	19	1030	23	420	21	620	18
1080	23	1220	30	800	20	1150	28
1000	19	1080	22	1140	24	970	20
1030	25	970	20	920	21	1060	24

(a) What is the mean of these predicted values? Compare it with the mean of the ACT scores.

(b) Compare the standard deviation of the predicted values with the standard deviation of the actual ACT scores. If least-squares regression is used to predict ACT scores for a large number of students such as these, the average predicted value will be accurate but the variability of the predicted scores will be too small.

(c) Find the SAT score for a student who is one standard deviation above the mean ($z = (x - \bar{x})/s = 1$). Find the predicted ACT score and standardize this score. (Use the means and standard deviations from this set of data for these calculations.)

(d) Repeat part (c) for a student whose SAT score is one standard deviation below the mean ($z = -1$).

(e) What do you conclude from parts (c) and (d)? Perform additional calculations for different z's if needed.

10.41   Refer to the previous two exercises. An alternative to the least-squares method is based on matching standardized scores. Specifically, we set

$$\frac{(y - \bar{y})}{s_y} = \frac{(x - \bar{x})}{s_x}$$

and solve for y. Let's use the notation $y = a_0 + a_1 x$ for this line. The slope is $a_1 = s_x/s_y$ and the intercept is $a_0 = \bar{y} - a_1\bar{x}$. Compare these expressions with the formulas for the least-squares slope and intercept (page 666).

(a) Using the data in Table 10.4, find the values of a_0 and a_1.

(b) Plot the data with the least-squares line and the new prediction line.

(c) Use the new line to find predicted ACT scores. Find the mean and the standard deviation of these scores. How do they compare with the mean and standard deviation of the ACT scores?

NOTES

1. The quote and the example are taken from J. V. G. A. Durnin and J. Womersley, "Body fat assessed from total body density and its estimation from skinfold thicknesses: measurements on 481 men and women aged 16 to 72 years," *British Journal of Nutrition*, 32 (1974), pp. 77–97. The data analyzed in the examples that follow have been generated by computer simulation in such a way that they give the same statistical results reported in the article.

2. Minitab does this directly in response to the PREDICT subcommand of the REGRESS command. Some software packages, such as SAS, use an indirect method: Define an additional observation with LSKIN equal to the value of x^* and DEN missing. The software ignores the additional case in computing the least-squares line but computes and prints the predicted value and associated statistics.

3. The data were provided by Linda McCabe and were collected as part of a large study of women's bone health and another study of calcium kinetics, both directed by Professor Connie Weaver of the Foods and Nutrition Department, Purdue University. Results related to this example are reported in Dorothy Teegarden et al., "Peak bone mass in young women," *Journal of Bone and Mineral Research*, 10 (1995), pp. 711–715.

4. This example is based on summaries in Charles Fombrun and Mark Shanley, "What's in a name? Reputation building and corporate strategy," *Academy of Management Journal*, 33 (1990), pp. 233–258.

5. These data were provided by Robert Dale of Purdue University.

6. This quantity is the estimated standard deviation of $\hat{y} - y$, not the estimated standard deviation of $\hat{y}$ alone.

7. The method is described in Chapter 10 of G. W. Snedecor and W. G. Cochran, *Statistical Methods*, 8th ed., Iowa State University Press, Ames, Iowa, 1989.

Prelude

In Chapter 10 we studied formal inference for the situation where there is one explanatory variable and one response variable. In this chapter we introduce the more complex setting in which several explanatory variables work together to explain the response. The descriptive tools we learned in Chapter 2—scatterplots, least-squares regression, and correlation— are essential preliminaries to inference and also provide a foundation for confidence intervals and significance tests. Here are some examples of questions we can answer from data using the methods in this chapter:

- We want to predict the college grade point average of newly admitted students. We have data on their high school grades in several subjects and their scores on the two parts of the SAT. How well can we predict college grades from this information? Do high school grades or SAT scores predict college grades more accurately?

- How do banks determine the interest rates that they will charge for a loan? To what extent do the rates depend on characteristics of the loan and characteristics of the borrower?

Clinton Hill, *Close Encounter*, 1987.

Multiple Regression

In Chapter 10 we presented methods for inference in the setting of a linear relationship between a response variable y and a *single* explanatory variable x. With multiple linear regression we use *more than one* explanatory variable to explain or predict a single response variable. Many of the ideas that we encountered in our study of simple linear regression carry over to this more complicated case. However, the introduction of several explanatory variables leads to many additional considerations. In this short chapter we cannot explore all of these issues. Rather, we will outline some basic facts about inference in the multiple regression setting and then illustrate multiple regression analysis with a case study.

Population multiple regression equation

The simple linear regression model assumes that the mean of the response variable y depends on the explanatory variable x according to a linear equation

$$\mu_y = \beta_0 + \beta_1 x$$

For any fixed value of x, the response y varies normally around this mean and has a standard deviation σ that is the same for all values of x.

In the multiple regression setting, the response variable y depends on not one but p explanatory variables. We will denote these explanatory variables by $x_1, x_2, \ldots, x_p$. The mean response is a linear function of the explanatory variables:

$$\mu_y = \beta_0 + \beta_1 x_1 + \beta_2 x_2 + \cdots + \beta_p x_p$$

population regression equation

This expression is the **population regression equation.** We do not observe the mean response because the observed values of y vary about their means. We can think of subpopulations of responses, each corresponding to a particular set of values for *all* of the explanatory variables $x_1, x_2, \ldots, x_p$. In each subpopulation, y varies normally with a mean given by the population regression equation. The regression model assumes that the standard deviation σ of the responses is the same in all subpopulations.

EXAMPLE 11.1

Our case study uses data collected at a large university on all first-year computer science majors in a particular year.[1] The purpose of the study was to attempt to predict success in the early university years. One measure of success was the cumulative grade point average (GPA) after three semesters. Among the explanatory variables recorded at the time the students enrolled in the university were average high school grades in mathematics (HSM), science (HSS), and English (HSE).

We will use high school grades to predict the response variable GPA. There are $p = 3$ explanatory variables: $x_1 = $ HSM, $x_2 = $ HSS, and $x_3 = $ HSE. The high school grades are coded on a scale from 1 to 10, with 10 corresponding to A, 9 to A−, 8 to B+, and so on. These grades define the subpopulations. For example, the straight-C students are the subpopulation defined by HSM = 4, HSS = 4, and HSE = 4.

One possible multiple regression model for the subpopulation mean GPAs is

$$\mu_{\text{GPA}} = \beta_0 + \beta_1 \text{HSM} + \beta_2 \text{HSS} + \beta_3 \text{HSE}$$

For the straight-C subpopulation of students, the model gives the subpopulation mean as

$$\mu_{\text{GPA}} = \beta_0 + 4\beta_1 + 4\beta_2 + 4\beta_3$$

Data for multiple regression

The data for a simple linear regression problem consist of observations (x_i, y_i) on the two variables. Because there are several explanatory variables in multiple regression, the notation needed to describe the data is more elaborate. In our case study, each observation or case consists of a value for the response variable and for each of the explanatory variables for one student. Call x_{ij} the value of the jth explanatory variable for the ith student. The data are then

$$\text{Student 1: } (x_{11}, x_{12}, \ldots, x_{1p}, y_1)$$
$$\text{Student 2: } (x_{21}, x_{22}, \ldots, x_{2p}, y_2)$$
$$\vdots$$
$$\text{Student } n: (x_{n1}, x_{n2}, \ldots, x_{np}, y_n)$$

Here, n is the number of cases and p is the number of explanatory variables. Data are often entered into computer regression programs in this format. Each row is a case and each column corresponds to a different variable. The data for Example 11.1, with several additional explanatory variables, appear in this format in the CSDATA data set described in the Data Appendix.

Multiple linear regression model

We combine the population regression equation and assumptions about variation to construct the multiple linear regression model. The subpopulation means describe the FIT part of our statistical model. The RESIDUAL part represents the variation of observations about the means. We will use the same notation for the residual that we used in the simple linear regression model. The symbol ϵ represents the deviation of an individual observation from its subpopulation mean. We assume that these deviations are normally distributed with mean 0 and an unknown standard deviation σ that does not depend on the values of the x variables.

Multiple Linear Regression Model

The **statistical model for multiple linear regression** is

$$y_i = \beta_0 + \beta_1 x_{i1} + \beta_2 x_{i2} + \cdots + \beta_p x_{ip} + \epsilon_i$$

for $i = 1, 2, \ldots, n$.

The **mean response μ_y** is a linear function of the explanatory variables:

$$\mu_y = \beta_0 + \beta_1 x_1 + \beta_2 x_2 + \cdots + \beta_p x_p$$

The **deviations ϵ_i** are independent and normally distributed with mean 0 and standard deviation σ. In other words, they are an SRS from the $N(0, \sigma)$ distribution.

The parameters of the model are $\beta_0, \beta_1, \beta_2, \ldots, \beta_p$, and σ.

The assumption that the subpopulation means are related to the regression coefficients β by the equation

$$\mu_y = \beta_0 + \beta_1 x_1 + \beta_2 x_2 + \cdots + \beta_p x_p$$

implies that we can estimate all subpopulation means from estimates of the β's. To the extent that this equation is accurate, we have a useful tool for describing how the mean of y varies with the x's.

Estimation of the multiple regression parameters

For simple linear regression we used the principle of least squares first described in Section 2.3 to obtain estimators of the intercept and slope of the regression line. For multiple regression the principle is the same but the details are more complicated. Let

$$b_0, b_1, b_2, \ldots, b_p$$

denote the estimators of the parameters

$$\beta_0, \beta_1, \beta_2, \ldots, \beta_p$$

For the ith observation the predicted response is

$$\hat{y}_i = b_0 + b_1 x_{i1} + b_2 x_{i2} + \cdots + b_p x_{ip}$$

residual

The ith **residual,** the difference between the observed and predicted response, is therefore

$$
\begin{aligned}
e_i &= \text{observed response} - \text{predicted response} \\
&= y_i - \hat{y}_i \\
&= y_i - b_0 - b_1 x_{i1} - b_2 x_{i2} - \cdots - b_p x_{ip}
\end{aligned}
$$

method of least squares

    The **method of least squares** chooses the values of the b's that make the sum of the squares of the residuals as small as possible. In other words, the parameter estimates $b_0, b_1, b_2, \ldots, b_p$ minimize the quantity

$$\sum (y_i - b_0 - b_1 x_{i1} - b_2 x_{i2} - \cdots - b_p x_{ip})^2$$

The formula for the least-squares estimates is complicated. We will be content to understand the principle on which they are based and to let software do the computations.

    The parameter σ^2 measures the variability of the responses about the population regression equation. As in the case of simple linear regression, we estimate σ^2 by an average of the squared residuals. The estimator is

$$
\begin{aligned}
s^2 &= \frac{\sum e_i^2}{n - p - 1} \\
&= \frac{\sum (y_i - \hat{y}_i)^2}{n - p - 1}
\end{aligned}
$$

degrees of freedom

The quantity $n - p - 1$ is the **degrees of freedom** associated with s^2. The degrees of freedom equal the sample size n minus $p + 1$, the number of β's we must estimate to fit the model. In the simple linear regression case there is just one explanatory variable, so $p - 1$ and the degrees of freedom are $n - 2$. To estimate σ we use

$$s = \sqrt{s^2}$$

Confidence intervals and significance tests for regression coefficients

We can obtain confidence intervals and significance tests for each of the regression coefficients β_j as we did in simple linear regression. The standard errors of the b's have more complicated formulas, but all are multiples of s. We again rely on statistical software for the calculations.

Confidence Intervals and Significance Tests for β_j

A **level C confidence interval** for β_j is

$$b_j \pm t^* \mathrm{SE}_{b_j}$$

where SE_{b_j} is the standard error of b_j and t^* is the value for the $t(n - p - 1)$ density curve with area C between $-t^*$ and t^*.

To test the hypothesis $H_0 : \beta_j = 0$, compute the **t statistic**

$$t = \frac{b_j}{\mathrm{SE}_{b_j}}$$

In terms of a random variable T having the $t(n - p - 1)$ distribution, the P-value for a test of H_0 against

$H_a : \beta_j > 0$ is $P(T \geq t)$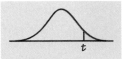

$H_a : \beta_j < 0$ is $P(T \leq t)$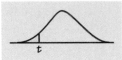

$H_a : \beta_j \neq 0$ is $2P(T \geq |t|)$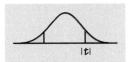

confidence intervals for mean response

prediction intervals

        Because regression is often used for prediction, we may wish to construct **confidence intervals for a mean response** and **prediction intervals** for a future observation from multiple regression models. The basic ideas are the same as in the simple linear regression case. In most software systems, the same commands that give confidence and prediction intervals for simple linear regression work for multiple regression. The only difference is that we specify a list of explanatory variables rather than a single variable. Modern software allows us to perform these rather complex calculations without an intimate knowledge of all of the computational details. This frees us to concentrate on the meaning and appropriate use of the results.

ANOVA table for multiple regression

In simple linear regression the F test from the ANOVA table is equivalent to the two-sided t test of the hypothesis that the slope of the regression line is 0. For multiple regression there is a corresponding **ANOVA F test,** but it tests the hypothesis that *all* of the regression coefficients (with the exception of the intercept) are 0. Here is the general form of the ANOVA table for multiple regression:

ANOVA F test

Source	Degrees of freedom	Sum of squares	Mean square	F
Model	p	$\sum(\hat{y}_i - \bar{y})^2$	SSM/DFM	MSM/MSE
Error	$n - p - 1$	$\sum(y_i - \hat{y}_i)^2$	SSE/DFE	
Total	$n - 1$	$\sum(y_i - \bar{y})^2$	SST/DFT	

The ANOVA table is similar to that for simple linear regression. The degrees of freedom for the model increase from 1 to p to reflect the fact that we now have p explanatory variables rather than just one. As a consequence, the degrees of freedom for error decrease by the same amount. The sums of squares represent sources of variation. Once again, both the sums of squares and their degrees of freedom add:

$$SST = SSM + SSE$$
$$DFT = DFM + DFE$$

The estimate of the variance σ^2 for our model is again given by the MSE in the ANOVA table. That is, $s^2 = $ MSE.

F statistic

The ratio MSM/MSE is an **F statistic** for testing the null hypothesis

$$H_0: \beta_1 = \beta_2 = \cdots = \beta_p = 0$$

against the alternative hypothesis

$$H_a: \text{at least one of the } \beta_j \text{ is not 0}$$

The null hypothesis says that none of the explanatory variables are predictors of the response variable when used in the form expressed by the multiple regression equation. The alternative states that at least one of them is linearly related to the response. As in simple linear regression, large values of F give evidence against H_0. When H_0 is true, F has the $F(p, n - p - 1)$ distribution. The degrees of freedom for the F distribution are those associated with the model and error in the ANOVA table.

Analysis of Variance F Test

In the multiple regression model, the hypothesis

$$H_0: \beta_1 = \beta_2 = \cdots = \beta_p = 0$$

is tested by the analysis of variance F statistic

$$F = \frac{\text{MSM}}{\text{MSE}}$$

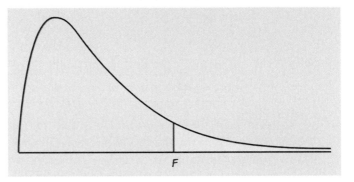

The P-value is the probability that a random variable having the $F(p, n - p - 1)$ distribution is greater than or equal to the calculated value of the F statistic.

Squared multiple correlation R^2

For simple linear regression we noted that the square of the sample correlation could be written as the ratio of SSM to SST and could be interpreted as the proportion of variation in y explained by x. A similar statistic is routinely calculated for multiple regression.

The Squared Multiple Correlation

The statistic

$$R^2 = \frac{\text{SSM}}{\text{SST}} = \frac{\sum (\hat{y}_i - \bar{y})^2}{\sum (y_i - \bar{y})^2}$$

is the proportion of the variation of the response variable y that is explained by the explanatory variables $x_1, x_2, \ldots, x_p$ in a multiple linear regression.

Often, R^2 is multiplied by 100 and expressed as a percent. The square root of R^2, called the **multiple correlation coefficient,** is the correlation between the observations y_i and the predicted values $\hat{y}_i$.

multiple correlation
coefficient

A case study: preliminary analysis

In this section we illustrate multiple regression by analyzing the data from the study described in Example 11.1. The response variable is the cumulative GPA after three semesters for a group of computer science majors at a large university. The explanatory variables previously mentioned are average high school grades, represented by HSM, HSS, and HSE. We also examine the SAT mathematics and SAT verbal scores as explanatory variables. We have data for $n = 224$ students in the study. We use SAS to illustrate the outputs that are given by most software.

The first step in the analysis is to carefully examine each of the variables. Means, standard deviations, and minimum and maximum values appear in Figure 11.1.

The minimum value for the SAT mathematics (SATM) variable appears to be rather extreme; it is $(595 - 300)/86 = 3.43$ standard deviations below the mean. We do not discard this case at this time but will take care in our subsequent analyses to see if it has an excessive influence on our results. The mean for the SATM score is higher than the mean for the verbal score (SATV), as we might expect for a group of computer science majors. The two standard deviations are about the same. The means of the three high school grade variables are similar, with the mathematics grades being a bit higher. The standard deviations for the high school grade variables are very close to each other. The mean GPA is 2.635 on a 4-point scale, with standard deviation 0.779.

Because the variables GPA, SATM, and SATV have many possible values, we could use stemplots or histograms to examine the shapes of their distributions. Normal quantile plots indicate whether or not the distributions look normal. It is important to note that the multiple regression model *does not* require any of these distributions to be normal. Only the deviations of the responses y from their means are assumed to be normal. The purpose of examining these plots is to understand something about each variable alone before attempting to use it in a complicated model. Extreme values of any variable should be noted and checked for accuracy. If found to be correct, the cases with these values should be carefully examined to see if they are truly

Variable	N	Mean	Std Dev	Minimum	Maximum
GPA	224	2.6352232	0.7793949	0.1200000	4.0000000
SATM	224	595.2857143	86.4014437	300.0000000	800.0000000
SATV	224	504.5491071	92.6104591	285.0000000	760.0000000
HSM	224	8.3214286	1.6387367	2.0000000	10.0000000
HSS	224	8.0892857	1.6996627	3.0000000	10.0000000
HSE	224	8.0937500	1.5078736	3.0000000	10.0000000

FIGURE 11.1 Descriptive statistics for the computer science student case study.

exceptional and perhaps do not belong in the same analysis with the other cases. When these data are examined in this way, no obvious problems are evident.

The high school grade variables HSM, HSS, and HSE have relatively few values and are best summarized by giving the relative frequencies for each possible value. The output in Figure 11.2 provides these summaries. The distributions are all skewed, with a large proportion of high grades (10 = A and 9 = A−). Again we emphasize that these distributions need not be normal.

HSM	Frequency	Percent	Cumulative Frequency	Cumulative Percent
2	1	0.4	1	0.4
3	1	0.4	2	0.9
4	4	1.8	6	2.7
5	6	2.7	12	5.4
6	23	10.3	35	15.6
7	28	12.5	63	28.1
8	36	16.1	99	44.2
9	59	26.3	158	70.5
10	66	29.5	224	100.0

HSM	Frequency	Percent	Cumulative Frequency	Cumulative Percent
3	1	0.4	1	0.4
4	7	3.1	8	3.6
5	9	4.0	17	7.6
6	24	10.7	41	18.3
7	42	18.8	83	37.1
8	31	13.8	114	50.9
9	50	22.3	164	73.2
10	60	26.8	224	100.0

HSM	Frequency	Percent	Cumulative Frequency	Cumulative Percent
3	1	0.4	1	0.4
4	4	1.8	5	2.2
5	5	2.2	10	4.5
6	23	10.3	33	14.7
7	43	19.2	76	33.9
8	49	21.9	125	55.8
9	42	23.2	177	79.0
10	47	21.0	224	100.0

FIGURE 11.2   The distributions of the high school grade variables.

Relationships between pairs of variables

The second step in our analysis is to examine the relationships between all pairs of variables. Scatterplots and correlations are our tools for studying two-variable relationships. The correlations appear in Figure 11.3. The output includes the P-value for the test of the null hypothesis that the population correlation is 0 versus the two-sided alternative for each pair. Thus, we see that the correlation between GPA and HSM is 0.44, with a P-value of 0.0001, whereas the correlation between GPA and SATV is 0.11, with a P-value of 0.087. The first is statistically significant by any reasonable standard, and the second is rather marginal.

The high school grades all have higher correlations with GPA than do the SAT scores. As we might expect, math grades have the highest correlation ($r = 0.44$), followed by science grades (0.33) and then English grades (0.29). The two SAT scores have a rather high correlation with each other (0.46), and the high school grades also correlate well with each other (0.45 to 0.58). The SAT mathematics score correlates well with HSM (0.45), less well with HSS (0.24), and rather poorly with HSE (0.11). The correlations of the SAT verbal score with the three high school grades are about equal, ranging from 0.22 to 0.26.

It is important to keep in mind that by examining pairs of variables we are seeking a better understanding of the data. The fact that the correlation of a particular explanatory variable with the response variable does not achieve

Pearson Correlation Coefficients / Prob > |R| under Ho: Rho=0 / N = 224

	GPA	SATM	SATV	HSM	HSS	HSE
GPA	1.00000	0.25171	0.11449	0.43650	0.32943	0.28900
	0.0	0.0001	0.0873	0.0001	0.0001	0.0001
SATM	0.25171	1.00000	0.46394	0.45351	0.24048	0.10828
	0.0001	0.0	0.0001	0.0001	0.0003	0.1060
SATV	0.11449	0.46394	1.00000	0.22112	0.26170	0.24371
	0.0873	0.0001	0.0	0.0009	0.0001	0.0002
HSM	0.43650	0.45351	0.22112	1.00000	0.57569	0.44689
	0.0001	0.0001	0.0009	0.0	0.0001	0.0001
HSS	0.32943	0.24048	0.26170	0.57569	1.00000	0.57937
	0.0001	0.0003	0.0001	0.0001	0.0	0.0001
HSE	0.28900	0.10828	0.24371	0.44689	0.57937	1.00000
	0.0001	0.1060	0.0002	0.0001	0.0001	0.0

FIGURE 11.3 Correlations among the case study variables.

statistical significance does not *necessarily* imply that it will not be a useful (and significant) predictor in a multiple regression.

Numerical summaries such as correlations are useful, but plots are generally more informative when seeking to understand data. Plots tell us whether the numerical summary gives a fair representation of the data. For a multiple regression, each pair of variables should be plotted. For the six variables in our case study, this means that we should examine 15 plots. In general there are $p + 1$ variables in a multiple regression analysis with p explanatory variables, so that $p(p + 1)/2$ plots are required. Multiple regression is a complicated procedure. If we do not do the necessary preliminary work, we are in serious danger of producing useless or misleading results. We leave the task of making these plots as an exercise.

Regression on high school grades

To explore the relationship between the explanatory variables and our response variable GPA, we run several multiple regressions. The explanatory variables fall into two classes. High school grades are represented by HSM, HSS, and HSE, and standardized tests are represented by the two SAT scores. We begin our analysis by using the high school grades to predict GPA. Figure 11.4 gives the multiple regression output.

```
Dependent Variable: GPA


                            Analysis of Variance

                         Sum of           Mean
Source          DF       Squares         Square      F Value     Prob>F

Model            3       27.71233        9.23744      18.861      0.0001
Error          220      107.75046        0.48977
C Total        223      135.46279


        Root MSE        0.69984      R-Square      0.2046
        Dep Mean        2.63522      Adj R-sq      0.1937
        C.V.           26.55711


                          Parameter Estimates

                     Parameter       Standard      T for H0:
Variable    DF       Estimate          Error     Parameter=0     Prob > |T|

INTERCEP     1       0.589877       0.29424324       2.005         0.0462
HSM          1       0.168567       0.03549214       4.749         0.0001
HSS          1       0.034316       0.03755888       0.914         0.3619
HSE          1       0.045102       0.03869585       1.166         0.2451
```

FIGURE 11.4 Multiple regression output for regression using high school grades to predict GPA.

ANOVA table

The output contains an **ANOVA table,** some additional descriptive statistics, and information about the parameter estimates. When examining any ANOVA table, it is a good idea to first verify the degrees of freedom. This ensures that we have not made some serious error in specifying the model for the program or in entering the data. Because there are $n = 224$ cases, we have DFT $= n - 1 = 223$. The three explanatory variables give DFM $= p = 3$ and DFE $= n - p - 1 = 223 - 3 = 220$.

F statistic

The ANOVA **F statistic** is 18.86, with a P-value of 0.0001. Under the null hypothesis

$$H_0: \beta_1 = \beta_2 = \beta_3 = 0$$

the F statistic has an $F(3, 220)$ distribution. According to this distribution, the chance of obtaining an F statistic of 18.86 or larger is 0.0001. We therefore conclude that at least one of the three regression coefficients for the high school grades is different from 0 in the population regression equation.

estimate of σ

R^2

In the descriptive statistics that follow the ANOVA table we find that Root MSE is 0.6998. This value is the square root of the MSE given in the ANOVA table and is s, the **estimate of the parameter σ** of our model. The value of R^2 is 0.20. That is, 20% of the observed variation in the GPA scores is explained by linear regression on high school grades. Although the P-value is very small, the model does not explain very much of the variation in GPA. Remember, a small P-value does not necessarily tell us that we have a large effect, particularly when the sample size is large.

fitted regression equation

From the Parameter Estimates section we obtain the **fitted regression equation**

$$\widehat{\text{GPA}} = 0.590 + 0.169\text{HSM} + 0.034\text{HSS} + 0.045\text{HSE}$$

t statistics

Recall that the **t statistics** for testing the regression coefficients are obtained by dividing the estimates by their standard errors. Thus, for the coefficient of HSM we obtain the t-value given in the output by calculating

$$t = \frac{b}{\text{SE}_b} = \frac{0.168567}{0.03549214} = 4.749$$

P-values

The **P-values** appear in the last column. Note that these P-values are for the two-sided alternatives. HSM has a P-value of 0.0001, and we conclude that the regression coefficient for this explanatory variable is significantly different from 0. The P-values for the other explanatory variables (0.36 for HSS and 0.25 for HSE) do not achieve statistical significance.

Interpretation of results

The significance tests for the individual regression coefficients seem to contradict the impression obtained by examining the correlations in Figure 11.3. In that display we see that the correlation between GPA and HSS is 0.33 and the correlation between GPA and HSE is 0.29. The P-values for both of these correlations are 0.0001. In other words, if we used HSS alone in a regression to predict GPA, or if we used HSE alone, we would obtain statistically significant regression coefficients.

This phenomenon is not unusual in multiple regression analysis. Part of the explanation lies in the correlations between HSM and the other two explanatory variables. These are rather high (at least compared with the other correlations in Figure 11.3). The correlation between HSM and HSS is 0.58, and that between HSM and HSE is 0.45. Thus, when we have a regression model that contains all three high school grades as explanatory variables, there is considerable overlap of the predictive information contained in these variables. *The significance tests for individual regression coefficients assess the significance of each predictor variable assuming that all other predictors are included in the regression equation.* Given that we use a model with HSM and HSS as predictors, the coefficient of HSE is not statistically significant. Similarly, given that we have HSM and HSE in the model, HSS does not have a significant regression coefficient. HSM, however, adds significantly to our ability to predict GPA even after HSS and HSE are already in the model.

Unfortunately, we cannot conclude from this analysis that the *pair* of explanatory variables HSS and HSE contribute nothing significant to our model for predicting GPA once HSM is in the model. The impact of relations among the several explanatory variables on fitting models for the response is the most important new phenomenon encountered in moving from simple linear regression to multiple regression. We can only hint at the many complicated problems that arise.

Residuals

As in simple linear regression, we should always examine the residuals as an aid to determining whether the multiple regression model is appropriate for the data. Because there are several explanatory variables, we must examine several residual plots. It is usual to plot the residuals versus the predicted values $\hat{y}$ and also versus each of the explanatory variables. Look for outliers, influential observations, evidence of a curved (rather than linear) relation, and anything else unusual. Again, we leave the task of making these plots as an exercise. The plots all appear to show more or less random noise around the center value of 0.

If the deviations ϵ in the model are normally distributed, the residuals should be normally distributed. Figure 11.5 presents a normal quantile plot of the residuals. The distribution appears to be approximately normal. There are many other specialized plots that help detect departures from the multiple regression model. Discussion of these, however, is more than we can undertake in this chapter.

Refining the model

Because the variable HSS has the largest P-value of the three explanatory variables (see Figure 11.4) and therefore appears to contribute the least to our explanation of GPA, we rerun the regression using only HSM and HSE as explanatory variables. The SAS output appears in Figure 11.6. The F statistic indicates that we can reject the null hypothesis that the regression coefficients

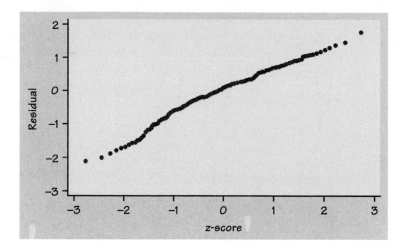

FIGURE 11.5 Normal quantile plot of the residuals from the high school grade model. There are no important deviations from normality.

for the two explanatory variables are both 0. The P-value is still 0.0001. The value of R^2 has dropped very slightly compared with our previous run, from 0.2046 to 0.2016. Dropping HSS from the model resulted in the loss of very little explanatory power. The measure s of variation about the fitted equation (Root MSE in the printout) is nearly identical for the two regressions,

Dependent Variable: GPA

Analysis of Variance

Source	DF	Sum of Squares	Mean Square	F Value	Prob>F
Model	2	27.30349	13.65175	27.894	0.0001
Error	221	108.15930	0.48941		
C Total	223	135.46279			

Root MSE	0.69958	R-Square	0.2016	
Dep Mean	2.63522	Adj R-sq	0.1943	
C.V.	26.54718			

Parameter Estimates

Variable	DF	Parameter Estimate	Standard Error	T for H0: Parameter=0	Prob > |T|
INTERCEP	1	0.624228	0.29172204	2.140	0.0335
HSM	1	0.182654	0.03195581	5.716	0.0001
HSE	1	0.060670	0.03472914	1.747	0.0820

FIGURE 11.6 Multiple regression output for regression using HSM and HSE to predict GPA.

another indication that dropping HSS loses very little. The t statistics for the individual regression coefficients indicate that HSM is still clearly significant ($P = 0.0001$), while the coefficient of HSE is larger than before (1.747 versus 1.166) and approaches the traditional 0.05 level of significance ($P = 0.082$).

Comparison of the fitted equations for the two multiple regression analyses tells us something more about the intricacies of this procedure. For the first run we have

$$\widehat{GPA} = 0.590 + 0.169 HSM + 0.034 HSS + 0.045 HSE$$

whereas the second gives us

$$\widehat{GPA} = 0.624 + 0.183 HSM + 0.061 HSE$$

Eliminating HSS from the model changes the regression coefficients for all of the remaining variables and the intercept. This phenomenon occurs quite generally in multiple regression. *Individual regression coefficients, their standard errors, and significance tests are meaningful only when interpreted in the context of the other explanatory variables in the model.*

Regression on SAT scores

We now turn to the problem of predicting GPA using the two SAT scores. Figure 11.7 gives the output. The fitted model is

$$\widehat{GPA} = 1.289 + 0.002283 SATM - 0.000025 SATV$$

The degrees of freedom are as expected: 2, 221, and 223. The F statistic is 7.476, with a P-value of 0.0007. We conclude that the regression coefficients for SATM and SATV are not both 0. Recall that we obtained the P-value 0.0001 when we used high school grades to predict GPA. Both multiple regression equations are highly significant, but this obscures the fact that the two models have quite different explanatory power. For the SAT regression, $R^2 = 0.0634$, whereas for the high school grades model even with only HSM and HSE (Figure 11.6), we have $R^2 = 0.2016$, a value more than three times as large. Stating that we have a statistically significant result is quite different from saying that an effect is large or important.

Further examination of the output in Figure 11.7 reveals that the coefficient of SATM is significant ($t = 3.44$, $P = 0.0007$), and that for SATV is not ($t = -0.04$, $P = 0.9684$). For a complete analysis we should carefully examine the residuals. Also, we might want to run the analysis with SATM as the only explanatory variable.

Regression using all variables

We have seen that either the high school grades or the SAT scores give a highly significant regression equation. The mathematics component of each of these

```
Dependent Variable: GPA

                          Analysis of Variance

                           Sum of          Mean
Source            DF       Squares        Square      F Value      Prob>F

Model              2       8.58384       4.29192        7.476      0.0007
Error            221     126.87895       0.57411
C Total          223     135.46279


        Root MSE          0.75770    R-Square       0.0634
        Dep Mean          2.63522    Adj R-sq       0.0549
        C.V.             28.75287


                          Parameter Estimates

                     Parameter        Standard     T for H0:
     Variable  DF     Estimate          Error     Parameter=0    Prob > |T|

     INTERCEP   1     1.288677       0.37603684        3.427       0.0007
     SATM       1     0.002283       0.00066291        3.444       0.0007
     SATV       1    -0.000024562    0.00061847       -0.040       0.9684
```

FIGURE 11.7 Multiple regression output for regression using SAT scores to predict GPA.

groups of explanatory variables appears to be the key predictor. Comparing the values of R^2 for the two models indicates that high school grades are better predictors than SAT scores. Can we get a better prediction equation using all of the explanatory variables together in one multiple regression?

To address this question we run the regression with all five explanatory variables. The output appears in Figure 11.8. The F statistic is 11.69, with a P-value of 0.0001, so at least one of our explanatory variables has a nonzero regression coefficient. This result is not surprising given that we have already seen that HSM and SATM are strong predictors of GPA. The value of R^2 is 0.2115, not much higher than the value of 0.2046 that we found for the high school grades regression.

Examination of the t statistics and the associated P-values for the individual regression coefficients reveals that HSM is the only one that is significant ($P = 0.0003$). That is, only HSM makes a significant contribution when it is added to a model that already has the other four explanatory variables. Once again it is important to understand that this result does not necessarily mean that the regression coefficients for the four other explanatory variables are *all* 0. Many statistical software packages provide the capability for testing whether a collection of regression coefficients in a multiple regression model are *all* 0. We use this approach to address two interesting questions about this set of data. We did not discuss such tests in the outline that opened this section, but the basic idea is quite simple.

```
Dependent Variable: GPA

                            Analysis of Variance

                        Sum of          Mean
Source          DF      Squares         Square      F Value      Prob>F

Model            5      28.64364        5.72873      11.691      0.0001
Error          218     106.81914        0.49000
C Total        223     135.46279


        Root MSE        0.70000      R-Square        0.2115
        Dep Mean        2.63522      Adj R-sq        0.1934
        C.V.           26.56311


                        Parameter Estimates

                    Parameter        Standard      T for H0:
   Variable   DF     Estimate          Error     Parameter=0     Prob > |T|

   INTERCEP   1      0.326719       0.39999643       0.817        0.4149
   SATM       1      0.000944       0.00068566       1.376        0.1702
   SATV       1     -0.000408       0.00059189      -0.689        0.4915
   HSM        1      0.145961       0.03926097       3.718        0.0003
   HSS        1      0.035905       0.03779841       0.950        0.3432
   HSE        1      0.055293       0.03956869       1.397        0.1637


Test: SAT      Numerator:        0.4657  DF:     2   F value:     0.9503
               Denominator:  0.489996    DF:   218   Prob>F:      0.3882


Test: HS       Numerator:        6.6866  DF:     3   F value:    13.6462
               Denominator:  0.489996    DF:   218   Prob>F:      0.0001
```

FIGURE 11.8   Multiple regression output for regression using all variables to predict GPA.

Figure 11.9 gives the Excel and Minitab multiple regression output for this problem. Although the format and organization of output differ among software packages, the basic results that we need are easy to find.

Test for a collection of regression coefficients

In the context of the multiple regression model with all five predictors, we ask first whether or not the coefficients for the two SAT scores are both 0. In other words, do the SAT scores add any significant predictive information to that already contained in the high school grades? To be fair, we also ask the complementary question—do the high school grades add any significant predictive information to that already contained in the SAT scores?

Excel

	A	B	C	D	E	F	G
1	SUMMARY OUTPUT						
2							
3	*Regression Statistics*						
4	Multiple R	0.459837234					
5	R Square	0.211450282					
6	Adjusted R Square	0.193364279					
7	Standard Error	0.699997195					
8	Observations	224					
9							
10	ANOVA						
11		*df*	*SS*	*MS*	*F*	*Signifcance F*	
12	Regression	5	28.64364489	5.728729	11.69138	5.06E-10	
13	Residual	218	106.8191439	0.489996			
14	Total	223	135.4627888				
15							
16		*Coefficients*	*Standard Error*	*t Stat*	*P-value*	*Lower 95%*	*upper 95%*
17	intercept	0.326718739	0.399996431	0.816804	0.414932	-0.461636967	1.115074446
18	HSM	0.14596108	0.039260974	3.717714	0.000256	0.068581358	0.223340801
19	HSS	0.03590532	0.037798412	0.949916	0.343207	-0.03859183	0.11040247
20	HSE	0.055292581	0.039568691	1.397382	0.163719	-0.022693622	0.133278785
21	SATM	0.000843593	0.000685657	1.376187	0.170176	-0.000407774	0.002294959
22	SATV	0.00040785	0.000591893	-0.68906	0.491518	-0.001574415	0.00075816

Sheet1 \ **Sheet2** / e data /

Ready   Sum=0   CARS

Minitab

```
The regression equation is
GPA = 0.327 + 0.146 HSM + 0.0359 HSS + 0.0553 HSE +0.000944 SATM -
0.000408 SATV

Predictor        Coef        StDev            T          P
Constant       0.3267      0.4000         0.82      0.415
HSM            0.14596     0.03926        3.72      0.000
HSS            0.03591     0.03780        0.95      0.343
HSE            0.05529     0.03957        1.40      0.164
SATM         0.0009436   0.0006857        1.38      0.170
SATV        -0.0004078   0.0005919       -0.69      0.492

S = 0.7000        R-Sq = 21.1%      R-Sq(adj) = 19.3%

Analysis of Variance

Source         DF          SS           MS         F         P
Regression      5      28.6436       5.7287     11.69     0.000
Error         218     106.8191       0.4900
Total         223     135.4628
```

FIGURE 11.9 Excel and Minitab multiple regression output for regression using all variables to predict GPA.

The answers are given in the last two parts of the output in Figure 11.8. For the first test we see that $F = 0.9503$. Under the null hypothesis that the two SAT coefficients are 0, this statistic has an $F(2, 218)$ distribution and the P-value is 0.39. We conclude that the SAT scores are not significant predictors of GPA in a regression that already contains the high school scores as predictor variables. Recall that the model with just SAT scores has a highly significant F statistic. We now see that whatever predictive information is in the SAT scores can also be found in the high school grades. In this sense, the SAT scores are superfluous.

The test statistic for the three high school grade variables is $F = 13.6462$. Under the null hypothesis that these three regression coefficients are 0, the statistic has an $F(3, 218)$ distribution and the P-value is 0.0001. We conclude that high school grades contain useful information for predicting GPA that is not contained in SAT scores.

Of course, our statistical analysis of these data does not imply that SAT scores are less useful than high school grades for predicting college grades for all groups of students. We have studied a select group of students—computer science majors—from a specific university. Generalizations to other situations are beyond the scope of inference based on these data alone.

Beyond the basics ▶ multiple logistic regression

Many studies have yes/no or success/failure response variables. A surgery patient lives or dies; a consumer does or does not purchase a product after viewing an advertisement. Because the response variable in a multiple regression is assumed to have a normal distribution, this methodology is not suitable for predicting such responses. However, there are models that apply the ideas of regression to response variables with only two possible outcomes.

logistic regression   One type of model that can be used is called **logistic regression.** We think in terms of a binomial model for the two possible values of the response variable and use one or more explanatory variables to explain the probability of success. Details are even more complicated than those for multiple regression but the fundamental ideas are very much the same. Here is an example.

EXAMPLE 11.2   A company uses many different mailed catalogs to sell different types of merchandise. One catalog that features home goods, such as bedspreads and pillows, was mailed to 200,000 people who were not current customers of the catalog.[2] The response variable is whether or not a purchase was made from this catalog. We use logistic regression[3] to model the probability of a purchase, say p, as a function of five explanatory variables. These are x_1, number of purchases within the last 24 months from a home gift catalog; x_2, the proportion of single people in the zip code area based on census data; x_3, number of credit cards; x_4, a variable that distinguishes apartment dwellers from those who live in single-family homes; and x_5, an indicator of whether or not the customer has ever made a purchase from a similar type of catalog.

Similar to the F test in multiple regression, there is a chi-square test for multiple logistic regression that tests the null hypothesis that *all* coefficients of the explanatory variables are zero. The value is $X^2 = 141.34$, and the degrees of freedom are the number of explanatory variables, 5 in this case. (Recall that we studied the chi-square distribution in Chapter 9, where we analyzed two-way tables of counts.) The P-value is reported by software as $P = 0.0001$. (You can verify that it is less than 0.0005 using Table F.) We conclude that not all of the explanatory variables have zero coefficients.

Interpretation of the coefficients is a little more difficult in multiple logistic regression because of the form of the model. For our example, the fitted model is

$$\log\left(\frac{p}{1-p}\right) = -4.80 + 0.11x_1 - 0.11x_2 + 0.14x_3 - 0.43x_4 + 0.22x_5$$

The expression $p/(1-p)$ is the odds of a successful outcome. Logistic regression models the "log odds" as a linear combination of the explanatory variables.

In place of the t tests for individual coefficients in multiple regression, chi-square tests, each with 1 degree of freedom, are used to test whether individual coefficients are zero. The P-values for each of these tests are very small, ranging from 0.0067 to 0.0001. We conclude that all are nonzero. The signs of the coefficients tell us whether the explanatory variable is positively or negatively related to the response in the context of the model. We see that having made purchases in the past and having more credit cards are associated with a higher chance of buying from this catalog, whereas people who live in an apartment and in a neighborhood with a high proportion of single people are less likely to purchase these home goods.

The predicted value for this model allows us to estimate the probability p that a customer will make a purchase from the catalog. In the future, this model will be applied to potential customers who have not yet received the catalog. It will be mailed to those with an estimated probability of making a purchase that is above some cutoff value. From the response rate needed for a mailing to be profitable, an appropriate cutoff can be determined.

SUMMARY

The statistical model for **multiple linear regression** with response variable y and p explanatory variables $x_1, x_2, \ldots, x_p$ is

$$y_i = \beta_0 + \beta_1 x_{i1} + \beta_2 x_{i2} + \cdots + \beta_p x_{ip} + \epsilon_i$$

where $i = 1, 2, \ldots, n$. The ϵ_i are assumed to be independent and normally distributed with mean 0 and standard deviation σ. The **parameters** of the model are $\beta_0, \beta_1, \beta_2, \ldots, \beta_p$, and σ.

The β's are estimated by $b_0, b_1, b_2, \ldots, b_p$, which are obtained by the **method of least squares.** The parameter σ is estimated by

$$s = \sqrt{\text{MSE}} = \sqrt{\frac{\sum e_i^2}{n - p - 1}}$$

where the e_i are the **residuals,**

$$e_i = y_i - \hat{y}_i$$

A **level C confidence interval** for β_j is

$$b_j \pm t^* \text{SE}_{b_j}$$

where t^* is the value for the $t(n - p - 1)$ density curve with area C between $-t^*$ and t^*.

The test of the hypothesis $H_0: \beta_j = 0$ is based on the **t statistic:**

$$t = \frac{b_j}{SE_{b_j}}$$

and the $t(n - p - 1)$ distribution.

The estimate b_j of β_j and the test and confidence interval for β_j are all based on a specific multiple linear regression model. The results of all of these procedures change if other explanatory variables are added to or deleted from the model.

The **ANOVA table** for a multiple linear regression gives the degrees of freedom, sum of squares, and mean squares for the model, error, and total sources of variation. The **ANOVA F statistic** is the ratio MSM/MSE and is used to test the null hypothesis

$$H_0: \beta_1 = \beta_2 = \cdots = \beta_p = 0$$

If H_0 is true, this statistic has an $F(p, n - p - 1)$ distribution.

The **squared multiple correlation** is given by the expression

$$R^2 = \frac{SSM}{SST}$$

and is interpreted as the proportion of the variability in the response variable y that is explained by the explanatory variables $x_1, x_2, \ldots, x_p$ in the multiple linear regression.

CHAPTER 11 EXERCISES

11.1   Banks charge different interest rates for different loans. A random sample of 2,229 loans made for the purchase of new automobiles was studied to identify variables that explain the interest rate charged. All of these loans were made directly by the bank. A multiple regression was run with interest rate as the response variable and 13 explanatory variables. (From Michael E. Staten et al., "Information costs and the organization of credit markets: a theory of indirect lending," *Economic Inquiry*, 28 (1990), pp. 508–529.)

(a) The F statistic reported is 71.34. State the null and alternative hypotheses for this statistic. Give the degrees of freedom and the P-value for this test. What do you conclude?

(b) The value of R^2 is 0.297. What percent of the variation in interest rates is explained by the 13 explanatory variables?

(c) The researchers report a t statistic for each of the regression coefficients. State the null and alternative hypotheses tested by each of these statistics. What are the degrees of freedom for these t statistics? What values of t will lead to rejection of the null hypothesis at the 5% level?

(d) The following table gives the explanatory variables and the t statistics (these are given without the sign, assuming that all tests are two-sided) for the regression coefficients. Which of the explanatory variables are significantly different from zero in this model?

Variable	b	t
Intercept	15.47	
Loan size (in dollars)	−0.0015	10.30
Length of loan (in months)	−0.906	4.20
Percent down payment	−0.522	8.35
Cosigner (0 = no, 1 = yes)	−0.009	3.02
Unsecured loan (0 = no, 1 = yes)	0.034	2.19
Total payments (borrower's monthly installment debt)	0.100	1.37
Total income (borrower's total monthly income)	−0.170	2.37
Bad credit report (0 = no, 1 = yes)	0.012	1.99
Young borrower (0 = older than 25, 1 = 25 or younger)	0.027	2.85
Male borrower (0 = female, 1 = male)	−0.001	0.89
Married (0 = no, 1 = yes)	−0.023	1.91
Own home (0 = no, 1 = yes)	−0.011	2.73
Years at current address	−0.124	4.21

(e) The signs of many of these coefficients are what we might expect before looking at the data. For example, the negative coefficient of loan size means that larger loans get a smaller interest rate. This is very reasonable. Examine the signs of each of the statistically significant coefficients and give a short explanation of what they mean.

11.2 Refer to the previous exercise. The researchers also looked at loans made indirectly, that is, through an auto dealer. They studied 5,664 indirect loans. Multiple regression was used to predict the interest rate using the same set of explanatory variables.

(a) The F statistic reported is 27.97. State the null and alternative hypotheses for this statistic. Give the degrees of freedom and the P-value for this test. What do you conclude?

(b) The value of R^2 is 0.141. What percent of the variation in interest rates is explained by the 13 explanatory variables? Compare this value with the percent explained for direct loans in the previous exercise.

(c) The researchers report a t statistic for each of the regression coefficients. State the null and alternative hypotheses tested by each of these statistics. What are the degrees of freedom for these t statistics? What values of t will lead to rejection of the null hypothesis at the 5% level?

(d) The following table gives the explanatory variables and the t statistics (these are given without the sign, assuming that all tests are two-sided) for the regression coefficients. Which of the explanatory variables are significantly different from zero in this model?

Variable	b	t
Intercept	15.89	
Loan size (in dollars)	−0.0029	17.40
Length of loan (in months)	−1.098	5.63
Percent down payment	−0.308	4.92
Cosigner (0 = no, 1 = yes)	−0.001	1.41
Unsecured loan (0 = no, 1 = yes)	0.028	2.83
Total payments (borrower's monthly installment debt)	−0.513	1.37
Total income (borrower's total monthly income)	0.078	0.75
Bad credit report (0 = no, 1 = yes)	0.039	1.76
Young borrower (0 = older than 25, 1 = 25 or younger)	−0.036	1.33
Male borrower (0 = female, 1 = male)	−0.179	1.03
Married (0 = no, 1 = yes)	−0.043	1.61
Own home (0 = no, 1 = yes)	−0.047	1.59
Years at current address	−0.086	1.73

(e) The signs of many of these coefficients are what we might expect before looking at the data. For example, the negative coefficient of loan size means that larger loans get a smaller interest rate. This is very reasonable. Examine the signs of each of the statistically significant coefficients and give a short explanation of what they mean.

11.3 Refer to the previous two exercises. The authors conclude that banks take higher risks with indirect loans because they do not take into account borrower characteristics when setting the loan rate. Explain how the results of the multiple regressions lead to this conclusion.

11.4 Refer to the educational data for 78 seventh-grade students given in Table 1.6 (page 33). We view GPA as the response variable. IQ, gender, and self-concept are the explanatory variables.

(a) Find the correlation between GPA and each of the explanatory variables. What percent of the total variation in student GPAs can be explained by the straight-line relationship for each of the explanatory variables?

(b) The importance of IQ in explaining GPA is not surprising. The purpose of the study is to assess the influence of self-concept on GPA. So we will include IQ in the regression model and ask, "How much does self-concept contribute to explaining GPA after the effect of IQ on GPA is taken into account?" Give a model that can be used to answer this question.

(c) Run the model and report the fitted regression equation. What percent of the variation in GPA is explained by the explanatory variables in your model?

(d) Translate the question of interest into appropriate null and alternative hypotheses about the model parameters. Give the value of the test statistic and its P-value. Write a short summary of your analysis with an emphasis on your conclusion.

11.5 One model for subpopulation means for the computer science study is described in Example 11.1 (page 712) as

$$\mu_{GPA} = \beta_0 + \beta_1 HSM + \beta_2 HSS + \beta_3 HSE$$

(a) Give the model for the subpopulation mean GPA for students having high school grade scores HSM = 9 (A−), HSS = 8 (B+), and HSE = 7 (B).

(b) Using the parameter estimates given in Figure 11.4 (page 722), calculate the estimate of this subpopulation mean. Then briefly explain in words what your numerical answer means.

11.6 Use the model given in the previous exercise to do the following:

(a) For students having high school grade scores HSM = 6 (B−), HSS = 7 (B), and HSE = 8 (B+), express the subpopulation mean in terms of the parameters β_j.

(b) Calculate the estimate of this subpopulation mean using the b_j given in Figure 11.4 (page 722). Briefly explain the meaning of the number you obtain.

11.7 Consider the regression problem of predicting GPA from the three high school grade variables. The computer output appears in Figure 11.4 (page 722).

(a) Use information given in the output to calculate a 95% confidence interval for the regression coefficient of HSM. Explain in words what this regression coefficient means in the context of this particular model.

(b) Do the same for the explanatory variable HSE.

11.8 Computer output for the regression run for predicting GPA from the high school grade variables HSM and HSE appears in Figure 11.6 (page 725). Answer (a) and (b) of the previous exercise for this regression. Why are the results for this exercise different from those that you calculated for the previous exercise?

11.9 Consider the regression problem of predicting GPA from the three high school grade variables. The computer output appears in Figure 11.4 (page 722).

(a) Write the estimated regression equation.

(b) What is the value of s, the estimate of σ?

(c) State the H_0 and H_a tested by the ANOVA F statistic for this problem. After stating the hypotheses in symbols, explain them in words.

(d) What is the distribution of the F statistic under H_0? What conclusion do you draw from the F test?

(e) What percent of the variation in GPA is explained by these three high school grade variables?

11.10 Figure 11.7 (page 727) gives the output for the regression analysis in which the two SAT scores are used to predict GPA. Answer the questions in the previous exercise for this analysis.

11.11   Multiple regressions are sometimes used in litigation. In the case of *Cargill, Inc. v. Hardin,* the prosecution charged that the cash price of wheat was manipulated in violation of the Commodity Exchange Act. In a statistical study conducted for this case, a multiple regression model was constructed to predict the price of wheat using three supply-and-demand explanatory variables. Data for 14 years were used to construct the regression equation, and a prediction for the suspect period was computed from this equation. The value of R^2 was 0.989. The predicted value was reported as $2.136 with a standard error of $0.013. Express the prediction as an interval. (The degrees of freedom were large for this analysis, so use 100 as the df to determine t^*.) The actual price for the period in question was $2.13. The judge in this case decided that the analysis provided evidence that the price was not artificially depressed, and the opinion was sustained by the court of appeals. Write a short summary of the results of the analysis that relate to the decision and explain why you agree or disagree with it. (A description of this case as well as other examples of the use of statistics in legal settings is given in Michael O. Finkelstein, *Quantitative Methods in Law,* Free Press, New York, 1978.)

The following eight exercises are related to the case study described in this chapter. They require use of the CSDATA data set described in the Data Appendix.

11.12   Use software to make a plot of GPA versus SATM. Do the same for GPA versus SATV. Describe the general patterns. Are there any unusual values?

11.13   Make a plot of GPA versus HSM. Do the same for the other two high school grade variables. Describe the three plots. Are there any outliers or influential points?

11.14   Regress GPA on the three high school grade variables. Calculate and store the residuals from this regression. Plot the residuals versus each of the three predictors and versus the predicted value of GPA. Are there any unusual points or patterns in these four plots?

11.15   Use the two SAT scores in a multiple regression to predict GPA. Calculate and store the residuals. Plot the residuals versus each of the explanatory variables and versus the predicted GPA. Describe the plots.

11.16   It appears that the mathematics explanatory variables are strong predictors of GPA in the computer science study. Run a multiple regression using HSM and SATM to predict GPA.

(a) Give the fitted regression equation.

(b) State the H_0 and H_a tested by the ANOVA F statistic, and explain their meaning in plain language. Report the value of the F statistic, its P-value, and your conclusion.

(c) Give 95% confidence intervals for the regression coefficients of HSM and SATM. Do either of these include the point 0?

(d) Report the t statistics and P-values for the tests of the regression coefficients of HSM and SATM. What conclusions do you draw from these tests?

(e) What is the value of s, the estimate of σ?

(f) What percent of the variation in GPA is explained by HSM and SATM in your model?

11.17 How well do verbal variables predict the performance of computer science students? Perform a multiple regression analysis to predict GPA from HSE and SATV. Summarize the results and compare them with those obtained in the previous exercise. In what ways do the regression results indicate that the mathematics variables are better predictors?

11.18 The variable SEX has the value 1 for males and 2 for females. Create a data set containing the values for males only. Run a multiple regression analysis for predicting GPA from the three high school grade variables for this group. Using the case study in the text as a guide, interpret the results and state what conclusions can be drawn from this analysis. In what way (if any) do the results for males alone differ from those for all students?

11.19 Refer to the previous exercise. Perform the analysis using the data for females only. Are there any important differences between female and male students in predicting GPA?

The following 3 exercises use the CONCEPT data set described in the Data Appendix.

11.20 Find the correlations between the response variable GPA and each of the explanatory variables IQ, AGE, SEX, SC, and C1 to C6. Of all the explanatory variables, IQ does the best job of explaining GPA in a simple linear regression with only one variable. How do you know this *without* doing all of the regressions? What percent of the variation in GPA can be explained by the straight-line relationship between GPA and IQ?

11.21 Let us look at the role of positive self-concept about one's physical appearance (variable C3). We include IQ in the model because it is a known important predictor of GPA.

(a) Report the fitted regression model

$$\widehat{GPA} = b_0 + b_1 IQ + b_2 C3$$

with its R^2, and report the t statistic for significance of self-concept about one's physical appearance with its P-value. Does C3 contribute significantly to explaining GPA when added to IQ? How much does adding C3 to the model raise R^2?

(b) Now start with the model that includes overall self-concept SC along with IQ. Does C3 help explain GPA significantly better? Does it raise R^2 enough to be of practical value? In answering these questions, report the fitted regression model

$$\widehat{GPA} = b_0 + b_1 IQ + b_2 C3 + b_3 SC$$

with its R^2 and the t statistic for significance of C3 with its P-value.

(c) Explain carefully in words why the coefficient b_2 for the variable C3 takes quite different values in the regressions of parts (a) and (b). Then explain simply how it can happen that C3 is a useful explanatory variable in part (a) but worthless in part (b).

11.22 A reasonable model explains GPA using IQ, C1 (behavior self-concept), and C5 (popularity self-concept). These three explanatory variables are all significant in the presence of the other two, and no other explanatory variable is significant when added to these three. (You do not have to verify these statements.) Let's use this model.

(a) What is the fitted regression model, its R^2, and the standard deviation s about the fitted model? What GPA does this model predict for a student with IQ 109, C1 score 13, and C5 score 8?

(b) If both C1 and C5 could be held constant, how much would GPA increase for each additional point of IQ score, according to the fitted model? Give a 95% confidence interval for the mean increase in GPA in the entire population if IQ could be increased by 1 point while holding C1 and C5 constant.

(c) Compute the residuals for this model. Plot the residuals against the predicted values and also make a normal quantile plot. Which observation (value of OBS) produces the most extreme residual? Circle this observation in both of your plots of the residuals. For which individual variables does this student have very unusual values? What are these values? (Look at all the variables, not just those in the model.)

(d) Repeat part (a) with this one observation removed. How much did this one student affect the fitted model and the prediction?

The following 9 exercises use the CHEESE data set described in the Data Appendix.

11.23 For each of the four variables in the CHEESE data set, find the mean, median, standard deviation, and interquartile range. Display each distribution by means of a stemplot and use a normal quantile plot to assess normality of the data. Summarize your findings. Note that when doing regressions with these data, we do not assume that these distributions are normal. Only the residuals from our model need be (approximately) normal. The careful study of each variable to be analyzed is nonetheless an important first step in any statistical analysis.

11.24 Make a scatterplot for each pair of variables in the CHEESE data set (you will have six plots). Describe the relationships. Calculate the correlation for each pair of variables and report the P-value for the test of zero population correlation in each case.

11.25 Perform a simple linear regression analysis using Taste as the response variable and Acetic as the explanatory variable. Be sure to examine the residuals carefully. Summarize your results. Include a plot of the data with the least-squares regression line. Plot the residuals versus each of the other two chemicals. Are any patterns evident? (The concentrations of the other chemicals are lurking variables for the simple linear regression.)

11.26   Repeat the analysis of Exercise 11.25 using Taste as the response variable and H2S as the explanatory variable. $Taste = b_0 + b_1 H2S$

11.27   Repeat the analysis of Exercise 11.25 using Taste as the response variable and Lactic as the explanatory variable.

11.28   Compare the results of the regressions performed in the three previous exercises. Construct a table with values of the F statistic, its P-value, R^2, and the estimate s of the standard deviation for each model. Report the three regression equations. Why are the intercepts in these three equations different?

11.29   Carry out a multiple regression using Acetic and H2S to predict Taste. Summarize the results of your analysis. Compare the statistical significance of Acetic in this model with its significance in the model with Acetic alone as a predictor (Exercise 11.25). Which model do you prefer? Give a simple explanation for the fact that Acetic alone appears to be a good predictor of Taste but with H2S in the model, it is not.   Is it significant to add H2S as a variable?

11.30   Carry out a multiple regression using H2S and Lactic to predict Taste. Comparing the results of this analysis with the simple linear regressions using each of these explanatory variables alone, it is evident that a better result is obtained by using both predictors in a model. Support this statement with explicit information obtained from your analysis.

11.31   Use the three explanatory variables Acetic, H2S, and Lactic in a multiple regression to predict Taste. Write a short summary of your results, including an examination of the residuals. Based on all of the regression analyses you have carried out on these data, which model do you prefer and why?

The following 7 exercises use the corn and soybean data given in Tables 10.2 (page 696) and 10.3 (page 703).

11.32   Run the simple linear regression using year to predict corn yield.

   (a) Summarize the results of your analysis, including the significance test results for the slope and R^2 for this model.

   (b) Analyze the residuals with a normal quantile plot. Is there any indication in the plot that the residuals are not normal?

   (c) Plot the residuals versus soybean yield. Does the plot indicate that soybean yield might be useful in a multiple linear regression with year to predict corn yield?

11.33   Run the simple linear regression using soybean yield to predict corn yield.

   (a) Summarize the results of your analysis, including the significance test results for the slope and R^2 for this model.

   (b) Analyze the residuals with a normal quantile plot. Is there any indication in the plot that the residuals are not normal?

   (c) Plot the residuals versus year. Does the plot indicate that year might be useful in a multiple linear regression with soybean yield to predict corn yield?

11.34 From the previous two exercises, we conclude that year *and* soybean yield may be useful together in a model for predicting corn yield. Run this multiple regression.

(a) Explain the results of the ANOVA F test. Give the null and alternative hypotheses, the test statistic with degrees of freedom, and the P-value. What do you conclude?

(b) What percent of the variation in corn yield is explained by these two variables? Compare it with the percent explained in the simple linear regression models of the previous two exercises.

(c) Give the fitted model. Why do the coefficients of year and soybean yield differ from those in the previous two exercises?

(d) Summarize the significance test results for the regression coefficients for year and soybean yield.

(e) Give a 95% confidence interval for each of these coefficients.

(f) Plot the residuals versus year and versus soybean yield. What do you conclude?

11.35 We need a new variable to model the curved relation that we see between corn yield and year in the residual plot of the last exercise. Let year2 $=$ (year $-$ 1976.5)^2. (When adding a quadratic (squared) term to a multiple regression model, it is a good idea to subtract the mean of the variable being squared before squaring. This avoids all sorts of messy problems that we cannot discuss here.)

(a) Run the multiple linear regression using year, year2, and soybean yield to predict corn yield. Give the fitted regression equation.

(b) Give the null and alternative hypotheses for the ANOVA F test. Report the results of this test, giving the test statistic, degrees of freedom, P-value, and conclusion.

(c) What percent of the variation in corn yield is explained by this multiple regression? Compare this with the model in the previous exercise.

(d) Summarize the results of the significance tests for the individual regression coefficients.

(e) Analyze the residuals and summarize your conclusions.

11.36 Run the model to predict corn yield using year and the squared term year2 defined in the previous exercise.

(a) Summarize the significance test results.

(b) The coefficient of year2 is not statistically significant in this run, but it was highly significant in the model analyzed in the previous exercise. Explain how this can happen.

(c) Obtain the fitted values for each year in the data set and use these to sketch the curve on a plot of the data. Plot the least-squares line on this graph for comparison. Describe the differences between the two regression functions. For what years do they give very similar fitted values; for what years are the differences between the two relatively large?

11.37 Use the simple linear regression model with corn yield as the response variable and year as the explanatory variable to predict the corn yield for the year 2006, and give the 95% prediction interval. Also use the multiple regression model where year and year2 are both explanatory variables to find another predicted value with the 95% interval. Explain why these two predicted values are so different.

11.38 Repeat the previous exercise doing the prediction for 1997, one year after the last available data point. Compare the results of this exercise with the previous one. Why are they different?

The following 3 exercises use the bank wages data given in Table 10.1 (page 695). For these exercises we code the size of the bank as 1 if it is large and 0 if it is small. There is one outlier in this data set. Delete it and use the remaining 59 observations in these exercises.

11.39 Use length of service (LOS) to predict wages with a simple linear regression. Write a short summary of your results and conclusions.

11.40 Predict wages using the size of the bank as the explanatory variable. Use the coded values 0 and 1 for this model.

    (a) Summarize the results of your analysis. Include a statement of all hypotheses, test statistics with degrees of freedom, P-values, and conclusions.

    (b) Calculate the t statistic for comparing the mean wages for the large and the small banks assuming equal standard deviations. Give the degrees of freedom. Verify that this t is the same as the t statistic for the coefficient of size in the regression. Explain why this makes sense.

    (c) Plot the residuals versus LOS. What do you conclude?

11.41 Use a multiple linear regression to predict wages from LOS and the size of the bank. Write a report summarizing your work. Include graphs and the results of significance tests.

NOTES

    1. Results of the study are reported in P. F. Campbell and G. P. McCabe, "Predicting the success of freshmen in a computer science major," *Communications of the ACM,* 27 (1984), pp. 1108–1113.

    2. This example describes an analysis of real data performed by Christine Smiley of Kestnbaum & Company, Chicago, Illinois.

    3. For more information on logistic regression see Chapter 14 of J. Neter et al., *Applied Linear Statistical Models,* 4th ed., Irwin, Chicago, 1996.

Prelude

Many of the most effective statistical studies are comparative, whether we compare the salaries of women and men in a sample of firms or the responses to two treatments in a clinical trial. We display comparisons with back-to-back stemplots or side-by-side boxplots, and we measure them with five-number summaries or means and standard deviations. Now we ask: "Is the difference between groups statistically significant?"

When only two groups are compared, Chapter 7 provides the tools we need. Two-sample t procedures compare the means of two normal populations, and we saw that these procedures, unlike comparisons of spread, are sufficiently robust to be widely useful. Now we will compare any number of means by techniques that generalize the two-sample t and share its robustness and usefulness.

- Do three methods for teaching reading in second grade differ in effectiveness? Do children taught by any of these methods learn significantly faster than children taught by the other methods?

- Careful observation of small samples of young children in Egypt, Kenya, and Mexico shows that the average food intake in the samples is different. Egyptian children average 1217 calories daily, while Mexican children receive 1119 and Kenyan children only 844 calories. Are these just chance differences, or are they statistically significant? Like the comparison of methods of teaching reading, this is a one-way design; countries or teaching methods are just lined up for comparison.

Clinton Hill, *Hotflash*, 1991.

One-Way Analysis of Variance

Which of four advertising offers mailed to sample households produces the highest sales in dollars? Which of ten brands of automobile tires wears longest? How long do cancer patients live under each of three therapies for their cancer? In each of these settings we wish to compare several treatments. In each case the data are subject to sampling variability—if we mailed the advertising offers to another set of households, we would get different data. We therefore pose the question for inference in terms of the *mean* response. We compare, for example, the mean tread lifetime of different brands of tires. In Chapter 7 we met procedures for comparing the means of two populations. We are now ready to extend those methods to problems involving more than two populations. The statistical methodology for comparing several means is called **analysis of variance,** or simply **ANOVA.**

ANOVA

one–way ANOVA

We will consider two ANOVA techniques. When there is only one way to classify the populations of interest, we use **one-way ANOVA** to analyze the data. For example, to compare the tread lifetimes of ten specific brands of tires we use one-way ANOVA. This chapter presents the details for one-way ANOVA.

In many practical situations there is more than one way to classify the populations. A mail-order firm might want to compare mailings that offer different discounts and also have different layouts. Will a lower price offered in a plain format draw more sales on the average than a higher price offered in a fancy brochure? Analyzing the effect of price and layout together requires **two-way ANOVA,** and adding yet more factors necessitates higher-way ANOVA techniques. Most of the new ideas in ANOVA with more than one factor already appear in two-way ANOVA, which we discuss in Chapter 13.

two–way ANOVA

Data for a one–way ANOVA

Do two population means differ? If we have random samples from the two populations, we compute a two-sample t statistic and its P-value to assess the statistical significance of the difference in the sample means. We compare several means in much the same way. Instead of a t statistic, ANOVA uses an F statistic and its P-value to evaluate the null hypothesis that all of several population means are equal.

In the sections that follow, we will examine the basic ideas and assumptions that are needed for the analysis of variance. Although the details differ, many of the concepts are similar to those discussed in the two-sample case.

One-way analysis of variance is a statistical method for comparing several population means. We draw a simple random sample (SRS) from each population and use the data to test the null hypothesis that the population means are all equal.

EXAMPLE 12.1

A medical researcher wants to compare the effectiveness of three different treatments to lower the cholesterol of patients with high blood cholesterol levels. He assigns 60 individuals at random to the three treatments (20 to each) and records the reduction in cholesterol for each patient.

EXAMPLE 12.2 | An ecologist is interested in comparing the concentration of the pollutant cadmium in five streams. She collects 50 water specimens from each stream and measures the concentration of cadmium in each specimen.

These two examples have several similarities. In both there is a single response variable measured on several units; the units are people in the first example and water specimens in the second. We expect the data to be approximately normal and will consider a transformation if they are not. We wish to compare several populations, three treatments in the first example and five streams in the second. But there is also an important difference between the examples. The first is an **experiment** in which patients are randomly assigned to treatments. The second example is an **observational study.** By taking water specimens at randomly chosen locations, the ecologist tries to ensure that her specimens from each stream are a random sample of all water specimens that could be obtained from that stream. In both cases we can use ANOVA to compare the mean responses. The same ANOVA methods apply to data from random samples and randomized experiments. Do keep the data-production method in mind when interpreting the results, however. A strong case for causation is best made by a randomized experiment. We will often use the term *groups* for the populations to be compared in a one-way ANOVA.

experiment
observational study

Comparing means

To assess whether several populations all have the same mean, we compare the means of samples drawn from each population. Figure 12.1 displays the sample means for Example 12.1. Treatment 2 had the highest average cholesterol reduction. But is the observed difference among the groups just the result of chance variation? We do not expect sample means to be equal even if the population means are all identical. The purpose of ANOVA is to assess whether the observed differences among sample means are *statistically*

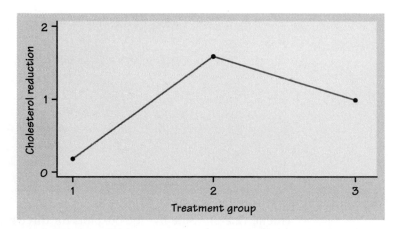

FIGURE 12.1 Mean serum cholesterol reduction in three groups.

significant. In other words, could a variation this large be plausibly due to chance, or is it good evidence for a difference among the population means? This question can't be answered from the sample means alone. Because the standard deviation of a sample mean $\bar{x}$ is the population standard deviation σ divided by $\sqrt{n}$, the answer depends upon both the variation within the groups of observations and the sizes of the samples.

Side-by-side boxplots help us see the within-group variation. Compare Figures 12.2(a) and 12.2(b). The sample medians are the same in both figures, but the large variation within the groups in Figure 12.2(a) suggests that the

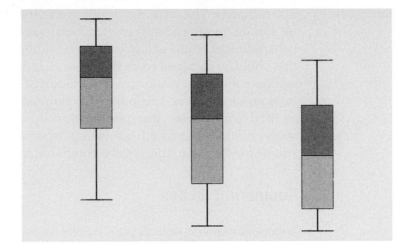

FIGURE 12.2(a)   Side-by-side boxplots for three groups with large within-group variation. The differences among centers may be just chance variation.

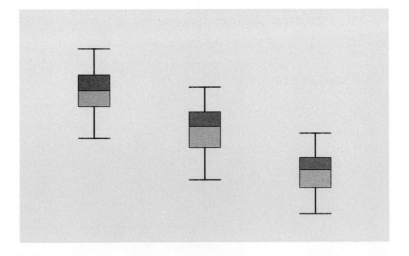

FIGURE 12.2(b)   Side-by-side boxplots for three groups with the same centers as in Figure 12.2(a) but with small within-group variation. The differences among centers are more likely to be significant.

differences among the sample medians could be due simply to chance variation. The data in Figure 12.2(b) are much more convincing evidence that the populations differ. Even the boxplots omit essential information, however. To assess the observed differences, we must also know how large the samples are. Nonetheless, boxplots are a good preliminary display of ANOVA data. (ANOVA compares means, and boxplots display medians. If the distributions are nearly symmetric, these two measures of center will be close together.)

The two–sample *t* statistic

Two-sample t statistics compare the means of two populations. If the two populations are assumed to have equal but unknown standard deviations and the sample sizes are both equal to n, the t statistic is (page 551)

$$t = \frac{\bar{x} - \bar{y}}{s_p \sqrt{\dfrac{1}{n} + \dfrac{1}{n}}} = \frac{\sqrt{\dfrac{n}{2}}(\bar{x} - \bar{y})}{s_p}$$

The square of this t statistic is

$$t^2 = \frac{\dfrac{n}{2}(\bar{x} - \bar{y})^2}{s_p^2}$$

If we use ANOVA to compare two populations, the ANOVA F statistic is exactly equal to this t^2. We can therefore learn something about how ANOVA works by looking carefully at the statistic in this form.

between–group variation

within–group variation

The numerator in the t^2 statistic measures the variation **between** the groups in terms of the difference between their sample means $\bar{x}$ and $\bar{y}$. It includes a factor for the common sample size n. The numerator can be large because of a large difference between the sample means or because the sample sizes are large. The denominator measures the variation **within** groups by s_p^2, the pooled estimator of the common variance. If the within-group variation is small, the same variation between the groups produces a larger statistic and a more significant result.

Although the general form of the F statistic is more complicated, the idea is the same. To assess whether several populations all have the same mean, we compare the variation *among* the means of several groups with the variation *within* groups. Because we are comparing variation, the method is called *analysis of variance*.

ANOVA hypotheses

ANOVA tests the null hypothesis that the population means are *all* equal. The alternative is that they are not all equal. This alternative could be true because all of the means are different, or simply because one of them differs from the rest. This is a more complex situation than comparing just two populations. If we reject the null hypothesis, we need to perform some further analysis to draw conclusions about which population means differ from which others.

The computations needed for an ANOVA are more lengthy than those for the t test. For this reason we generally use computer programs to perform the calculations. Automating the calculations frees us from the burden of arithmetic and allows us to concentrate on interpretation. The following example illustrates the practical use of ANOVA in analyzing data. Later we will explore the technical details.

EXAMPLE 12.3

In the computer science department of a large university, many students change their major after the first year. A detailed study[1] of the 256 students enrolled as first-year computer science majors in one year was undertaken to help understand this phenomenon. Students were classified on the basis of their status at the beginning of their second year, and several variables measured at the time of their entrance to the university were obtained. Here are summary statistics for the SAT mathematics scores:

Second-year major	n	$\overline{x}$	s
Computer science	103	619	86
Engineering and other sciences	31	629	67
Other	122	575	83

Figure 12.3 gives side-by-side boxplots of the SAT data. Compare Figure 12.3 with the plot of the mean scores in Figure 12.4. The means appear to be different, but there is a large amount of overlap in the three distributions. When we perform an ANOVA on these data, we ask a question about the group means. The null hypothesis is that the population mean SAT scores for the three groups are equal, and the alternative is that they are not all equal. The report on the study states that the ANOVA F statistic is 10.35 with $P < 0.001$. There is very strong evidence that the three groups of students do not all have the same mean SAT mathematics scores.

Although we rejected the null hypothesis, we have not demonstrated that all three population means are different. Inspection of Figures 12.3 and 12.4 suggests that the means of the first two groups are nearly equal and that the third group differs from these two. We need to do some additional analysis to clarify the situation. The researchers expected the mathematics scores of the engineering and other sciences group to be similar to those of the computer science group. They suspected that the students in the third (other major) group would have lower mathematics scores, because these students chose to transfer to programs of study that require less mathematics. The researchers had this specific comparison in mind before seeing the data, so they used

contrasts

contrasts to assess the significance of this specific relation among the means. If no specific relations among the means are in mind before looking at the data, we instead use a **multiple comparisons**

multiple comparisons procedure to determine which pairs of population means differ significantly. In later sections we will explore both contrasts and multiple comparisons in detail.

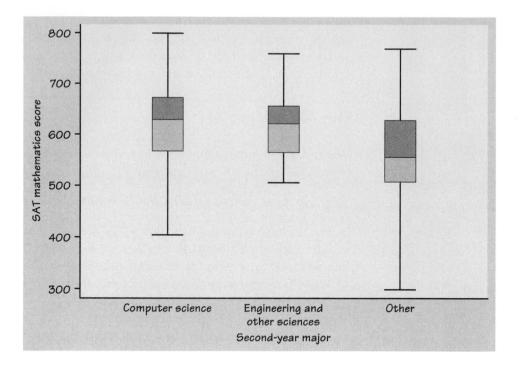

FIGURE 12.3 Side-by-side boxplots of SAT mathematics scores for the change-of-majors study.

A purist might argue that ANOVA is inappropriate in Example 12.3. *All* first-year computer science majors in this university for the year studied were included in the analysis. There was no random sampling from larger populations. On the other hand, we can think of these three groups of students as samples from three populations of students in similar circumstances. They

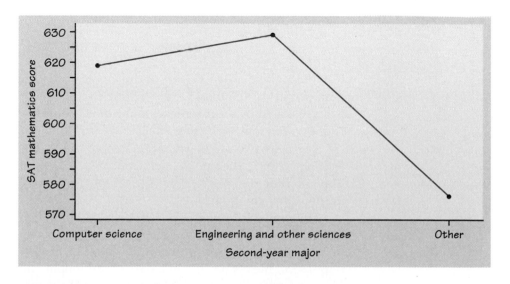

FIGURE 12.4 SAT mathematics score means for the change-of-majors study.

may be representative of students in the next few years at the same university or other universities with similar programs. Judgments such as this are very important and must be made not by statisticians but by people who are knowledgeable about the subject of the study.

The ANOVA model

When analyzing data, the following equation reminds us that we look for an overall pattern and deviations from it:

$$\text{DATA} = \text{FIT} + \text{RESIDUAL}$$

In the regression model of Chapter 10, the FIT was the population regression line and the RESIDUAL represented the deviations of the data from this line. We now apply this framework to describe the statistical models used in ANOVA. These models provide a convenient way to summarize the assumptions that are the foundation for our analysis. They also give us the necessary notation to describe the calculations needed.

First, recall the statistical model for a random sample of observations from a single normal population with mean μ and standard deviation σ. If the observations are

$$x_1, x_2, \ldots, x_n$$

we can describe this model by saying that the x_j are an SRS from the $N(\mu, \sigma)$ distribution. Another way to describe the same model is to think of the x's varying about their population mean. To do this, write each observation x_j as

$$x_j = \mu + \epsilon_j$$

The ϵ_j are then an SRS from the $N(0, \sigma)$ distribution. Because μ is unknown, the ϵ's cannot actually be observed. This form more closely corresponds to our

$$\text{DATA} = \text{FIT} + \text{RESIDUAL}$$

way of thinking. The FIT part of the model is represented by μ. It is the systematic part of the model, like the line in a regression. The RESIDUAL part is represented by ϵ_j. It represents the deviations of the data from the fit and is due to random, or chance, variation.

There are two unknown parameters in this statistical model: μ and σ. We estimate μ by $\overline{x}$, the sample mean, and σ by s, the sample standard deviation. The differences $e_j = x_j - \overline{x}$ are the sample residuals and correspond to the ϵ_j in the statistical model.

The model for one-way ANOVA is very similar. We take random samples from each of I different populations. The sample size is n_i for the ith population. Let x_{ij} represent the jth observation from the ith population. The I population means are the FIT part of the model and are represented by μ_i. The random variation, or RESIDUAL, part of the model is represented by the deviations ϵ_{ij} of the observations from the means.

> ### The One-Way ANOVA Model
>
> The **one-way ANOVA model** is
>
> $$x_{ij} = \mu_i + \epsilon_{ij}$$
>
> for $i = 1, \ldots, I$ and $j = 1, \ldots, n_i$. The ϵ_{ij} are assumed to be from an $N(0, \sigma)$ distribution. The **parameters of the model** are the population means $\mu_1, \mu_2, \ldots, \mu_I$ and the common standard deviation σ.

Note that the sample sizes n_i may differ, but the standard deviation σ is assumed to be the same in all of the populations. Figure 12.5 pictures this model for $I = 3$. The three population means μ_i are different, but the shapes of the three normal distributions are the same, reflecting the assumption that all three populations have the same standard deviation.

EXAMPLE 12.4

A survey of college students attempted to determine how much money they spend per year on textbooks and to compare students by year of study. From lists of students provided by the registrar, SRSs of size 50 were chosen from each of the four classes (freshmen, sophomores, juniors, and seniors). The students selected were asked how much they spent on textbooks during the current semester.

There are $I = 4$ populations. The population means μ_1, μ_2, μ_3, and μ_4 are the average amounts spent on textbooks by *all* freshmen, sophomores, juniors, and seniors at this college for this semester. The sample sizes n_i are 50, 50, 50, and 50.

Suppose the first freshman sampled is Eve Brogden. The observation $x_{1,1}$ is the amount spent by Eve. The data for the other freshmen sampled are denoted by $x_{1,2}, x_{1,3}, \ldots, x_{1,50}$. Similarly, the data for the other groups have a first subscript indicating the group and a second subscript indicating the student in the group.

According to our model, Eve's spending is $x_{1,1} = \mu_1 + \epsilon_{1,1}$, where μ_1 is the average for *all* of the students in the freshman class and $\epsilon_{1,1}$ is the chance variation due to Eve's specific needs. We are assuming that the ϵ_{ij} are independent and normally distributed (at least approximately) with mean 0 and standard deviation σ.

Estimates of population parameters

The unknown parameters in the statistical model for ANOVA are the I population means μ_i and the common population standard deviation σ. To estimate μ_i we use the sample mean for the ith group

$$\overline{x}_i = \frac{1}{n_i} \sum_{j=1}^{n_i} x_{ij}$$

The residuals $e_{ij} = x_{ij} - \overline{x}_i$ reflect the variation about the sample means that we see in the data.

The ANOVA model assumes that the population standard deviations are all equal. If we have unequal standard deviations, we generally try to transform the data so that they are approximately equal. We might, for example, work with $\sqrt{x_{ij}}$ or $\log x_{ij}$. Fortunately, we can often find a transformation that *both* makes the group standard deviations more nearly equal and also makes the distributions of observations in each group more nearly normal. If the

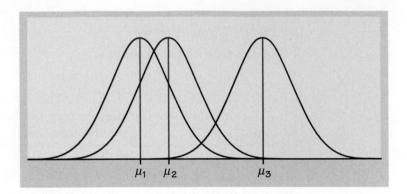

FIGURE 12.5 Model for one-way ANOVA with three groups. The three populations have normal distributions with the same standard deviation.

standard deviations are markedly different and cannot be made similar by a transformation, inference requires different methods.

Unfortunately, formal tests for the equality of standard deviations in several groups share the lack of robustness against nonnormality that we noted in Chapter 7 for the case of two groups. Because ANOVA procedures are not extremely sensitive to unequal standard deviations, we do *not* recommend a formal test of equality of standard deviations as a preliminary to the ANOVA. Instead, we will use the following rule of thumb.

Rule for Examining Standard Deviations in ANOVA

If the largest standard deviation is less than twice the smallest standard deviation, we can use methods based on the assumption of equal standard deviations and our results will still be approximately correct.[2]

When we assume that the population standard deviations are equal, each sample standard deviation is an estimate of σ. To combine these into a single estimate, we use a generalization of the pooling method introduced in Chapter 7.

Pooled Estimator of σ

Suppose we have sample variances $s_1^2, s_2^2, \ldots, s_I^2$ from I independent SRSs of sizes $n_1, n_2, \ldots, n_I$ from populations with common variance σ^2. The **pooled sample variance**

$$s_p^2 = \frac{(n_1 - 1)s_1^2 + (n_2 - 1)s_2^2 + \cdots + (n_I - 1)s_I^2}{(n_1 - 1) + (n_2 - 1) + \cdots + (n_I - 1)}$$

is an unbiased estimator of σ^2. The **pooled standard deviation**

$$s_p = \sqrt{s_p^2}$$

is the estimate of σ.

Pooling gives more weight to groups with larger sample sizes. If the sample sizes are equal, s_p^2 is just the average of the I sample variances. Note that s_p is *not* the average of the I sample standard deviations.

EXAMPLE 12.5

In the change-of-majors study of Example 12.3 there are $I = 3$ groups and the sample sizes are $n_1 = 103, n_2 = 31$, and $n_3 = 122$. The sample means for the SAT mathematics scores are $\bar{x}_1 = 619, \bar{x}_2 = 629$, and $\bar{x}_3 = 575$. The sample standard deviations are $s_1 = 86, s_2 = 67$, and $s_3 = 83$.

Because the largest standard deviation (86) is less than twice the smallest ($2 \times 67 = 134$), our rule of thumb indicates that we can use the assumption of equal population standard deviations.

The pooled variance estimate is

$$s_p^2 = \frac{(n_1 - 1)s_1^2 + (n_2 - 1)s_2^2 + (n_3 - 1)s_3^2}{(n_1 - 1) + (n_2 - 1) + (n_3 - 1)}$$

$$= \frac{(102)(86)^2 + (30)(67)^2 + (121)(83)^2}{102 + 30 + 121}$$

$$= \frac{1,722,631}{253} = 6809$$

The pooled standard deviation is

$$s_p = \sqrt{6809} = 82.5$$

This is our estimate of the common standard deviation σ of the SAT mathematics scores in the three populations of students.

Testing hypotheses in one-way ANOVA

Comparison of several means is accomplished by using an F statistic to compare the variation among groups with the variation within groups. We now show how the F statistic expresses this comparison. Calculations are organized in an **ANOVA table,** which contains numerical measures of the variation among groups and within groups.

ANOVA table

First we must specify our hypotheses for one-way ANOVA. As usual, I represents the number of populations to be compared.

Hypotheses for One-Way ANOVA

The **null and alternative hypotheses** for one-way ANOVA are

$$H_0 : \mu_1 = \mu_2 = \ldots = \mu_I$$
$$H_a : \text{not all of the } \mu_i \text{ are equal}$$

The example and discussion that follow illustrate how to do a one-way ANOVA. The calculations are generally performed using statistical software on a computer, so we focus on interpretation of the output.

TABLE 12.1   Pretest reading scores

Group	Subject	Score	Group	Subject	Score	Group	Subject	Score
Basal	1	4	DRTA	23	7	Strat	45	11
Basal	2	6	DRTA	24	7	Strat	46	7
Basal	3	9	DRTA	25	12	Strat	47	4
Basal	4	12	DRTA	26	10	Strat	48	7
Basal	5	16	DRTA	27	16	Strat	49	7
Basal	6	15	DRTA	28	15	Strat	50	6
Basal	7	14	DRTA	29	9	Strat	51	11
Basal	8	12	DRTA	30	8	Strat	52	14
Basal	9	12	DRTA	31	13	Strat	53	13
Basal	10	8	DRTA	32	12	Strat	54	9
Basal	11	13	DRTA	33	7	Strat	55	12
Basal	12	9	DRTA	34	6	Strat	56	13
Basal	13	12	DRTA	35	8	Strat	57	4
Basal	14	12	DRTA	36	9	Strat	58	13
Basal	15	12	DRTA	37	9	Strat	59	6
Basal	16	10	DRTA	38	8	Strat	60	12
Basal	17	8	DRTA	39	9	Strat	61	6
Basal	18	12	DRTA	40	13	Strat	62	11
Basal	19	11	DRTA	41	10	Strat	63	14
Basal	20	8	DRTA	42	8	Strat	64	8
Basal	21	7	DRTA	43	8	Strat	65	5
Basal	22	9	DRTA	44	10	Strat	66	8

EXAMPLE 12.6

A study of reading comprehension in children compared three methods of instruction.[3] As is common in such studies, several pretest variables were measured before any instruction was given. One purpose of the pretest was to see if the three groups of children were similar in their comprehension skills. One of the pretest variables was an "intruded sentences" measure, which measures one type of reading comprehension skill. The data for the 22 subjects in each group are given in Table 12.1. The three methods of instruction are called basal, DRTA, and strategies. We use Basal, DRTA, and Strat as values for the categorical variable indicating which method each student received.

Before proceeding with ANOVA, several important preliminaries need attention. Look at the normal quantile plots for the three groups in Figures 12.6(a), 12.6(b), and 12.6(c). The data look reasonably normal; we have no evidence to doubt our assumption that our residuals are normal. Although there are a small number of possible values for the score (the values are all integers between 4 and 17), this should not cause difficulties in using ANOVA. Side-by-side boxplots do not give a particularly good picture of these data. This is a consequence of the relatively small number of observations in each group (22) combined with the fact that the variable has only a few possible values.

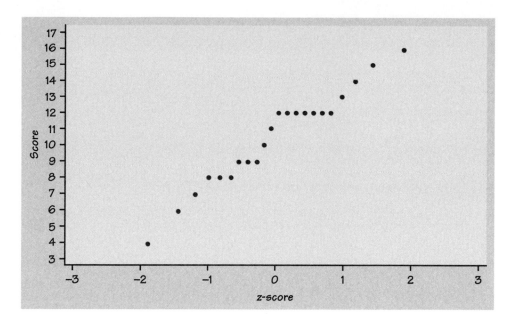

FIGURE 12.6(a) Normal quantile plot of the pretest scores in the Basal group of the reading comprehension study.

Figure 12.7 gives the summary statistics generated by SAS. Because the largest standard deviation (3.34) is less than twice the smallest ($2 \times 2.69 = 5.38$), our rule of thumb tells us that we need not be concerned about violating the assumption that the three populations have the same standard deviation.

Because the data look reasonably normal and the assumption of equal standard deviations can be used, we proceed with the analysis of variance.

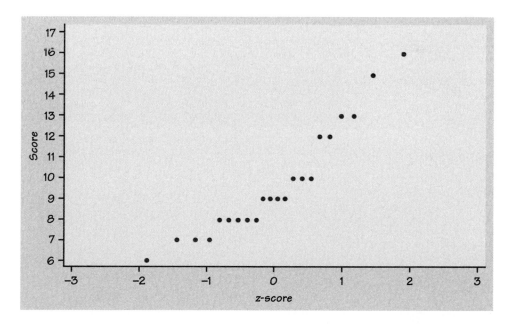

FIGURE 12.6(b) Normal quantile plot of the pretest scores in the DRTA group of the reading comprehension study.

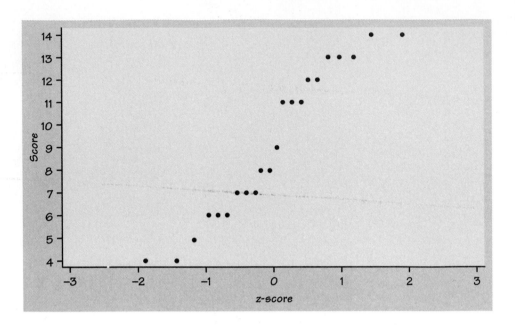

FIGURE 12.6(c)   Normal quantile plot of the pretest scores in the Strat group of the reading comprehension study.

The results produced by SAS are shown in Figure 12.8. The pooled standard deviation s_p is given as Root MSE and has the value 3.01. The calculated value of the F statistic appears under the heading F Value, and its P-value is under the heading Pr > F. The value of F is 1.13, with a P-value of 0.3288. That is, an F of 1.13 or larger would occur about 33% of the time by chance when the population means are equal. We have no evidence to reject the null hypothesis that the three populations have equal means.

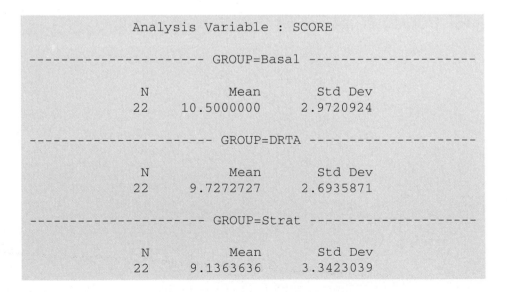

FIGURE 12.7   Summary statistics for the pretest scores in the three groups of the reading comprehension study.

```
                    General Linear Models Procedure

Dependent Variable: SCORE


                                 Sum of           Mean
Source                 DF        Squares          Square    F Value    Pr > F
Model                   2       20.575758       10.287879      1.13    0.3288
Error                  63      572.454545        9.086580
Corrected Total        65      593.030303


               R-Square              C.V.        Root MSE          SCORE Mean
               0.034696          30.79723          3.0144             9.7879
```

FIGURE 12.8 Analysis of variance output for the pretest scores in the reading comprehension study.

The ANOVA table

ANOVA table

The information in an analysis of variance is organized in an **ANOVA table.** In the computer output in Figure 12.8, the columns of this table are labeled Source, DF, Sum of Squares, Mean Square, F Value, and Pr > F. The rows are labeled Model, Error, and Corrected Total. These are the three sources of variation in the one-way ANOVA.

The Model row in the table corresponds to the FIT term in our DATA = FIT + RESIDUAL way of thinking. It gives information related to the variation

variation among groups

among group means. In writing ANOVA tables we will use the generic label "groups" or some other term that describes the factor being studied for this row.

The Error row in the table corresponds to the RESIDUAL term in our

variation within groups

DATA = FIT + RESIDUAL. It gives information related to the variation **within** groups. The term "error" is most appropriate for experiments in the physical sciences where the observations within a group differ because of measurement error. In business and the biological and social sciences, on the other hand, the within-group variation is often due to the fact that not all firms or plants or people are the same. This sort of variation is not due to errors and is better described as "residual." Finally, the Corrected Total row in the table corresponds to the DATA term in our DATA = FIT + RESIDUAL framework. Often this row is simply labeled Total. So, for analysis of variance,

$$\text{DATA} = \text{FIT} + \text{RESIDUAL}$$

translates into

$$\text{total} = \text{model} + \text{residual}$$

sum of squares

As you might expect, each **sum of squares** is a sum of squared deviations. We use SSG, SSE, and SST for the entries in this column corresponding to groups, error, and total. Each sum of squares measures a different type of variation. SST measures variation of the data around the overall mean, $x_{ij} - \overline{x}$. Variation of the group means around the overall mean $\overline{x}_i - \overline{x}$ is measured by SSG. Finally, SSE measures variation of each observation around its group mean, $x_{ij} - \overline{x}_i$.

The Sum of Squares column in Figure 12.8 gives the values for these three sums of squares. The output gives many more digits than we need. Rounded-off values are SSG = 21, SSE = 572, and SST = 593. In this example it appears that most of the variation is coming from Error, that is, from within groups. You can verify that SST = SSG + SSE for this example. This fact is true in general. The total variation is always equal to the among-group variation plus the within-group variation.

degrees of freedom

Associated with each sum of squares is a quantity called the **degrees of freedom.** Because SST measures variation of all N observations around the overall mean, its degrees of freedom are DFT = $N - 1$. This is the same as the degrees of freedom for the ordinary sample variance. Similarly, because SSG measures the variation of the I sample means around the overall mean, its degrees of freedom are DFG = $I - 1$. Finally, SSE is the sum of squares of the deviations $x_{ij} - \overline{x}_i$. Here we have N observations being compared with I sample means and DFE = $N - I$.

In our example, we have $I = 3$ and $N = 66$. Therefore,

$$\text{DFT} = N - 1 = 66 - 1 = 65$$
$$\text{DFG} = I - 1 = 3 - 1 = 2$$
$$\text{DFE} = N - I = 66 - 3 = 63$$

These are the entries in the DF column of Figure 12.8. Note that the degrees of freedom add in the same way that the sums of squares add. That is, DFT = DFG + DFE.

mean square

For each source of variation, the **mean square** is the sum of squares divided by the degrees of freedom. You can verify this by doing the divisions for the values given on the output in Figure 12.8.

Sums of Squares, Degrees of Freedom, and Mean Squares

Sums of squares represent variation present in the data. They are calculated by summing squared deviations. In the one-way ANOVA there are three **sources of variation:** groups, error, and total. The sums of squares are related by the formula

$$\text{SST} = \text{SSG} + \text{SSE}$$

Thus the total variation is composed of two parts, one due to groups and one due to error.

Degrees of freedom are related to the deviations that are used in the sums of squares. The degrees of freedom are related in the same way as the sums of squares:

$$\text{DFT} = \text{DFG} + \text{DFE}$$

To calculate each **mean square,** divide the corresponding sum of squares by its degrees of freedom.

The pooled variance for Example 12.6 is $572.45/63 = 9.09$, the mean square in the error row in Figure 12.8. It is true in general that

$$s_p^2 = \text{MSE} = \frac{\text{SSE}}{\text{DFE}}$$

That is, the error mean square is an estimate of the within-group variance, σ^2. The output gives s_p under the heading Root MSE.

The F test

If H_0 is true, there are no differences among the group means. The ratio MSG/MSE is a statistic that is approximately 1 if H_0 is true and tends to be larger if H_a is true. This is the ANOVA F statistic. In our example, MSG = 10.29 and MSE = 9.09, so the ANOVA F statistic is

$$F = \frac{\text{MSG}}{\text{MSE}} = \frac{10.29}{9.09} = 1.13$$

When H_0 is true, the F statistic has an F distribution that depends upon two numbers: the *degrees of freedom for the numerator* and the *degrees of freedom for the denominator*. These degrees of freedom are those associated with the mean squares in the numerator and denominator of the F statistic. For one way ANOVA, the degrees of freedom for the numerator are DFG = $I - 1$ and the degrees of freedom for the denominator are DFE = $N - I$. We use the notation $F(I - 1, N - I)$ for this distribution.

The ANOVA F Test

To test the null hypothesis in a one-way ANOVA, calculate the **F statistic**

$$F = \frac{\text{MSG}}{\text{MSE}}$$

When H_0 is true, the F statistic has the $F(I - 1, N - I)$ distribution. When H_a is true, the F statistic tends to be large. We reject H_0 in favor of H_a if the F statistic is sufficiently large.

The **P-value** of the F test is the probability that a random variable having the $F(I - 1, N - I)$ distribution is greater than or equal to the calculated value of the F statistic.

Tables of F critical values are available for use when software does not give the P-value. Table E in the back of the book contains the F critical values for probabilities $p = 0.100$, 0.050, 0.025, 0.010, and 0.001. For one-way ANOVA we use critical values from the table corresponding to $I - 1$ degrees of freedom in the numerator and $N - I$ degrees of freedom in the denominator. We have already seen several examples where the F statistic and its P-value were used to choose between H_0 and H_a.

EXAMPLE 12.7

In the study of computer science students in Example 12.3, $F = 10.35$. There were three populations, so the degrees of freedom in the numerator are DFG $= I - 1 = 2$. For this example the degrees of freedom in the denominator are DFE $= N - I = 256 - 3 = 253$. In Table E we first find the column corresponding to 2 degrees of freedom in the numerator. For the degrees of freedom in the denominator (DFD in the table), we see that there are entries for 200 and 1000. These entries are very close. To be conservative we use critical values corresponding to 200 degrees of freedom in the denominator since these are slightly larger. Because 10.35 is larger than all of the tabulated values, we reject H_0 and conclude that the differences in means are statistically significant with $P < 0.001$.

p	Critical value
0.100	2.33
0.050	3.04
0.025	3.76
0.010	4.71
0.001	7.15

EXAMPLE 12.8

In the comparison of teaching methods of Example 12.6, $F = 1.13$. Because we are comparing three populations, the degrees of freedom in the numerator are DFG $= I - 1 = 2$, and the degrees of freedom in the denominator are DFE $= N - I = 66 - 3 = 63$. Table E has entries for 2 degrees of freedom in the numerator. We do not find entries for 63 degrees of freedom in the denominator, so we use the slightly larger values corresponding to 60.

p	Critical value
0.100	2.39
0.050	3.15
0.025	3.93
0.010	4.98
0.001	7.77

Because 1.13 is less than any of the tabulated values, we do not have strong evidence in the data against the hypothesis of equal means. We conclude that the differences in sample means are not significant at the 10% level. Statistical software (Figure 12.8) gives the exact value $P = 0.3288$.

Remember that the F test is always one-sided because any differences among the group means tend to make F large. The ANOVA F test shares the robustness of the two-sample t test. It is relatively insensitive to moderate nonnormality and unequal variances, especially when the sample sizes are similar.

The following display shows the general form of a one-way ANOVA table with the F statistic. The formulas in the sum of squares column can be used for calculations in small problems. There are other formulas that are more efficient for hand or calculator use, but ANOVA calculations are usually done by computer software.

Source	Degrees of freedom	Sum of squares	Mean square	F
Groups	$I - 1$	$\sum_{\text{groups}} n_i (\bar{x}_i - \bar{x})^2$	SSG/DFG	MSG/MSE
Error	$N - I$	$\sum_{\text{groups}} (n_i - 1)s_i^2$	SSE/DFE	
Total	$N - 1$	$\sum_{\text{obs}} (x_{ij} - \bar{x})^2$	SST/DFT	

EXAMPLE 12.9

In Example 12.8 we could not reject H_0 that the three groups of students in Example 12.6 had equal mean scores on a pretest of reading comprehension. We now change the data in a way that favors H_a. Add one point to the score of each student in the Basal group, and subtract one point from each of the observations in the Strat group. Leave the observations in the DRTA group as they were. Call the new variable SCOREX. The summary statistics for SCOREX appear in Figure 12.9. Compared with the original means, given in Figure 12.7, the SCOREX means are farther apart. Note that we have not changed any of the standard deviations.

The ANOVA results appear in Figure 12.10. Compare this table with the ANOVA for the original variable SCORE given in Figure 12.8. The degrees of freedom are the same. The values of SSG and SST have increased, but SSE has not changed. In constructing SCOREX, we have added more total variation to the data, and all of this added variation went into the variation between groups. The increase in SSG is reflected in MSG. The new value is about six times the old. Because MSE has not changed, the F statistic is also about six times as large. The P-value has decreased dramatically and we now have evidence to reject H_0 in favor of H_a.

coefficient of determination

One other item in the computer output for ANOVA is worth noting. For an analysis of variance, we define the **coefficient of determination** as

$$R^2 = \frac{\text{SSG}}{\text{SST}}$$

The coefficient of determination plays the same role as the squared multiple correlation R^2 in a multiple regression. In Figure 12.8, R^2 is listed under the heading R-Square. The value is 0.035. This result says that the FIT part of the model (that is, differences among the means of the groups) accounts for 3.5% of the total variation in the data. Compare this value with that for the changed data described in Example 12.9 and given in Figure 12.10. The change has increased the coefficient of determination to 17.9%.

```
                    Analysis Variable : SCOREX

--------------------- GROUP=Basal ---------------------

                N          Mean        Std Dev
               22     11.5000000      2.9720924

--------------------- GROUP=DRTA ---------------------

                N          Mean        Std Dev
               22      9.7272727      2.6935871

--------------------- GROUP=Strat ---------------------

                N          Mean        Std Dev
               22      8.1363636      3.3423039
```

FIGURE 12.9 Summary statistics for the adjusted scores SCOREX in the reading comprehension study.

Contrasts

The ANOVA F test gives a general answer to a general question: Are the differences among observed group means significant? Unfortunately, a small P-value simply tells us that the group means are not all the same. It does not tell us specifically which means differ from each other. Plotting and inspecting the means give us some indication of where the differences lie, but we would like to supplement inspection with formal inference.

In the ideal situation, specific questions regarding comparisons among the means are posed before the data are collected. We can answer specific questions of this kind and attach a level of confidence to the answers we give. We now explore these ideas through an example.

```
                    General Linear Models Procedure

Dependent Variable: SCOREX

                              Sum of         Mean
Source                 DF    Squares       Square   F Value    Pr > F
Model                   2   124.57576     62.28788     6.85    0.0020
Error                  63   572.45455      9.08658
Corrected Total        65   697.03030

              R-Square          C.V.      Root MSE          SCOREX Mean
              0.178724      30.79723        3.0144               9.7879
```

FIGURE 12.10 Analysis of variance output for SCOREX in the reading comprehension study.

EXAMPLE 12.10

Example 12.6 describes a randomized comparative experiment to compare three methods for teaching reading. We analyzed the pretest scores and found no reason to reject H_0 that the three groups had similar population means on this measure. This was the desired outcome. We now turn to the response variable, a measure of reading comprehension called COMP that was measured by a test taken after the instruction was completed.

Figure 12.11 gives the summary statistics for COMP computed by SAS. Side-by-side boxplots appear in Figure 12.12, and Figure 12.13 plots the group means. The ANOVA results generated by SAS are given in Figure 12.14, and a normal quantile plot of the residuals appears in Figure 12.15.
The ANOVA null hypothesis is

$$H_0: \mu_B = \mu_D = \mu_S$$

where the subscripts correspond to the group labels Basal, DRTA, and Strat. Figure 12.14 shows that $F = 4.48$ with degrees of freedom 2 and 63. The P-value is 0.0152. We have good evidence against H_0.
What can the researchers conclude from this analysis? The alternative hypothesis is true if $\mu_B \neq \mu_D$ or if $\mu_B \neq \mu_S$ or if $\mu_D \neq \mu_S$ or if any combination of these statements is true. We would like to be more specific.

The researchers were investigating a specific theory about reading comprehension. The instruction for the Basal group was the standard method commonly used in schools. The DRTA and Strat groups received innovative methods of teaching that were designed to increase the reading comprehension of the children. The DRTA and Strat methods were not identical, but they both involved teaching the students to use similar comprehension strategies in their reading.

The ANOVA F test demonstrates that there is some difference among the mean responses, but it does not directly answer the researchers' questions.

```
               Analysis Variable : COMP

------------------------ GROUP=Basal ------------------------


           N          Mean          Std Dev
          22     41.0454545        5.6355781

------------------------ GROUP=DRTA -------------------------


           N          Mean          Std Dev
          22     46.7272727        7.3884196

------------------------ GROUP=Strat ------------------------


           N          Mean          Std Dev
          22     44.2727273        5.7667505
```

FIGURE 12.11 Summary statistics for the comprehension scores in the three groups of the reading comprehension study.

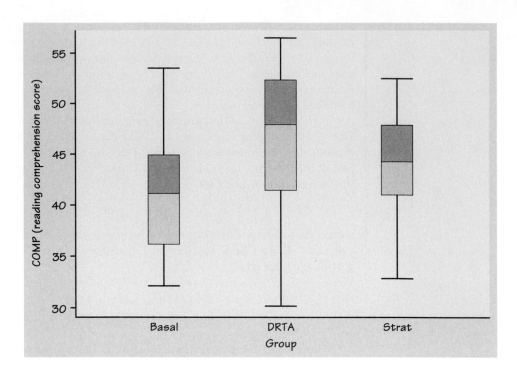

FIGURE 12.12   Side-by-side boxplots of the comprehension scores for the reading comprehension study.

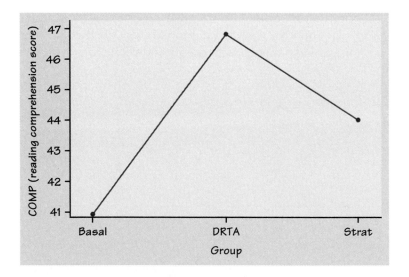

FIGURE 12.13   Comprehension score group means for the reading comprehension study.

```
                    General Linear Models Procedure

Dependent Variable: COMP

                              Sum of          Mean
Source              DF        Squares        Square    F Value    Pr > F
Model                2     357.30303     178.65152       4.48    0.0152
Error               63    2511.68182      39.86797
Corrected Total     65    2868.98485

              R-Square          C.V.      Root MSE         COMP Mean
              0.124540      14.34531        6.3141            44.015


Contrast            DF   Contrast SS  Mean Square    F Value    Pr > F
B vs D and S         1     291.03030    291.03030       7.30    0.0088
D vs S               1      66.27273     66.27273       1.65    0.2020
```

FIGURE 12.14 Analysis of variance and contrasts output for the comprehension scores in the reading comprehension study.

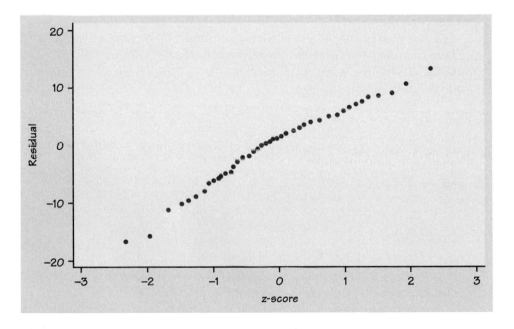

FIGURE 12.15 Normal quantile plot of the residuals for the pretest scores in the reading comprehension study.

EXAMPLE 12.11   The primary question can be formulated as a null hypothesis:

$$H_{01}: \frac{1}{2}(\mu_D + \mu_S) = \mu_B$$

with the alternative

$$H_{a1}: \frac{1}{2}(\mu_D + \mu_S) > \mu_B$$

The hypothesis H_{01} compares the average of the two innovative methods (DRTA and Strat) with the standard method (Basal). The alternative is one-sided because the researchers are interested in demonstrating that the new methods are better than the old. We use the subscripts 1 and 2 to distinguish the two sets of hypotheses.

EXAMPLE 12.12   A secondary question involves a comparison of the two new methods. We formulate this as the hypothesis that the methods DRTA and Strat are equally effective:

$$H_{02}: \mu_D = \mu_S$$

versus the alternative

$$H_{a2}: \mu_D \neq \mu_S$$

contrast ψ

Each of H_{01} and H_{02} says that a combination of population means is 0. These combinations of means are called **contrasts.** We use ψ, the Greek letter psi, for contrasts among population means. The two contrasts that arise from our two null hypotheses are

$$
\begin{aligned}
\psi_1 &= -\mu_B + \frac{1}{2}(\mu_D + \mu_S) \\
&= (-1)\mu_B + (0.5)\mu_D + (0.5)\mu_S
\end{aligned}
$$

and

$$\psi_2 = \mu_D - \mu_S$$

In each case, the value of the contrast is 0 when H_0 is true. Note that we have chosen to define the contrasts so that they will be positive when the alternative of interest (what we expect) is true. Whenever possible, this is a good idea because it makes some computations easier.

sample contrast

A contrast expresses an effect in the population as a combination of population means. To estimate the contrast, form the corresponding **sample contrast** by using sample means in place of population means. Under the ANOVA assumptions, a sample contrast is a linear combination of independent normal variables and therefore has a normal distribution. We can obtain the standard error of a contrast by using the rules for variances given in Section 4.4. Inference is based on t statistics. Here are the details.

Contrasts

A **contrast** is a combination of population means of the form

$$\psi = \sum a_i \mu_i$$

where the coefficients a_i have sum 0. The corresponding **sample contrast** is

$$c = \sum a_i \overline{x}_i$$

The **standard error of c** is

$$SE_c = s_p \sqrt{\sum \frac{a_i^2}{n_i}}$$

To test the null hypothesis

$$H_0 \colon \psi = 0$$

use the *t* **statistic**

$$t = \frac{c}{SE_c}$$

with degrees of freedom DFE that are associated with s_p. The alternative hypothesis can be one-sided or two-sided.

A **level *C* confidence interval for** ψ is

$$c \pm t^* SE_c$$

where t^* is the value for the t (DFE) density curve with area C between $-t^*$ and t^*.

Because each $\overline{x}_i$ estimates the corresponding μ_i, the addition rule for means tells us that the mean μ_c of the sample contrast c is ψ. In other words, c is an unbiased estimator of ψ. Testing the hypothesis that a contrast is 0 assesses the significance of the effect measured by the contrast. It is often more informative to estimate the size of the effect using a confidence interval for the population contrast.

EXAMPLE 12.13   In our example the coefficients in the contrasts are $a_1 = -1$, $a_2 = 0.5$, $a_3 = 0.5$ for ψ_1 and $a_1 = 0$, $a_2 = 1$, $a_3 = -1$ for ψ_2, where the subscripts 1, 2, and 3 correspond to B, D, and S. In each case the sum of the a_i is 0. We look at inference for each of these contrasts in turn.

EXAMPLE 12.14　The sample contrast that estimates ψ_1 is

$$c_1 = \bar{x}_B + \frac{1}{2}(\bar{x}_D + \bar{x}_S)$$

$$= -41.05 + \frac{1}{2}(46.73 + 44.27) = 4.45$$

with standard error

$$SE_{c_1} = 6.314\sqrt{\frac{(-1)^2}{22} + \frac{(0.5)^2}{22} + \frac{(0.5)^2}{22}}$$

$$= 1.65$$

The t statistic for testing $H_{01}: \psi_1 = 0$ versus $H_{a1}: \psi_1 > 0$ is

$$t = \frac{c_1}{SE_{c_1}} = \frac{4.45}{1.65} = 2.70$$

Because s_p has 63 degrees of freedom, software using the $t(63)$ distribution gives the one-sided P-value as 0.0044. If we used Table D, we would conclude that $P < 0.005$. The P-value is small, so there is strong evidence against H_{01}. The researchers have shown that the new methods produce higher mean scores than the old. The size of the improvement can be described by a confidence interval. To find the 95% confidence interval for ψ_1, we combine the estimate with its margin of error:

$$c_1 \pm t^*SE_{c_1} = 4.45 \pm (2.00)(1.65)$$

$$= 4.45 \pm 3.30$$

The interval is (1.15, 7.75). We are 95% confident that the mean improvement obtained by using the innovative methods as compared with the old method is between 1.15 and 7.75 points.

EXAMPLE 12.15　The second sample contrast, which compares the two new methods, is

$$c_2 = 46.73 - 44.27 = 2.46$$

with standard error

$$SE_{c_2} = 6.314\sqrt{\frac{(1)^2}{22} + \frac{(-1)^2}{22}}$$

$$= 1.90$$

The t statistic for assessing the significance of this contrast is

$$t = \frac{2.46}{1.90} = 1.29$$

The P-value for the two-sided alternative is 0.2020. We conclude that either the two new methods have the same population means or the sample sizes are not sufficiently large to distinguish them. A confidence interval helps clarify this statement. To find the 95% confidence interval for ψ_2, we combine the estimate with its margin of error:

$$c_2 \pm t^*SE_{c_2} = 2.46 \pm (2.00)(1.90)$$

$$= 2.46 \pm 3.80$$

The interval is $(-1.34, 6.26)$. With 95% confidence we state that the difference between the population means for the two new methods is between -1.34 and 6.26.

Many statistical software packages report the test statistics associated with contrasts as F statistics rather than t statistics. These F statistics are the squares of the t statistics described above. SAS output for contrasts is given at the bottom of Figure 12.14. The F statistics 7.30 and 1.66 for our two contrasts appear in the F Value column. These are the squares of the t-values 2.70 and 1.29 (up to roundoff error). The degrees of freedom for each F are 1 and 63. The P-values are correct for two-sided alternative hypotheses. They are 0.0088 and 0.2020. To convert the computer-generated results to apply to our one-sided hypothesis concerning ψ_1, simply divide the reported P-value by 2 after checking that the value of c is in the direction of H_a (that is, that c is positive).

Questions about population means are expressed as hypotheses about contrasts. A contrast should express a specific question that we have in mind when designing the study. When contrasts are formulated before seeing the data, *inference about contrasts is valid whether or not the ANOVA H_0 of equality of means is rejected.* Because the F test answers a very general question, it is less powerful than tests for contrasts designed to answer specific questions. Specifying the important questions before the analysis is undertaken enables us to use this powerful statistical technique.

Multiple comparisons

multiple comparisons

In many studies, specific questions cannot be formulated in advance of the analysis. If H_0 is not rejected, we conclude that the population means are indistinguishable on the basis of the data given. On the other hand, if H_0 is rejected, we would like to know which pairs of means differ. **Multiple comparisons** methods address this issue. It is important to keep in mind that multiple comparisons methods are used only *after rejecting* the ANOVA H_0.

EXAMPLE 12.16

Return once more to the reading comprehension data given in Example 12.10. There are three pairs of population means, comparing groups 1 and 2, 1 and 3, and 2 and 3. We can write a t statistic for each of these pairs. For example, the statistic

$$t_{12} = \frac{\bar{x}_1 - \bar{x}_2}{s_p \sqrt{\frac{1}{n_1} + \frac{1}{n_2}}} = \frac{41.05 - 46.73}{6.31 \sqrt{\frac{1}{22} + \frac{1}{22}}} = -2.99$$

compares populations 1 and 2. The subscripts on t specify which groups are compared.

The t statistics for the other two pairs are

$$t_{13} = \frac{\bar{x}_1 - \bar{x}_3}{s_p \sqrt{\frac{1}{n_1} + \frac{1}{n_3}}} = \frac{41.05 - 44.27}{6.31 \sqrt{\frac{1}{22} + \frac{1}{22}}} = -1.69$$

and

$$t_{23} = \frac{\bar{x}_2 - \bar{x}_3}{s_p \sqrt{\frac{1}{n_2} + \frac{1}{n_3}}} = \frac{46.73 - 44.27}{6.31 \sqrt{\frac{1}{22} + \frac{1}{22}}} = 1.29$$

We performed the last calculation when we analyzed the contrast ψ_2 in the previous section, because that contrast was $\mu_2 - \mu_3$. These t statistics are very similar to the pooled two-sample t statistic for comparing two population means that we saw in Chapter 7 (page 550). The difference is that we now have more than two populations, so each statistic uses the pooled estimator s_p from all groups rather than the pooled estimator from just the two groups being compared. This additional information about the common σ increases the power of the tests. The degrees of freedom for all of these statistics are DFE = 63, those associated with s_p.

Because we do not have any specific ordering of the means in mind as an alternative to equality, we must use a two-sided approach to the problem of deciding which pairs of means are significantly different.

Multiple Comparisons

To perform a **multiple comparisons procedure,** compute t **statistics** for all pairs of means using the formula

$$t_{ij} = \frac{\overline{x}_i - \overline{x}_j}{s_p \sqrt{\frac{1}{n_i} + \frac{1}{n_j}}}$$

If

$$|t_{ij}| \geq t^{**}$$

we declare that the population means μ_i and μ_j are different. Otherwise, we conclude that the data do not distinguish between them. The value of t^{**} depends upon which multiple comparisons procedure we choose.

One obvious choice for t^{**} is the upper $\alpha/2$ critical value for the t (DFE) distribution. This choice simply carries out as many separate significance tests of fixed level α as there are pairs of means to be compared. The proce-

LSD method

dure based on this choice is called the **least-significant differences** method, or simply LSD. LSD has some undesirable properties, particularly if the number of means being compared is large. Suppose, for example, that there are $I = 20$ groups and we use LSD with $\alpha = 0.05$. There are 190 different pairs of means. If we perform 190 t tests, each with an error rate of 5%, our overall error rate will be unacceptably large. We expect about 5% of the 190 to be significant even if the corresponding population means are the same. Since 5% of 190 is 9.5, we expect 9 or 10 false rejections.

The LSD procedure fixes the probability of a false rejection for each single pair of means being compared. It does not control the overall probability of *some* false rejection among all pairs. Other choices of t^{**} control possible errors in other ways. The choice of t^{**} is therefore a complex problem and a detailed discussion of it is beyond the scope of this text. Many choices for t^{**} are used in practice. One major statistical package allows selection from a list of over a dozen choices.

Bonferroni method

We will discuss only one of these, called the **Bonferroni method.** Use of this procedure with $\alpha = 0.05$, for example, guarantees that the probability of *any* false rejection among all comparisons made is no greater than 0.05. This is much stronger protection than controlling the probability of a false rejection at 0.05 for *each separate* comparison.

Many computer programs display the output for multiple comparisons procedures as a table of the group means along with the smallest difference between two means that is statistically significant, called the **minimum significant difference,** or MSD. The MSD depends on the type of multiple comparison we choose. Any two sample means that differ by more than the MSD are significantly different. That is, we conclude that the corresponding population means are different. In our notation, the MSD is

minimum significant difference, MSD

$$\text{MSD} = t^{**} s_p \sqrt{\frac{1}{n_i} + \frac{1}{n_j}}$$

EXAMPLE 12.17

We apply the Bonferroni multiple comparisons procedure with $\alpha = 0.05$ to the data from the reading comprehension study in Example 12.10. The value of t^{**} for this procedure (from software or special tables) is 2.46. Of the statistics $t_{12} = -2.99$, $t_{13} = -1.69$, and $t_{23} = 1.29$ calculated in the beginning of this section, only t_{12} is significant.

The output generated by SAS for Bonferroni comparisons appears in Figure 12.16. The critical value t^{**} is given on the output as Critical Value of T. The MSD is 4.6825. The output shows which group means *are not* significantly different. The letters A under the heading Bon Grouping extend from the DRTA group to the Strat group. This means that these two groups are not distinguishable. Similarly, the letters B extend from the Strat group to the Basal group, indicating that these two groups also cannot be distinguished. There is no collection of letters that connects the DRTA and Basal groups. This means that we can declare these two means to be significantly different.

This conclusion appears to be illogical. If μ_1 is the same as μ_3, and μ_2 is the same as μ_3, doesn't it follow that μ_1 is the same as μ_2? Logically, the answer must be yes.

This apparent contradiction points out dramatically the nature of the conclusions of statistical tests of significance. Some of the difficulty can be resolved by noting the choice of words used above. In describing the inferences, we talk about being able or unable to distinguish between pairs of means. In making the logical statements, we say, "is the same as." There is a big difference between the two modes of thought. Statistical tests ask, "Do we have adequate evidence to distinguish two means?" It is not illogical to conclude that we have evidence to distinguish μ_1 from μ_3, but not μ_1 from μ_2 or μ_2 from μ_3. It is very unlikely that any two methods of teaching reading comprehension would give *exactly* the same population means, but the data can fail to provide good evidence of a difference.

One way to deal with these difficulties of interpretation is to give confidence intervals for the differences. The intervals remind us that the differences are not known exactly. We want to give **simultaneous confidence intervals,** that is, intervals for *all* differences among the population means at once. Again, we must face the problem that there are many competing procedures—in this case, many methods of obtaining simultaneous intervals.

simultaneous confidence intervals

```
                     General Linear Models Procedure

             Bonferroni (Dunn) T tests for variable: COMP

NOTE: This test controls the type I experimentwise error rate,
      but generally has a higher type II error rate than REGWQ.

                Alpha= 0.05   df= 63   MSE= 39.86797
                     Critical Value of T= 2.46
                Minimum Significant Difference= 4.6825

Means with the same letter are not significantly different.

         Bon Grouping                    Mean      N   GROUP

                         A             46.727     22   DRTA
                         A
                 B       A             44.273     22   Strat
                 B
                 B                     41.045     22   Basal
```

FIGURE 12.16 Bonferroni multiple comparisons output for the comprehension scores in the reading comprehension study.

Simultaneous Confidence Intervals for Differences Between Means

Simultaneous confidence intervals for all differences $\mu_i - \mu_j$ between population means have the form

$$(\bar{x}_i - \bar{x}_j) \pm t^{**} s_p \sqrt{\frac{1}{n_i} + \frac{1}{n_j}}$$

The critical values t^{**} are the same as those used for the multiple comparisons procedure chosen.

The confidence intervals generated by a particular choice of t^{**} are closely related to the multiple comparisons results for that same method. If one of the confidence intervals includes the value 0, then that pair of means will not be declared significantly different, and vice versa.

EXAMPLE 12.18 For simultaneous 95% Bonferroni confidence intervals SAS gives the output in Figure 12.17 for the data in Example 12.10. We are 95% confident that *all three* intervals simultaneously contain the true values of the population mean differences.

```
              Bonferroni (Dunn) T tests for variable: COMP

    NOTE: This test controls the type I experimentwise error rate but
          generally has a higher type II error rate than Tukey's for
          all pairwise comparisons.

        Alpha= 0.05  Confidence= 0.95  df= 63  MSE= 39.86797
                    Critical Value of T= 2.45958
                 Minimum Significant Difference= 4.6825

Comparisons significant at the 0.05 level are indicates by '***'.

                            Simultaneous              Simultaneous
                               Lower     Difference      Upper
                GROUP        Confidence   Between      Confidence
              Comparison       Limit       Means         Limit

          DRTA   - Strat      -2.228       2.455         7.137
          DRTA   - Basal       0.999       5.682        10.364    ***


          Strat  - DRTA       -7.137      -2.455         2.228
          Strat  - Basal      -1.455       3.227         7.910


          Basal  - DRTA      -10.364      -5.682        -0.999    ***
          Basal  - Strat      -7.910      -3.227         1.455
```

FIGURE 12.17   Bonferroni simultaneous confidence intervals output for the comprehension scores in the reading comprehension study.

Software

We have used SAS to illustrate the analysis of the reading comprehension data. Other statistical software gives similar output and you should be able to read it without any difficulty.

EXAMPLE 12.19 | Research suggests that customers think that a product is of high quality if it is heavily advertised. An experiment designed to explore this idea collected quality ratings (on a 1 to 7 scale) of a new line of take-home refrigerated entrées based on reading a magazine ad. Three groups were compared. The first group's ad included information that would undermine (U) the expected positive association between quality and advertising; the second group's ad contained information that would affirm (A) the association; and the third group was a control (C).[4] The data are given in Table 12.2. Outputs from SAS, Excel, Minitab, and the TI-83 are given in Figure 12.18.

TABLE 12.2	Quality ratings in three groups		
Group	Quality ratings		
Undermine (n = 55)	6 5 5 5 4 5 4 6 5 5 5 5 3 3 5 4 5 5 5 4 5 4 4 5 4 4 4 4 5 5 5 4 5 5 5 4 4 5 5 4 5 5 4 4 5 4 3 4 5 5 5 3 4 4 4 4		
Affirm (n = 36)	4 6 4 6 5 5 5 6 4 5 5 5 4 6 6 5 5 7 4 6 6 4 5 4 5 5 6 4 5 5 4 6 4 6 5 5		
Control (n = 36)	5 4 5 6 5 7 5 6 7 5 7 5 4 5 4 4 6 6 5 6 5 5 4 5 5 6 6 6 5 6 6 7 6 6 5 5		

SAS

```
General Linear Models Procedure

Dependent Variable: QUALITY
                             Sum of          Mean
Source            DF         Squares         Square    F Value    Pr > F
Model              2        18.828255       9.414127    15,28     0.0001
Error            124        76.384343       0.616003
Corrected Total  126        95.212598

           R-Square           C.V.        Root MSE        QUALITY Mean
           0.197750         15.94832        0.7849            4.9213

Level of        -----------QUALITY-----------
      GROUP       N         Mean                  SD

      affirm     36       5.05555556          0.82615960
      control    36       5.41666667          0.87423436
      under      55       4.50909091          0.69048365
```

Excel

	A	B	C	D	E	F	G
1	Anova Single Factor						
2							
3	SUMMARY						
4	Groups	Count	Sum	Average	Variance		
5	Column 1	55	248	4.509091	0.476768		
6	Column 2	36	182	5.055556	0.68254		
7	Column 3	36	195	5.416667	0.764286		
8							
9							
10	ANOVA						
11	Source of Variation	SS	df	MS	F	P-value	F crit
12	Between Groups	18.82825	2	9.414127	15.28261	1.17E-06	3.069289
13	Within Groups	76.38434	124	0.616003			
14							
15	Total	95.2126	126				

| Sheet17 \ **Sheet18** / Sheet 1 / Sheet 2 / Sheet 3 / |

Ready   Sum=0

FIGURE 12.18(a)   SAS and Excel output for the advertising study.

Minitab

```
One-Way Analysis of Variance

Analysis of Variance
Source      DF          SS          MS          F           P
Factor       2      18.828       9.414      15.28      0.0000
Error      124      76.384       0.616
Total      126      95.123
                                         Individual 95% CIs For Mean
                                         Based on Pooled StDev
Level        N        Mean       StDev   ---+---------+---------+---------+---
Undermin    55      4.5091      0.6905   (-----*----)
Affirm      36      5.0556      0.8262                (-----*------)
Control     36      5.4167      0.8742                             (-----*------)
                                         ---+---------+---------+---------+---
Pooled StDev =      0.7849               4.40      4.80      5.20      5.60
```

TI-83

```
One-way ANOVA
 F=15.28260579
 p=1.1673896E-6
 Factor
  df=2
  SS=18.828255
  MS=9.4141275
 Error
  df=124
  SS=76.3843434
  MS=.61600277
 Sxp=.784858439
```

FIGURE 12.18(b)   Minitab and TI-83 output for the advertising study.

Power*

Recall that the power of a test is the probability of rejecting H_0 when H_a is in fact true. Power measures how likely a test is to detect a specific alternative. When planning a study in which ANOVA will be used for the analysis, it is important to perform power calculations to check that the sample sizes are

---

*This section is optional.

adequate to detect differences among means that are judged to be important. Power calculations also help evaluate and interpret the results of studies in which H_0 was not rejected. We sometimes find that the power of the test was so low against reasonable alternatives that there was little chance of obtaining a significant F.

In Chapter 7 we found the power for the two-sample t test. One-way ANOVA is a generalization of the two-sample t test, so it is not surprising that the procedure for calculating power is quite similar. Here are the steps that are needed:

1. Specify
    (a) an alternative (H_a) that you consider important; that is, values for the true population means $\mu_1, \mu_2, \ldots, \mu_I$;
    (b) sample sizes $n_1, n_2, \ldots, n_I$; usually these will all be equal to the common value n;
    (c) a level of significance α, usually equal to 0.05; and
    (d) a guess at the standard deviation σ.

2. Find the degrees of freedom DFG $= I - 1$ and DFE $= N - I$ and the critical value that will lead to rejection of H_0. This value, which we denote by F^*, is the upper α critical value for the F (DFG, DFE) distribution.

noncentrality parameter
3. Calculate the **noncentrality parameter**[5]

$$\lambda = \frac{\sum n_i (\mu_i - \overline{\mu})^2}{\sigma^2}$$

where $\overline{\mu}$ is a weighted average of the group means,

$$\overline{\mu} = \sum w_i \mu_i$$

and the weights are proportional to the sample sizes,

$$w_i = \frac{n_i}{\sum n_i} = \frac{n_i}{N}$$

4. Find the power, which is the probability of rejecting H_0 when the alternative hypothesis is true, that is, the probability that the observed F is greater than F^*. Under H_a, the F statistic has a distribution known as the **noncentral F distribution.** Using the SAS function for this distribution, we find the power as

noncentral F distribution

$$\text{Power} = 1 - \text{PROBF}(F^*, \text{DFG}, \text{DFE}, \lambda)$$

Note that if the n_i are all equal to the common value n, then $\overline{\mu}$ is the ordinary average of the μ_i and

$$\lambda = \frac{n \sum (\mu_i - \overline{\mu})^2}{\sigma^2}$$

If the means are all equal (the ANOVA H_0), then $\lambda = 0$. The noncentrality parameter measures how unequal the given set of means is. Large λ points to an alternative far from H_0, and we expect the ANOVA F test to have high power. Software makes calculation of the power quite easy, but tables and charts are also available.

EXAMPLE 12.20

Suppose that a study on reading comprehension similar to the one described in Example 12.10 has 10 students in each group. How likely is this study to detect differences in the mean responses similar to those observed in the actual study? Based on the results of the actual study, we will use $\mu_1 = 41$, $\mu_2 = 47$, $\mu_3 = 44$, and $\sigma = 7$ in a calculation of power. The n_i are equal, so $\overline{\mu}$ is simply the average of the μ_i:

$$\overline{\mu} = \frac{41 + 47 + 44}{3} = 44$$

The noncentrality parameter is therefore

$$\lambda = \frac{n \sum (\mu_i - \overline{\mu})^2}{\sigma^2}$$
$$= \frac{(10)[(41 - 44)^2 + (47 - 44)^2 + (44 - 44)^2]}{49}$$
$$= \frac{(10)(18)}{49} = 3.67$$

Because there are three groups with 10 observations per group, DFG $= 2$ and DFE $= 27$. The critical value for $\alpha = 0.05$ is $F^* = 3.35$. The power is therefore

$$1 - \text{PROBF}(3.35, 2, 27, 3.67) = 0.3486$$

The chance that we reject the ANOVA H_0 at the 5% significance level is only about 35%.

If the assumed values of the μ_i in this example describe differences among the groups that the experimenter wants to detect, then we would want to use more than 10 subjects per group. Although H_0 is assumed to be false, the chance of rejecting it is only about 35%. This chance can be increased to acceptable levels by increasing the sample sizes.

EXAMPLE 12.21

To decide on an appropriate sample size for the experiment described in the previous example, we repeat the power calculation for different values of n, the number of subjects in each group. Here are the results:

n	DFG	DFE	F^*	λ	Power
20	2	57	3.16	7.35	0.65
30	2	87	3.10	10.02	0.80
40	2	117	3.07	14.69	0.93
50	2	147	3.06	18.37	0.97
100	2	297	3.03	36.73	≈ 1

With $n = 40$ the experimenters have a 93% chance of rejecting H_0 with $\alpha = 0.05$ and thereby demonstrating that the groups have different means. In the long run, 93 out of every 100 such experiments would reject H_0 at the $\alpha = 0.05$ level of significance. Using 50 subjects per group increases their chance of finding significance to 97%. With 100 subjects per group, they are virtually certain to reject H_0. The exact power for $n = 100$ is 0.99989. In most cases the additional cost of increasing the sample size from 50 to 100 subjects per group would not be justified by the relatively small increase in the chance of obtaining statistically significant results.

SUMMARY

One-way analysis of variance (ANOVA) is used to compare several population means based on independent SRSs from each population. The populations are assumed to be normal with possibly different means and the same standard deviation.

To do an analysis of variance, first compute sample means and standard deviations for all groups. Side-by-side boxplots give an overview of the data. Examine normal quantile plots (either for each group separately or for the residuals) to detect outliers or extreme deviations from normality. Compute the ratio of the largest to the smallest sample standard deviation. If this ratio is less than 2 and the normal quantile plots are satisfactory, ANOVA can be performed.

The **null hypothesis** is that the population means are *all* equal. The **alternative hypothesis** is true if there are *any* differences among the population means.

ANOVA is based on separating the total variation observed in the data into two parts: variation **among group means** and variation **within groups.** If the variation among groups is large relative to the variation within groups, we have evidence against the null hypothesis.

An **analysis of variance table** organizes the ANOVA calculations. **Degrees of freedom, sums of squares,** and **mean squares** appear in the table. The **F statistic** and its ***P*-value** are used to test the null hypothesis.

Specific questions formulated before examination of the data can be expressed as **contrasts.** Tests and confidence intervals for contrasts provide answers to these questions.

If no specific questions are formulated before examination of the data and the null hypothesis of equality of population means is rejected, **multiple comparisons** methods are used to assess the statistical significance of the differences between pairs of means.

The **power** of the F test depends upon the sample sizes, the variation among population means, and the within-group standard deviation.

CHAPTER 12 EXERCISES

12.1 Does bread lose its vitamins when stored? Small loaves of bread were prepared with flour that was fortified with a fixed amount of vitamins. After baking, the vitamin C content of two loaves was measured. Another two loaves were baked at the same time, stored for one day, and then the vitamin C content was measured. In a similar manner, two loaves were stored for three, five, and seven days before measurements were taken. The units are milligrams per hundred grams of flour (mg/100 g). Here are the data (provided by Helen Park; see H. Park et al., "Fortifying bread with each of three antioxidants," *Cereal Chemistry*, 74 (1997), pp. 202–206):

Condition	Vitamin C (mg/100 g)	
Immediately after baking	47.62	49.79
One day after baking	40.45	43.46
Three days after baking	21.25	22.34
Five days after baking	13.18	11.65
Seven days after baking	8.51	8.13

(a) Give a table with sample size, mean, standard deviation, and standard error for each condition.

(b) Perform a one-way ANOVA for these data. Be sure to state your hypotheses, the test statistic with degrees of freedom, and the *P*-value.

(c) Summarize the data and the means with a plot. Use the plot and the ANOVA results to write a short summary of your conclusions.

12.2 Refer to the previous exercise. Measurements of the amounts of vitamin A (β-carotene) and vitamin E in each loaf are given below. Analyze the data for each of these vitamins and write a report summarizing what happens to the three vitamins in bread when it is stored.

Condition	Vitamin A (mg/100 g)		Vitamin E (mg/100 g)	
Immediately after baking	3.36	3.34	94.6	96.0
One day after baking	3.28	3.20	95.7	93.2
Three days after baking	3.26	3.16	97.4	94.3
Five days after baking	3.25	3.36	95.0	97.7
Seven days after baking	3.01	2.92	92.3	95.1

12.3 If a supermarket product is offered at a reduced price frequently, do customers expect the price of the product to be lower in the future? This question was examined by researchers in a study conducted on students enrolled in an introductory management course at a large midwestern university. For 10 weeks 160 subjects received information about the products. The treatment conditions corresponded to the number of promotions (1, 3, 5, or 7) that were described during this 10-week period.

TABLE 12.3   Price promotion data

Number of promotions	Expected price (dollars)									
1	3.78	3.82	4.18	4.46	4.31	4.56	4.36	4.54	3.89	4.13
	3.97	4.38	3.98	3.91	4.34	4.24	4.22	4.32	3.96	4.73
	3.62	4.27	4.79	4.58	4.46	4.18	4.40	4.36	4.37	4.23
	4.06	3.86	4.26	4.33	4.10	3.94	3.97	4.60	4.50	4.00
3	4.12	3.91	3.96	4.22	3.88	4.14	4.17	4.07	4.16	4.12
	3.84	4.01	4.42	4.01	3.84	3.95	4.26	3.95	4.30	4.33
	4.17	3.97	4.32	3.87	3.91	4.21	3.86	4.14	3.93	4.08
	4.07	4.08	3.95	3.92	4.36	4.05	3.96	4.29	3.60	4.11
5	3.32	3.86	4.15	3.65	3.71	3.78	3.93	3.73	3.71	4.10
	3.69	3.83	3.58	4.08	3.99	3.72	4.41	4.12	3.73	3.56
	3.25	3.76	3.56	3.48	3.47	3.58	3.76	3.57	3.87	3.92
	3.39	3.54	3.86	3.77	4.37	3.77	3.81	3.71	3.58	3.69
7	3.45	3.64	3.37	3.27	3.58	4.01	3.67	3.74	3.50	3.60
	3.97	3.57	3.50	3.81	3.55	3.08	3.78	3.86	3.29	3.77
	3.25	3.07	3.21	3.55	3.23	2.97	3.86	3.14	3.43	3.84
	3.65	3.45	3.73	3.12	3.82	3.70	3.46	3.73	3.79	3.94

Students were randomly assigned to the four groups. Table 12.3 gives the data. (Based on M. U. Kalwani and C. K. Yim, "Consumer price and promotion expectations: an experimental study," *Journal of Marketing Research*, 29 (1992), pp. 90–100.)

(a) Make a normal quantile plot for the data in each of the four treatment groups. Summarize the information in the plots and draw a conclusion regarding the normality of these data.

(b) Summarize the data with a table containing the sample size, mean, standard deviation, and standard error for each group.

(c) Is the assumption of equal standard deviations reasonable here? Explain why or why not.

(d) Run the one-way ANOVA. Give the hypotheses tested, the test statistic with degrees of freedom, and the *P*-value. Summarize your conclusion.

12.4   Refer to the previous exercise. Use the Bonferroni or another multiple comparisons procedure to compare the group means. Give a summary of the results and support your conclusions with a graph of the means.

12.5   Do piano lessons improve the spatial-temporal reasoning of preschool children? In Exercise 7.51 we examined this question by comparing the change scores (after treatment minus before treatment) of 34 children who took piano lessons with the scores of 44 children who did not. The

TABLE 12.4 Piano lesson data

Lessons	Scores									
Piano	2	5	7	−2	2	7	4	1	0	7
	3	4	3	4	9	4	5	2	9	6
	0	3	6	−1	3	4	6	7	−2	7
	−3	3	4	4						
Singing	1	−1	0	1	−4	0	0	1	0	−1
Computer	0	1	1	−3	−2	4	−1	2	4	2
	2	2	−3	−3	0	2	0	−1	3	−1
None	5	−1	7	0	4	0	2	1	−6	0
	2	−1	0	−2						

3.3621

latter group actually contained three groups of children: 10 were given singing lessons, 20 had some computer instruction, and 14 received no extra lessons. The data appear in Table 12.4.

(a) Make a table giving the sample size, the mean, the standard deviation, and the standard error for each group.

(b) Analyze the data using a one-way analysis of variance. State the null and alternative hypotheses, the test statistic with degrees of freedom, the P-value, and your conclusion.

12.6 Refer to the previous exercise. Use the Bonferroni or another multiple comparisons procedure to compare the group means. Give a summary of the results and support your conclusions with a graph of the means.

12.7 The researchers in Exercise 12.5 based their research on a biological argument for a causal link between music and spatial-temporal reasoning. Therefore, it is natural to test the contrast that compares the mean of the piano lesson group with the average of the three other means. Construct this contrast, perform the significance test, and summarize the results. Note that this is not the test that we performed in Exercise 7.51, where we did not differentiate among the three groups who did not receive piano instruction.

12.8 For each of the following situations, identify the response variable and the populations to be compared, and give I, the n_i, and N.

(a) To compare four varieties of tomato plants, 12 plants of each variety are grown and the yield in pounds of tomatoes is recorded.

(b) A marketing experiment compares five different types of packaging for a laundry detergent. Each package is shown to 40 different potential consumers, who rate the attractiveness of the product on a 1 to 10 scale.

(c) To compare the effectiveness of three different weight-loss programs, 20 people are randomly assigned to each. At the end of the program, the weight loss for each of the participants is recorded.

12.9 For each of the following situations, identify the response variable and the populations to be compared, and give I, the n_i, and N.

(a) In a study on smoking, subjects are classified as nonsmokers, moderate smokers, or heavy smokers. A sample of size 100 is drawn from each group. Each person is asked to report the number of hours of sleep he or she gets on a typical night.

(b) The strength of concrete depends upon the formula used to prepare it. One study compared four different mixtures. Five batches of each mixture were prepared, and the strength of the concrete made from each batch was measured.

(c) Which of three methods of teaching sign language is most effective? Twenty students are randomly assigned to each of the methods, and their scores on a final exam are recorded.

12.10 How do nematodes (microscopic worms) affect plant growth? A botanist prepares 16 identical planting pots and then introduces different numbers of nematodes into the pots. A tomato seedling is transplanted into each plot. Here are data on the increase in height of the seedlings (in centimeters) 16 days after planting (data provided by Matthew Moore):

Nematodes	Seedling growth			
0	10.8	9.1	13.5	9.2
1,000	11.1	11.1	8.2	11.3
5,000	5.4	4.6	7.4	5.0
10,000	5.8	5.3	3.2	7.5

(a) Make a table of means and standard deviations for the four treatments, and plot the means. What does the plot of the means show?

(b) State H_0 and H_a for an ANOVA on these data, and explain in words what ANOVA tests in this setting.

(c) Using computer software, run the ANOVA. What are the F statistic and its P-value? Give the values of s_p and R^2. Report your conclusion.

12.11 The presence of harmful insects in farm fields is detected by erecting boards covered with a sticky material and then examining the insects trapped on the board. To investigate which colors are most attractive to cereal leaf beetles, researchers placed six boards of each of four colors in a field of oats in July. The table below gives data on the number of cereal leaf beetles trapped. (Modified from M. C. Wilson and R. E. Shade, "Relative attractiveness of various luminescent colors to the cereal leaf beetle and the meadow spittlebug," *Journal of Economic Entomology,* 60 (1967), pp. 578–580.)

Color	Insects trapped
Lemon yellow	45 59 48 46 38 47
White	21 12 14 17 13 17
Green	37 32 15 25 39 41
Blue	16 11 20 21 14 7

(a) Make a table of means and standard deviations for the four colors, and plot the means.

(b) State H_0 and H_a for an ANOVA on these data, and explain in words what ANOVA tests in this setting.

(c) Using computer software, run the ANOVA. What are the F statistic and its P-value? Give the values of s_p and R^2. What do you conclude?

12.12 Return to the nematode problem in Exercise 12.10.

(a) Define the contrast that compares the 0 treatment (the control group) with the average of the other three.

(b) State H_0 and H_a for using this contrast to test whether or not the presence of nematodes causes decreased growth in tomato seedlings.

(c) Perform the significance test and give the P-value. Do you reject H_0?

(d) Define the contrast that compares the 0 treatment with the treatment with 10,000 nematodes. This contrast is a measure of the decrease in growth due to having a very large nematode infestation. Give a 95% confidence interval for this decrease in growth.

12.13 Return to the color attractiveness problem in Exercise 12.11. For the Bonferroni procedure with $\alpha = 0.05$, the value of t^{**} is 2.61. Use this multiple comparisons procedure to decide which pairs of colors are significantly different. Summarize your results. Which color would you recommend for attracting cereal leaf beetles?

12.14 In large classes instructors sometimes use different forms of an examination. When average scores for the different forms are calculated, students who received the form with the lowest average score may complain that their examination was more difficult than the others. Analysis of variance can help determine whether the variation in mean scores is larger than would be expected by chance. One such class used three forms. Summary statistics were as follows. (Data provided by Peter Georgeoff of the Purdue University Department of Educational Studies.)

Form	n	$\overline{x}$	s	Min.	Q_1	Median	Q_3	Max.
1	79	31.78	4.45	18	29	32	35	42
2	81	32.88	4.40	20	30	33	36	42
3	81	34.47	4.29	24	32	35	38	46

Here is the SAS output for a one-way ANOVA run on the exam scores:

```
                           Sum of            Mean
Source                DF     Squares        Square    F Value     Pr > F
Model                  2   292.01871     146.00936       7.61     0.0006
Error                238  4566.28004      19.18605
Corrected Total      240  4858.29876


            R-Square             C.V.      Root MSE            SCORE Mean
            0.060107         13.25164        4.3802                33.054


       Bonferroni (Dunn) T tests for variable: SCORE

NOTE: This test controls the type I experimentwise error rate but
      generally has a higher type II error rate than Tukey's for
      all pairwise comparisons.

   Alpha= 0.05  Confidence= 0.95  df= 238  MSE= 19.18605
                 Critical value of T= 2.41102


Comparisons significant at the 0.05 level are indicated by '***'.

                     Simultaneous                 Simultaneous
                        Lower     Difference          Upper
            FORM      Confidence   Between        Confidence
         Comparison     Limit       Means            Limit

         3    - 2      -0.0669      1.5926          3.2521
         3    - 1       1.0144      2.6843          4.3543      ***

         2    - 3      -3.2521     -1.5926          0.0669
         2    - 1      -0.5782      1.0917          2.7617

         1    - 3      -4.3543     -2.6843         -1.0144      ***
         1    - 2      -2.7617     -1.0917          0.5782
```

(a) Compare the distributions of exam scores for the three forms with side-by-side boxplots. Give a short summary of the information contained in these plots.

(b) Summarize and interpret the results of the ANOVA, including the multiple comparisons procedure.

12.15   The presence of lead in the soil of forests is an important ecological concern. One source of lead contamination is the exhaust from automobiles. In recent years this source has been greatly reduced by the elimination of lead from gasoline. Can the effects be seen in our forests? The Hubbard Brook Experimental Forest in West Thornton, New Hampshire, is the site of an ongoing study of the forest floor. Lead measurements of samples taken from this forest are available for several years. The variable of interest is lead concentration recorded as milligrams per square meter. Because the data are strongly skewed to the right, logarithms of the concentrations were analyzed. Here are some summary statistics for 5 years (data provided by Tom Siccama of the Yale University School of Forestry and Environmental Studies):

Year	n	$\bar{x}$	s	Min.	Q_1	Median	Q_3	Max.
76	59	6.80	.58	5.74	6.33	6.73	7.32	8.05
77	58	6.75	.68	3.95	6.39	6.80	7.23	8.10
78	58	6.76	.50	5.01	6.50	6.78	7.10	7.66
82	68	6.50	.55	5.15	6.11	6.53	6.83	7.82
87	70	6.40	.68	4.38	6.09	6.46	6.85	8.15

Here is the SAS output for a one-way ANOVA run on the logs of the lead concentrations:

```
                       Sum of         Mean
Source          DF     Squares        Square    F Value     Pr > F
Model            4    8.4437799     2.1109450     5.75      0.0002
Error          308  113.1440666     0.3673509
Corrected Total 312  121.5878465

          R-Square            C.V.       Root MSE           LLEAD Mean
          0.069446         9.143762       0.6061               6.6285


        Bonferroni (Dunn) T tests for variable: LLEAD

NOTE: This test controls the type I experimentwise error rate but
      generally has a higher type II error rate than Tukey's for
      all pairwise comparisons.

   Alpha= 0.05  Confidence= 0.95  df= 308  MSE= 0.367351
             Critical value of T= 2.82740


Comparisons significant at the 0.05 level are indicated by '***'.

                   Simultaneous                 Simultaneous
                      Lower      Difference         Upper
           YEAR     Confidence    Between        Confidence
        Comparison    Limit        Means            Limit


        76  - 78     -0.2745      0.0424          0.3592
        76  - 77     -0.2687      0.0482          0.3650
        76  - 82     -0.0046      0.3003          0.6052
        76  - 87      0.0995      0.4024          0.7052      ***
```

78	− 76	−0.3592	−0.0424	0.2745
78	− 77	−0.3124	0.0058	0.3240
78	− 82	−0.0484	0.2579	0.5642
78	− 87	0.0557	0.3600	0.6643 ★★★
77	− 76	−0.3650	−0.0482	0.2687
77	− 78	−0.3240	−0.0058	0.3124
77	− 82	−0.0542	0.2521	0.5584
77	− 87	0.0499	0.3542	0.6585 ★★★
82	− 76	−0.6052	−0.3003	0.0046
82	− 78	−0.5642	−0.2579	0.0484
82	− 77	−0.5584	−0.2521	0.0542
82	− 87	−0.1897	0.1021	0.3938
87	− 76	−0.7052	−0.4024	−0.0995 ★★★
87	− 78	−0.6643	−0.3600	−0.0557 ★★★
87	− 77	−0.6585	−0.3542	−0.0499 ★★★
87	− 82	−0.3938	−0.1021	0.1897

(a) Display the data with side-by-side boxplots. Describe the major features of the data.

(b) Summarize the ANOVA results. Do the data suggest that the (log) concentration of lead in the Hubbard Forest floor is decreasing?

12.16 A randomized comparative experiment compares three programs designed to help people lose weight. There are 20 subjects in each program. The sample standard deviations for the amount of weight lost (in pounds) are 5.2, 8.9, and 10.1. Can you use the assumption of equal standard deviations to analyze these data? Compute the pooled variance and find s_p.

12.17 A study of physical fitness collected data on the weight (in kilograms) of men in four different age groups. The sample sizes for the groups were 92, 34, 35, and 24. The sample standard deviations for the groups were 12.2, 10.4, 9.2, and 11.7. Can you use the assumption of equal standard deviations to analyze these data? Compute the pooled variance and find s_p.

12.18 For each part of Exercise 12.8, outline the ANOVA table, giving the sources of variation and the degrees of freedom. (Do not compute the numerical values for the sums of squares and mean squares.)

12.19 For each part of Exercise 12.9, outline the ANOVA table, giving the sources of variation and the degrees of freedom. (Do not compute the numerical values for the sums of squares and mean squares.)

12.20 Return to the change-of-majors study described in Example 12.3 (page 748).

(a) State H_0 and H_a for ANOVA.

(b) Outline the ANOVA table, giving the sources of variation and the degrees of freedom.

(c) What is the distribution of the F statistic under the assumption that H_0 is true?

(d) Using Table E, find the critical value for an $\alpha = 0.05$ test.

12.21 Return to the survey of college students described in Example 12.4 (page 751).

(a) State H_0 and H_a for ANOVA.

(b) Outline the ANOVA table, giving the sources of variation and the degrees of freedom.

(c) What is the distribution of the F statistic under the assumption that H_0 is true?

(d) Using Table E, find the critical value for an $\alpha = 0.05$ test.

12.22 A study of the effects of exercise on physiological and psychological variables compared four groups of male subjects. The treatment group (T) consisted of 10 participants in an exercise program. A control group (C) of 5 subjects volunteered for the program but were unable to attend for various reasons. Subjects in the other two groups were selected to be similar to those in the first two groups in age and other characteristics. These were 11 joggers (J) and 10 sedentary people (S) who did not regularly exercise. (Data provided by Dennis Lobstein, from his Ph.D. dissertation, "A multivariate study of exercise training effects on beta-endorphin and emotionality in psychologically normal, medically healthy men," Purdue University, 1983.) One of the variables measured at the end of the program was a physical fitness score. Part of the ANOVA table used to analyze these data is given below.

Source	Degrees of freedom	Sum of squares	Mean square	F
Groups	3	104,855.87		
Error	32	70,500.59		
Total				

(a) Fill in the missing entries in the ANOVA table.

(b) State H_0 and H_a for this experiment.

(c) What is the distribution of the F statistic under the assumption that H_0 is true? Using Table E, give an approximate P-value for the ANOVA test. Write a brief conclusion.

(d) What is s_p^2, the estimate of the within-group variance? What is s_p?

12.23 Another variable measured in the experiment described in the previous exercise was a depression score. Higher values of this score indicate more depression. Part of the ANOVA table for these data appears below.

Source	Degrees of freedom	Sum of squares	Mean square	F
Groups	3		158.96	
Error	32		62.81	
Total				

(a) Fill in the missing entries in the ANOVA table.

(b) State H_0 and H_a for this experiment.

(c) What is the distribution of the F statistic under the assumption that H_0 is true? Using Table E, give an approximate P-value for the ANOVA test. What do you conclude?

(d) What is s_p^2, the estimate of the within-group variance? What is s_p?

12.24   The weight gain of women during pregnancy has an important effect on the birth weight of their children. If the weight gain is not adequate, the infant is more likely to be small and will tend to be less healthy. In a study conducted in three countries, weight gains (in kilograms) of women during the third trimester of pregnancy were measured. The results are summarized in the following table. (Data taken from Collaborative Research Support Program in Food Intake and Human Function, *Management Entity Final Report*, University of California, Berkeley, 1988.)

Country	n	$\bar{x}$	s
Egypt	46	3.7	2.5
Kenya	111	3.1	1.8
Mexico	52	2.9	1.8

(a) Find the pooled estimate of the within-country variance s_p^2. What entry in the ANOVA table gives this quantity?

(b) The sum of squares for countries (groups) is 17.22. Use this information and that given above to complete all entries in an ANOVA table.

(c) State H_0 and H_a for this study.

(d) What is the distribution of the F statistic under the assumption that H_0 is true? Use Table E to find an approximate P-value for the significance test. Report your conclusion.

(e) Calculate R^2, the coefficient of determination.

12.25   In another part of the study described in the previous exercise, measurements of food intake in kilocalories were taken on many individuals several times during the period of a year. From these data, average daily food intake values were computed for each individual. The results for toddlers aged 18 to 30 months are summarized in the following table:

Country	n	$\bar{x}$	s
Egypt	88	1217	327
Kenya	91	844	184
Mexico	54	1119	285

(a) Find the pooled estimate of the within-country variance s_p^2. What entry in the ANOVA table gives this quantity?

(b) The sum of squares for countries (groups) is 6,572,551. Use this information and that given above to complete all entries in an ANOVA table.

(c) State H_0 and H_a for this study.

(d) What is the distribution of the F statistic under the assumption that H_0 is true? Use Table E to find an approximate P-value for the significance test. Report your conclusion.

(e) Calculate R^2, the coefficient of determination.

12.26 Refer to the change-of-majors study described in Example 12.3 (page 748). Let μ_1, μ_2, and μ_3 represent the population mean SAT mathematics scores for the three groups.

(a) Because the first two groups (computer science, engineering and other sciences) are majoring in areas that require mathematics skills, it is natural to compare the average of these two with the third group. Write an expression in terms of the μ_i for a contrast ψ_1 that represents this comparison.

(b) Let ψ_2 be a contrast which compares the first two groups. Write an expression in terms of the μ_i for ψ_2 that represents this comparison.

12.27 Return to the survey of college students described in Example 12.4 (page 751). Let μ_1, μ_2, μ_3, and μ_4 represent the population mean expenditures on textbooks for the freshmen, sophomores, juniors, and seniors.

(a) Because juniors and seniors take higher-level courses, which might use more expensive textbooks, we want to compare the average of the freshmen and sophomores with the average of the juniors and seniors. Write a contrast that expresses this comparison.

(b) Write a contrast for comparing the freshmen with the sophomores.

(c) Write a contrast for comparing the juniors with the seniors.

12.28 Return to the SAT mathematics scores for the change-of-majors study in Example 12.3 (page 748). Answer the following questions for the two contrasts that you defined in Exercise 12.26.

(a) For each contrast give H_0 and an appropriate H_a. In choosing the alternatives you should use information given in the description of the problem, but you may not consider any impressions obtained by inspection of the sample means.

(b) Find the values of the corresponding sample contrasts.

(c) Using the value $s_p = 82.5$, calculate the standard errors for the sample contrasts.

(d) Give the test statistics and approximate P-values for the two significance tests. What do you conclude?

(e) Compute 95% confidence intervals for the two contrasts.

12.29　The following table presents data on the high school mathematics grades of the students in the change-of-majors study described in Example 12.3 (page 748). The sample sizes in this exercise are different from those in Example 12.3 because complete information was not available on all students. The values have been coded so that 10 = A, 9 = A−, and so on.

Second-year major	n	$\bar{x}$	s
Computer science	90	8.77	1.41
Engineering and other sciences	28	8.75	1.46
Other	106	7.83	1.74

Answer the following questions for the two contrasts that you defined in (a) and (b) of Exercise 12.26.

(a) For each contrast state H_0 and an appropriate H_a. In choosing the alternatives you should use information given in the description of the problem, but you may not consider any impressions obtained by inspection of the sample means.

(b) Find the values of the corresponding sample contrasts c_1 and c_2.

(c) Using the value $s_p = 1.581$, calculate the standard errors s_{c_1} and s_{c_2} for the sample contrasts.

(d) Give the test statistics and approximate P-values for the two significance tests. What do you conclude?

(e) Compute 95% confidence intervals for the two contrasts.

12.30　In the exercise program study described in Exercise 12.22, the summary statistics for physical fitness scores are as follows:

Group	n	$\bar{x}$	s
Treatment (T)	10	291.91	38.17
Control (C)	5	308.97	32.07
Joggers (J)	11	366.87	41.19
Sedentary (S)	10	226.07	63.53

The researchers wanted to address the following questions for the physical fitness scores. In these questions "better" means a higher fitness score. (1) Is

T better than C? (2) Is T better than the average of C and S? (3) Is J better than the average of the other three groups?

(a) For each of the three questions, define an appropriate contrast. Translate the questions into null and alternative hypotheses about these contrasts.

(b) Test your hypotheses and give approximate P-values. Summarize your conclusions. Do you think that the use of contrasts in this way gives an adequate summary of the results?

(c) You found that the groups differ significantly in the physical fitness scores. Does this study allow conclusions about causation—for example, that a sedentary lifestyle causes people to be less physically fit? Explain your answer.

12.31 Exercise 12.23 gives the ANOVA table for depression scores from the exercise program study described in Exercise 12.22. Here are the summary statistics for the depression scores:

Group	n	$\bar{x}$	s
Treatment (T)	10	51.90	6.42
Control (C)	5	57.40	10.46
Joggers (J)	11	49.73	6.27
Sedentary (S)	10	58.20	9.49

In planning the experiment, the researchers wanted to address the following questions for the depression scores. In these questions "better" means a lower depression score. (1) Is T better than C? (2) Is T better than the average of C and S? (3) Is J better than the average of the other three groups?

(a) For each of the three questions, define an appropriate contrast. Translate the questions into null and alternative hypotheses about these contrasts.

(b) Test your hypotheses and give approximate P-values. Summarize your conclusions. Do you think that the use of contrasts in this way gives an adequate summary of the results?

(c) You found that the groups differ significantly in the depression scores. Does this study allow conclusions about causation—for example, that a sedentary lifestyle causes people to be more depressed? Explain your answer.

12.32 Exercise 12.24 gives data on the weight gains of pregnant women in Egypt, Kenya, and Mexico. Computer software gives the critical value for the Bonferroni multiple comparisons procedure with $\alpha = 0.05$ as $t^{**} = 2.41$. Explain in plain language what $\alpha = 0.05$ means in the Bonferroni procedure. Use this procedure to compare the mean weight gains for the three countries. Summarize your conclusions.

12.33 Exercise 12.25 gives summary statistics for the food intake values for toddlers in Egypt, Kenya, and Mexico. Computer software gives the critical value for the Bonferroni multiple comparisons procedure with $\alpha = 0.05$ as $t^{**} = 2.41$. Explain in plain language what $\alpha = 0.05$ means in the Bonferroni procedure. Use this procedure to compare the toddler food intake means for the three countries. What do you conclude?

12.34 Refer to the physical fitness scores for the four groups in the exercise program study discussed in Exercises 12.22 and 12.30. Computer software gives the critical value for the Bonferroni multiple comparisons procedure with $\alpha = 0.05$ as $t^{**} = 2.53$. Use this procedure to compare the mean fitness scores for the four groups. Summarize your conclusions.

12.35 Refer to the depression scores for the four groups in the exercise program study discussed in Exercises 12.23 and 12.31. Computer software gives the critical value for the Bonferroni multiple comparisons procedure with $\alpha = 0.05$ as $t^{**} = 2.53$. Use this procedure to compare the mean depression scores for the four groups. Summarize your conclusions.

12.36 You are planning a study of the weight gains of pregnant women during the third trimester of pregnancy similar to that described in Exercise 12.24. The standard deviations given in that exercise range from 1.8 to 2.5. To perform power calculations, assume that the standard deviation is $\sigma = 2.3$. You have three groups, each with n subjects, and you would like to reject the ANOVA H_0 when the alternative $\mu_1 = 2.5$, $\mu_2 = 3.0$, and $\mu_3 = 3.5$ is true. Use software to make a table of powers against this alternative (similar to the table in Example 12.21) for the following numbers of women in each group: $n = 50, 100, 150, 175$, and 200. What sample size would you choose for your study?

12.37 Repeat the previous exercise for the alternative $\mu_1 = 2.7$, $\mu_2 = 3.0$, and $\mu_3 = 3.3$.

12.38 Analysis of variance methods are often used in clinical trials where the goal is to assess the effectiveness of one or more treatments for a particular medical condition. One such study compared three treatments for dandruff and a placebo. The treatments were 1% pyrithione zinc shampoo (PyrI), the same shampoo but with instructions to shampoo two times (PyrII), 2% ketoconazole shampoo (Keto), and a placebo shampoo (Placebo). After six weeks of treatment, eight sections of the scalp were examined and given a score that measured the amount of scalp flaking on a 0 to 10 scale. The response variable was the sum of these eight scores. An analysis of the baseline flaking measure indicated that randomization of patients to treatments was successful in that no differences were found between the groups. At baseline there were 112 subjects in each of the three treatment groups and 28 subjects in the Placebo group. During the clinical trial, 3 dropped out from the PyrII group and 6 from the Keto group. No patients dropped out of the other two groups. The data are given in the DANDRUFF data set described in the Data Appendix.

(a) Find the mean, standard deviation, and standard error for the subjects in each group. Summarize these, along with the sample sizes, in a table and make a graph of the means.

(b) Run the analysis of variance on these data. Write a short summary of the results and your conclusion. Be sure to include the hypotheses tested, the test statistic with degrees of freedom, and the *P*-value.

12.39 Refer to the previous exercise.

(a) Plot the residuals versus case number (the first variable in the data set). Describe the plot. Is there any pattern that would cause you to question the assumption that the data are independent?

(b) Examine the standard deviations for the four treatment groups. Is there a problem with the assumption of equal standard deviations for ANOVA in this data set? Explain your answer.

(c) Prepare normal quantile plots for each treatment group. What do you conclude from these plots?

(d) Obtain the residuals from the analysis of variance and prepare a normal quantile plot of these. What do you conclude?

12.40 Refer to Exercise 12.38. Use the Bonferroni or another multiple comparisons procedure that your software provides to compare the individual group means in the dandruff study. Write a short summary of your conclusions.

12.41 Refer to Exercise 12.38. There are several natural contrasts in this experiment that describe comparisons of interest to the experimenters. They are (1) Placebo versus the average of the three treatments; (2) Keto versus the average of the two Pyr treatments; and (3) PyrI versus PyrII.

(a) Express each of these three contrasts in terms of the means (μ's) of the treatments.

(b) Give estimates with standard errors for each of the contrasts.

(c) Perform the significance tests for the contrasts. Summarize the results of your tests and your conclusions.

12.42 Refer to Exercise 12.1, where we studied the effects of storage on the vitamin C content of bread. In this experiment 64 mg/100 g of vitamin C was added to the flour that was used to make each loaf.

(a) Convert the vitamin C amounts (mg/100 g) to percents of the amounts originally in the loaves by dividing the amounts in Exercise 12.1 by 64 and multiplying by 100. Calculate the transformed means, standard deviations, and standard errors and summarize them with the sample sizes in a table.

(b) Explain how you could have calculated the table entries directly from the table you gave in part (a) of Exercise 12.1.

(c) Analyze the percents using analysis of variance. Compare the test statistic, degrees of freedom, *P*-value, and conclusion you obtain here with the corresponding values that you found in Exercise 12.1.

12.43 Refer to the previous exercise and Exercise 12.2. The flour used to make the loaves contained 5 mg/100 g of vitamin A and 100 mg/100 g of vitamin E.

Summarize the effects of transforming the data to percents for the three vitamins.

12.44   Refer to the previous exercise. Can you suggest a general conclusion regarding what happens to the test statistic, degrees of freedom, P-value, and conclusion when you perform analysis of variance on data that have been transformed by multiplying the raw data by a constant and then adding another constant? (That is, if y is the original data, we analyze y^*, where $y^* = a + by$ and a and $b \neq 0$ are constants.)

12.45   Example 12.6 (page 754) describes an experiment to compare three methods of teaching reading. One of the pretest measures is analyzed in that example and one of the posttest measures is analyzed in Example 12.10 (page 763). In the actual study there were two pretest and three posttest measures. The data are given in the READING data set described in the Data Appendix. The pretest variable analyzed in Example 12.6 is Pre1 and the posttest variable analyzed in Example 12.10 is Post3. For this exercise, use the data for the pretest variable Pre2.

(a) Summarize the data with a table giving the sample sizes, means, and standard deviations. Plot the means.

(b) Examine the distribution of the scores in the three groups by constructing normal quantile plots. Do the data look normal?

(c) What is the ratio of the largest to the smallest standard deviation? Is it reasonable to proceed with an ANOVA?

(d) Run a one-way ANOVA on these data. State H_0 and H_a for the ANOVA significance test. Report the value of the F statistic and its P-value. Do you have strong evidence against H_0?

(e) Summarize your conclusions from this analysis.

12.46   Refer to the previous exercise. To examine the effect on the ANOVA of changing the means, you will do an analysis similar to that described in Example 12.9. Use the pretest variable Pre2 for this exercise. For each of the observations in the Basal group add one point to the Pre2. For each of the observations in the Strat group, subtract one point. Leave the observations in the DRTA group unchanged. Call the new variable Pre2X.

(a) Give the sample means and standard deviations for Pre2X in each group. (You can find these from the means and standard deviations for Pre2.)

(b) Run the ANOVA on the variable Pre2X.

(c) Compare the results of this analysis with the ANOVA for Pre2.

(d) Summarize your conclusions.

12.47   Example 12.10 analyzed the scores for the third posttest variable in the reading comprehension study. For this exercise you will analyze the first posttest variable, given as Post1 in the READING data set.

(a) Summarize the data in a table giving the sample sizes, means, and standard deviations. Plot the means.

(b) Examine the distribution of the scores in the three groups by constructing normal quantile plots. Do the data look normal?

(c) What is the ratio of the largest to the smallest standard deviation? Is it reasonable to proceed with the ANOVA?

(d) Run the ANOVA and summarize the results.

(e) Following the procedure given in the text in the section on contrasts, analyze the contrast for comparing the Basal group with the average of the other two groups. Test the one-sided H_a and give a 95% confidence interval.

(f) Analyze the contrast that compares the DRTA and the Strat groups. Use a two-sided test and give a 95% confidence interval.

(g) Based on your work in (a), (d), (e), and (f), write a brief nontechnical summary of your conclusions.

12.48 Example 12.10 analyzed the scores for the third posttest variable in the reading comprehension study. For this exercise you will analyze the second posttest variable, given as Post2 in the READING data set.

(a) Summarize the data in a table of the sample sizes, means, and standard deviations. Plot the means.

(b) Examine the distribution of the scores in the three groups by constructing normal quantile plots. Do the data look normal?

(c) What is the ratio of the largest to the smallest standard deviation? Is it reasonable to proceed with the ANOVA?

(d) Run the ANOVA and summarize the results.

(e) Following the procedure given in the text in the section on contrasts, analyze the contrast for comparing the Basal group with the average of the other two groups. Test the one-sided H_a and give a 95% confidence interval.

(f) Analyze the contrast that compares the DRTA and the Strat groups. Use a two-sided test and give a 95% confidence interval.

(g) Based on your work in (a), (d), (e), and (f), write a brief nontechnical summary of your conclusions.

12.49 Return to the nematode experiment described in Exercise 12.10 (page 782). Suppose that when entering the data into the computer, you accidentally entered the first observation as 108 rather than 10.8.

(a) Run the ANOVA with the incorrect observation. Summarize the results.

(b) Compare this run with the results obtained with the correct data set. What does this illustrate about the effect of outliers in an ANOVA?

(c) Compute a table of means and standard deviations for each of the four treatments using the incorrect data. How would this table have helped you to detect the incorrect observation?

12.50 Refer to the color attractiveness experiment described in Exercise 12.11 (page 782). Suppose that when entering the data into the computer, you accidentally entered the first observation as 450 rather than 45.

(a) Run the ANOVA with the incorrect observation. Summarize the results.

(b) Compare this run with the results obtained with the correct data set. What does this illustrate about the effect of outliers in an ANOVA?

(c) Compute a table of means and standard deviations for each of the four treatments using the incorrect data. How would this table have helped you to detect the incorrect observation?

12.51 With small numbers of observations in each group, it can be difficult to detect deviations from normality and violations of the equal standard deviations assumption for ANOVA. Return to the nematode experiment described in Exercise 12.10 (page 782). The log transformation is often used for variables such as the growth of plants. In many cases this will tend to make the standard deviations more similar across groups and to make the data within each group look more normal. Rerun the ANOVA using the logarithms of the recorded values. Answer the questions given in Exercise 12.10. Compare these results with those obtained by analyzing the raw data.

12.52 Refer to the color attractiveness experiment described in Exercise 12.11 (page 782). The square root transformation is often used for variables that are counts, such as the number of insects trapped in this example. In many cases data transformed in this way will conform more closely to the assumptions of normality and equal standard deviations. Rerun the ANOVA using the square roots of the original counts of insects. Answer the questions given in Exercise 12.11. Compare these results with those obtained by analyzing the raw data.

12.53 You are planning a study of the SAT mathematics scores of four groups of students. From Example 12.3, we know that the standard deviations of the three groups considered in that study were 86, 67, and 83. In Example 12.5, we found the pooled standard deviation to be 82.5. Since the power of the F test decreases as the standard deviation increases, use $\sigma = 90$ for the calculations in this exercise. This choice will lead to sample sizes that are perhaps a little larger than we need but will prevent us from choosing sample sizes that are too small to detect the effects of interest. You would like to conclude that the population means are different when $\mu_1 = 620$, $\mu_2 = 600$, $\mu_3 = 580$, and $\mu_4 = 560$.

(a) Pick several values for n (the number of students that you will select from each group) and calculate the power of the ANOVA F test for each of your choices.

(b) Plot the power versus the sample size. Describe the general shape of the plot.

(c) What choice of n would you choose for your study? Give reasons for your answer.

12.54 Refer to the previous exercise. Repeat all parts for the alternative $\mu_1 = 610$, $\mu_2 = 600$, $\mu_3 = 590$, and $\mu_4 - 580$.

12.55 Refer to the price promotion study that we examined in Exercise 12.3. The explanatory variable in this study is the number of price promotions in a ten-day period with possible values 1, 3, 5, and 7. When using analysis of variance we treat the explanatory variable as categorical. An alternative analysis is to use simple linear regression. Perform this analysis and summarize the results. Plot the residuals from the regression model versus the number of promotions. What do you conclude?

NOTES

1. Results of the study are reported in P. F. Campbell and G. P. McCabe, "Predicting the success of freshmen in a computer science major," *Communications of the ACM,* 27 (1984), pp. 1108–1113.

2. This rule is intended to provide a general guideline for deciding when serious errors may result by applying ANOVA procedures. When the sample sizes in each group are very small, the sample variances will tend to vary much more than when the sample sizes are large. In this case, the rule may be a little too conservative. For unequal sample sizes, particular difficulties can arise when a relatively small sample size is associated with a population having a relatively large standard deviation. Careful judgment is needed in all cases. By considering P-values rather than fixed level α testing, judgments in ambiguous cases can more easily be made; for example, if the P-value is very small, say 0.001, then it is probably safe to reject the H_0 even if there is a fair amount of variation in the sample standard deviations.

3. This example is based on data provided from a study conducted by Jim Baumann and Leah Jones of the Purdue University School of Education.

4. This example is based on Amna Kirmani and Peter Wright, "Money talks: perceived advertising expense and expected product quality," *Journal of Consumer Research,* 16 (1989), pp. 344–353.

5. Several different definitions for the noncentrality parameter of the noncentral F distribution are in use. When $I = 2$, the λ defined here is equal to the square of the noncentrality parameter δ that we used for the two-sample t test in Chapter 7. Many authors prefer $\phi = \sqrt{\lambda/I}$. We have chosen to use λ because it is the form needed for the SAS function PROBF.

Prelude

The *t* procedures of Chapter 7 compare the means of two populations. We generalized these in Chapter 12 so that we could compare the means of several populations. In this chapter we classify populations according to two categorical variables and compare the means.

- The Bureau of Labor Statistics estimates earnings of United States workers classified by age and by gender. Are there systematic differences in wages across age groups? Do men earn more than women? For all age groups or only for some? This two-way design classifies workers both by age and by gender, leading to more complex questions about the relation between age and gender in their influence on wages.

- Do left-handed people live shorter lives than right-handed people? Because women live longer than men, a two-way design examined the relationship between handedness and gender.

- How does the frequency that a supermarket product is promoted at a discount affect the price that customers expect to pay for the product? Does the percent reduction also affect this expectation?

Two–Way Analysis of Variance

Two-way ANOVA compares the means of populations that are classified in two ways or the mean responses in two-factor experiments. Many of the key concepts are similar to those of one-way ANOVA, but the presence of more than one factor also introduces some new ideas. We once more assume that the data are approximately normal and that groups have possibly different means but the same standard deviation; we again pool to estimate the variance; and we again use F statistics for significance tests. The major difference between one-way and two-way ANOVA is in the FIT part of the model. We will carefully study this term, and we will find much that is both new and useful.

Advantages of two–way ANOVA

In one-way ANOVA, we classify populations according to one categorical variable, or factor. In the two-way ANOVA model there are two factors, each with its own number of levels. When we are interested in the effects of two factors, a two-way design offers great advantages over several single-factor studies. Several examples will illustrate these advantages.

EXAMPLE 13.1

In an experiment on the influence of dietary minerals on blood pressure, rats receive diets prepared with varying amounts of calcium and varying amounts of magnesium, but with all other ingredients of the diets the same. Calcium and magnesium are the two factors in this experiment.

As is common in such experiments, high, normal, and low values for each of the two minerals were selected for study. So there are three levels for each of the factors and a total of nine diets, or treatments. The following diagram summarizes the factors and levels for this experiment:

	Calcium		
Magnesium	L	M	H
L	1	2	3
M	4	5	6
H	7	8	9

For example, Diet 2 contains magnesium at its low level combined with the normal (medium) level of calcium. Each diet is fed to 9 rats, giving a total of 81 rats used in this experiment. The response variable is the blood pressure of a rat after some time on the diet. (This experiment was conducted by Connie Weaver, Department of Foods and Nutrition, Purdue University.)

The researcher in this example was primarily interested in the effect of calcium on the blood pressure. She could have conducted a single-factor experiment with magnesium set at its normal value and three levels of calcium, using one-way ANOVA to analyze the responses. Had she then decided to study the effect of magnesium, she could have performed another experiment with calcium set at its normal value and three levels of magnesium. Suppose that the budget allowed for only 80 to 90 rats in all. For the first experiment, the researcher might assign 15 rats to each of the three calcium diets. Similarly, she could assign 15 rats to each of the three magnesium diets in the second experiment. This would require a total of 90 rats, 9 more than the two-factor design in Example 13.1.

Let us compare the two-way experiment with the two one-way experiments. In the two-way experiment 27 rats are assigned to each of the three calcium diets and 27 rats to each of the three magnesium diets. In the one-way experiments there are only 15 rats per diet. Therefore, the two-way experiment provides more information to estimate the effects of both calcium and magnesium. This fact can be expressed in terms of standard deviations. If σ is the standard deviation of blood pressure for rats fed the same diet, then the standard deviation of a sample mean for n rats is $\sigma/\sqrt{n}$. For the two-way experiment this is $\sigma/\sqrt{27} = 0.19\sigma$, whereas for the two one-way experiments we have $\sigma/\sqrt{15} = 0.26\sigma$. The sample mean responses for any one level of either calcium or magnesium are less variable in the two-way design.

When the actual experiment was analyzed using two-way ANOVA, no effect of magnesium on blood pressure was found. On the other hand, the data clearly demonstrated that the high-calcium diet had the beneficial effect of lowering the average blood pressure. Means for the three calcium levels were reported with standard errors based on 27 rats per group.

The conclusions about the effect of calcium on blood pressure are also more general in the two-way experiment. In the one-way experiments, calcium is varied only for the normal level of magnesium. All levels of calcium are combined with all levels of magnesium in the two-way design, so that conclusions about calcium are not confined to one level of magnesium. The same increased generality also applies to conclusions about magnesium. So the conclusions are both more precise and more general, yet the two-way design requires fewer rats. Experiments with several factors are an efficient use of resources.

EXAMPLE 13.2

A group of students and administrators concerned about abuse of alcohol is planning an education and awareness program for the student body at a large university. To aid in the design of their program they decide to gather information on the attitudes of undergraduate students toward alcohol.

Initially, they consider taking an SRS of all the students on the campus. They soon realize that the attitudes of freshmen, sophomores, juniors, and seniors may be quite different. A stratified sample is therefore more appropriate. The study planners decide to use year in school as a factor with four levels and to take separate random samples from each class of students.

Further discussion suggests that men and women may have different attitudes toward alcohol. The team therefore decides to add gender as a second factor. The final sampling design has two factors: year with four levels, and gender with two levels. An SRS of size 50 is drawn from each of the eight populations. The eight groups for this study are shown in the following diagram:

Gender	Fr.	So.	Jr.	Sr.
Male	1	2	3	4
Female	5	6	7	8

(Year spans the Fr., So., Jr., Sr. columns.)

Example 13.2 illustrates another advantage of two-way designs analyzed by two-way ANOVA. In the one-way design with year as the only factor, all four random samples would include both men and women. If a gender difference exists, the one-way ANOVA would assign this variation to the RESIDUAL (within groups) part of the model. In the two-way ANOVA, gender is included as a factor, and therefore this variation is included in the FIT part of the model. Whenever we can move variation from RESIDUAL to FIT, we reduce the σ of our model and increase the power of our tests.

EXAMPLE 13.3

A textile researcher is interested in how four different colors of dye affect the durability of fabrics. Because the effects of the dyes may be different for different types of cloth, he applies each dye to five different kinds of cloth. The two factors in this experiment are dyes with four levels and cloth types with five levels. Six fabric specimens are dyed for each of the 20 dye-cloth combinations, and all 120 specimens are tested for durability.

The researcher in Example 13.3 could have used only one type of cloth and compared the four dyes using a one-way analysis. However, based on knowledge of the chemistry of textiles and dyes, he suspected that some of the dyes might have a chemical reaction with some types of cloth. Such a reaction could have a strong negative effect on the durability of the cloth. He wanted his results to be useful to manufacturers who would use the dyes on a variety of cloths. He chose several cloth types for the study to represent the different kinds that would be used with these dyes.

If one or more specific dye-cloth combinations produced exceptionally bad or exceptionally good durability measures, the experiment should discover this combined effect. Effects of the dyes that may be different for different types of cloth are represented in the FIT part of a two-way model as **interactions.** In contrast, the average values for dyes and cloths are represented as **main effects.** The two-way model represents FIT as the sum of a main effect for each of the two factors *and* an interaction. One-way designs that vary a single factor and hold other factors fixed cannot discover interactions. We will discuss interactions more fully in a later section.

interactions
main effects

These examples illustrate several reasons why two-way designs are preferable to one-way designs.

Advantages of Two-Way ANOVA

1. It is more efficient to study two factors simultaneously rather than separately.

2. We can reduce the residual variation in a model by including a second factor thought to influence the response.

3. We can investigate interactions between factors.

These considerations also apply to study designs with more than two factors. We will be content to explore only the two-way case. The choice of sampling or experimental design is fundamental to any statistical study. Factors and levels must be carefully selected by an individual or team who understands both the statistical models and the issues that the study will address.

The two-way ANOVA model

When discussing two-way models in general, we will use the labels A and B for the two factors. For particular examples and when using statistical software, it is better to use names for these categorical variables that suggest their meaning. Thus, in Example 13.1 we would say that the factors are calcium and magnesium. The numbers of levels of the factors are often used to describe the model. Again using Example 13.1 as an illustration, we would call this a 3×3 ANOVA. Similarly, Example 13.2 illustrates a 2×4 ANOVA. In general, factor A will have I levels and factor B will have J levels. Therefore, we call the general two-way problem an $I \times J$ ANOVA.

In a two-way design every level of A appears in combination with every level of B, so that $I \times J$ groups are compared. The sample size for level i of factor A and level j of factor B is n_{ij}.[1]

The total number of observations is

$$N = \sum n_{ij}$$

Assumptions for Two-Way ANOVA

We have independent SRSs of size n_{ij} from each of $I \times J$ normal populations. The population means μ_{ij} may differ, but all populations have the same standard deviation σ. The μ_{ij} and σ are unknown parameters.

Let x_{ijk} represent the kth observation from the population having factor A at level i and factor B at level j. The statistical model is

$$x_{ijk} = \mu_{ij} + \epsilon_{ijk}$$

for $i = 1, \ldots, I$ and $j = 1, \ldots, J$ and $k = 1, \ldots, n_{ij}$, where the deviations ϵ_{ijk} are from an $N(0, \sigma)$ distribution.

The FIT part of the model is the means μ_{ij}, and the RESIDUAL part is the deviations ϵ_{ijk} of the individual observations from their group means. To estimate a population mean μ_{ij} we use the sample mean of the observations in the samples from this group:

$$\overline{x}_{ij} = \frac{1}{n_{ij}} \sum_k x_{ijk}$$

The k below the $\sum$ means that we sum the n_{ij} observations that belong to the (i,j)th sample.

The RESIDUAL part of the model contains the unknown σ. We calculate the sample variances for each SRS and pool these to estimate σ^2:

$$s_p^2 = \frac{\sum(n_{ij} - 1)s_{ij}^2}{\sum(n_{ij} - 1)}$$

Just as in one-way ANOVA, the numerator in this fraction is SSE and the denominator is DFE. Also as in the one-way analysis, DFE is the total number of observations minus the number of groups. That is, DFE $= N - IJ$. The estimator of σ is s_p.

Main effects and interactions

In this section we will explore in detail the FIT part of the two-way ANOVA, which is represented in the model by the population means μ_{ij}. The two-way design gives some structure to the set of means μ_{ij}.

Because we have independent samples from each of $I \times J$ groups, we can think of the problem initially as a one-way ANOVA with IJ groups. Each population mean μ_{ij} is estimated by the corresponding sample mean $\overline{x}_{ij}$. We can calculate sums of squares and degrees of freedom as in one-way ANOVA. In accordance with the conventions used by many computer software packages, we use the term *model* when discussing the sums of squares and degrees of freedom calculated as in one-way ANOVA with IJ groups. Thus, SSM is a model sum of squares constructed from deviations of the form $\overline{x}_{ij} - \overline{x}$, where $\overline{x}$ is the average of all of the observations and $\overline{x}_{ij}$ is the mean of the (i,j)th group. Similarly, DFM is simply $IJ - 1$.

In two-way ANOVA, the terms SSM and DFM are broken down into terms corresponding to a main effect for A, a main effect for B, and an AB interaction. Each of SSM and DFM is then a sum of terms:

$$\text{SSM} = \text{SSA} + \text{SSB} + \text{SSAB}$$

and

$$\text{DFM} = \text{DFA} + \text{DFB} + \text{DFAB}$$

The term SSA represents variation among the means for the different levels of the factor A. Because there are I such means, DFA $= I - 1$. Similarly, SSB represents variation among the means for the different levels of the factor B, with DFB $= J - 1$ degrees of freedom.

Interactions are a bit more involved. We can see that SSAB, which is SSM $-$ SSA $-$ SSB, represents the variation in the model that is not accounted for by the main effects. By subtraction we see that its degrees of freedom are

$$\text{DFAB} = (IJ - 1) - (I - 1) - (J - 1) = (I - 1)(J - 1)$$

There are many kinds of interactions. The easiest way to study them is through examples.

EXAMPLE 13.4   The Bureau of Labor Statistics collects data on earnings of workers in the United States classified according to various characteristics. Here are the weekly earnings in dollars of men and women in two age groups who were working full-time in the first quarter of 1997:[2]

Age	Women	Men	Mean
16–19	239	264	251.5
20–24	302	333	317.5
Mean	270.5	298.5	284.5

The table includes averages of the means in the rows and columns. For example, the first entry in the far right margin is the average of the salaries for women and men aged 16 to 19:

$$\frac{239 + 264}{2} = 251.5$$

Similarly, the average of women's salaries in the two age groups is

$$\frac{239 + 302}{2} = 270.5$$

marginal means   These averages are called **marginal means** (because of their location at the *margins* of such tabulations).

Figure 13.1 is a plot of the group means. From the plot it is clear that men earn more than women. In statistical language, there is a main effect for gender. We also see that workers aged 20 to 24 earn more than those aged 16 to 19. This is the main effect of age. These main effects can be described by

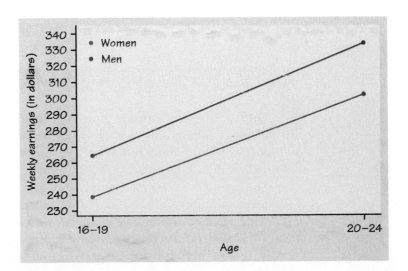

FIGURE 13.1   Plot of median weekly earnings of men and women for two age groups, Example 13.4.

differences between the marginal means. For example, the mean for men is $298.50, which is $28 higher than $270.50, the mean for women.

What about the interaction between gender and age? An interaction is present if the main effects provide an incomplete description of the data. The marginal means for gender show that the salary difference for men versus women is $28. This is the main effect of gender. The body of the table shows that the gender salary differences for the two age groups are $25 and $31. Although the difference is a bit higher for the older group, most of the information in the table concerning the gender difference is summarized by the main effect, the marginal difference of $28. Similarly, the main effect of age, $251.50 versus $317.50 summarizes the fact that the older group has higher earnings. The difference is similar for men and women. In this example, the interaction is small.

The lack of interaction can be seen in Figure 13.1. The line connecting the earnings for women and the line connecting the earnings for men are roughly parallel. This happens because the difference between the lines is the difference between the men's and women's salaries for each age group. As we have seen, this difference is approximately the same for the two groups.

In the next example, we add four more age groups to our data. We will see that in this case a very strong interaction is present.

EXAMPLE 13.5 The survey described in the previous example also gave weekly earnings for groups of older workers. Adding these gives the following table:

Age	Women	Men	Mean
16–19	239	264	251.5
20–24	302	333	317.5
25–34	422	514	468.0
35–44	475	639	557.0
45–54	496	692	594.0
55–64	426	660	543.0
Mean	393	517	455.0

Including the additional age groups changes the marginal means for gender and the overall mean of all groups. Figure 13.2 is a plot of the group means.

The two lines in Figure 13.2 are clearly not parallel. The gender difference in earnings depends on which age group we examine. The difference increases with age. This is an interaction between gender and age. The marginal means alone do not provide an adequate summary of the information in this table.

The presence of an interaction does not necessarily mean that the main effects are uninformative. It is still clear that men earn more than women, and this is evident in the marginal means for the two genders. Similarly, the marginal means for age show that earnings increase with age until the last group, where there is a decrease. We see this pattern in the plot for both men and women.

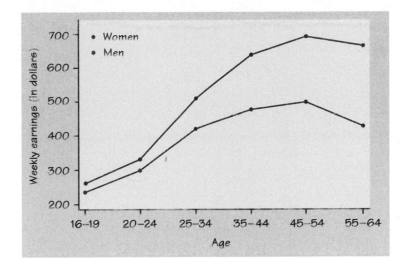

FIGURE 13.2   Plot of median weekly earnings of men and women for six age groups, Example 13.5.

Interactions come in many shapes and forms. When we find them, a careful examination of the means is needed to properly interpret the data. Simply stating that interactions are significant tells us little. Plots of the group means are very helpful.

EXAMPLE 13.6   A study of the energy expenditure of farmers in Burkina Faso (formerly known as Upper Volta) collected data during the wet and dry seasons.[3] The farmers grow millet during the wet season. In the dry season, there is relatively little activity because the ground is too hard to grow crops. The mean energy expended (in calories) by men and women in Burkina Faso during the wet and dry seasons is given in the following table:

Season	Men	Women	Mean
Dry	2310	2320	2315
Wet	3460	2890	3175
Mean	2885	2605	2745

Figure 13.3 is the plot of the group means.

During the dry season both men and women use about the same number of calories. When the wet season arrives, both genders use more energy. The men, who by social convention do most of the field work, expend much more energy than the women. The amount of energy used by the men in the wet season is very high by any reasonable standard. Such values are sometimes found in developed countries for people engaged in coal mining and lumberjacking.

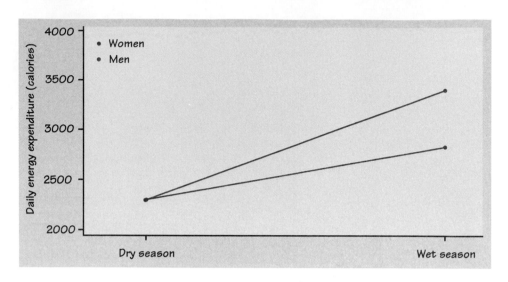

FIGURE 13.3 Plot of mean energy expenditures of farmers in Burkina Faso by sex and season, Example 13.6.

In a statistical analysis, the pattern of means shown in Figure 13.3 produced significant main effects for season and gender in addition to a season-by-gender interaction. The main effects record that men use more energy than women and that the wet season is associated with greater activity than the dry season. This clearly does not tell the whole story. We need to discuss the men and women in each of the two seasons to fully understand how energy is used in Burkina Faso.

A different kind of interaction is present in the next example. Here, we must be very cautious in our interpretation of the main effects since one of them can lead to a distorted conclusion.

EXAMPLE 13.7   An experiment to study how noise affects the performance of children tested second-grade hyperactive children and a control group of second graders who were not hyperactive.[4] One of the tasks involved solving math problems. The children solved problems under both high-noise and low-noise conditions. Here are the mean scores:

Group	High noise	Low noise	Mean
Control	214	170	192
Hyperactive	120	140	130
Mean	167	155	161

The means are plotted in Figure 13.4. In the analysis of this experiment, both of the main effects and the interaction were statistically significant. How are we to interpret these results?

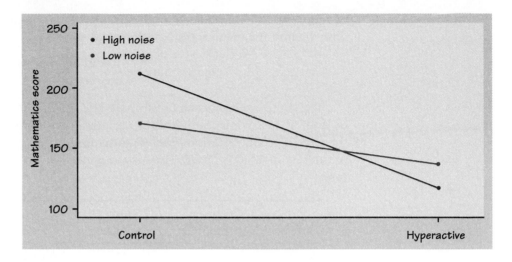

FIGURE 13.4 Plot of mean math scores of control and hyperactive children under low and high noise conditions, Example 13.7.

What catches our eye in the plot is that the lines cross. The hyperactive children did better in low noise than in high, but the control children performed better in high noise than in low. This interaction between noise and type of children is an important result. Which noise level is preferable depends on which type of children are in the class. The main effect for type of children is also straightforward: the normal children did better than the hyperactive children at both levels of noise. However, the main effect for noise could easily be misinterpreted if we do not look at the means carefully. Under high-noise conditions the marginal mean was 167, compared with a low-noise mean of 155. This suggests that high noise is, on the average, a more favorable condition than low noise. In fact, high noise is more favorable only for the control group. The hyperactive children performed better under the low-noise condition. Because of the interaction, we must know which type of children we have in order to say which noise level is better for working math problems.

The ANOVA table for two-way ANOVA

Two-way ANOVA is the statistical analysis for a two-way design with a quantitative response variable. The results of a two-way ANOVA are summarized in an ANOVA table based on splitting the total variation SST and the total degrees of freedom DFT among the two main effects and the interaction. Both the sums of squares (which measure variation) and the degrees of freedom add:

$$SST = SSA + SSB + SSAB + SSE$$
$$DFT = DFA + DFB + DFAB + DFE$$

The sums of squares are always calculated in practice by statistical software. When the n_{ij} are not all equal, some methods of analysis can give sums of squares that do not add. From each sum of squares and its degrees of freedom we find the mean square in the usual way:

$$\text{mean square} = \frac{\text{sum of squares}}{\text{degrees of freedom}}$$

The significance of each of the main effects and the interaction is assessed by an F statistic that compares the variation due to the effect of interest with the within-group variation. Each F statistic is the mean square for the source of interest divided by MSE. Here is the general form of the two-way ANOVA table:

Source	Degrees of freedom	Sum of squares	Mean square	F
A	$I - 1$	SSA	SSA/DFA	MSA/MSE
B	$J - 1$	SSB	SSB/DFB	MSB/MSE
AB	$(I - 1)(J - 1)$	SSAB	SSAB/DFAB	MSAB/MSE
Error	$N - IJ$	SSE	SSE/DFE	
Total	$N - 1$	SST	SST/DFT	

There are three null hypotheses in two-way ANOVA, with an F test for each. We can test for significance of the main effect of A, the main effect of B, and the AB interaction. It is generally good practice to examine the test for interaction first, since the presence of a strong interaction may influence the interpretation of the main effects. Be sure to plot the means as an aid to interpreting the results of the significance tests.

Significance Tests in Two-Way ANOVA

To test the main effect of A, use the F statistic

$$F_A = \frac{\text{MSA}}{\text{MSE}}$$

To test the main effect of B, use the F statistic

$$F_B = \frac{\text{MSB}}{\text{MSE}}$$

To test the interaction of A and B, use the F statistic

$$F_{AB} = \frac{\text{MSAB}}{\text{MSE}}$$

If the effect being tested is zero, the calculated F statistic has an F distribution with numerator degrees of freedom corresponding to the effect and denominator degrees of freedom equal to DFE. Large values of the F statistic lead to rejection of the null hypothesis. The P-value is the probability that a random variable having the corresponding F distribution is greater than or equal to the calculated value.

The following example illustrates how to do a two-way ANOVA. As with the one-way ANOVA, we focus our attention on interpretation of the computer output.

EXAMPLE 13.8

A study of cardiovascular risk factors compared runners who averaged at least 15 miles per week with a control group described as "generally sedentary." Both men and women were included in the study.[5] The design is a 2×2 ANOVA with the factors group and gender. There were 200 subjects in each of the four combinations. One of the variables measured was the heart rate after 6 minutes of exercise on a treadmill. SAS computer analysis produced the outputs in Figure 13.5 and Figure 13.6.

We begin with the usual preliminary examination. From Figure 13.5 we see that the ratio of the largest to the smallest standard deviation is less than 2. Therefore, we are not concerned about violating the assumption of equal

```
Analysis Variable : HR
FIGURE 13.5 Summary statistics for heart rates in the
four groups of the 2 x 2 ANOVA.

Analysis Variable : HR


 GROUP=Control GENDER=Female
   N            Mean          Std Dev       Std Error
----  --------------------------------------------
 200         148.00            16.27            1.15
----  --------------------------------------------


 GROUP=Control GENDER=Male
   N            Mean          Std Dev       Std Error
------------------------------------------------
 200         130.00            17.10            1.21
------------------------------------------------


 GROUP=Runners GENDER=Female
   N            Mean          Std Dev       Std Error
------------------------------------------------
 200         115.99            15.97            1.13
------------------------------------------------


 GROUP=Runners GENDER=Male
   N            Mean          Std Dev       Std Error
------------------------------------------------
 200         103.98            12.50            0.88
------------------------------------------------
```

FIGURE 13.5 Summary statistics for heart rates in the four groups of a 2×2 ANOVA, Example 13.8.

```
General Linear Models Procedure
Dependent Variable: HR
                                  Sum of        Mean
Source                   DF      Squares      Square   F Value    Pr > F
Model                     3   215256.09    71752.03    296.35    0.0001
Error                   796   192729.83      242.12
Corrected Total         799   407985.92

                 R-square          C.V.    Root MSE              HR Mean
                 0.527607      12.49924      15.560               124.49

Source                   DF    Type I SS  Mean Square  F Value    Pr > F
GROUP                     1   168432.08   168432.08    695.65    0.0001
GENDER                    1    45030.00    45030.00    185.98    0.0001
GROUP*GENDER              1     1794.01     1794.01      7.41    0.0066
```

FIGURE 13.6 Two-way analysis of variance output for heart rates, Example 13.8.

population standard deviations. Normal quantile plots (not shown) do not reveal any outliers, and the data appear to be reasonably normal.

The ANOVA table at the top of the output in Figure 13.6 is in effect a one-way ANOVA with four groups: female control, female runner, male control, and male runner. In this analysis Model has 3 degrees of freedom, and Error has 796 degrees of freedom. The F test and its associated P-value for this analysis refer to the hypothesis that all four groups have the same population mean. We are interested in the main effects and interaction, so we ignore this test.

The sums of squares for the group and gender main effects and the group-by-gender interaction appear at the bottom of Figure 13.6 under the heading Type I SS. These sum to the sum of squares for Model. Similarly, the degrees of freedom for these sums of squares sum to the degrees of freedom for Model. Two-way ANOVA splits the variation among the means (expressed by the Model sum of squares) into three parts that reflect the two-way layout.

Because the degrees of freedom are all 1 for the main effects and the interaction, the mean squares are the same as the sums of squares. The F statistics for the three effects appear in the column labeled F Value, and the P-values are under the heading Pr > F. For the group main effect, we verify the calculation of F as follows:

$$F = \frac{\text{MSG}}{\text{MSE}} = \frac{168,432}{242.12} = 695.65$$

All three effects are statistically significant. The group effect has the largest F, followed by the gender effect and the group-by-gender interaction. To interpret these results, we examine the plot of means with bars indicating

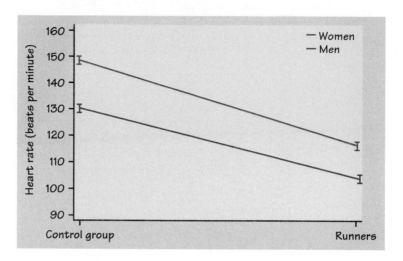

FIGURE 13.7   Plot of the group means with standard errors for heart rates in the 2×2 ANOVA, Example 13.8.

one standard error in Figure 13.7. Note that the standard errors are quite small due to the large sample sizes. The significance of the main effect for group is due to the fact that the controls have higher average heart rates than the runners for both genders. This is the largest effect evident in the plot.

The significance of the main effect for gender is due to the fact that the females have higher heart rates than men in both groups. The differences are not as large as those for the group effect, and this is reflected in the smaller value of the F statistic.

The analysis indicates that a complete description of the average heart rates requires consideration of the interaction in addition to the main effects. The two lines in the plot are not parallel. This interaction can be described in two ways. The female-male difference in average heart rates is greater for the controls than for the runners. Alternatively, the difference in average heart rates between controls and runners is greater for women than for men. As the plot suggests, the interaction is not large. It is statistically significant because there were 800 subjects in the study.

Two-way ANOVA output for the other software is similar to that given by SAS. Figure 13.8 gives the analysis of the heart rate data using Excel and Minitab.

SUMMARY

Two-way analysis of variance is used to compare population means when populations are classified according to two factors.

ANOVA assumes that the populations are normal with possibly different means and the same standard deviation and that independent SRSs are drawn from each population.

Excel

	E	F	G	H	I	J	K	L	M
1		Anova: Two Factor With Replication							
2									
3		SUMMARY	Control	Runners	Total				
4		*Female*							
5		Count	200	200	400				
6		Sum	29600	23197	52797				
7		Average	148	115.985	131.9925				
8		Variance	264.7437	255.0902	516.1478				
9									
10		*Male*							
11		Count	200	200	400				
12		Sum	26000	20795	46795				
13		Average	130	103.975	116.9875				
14		Variance	292.4221	156.2356.2	393.5161				
15									
16		*Total*							
17		Count	400	400					
18		Sum	55600	43992					
19		Average	139	109.98					
20		Variance	359.0877	241.2978					
21									
22									
23		ANOVA							
24		*Source of Variation*	SS	df	MS	F	P-value	F crit	
25		Sample	45030.01	1	45030.01	185.9799	3.29E-38	3.85316	
26		Columns	168432.1	1	168432.1	695.647	1.10E-110	3.85316	
27		Interaction	1794.005	1	1794.005	7.409481	0.00663	3.85316	
28		Within	192729.8	796	242.1229				
29									
30		Total	407985.9	799					
31									

hrate1 / Ready | Sum=0 | NUM

Minitab

```
Analysis of Variance for HR

Source         DF          SS          MS

Group           1      168432      168432

Gender          1       45030       45030

Interaction     1        1794        1794

Error         796      192730         242

Total         799      407986
```

FIGURE 13.8 Excel and Minitab two-way analysis of variance output for the heart rate study, Example 13.8.

As with one-way ANOVA, preliminary analysis includes examination of means, standard deviations, and normal quantile plots. **Marginal means** are calculated by taking averages of the cell means across rows and columns. Pooling is used to estimate the within-group variance.

ANOVA separates the total variation into parts for the **model** and **error.** The model variation is separated into parts for each of the **main effects** and the **interaction.**

The calculations are organized into an **ANOVA table.** F statistics and P-values are used to test hypotheses about the main effects and the interaction.

Careful inspection of the means is necessary to interpret significant main effects and interactions. Plots are a useful aid.

CHAPTER 13 EXERCISES

13.1 Each of the following situations is a two-way study design. For each case, identify the response variable and both factors, and state the number of levels for each factor (I and J) and the total number of observations (N).

(a) A study of the productivity of tomato plants compares five varieties of tomatoes and two types of fertilizer. Four plants of each variety are grown with each type of fertilizer. The yield in pounds of tomatoes is recorded for each plant.

(b) A marketing experiment compares six different types of packaging for a laundry detergent. A survey is conducted to determine the attractiveness of the packaging in six U.S. cities. Each type of packaging is shown to 50 different consumers in each city, who rate the attractiveness of the product on a 1 to 10 scale.

(c) To compare the effectiveness of four different weight-loss programs, 10 men and 10 women are randomly assigned to each. At the end of the program, the weight loss for each of the participants is recorded.

13.2 Each of the following situations is a two-way study design. For each case, identify the response variable and both factors, and state the number of levels for each factor (I and J) and the total number of observations (N).

(a) A study of smoking classifies subjects as nonsmokers, moderate smokers, or heavy smokers. Samples of 120 men and 120 women are drawn from each group. Each person reports the number of hours of sleep he or she gets on a typical night.

(b) The strength of concrete depends upon the formula used to prepare it. An experiment compares four different mixtures. Six specimens of concrete are poured from each mixture. Two of these specimens are subjected to 0 cycles of freezing and thawing, two are subjected to 100 cycles, and two specimens are subjected to 500 cycles. The strength of each specimen is then measured.

(c) Three methods for teaching sign language are to be compared. Seven students in special education and seven students in linguistics are randomly assigned to each of the methods and the scores on a final exam are recorded.

13.3 For each part of Exercise 13.1, outline the ANOVA table, giving the sources of variation and the degrees of freedom. (Do not compute the numerical values for the sums of squares and mean squares.)

13.4 For each part of Exercise 13.2, outline the ANOVA table, giving the sources of variation and the degrees of freedom. (Do not compute the numerical values for the sums of squares and mean squares.)

13.5 In the course of a clinical trial of measures to prevent coronary heart disease, blood pressure measurements were taken on 12,866 men. Individuals were classified by age group and race. The means for systolic blood pressure are given in the following table. (Data from W. M. Smith et al., "The multiple risk factor intervention trial," in H. M. Perry, Jr., and W. M. Smith (eds.), *Mild Hypertension: To Treat or Not to Treat,* New York Academy of Sciences, 1978, pp. 293–308.)

	35–39	40–44	45–49	50–54	55–59
White	131.0	132.3	135.2	139.4	142.0
Nonwhite	132.3	134.2	137.2	141.3	144.1

(a) Plot the group means, with age on the x axis and blood pressure on the y axis. For each racial group connect the points for the different ages.

(b) Describe the patterns you see. Does there appear to be a difference between the two racial groups? Does systolic blood pressure appear to vary with age? If so, how does it vary? Is there an interaction?

(c) Compute the marginal means. Then find the differences between the white and nonwhite mean blood pressures for each age group. Use this information to summarize numerically the patterns in the plot.

13.6 The means for diastolic blood pressure recorded in the clinical trial described in the previous exercise are:

	35–39	40–44	45–49	50–54	55–59
White	89.4	90.2	90.9	91.6	91.4
Nonwhite	91.2	93.1	93.3	94.5	93.5

(a) Plot the group means with age on the x axis and blood pressure on the y axis. For each racial group connect the points for the different ages.

(b) Describe the patterns you see. Does there appear to be a difference between the two racial groups? Does diastolic blood pressure appear to vary with age? If so, how does it vary? Is there an interaction between race and age?

(c) Compute the marginal means. Find the differences between the white and nonwhite mean blood pressures for each age group. Use this information to summarize numerically the patterns in the plot.

13.7 The amount of chromium in the diet has an effect on the way the body processes insulin. In an experiment designed to study this phenomenon, four diets were fed to male rats. There were two factors. Chromium had two levels: low (L) and normal (N). The rats were allowed to eat as much as they wanted (M) or the total amount that they could eat was restricted (R). We call the second factor Eat. One of the variables measured was the amount of an enzyme called GITH. The means for this response variable appear in the following table. (Data provided by Julie Hendricks and V. J. K. Liu of the Department of Foods and Nutrition, Purdue University.)

	Eat	
Chromium	M	R
L	4.545	5.175
N	4.425	5.317

(a) Make a plot of the mean GITH for these diets, with the factor Chromium on the x axis and GITH on the y axis. For each Eat group connect the points for the two Chromium means.

(b) Describe the patterns you see. Does the amount of chromium in the diet appear to affect the GITH mean? Does restricting the diet as compared with letting the rats eat as much as they want appear to have an effect? Is there an interaction?

(c) Compute the marginal means. Compute the differences between the M and R diets for each level of Chromium. Use this information to summarize numerically the patterns in the plot.

13.8 The Chapin Social Insight Test measures how well people can appraise others and predict what they may say or do. A study administered this test to different groups of people and compared the mean scores. Some of the results are given in the table below. Means for males and females who were psychology graduate students (PG) and liberal arts undergraduates (LU) are presented. The two factors are labeled Gender and Group. (This example is based on results reported in H. G. Gough, *The Chapin Social Insight Test*, Consulting Psychologists Press, Palo Alto, Calif., 1968.)

	Group	
Gender	PG	LU
Males	27.56	25.34
Females	29.25	24.94

Plot the means and describe the essential features of the data in terms of main effects and interactions.

13.9 The change-of-majors study described in Example 12.3 (page 748) classified students into one of three groups depending upon their major in the sophomore year. In this exercise we also consider the gender of the students. There are now two factors: major with three levels, and gender with two levels. The mean SAT mathematics scores for the six groups appear in the following table. For convenience we use the labels CS for computer science majors, EO for engineering and other science majors, and O for other majors.

Gender	Major		
	CS	EO	O
Males	628	618	589
Females	582	631	543

Describe the main effects and interaction using appropriate graphs and calculations.

13.10 The mean high school mathematics grades for the students in the previous exercise are summarized in the following table. The grades have been coded so that 10 = A, 9 = A−, etc.

Gender	Major		
	CS	EO	O
Males	8.68	8.35	7.65
Females	9.11	9.36	8.04

Summarize the results of this study using appropriate plots and calculations to describe the main effects and interaction.

13.11 A large research project studied the physical properties of wood materials constructed by bonding together small flakes of wood. Different species of trees were used, and the flakes were made in different sizes. One of the physical properties measured was the tension modulus of elasticity in the direction perpendicular to the alignment of the flakes, in pounds per square inch (psi). Some of the data are given in the following table. The sizes of the flakes are S1 = 0.015 inches by 2 inches and S2 = 0.025 inches by 2 inches. (Data provided by Michael Hunt and Bob Lattanzi of the Purdue University Forestry Department.)

Species	Size of flakes	
	S1	S2
Aspen	308	278
	428	398
	426	331
Birch	214	534
	433	512
	231	320
Maple	272	158
	376	503
	322	220

(a) Compute means and standard deviations for the three observations in each species-size group. Find the marginal mean for each species and for each size of flakes. Display the means and marginal means in a table.

(b) Plot the means of the six groups. Put species on the x axis and modulus of elasticity on the y axis. For each size connect the three points corresponding to the different species. Describe the patterns you see. Do the species appear to be different? What about the sizes? Does there appear to be an interaction?

(c) Run a two-way ANOVA on these data. Summarize the results of the significance tests. What do these results say about the impressions that you described in part (b) of this exercise?

13.12 Refer to the previous exercise. Another of the physical properties measured was the strength, in kilopounds per square inch (ksi), in the direction perpendicular to the alignment of the flakes. Some of the data are given in the following table. The sizes of the flakes are S1 = 0.015 inches by 2 inches and S2 = 0.025 inches by 2 inches.

	Size of flakes	
Species	S1	S2
Aspen	1296	1472
	1997	1441
	1686	1051
Birch	903	1422
	1246	1376
	1355	1238
Maple	1211	1440
	1827	1238
	1541	748

(a) Compute means and standard deviations for the three observations in each species-size group. Find the marginal means for the species and for the flake sizes. Display the means and marginal means in a table.

(b) Plot the means of the six groups. Put species on the x axis and strength on the y axis. For each size connect the three points corresponding to the different species. Describe the patterns you see. Do the species appear to be different? What about the sizes? Does there appear to be an interaction?

(c) Run a two-way ANOVA on these data. Summarize the results of the significance tests. What do these results say about the impressions that you described in part (b) of this exercise?

13.13 One step in the manufacture of large engines requires that holes of very precise dimensions be drilled. The tools that do the drilling are regularly

Tool	Time	Diameter		
1	1	25.030	25.030	25.032
1	2	25.028	25.028	25.028
1	3	25.026	25.026	25.026
2	1	25.016	25.018	25.016
2	2	25.022	25.020	25.018
2	3	25.016	25.016	25.016
3	1	25.005	25.008	25.006
3	2	25.012	25.012	25.014
3	3	25.010	25.010	25.008
4	1	25.012	25.012	25.012
4	2	25.018	25.020	25.020
4	3	25.010	25.014	25.018
5	1	24.996	24.998	24.998
5	2	25.006	25.006	25.006
5	3	25.000	25.002	24.999

TABLE 13.1 Tool diameter data

examined and are adjusted to ensure that the holes meet the required specifications. Part of the examination involves measurement of the diameter of the drilling tool. A team studying the variation in the sizes of the drilled holes selected this measurement procedure as a possible cause of variation in the drilled holes. They decided to use a designed experiment as one part of this examination. Some of the data are given in Table 13.1. The diameters in millimeters (mm) of five tools were measured by the same operator at three times (8:00 A.M., 11:00 A.M., and 3:00 P.M.). Three measurements were taken on each tool at each time. The person taking the measurements could not tell which tool was being measured and the measurements were taken in random order. (From Renée A. Jones and Regina P. Becker "Analysis of process variation," Purdue Technical Assistance Program Report TAP961210, 1996.)

(a) Make a table of means and standard deviations for each of the 5 × 3 combinations of the two factors.

(b) Plot the means and describe how the means vary with tool and time. Note that we expect the tools to have slightly different diameters. These will be adjusted as needed. It is the process of measuring the diameters that is important.

(c) Use a two-way ANOVA to analyze these data. Report the test statistics, degrees of freedom, and P-values for the significance tests.

(d) Write a short report summarizing your results.

13.14 Refer to the previous exercise. Multiply each measurement by 0.04 to convert from millimeters to inches. Redo the plots and rerun the ANOVA using the transformed measurements. Summarize what parts of the analysis have changed and what parts have remained the same.

13.15 How does the frequency that a supermarket product is promoted at a discount affect the price that customers expect to pay for the product? Does the percent reduction also affect this expectation? These questions were examined by researchers in a study conducted on students enrolled in an introductory management course at a large midwestern university. For 10 weeks 160 subjects received information about the products. The treatment conditions corresponded to the number of promotions (1, 3, 5, or 7) that were described during this 10-week period and the percent that the product was discounted (10%, 20%, 30%, and 40%). Ten students were randomly assigned to each of the 4 × 4 = 16 treatments. Table 13.2 gives the data. (Based on M. U. Kalwani and C. K. Yim, "Consumer price and promotion expectations: an experimental study," *Journal of Marketing Research*, 29 (1992), pp. 90–100.)

TABLE 13.2 Expected price data

Number of promotions	Percent discount	Expected price									
1	10	4.10	4.50	4.47	4.42	4.56	4.69	4.42	4.17	4.31	4.59
1	20	3.57	3.77	3.90	4.49	4.00	4.66	4.48	4.64	4.31	4.43
1	30	4.94	4.59	4.58	4.48	4.55	4.53	4.59	4.66	4.73	5.24
1	40	5.19	4.88	4.78	4.89	4.69	4.96	5.00	4.93	5.10	4.78
3	10	4.07	4.13	4.25	4.23	4.57	4.33	4.17	4.47	4.60	4.02
3	20	4.20	3.94	4.20	3.88	4.35	3.99	4.01	4.22	3.70	4.48
3	30	4.88	4.80	4.46	4.73	3.96	4.42	4.30	4.68	4.45	4.56
3	40	4.90	5.15	4.68	4.98	4.66	4.46	4.70	4.37	4.69	4.97
5	10	3.89	4.18	3.82	4.09	3.94	4.41	4.14	4.15	4.06	3.90
5	20	3.90	3.77	3.86	4.10	4.10	3.81	3.97	3.67	4.05	3.67
5	30	4.11	4.35	4.17	4.11	4.02	4.41	4.48	3.76	4.66	4.44
5	40	4.31	4.36	4.75	4.62	3.74	4.34	4.52	4.37	4.40	4.52
7	10	3.56	3.91	4.05	3.91	4.11	3.61	3.72	3.69	3.79	3.45
7	20	3.45	4.06	3.35	3.67	3.74	3.80	3.90	4.08	3.52	4.03
7	30	3.89	4.45	3.80	4.15	4.41	3.75	3.98	4.07	4.21	4.23
7	40	4.04	4.22	4.39	3.89	4.26	4.41	4.39	4.52	3.87	4.70

(a) Summarize the means and standard deviations in a table and plot the means. Summarize main features of the plot.

(b) Analyze the data with a two-way ANOVA. Report the results of this analysis.

(c) Using your plot and the ANOVA results, prepare a short report explaining how the expected price depends on the number of promotions and the percent of the discount.

13.16   Refer to the previous exercise. Rerun the analysis as a one-way ANOVA with $4 \times 4 = 16$ treatments. Summarize the results of this analysis. Use the Bonferroni multiple comparisons procedure to describe combinations of number of promotions and percent discounts that are similar or different.

13.17   Return to the chromium-insulin experiment described in Exercise 13.7. Here is part of the ANOVA table for these data:

Source	Degrees of freedom	Sum of squares	Mean square	F
A (Chromium)		0.00121		
B (Eat)		5.79121		
AB		0.17161		
Error		1.08084		
Total				

(a) In all, 40 rats were used in this experiment. Fill in the missing values in the ANOVA table.

(b) What is the value of the F statistic to test the null hypothesis that there is no interaction? What is its distribution when the null hypothesis is true? Using Table E, find an approximate P-value for this test.

(c) Answer the questions in part (b) for the main effect of Chromium and the main effect of Eat.

(d) What is s_p^2, the within-group variance? What is s_p?

(e) Using what you have learned in this exercise and your answer to Exercise 13.7, summarize the results of this experiment.

13.18   Return to the Chapin Social Insight Test study described in Exercise 13.8. Part of the ANOVA table for these data is given below:

Source	Degrees of freedom	Sum of squares	Mean square	F
A (Gender)		62.40		
B (Group)		1,599.03		
AB				
Error		13,633.29		
Total		15,458.52		

(a) There were 150 individuals tested in each of the groups. Fill in the missing values in the ANOVA table.

(b) What is the value of the F statistic to test the null hypothesis that there is no interaction? What is its distribution when the null hypothesis is true? Using Table E, find an approximate P-value for this test.

(c) Answer the questions in part (b) for the main effect of Gender and the main effect of Group.

(d) What is s_p^2, the within-group variance? What is s_p?

(e) Using what you have learned in this exercise and your answer to Exercise 13.8, summarize the results of this study.

13.19 Do left-handed people live shorter lives than right-handed people? A study of this question examined a sample of 949 death records and contacted next of kin to determine handedness. Note that there are many possible definitions of "left-handed." The researchers examined the effects of different definitions on the results of their analysis and found that their conclusions were not sensitive to the exact definition used. For the results presented here, people were defined to be right-handed if they wrote, drew, and threw a ball with the right hand. All others were defined to be left-handed. People were classified by gender (female or male) and handedness (left or right), and a 2 × 2 ANOVA was run with the age at death as the response variable. The F statistics were 22.36 (handedness), 37.44 (gender), and 2.10 (interaction). The following marginal mean ages at death (in years) were reported: 77.39 (females), 71.32 (males), 75.00 (right-handed), and 66.03 (left-handed). (For a summary of this study and other research in this area see Stanley Coren and Diane F. Halpern, "Left-handedness: a marker for decreased survival fitness," *Psychological Bulletin*, 109 (1991), pp. 90–106.)

(a) For each of the F statistics given above find the degrees of freedom and an approximate P-value. Summarize the results of these tests.

(b) Using the information given, write a short summary of the results of the study.

13.20 Scientists believe that exposure to the radioactive gas radon is associated with some types of cancers in the respiratory system. Radon from natural sources is present in many homes in the United States. A group of researchers decided to study the problem in dogs because dogs get similar types of cancers and are exposed to environments similar to their owners. Radon detectors are available for home monitoring but the researchers wanted to obtain actual measures of the exposure of a sample of dogs. To do this they placed the detectors in holders and attached them to the collars of the dogs. One problem was that the holders might in some way affect the radon readings. The researchers therefore devised a laboratory experiment to study the effects of the holders. Detectors from four series of production were available, so they used a two-way ANOVA design (Series with 4 levels and Holder with 2, representing the presence or absence of a holder). All detectors were exposed to the same radon source and the radon measure in picocuries per liter was recorded. (Data provided by Neil Zimmerman of the Purdue University School of Health Sciences.) Here is the SAS output:

```
                                Sum of         Mean
Source                 DF       Squares       Square   F Value   Pr > F
Model                   7    15612.073      2230.296      3.82   0.0017
Error                  61    35608.165       583.740
Corrected Total        68    51220.238


               R-Square          C.V.      Root MSE           RADON Mean
               0.304803      7.837457        24.161               308.27


Source                 DF    Type I SS  Mean Square   F Value   Pr > F
SERIES                  3    12293.636     4097.879      7.02   0.0004
HOLDER                  1     1144.666     1144.666      1.96   0.1665
SERIES*HOLDER           3     2173.771      724.590      1.24   0.3026
```

(a) Summarize the results of the ANOVA. Is there evidence to suggest that the holders make a difference in the radon readings?

(b) The mean radon readings for the four series were 330, 303, 302, and 295. The results of the significance test for Series were of great concern to the researchers. Explain why.

13.21 Refer to the data given for the change-of-majors study in the data set MAJORS described in the Data Appendix. Analyze the data for SATV, the SAT verbal score. Your analysis should include a table of sample sizes, means, and standard deviations; normal quantile plots; a plot of the means; and a two-way ANOVA using sex and major as the factors. Write a short summary of your conclusions.

13.22 Refer to the data given for the change-of-majors study in the data set MAJORS described in the Data Appendix. Analyze the data for HSS, the high school science grades. Your analysis should include a table of sample sizes, means, and standard deviations; normal quantile plots; a plot of the means; and a two-way ANOVA using sex and major as the factors. Write a short summary of your conclusions.

13.23 Refer to the data given for the change-of-majors study in the data set MAJORS described in the Data Appendix. Analyze the data for HSE, the high school English grades. Your analysis should include a table of sample sizes, means, and standard deviations; normal quantile plots; a plot of the means; and a two-way ANOVA using sex and major as the factors. Write a short summary of your conclusions.

13.24 Refer to the data given for the change-of-majors study in the data set MAJORS described in the Data Appendix. Analyze the data for GPA, the college grade point average. Your analysis should include a table of sample sizes, means, and standard deviations; normal quantile plots; a plot of the means; and a two-way ANOVA using sex and major as the factors. Write a short summary of your conclusions.

NOTES

1. We present the two-way ANOVA model and analysis for the general case in which the sample sizes may be unequal. If the sample sizes vary a great deal, serious complications can arise. There is no longer a single standard ANOVA analysis. Most computer packages offer several options for the computation of the ANOVA table when cell counts are unequal. When the counts are approximately equal, all methods give essentially the same results.

2. The data in Examples 13.4 and 13.5 are from http://stats.bls.gov, the U.S. Bureau of Labor Statistics Web site.

3. This example is based on a study described in P. Payne, "Nutrition adaptation in man: social adjustments and their nutritional implications," in K. Blaxter and J. C. Waterlow (eds.), *Nutrition Adaptation in Man*, Libbey, London, 1985.

4. Example 13.7 is based on a study described in S. S. Zentall and J. H. Shaw, "Effects of classroom noise on performance and activity of second-grade hyperactive and control children," *Journal of Educational Psychology*, 72 (1980), pp. 830–840.

5. This example is based on a study described in P. D. Wood et al., "Plasma lipoprotein distributions in male and female runners," in P. Milvey (ed.), *The Marathon: Physiological, Medical, Epidemiological, and Psychological Studies*, New York Academy of Sciences, 1977.

DATA APPENDIX

Some of the computer exercises in the text refer to seven relatively large data sets that are on the CD that accompanies this text. The CD also contains data from many other exercises and examples.

Background information for each of the seven data sets is presented below. For each, the first five cases are given here.

1 Cheese

Text Reference:
Exercises 1.124 and
11.23–11.31

As cheddar cheese matures, a variety of chemical processes take place. The taste of matured cheese is related to the concentration of several chemicals in the final product. In a study of cheddar cheese from the LaTrobe Valley of Victoria, Australia, samples of cheese were analyzed for their chemical composition and were subjected to taste tests.

Data for one type of cheese manufacturing process appear below. The variable "Case" is used to number the observations from 1 to 30. "Taste" is the response variable of interest. The taste scores were obtained by combining the scores from several tasters.

Three of the chemicals whose concentrations were measured were acetic acid, hydrogen sulfide, and lactic acid. For acetic acid and hydrogen sulfide (natural) log transformations were taken. Thus the explanatory variables are the transformed concentrations of acetic acid ("Acetic") and hydrogen sulfide ("H2S") and the untransformed concentration of lactic acid ("Lactic"). These data are based on experiments performed by G. T. Lloyd and E. H. Ramshaw of the CSIRO Division of Food Research, Victoria, Australia. Some results of the statistical analyses of these data are given in G. P. McCabe, L. McCabe, and A. Miller, "Analysis of taste and chemical composition of cheddar cheese, 1982–83 experiments," CSIRO Division of Mathematics and Statistics Consulting Report VT85/6, and in I. Barlow et al., "Correlations and changes in flavour and chemical parameters of cheddar cheeses during maturation," *Australian Journal of Dairy Technology*, 44 (1989), pp. 7–18.

Case	Taste	Acetic	H2S	Lactic
01	12.3	4.543	3.135	0.86
02	20.9	5.159	5.043	1.53
03	39.0	5.366	5.438	1.57
04	47.9	5.759	7.496	1.81
05	5.6	4.663	3.807	0.99

2 CSDATA

Text Reference:
Exercises 1.125, 2.123,
3.64, 7.113, and
11.12–11.19

The computer science department of a large university was interested in understanding why a large proportion of their first-year students failed to graduate as computer science majors. An examination of records from the registrar indicated that most of the attrition occurred during the first three semesters. Therefore, they decided to study all first-year students entering their program in a particular year and to follow their progress for the first three semesters.

The variables studied included the grade point average after three semesters and a collection of variables that would be available as students entered their program. These included standardized tests such as the SATs and high school grades in various subjects. The individuals who conducted the study were also interested in examining differences between men and women in this program. Therefore, sex was included as a variable.

Data on 224 students who began study as computer science majors in a particular year were analyzed. A few exceptional cases were excluded, such as students who did not have complete data available on the variables of interest (a few students were admitted who did not take the SATs). Data for the first 5 students appear below. There are eight variables for each student. OBS is a variable used to identify the student. The data files kept by the registrar identified students by social security number, but for this study they were simply given a number from 1 to 224. The grade point average after three semesters is the variable GPA. This university uses a four-point scale, with A corresponding to 4, B to 3, C to 2, etc. A straight A student has a 4.00 GPA.

The high school grades included in the data set are the variables HSM, HSS, and HSE. These correspond to average high school grades in math, science, and English. High schools use different grading systems (some high schools have a grade higher than A for honors courses), so the university's task in constructing these variables is not easy. The researchers were willing to accept the university's judgment and used its values. High school grades were recorded on a scale from 1 to 10, with 10 corresponding to A, 9 to A−, 8 to B+, etc.

The SAT scores are SATM and SATV, corresponding to the mathematics and verbal parts of the SAT. Gender was recorded as 1 for men and 2 for women. This is an arbitrary code. For software packages that can use alphanumeric variables (values do not have to be numbers), it is more convenient to use M and F or Men and Women as values for the sex variable. With this kind of user-friendly capability, you do not have to remember who are the 1's and who are the 2's.

Results of the study are reported in P. F. Campbell and G. P. McCabe, "Predicting the success of freshmen in a computer science major," *Communications of the ACM*, 27 (1984), pp. 1108–1113. The table below gives data for the first 5 students.

OBS	GPA	HSM	HSS	HSE	SATM	SATV	SEX
001	3.32	10	10	10	670	600	1
002	2.26	6	8	5	700	640	1
003	2.35	8	6	8	640	530	1
004	2.08	9	10	7	670	600	1
005	3.38	8	9	8	540	580	1

3   WOOD

Text Reference:
Exercises 2.124–2.126,
7.114, 10.30, and
10.31

The WOOD data set comes from a study of the strength of wood products conducted by Mike Hunt and Mike Triche of the Purdue University Department of Forestry. These researchers measured the modulus of elasticity for each of 50 strips of yellow poplar wood two times. The units are millions of pounds per square inch. The data for the first 5 strips appear below. The variable S identifies the strip. T1 and T2 are the measures of the modulus of elasticity at the two times.

S	T1	T2
01	1.58	1.55
02	1.53	1.54
03	1.38	1.37
04	1.24	1.25
05	1.59	1.57

4   Reading

Text Reference:
Exercises 7.115, 12.45–
12.48 and Examples
12.6 and 12.8–12.18

Jim Baumann and Leah Jones of the Purdue University School of Education conducted a study to compare three methods of teaching reading comprehension. The
66 students who participated in the study were randomly assigned to the methods
(22 to each). The standard practice of comparing new methods with a traditional one
was used in this study. The traditional method is called Basal and the two innovative
methods are called DRTA and Strat.

In the data set the variable Subject is used to identify the individual students. The
values are 1 to 66. The method of instruction is indicated by the variable Group, with
values B, D, and S, corresponding to Basal, DRTA, and Strat. Two pretests and three
posttests were given to all students. These are the variables Pre1, Pre2, Post1, Post2,
and Post3. Data for the first 5 subjects are given below.

Subject	Group	Pre1	Pre2	Post1	Post2	Post3
01	B	4	3	5	4	41
02	B	6	5	9	5	41
03	B	9	4	5	3	43
04	B	12	6	8	5	46
05	B	16	5	10	9	46

5   Majors

Text Reference:
Exercises 13.21–13.24

See the description of the CSDATA data set for background information on the study
for this data set. In this data file, the variables described for CSDATA are given with an
additional variable "Maj" that specifies the student's major field of study at the end of
three semesters. The codes 1, 2, and 3 correspond to Computer Science, Engineering
and Other Sciences, and Other. All available data were used in the analyses performed,
which resulted in sample sizes that were unequal in the six sex-by-major groups.

For a one-way ANOVA this causes no particular problems. However, for a two-
way ANOVA several complications arise when the sample sizes are unequal. A detailed
discussion of these complications is beyond the scope of this text. To avoid these difficulties and still use these interesting data, simulated data based on the results of this
study are given on the data disk. ANOVA based on these simulated data gives the same
qualitative conclusions as those obtained with the original data. Here are the first 5
cases:

OBS	SEX	Maj	SATM	SATV	HSM	HSS	HSE	GPA
001	1	1	640	530	8	6	8	2.35
002	1	1	670	600	9	10	7	2.08
003	1	1	600	400	8	8	7	3.21
004	1	1	570	480	7	7	6	2.34
005	1	1	510	530	6	8	8	1.40

6   CONCEPT

Text Reference:
Exercises 11.20–11.22

Darlene Gordon of the Purdue University School of Education provided the data in
the CONCEPT data set. The data were collected on 78 seventh-grade students in a
rural midwestern school. The research concerned the relationship between the students' "self-concept" and their academic performance. The variables are OBS, a subject identification number; GPA, grade point average; IQ, score on an IQ test; AGE, age
in years; SEX, female (1) or male (2); SC, overall score on the Piers-Harris Children's

Self-Concept Scale; and C1 to C6, "cluster scores" for specific aspects of self-concept: C1 = behavior, C2 = school status, C3 = physical appearance, C4 = anxiety, C5 = popularity, and C6 = happiness. Here are the first 5 cases:

OBS	GPA	IQ	AGE	SEX	SC	C1	C2	C3	C4	C5	C6
001	7.940	111	13	2	67	15	17	13	13	11	9
002	8.292	107	12	2	43	12	12	7	7	6	6
003	4.643	100	13	2	52	11	10	5	8	9	7
004	7.470	107	12	2	66	14	15	11	11	9	9
005	8.882	114	12	1	58	14	15	10	12	11	6

7 DANDRUFF

Text Reference:
Exercises 12.38–12.41

The DANDRUFF data set is based on W. L. Billhimer et al. "Results of a clinical trial comparing 1% pyrithione zinc and 2% ketoconazole shampoos," *Cosmetic Dermatology*, 9 (1996), pp. 34–39. The study reported in this paper is a clinical trial that compared three treatments for dandruff and a placebo. The treatments were 1% pyrithione zinc shampoo (PyrI), the same shampoo but with instructions to shampoo two times (PyrII), 2% ketoconazole shampoo (Keto), and a placebo shampoo (Placebo). After six weeks of treatment, eight sections of the scalp were examined and given a score that measured the amount of scalp flaking on a 0 to 10 scale. The response variable was the sum of these eight scores. An analysis of the baseline flaking measure indicated that randomization of patients to treatments was successful in that no differences were found between the groups. At baseline there were 112 subjects in each of the three treatment groups and 28 subjects in the Placebo group. During the clinical trial 3 dropped out from the PyrII group and 6 from the Keto group. No patients dropped out of the other two groups. Summary statistics given in the paper were used to generate random data that give the same conclusions. Here are the first 5 cases:

OBS	Treatment	Flaking
001	PyrI	17
002	PyrI	16
003	PyrI	18
004	PyrI	17
005	PyrI	18

TABLES

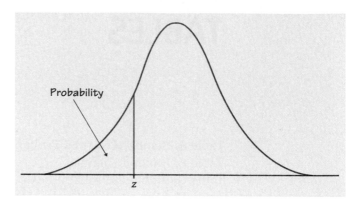

Table entry for z is the probability lying below z.

TABLE A Standard normal probabilities

z	.00	.01	.02	.03	.04	.05	.06	.07	.08	.09
−3.4	.0003	.0003	.0003	.0003	.0003	.0003	.0003	.0003	.0003	.0002
−3.3	.0005	.0005	.0005	.0004	.0004	.0004	.0004	.0004	.0004	.0003
−3.2	.0007	.0007	.0006	.0006	.0006	.0006	.0006	.0005	.0005	.0005
−3.1	.0010	.0009	.0009	.0009	.0008	.0008	.0008	.0008	.0007	.0007
−3.0	.0013	.0013	.0013	.0012	.0012	.0011	.0011	.0011	.0010	.0010
−2.9	.0019	.0018	.0018	.0017	.0016	.0016	.0015	.0015	.0014	.0014
−2.8	.0026	.0025	.0024	.0023	.0023	.0022	.0021	.0021	.0020	.0019
−2.7	.0035	.0034	.0033	.0032	.0031	.0030	.0029	.0028	.0027	.0026
−2.6	.0047	.0045	.0044	.0043	.0041	.0040	.0039	.0038	.0037	.0036
−2.5	.0062	.0060	.0059	.0057	.0055	.0054	.0052	.0051	.0049	.0048
−2.4	.0082	.0080	.0078	.0075	.0073	.0071	.0069	.0068	.0066	.0064
−2.3	.0107	.0104	.0102	.0099	.0096	.0094	.0091	.0089	.0087	.0084
−2.2	.0139	.0136	.0132	.0129	.0125	.0122	.0119	.0116	.0113	.0110
−2.1	.0179	.0174	.0170	.0166	.0162	.0158	.0154	.0150	.0146	.0143
−2.0	.0228	.0222	.0217	.0212	.0207	.0202	.0197	.0192	.0188	.0183
−1.9	.0287	.0281	.0274	.0268	.0262	.0256	.0250	.0244	.0239	.0233
−1.8	.0359	.0351	.0344	.0336	.0329	.0322	.0314	.0307	.0301	.0294
−1.7	.0446	.0436	.0427	.0418	.0409	.0401	.0392	.0384	.0375	.0367
−1.6	.0548	.0537	.0526	.0516	.0505	.0495	.0485	.0475	.0465	.0455
−1.5	.0668	.0655	.0643	.0630	.0618	.0606	.0594	.0582	.0571	.0559
−1.4	.0808	.0793	.0778	.0764	.0749	.0735	.0721	.0708	.0694	.0681
−1.3	.0968	.0951	.0934	.0918	.0901	.0885	.0869	.0853	.0838	.0823
−1.2	.1151	.1131	.1112	.1093	.1075	.1056	.1038	.1020	.1003	.0985
−1.1	.1357	.1335	.1314	.1292	.1271	.1251	.1230	.1210	.1190	.1170
−1.0	.1587	.1562	.1539	.1515	.1492	.1469	.1446	.1423	.1401	.1379
−0.9	.1841	.1814	.1788	.1762	.1736	.1711	.1685	.1660	.1635	.1611
−0.8	.2119	.2090	.2061	.2033	.2005	.1977	.1949	.1922	.1894	.1867
−0.7	.2420	.2389	.2358	.2327	.2296	.2266	.2236	.2206	.2177	.2148
−0.6	.2743	.2709	.2676	.2643	.2611	.2578	.2546	.2514	.2483	.2451
−0.5	.3085	.3050	.3015	2981	2946	2912	.2877	.2843	.2810	.2776
−0.4	.3446	.3409	.3372	.3336	.3300	.3264	.3228	.3192	.3156	.3121
−0.3	.3821	.3783	.3745	.3707	.3669	.3632	.3594	.3557	.3520	.3483
−0.2	.4207	.4168	.4129	.4090	.4052	.4013	.3974	.3936	.3897	.3859
−0.1	.4602	.4562	.4522	.4483	.4443	.4404	.4364	.4325	.4286	.4247
−0.0	.5000	.4960	.4920	.4880	.4840	.4801	.4761	.4721	.4681	.4641

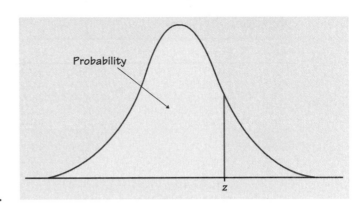

Table entry for z is the
probability lying below z.

TABLE A Standard normal probabilities (*Continued*)

z	.00	.01	.02	.03	.04	.05	.06	.07	.08	.09
0.0	.5000	.5040	.5080	.5120	.5160	.5199	.5239	.5279	.5319	.5359
0.1	.5398	.5438	.5478	.5517	.5557	.5596	.5636	.5675	.5714	.5753
0.2	.5793	.5832	.5871	.5910	.5948	.5987	.6026	.6064	.6103	.6141
0.3	.6179	.6217	.6255	.6293	.6331	.6368	.6406	.6443	.6480	.6517
0.4	.6554	.6591	.6628	.6664	.6700	.6736	.6772	.6808	.6844	.6879
0.5	.6915	.6950	.6985	.7019	.7054	.7088	.7123	.7157	.7190	.7224
0.6	.7257	.7291	.7324	.7357	.7389	.7422	.7454	.7486	.7517	.7549
0.7	.7580	.7611	.7642	.7673	.7704	.7734	.7764	.7794	.7823	.7852
0.8	.7881	.7910	.7939	.7967	.7995	.8023	.8051	.8078	.8106	.8133
0.9	.8159	.8186	.8212	.8238	.8264	.8289	.8315	.8340	.8365	.8389
1.0	.8413	.8438	.8461	.8485	.8508	.8531	.8554	.8577	.8599	.8621
1.1	.8643	.8665	.8686	.8708	.8729	.8749	.8770	.8790	.8810	.8830
1.2	.8849	.8869	.8888	.8907	.8925	.8944	.8962	.8980	.8997	.9015
1.3	.9032	.9049	.9066	.9082	.9099	.9115	.9131	.9147	.9162	.9177
1.4	.9192	.9207	.9222	.9236	.9251	.9265	.9279	.9292	.9306	.9319
1.5	.9332	.9345	.9357	.9370	.9382	.9394	.9406	.9418	.9429	.9441
1.6	.9452	.9463	.9474	.9484	.9495	.9505	.9515	.9525	.9535	.9545
1.7	.9554	.9564	.9573	.9582	.9591	.9599	.9608	.9616	.9625	.9633
1.8	.9641	.9649	.9656	.9664	.9671	.9678	.9686	.9693	.9699	.9706
1.9	.9713	.9719	.9726	.9732	.9738	.9744	.9750	.9756	.9761	.9767
2.0	.9772	.9778	.9783	.9788	.9793	.9798	.9803	.9808	.9812	.9817
2.1	.9821	.9826	.9830	.9834	.9838	.9842	.9846	.9850	.9854	.9857
2.2	.9861	.9864	.9868	.9871	.9875	.9878	.9881	.9884	.9887	.9890
2.3	.9893	.9896	.9898	.9901	.9904	.9906	.9909	.9911	.9913	.9916
2.4	.9918	.9920	.9922	.9925	.9927	.9929	.9931	.9932	.9934	.9936
2.5	.9938	.9940	.9941	.9943	.9945	.9946	.9948	.9949	.9951	.9952
2.6	.9953	.9955	.9956	.9957	.9959	.9960	.9961	.9962	.9963	.9964
2.7	.9965	.9966	.9967	.9968	.9969	.9970	.9971	.9972	.9973	.9974
2.8	.9974	.9975	.9976	.9977	.9977	.9978	.9979	.9979	.9980	.9981
2.9	.9981	.9982	.9982	.9983	.9984	.9984	.9985	.9985	.9986	.9986
3.0	.9987	.9987	.9987	.9988	.9988	.9989	.9989	.9989	.9990	.9990
3.1	.9990	.9991	.9991	.9991	.9992	.9992	.9992	.9992	.9993	.9993
3.2	.9993	.9993	.9994	.9994	.9994	.9994	.9994	.9995	.9995	.9995
3.3	.9995	.9995	.9995	.9996	.9996	.9996	.9996	.9996	.9996	.9997
3.4	.9997	.9997	.9997	.9997	.9997	.9997	.9997	.9997	.9997	.9998

TABLE B Random digits

Line								
101	19223	95034	05756	28713	96409	12531	42544	82853
102	73676	47150	99400	01927	27754	42648	82425	36290
103	45467	71709	77558	00095	32863	29485	82226	90056
104	52711	38889	93074	60227	40011	85848	48767	52573
105	95592	94007	69971	91481	60779	53791	17297	59335
106	68417	35013	15529	72765	85089	57067	50211	47487
107	82739	57890	20807	47511	81676	55300	94383	14893
108	60940	72024	17868	24943	61790	90656	87964	18883
109	36009	19365	15412	39638	85453	46816	83485	41979
110	38448	48789	18338	24697	39364	42006	76688	08708
111	81486	69487	60513	09297	00412	71238	27649	39950
112	59636	88804	04634	71197	19352	73089	84898	45785
113	62568	70206	40325	03699	71080	22553	11486	11776
114	45149	32992	75730	66280	03819	56202	02938	70915
115	61041	77684	94322	24709	73698	14526	31893	32592
116	14459	26056	31424	80371	65103	62253	50490	61181
117	38167	98532	62183	70632	23417	26185	41448	75532
118	73190	32533	04470	29669	84407	90785	65956	86382
119	95857	07118	87664	92099	58806	66979	98624	84826
120	35476	55972	39421	65850	04266	35435	43742	11937
121	71487	09984	29077	14863	61683	47052	62224	51025
122	13873	81598	95052	90908	73592	75186	87136	95761
123	54580	81507	27102	56027	55892	33063	41842	81868
124	71035	09001	43367	49497	72719	96758	27611	91596
125	96746	12149	37823	71868	18442	35119	62103	39244
126	96927	19931	36089	74192	77567	88741	48409	41903
127	43909	99477	25330	64359	40085	16925	85117	36071
128	15689	14227	06565	14374	13352	49367	81982	87209
129	36759	58984	68288	22913	18638	54303	00795	08727
130	69051	64817	87174	09517	84534	06489	87201	97245
131	05007	16632	81194	14873	04197	85576	45195	96565
132	68732	55259	84292	08796	43165	93739	31685	97150
133	45740	41807	65561	33302	07051	93623	18132	09547
134	27816	78416	18329	21337	35213	37741	04312	68508
135	66925	55658	39100	78458	11206	19876	87151	31260
136	08421	44753	77377	28744	75592	08563	79140	92454
137	53645	66812	61421	47836	12609	15373	98481	14592
138	66831	68908	40772	21558	47781	33586	79177	06928
139	55588	99404	70708	41098	43563	56934	48394	51719
140	12975	13258	13048	45144	72321	81940	00360	02428
141	96767	35964	23822	96012	94591	65194	50842	53372
142	72829	50232	97892	63408	77919	44575	24870	04178
143	88565	42628	17797	49376	61762	16953	88604	12724
144	62964	88145	83083	69453	46109	59505	69680	00900
145	19687	12633	57857	95806	09931	02150	43163	58636
146	37609	59057	66967	83401	60705	02384	90597	93600
147	54973	86278	88737	74351	47500	84552	19909	67181
148	00694	05977	19664	65441	20903	62371	22725	53340
149	71546	05233	53946	68743	72460	27601	45403	88692
150	07511	88915	41267	16853	84569	79367	32337	03316

TABLE B Random digits (*Continued*)

Line								
151	03802	29341	29264	80198	12371	13121	54969	43912
152	77320	35030	77519	41109	98296	18984	60869	12349
153	07886	56866	39648	69290	03600	05376	58958	22720
154	87065	74133	21117	70595	22791	67306	28420	52067
155	42090	09628	54035	93879	98441	04606	27381	82637
156	55494	67690	88131	81800	11188	28552	25752	21953
157	16698	30406	96587	65985	07165	50148	16201	86792
158	16297	07626	68683	45335	34377	72941	41764	77038
159	22897	17467	17638	70043	36243	13008	83993	22869
160	98163	45944	34210	64158	76971	27689	82926	75957
161	43400	25831	06283	22138	16043	15706	73345	26238
162	97341	46254	88153	62336	21112	35574	99271	45297
163	64578	67197	28310	90341	37531	63890	52630	76315
164	11022	79124	49525	63078	17229	32165	01343	21394
165	81232	43939	23840	05995	84589	06788	76358	26622
166	36843	84798	51167	44728	20554	55538	27647	32708
167	84329	80081	69516	78934	14293	92478	16479	26974
168	27788	85789	41592	74472	96773	27090	24954	41474
169	99224	00850	43737	75202	44753	63236	14260	73686
170	38075	73239	52555	46342	13365	02182	30443	53229
171	87368	49451	55771	48343	51236	18522	73670	23212
172	40512	00681	44282	47178	08139	78693	34715	75606
173	81636	57578	54286	27216	58758	80358	84115	84568
174	26411	94292	06340	97762	37033	85968	94165	46514
175	80011	09937	57195	33906	94831	10056	42211	65491
176	92813	87503	63494	71379	76550	45984	05481	50830
177	70348	72871	63419	57363	29685	13090	18763	31714
178	24005	52114	26224	39078	80798	15220	13186	00976
179	85063	55810	10470	08029	30025	29734	61181	72090
180	11532	73186	92541	06915	72954	10167	12142	26492
181	59618	03914	05208	84088	20426	39004	84582	87317
182	92965	50837	39921	84661	82514	81899	24565	60874
183	85116	27684	14597	85747	01596	25889	41998	15635
184	15106	10411	90221	49377	44369	28185	80959	76355
185	03638	31589	07871	25792	85823	55400	56026	12193
186	97971	48932	45792	63993	95635	28753	46069	84635
187	49345	18305	76213	82390	77412	97401	50650	71755
188	87370	88099	89695	87633	76987	85503	26257	51736
189	88296	95670	74932	65317	93848	43988	47597	83044
190	79485	92200	99401	54473	34336	82786	05457	60343
191	40830	24979	23333	37619	56227	95941	59494	86539
192	32006	76302	81221	00693	95197	75044	46596	11628
193	37569	85187	44692	50706	53161	69027	88389	60313
194	56680	79003	23361	67094	15019	63261	24543	52884
195	05172	08100	22316	54495	60005	29532	18433	18057
196	74782	27005	03894	98038	20627	40307	47317	92759
197	85288	93264	61409	03404	09649	55937	60843	66167
198	68309	12060	14762	58002	03716	81968	57934	32624
199	26461	88346	52430	60906	74216	96263	69296	90107
200	42672	67680	42376	95023	82744	03971	96560	55148

TABLE C   Binomial probabilities

Entry is $P(X = k) = \binom{n}{k} p^k (1-p)^{n-k}$

n	k	.01	.02	.03	.04	.05	.06	.07	.08	.09
						p				
2	0	.9801	.9604	.9409	.9216	.9025	.8836	.8649	.8464	.8281
	1	.0198	.0392	.0582	.0768	.0950	.1128	.1302	.1472	.1638
	2	.0001	.0004	.0009	.0016	.0025	.0036	.0049	.0064	.0081
3	0	.9703	.9412	.9127	.8847	.8574	.8306	.8044	.7787	.7536
	1	.0294	.0576	.0847	.1106	.1354	.1590	.1816	.2031	.2236
	2	.0003	.0012	.0026	.0046	.0071	.0102	.0137	.0177	.0221
	3				.0001	.0001	.0002	.0003	.0005	.0007
4	0	.9606	.9224	.8853	.8493	.8145	.7807	.7481	.7164	.6857
	1	.0388	.0753	.1095	.1416	.1715	.1993	.2252	.2492	.2713
	2	.0006	.0023	.0051	.0088	.0135	.0191	.0254	.0325	.0402
	3			.0001	.0002	.0005	.0008	.0013	.0019	.0027
	4									.0001
5	0	.9510	.9039	.8587	.8154	.7738	.7339	.6957	.6591	.6240
	1	.0480	.0922	.1328	.1699	.2036	.2342	.2618	.2866	.3086
	2	.0010	.0038	.0082	.0142	.0214	.0299	.0394	.0498	.0610
	3		.0001	.0003	.0006	.0011	.0019	.0030	.0043	.0060
	4						.0001	.0001	.0002	.0003
	5									
6	0	.9415	.8858	.8330	.7828	.7351	.6899	.6470	.6064	.5679
	1	.0571	.1085	.1546	.1957	.2321	.2642	.2922	.3164	.3370
	2	.0014	.0055	.0120	.0204	.0305	.0422	.0550	.0688	.0833
	3		.0002	.0005	.0011	.0021	.0036	.0055	.0080	.0110
	4					.0001	.0002	.0003	.0005	.0008
	5									
	6									
7	0	.9321	.8681	.8080	.7514	.6983	.6485	.6017	.5578	.5168
	1	.0659	.1240	.1749	.2192	.2573	.2897	.3170	.3396	.3578
	2	.0020	.0076	.0162	.0274	.0406	.0555	.0716	.0886	.1061
	3		.0003	.0008	.0019	.0036	.0059	.0090	.0128	.0175
	4				.0001	.0002	.0004	.0007	.0011	.0017
	5								.0001	.0001
	6									
	7									
8	0	.9227	.8508	.7837	.7214	.6634	.6096	.5596	.5132	.4703
	1	.0746	.1389	.1939	.2405	.2793	.3113	.3370	.3570	.3721
	2	.0026	.0099	.0210	.0351	.0515	.0695	.0888	.1087	.1288
	3	.0001	.0004	.0013	.0029	.0054	.0089	.0134	.0189	.0255
	4			.0001	.0002	.0004	.0007	.0013	.0021	.0031
	5							.0001	.0001	.0002
	6									
	7									
	8									

TABLE C (*Continued*)

Entry is $P(X = k) = \binom{n}{k} p^k (1 - p)^{n-k}$

		p								
n	k	.10	.15	.20	.25	.30	.35	.40	.45	.50
2	0	.8100	.7225	.6400	.5625	.4900	.4225	.3600	.3025	.2500
	1	.1800	.2550	.3200	.3750	.4200	.4550	.4800	.4950	.5000
	2	.0100	.0225	.0400	.0625	.0900	.1225	.1600	.2025	.2500
3	0	.7290	.6141	.5120	.4219	.3430	.2746	.2160	.1664	.1250
	1	.2430	.3251	.3840	.4219	.4410	.4436	.4320	.4084	.3750
	2	.0270	.0574	.0960	.1406	.1890	.2389	.2880	.3341	.3750
	3	.0010	.0034	.0080	.0156	.0270	.0429	.0640	.0911	.1250
4	0	.6561	.5220	.4096	.3164	.2401	.1785	.1296	.0915	.0625
	1	.2916	.3685	.4096	.4219	.4116	.3845	.3456	.2995	.2500
	2	.0486	.0975	.1536	.2109	.2646	.3105	.3456	.3675	.3750
	3	.0036	.0115	.0256	.0469	.0756	.1115	.1536	.2005	.2500
	4	.0001	.0005	.0016	.0039	.0081	.0150	.0256	.0410	.0625
5	0	.5905	.4437	.3277	.2373	.1681	.1160	.0778	.0503	.0313
	1	.3280	.3915	.4096	.3955	.3602	.3124	.2592	.2059	.1563
	2	.0729	.1382	.2048	.2637	.3087	.3364	.3456	.3369	.3125
	3	.0081	.0244	.0512	.0879	.1323	.1811	.2304	.2757	.3125
	4	.0004	.0022	.0064	.0146	.0284	.0488	.0768	.1128	.1562
	5		.0001	.0003	.0010	.0024	.0053	.0102	.0185	.0312
6	0	.5314	.3771	.2621	.1780	.1176	.0754	.0467	.0277	.0156
	1	.3543	.3993	.3932	.3560	.3025	.2437	.1866	.1359	.0938
	2	.0984	.1762	.2458	.2966	.3241	.3280	.3110	.2780	.2344
	3	.0146	.0415	.0819	.1318	.1852	.2355	.2765	.3032	.3125
	4	.0012	.0055	.0154	.0330	.0595	.0951	.1382	.1861	.2344
	5	.0001	.0004	.0015	.0044	.0102	.0205	.0369	.0609	.0937
	6			.0001	.0002	.0007	.0018	.0041	.0083	.0156
7	0	.4783	.3206	.2097	.1335	.0824	.0490	.0280	.0152	.0078
	1	.3720	.3960	.3670	.3115	.2471	.1848	.1306	.0872	.0547
	2	.1240	.2097	.2753	.3115	.3177	.2985	.2613	.2140	.1641
	3	.0230	.0617	.1147	.1730	.2269	.2679	.2903	.2918	.2734
	4	.0026	.0109	.0287	.0577	.0972	.1442	.1935	.2388	.2734
	5	.0002	.0012	.0043	.0115	.0250	.0466	.0774	.1172	.1641
	6		.0001	.0004	.0013	.0036	.0084	.0172	.0320	.0547
	7				.0001	.0002	.0006	.0016	.0037	.0078
8	0	.4305	.2725	.1678	.1001	.0576	.0319	.0168	.0084	.0039
	1	.3826	.3847	.3355	.2670	.1977	.1373	.0896	.0548	.0313
	2	.1488	.2376	.2936	.3115	.2965	.2587	.2090	.1569	.1094
	3	.0331	.0839	.1468	.2076	.2541	.2786	.2787	.2568	.2188
	4	.0046	.0185	.0459	.0865	.1361	.1875	.2322	.2627	.2734
	5	.0004	.0026	.0092	.0231	.0467	.0808	.1239	.1719	.2188
	6		.0002	.0011	.0038	.0100	.0217	.0413	.0703	.1094
	7			.0001	.0004	.0012	.0033	.0079	.0164	.0312
	8					.0001	.0002	.0007	.0017	.0039

TABLE C (*Continued*)

						p				
n	k	.01	.02	.03	.04	.05	.06	.07	.08	.09
9	0	.9135	.8337	.7602	.6925	.6302	.5730	.5204	.4722	.4279
	1	.0830	.1531	.2116	.2597	.2985	.3292	.3525	.3695	.3809
	2	.0034	.0125	.0262	.0433	.0629	.0840	.1061	.1285	.1507
	3	.0001	.0006	.0019	.0042	.0077	.0125	.0186	.0261	.0348
	4			.0001	.0003	.0006	.0012	.0021	.0034	.0052
	5						.0001	.0002	.0003	.0005
	6									
	7									
	8									
	9									
10	0	.9044	.8171	.7374	.6648	.5987	.5386	.4840	.4344	.3894
	1	.0914	.1667	.2281	.2770	.3151	.3438	.3643	.3777	.3851
	2	.0042	.0153	.0317	.0519	.0746	.0988	.1234	.1478	.1714
	3	.0001	.0008	.0026	.0058	.0105	.0168	.0248	.0343	.0452
	4			.0001	.0004	.0010	.0019	.0033	.0052	.0078
	5					.0001	.0001	.0003	.0005	.0009
	6									.0001
	7									
	8									
	9									
	10									
12	0	.8864	.7847	.6938	.6127	.5404	.4759	.4186	.3677	.3225
	1	.1074	.1922	.2575	.3064	.3413	.3645	.3781	.3837	.3827
	2	.0060	.0216	.0438	.0702	.0988	.1280	.1565	.1835	.2082
	3	.0002	.0015	.0045	.0098	.0173	.0272	.0393	.0532	.0686
	4		.0001	.0003	.0009	.0021	.0039	.0067	.0104	.0153
	5				.0001	.0002	.0004	.0008	.0014	.0024
	6							.0001	.0001	.0003
	7									
	8									
	9									
	10									
	11									
	12									
15	0	.8601	.7386	.6333	.5421	.4633	.3953	.3367	.2863	.2430
	1	.1303	.2261	.2938	.3388	.3658	.3785	.3801	.3734	.3605
	2	.0092	.0323	.0636	.0988	.1348	.1691	.2003	.2273	.2496
	3	.0004	.0029	.0085	.0178	.0307	.0468	.0653	.0857	.1070
	4		.0002	.0008	.0022	.0049	.0090	.0148	.0223	.0317
	5			.0001	.0002	.0006	.0013	.0024	.0043	.0069
	6						.0001	.0003	.0006	.0011
	7								.0001	.0001
	8									
	9									
	10									
	11									
	12									
	13									
	14									
	15									

TABLE C (*Continued*)

				p				
.10	.15	.20	.25	.30	.35	.40	.45	.50
.3874	.2316	.1342	.0751	.0404	.0207	.0101	.0046	.0020
.3874	.3679	.3020	.2253	.1556	.1004	.0605	.0339	.0176
.1722	.2597	.3020	.3003	.2668	.2162	.1612	.1110	.0703
.0446	.1069	.1762	.2336	.2668	.2716	.2508	.2119	.1641
.0074	.0283	.0661	.1168	.1715	.2194	.2508	.2600	.2461
.0008	.0050	.0165	.0389	.0735	.1181	.1672	.2128	.2461
.0001	.0006	.0028	.0087	.0210	.0424	.0743	.1160	.1641
		.0003	.0012	.0039	.0098	.0212	.0407	.0703
			.0001	.0004	.0013	.0035	.0083	.0176
					.0001	.0003	.0008	.0020

.10	.15	.20	.25	.30	.35	.40	.45	.50
.3487	.1969	.1074	.0563	.0282	.0135	.0060	.0025	.0010
.3874	.3474	.2684	.1877	.1211	.0725	.0403	.0207	.0098
.1937	.2759	.3020	.2816	.2335	.1757	.1209	.0763	.0439
.0574	.1298	.2013	.2503	.2668	.2522	.2150	.1665	.1172
.0112	.0401	.0881	.1460	.2001	.2377	.2508	.2384	.2051
.0015	.0085	.0264	.0584	.1029	.1536	.2007	.2340	.2461
.0001	.0012	.0055	.0162	.0368	.0689	.1115	.1596	.2051
	.0001	.0008	.0031	.0090	.0212	.0425	.0746	.1172
		.0001	.0004	.0014	.0043	.0106	.0229	.0439
				.0001	.0005	.0016	.0042	.0098
						.0001	.0003	.0010

.10	.15	.20	.25	.30	.35	.40	.45	.50
.2824	.1422	.0687	.0317	.0138	.0057	.0022	.0008	.0002
.3766	.3012	.2062	.1267	.0712	.0368	.0174	.0075	.0029
.2301	.2924	.2835	.2323	.1678	.1088	.0639	.0339	.0161
.0852	.1720	.2362	.2581	.2397	.1954	.1419	.0923	.0537
.0213	.0683	.1329	.1936	.2311	.2367	.2128	.1700	.1208
.0038	.0193	.0532	.1032	.1585	.2039	.2270	.2225	.1934
.0005	.0040	.0155	.0401	.0792	.1281	.1766	.2124	.2256
	.0006	.0033	.0115	.0291	.0591	.1009	.1489	.1934
	.0001	.0005	.0024	.0078	.0199	.0420	.0762	.1208
		.0001	.0004	.0015	.0048	.0125	.0277	.0537
				.0002	.0008	.0025	.0068	.0161
					.0001	.0003	.0010	.0029
							.0001	.0002

.10	.15	.20	.25	.30	.35	.40	.45	.50
.2059	.0874	.0352	.0134	.0047	.0016	.0005	.0001	.0000
.3432	.2312	.1319	.0668	.0305	.0126	.0047	.0016	.0005
.2669	.2856	.2309	.1559	.0916	.0476	.0219	.0090	.0032
.1285	.2184	.2501	.2252	.1700	.1110	.0634	.0318	.0139
.0428	.1156	.1876	.2252	.2186	.1792	.1268	.0780	.0417
.0105	.0449	.1032	.1651	.2061	.2123	.1859	.1404	.0916
.0019	.0132	.0430	.0917	.1472	.1906	.2066	.1914	.1527
.0003	.0030	.0138	.0393	.0811	.1319	.1771	.2013	.1964
	.0005	.0035	.0131	.0348	.0710	.1181	.1647	.1964
	.0001	.0007	.0034	.0116	.0298	.0612	.1048	.1527
		.0001	.0007	.0030	.0096	.0245	.0515	.0916
			.0001	.0007	.0024	.0074	.0191	.0417
				.0001	.0004	.0016	.0052	.0139
					.0001	.0003	.0010	.0032
							.0001	.0005

TABLE C (*Continued*)

						p				
n	k	.01	.02	.03	.04	.05	.06	.07	.08	.09
20	0	.8179	.6676	.5438	.4420	.3585	.2901	.2342	.1887	.1516
	1	.1652	.2725	.3364	.3683	.3774	.3703	.3526	.3282	.3000
	2	.0159	.0528	.0988	.1458	.1887	.2246	.2521	.2711	.2818
	3	.0010	.0065	.0183	.0364	.0596	.0860	.1139	.1414	.1672
	4		.0006	.0024	.0065	.0133	.0233	.0364	.0523	.0703
	5			.0002	.0009	.0022	.0048	.0088	.0145	.0222
	6				.0001	.0003	.0008	.0017	.0032	.0055
	7						.0001	.0002	.0005	.0011
	8								.0001	.0002
	9									
	10									
	11									
	12									
	13									
	14									
	15									
	16									
	17									
	18									
	19									
	20									

						p				
n	k	.10	.15	.20	.25	.30	.35	.40	.45	.50
20	0	.1216	.0388	.0115	.0032	.0008	.0002	.0000	.0000	.0000
	1	.2702	.1368	.0576	.0211	.0068	.0020	.0005	.0001	.0000
	2	.2852	.2293	.1369	.0669	.0278	.0100	.0031	.0008	.0002
	3	.1901	.2428	.2054	.1339	.0716	.0323	.0123	.0040	.0011
	4	.0898	.1821	.2182	.1897	.1304	.0738	.0350	.0139	.0046
	5	.0319	.1028	.1746	.2023	.1789	.1272	.0746	.0365	.0148
	6	.0089	.0454	.1091	.1686	.1916	.1712	.1244	.0746	.0370
	7	.0020	.0160	.0545	.1124	.1643	.1844	.1659	.1221	.0739
	8	.0004	.0046	.0222	.0609	.1144	.1614	.1797	.1623	.1201
	9	.0001	.0011	.0074	.0271	.0654	.1158	.1597	.1771	.1602
	10		.0002	.0020	.0099	.0308	.0686	.1171	.1593	.1762
	11			.0005	.0030	.0120	.0336	.0710	.1185	.1602
	12			.0001	.0008	.0039	.0136	.0355	.0727	.1201
	13				.0002	.0010	.0045	.0146	.0366	.0739
	14					.0002	.0012	.0049	.0150	.0370
	15						.0003	.0013	.0049	.0148
	16							.0003	.0013	.0046
	17								.0002	.0011
	18									.0002
	19									
	20									

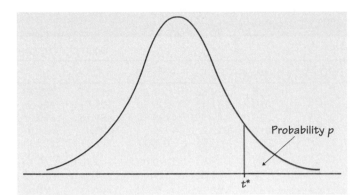

Table entry for p and C is the point t^* with probability p lying above it and probability C lying between $-t^*$ and t^*.

Probability p

t^*

TABLE D t distribution critical values

df	\multicolumn{12}{c}{Tail probability p}
	.25	.20	.15	.10	.05	.025	.02	.01	.005	.0025	.001	.0005
1	1.000	1.376	1.963	3.078	6.314	12.71	15.89	31.82	63.66	127.3	318.3	636.6
2	0.816	1.061	1.386	1.886	2.920	4.303	4.849	6.965	9.925	14.09	22.33	31.60
3	0.765	0.978	1.250	1.638	2.353	3.182	3.482	4.541	5.841	7.453	10.21	12.92
4	0.741	0.941	1.190	1.533	2.132	2.776	2.999	3.747	4.604	5.598	7.173	8.610
5	0.727	0.920	1.156	1.476	2.015	2.571	2.757	3.365	4.032	4.773	5.893	6.869
6	0.718	0.906	1.134	1.440	1.943	2.447	2.612	3.143	3.707	4.317	5.208	5.959
7	0.711	0.896	1.119	1.415	1.895	2.365	2.517	2.998	3.499	4.029	4.785	5.408
8	0.706	0.889	1.108	1.397	1.860	2.306	2.449	2.896	3.355	3.833	4.501	5.041
9	0.703	0.883	1.100	1.383	1.833	2.262	2.398	2.821	3.250	3.690	4.297	4.781
10	0.700	0.879	1.093	1.372	1.812	2.228	2.359	2.764	3.169	3.581	4.144	4.587
11	0.697	0.876	1.088	1.363	1.796	2.201	2.328	2.718	3.106	3.497	4.025	4.437
12	0.695	0.873	1.083	1.356	1.782	2.179	2.303	2.681	3.055	3.428	3.930	4.318
13	0.694	0.870	1.079	1.350	1.771	2.160	2.282	2.650	3.012	3.372	3.852	4.221
14	0.692	0.868	1.076	1.345	1.761	2.145	2.264	2.624	2.977	3.326	3.787	4.140
15	0.691	0.866	1.074	1.341	1.753	2.131	2.249	2.602	2.947	3.286	3.733	4.073
16	0.690	0.865	1.071	1.337	1.746	2.120	2.235	2.583	2.921	3.252	3.686	4.015
17	0.689	0.863	1.069	1.333	1.740	2.110	2.224	2.567	2.898	3.222	3.646	3.965
18	0.688	0.862	1.067	1.330	1.734	2.101	2.214	2.552	2.878	3.197	3.611	3.922
19	0.688	0.861	1.066	1.328	1.729	2.093	2.205	2.539	2.861	3.174	3.579	3.883
20	0.687	0.860	1.064	1.325	1.725	2.086	2.197	2.528	2.845	3.153	3.552	3.850
21	0.686	0.859	1.063	1.323	1.721	2.080	2.189	2.518	2.831	3.135	3.527	3.819
22	0.686	0.858	1.061	1.321	1.717	2.074	2.183	2.508	2.819	3.119	3.505	3.792
23	0.685	0.858	1.060	1.319	1.714	2.069	2.177	2.500	2.807	3.104	3.485	3.768
24	0.685	0.857	1.059	1.318	1.711	2.064	2.172	2.492	2.797	3.091	3.467	3.745
25	0.684	0.856	1.058	1.316	1.708	2.060	2.167	2.485	2.787	3.078	3.450	3.725
26	0.684	0.856	1.058	1.315	1.706	2.056	2.162	2.479	2.779	3.067	3.435	3.707
27	0.684	0.855	1.057	1.314	1.703	2.052	2.158	2.473	2.771	3.057	3.421	3.690
28	0.683	0.855	1.056	1.313	1.701	2.048	2.154	2.467	2.763	3.047	3.408	3.674
29	0.683	0.854	1.055	1.311	1.699	2.045	2.150	2.462	2.756	3.038	3.396	3.659
30	0.683	0.854	1.055	1.310	1.697	2.042	2.147	2.457	2.750	3.030	3.385	3.646
40	0.681	0.851	1.050	1.303	1.684	2.021	2.123	2.423	2.704	2.971	3.307	3.551
50	0.679	0.849	1.047	1.299	1.676	2.009	2.109	2.403	2.678	2.937	3.261	3.496
60	0.679	0.848	1.045	1.296	1.671	2.000	2.099	2.390	2.660	2.915	3.232	3.460
80	0.678	0.846	1.043	1.292	1.664	1.990	2.088	2.374	2.639	2.887	3.195	3.416
100	0.677	0.845	1.042	1.290	1.660	1.984	2.081	2.364	2.626	2.871	3.174	3.390
1000	0.675	0.842	1.037	1.282	1.646	1.962	2.056	2.330	2.581	2.813	3.098	3.300
z^*	0.674	0.841	1.036	1.282	1.645	1.960	2.054	2.326	2.576	2.807	3.091	3.291
	50%	60%	70%	80%	90%	95%	96%	98%	99%	99.5%	99.8%	99.9%

\multicolumn{13}{c}{Confidence level C}

TABLE E *F* critical values

DFD	p	\multicolumn{9}{c}{Degrees of freedom in the numerator}
		1	2	3	4	5	6	7	8	9
1	.100	39.86	49.50	53.59	55.83	57.24	58.20	58.91	59.44	59.86
	.050	161.45	199.50	215.71	224.58	230.16	233.99	236.77	238.88	240.54
	.025	647.79	799.50	864.16	899.58	921.85	937.11	948.22	956.66	963.28
	.010	4052.2	4999.5	5403.4	5624.6	5763.6	5859.0	5928.4	5981.1	6022.5
	.001	405284	500000	540379	562500	576405	585937	592873	598144	602284
2	.100	8.53	9.00	9.16	9.24	9.29	9.33	9.35	9.37	9.38
	.050	18.51	19.00	19.16	19.25	19.30	19.33	19.35	19.37	19.38
	.025	38.51	39.00	39.17	39.25	39.30	39.33	39.36	39.37	39.39
	.010	98.50	99.00	99.17	99.25	99.30	99.33	99.36	99.37	99.39
	.001	998.50	999.00	999.17	999.25	999.30	999.33	999.36	999.37	999.39
3	.100	5.54	5.46	5.39	5.34	5.31	5.28	5.27	5.25	5.24
	.050	10.13	9.55	9.28	9.12	9.01	8.94	8.89	8.85	8.81
	.025	17.44	16.04	15.44	15.10	14.88	14.73	14.62	14.54	14.47
	.010	34.12	30.82	29.46	28.71	28.24	27.91	27.67	27.49	27.35
	.001	167.03	148.50	141.11	137.10	134.58	132.85	131.58	130.62	129.86
4	.100	4.54	4.32	4.19	4.11	4.05	4.01	3.98	3.95	3.94
	.050	7.71	6.94	6.59	6.39	6.26	6.16	6.09	6.04	6.00
	.025	12.22	10.65	9.98	9.60	9.36	9.20	9.07	8.98	8.90
	.010	21.20	18.00	16.69	15.98	15.52	15.21	14.98	14.80	14.66
	.001	74.14	61.25	56.18	53.44	51.71	50.53	49.66	49.00	48.47
5	.100	4.06	3.78	3.62	3.52	3.45	3.40	3.37	3.34	3.32
	.050	6.61	5.79	5.41	5.19	5.05	4.95	4.88	4.82	4.77
	.025	10.01	8.43	7.76	7.39	7.15	6.98	6.85	6.76	6.68
	.010	16.26	13.27	12.06	11.39	10.97	10.67	10.46	10.29	10.16
	.001	47.18	37.12	33.20	31.09	29.75	28.83	28.16	27.65	27.24
6	.100	3.78	3.46	3.29	3.18	3.11	3.05	3.01	2.98	2.96
	.050	5.99	5.14	4.76	4.53	4.39	4.28	4.21	4.15	4.10
	.025	8.81	7.26	6.60	6.23	5.99	5.82	5.70	5.60	5.52
	.010	13.75	10.92	9.78	9.15	8.75	8.47	8.26	8.10	7.98
	.001	35.51	27.00	23.70	21.92	20.80	20.03	19.46	19.03	18.69
7	.100	3.59	3.26	3.07	2.96	2.88	2.83	2.78	2.75	2.72
	.050	5.59	4.74	4.35	4.12	3.97	3.87	3.79	3.73	3.68
	.025	8.07	6.54	5.89	5.52	5.29	5.12	4.99	4.90	4.82
	.010	12.25	9.55	8.45	7.85	7.46	7.19	6.99	6.84	6.72
	.001	29.25	21.69	18.77	17.20	16.21	15.52	15.02	14.63	14.33
8	.100	3.46	3.11	2.92	2.81	2.73	2.67	2.62	2.59	2.56
	.050	5.32	4.46	4.07	3.84	3.69	3.58	3.50	3.44	3.39
	.025	7.57	6.06	5.42	5.05	4.82	4.65	4.53	4.43	4.36
	.010	11.26	8.65	7.59	7.01	6.63	6.37	6.18	6.03	5.91
	.001	25.41	18.49	15.83	14.39	13.48	12.86	12.40	12.05	11.77
9	.100	3.36	3.01	2.81	2.69	2.61	2.55	2.51	2.47	2.44
	.050	5.12	4.26	3.86	3.63	3.48	3.37	3.29	3.23	3.18
	.025	7.21	5.71	5.08	4.72	4.48	4.32	4.20	4.10	4.03
	.010	10.56	8.02	6.99	6.42	6.06	5.80	5.61	5.47	5.35
	.001	22.86	16.39	13.90	12.56	11.71	11.13	10.70	10.37	10.11
10	.100	3.29	2.92	2.73	2.61	2.52	2.46	2.41	2.38	2.35
	.050	4.96	4.10	3.71	3.48	3.33	3.22	3.14	3.07	3.02
	.025	6.94	5.46	4.83	4.47	4.24	4.07	3.95	3.85	3.78
	.010	10.04	7.56	6.55	5.99	5.64	5.39	5.20	5.06	4.94
	.001	21.04	14.91	12.55	11.28	10.48	9.93	9.52	9.20	8.96

Degrees of freedom in the denominator

TABLE E (*Continued*)

			Degrees of freedom in the numerator							
10	12	15	20	25	30	40	50	60	120	1000
60.19	60.71	61.22	61.74	62.05	62.26	62.53	62.69	62.79	63.06	63.30
241.88	243.91	245.95	248.01	249.26	250.10	251.14	251.77	252.20	253.25	254.19
968.63	976.71	984.87	993.10	998.08	1001.4	1005.6	1008.1	1009.8	1014.0	1017.7
6055.8	6106.3	6157.3	6208.7	6239.8	6260.6	6286.8	6302.5	6313.0	6339.4	6362.7
605621	610668	615764	620908	624017	626099	628712	630285	631337	633972	636301
9.39	9.41	9.42	9.44	9.45	9.46	9.47	9.47	9.47	9.48	9.49
19.40	19.41	19.43	19.45	19.46	19.46	19.47	19.48	19.48	19.49	19.49
39.40	39.41	39.43	39.45	39.46	39.46	39.47	39.48	39.48	39.49	39.50
99.40	99.42	99.43	99.45	99.46	99.47	99.47	99.48	99.48	99.49	99.50
999.40	999.42	999.43	999.45	999.46	999.47	999.47	999.48	999.48	999.49	999.50
5.23	5.22	5.20	5.18	5.17	5.17	5.16	5.15	5.15	5.14	5.13
8.79	8.74	8.70	8.66	8.63	8.62	8.59	8.58	8.57	8.55	8.53
14.42	14.34	14.25	14.17	14.12	14.08	14.04	14.01	13.99	13.95	13.91
27.23	27.05	26.87	26.69	26.58	26.50	26.41	26.35	26.32	26.22	26.14
129.25	128.32	127.37	126.42	125.84	125.45	124.96	124.66	124.47	123.97	123.53
3.92	3.90	3.87	3.84	3.83	3.82	3.80	3.80	3.79	3.78	3.76
5.96	5.91	5.86	5.80	5.77	5.75	5.72	5.70	5.69	5.66	5.63
8.84	8.75	8.66	8.56	8.50	8.46	8.41	8.38	8.36	8.31	8.26
14.55	14.37	14.20	14.02	13.91	13.84	13.75	13.69	13.65	13.56	13.47
48.05	47.41	46.76	46.10	45.70	45.43	45.09	44.88	44.75	44.40	44.09
3.30	3.27	3.24	3.21	3.19	3.17	3.16	3.15	3.14	3.12	3.11
4.74	4.68	4.62	4.56	4.52	4.50	4.46	4.44	4.43	4.40	4.37
6.62	6.52	6.43	6.33	6.27	6.23	6.18	6.14	6.12	6.07	6.02
10.05	9.89	9.72	9.55	9.45	9.38	9.29	9.24	9.20	9.11	9.03
26.92	26.42	25.91	25.39	25.08	24.87	24.60	24.44	24.33	24.06	23.82
2.94	2.90	2.87	2.84	2.81	2.80	2.78	2.77	2.76	2.74	2.72
4.06	4.00	3.94	3.87	3.83	3.81	3.77	3.75	3.74	3.70	3.67
5.46	5.37	5.27	5.17	5.11	5.07	5.01	4.98	4.96	4.90	4.86
7.87	7.72	7.56	7.40	7.30	7.23	7.14	7.09	7.06	6.97	6.89
18.41	17.99	17.56	17.12	16.85	16.67	16.44	16.31	16.21	15.98	15.77
2.70	2.67	2.63	2.59	2.57	2.56	2.54	2.52	2.51	2.49	2.47
3.64	3.57	3.51	3.44	3.40	3.38	3.34	3.32	3.30	3.27	3.23
4.76	4.67	4.57	4.47	4.40	4.36	4.31	4.28	4.25	4.20	4.15
6.62	6.47	6.31	6.16	6.06	5.99	5.91	5.86	5.82	5.74	5.66
14.08	13.71	13.32	12.93	12.69	12.53	12.33	12.20	12.12	11.91	11.72
2.54	2.50	2.46	2.42	2.40	2.38	2.36	2.35	2.34	2.32	2.30
3.35	3.28	3.22	3.15	3.11	3.08	3.04	3.02	3.01	2.97	2.93
4.30	4.20	4.10	4.00	3.94	3.89	3.84	3.81	3.78	3.73	3.68
5.81	5.67	5.52	5.36	5.26	5.20	5.12	5.07	5.03	4.95	4.87
11.54	11.19	10.84	10.48	10.26	10.11	9.92	9.80	9.73	9.53	9.36
2.42	2.38	2.34	2.30	2.27	2.25	2.23	2.22	2.21	2.18	2.16
3.14	3.07	3.01	2.94	2.89	2.86	2.83	2.80	2.79	2.75	2.71
3.96	3.87	3.77	3.67	3.60	3.56	3.51	3.47	3.45	3.39	3.34
5.26	5.11	4.96	4.81	4.71	4.65	4.57	4.52	4.48	4.40	4.32
9.89	9.57	9.24	8.90	8.69	8.55	8.37	8.26	8.19	8.00	7.84
2.32	2.28	2.24	2.20	2.17	2.16	2.13	2.12	2.11	2.08	2.06
2.98	2.91	2.85	2.77	2.73	2.70	2.66	2.64	2.62	2.58	2.54
3.72	3.62	3.52	3.42	3.35	3.31	3.26	3.22	3.20	3.14	3.09
4.85	4.71	4.56	4.41	4.31	4.25	4.17	4.12	4.08	4.00	3.92
8.75	8.45	8.13	7.80	7.60	7.47	7.30	7.19	7.12	6.94	6.78

TABLE E *(Continued)*

<table>
<thead>
<tr><th colspan="2"></th><th colspan="9">Degrees of freedom in the numerator</th></tr>
<tr><th>DFD</th><th>p</th><th>1</th><th>2</th><th>3</th><th>4</th><th>5</th><th>6</th><th>7</th><th>8</th><th>9</th></tr>
</thead>
<tbody>
<tr><td rowspan="5">11</td><td>.100</td><td>3.23</td><td>2.86</td><td>2.66</td><td>2.54</td><td>2.45</td><td>2.39</td><td>2.34</td><td>2.30</td><td>2.27</td></tr>
<tr><td>.050</td><td>4.84</td><td>3.98</td><td>3.59</td><td>3.36</td><td>3.20</td><td>3.09</td><td>3.01</td><td>2.95</td><td>2.90</td></tr>
<tr><td>.025</td><td>6.72</td><td>5.26</td><td>4.63</td><td>4.28</td><td>4.04</td><td>3.88</td><td>3.76</td><td>3.66</td><td>3.59</td></tr>
<tr><td>.010</td><td>9.65</td><td>7.21</td><td>6.22</td><td>5.67</td><td>5.32</td><td>5.07</td><td>4.89</td><td>4.74</td><td>4.63</td></tr>
<tr><td>.001</td><td>19.69</td><td>13.81</td><td>11.56</td><td>10.35</td><td>9.58</td><td>9.05</td><td>8.66</td><td>8.35</td><td>8.12</td></tr>

<tr><td rowspan="5">12</td><td>.100</td><td>3.18</td><td>2.81</td><td>2.61</td><td>2.48</td><td>2.39</td><td>2.33</td><td>2.28</td><td>2.24</td><td>2.21</td></tr>
<tr><td>.050</td><td>4.75</td><td>3.89</td><td>3.49</td><td>3.26</td><td>3.11</td><td>3.00</td><td>2.91</td><td>2.85</td><td>2.80</td></tr>
<tr><td>.025</td><td>6.55</td><td>5.10</td><td>4.47</td><td>4.12</td><td>3.89</td><td>3.73</td><td>3.61</td><td>3.51</td><td>3.44</td></tr>
<tr><td>.010</td><td>9.33</td><td>6.93</td><td>5.95</td><td>5.41</td><td>5.06</td><td>4.82</td><td>4.64</td><td>4.50</td><td>4.39</td></tr>
<tr><td>.001</td><td>18.64</td><td>12.97</td><td>10.80</td><td>9.63</td><td>8.89</td><td>8.38</td><td>8.00</td><td>7.71</td><td>7.48</td></tr>

<tr><td rowspan="5">13</td><td>.100</td><td>3.14</td><td>2.76</td><td>2.56</td><td>2.43</td><td>2.35</td><td>2.28</td><td>2.23</td><td>2.20</td><td>2.16</td></tr>
<tr><td>.050</td><td>4.67</td><td>3.81</td><td>3.41</td><td>3.18</td><td>3.03</td><td>2.92</td><td>2.83</td><td>2.77</td><td>2.71</td></tr>
<tr><td>.025</td><td>6.41</td><td>4.97</td><td>4.35</td><td>4.00</td><td>3.77</td><td>3.60</td><td>3.48</td><td>3.39</td><td>3.31</td></tr>
<tr><td>.010</td><td>9.07</td><td>6.70</td><td>5.74</td><td>5.21</td><td>4.86</td><td>4.62</td><td>4.44</td><td>4.30</td><td>4.19</td></tr>
<tr><td>.001</td><td>17.82</td><td>12.31</td><td>10.21</td><td>9.07</td><td>8.35</td><td>7.86</td><td>7.49</td><td>7.21</td><td>6.98</td></tr>

<tr><td rowspan="5">14</td><td>.100</td><td>3.10</td><td>2.73</td><td>2.52</td><td>2.39</td><td>2.31</td><td>2.24</td><td>2.19</td><td>2.15</td><td>2.12</td></tr>
<tr><td>.050</td><td>4.60</td><td>3.74</td><td>3.34</td><td>3.11</td><td>2.96</td><td>2.85</td><td>2.76</td><td>2.70</td><td>2.65</td></tr>
<tr><td>.025</td><td>6.30</td><td>4.86</td><td>4.24</td><td>3.89</td><td>3.66</td><td>3.50</td><td>3.38</td><td>3.29</td><td>3.21</td></tr>
<tr><td>.010</td><td>8.86</td><td>6.51</td><td>5.56</td><td>5.04</td><td>4.69</td><td>4.46</td><td>4.28</td><td>4.14</td><td>4.03</td></tr>
<tr><td>.001</td><td>17.14</td><td>11.78</td><td>9.73</td><td>8.62</td><td>7.92</td><td>7.44</td><td>7.08</td><td>6.80</td><td>6.58</td></tr>

<tr><td rowspan="5">15</td><td>.100</td><td>3.07</td><td>2.70</td><td>2.49</td><td>2.36</td><td>2.27</td><td>2.21</td><td>2.16</td><td>2.12</td><td>2.09</td></tr>
<tr><td>.050</td><td>4.54</td><td>3.68</td><td>3.29</td><td>3.06</td><td>2.90</td><td>2.79</td><td>2.71</td><td>2.64</td><td>2.59</td></tr>
<tr><td>.025</td><td>6.20</td><td>4.77</td><td>4.15</td><td>3.80</td><td>3.58</td><td>3.41</td><td>3.29</td><td>3.20</td><td>3.12</td></tr>
<tr><td>.010</td><td>8.68</td><td>6.36</td><td>5.42</td><td>4.89</td><td>4.56</td><td>4.32</td><td>4.14</td><td>4.00</td><td>3.89</td></tr>
<tr><td>.001</td><td>16.59</td><td>11.34</td><td>9.34</td><td>8.25</td><td>7.57</td><td>7.09</td><td>6.74</td><td>6.47</td><td>6.26</td></tr>

<tr><td rowspan="5">16</td><td>.100</td><td>3.05</td><td>2.67</td><td>2.46</td><td>2.33</td><td>2.24</td><td>2.18</td><td>2.13</td><td>2.09</td><td>2.06</td></tr>
<tr><td>.050</td><td>4.49</td><td>3.63</td><td>3.24</td><td>3.01</td><td>2.85</td><td>2.74</td><td>2.66</td><td>2.59</td><td>2.54</td></tr>
<tr><td>.025</td><td>6.12</td><td>4.69</td><td>4.08</td><td>3.73</td><td>3.50</td><td>3.34</td><td>3.22</td><td>3.12</td><td>3.05</td></tr>
<tr><td>.010</td><td>8.53</td><td>6.23</td><td>5.29</td><td>4.77</td><td>4.44</td><td>4.20</td><td>4.03</td><td>3.89</td><td>3.78</td></tr>
<tr><td>.001</td><td>16.12</td><td>10.97</td><td>9.01</td><td>7.94</td><td>7.27</td><td>6.80</td><td>6.46</td><td>6.19</td><td>5.98</td></tr>

<tr><td rowspan="5">17</td><td>.100</td><td>3.03</td><td>2.64</td><td>2.44</td><td>2.31</td><td>2.22</td><td>2.15</td><td>2.10</td><td>2.06</td><td>2.03</td></tr>
<tr><td>.050</td><td>4.45</td><td>3.59</td><td>3.20</td><td>2.96</td><td>2.81</td><td>2.70</td><td>2.61</td><td>2.55</td><td>2.49</td></tr>
<tr><td>.025</td><td>6.04</td><td>4.62</td><td>4.01</td><td>3.66</td><td>3.44</td><td>3.28</td><td>3.16</td><td>3.06</td><td>2.98</td></tr>
<tr><td>.010</td><td>8.40</td><td>6.11</td><td>5.19</td><td>4.67</td><td>4.34</td><td>4.10</td><td>3.93</td><td>3.79</td><td>3.68</td></tr>
<tr><td>.001</td><td>15.72</td><td>10.66</td><td>8.73</td><td>7.68</td><td>7.02</td><td>6.56</td><td>6.22</td><td>5.96</td><td>5.75</td></tr>

<tr><td rowspan="5">18</td><td>.100</td><td>3.01</td><td>2.62</td><td>2.42</td><td>2.29</td><td>2.20</td><td>2.13</td><td>2.08</td><td>2.04</td><td>2.00</td></tr>
<tr><td>.050</td><td>4.41</td><td>3.55</td><td>3.16</td><td>2.93</td><td>2.77</td><td>2.66</td><td>2.58</td><td>2.51</td><td>2.46</td></tr>
<tr><td>.025</td><td>5.98</td><td>4.56</td><td>3.95</td><td>3.61</td><td>3.38</td><td>3.22</td><td>3.10</td><td>3.01</td><td>2.93</td></tr>
<tr><td>.010</td><td>8.29</td><td>6.01</td><td>5.09</td><td>4.58</td><td>4.25</td><td>4.01</td><td>3.84</td><td>3.71</td><td>3.60</td></tr>
<tr><td>.001</td><td>15.38</td><td>10.39</td><td>8.49</td><td>7.46</td><td>6.81</td><td>6.35</td><td>6.02</td><td>5.76</td><td>5.56</td></tr>

<tr><td rowspan="5">19</td><td>.100</td><td>2.99</td><td>2.61</td><td>2.40</td><td>2.27</td><td>2.18</td><td>2.11</td><td>2.06</td><td>2.02</td><td>1.98</td></tr>
<tr><td>.050</td><td>4.38</td><td>3.52</td><td>3.13</td><td>2.90</td><td>2.74</td><td>2.63</td><td>2.54</td><td>2.48</td><td>2.42</td></tr>
<tr><td>.025</td><td>5.92</td><td>4.51</td><td>3.90</td><td>3.56</td><td>3.33</td><td>3.17</td><td>3.05</td><td>2.96</td><td>2.88</td></tr>
<tr><td>.010</td><td>8.18</td><td>5.93</td><td>5.01</td><td>4.50</td><td>4.17</td><td>3.94</td><td>3.77</td><td>3.63</td><td>3.52</td></tr>
<tr><td>.001</td><td>15.08</td><td>10.16</td><td>8.28</td><td>7.27</td><td>6.62</td><td>6.18</td><td>5.85</td><td>5.59</td><td>5.39</td></tr>

<tr><td rowspan="5">20</td><td>.100</td><td>2.97</td><td>2.59</td><td>2.38</td><td>2.25</td><td>2.16</td><td>2.09</td><td>2.04</td><td>2.00</td><td>1.96</td></tr>
<tr><td>.050</td><td>4.35</td><td>3.49</td><td>3.10</td><td>2.87</td><td>2.71</td><td>2.60</td><td>2.51</td><td>2.45</td><td>2.39</td></tr>
<tr><td>.025</td><td>5.87</td><td>4.46</td><td>3.86</td><td>3.51</td><td>3.29</td><td>3.13</td><td>3.01</td><td>2.91</td><td>2.84</td></tr>
<tr><td>.010</td><td>8.10</td><td>5.85</td><td>4.94</td><td>4.43</td><td>4.10</td><td>3.87</td><td>3.70</td><td>3.56</td><td>3.46</td></tr>
<tr><td>.001</td><td>14.82</td><td>9.95</td><td>8.10</td><td>7.10</td><td>6.46</td><td>6.02</td><td>5.69</td><td>5.44</td><td>5.24</td></tr>
</tbody>
</table>

Degrees of freedom in the denominator

TABLE E (*Continued*)

	Degrees of freedom in the numerator										
10	12	15	20	25	30	40	50	60	120	1000
2.25	2.21	2.17	2.12	2.10	2.08	2.05	2.04	2.03	2.00	1.98
2.85	2.79	2.72	2.65	2.60	2.57	2.53	2.51	2.49	2.45	2.41
3.53	3.43	3.33	3.23	3.16	3.12	3.06	3.03	3.00	2.94	2.89
4.54	4.40	4.25	4.10	4.01	3.94	3.86	3.81	3.78	3.69	3.61
7.92	7.63	7.32	7.01	6.81	6.68	6.52	6.42	6.35	6.18	6.02
2.19	2.15	2.10	2.06	2.03	2.01	1.99	1.97	1.96	1.93	1.91
2.75	2.69	2.62	2.54	2.50	2.47	2.43	2.40	2.38	2.34	2.30
3.37	3.28	3.18	3.07	3.01	2.96	2.91	2.87	2.85	2.79	2.73
4.30	4.16	4.01	3.86	3.76	3.70	3.62	3.57	3.54	3.45	3.37
7.29	7.00	6.71	6.40	6.22	6.09	5.93	5.83	5.76	5.59	5.44
2.14	2.10	2.05	2.01	1.98	1.96	1.93	1.92	1.90	1.88	1.85
2.67	2.60	2.53	2.46	2.41	2.38	2.34	2.31	2.30	2.25	2.21
3.25	3.15	3.05	2.95	2.88	2.84	2.78	2.74	2.72	2.66	2.60
4.10	3.96	3.82	3.66	3.57	3.51	3.43	3.38	3.34	3.25	3.18
6.80	6.52	6.23	5.93	5.75	5.63	5.47	5.37	5.30	5.14	4.99
2.10	2.05	2.01	1.96	1.93	1.91	1.89	1.87	1.86	1.83	1.80
2.60	2.53	2.46	2.39	2.34	2.31	2.27	2.24	2.22	2.18	2.14
3.15	3.05	2.95	2.84	2.78	2.73	2.67	2.64	2.61	2.55	2.50
3.94	3.80	3.66	3.51	3.41	3.35	3.27	3.22	3.18	3.09	3.02
6.40	6.13	5.85	5.56	5.38	5.25	5.10	5.00	4.94	4.77	4.62
2.06	2.02	1.97	1.92	1.89	1.87	1.85	1.83	1.82	1.79	1.76
2.54	2.48	2.40	2.33	2.28	2.25	2.20	2.18	2.16	2.11	2.07
3.06	2.96	2.86	2.76	2.69	2.64	2.59	2.55	2.52	2.46	2.40
3.80	3.67	3.52	3.37	3.28	3.21	3.13	3.08	3.05	2.96	2.88
6.08	5.81	5.54	5.25	5.07	4.95	4.80	4.70	4.64	4.47	4.33
2.03	1.99	1.94	1.89	1.86	1.84	1.81	1.79	1.78	1.75	1.72
2.49	2.42	2.35	2.28	2.23	2.19	2.15	2.12	2.11	2.06	2.02
2.99	2.89	2.79	2.68	2.61	2.57	2.51	2.47	2.45	2.38	2.32
3.69	3.55	3.41	3.26	3.16	3.10	3.02	2.97	2.93	2.84	2.76
5.81	5.55	5.27	4.99	4.82	4.70	4.54	4.45	4.39	4.23	4.08
2.00	1.96	1.91	1.86	1.83	1.81	1.78	1.76	1.75	1.72	1.69
2.45	2.38	2.31	2.23	2.18	2.15	2.10	2.08	2.06	2.01	1.97
2.92	2.82	2.72	2.62	2.55	2.50	2.44	2.41	2.38	2.32	2.26
3.59	3.46	3.31	3.16	3.07	3.00	2.92	2.87	2.83	2.75	2.66
5.58	5.32	5.05	4.78	4.60	4.48	4.33	4.24	4.18	4.02	3.87
1.98	1.93	1.89	1.84	1.80	1.78	1.75	1.74	1.72	1.69	1.66
2.41	2.34	2.27	2.19	2.14	2.11	2.06	2.04	2.02	1.97	1.92
2.87	2.77	2.67	2.56	2.49	2.44	2.38	2.35	2.32	2.26	2.20
3.51	3.37	3.23	3.08	2.98	2.92	2.84	2.78	2.75	2.66	2.58
5.39	5.13	4.87	4.59	4.42	4.30	4.15	4.06	4.00	3.84	3.69
1.96	1.91	1.86	1.81	1.78	1.76	1.73	1.71	1.70	1.67	1.64
2.38	2.31	2.23	2.16	2.11	2.07	2.03	2.00	1.98	1.93	1.88
2.82	2.72	2.62	2.51	2.44	2.39	2.33	2.30	2.27	2.20	2.14
3.43	3.30	3.15	3.00	2.91	2.84	2.76	2.71	2.67	2.58	2.50
5.22	4.97	4.70	4.43	4.26	4.14	3.99	3.90	3.84	3.68	3.53
1.94	1.89	1.84	1.79	1.76	1.74	1.71	1.69	1.68	1.64	1.61
2.35	2.28	2.20	2.12	2.07	2.04	1.99	1.97	1.95	1.90	1.85
2.77	2.68	2.57	2.46	2.40	2.35	2.29	2.25	2.22	2.16	2.09
3.37	3.23	3.09	2.94	2.84	2.78	2.69	2.64	2.61	2.52	2.43
5.08	4.82	4.56	4.29	4.12	4.00	3.86	3.77	3.70	3.54	3.40

TABLE E (*Continued*)

DFD	p	\multicolumn{9}{c}{Degrees of freedom in the numerator}
		1	2	3	4	5	6	7	8	9
21	.100	2.96	2.57	2.36	2.23	2.14	2.08	2.02	1.98	1.95
	.050	4.32	3.47	3.07	2.84	2.68	2.57	2.49	2.42	2.37
	.025	5.83	4.42	3.82	3.48	3.25	3.09	2.97	2.87	2.80
	.010	8.02	5.78	4.87	4.37	4.04	3.81	3.64	3.51	3.40
	.001	14.59	9.77	7.94	6.95	6.32	5.88	5.56	5.31	5.11
22	.100	2.95	2.56	2.35	2.22	2.13	2.06	2.01	1.97	1.93
	.050	4.30	3.44	3.05	2.82	2.66	2.55	2.46	2.40	2.34
	.025	5.79	4.38	3.78	3.44	3.22	3.05	2.93	2.84	2.76
	.010	7.95	5.72	4.82	4.31	3.99	3.76	3.59	3.45	3.35
	.001	14.38	9.61	7.80	6.81	6.19	5.76	5.44	5.19	4.99
23	.100	2.94	2.55	2.34	2.21	2.11	2.05	1.99	1.95	1.92
	.050	4.28	3.42	3.03	2.80	2.64	2.53	2.44	2.37	2.32
	.025	5.75	4.35	3.75	3.41	3.18	3.02	2.90	2.81	2.73
	.010	7.88	5.66	4.76	4.26	3.94	3.71	3.54	3.41	3.30
	.001	14.20	9.47	7.67	6.70	6.08	5.65	5.33	5.09	4.89
24	.100	2.93	2.54	2.33	2.19	2.10	2.04	1.98	1.94	1.91
	.050	4.26	3.40	3.01	2.78	2.62	2.51	2.42	2.36	2.30
	.025	5.72	4.32	3.72	3.38	3.15	2.99	2.87	2.78	2.70
	.010	7.82	5.61	4.72	4.22	3.90	3.67	3.50	3.36	3.26
	.001	14.03	9.34	7.55	6.59	5.98	5.55	5.23	4.99	4.80
25	.100	2.92	2.53	2.32	2.18	2.09	2.02	1.97	1.93	1.89
	.050	4.24	3.39	2.99	2.76	2.60	2.49	2.40	2.34	2.28
	.025	5.69	4.29	3.69	3.35	3.13	2.97	2.85	2.75	2.68
	.010	7.77	5.57	4.68	4.18	3.85	3.63	3.46	3.32	3.22
	.001	13.88	9.22	7.45	6.49	5.89	5.46	5.15	4.91	4.71
26	.100	2.91	2.52	2.31	2.17	2.08	2.01	1.96	1.92	1.88
	.050	4.23	3.37	2.98	2.74	2.59	2.47	2.39	2.32	2.27
	.025	5.66	4.27	3.67	3.33	3.10	2.94	2.82	2.73	2.65
	.010	7.72	5.53	4.64	4.14	3.82	3.59	3.42	3.29	3.18
	.001	13.74	9.12	7.36	6.41	5.80	5.38	5.07	4.83	4.64
27	.100	2.90	2.51	2.30	2.17	2.07	2.00	1.95	1.91	1.87
	.050	4.21	3.35	2.96	2.73	2.57	2.46	2.37	2.31	2.25
	.025	5.63	4.24	3.65	3.31	3.08	2.92	2.80	2.71	2.63
	.010	7.68	5.49	4.60	4.11	3.78	3.56	3.39	3.26	3.15
	.001	13.61	9.02	7.27	6.33	5.73	5.31	5.00	4.76	4.57
28	.100	2.89	2.50	2.29	2.16	2.06	2.00	1.94	1.90	1.87
	.050	4.20	3.34	2.95	2.71	2.56	2.45	2.36	2.29	2.24
	.025	5.61	4.22	3.63	3.29	3.06	2.90	2.78	2.69	2.61
	.010	7.64	5.45	4.57	4.07	3.75	3.53	3.36	3.23	3.12
	.001	13.50	8.93	7.19	6.25	5.66	5.24	4.93	4.69	4.50
29	.100	2.89	2.50	2.28	2.15	2.06	1.99	1.93	1.89	1.86
	.050	4.18	3.33	2.93	2.70	2.55	2.43	2.35	2.28	2.22
	.025	5.59	4.20	3.61	3.27	3.04	2.88	2.76	2.67	2.59
	.010	7.60	5.42	4.54	4.04	3.73	3.50	3.33	3.20	3.09
	.001	13.39	8.85	7.12	6.19	5.59	5.18	4.87	4.64	4.45
30	.100	2.88	2.49	2.28	2.14	2.05	1.98	1.93	1.88	1.85
	.050	4.17	3.32	2.92	2.69	2.53	2.42	2.33	2.27	2.21
	.025	5.57	4.18	3.59	3.25	3.03	2.87	2.75	2.65	2.57
	.010	7.56	5.39	4.51	4.02	3.70	3.47	3.30	3.17	3.07
	.001	13.29	8.77	7.05	6.12	5.53	5.12	4.82	4.58	4.39

Degrees of freedom in the denominator

TABLE E (*Continued*)

	Degrees of freedom in the numerator										
10	12	15	20	25	30	40	50	60	120	1000
1.92	1.87	1.83	1.78	1.74	1.72	1.69	1.67	1.66	1.62	1.59
2.32	2.25	2.18	2.10	2.05	2.01	1.96	1.94	1.92	1.87	1.82
2.73	2.64	2.53	2.42	2.36	2.31	2.25	2.21	2.18	2.11	2.05
3.31	3.17	3.03	2.88	2.79	2.72	2.64	2.58	2.55	2.46	2.37
4.95	4.70	4.44	4.17	4.00	3.88	3.74	3.64	3.58	3.42	3.28
1.90	1.86	1.81	1.76	1.73	1.70	1.67	1.65	1.64	1.60	1.57
2.30	2.23	2.15	2.07	2.02	1.98	1.94	1.91	1.89	1.84	1.79
2.70	2.60	2.50	2.39	2.32	2.27	2.21	2.17	2.14	2.08	2.01
3.26	3.12	2.98	2.83	2.73	2.67	2.58	2.53	2.50	2.40	2.32
4.83	4.58	4.33	4.06	3.89	3.78	3.63	3.54	3.48	3.32	3.17
1.89	1.84	1.80	1.74	1.71	1.69	1.66	1.64	1.62	1.59	1.55
2.27	2.20	2.13	2.05	2.00	1.96	1.91	1.88	1.86	1.81	1.76
2.67	2.57	2.47	2.36	2.29	2.24	2.18	2.14	2.11	2.04	1.98
3.21	3.07	2.93	2.78	2.69	2.62	2.54	2.48	2.45	2.35	2.27
4.73	4.48	4.23	3.96	3.79	3.68	3.53	3.44	3.38	3.22	3.08
1.88	1.83	1.78	1.73	1.70	1.67	1.64	1.62	1.61	1.57	1.54
2.25	2.18	2.11	2.03	1.97	1.94	1.89	1.86	1.84	1.79	1.74
2.64	2.54	2.44	2.33	2.26	2.21	2.15	2.11	2.08	2.01	1.94
3.17	3.03	2.89	2.74	2.64	2.58	2.49	2.44	2.40	2.31	2.22
4.64	4.39	4.14	3.87	3.71	3.59	3.45	3.36	3.29	3.14	2.99
1.87	1.82	1.77	1.72	1.68	1.66	1.63	1.61	1.59	1.56	1.52
2.24	2.16	2.09	2.01	1.96	1.92	1.87	1.84	1.82	1.77	1.72
2.61	2.51	2.41	2.30	2.23	2.18	2.12	2.08	2.05	1.98	1.91
3.13	2.99	2.85	2.70	2.60	2.54	2.45	2.40	2.36	2.27	2.18
4.56	4.31	4.06	3.79	3.63	3.52	3.37	3.28	3.22	3.06	2.91
1.86	1.81	1.76	1.71	1.67	1.65	1.61	1.59	1.58	1.54	1.51
2.22	2.15	2.07	1.99	1.94	1.90	1.85	1.82	1.80	1.75	1.70
2.59	2.49	2.39	2.28	2.21	2.16	2.09	2.05	2.03	1.95	1.89
3.09	2.96	2.81	2.66	2.57	2.50	2.42	2.36	2.33	2.23	2.14
4.48	4.24	3.99	3.72	3.56	3.44	3.30	3.21	3.15	2.99	2.84
1.85	1.80	1.75	1.70	1.66	1.64	1.60	1.58	1.57	1.53	1.50
2.20	2.13	2.06	1.97	1.92	1.88	1.84	1.81	1.79	1.73	1.68
2.57	2.47	2.36	2.25	2.18	2.13	2.07	2.03	2.00	1.93	1.86
3.06	2.93	2.78	2.63	2.54	2.47	2.38	2.33	2.29	2.20	2.11
4.41	4.17	3.92	3.66	3.49	3.38	3.23	3.14	3.08	2.92	2.78
1.84	1.79	1.74	1.69	1.65	1.63	1.59	1.57	1.56	1.52	1.48
2.19	2.12	2.04	1.96	1.91	1.87	1.82	1.79	1.77	1.71	1.66
2.55	2.45	2.34	2.23	2.16	2.11	2.05	2.01	1.98	1.91	1.84
3.03	2.90	2.75	2.60	2.51	2.44	2.35	2.30	2.26	2.17	2.08
4.35	4.11	3.86	3.60	3.43	3.32	3.18	3.09	3.02	2.86	2.72
1.83	1.78	1.73	1.68	1.64	1.62	1.58	1.56	1.55	1.51	1.47
2.18	2.10	2.03	1.94	1.89	1.85	1.81	1.77	1.75	1.70	1.65
2.53	2.43	2.32	2.21	2.14	2.09	2.03	1.99	1.96	1.89	1.82
3.00	2.87	2.73	2.57	2.48	2.41	2.33	2.27	2.23	2.14	2.05
4.29	4.05	3.80	3.54	3.38	3.27	3.12	3.03	2.97	2.81	2.66
1.82	1.77	1.72	1.67	1.63	1.61	1.57	1.55	1.54	1.50	1.46
2.16	2.09	2.01	1.93	1.88	1.84	1.79	1.76	1.74	1.68	1.63
2.51	2.41	2.31	2.20	2.12	2.07	2.01	1.97	1.94	1.87	1.80
2.98	2.84	2.70	2.55	2.45	2.39	2.30	2.25	2.21	2.11	2.02
4.24	4.00	3.75	3.49	3.33	3.22	3.07	2.98	2.92	2.76	2.61

TABLE E *(Continued)*

| DFD | p | \multicolumn{9}{c}{Degrees of freedom in the numerator} |
		1	2	3	4	5	6	7	8	9
40	.100	2.84	2.44	2.23	2.09	2.00	1.93	1.87	1.83	1.79
	.050	4.08	3.23	2.84	2.61	2.45	2.34	2.25	2.18	2.12
	.025	5.42	4.05	3.46	3.13	2.90	2.74	2.62	2.53	2.45
	.010	7.31	5.18	4.31	3.83	3.51	3.29	3.12	2.99	2.89
	.001	12.61	8.25	6.59	5.70	5.13	4.73	4.44	4.21	4.02
50	.100	2.81	2.41	2.20	2.06	1.97	1.90	1.84	1.80	1.76
	.050	4.03	3.18	2.79	2.56	2.40	2.29	2.20	2.13	2.07
	.025	5.34	3.97	3.39	3.05	2.83	2.67	2.55	2.46	2.38
	.010	7.17	5.06	4.20	3.72	3.41	3.19	3.02	2.89	2.78
	.001	12.22	7.96	6.34	5.46	4.90	4.51	4.22	4.00	3.82
60	.100	2.79	2.39	2.18	2.04	1.95	1.87	1.82	1.77	1.74
	.050	4.00	3.15	2.76	2.53	2.37	2.25	2.17	2.10	2.04
	.025	5.29	3.93	3.34	3.01	2.79	2.63	2.51	2.41	2.33
	.010	7.08	4.98	4.13	3.65	3.34	3.12	2.95	2.82	2.72
	.001	11.97	7.77	6.17	5.31	4.76	4.37	4.09	3.86	3.69
100	.100	2.76	2.36	2.14	2.00	1.91	1.83	1.78	1.73	1.69
	.050	3.94	3.09	2.70	2.46	2.31	2.19	2.10	2.03	1.97
	.025	5.18	3.83	3.25	2.92	2.70	2.54	2.42	2.32	2.24
	.010	6.90	4.82	3.98	3.51	3.21	2.99	2.82	2.69	2.59
	.001	11.50	7.41	5.86	5.02	4.48	4.11	3.83	3.61	3.44
200	.100	2.73	2.33	2.11	1.97	1.88	1.80	1.75	1.70	1.66
	.050	3.89	3.04	2.65	2.42	2.26	2.14	2.06	1.98	1.93
	.025	5.10	3.76	3.18	2.85	2.63	2.47	2.35	2.26	2.18
	.010	6.76	4.71	3.88	3.41	3.11	2.89	2.73	2.60	2.50
	.001	11.15	7.15	5.63	4.81	4.29	3.92	3.65	3.43	3.26
1000	.100	2.71	2.31	2.09	1.95	1.85	1.78	1.72	1.68	1.64
	.050	3.85	3.00	2.61	2.38	2.22	2.11	2.02	1.95	1.89
	.025	5.04	3.70	3.13	2.80	2.58	2.42	2.30	2.20	2.13
	.010	6.66	4.63	3.80	3.34	3.04	2.82	2.66	2.53	2.43
	.001	10.89	6.96	5.46	4.65	4.14	3.78	3.51	3.30	3.13

Degrees of freedom in the denominator

TABLE E (*Continued*)

| | Degrees of freedom in the numerator | | | | | | | | | |
10	12	15	20	25	30	40	50	60	120	1000
1.76	1.71	1.66	1.61	1.57	1.54	1.51	1.48	1.47	1.42	1.38
2.08	2.00	1.92	1.84	1.78	1.74	1.69	1.66	1.64	1.58	1.52
2.39	2.29	2.18	2.07	1.99	1.94	1.88	1.83	1.80	1.72	1.65
2.80	2.66	2.52	2.37	2.27	2.20	2.11	2.06	2.02	1.92	1.82
3.87	3.64	3.40	3.14	2.98	2.87	2.73	2.64	2.57	2.41	2.25
1.73	1.68	1.63	1.57	1.53	1.50	1.46	1.44	1.42	1.38	1.33
2.03	1.95	1.87	1.78	1.73	1.69	1.63	1.60	1.58	1.51	1.45
2.32	2.22	2.11	1.99	1.92	1.87	1.80	1.75	1.72	1.64	1.56
2.70	2.56	2.42	2.27	2.17	2.10	2.01	1.95	1.91	1.80	1.70
3.67	3.44	3.20	2.95	2.79	2.68	2.53	2.44	2.38	2.21	2.05
1.71	1.66	1.60	1.54	1.50	1.48	1.44	1.41	1.40	1.35	1.30
1.99	1.92	1.84	1.75	1.69	1.65	1.59	1.56	1.53	1.47	1.40
2.27	2.17	2.06	1.94	1.87	1.82	1.74	1.70	1.67	1.58	1.49
2.63	2.50	2.35	2.20	2.10	2.03	1.94	1.88	1.84	1.73	1.62
3.54	3.32	3.08	2.83	2.67	2.55	2.41	2.32	2.25	2.08	1.92
1.66	1.61	1.56	1.49	1.45	1.42	1.38	1.35	1.34	1.28	1.22
1.93	1.85	1.77	1.68	1.62	1.57	1.52	1.48	1.45	1.38	1.30
2.18	2.08	1.97	1.85	1.77	1.71	1.64	1.59	1.56	1.46	1.36
2.50	2.37	2.22	2.07	1.97	1.89	1.80	1.74	1.69	1.57	1.45
3.30	3.07	2.84	2.59	2.43	2.32	2.17	2.08	2.01	1.83	1.64
1.63	1.58	1.52	1.46	1.41	1.38	1.34	1.31	1.29	1.23	1.16
1.88	1.80	1.72	1.62	1.56	1.52	1.46	1.41	1.39	1.30	1.21
2.11	2.01	1.90	1.78	1.70	1.64	1.56	1.51	1.47	1.37	1.25
2.41	2.27	2.13	1.97	1.87	1.79	1.69	1.63	1.58	1.45	1.30
3.12	2.90	2.67	2.42	2.26	2.15	2.00	1.90	1.83	1.64	1.43
1.61	1.55	1.49	1.43	1.38	1.35	1.30	1.27	1.25	1.18	1.08
1.84	1.76	1.68	1.58	1.52	1.47	1.41	1.36	1.33	1.24	1.11
2.06	1.96	1.85	1.72	1.64	1.58	1.50	1.45	1.41	1.29	1.13
2.34	2.20	2.06	1.90	1.79	1.72	1.61	1.54	1.50	1.35	1.16
2.99	2.77	2.54	2.30	2.14	2.02	1.87	1.77	1.69	1.49	1.22

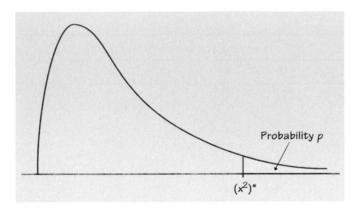

Table entry for p is the point $(X^2)^*$ with probability p lying above it.

TABLE F X^2 distribution critical values

						Tail probability p						
df	.25	.20	.15	.10	.05	.025	.02	.01	.005	.0025	.001	.0005
1	1.32	1.64	2.07	2.71	3.84	5.02	5.41	6.63	7.88	9.14	10.83	12.12
2	2.77	3.22	3.79	4.61	5.99	7.38	7.82	9.21	10.60	11.98	13.82	15.20
3	4.11	4.64	5.32	6.25	7.81	9.35	9.84	11.34	12.84	14.32	16.27	17.73
4	5.39	5.99	6.74	7.78	9.49	11.14	11.67	13.28	14.86	16.42	18.47	20.00
5	6.63	7.29	8.12	9.24	11.07	12.83	13.39	15.09	16.75	18.39	20.51	22.11
6	7.84	8.56	9.45	10.64	12.59	14.45	15.03	16.81	18.55	20.25	22.46	24.10
7	9.04	9.80	10.75	12.02	14.07	16.01	16.62	18.48	20.28	22.04	24.32	26.02
8	10.22	11.03	12.03	13.36	15.51	17.53	18.17	20.09	21.95	23.77	26.12	27.87
9	11.39	12.24	13.29	14.68	16.92	19.02	19.68	21.67	23.59	25.46	27.88	29.67
10	12.55	13.44	14.53	15.99	18.31	20.48	21.16	23.21	25.19	27.11	29.59	31.42
11	13.70	14.63	15.77	17.28	19.68	21.92	22.62	24.72	26.76	28.73	31.26	33.14
12	14.85	15.81	16.99	18.55	21.03	23.34	24.05	26.22	28.30	30.32	32.91	34.82
13	15.98	16.98	18.20	19.81	22.36	24.74	25.47	27.69	29.82	31.88	34.53	36.48
14	17.12	18.15	19.41	21.06	23.68	26.12	26.87	29.14	31.32	33.43	36.12	38.11
15	18.25	19.31	20.60	22.31	25.00	27.49	28.26	30.58	32.80	34.95	37.70	39.72
16	19.37	20.47	21.79	23.54	26.30	28.85	29.63	32.00	34.27	36.46	39.25	41.31
17	20.49	21.61	22.98	24.77	27.59	30.19	31.00	33.41	35.72	37.95	40.79	42.88
18	21.60	22.76	24.16	25.99	28.87	31.53	32.35	34.81	37.16	39.42	42.31	44.43
19	22.72	23.90	25.33	27.20	30.14	32.85	33.69	36.19	38.58	40.88	43.82	45.97
20	23.83	25.04	26.50	28.41	31.41	34.17	35.02	37.57	40.00	42.34	45.31	47.50
21	24.93	26.17	27.66	29.62	32.67	35.48	36.34	38.93	41.40	43.78	46.80	49.01
22	26.04	27.30	28.82	30.81	33.92	36.78	37.66	40.29	42.80	45.20	48.27	50.51
23	27.14	28.43	29.98	32.01	35.17	38.08	38.97	41.64	44.18	46.62	49.73	52.00
24	28.24	29.55	31.13	33.20	36.42	39.36	40.27	42.98	45.56	48.03	51.18	53.48
25	29.34	30.68	32.28	34.38	37.65	40.65	41.57	44.31	46.93	49.44	52.62	54.95
26	30.43	31.79	33.43	35.56	38.89	41.92	42.86	45.64	48.29	50.83	54.05	56.41
27	31.53	32.91	34.57	36.74	40.11	43.19	44.14	46.96	49.64	52.22	55.48	57.86
28	32.62	34.03	35.71	37.92	41.34	44.46	45.42	48.28	50.99	53.59	56.89	59.30
29	33.71	35.14	36.85	39.09	42.56	45.72	46.69	49.59	52.34	54.97	58.30	60.73
30	34.80	36.25	37.99	40.26	43.77	46.98	47.96	50.89	53.67	56.33	59.70	62.16
40	45.62	47.27	49.24	51.81	55.76	59.34	60.44	63.69	66.77	69.70	73.40	76.09
50	56.33	58.16	60.35	63.17	67.50	71.42	72.61	76.15	79.49	82.66	86.66	89.56
60	66.98	68.97	71.34	74.40	79.08	83.30	84.58	88.38	91.95	95.34	99.61	102.7
80	88.13	90.41	93.11	96.58	101.9	106.6	108.1	112.3	116.3	120.1	124.8	128.3
100	109.1	111.7	114.7	118.5	124.3	129.6	131.1	135.8	140.2	144.3	149.4	153.2

SOLUTIONS TO ODD-NUMBERED EXERCISES

Chapter 1

1.1 (a) Categorical. (b) Quantitative. (c) Categorical. (d) Categorical. (e) Quantitative. (f) Quantitative.

1.3 The individuals are vehicles (or "cars"). Variables: vehicle type (categorical), where made (categorical), city MPG (quantitative), and highway MPG (quantitative).

1.5 A tape measure (the measuring instrument) can be used to measure (in units of inches or centimeters) various lengths such as the longest single hair or the length of hair on sides or back or front.

1.7 (a) The number of deaths would tend to rise with the increasing population, even if cancer treatments become more effective over time: Since there are more people, there are more potential cases of cancer. (b) If treatments for other diseases are also improving, people who might have died from other causes would instead live long enough to succumb to cancer. Even if treatments were becoming *less* effective, many forms of cancer are detected earlier, as better tests are developed. In measuring five-year survival rates for (c), if we can detect cancer (say) one year earlier than was previously possible, then effectively, each patient lives one year longer after the cancer is detected, thus raising the five-year survival rate.

1.9 Using the proportion or percentage of repairs, Brand A is more reliable: $2942/13,376 \doteq 0.22 = 22\%$ for Brand A, and $192/480 = 0.4 = 40\%$ for Brand B.

1.11 Possible answers are total profits, number of employees, total value of stock, and total assets.

1.13 (a) Note that since the variable is nominal, not ordinal, the bars may be drawn in any order. (b) A pie chart would not be appropriate, since the different entries in the table do not represent parts of a single whole.

1.15 Fig. 1.10(a) is strongly skewed to the right with a peak at 0; Fig. 1.10(b) is somewhat symmetric with a central peak at 4. The peak is at the lowest value in Fig. 1.10(a) and is at a central value in Fig. 1.10(b).

1.17 There are two peaks. Most ACT states are in the upper portion of the distribution.

1.19 Left-skewed because newer coins are more common than older coins.

1.21 Because the data are not very spread out, split stems may be useful (e.g., one stem "10" for leaves 0–4, and another for 5–9). There does not seem to be a substantial difference between the two groups; this is supported by the fact that the medians are 111.5 (calcium) and 112 (placebo)—almost identical.

1.23 The distribution is slightly left-skewed. Preferences may vary.

1.25 (a) The first stemplot has the advantage of being compact, while the split stems suggest that there may be a second peak. (b) The distribution is roughly symmetric, with center around 12.6 or 12.7 percent. Alaska and Florida appear to be outliers; Alaska is low presumably because of its less attractive climate, while Florida is high because many retirees move there.

1.27 Either a stemplot or a histogram would be appropriate. There are no outliers, although the distribution is clearly right-skewed.

1.29 The distribution is fairly symmetric with center around 110 (clearly above 100). IQs range from the low 70s to the high 130s, with a "gap" in the low 80s.

1.31 (a) For 1950: $29.3/151.1 = 19.4\%$, $21.8/151.1 = 14.4\%$, etc. For 2075: $34.9/310.6 = 11.2\%$, $35.7/310.6 = 11.5\%$, etc. (b) Children (under 10) represent the single largest group in the population; about one out of five Americans was under 10 in 1950. There is a slight dip in the 10–19 age bracket, then the percentages trail off gradually after that. **(c)** The projections show a much greater proportion in the higher age brackets—there is now a gradual rise in the proportion up to ages 40–49, followed by the expected decline in the proportion of "senior citizens."

1.33 A class that is $20,000 wide should have bars one-fourth as tall as the bars for the $5000-wide classes.

1.35 **(a)** The death rate decreases fairly steadily over time. **(b)** The drop from the mid-1970s to the mid-1980s appears to be part of the overall trend; there is no particular suggestion in the plot that the decrease is any greater during that time, and thus no evidence that the lower speed limits saved lives. **(c)** A histogram *could* be made but would probably not be very useful. We want to see how these numbers change over time, not the number of times that, e.g., the death rate was between 4.5 and 5.0.

1.37 In his first five years, Ruth had few home runs. After that, until the last year of his career, his home run output fluctuated but was consistently high (25 or more).

1.39 **(a)** Weights are generally lower for toddlers with summer and late fall birthdays (June to November), and highest for toddlers with December to March birthdays. **(b)** Toddlers with summer birthdays appear to be slightly taller than those with winter birthdays. **(c)** Both plots have extremes in the summer and winter, but they are opposite: when one is high, the other is low. As a crude summary, the two plots together suggest that summer two-year-olds are tall and skinny, while winter two-year-olds are short and heavy.

1.41 **(a)** For best results, split stems five ways. The mean is 516.3; the median is 516.5. These are similar because the distribution is fairly symmetrical. **(b)** $s = 44.2$. Because the distribution is fairly symmetric, $\bar{x}$ and s are appropriate.

1.43 **(b)** Women: $\bar{x} = 141.06$, $M = 138.5$. Men: $\bar{x} = 121.25$, $M = 114.5$. The right skewness makes $\bar{x} > M$ in both cases. **(c)** Women: 101, 126, 138.5, 154, 200. Men: 70, 98, 114.5, 143, 187. The IQR for the women is 28, so 200 is an outlier. **(d)** All the displays and descriptions reveal that women generally score higher than men. The men's scores ($IQR = 45$) are more spread out than the women's; this is fairly clear from the boxplot, but not so obvious from the stemplot.

1.45 **(a)** Control: $\bar{x} = 366.3$ grams and $s = 50.8$ grams. Experimental: $\bar{x} = 402.95$ grams and $s = 42.73$ grams. **(b)** Both distributions appear to be relatively symmetric, with no outliers—which makes $\bar{x}$ and s appropriate descriptions.

1.47 Distribution of wealth will be skewed right, so $M = \$800,000$ and $\bar{x} = \$2.2$ million.

1.49 $\bar{x} = \$62,500$ and $M = \$25,000$. Seven of the employees earned less than the mean.

1.51 The mean rises to $87,500, while the median is unchanged.

1.53 **(a)** A stemplot with split stems shows a peak in the high 80s and suggests a slight skew to the right. (Without split stems, the skewness is not very apparent.) **(b)** $\bar{x} = 89.67$, $s^2 = 61.3089$, and $s = 7.83$. **(c)** $M = 88.5$, $Q_1 = 84.5$, $Q_3 = 93$, and $IQR = 8.5$. There are no outliers. **(d)** The slight skewness suggests that the quartiles should be used.

1.55 **(a)** 1, 1, 1, 1 is one answer. **(b)** 0, 0, 10, 10. **(c)** Any collection of equal numbers has variance 0, so (a) has 11 correct answers. The answer to (b) is unique.

1.57 The stemplot shows a low outlier of 4.9% (Alaska) and a high outlier of 18.6% (Florida). Because of the outliers, the five-number summary is a good choice: 4.9%, 11.4%, 12.6%, 13.9%, 18.6%.

1.59 The survival times are skewed to the right, so use the five-number summary: 43, 82.5, 102.5, 151.5, 598. Half the guinea pigs lived less than 102.5 days; typical lifetimes were 82.5 to 151.5 days. The longest-lived guinea pig died just short of 600 days, while one guinea pig lived only 43 days.

1.61 The logical number to choose as the 10th percentile is 10.55%. Likewise, the 90th percentile is 14.85%. The top 10% are Iowa (15.2%), West Virginia (15.3%), Rhode Island (15.7%), Pennsylvania (15.9%), and Florida (18.6%). The bottom 10% are Alaska (4.9%), Utah (8.8%), Colorado and Georgia (10%), and Texas (10.2%).

1.63 Five-number summaries: Beef—253, 320.5, 380.5, 478, 645. Meat—144, 379, 405, 501, 545. Poultry—357, 379, 430, 535, 588. Overall, beef hot dogs have less sodium (except for the one with the most sodium: 645 mg). Even if we ignore the low outlier among meat hot dogs, meat holds a slight edge over poultry.

1.65 (a) $a = 0, b = 746$. (b) $a = 0, b = 1/0.62 \doteq 1.61$. (c) $a = -98.6, b = 1$. (d) $a = 0$, $b = 100/30 = 10/3$.

1.67 Variance is changed by a factor of $2.54^2 = 6.4516$.

1.69 (a) The curve forms a 1×1 square, which has area 1. (b) $P = 0.25$. (c) $P = 0.8$.

1.71 $\bar{x} = M = 0.5, Q_1 = 0.25$, and $Q_3 = 0.75$.

1.73 Mark the center of the curve at 1.45; measure one standard deviation and mark 1.05 and 1.85, etc.

1.75 Using the 68–95–99.7 rule: 1.45 ± 0.80, or 0.65 to 2.25 grams per mile. Using Table A values: 1.45 ± 0.784, or 0.666 to 2.234 grams per mile.

1.77 (a) 266 ± 32, or 234 to 298 days. (b) Less than 234 days; longer than 298 days.

1.79 Eleanor: $z = 1.8$. Gerald: $z = 1.5$. Eleanor's score is higher.

1.81 (a) 0.9978. (b) 0.0022. (c) 0.9515. (d) 0.9493.

1.83 (a) -0.67 or -0.68. (b) 0.25.

1.85 SAT scores of 800+ correspond to $z > 3$; this is 0.15% (using the 68–95–99.7 rule).

1.87 (a) About 0.0655. (b) About 0.3336. (c) About $0.9936 - 0.3336 = 0.6600$.

1.89 Table A gives 0.0668.

1.91 Sarah's z-score is 1, while her mother's z-score is 1.2, so Sarah's mother scored relatively higher. But Sarah had the higher raw score, so she does stand higher in the variable measured. Sarah scored at the 84th percentile (0.8413). Her mother scored at the 88.5th percentile (0.8849).

1.93 (a) 50%. (b) 0.0918, or 9.18%. (c) 0.38%. (d) 40.82%.

1.95 (a) $Q_1 \doteq -0.67$, while $Q_3 \doteq 0.67$. (A more accurate value is ± 0.675.) (b) $Q_1 = 89.95$ (89.875 using 0.675), and $Q_3 = 110.05$ (or 110.125) (c) $IQR = 1.34$ (or 1.35). (d) Suspected outliers are below -2.68 (or -2.7), and above 2.68 (or 2.7). This percentage is $2 \times 0.0037 = 0.74\%$ (or $2 \times 0.0035 = 0.70\%$).

1.97 The plot does not suggest any major deviations from normality, except that the tails are less extreme than would be expected. This means extremely high and extremely low scores are "stacked up"—no one scored below 14 or above 54.

1.99 The plot is reasonably close to a line, apart from the stairstep appearance produced by granularity—presumably due to limited accuracy of the measuring instrument.

1.101 The plot suggests that the distribution is normal.

1.103 The boys' distribution seems to have two peaks: one in the low 50s and another in the high 60s/low 70s. For boys: $\bar{x} = 57.91$, Min $= 20, Q_1 = 51, M = 59, Q_3 = 67$, Max $= 80$. For girls: $\bar{x} = 55.52$, Min $= 21, Q_1 = 49, M = 60, Q_3 = 64$, Max $= 72$. Since both distributions are slightly skewed to the left, the five-number summaries may be more appropriate. The stemplot and five-number summaries suggest that the girls' scores do not seem to extend as high as the highest boys' scores.

1.105 Histograms will vary slightly but should suggest the density curve shown. Compared to a normal distribution, the uniform distribution does not extend as low or as high.

1.107 (a) Since a person cannot choose the day which he or she has a heart attack, one would expect that all days are "equally likely"—no day is favored over any other. (b) Monday through Thursday are fairly similar, but there is a pronounced peak on Friday, and lows on Saturday and Sunday. Patients do have some choice about when they leave the hospital, and many probably choose to leave on Friday.

1.109 The 1940 distribution is skewed to the left, while the 1980 distribution is fairly symmetric and considerably less spread out than the 1940 distribution. There are few low percentages in the 1980s, reflecting increased voting by blacks.

1.111 Either a bar chart or a pie chart would be appropriate. An "other methods" category (with 7.9%) is needed so that the total is 100%.

1.113 (a) $x_{\text{new}} = -50 + 2x$. (b) $x_{\text{new}} = -49.09 + (20/11)x$. (c) David: $x_{\text{new}} = 106$. Nancy: $x_{\text{new}} = 92.72$. David scored relatively higher. (d) David's score is $z = 0.3$ standard deviations above the mean, so about 61.79% of third graders score below 78. For Nancy, $z = -0.36$, so about 35.94% of sixth graders score below 78.

1.115 $z \doteq 2.33$ so about 0.99% score above 135; about 12 or 13 of the 1300 students.

1.117 The plot indicates that the data are approximately normally distributed; the mean and standard deviation are good measures for normal distributions. The mean is $35.\overline{09}$, and the standard deviation is 11.19.

1.119   (a) Software is the best way to produce these. If done by hand, the five-number summaries are QB:180, 185, 202, 207.5, 210; RB: 170, 190, 211, 225, 230; OL: 235, 275, 295, 295, 335; WR: 155, 179, 182.5, 189, 202; TE: 230, 235, 242.5, 256, 260; K: 160, 167.5, 175, 184, 193; DB: 170, 176, 190, 193, 195; LB: 205, 215, 220, 230, 237; DL: 220, 240, 245, 265, 285. (b) The heaviest players are on the offensive line, followed by defensive linemen and tight ends. The lightest players overall are kickers, followed by wide receivers and defensive backs. (c) The $(1.5 \times IQR)$ outlier test reveals outliers in OL and WR positions. Specifically, the outliers are the lightest (235 lb) and heaviest (335 lb) offensive linemen and the lightest (155 lb) wide receiver.

1.121   The distribution is strongly right-skewed, so use the five-number summary : 109,000, 205,000, 1,250,290, 2,300,000, 9,237,500. The highest salary is an outlier.

1.123   The distribution is strongly right-skewed, so use the five-number summary: 1232, 44,193, 150,816, 533,392, 9,127,751. The outlier test says that any population over 1,267,190.5 is an outlier, which would give seven outliers, but viewing a stemplot, one is inclined to take only the three largest counties as outliers. One division would be to include all of the top 25% of the counties (i.e., the counties with population over $Q_3 = 533,392$), some of the middle half (those with population between Q_1 and Q_3), and a small fraction of the bottom 25%.

1.125   Men seem to have higher SATM scores than women; each number in the five-number summary is about 40 to 50 points higher than the corresponding number for women. Women generally have higher GPAs than men, but the difference is less striking; in fact, the men's median is slightly higher. All four of the distributions seem reasonably close to normal.

Chapter 2

2.1   (a) Time spent studying is explanatory; grade is the response. (b) Explore the relationship; there is no reason to view one or the other as explanatory. (c) Rainfall is explanatory; crop yield is the response. (d) Explore the relationship. (e) The father's class is explanatory; the son's class is the response.

2.3   (a) The two variables are negatively related; the plot shows a clear curve, with an outlier (one car with high nitrogen oxides). (b) No; high carbon monoxide is associated with low nitrogen oxides, and vice versa.

2.5   (b) There is a fairly strong linear relationship. (c) The association is negative: countries with high wine consumption have fewer heart disease deaths, while low wine consumption tends to go with more deaths from heart disease. This does not prove causation; there may be some other reason for the link.

2.7   (a) Flow rate is explanatory. (b) As the flow rate increases, the amount of eroded soil increases. Yes, the pattern is approximately linear; the association is positive.

2.9   (a) Franklin's point is in the lower left corner of the graph. (b) There is a moderately strong positive linear association. There are no really extreme observations, though Bank 9 did rather well. Franklin does not look out of place.

2.11   (b) There is no clear relationship. (c) Generally, those with short incubation periods are more likely to die. (d) Person 6—the 17-year-old with a short incubation (20 hours) who survived—merits extra attention. He or she is also the youngest in the group by far. Among the other survivors, one (person 17) had an incubation period of 28 hours, and the rest had incubation periods of 43 hours or more.

2.13   (a) $\bar{x} = \$177,330$. The distribution is right-skewed, so the five-number summary is more appropriate: $113,000, $149,000, $174,900, $188,000, $327,500. (b) There is a weak positive relationship. (c) The five most expensive houses. The prices are outliers, and their points on the scatterplot lie above the general pattern.

2.15   (a) Means: 10.65, 10.43, 5.60, and 5.45. (b) Growth drops substantially between 1000 and 5000 nematodes.

2.17   (a) Means: 1520, 1707, 1540, 1816. (b) Against: Pecking order 1 had the lowest mean weight, while 4 was the heaviest on average.

2.19   (a) It appears that the correlation for men will be slightly smaller, since the men's points are more scattered. (b) Women: $r = 0.876$. Men: $r = 0.592$. (c) Women:

$\bar{x} = 43.03$ kg. Men: $\bar{x} = 53.10$ kg. This has no effect on the correlation. **(d)** The correlations would remain the same.

2.21 $r = 0.550$; without player 7, $r^* = 0.661$. Without player 7, the pattern of the scatterplot is more linear.

2.23   **(c)** $r_1 = r_2 = 0.253$, since r does not depend on the scale (units) of x and y.

2.25   The plot shows a relatively strong negative association; hence r is negative and large (close to -1). The correlation r does not describe the curve of the plot or the different patterns observed within the ACT and SAT states.

2.27   **(a)** Divide speed by 1.609; multiply fuel consumption by 1.609 and divide by 0.03785. The transformed data has the same correlation as the original data ($r = -0.172$). **(b)** The new correlation is $r^* = -0.043$; the new plot is even less linear than the first.

2.29   The plot shows a weak positive association; it is fairly linear. $r = 0.542$; there is some tendency for GPAs and self-concept scores to be high (or low) together.

2.31   The person who wrote the article interpreted a correlation close to 0 as if it were a correlation close to -1. Prof. McDaniel's findings mean there is little linear association between research and teaching—for example, knowing a professor is a good researcher gives little information about whether she is a good or bad teacher.

2.33   **(b)** When $x = 20$, $y = \$2500$. **(c)** $y = 500 + 200x$.

2.35   **(a)** $y = 100 + 40x$; the slope is 40 g/week. **(c)** When $x = 104$, $y = 4260$ grams, or about 9.4 pounds. Like humans, rats do not grow at a constant rate throughout their entire life.

2.37   **(b)** Initial pH: 5.4247; final pH: 4.6350. **(c)** The slope is -0.0053; the pH decreased by 0.0053 units per week (on the average).

2.39   **(a)** A moderately strong positive linear relationship. **(b)** $r = 0.9414$; about 88.6% of the variation in manatee deaths is explained by powerboat registrations, so predictions are reasonably accurate. **(c)** $\hat{y} \doteq -41.4 + 0.125x$; when $x = 716$, $y \doteq 48$ dead manatees are predicted. **(d)** When $x = 2000$, $y \doteq 208$; extrapolation makes this prediction unreliable. **(e)** Two of the points (those for 1992 and 1993) lie below the overall pattern (i.e., the actual manatee deaths were less than we might suspect), but otherwise there is no strong indication that the measures succeeded. **(f)** The mean for those years was 42—less than our predicted mean of 48 (which *might* suggest that the measures taken showed some results).

2.41   **(b)** $r = 0.507$ and $r^2 = 0.257 = 25.7\%$. There is a positive association between U.S. and overseas returns, but it is not very strong: knowing the U.S. return only accounts for about 26% of the variation in the overseas returns. **(c)** $\hat{y} = 5.64 + 0.692x$. **(d)** $\hat{y} = 12.6\%$; the error is $32.9\% - 12.6\% = 20.3\%$. Since the correlation is so low, the predictions will not be very reliable.

2.43   **(a)** $b = 0.16$; $a = 30.2$. **(b)** $\hat{y} = 78.2$. **(c)** Only 36% of the variability in y is accounted for by the regression, so $\hat{y} = 78.2$ could be quite different from the real score.

2.45 $r = 0.9670$, so $r^2 = 93.5\%$ of the variation in erosion is explained by the relationship between flow rate and erosion.

2.47   **(a)** Male height on female height: $\hat{y}_1 = 24 + 0.6818x$. Female height on male height: $\hat{y}_2 = 33.66 + 0.4688x$. The two slopes multiply to give $r^2 = 0.3196$, since the standard deviations cancel out. **(b)** Since regression lines always pass through $(\bar{x}, \bar{y})$, they intersect at $(66, 69)$—the first coordinate is the mean female height, and the second is the mean male height. **(c)** Slope b unchanged, intercept a multiplied by 2.54.

2.49   **(a)** $\hat{y} = 113 + 26.9x$. For every 1 kg increase in lean body mass, the metabolic rate rises by about 26.9. **(b)** $\bar{x} = 46.74$ kg, $s_x = 8.28$ kg; $\bar{y} = 1369.5$ cal, $s_y = 257.5$ cal; $r = 0.865$ (no units); $b = 26.9$ cal/kg, and $a = 113$ cal. **(c)** $\bar{x} = 102.83$ lb, $s_x = 18.23$ lb; $\bar{y}$, s_y, a, and r are unchanged; $b = 12.2$ cal/lb; $\hat{y} = 113 + 12.2x$.

2.51   **(b)** No; the pattern is not linear. **(c)** The sum is 0.01. The first and last residuals are negative, and those in the middle are positive.

2.53   **(b)** The first omits the outlier; it lies closer to the pattern of the other points (and farther from the omitted point).

2.55   **(a)** The relationship seems linear. **(b)** $\hat{y} = 1.77 + 0.0803x$. **(c)** The residuals are $0.011, -0.001, -0.001, -0.011, -0.009, 0.003, 0.009$. These add to 0.001. Results will vary. **(d)** Residuals are positive for low and high speeds, and negative for moderate speeds; this suggests that a curve (like a parabola) may be a better fit. We cannot plot residuals vs. time of observation since we do not have that information.

2.57 **(a)** In 1986, the overseas return was 69.4%—much higher than expected. The residual is 50.9%. The original equation was $\hat{y} = 5.64 + 0.692x$; without this point, it is $\hat{y} = 4.13 + 0.653x$. This is not much of a change; the point is not influential. **(b)** There is no obvious pattern to the residual plot.

2.59 **(b)** Sea scallops are relatively expensive; the point for scallops lies far away from the rest of the points, though it does not deviate greatly from the pattern. The fitted line changes slightly without that point. Lobsters might also be seen as outliers, though they are not as separated from the pack. **(c)** $r = 0.967$; 93.5% of the variation in 1980 prices is explained by 1970 prices. **(d)** Without scallops, $r^* = 0.940$; without scallops and lobsters, $r = 0.954$. The correlation drops slightly since, in the absence of the outlier(s), the scatter of the data is less, so the scatter about a line is (relatively) greater. **(e)** Yes; the plot suggests a linear relationship.

2.61 **(a)** $\hat{y} = 11.06 - 0.01466x$, but the plot does not suggest a linear relationship. Moral: Check to see if a line is an appropriate model for the data. **(b)** $\hat{y} = 2.72 + 0.166x$, but there is an influential point: Northern Ireland. Moral: Look for outliers and influential points.

2.63 **(a)** U.S.: -26.4%, 5.1%, 17.5%, 23.6%, 37.6%. Overseas: -23.2%, 2.5%, 12.0%, 29.6%, 69.4%. **(b)** Either answer is defensible: One-fourth of the time, overseas stocks did better than 29.6% (vs. 23.6% for U.S. stocks). On the other hand, half the time, U.S. stocks returned 17.5% or more (vs. 12% for overseas stocks). **(c)** Overseas stocks are more volatile—$Q_3 - Q_1 = 27.1\%$ for overseas stocks, about 50% larger than the U.S. *IQR* of 18.5%. The boxplot shows a lot more spread. Also, the low U.S. return (-26.4%) is an outlier; not so with the overseas stocks.

2.65 **(b)** The right-hand points lie below the left-hand points. (This means the right-hand times are shorter, so the subject is right-handed.) There is no striking pattern for the left-hand points; the pattern for right-hand points is obscured since they are squeezed at the bottom of the plot. **(c)** Right hand: $\hat{y} = 99.4 + 0.0283x$ ($r = 0.305$, $r^2 = 9.3\%$). Left hand: $\hat{y} = 172 + 0.262x$ ($r = 0.318$, $r^2 = 10.1\%$). The left-hand regression is slightly better, but neither is very good: distance only accounts for 9.3% (right) and 10.1% (left) of the variation in time. **(d)** Neither plot shows a pattern.

2.67 $r = 0.999$. With individual runners, the correlation would be smaller (closer to 0), since using data from individual runners would increase the "scatter" on the scatterplot, thus decreasing the strength of the relationship.

2.69 1 hour: $2^4 = 16$; 5 hours: $2^{20} = 1,048,576$.

2.71 **(a)** After 1 year: $500 \times 1.075 = \$537.50$. After 2 years: $500 \times 1.075^2 \doteq \$577.81$. Continue this pattern for each year; after 10 years: $500 \times 1.075^{10} \doteq \$1030.52$. **(c)** Rounded to three decimal places, the logarithms are 2.730, 2.762, 2.793, ..., 3.013.

2.73 **(a)** If the investment was made at the *beginning* of 1970: $1000 \times 1.1134^{26} \doteq \$16,327.95$. If the investment was made at the *end* of 1970: $1000 \times 1.1134^{25} \doteq \$14,664.94$. **(b)** $1000 \times 1.0562^{25} = \$3923.32$.

2.75 **(a)** There is a slight curve to the graph, suggesting exponential growth. **(b)** The logarithms (to three decimals) are 4.156, 4.339, 4.656, 4.999, 5.281, 5.496, 5.626, which gives $\hat{y} = -98.753 + 0.052501x$. Growth was faster from about 1965 to 1985, when the points in the scatterplot rise faster than the line. **(c)** Log $495,710 = 5.695$, which is less than $\hat{y} \doteq 5.829$, so the actual spending was less than predicted. Note: This logarithm corresponds to a predicted value of spending around $673,000.

2.77 **(b)** The logarithm plot has a greater slope—representing a faster growth rate—up to 1880 than after 1880. **(c)** $\hat{y} = -15.3815 + 0.0090210x$. When $x = 1997$, $\hat{y} = 2.633$, which corresponds to a population of 429.5 million. (Computation with "exact" values gives 429.9 million.) The actual value is much smaller than the predicted value.

2.79 For 1945, $\log y = 1.60865$, so $\hat{y} = 10^{1.60865} = 40.61$. For 1995, $\log y = 2.43715$, so $\hat{y} = 10^{2.43715} = 273.6$.

2.81 **(a)** Adding across the bottom (total) row: 14,050 thousand, or 14,050,000. **(b)** 37.9%. **(c)** 55.8%, 20.5%, 54.5%, 8.7%. **(d)** The 18 to 21 group constitutes the majority of full-time students at both 2- and 4-year institutions but makes up a much smaller proportion of part-time students.

2.83 **(a)** Adding across the rows, we find 246 (thousand), 5330, 5856, and 2618, respectively. Dividing by 14,050 gives 1.8%, 37.9%, 41.7%, and 18.6%. **(b)** 2.8%, 20.5%, 45.3%, 31.4%. **(c)** Two-year part-time students are more likely to be older than undergraduates in general. They are also slightly more likely to be under 18 and considerably less likely to be 18 to 21.

2.85 See 2.83 for the 2-year part-time distribution. For the others, divide the numbers in each column by the total at the bottom of each column; this gives 1.8%, 55.8%, 31.4%, 11.0% for 2-year full-time; 1.2%, 54.5%, 37.8%, 6.5% for 4-year full-time; 1.4%, 8.7%, 54.0%, 35.8% for 4-year part-time. The biggest difference is that full-time students tend to be younger—both 2- and 4-year full-time students have similar age distributions. Meanwhile, part-time students have similar distributions at both types of institutions and are predominantly over 21, with about one-third over 35.

2.87 **(a)** 59% did not respond. **(b)** 37.5% of small businesses, 59.5% of medium-sized businesses, and 80% of large businesses did not respond. Generally, the larger the business, the less likely it is to respond. **(d)** Small: 50.8%. Medium: 32.9%. Large: 16.3%. **(e)** No. Over half of respondents were small businesses, while less than 1/6 of responses came from large businesses.

2.89 58.3% of desipramine users did not have a relapse, compared to 25.0% of lithium users and 16.7% of those who received placebos. Desipramine seems to be effective.

2.91 **(a)** The sum is 58,929; the difference is due to roundoff error. **(b)** Divide each column total (from the bottom row) by 99,588 to obtain 19.4%, 59.2%, 11.1%, and 10.3%. **(c)** 18 to 24 years: 73.6%, 24.1%, 0.2%, 2.1%. 40 to 64 years: 6.3%, 72.7%, 6.0%, 15.0%. Among the younger women, almost three-fourths have not yet married, and those that are married have had little time to become widowed or divorced. Most of the older group are or have been married—only about 6% are still single. **(d)** 48.1% of never-married women are 18–24, 36.0% are 25–39, 11.9% are 40–64, and 4.0% are 65 or older. The target ages should be under 39.

2.93 **(a)** Males: 490 admitted, 210 refused. Females: 280 admitted, 220 refused. **(b)** 70% of male applicants are admitted, while only 56% of females are admitted. **(c)** 80% of male business school applicants are admitted, compared with 90% of females; in the law school, 10% of males are admitted, compared with 33.3% of females. **(d)** A majority (6/7) of male applicants apply to the business school, which admits 83% of all applicants. Meanwhile, a majority (3/5) of women apply to the law school, which only admits 27.5% of its applicants.

2.95 The key is to be sure that the three-way table has a lower percentage of overweight people among the smokers than among the nonsmokers. As one example, take 7 smokers, with 1 overweight person who dies early and 6 not overweight (of whom 4 die early). Then take 13 nonsmokers, with 9 overweight (of whom 3 die early) and 4 not overweight (of whom 1 dies early). Then overall, 4 of 10 overweight people die early, and 5 of 10 nonoverweight people die early.

2.97 Both reading ability and shoe size tend to increase with age—the lurking variable z.

2.99 No. The high death rate for C may occur because C is the anesthetic of choice in serious operations or for patients in poor condition.

2.101 Patients suffering from more serious illnesses are more likely to go to larger hospitals (which may have more or better facilities) for treatment. They are also likely to require more time to recuperate after treatment.

2.103 In this case, there may be a causative effect, but in the direction opposite to the one suggested: people who are overweight are more likely to be on diets and so choose artificial sweeteners over sugar.

2.105 The explanatory variable is whether or not a student has taken at least two years of foreign language, and the score on the test is the response. The lurking variable is the students' English skills *before* taking (or not taking) the foreign language: students who have a good command of English early in their high school career are more likely to choose (or be advised to choose) to take a foreign language.

2.107 We need information on the type of surgery and on the age, sex, and condition of the patients.

2.109 **(a)** Yes. The two lines appear to fit the data well. There do not appear to be any outliers or influential points. **(b)** Before: 0.189 hundred ft^3. After: 0.157 hundred ft^3. **(c)** Before: $\hat{y} = 7.704$. After: $\hat{y} = 6.348$. This amounts to $1.017 per day, or $31.53 for the month.

2.111 **(a)** Explanatory: weeds per meter; response: corn yield. **(b)** The stemplots give some evidence that yield decreases when there is more lamb's-quarter. **(c)** $\hat{y} = 166 - 1.10x$. Each additional plant per meter decreases yield by about 1.1 bushels/acre. **(d)** $\hat{y} = 159.4$ bushels/acre.

2.113 **(b)** The plot shows a negative association with no outliers. **(c)** $\hat{y} = 488 - 20.7x$; it is not a good match because the scatterplot does not suggest a straight line. **(d)** Length 5 to 9 inches: $\hat{y} = 668 - 46.9x$. Length 9 to 14 inches: $\hat{y} = 283 - 3.37x$. These two lines together describe the data fairly well. One might ask why strength decreases so rapidly with increasing length at first, then almost levels off.

2.115 **(a)** $11/20 = 55.0\%$, $68/91 \doteq 74.7\%$, and $3/8 = 37.5\%$. Some (but not too much) time spent in extracurricular activities seems to be beneficial. **(b)** No. There may be a lurking variable that affects both—e.g., a personality trait that "causes" students to do well and also to participate in extracurricular activities in moderation.

2.117 Some departments pay higher salaries than others; if women are concentrated in the lower-paying disciplines, their overall median salary will be lower than that of men even if all salaries in each department are identical.

2.119 **(a)** $68,838/109,672 \doteq 62.8\%$, and similarly we get 61.9%, 60.9%, 55.2%, 53.5%, 52.8%, 53.3%, 50.3%, 55.1%, and 49.1%. There is a fairly steady decline in participation, with a noticeably large drop after the 1960s. **(b)** More college students became eligible to vote after 1970, and that group may be more likely to miss an election, either because of apathy or because they are away from home when elections occur.

2.121 **(a)** Both distributions are skewed to the right. For pay: 26.0, 30.8, 32.6, 39.1, 50.0. There are no outliers. For spending: 3.67, 4.95, 5.66, 6.52, 9.93. Spending over $8.875 thousand per student qualifies as an outlier; these states are New York, New Jersey, and Alaska. **(b)** There is a moderate positive association. This makes sense since money spent for teacher salaries is part of the education budget; more money spent per pupil would typically translate to more money spent overall. **(c)** $\hat{y} = 14.1 + 3.53x$. For each additional $1000 spent per student, teacher salaries increase by about $3530. Regression of pay on spending explains about 62.9% of the variation in spending. **(d)** The residuals for the three states are small. Without those states, $\hat{y} = 12.5 + 3.82x$, which has a slightly greater slope than before—so the three points are *somewhat* influential.

2.123 $\hat{y} = 1.28 + 0.00227x$ and $r = 0.252$; the regression explains $r^2 = 6.3\%$ of the variation in GPA. By itself, SATM does not give reliable predictions of GPA.

2.125 **(a)** The new point has a very large residual. **(b)** The distributions of T1 and T2 do not appear to have any outliers. **(c)** With the new point, $\hat{y} = -0.017 + 1.011x$; without it, $\hat{y} = -0.033 + 1.018x$. The two lines are very similar, so the point is not influential. **(d)** With the new point, $r = 0.966$; without it, $r = 0.996$.

Chapter 3

3.1 One observation could have many explanations. Were the sunflowers in the sun and the okra in the shade? Were the sunflowers upwind from the okra? Repeated trials in controlled conditions are needed.

3.3 It is an observational study: information is gathered without imposing any treatment. A voter's gender is explanatory, and political party is the response.

3.5 **(a)** Which surgery was performed is explanatory, and survival time is the response. **(b)** This study uses available data; it is not an experiment because the study itself imposes no treatment on the subjects. **(c)** Any conclusions drawn from this study would have to be viewed with suspicion, because doctors may recommend treatment based on the patient's condition.

3.7 Yes. The treatment was walking briskly on the treadmill. The fact that eating was not recorded limits the conclusions that can be drawn. The explanatory variable was time after exercise, and the response variable was the metabolic rate.

3.9 Subjects: 300 sickle cell patients. Factor: drug given. Treatments: hydroxyurea and placebo. Response variable: number of pain episodes.

3.11 Subjects: students. Factors: length of commercial and number of repetitions. Treatments: 30 seconds repeated 1, 3, or 5 times; and 90 seconds repeated 1, 3, or 5 times. Response variables: recollection of ad, attitude about camera, and intention to buy camera.

3.13 **(a)** Make a diagram similar to Fig. 3.2, with "15 rats" replaced with "150 patients," "new/standard diet" replaced with "hydroxyurea/placebo," and "Compare weight gain" replaced with "Observe pain episodes."**(b)** A placebo allows researchers to control for the relief subjects might experience due to the psychological effect of taking a drug.

3.15 **(a)** The response variable is the company chosen. Make a diagram similar to Fig. 3.2. **(b)** If we assign labels 01 to 40, then choose two digits at a time from the table, we get subjects 29, 07, 34, 22, 10, 25, 13, 38, 15, 05, 09, 08, 27, 23, 30, 28, 18, 03, 01, and 36 for the child-care brochure. (It is best to list the subjects' *names* rather than just their numbers; they are omitted here for space reasons.)

3.17 Make a diagram similar to Fig. 3.2. Assign labels 01 to 20, then choose two digits at a time. Use Method A in plots 19, 06, 09, 10, 16, 01, 08, 20, 02, and 07.

3.19 If this year is considerably different in some way from last year, we cannot compare electricity consumption over the two years. The possible differences between the two years would confound the effects of the treatments.

3.21 Make a diagram similar to Figures 3.2 and 3.3, but with 4 branches (one for each color) rather than 2 or 3. Assign labels 01 to 16, then choose two digits at a time from the table: use blue on poles 04, 09, 14, and 03; green on 10, 06, 11, and 16; white on 02, 07, 13, and 15; and yellow on the rest.

3.23 Since there are 6 treatments, we can assign 16 students to each treatment and have 4 left over. These four can either be ignored, or each can be randomly assigned to one of the six groups. Make a diagram similar to Figures 3.2 and 3.3, but with 6 branches rather than 2 or 3.

3.25 For each person, flip the coin to decide which hand they should use first.

3.27 **(a)** Ordered by increasing weight, the five blocks are (1) Williams–22, Festinger–24, Hernandez–25, and Moses–25; (2) Santiago–27, Kendall–28, Mann–28, and Smith–29; (3) Brunk–30, Obrach–30, Rodriquez–30, and Loren–32; (4) Jackson–33, Stall–33, Brown–34, and Dixon–34; (5) Birnbaum–35, Tran–35, Nevesky–39, and Wilansky–42. **(b)** The exact randomization will vary with the starting line in Table B. Different methods are possible; perhaps the simplest is to number from 1 to 4 within each block, then assign the members of block 1 to a weight-loss treatment, then assign block 2, etc.

3.29 **(a)** Make a diagram similar to Figures 3.2 and 3.3, but with 4 branches rather than 2 or 3. **(b)** Practically, it may take a long time until enough claims have been filed to have the information we need. Ethically, this outline suggests that we assign subjects to an insurance plan; some might object to that. Other answers are possible.

3.31 **(a)** False. Such regularity holds only in the long run. **(b)** True. All pairs of digits (there are 100, from 00 to 99) are equally likely. **(c)** False. Four random digits have chance 1/10,000 to be 0000, so this sequence will occasionally occur.

3.33 The population is "employed adult women." The sample is the 48 women who return the questionnaires. 52% did not respond.

3.35 **(a)** Adult U.S. residents. **(b)** U.S. households. **(c)** All regulators from the supplier *or* the regulators in the last shipment.

3.37 Beginning with Agarwal and going down the columns, assign labels 01 to 28. We select 04–Bowman, 10–Frank, 17–Liang, 19–Naber, 12–Goel, 13–Gupta.

3.39 Taking three-digit numbers from Table B gives 214, 313, 409, 306, 511.

3.41 **(a)** We will choose one of the first 40 at random and then the addresses 40, 80, 120, and 160 places down the list from it. The addresses selected are 35, 75, 115, 155, 195. **(b)** All addresses are equally likely—each has chance 1/40 of being selected. To see this, note that each of the first 40 has chance 1/40 since one is chosen at random. Because each address in the second 40 is chosen exactly when the corresponding address in the first 40 is, each of the second 40 also has chance 1/40. And so on. This is not an SRS because the only possible samples have exactly 1 address from the first 25, 1 address from the second 25, and so on. An SRS could contain any 5 of the 200 addresses in the population.

3.43 Give each name on the alphabetized lists a number; 001 to 500 for females and 0001 to 2000 for males. The first five females are 138, 159, 052, 087, and 359. The first five men are 1369, 0815, 0727, 1025, and 1868.

3.45 **(a)** Households without telephones or with unlisted numbers. Such households would likely be made up of poor individuals (who cannot afford a phone), those who choose not to have phones, and those who do not wish to have their phone number published. **(b)** Those with unlisted numbers would be included in the sampling frame when a random digit dialer is used.

3.47 Voluntary response is the big reason. Opponents of gun control usually feel more strongly than supporters and so are more likely to call. The sampling method also reduces response from poorer people by requiring a phone and willingness to pay for the call.

3.49 Form A would draw the higher response favoring the ban. It is phrased to produce a negative reaction: "giving huge sums of money" vs. "contributing," and giving "to candidates" rather than "to campaigns." Also, form B presents both sides of the issue, allowing for special interest groups to have "a right to contribute."

3.51 6.2% is a statistic.

3.53 43 is a statistic; 52% is a parameter.

3.55 **(a)** High variability, high bias. **(b)** Low variability, low bias. **(c)** High variability, low bias. **(d)** Low variability, high bias.

3.57 For this exercise we assume the population proportions in all states are about the same. **(a)** No. The precision of an SRS of size 2000 is the same no matter what the population size (as long as the population is about 10 times the size of the sample or larger). **(b)** Yes. The sample sizes will vary from 32,000 (California) to 485 (Wyoming), so the precision will also vary (larger samples are less variable).

3.59 **(a)** Answers will vary. If, e.g., 8 heads are observed, then $\hat{p} = 8/20 = 0.4 = 40\%$. **(b)** Note that all the leaves in the stemplot should be either 0 or 5.

3.61 **(a)** We let the digits 0 and 1 represent the presence of eggs, while the other digits represent the absence of eggs. Use ten digits in each sample (one for each square yard). Answers will vary with the line chosen from Table B. **(b)** To make the stemplot, view each $\hat{p}$-value as having a 0 in the second place after the decimal—e.g., $\hat{p} = 0.20$ rather than just $\hat{p} = 0.2$—and use 0 for the leaf.

3.63 **(a)** $p = 27/95 \doteq 0.2842$. **(b)** Assign labels 01 through 95 to the players, then take digits two at a time. Answers will vary with the starting line in Table B; if the sample contains, say, 3 offensive backs, then $\hat{p} = 3/15 = 0.2$. **(c)** The distribution should look *roughly* normal, centered at p. **(d)** The mean should be fairly close to p; the lack of bias is (should be) illustrated by the clustering around p in the histogram.

3.65 An observational study—no treatment was imposed.

3.67 For each taster flip a coin. If heads, taste Pepsi, then Coke. If tails, do the opposite.

3.69 An example: You want to compare how long it takes to walk to class by two different routes. The experiment will take 20 days. The days are labeled from 01 to 20. Using Table B, the first 10 numbers between 01 and 20 will be assigned to route A; the others will be assigned to route B. Take the designated route on each day and record the time to get to class.

3.71 **(a)** Make a diagram similar to Fig. 3.2. **(b)** The patients are numbered from 01 to 30. Those receiving the beta-blockers are 21, 18, 23, 19, 10, 08, 03, 25, 06, 11, 15, 27, 13, 24, and 28.

3.73   Use a stratified sample with 50 faculty members selected from each class. This plan neglects the particular college where an individual is employed. An alternative scheme would be to sample colleges, and then faculty within those colleges.

3.75   **(a)** Use a block design, similar to Fig. 3.4, with black men as one block and white men as the other. **(b)** A larger group gives more information—when more subjects are involved, the random differences between individuals have less influence, and we can expect our sample to better represent the population.

3.77   **(a)** Make a diagram similar to Fig. 3.2. **(b)** Have each subject do the task twice, once under each temperature condition, randomly choosing which temperature comes first. Compute the difference in each subject's performances at the two temperatures.

3.79   The 1128 letters are a voluntary response sample, which do not necessarily reflect the opinions of her constituents, since persons with strong opinions on the subject are more likely to take the time to write.

3.81   Results will vary with the lines chosen in Table B, but probability computations reveal that about 95% of samples will have 3 to 7 defective rats in each sample.

3.83   Each histogram should be centered near 0.6, with the spread decreasing as the sample size increases. In particular, with $n = 50$, most $\hat{p}$-values wil be between 0.4 and 0.8; with $n = 200$, between 0.5 and 0.7; and with $n = 800$, between 0.55 and 0.65.

Chapter 4

4.1   Long trials of this experiment often approach 40% heads.

4.3   **(a)** We expect probability 1/2. **(b)** The theoretical probability is 2/3.

4.7   Theoretical probabilities: $1/16, 4/16 = 1/4, 6/16 = 3/8, 4/16 = 1/4, 1/16$.

4.9   **(a)** Most answers will be between 35% and 65%.

4.11   **(a)** 0. **(b)** 1. **(c)** 0.01. **(d)** 0.6 or 0.99.

4.13   **(a)** $S = \{$all numbers between 0 and 24$\}$. **(b)** $S = \{0, 1, 2, \ldots, 11\,000\}$. **(c)** $S = \{0, 1, 2, \ldots, 12\}$. **(d)** $S = \{$all numbers greater than or equal to 0$\}$ or $S = \{0, 0.01, 0.02, 0.03, \ldots\}$. **(e)** $S = \{$all positive and negative numbers$\}$. Note that the rats can lose weight.

4.15   **(a)** $P(\text{type AB}) = 0.04$. **(b)** $P(\text{type O or B}) = 0.49 + 0.20 = 0.69$.

4.17   Model 1: Legitimate. Model 2: Legitimate. Model 3: Probabilities have sum 6/7. Model 4: Probabilities cannot be negative.

4.19   No. The probabilities he describes are 0.1, 0.1, 0.3, and 0.6, which add up to 1.1.

4.21 $P(\text{either CV disease or cancer}) = 0.45 + 0.22 = 0.67$; $P(\text{other cause}) = 0.33$.

4.23   **(a)** The sum is 1. **(b)** 0.59. **(c)** 0.64. **(d)** 0.1681.

4.25   **(a)** The probabilities sum to 1. **(b)** 0.43. **(c)** 0.96. **(d)** 0.28. **(e)** 0.72.

4.27   **(a)** There are 10 pairs. Just using initials: { (A,D), (A,J), (A,S), (A,R), (D,J), (D,S), (D,R), (J,S), (J,R), (S,R)}. **(b)** $1/10$. **(c)** $4/10$. **(d)** $3/10$.

4.29 $(1 - 0.05)^{12} = (0.95)^{12} \doteq 0.540$.

4.31   **(a)** $P(A) \doteq 0.230$ since there are 38,225 (thousand) people who have completed 4+ years of college out of 166,438 (thousand). **(b)** 0.313. **(c)** 0.048; A and B are not independent since $P(A \text{ and } B) \neq P(A)P(B)$.

4.33   Look at the first five rolls in each sequence. All have one G and four R's, so those probabilities are the same. In the first sequence, you win regardless of the sixth roll; for the second, you win if the sixth roll is G, and for the third sequence, you win if it is R.

4.35   **(a)** 0.275. **(b)** 0.35. **(c)** 0.545.

4.37 $2/6 = 1/3$.

4.39   **(a)** 1%. **(b)** All probabilities are between 0 and 1; the probabilities add to 1. **(c)** 0.94. **(d)** 0.86. **(e)** Write either $X \geq 4$ or $X > 3$. The probability is 0.06.

4.41   **(a)** 75.2%. **(b)** All probabilities are between 0 and 1; the probabilities add to 1. **(c)** 0.983. **(d)** 0.976. **(e)** Write either $X \geq 9$ or $X > 8$. The probability is 0.931.

4.43   **(a)** 0.144. **(b)** The possible combinations are SSS, SSO, SOS, OSS, SOO, OSO, OOS, OOO (S = support, O = oppose). $P(\text{SSS}) = 0.216$; $P(\text{SSO}) = P(\text{SOS}) = P(\text{OSS}) = 0.144$; $P(\text{SOO}) = P(\text{OSO}) = P(\text{OOS}) = 0.096$; and $P(\text{OOO}) = 0.064$.

4.45 **(c)** $P(X = 0) = 0.216$; $P(X = 1) = 0.432$; $P(X = 2) = 0.288$; $P(X = 3) = 0.064$.
(d) Write either $X \geq 2$ or $X > 1$. The probability is 0.352.

4.45 **(a)** 0.49. **(b)** 0.73. **(c)** 0.73. **(d)** 0.2. **(e)** 0.5. **(f)** 0.

4.47 **(a)** The area of a triangle is $(1/2)bh = (1/2)(2)(1) = 1$. **(b)** 0.5. **(c)** 0.125.

4.49 **(a)** $P(\hat{p} \geq 0.16) = 0.1379$. **(b)** $P(0.14 \leq \hat{p} \leq 0.16) = 0.7242$.

4.51 $\mu = (0)(0.10) + (1)(0.15) + (2)(0.30) + (3)(0.30) + (4)(0.15) = 2.25$.

4.53 The mean μ of the company's "winnings" (premiums) and their "losses" (insurance claims) is positive. Even though the company will lose a large amount of money on a small number of policyholders who die, it will gain a small amount on the majority. The law of large numbers says that the average "winnings" minus "losses" should be close to μ, and overall the company will almost certainly show a profit.

4.55 **(a)** Independent: Weather conditions a year apart should be independent. **(b)** Not independent: Weather patterns tend to persist for several days; today's weather tells us something about tomorrow's. **(c)** Not independent: The two locations are very close together and would likely have similar weather conditions.

4.57 **(a)** The wheel is not affected by its past outcomes—it has no memory; outcomes are independent. So on any one spin, black and red remain equally likely. **(b)** Removing a card changes the composition of the remaining deck, so successive draws are not independent. If you hold 5 red cards, the deck now contains 5 fewer red cards, so your chance of another red decreases.

4.59 **(a)** The total mean is $11 + 20 = 31$ seconds. **(b)** No. Changing the standard deviations does not affect the means. **(c)** No. The total mean does not depend on dependence or independence of the two variables.

4.61 In 4.51, we had $\mu = 2.25$, so $\sigma_X^2 = 1.3875$, and $\sigma_X \doteq 1.178$.

4.63 $\mu_H = 2.6$ and $\mu_F = 3.14$ persons, $\sigma_H^2 = 2.02$ and $\sigma_F^2 = 1.5604$, $\sigma_H \doteq 1.421$ and $\sigma_F \doteq 1.249$ persons. Since families must include at least two people, it is not too surprising that the average family is slightly larger than the average household. For large family/household sizes, the differences between the distributions are small.

4.65 $\sigma_{\text{total}}^2 = 20$, so $\sigma_{\text{total}} \doteq 4.472$.

4.67 **(a)** $\sigma_Y^2 = 19{,}225$ and $\sigma_Y \doteq 138.65$ units. **(b)** $\sigma_{X+Y}^2 = 7{,}819{,}225$, so $\sigma_{X+Y} \doteq 2796.29$ units. **(c)** $\sigma_{2000X+3500Y} \doteq \$5{,}606{,}738$.

4.69 **(a)** $\mu_X = 550°$ Celsius; $\sigma_X^2 = 32.5$, so $\sigma_X \doteq 5.701°$C. **(b)** Mean: $0°$C; standard deviation: $5.701°$C. **(c)** $\mu_Y = (9/5)\mu_X + 32 = 1022°$F, and $\sigma_Y = (9/5)\sigma_X \doteq 10.26°$F.

4.71 $\sigma_X^2 = 94{,}236{,}826.64$, so that $\sigma_X \doteq \$9707.57$.

4.73 **(a)** $\mu_A = 300$ people. **(b)** There is no difference (except in the phrasing). **(c)** No. The choice seems to be based on how the options "sound."

4.75 $P(A \text{ or } B) = P(A) + P(B) - P(A \text{ and } B) = 0.125 + 0.237 - 0.077 = 0.285$.

4.77 **(a)** Household is both prosperous and educated. $P(A \text{ and } B) = 0.077$. **(b)** Household is prosperous but not educated. $P(A \text{ and } B^c) = 0.048$. **(c)** Household is not prosperous but is educated. $P(A^c \text{ and } B) = 0.160$. **(d)** Household is neither prosperous nor educated. $P(A^c \text{ and } B^c) = 0.715$.

4.79 **(a)** $18{,}262/99{,}585 \doteq 0.1834$. **(b)** $7{,}767/18{,}262 \doteq 0.4253$. **(c)** $7{,}767/99{,}585 \doteq 0.0780$. **(d)** $P(\text{over 65 and married}) = P(\text{over 65})P(\text{married} \mid \text{over 65}) = (0.1834)(0.4253)$.

4.81 **(a)** $3{,}046/58{,}929 \doteq 0.0517$. **(b)** "...married...age 18 to 24." **(c)** "0.0517 is the proportion of women who are *age 18 to 24* among those women who are *married*."

4.83 **(a)** $770/1626 \doteq 0.4736$. **(b)** $529/770 \doteq 0.6870$. **(c)** Using the multiplication rule: $P(\text{male and bachelor's degree}) = (0.4736)(0.6870) = 0.3254$. Directly: $529/1626 \doteq 0.3253$.

4.85 **(a)** 15% drink only cola. **(b)** 20% drink none of these.

4.87 If $F = \{\text{dollar falls}\}$ and $R = \{\text{renegotiation demanded}\}$, then $P(F \text{ and } R) = 0.32$.

4.89 $P(R) = 0.44$.

4.91 $P(\text{correct}) = P(\text{knows answer}) + P(\text{doesn't know but guesses correctly}) = 0.8$.

4.93 $P(\text{knows the answer} \mid \text{gives the correct answer}) = 0.75/0.80 = 15/16 = 0.9375$.

4.95 **(a)** The rat is in State A after Trials 1,2, and 3, and then changes to State B after Trial 4. **(b)** $P(X = 4) = 0.1024$. **(c)** $P(X = x) = (0.8)^{x-1}(0.2)$ for any $x \geq 1$—the rat fails to learn from the first $x - 1$ shocks, then learns from the last shock.

4.97 With $C = \{\text{building a plant is more profitable}\}$, we have $P(C) = 0.4906$ and $P(C^c) = 0.5094$. Contracting with a Hong Kong factory has a slight edge.

4.99 (a) $\mu_X = 3$ million dollars. $\sigma_X^2 = 503.375$, so $\sigma_X \doteq 22.436$ million dollars. (b) $\mu_Y = 0.9\mu_X - 0.225$ million dollars, and $\sigma_Y = 0.9\sigma_X \doteq 20.192$ million dollars.

4.101 (a) Asian stochastic beetle: $\mu = 1.3$ females. Benign boiler beetle: $\mu = 0.8$ females. (b) When a large population of beetles is considered, each generation of Asian stochastic beetles will contain close to 1.3 times as many females as the preceding generation. So the population will grow steadily. Each generation of benign boiler beetles, on the other hand, contains only about 80% as many females as the preceding generation.

4.103 (a) $S = \{3, 4, 5, \ldots, 18\}$ (these are not equally likely). (b) $\{X = 5\}$ means that the three dice come up $(1, 1, 3)$, $(1, 3, 1)$, $(3, 1, 1)$, $(1, 2, 2)$, $(2, 1, 2)$, or $(2, 2, 1)$. Each of these possibilities has probability $(1/6)^3$, so $P(X = 5) = 1/36$. (c) $\mu_{X_1} = \mu_{X_2} = \mu_{X_3} = 3.5$, and $\sigma_{X_i}^2 = 2.916$. Since the three rolls of the dice are independent, $\mu_X = \mu_{X_1} + \mu_{X_2} + \mu_{X_3} = 10.5$ and $\sigma_X^2 = \sigma_{X_1}^2 + \sigma_{X_2}^2 + \sigma_{X_3}^2 = 8.75$, so that $\sigma_X \doteq 2.958$.

4.105 $P(\text{roll a Venus}) = 24(0.4)(0.4)(0.1)(0.1) = 0.0384$.

4.107 (a) $P(X \geq 50) = 0.14 + 0.05 = 0.19$. (b) $P(X \geq 100 \mid X \geq 50) = 0.05/0.19 = 5/19$.

4.109 (a) 0.57. (b) 0.0481. (c) 0.0962. (d) 0.6864.

4.111 The response will be "No" with probability $0.35 = (0.5)(0.7)$. If the probability of plagiarism were 0.2, then $P(\text{student answers "No"}) = 0.4 - (0.5)(0.8)$. If 39% of students surveyed answered "No," then we estimate that $2 \times 39\% = 78\%$ have *not* plagiarized, so about 22% have plagiarized.

4.113 (a) The *exact* probabilities (to three decimals) are 0.001, 0.140, 0.352, 0.263, 0.136, 0.063, 0.027, 0.012, 0.005, 0.002, 0.001. [These are $P(X = 0)$, $P(X = 1)$, etc.] Actual simulation results will vary but should be similar. (b) This probability is about 0.508. Based on 50 simulated trials, most answers will be between 0.30 and 0.72. (c) Both computed means should be the same.

Chapter 5

5.1 (a) It may be binomial if we assume that there are no twins or other multiple births among the next 20, and that for all births, the probability that the baby is female is the same. (b) No. The number of observations is not fixed. (c) No. It is not reasonable to assume that the opinions of a husband and wife are independent.

5.3 (a) Yes. It is reasonable to assume that the results for the 50 students are independent, and each has the same chance of passing. (b) No. Since the student receives instruction after incorrect answers, her probability of success is likely to increase. (c) No. Temperature may affect the outcome of the test.

5.5 (a) There are 150 independent observations, each with probability of "success" (response) $p = 0.5$. (b) $\mu = np = (150)(0.5) = 75$. (c) Software gives 0.2312. (d) Use $n = 200$.

5.7 (a) $\hat{p} = 0.86$ (86%). (b) Software gives 0.1239. (c) Even when the claim is correct, there will be some variation in sample proportions. In particular, in about 12% of samples we can expect to observe 86 or fewer orders shipped on time.

5.9 (a) Find $P(0.41 \leq \hat{p} \leq 0.47) = P(123 \leq X \leq 141)$. Software gives 0.7309. (b) For $n = 600$, software gives 0.8719. For $n = 1200$, software gives 0.9663. Larger sample sizes are more likely to produce values of $\hat{p}$ close to the true value of p.

5.11 (a) 0.2816. (b) 0.5256. (c) 0.5256.

5.13 (a) $n = 4$ and $p = 0.25$. (b) $P(X = 0) = 0.3164$; $P(X = 1) = 0.4219$; $P(X = 2) = 0.2109$; $P(X = 3) = 0.0469$; $P(X = 4) = 0.0039$. (c) $\mu = np = 1$.

5.15 (a) $p = 0.25$. (b) 0.0139. (c) $\mu = 5$, $\sigma \doteq 1.9365$. (d) No. The trials would not be independent, since the subject may alter his or her strategy based on this information.

5.17 (a) The probability that all are assessed as truthful is 0.0687; the probability that at least one is reported to be a liar is 0.9313. (b) $\mu = 2.4$, $\sigma \doteq 1.3856$. (c) $P(X < \mu) = 0.5583$.

5.19 (a) $\mu = 180$ and $\sigma \doteq 12.5857$. (b) 0.2134, or 0.2252 (with continuity correction).

5.21 (a) $P(X \leq 70) = 0.1241$, or 0.1493 (with continuity correction). (b) $P(X \leq 175) = 0.0339$, or 0.0398 (with continuity correction). (c) 400 (with $n = 100$, $\sigma \doteq 0.0458$;

with $n = 400$, $\sigma \doteq 0.0229$). **(d)** Yes. Regardless of p, n must be quadrupled to cut σ in half.

5.23   **(a)** $\binom{n}{n} = \dfrac{n!}{n!0!} = 1$. The only way to distribute n successes among n observations is for all observations to be successes. **(b)** $\binom{n}{n-1} = \dfrac{n}{(n-1)!1!} = \dfrac{n \times (n-1)!}{(n-1)!} = n$. To distribute $n-1$ successes among n observations, the one failure must be either observation $1, 2, 3, \ldots, n-1$, or n. **(c)** $\binom{n}{k} = \dfrac{n!}{k!(n-k)!} = \dfrac{n!}{(n-k)![n-(n-k)]!} = \binom{n}{n-k}$. Distributing k successes is equivalent to distributing $n-k$ failures.

5.25   **(a)** $P(X < 0) = P(Z < 0.1346) = 0.5517$. **(b)** The mean is -3.5%. The standard deviation is 11.628%. **(c)** $P(\text{average return} < 0) = P(Z < 0.3010) = 0.6179$. Averages of several observations are more likely to be close to μ than an individual observation.

5.27   **(a)** $N(123 \text{ mg}, \sigma/\sqrt{3} \doteq 0.0462)$. **(b)** $P(Z > 21.65)$—essentially 0.

5.29   **(a)** $P(X < 295) = 0.1587$. **(b)** $\bar{x}$ has the $N(298 \text{ ml}, \sigma/\sqrt{6})$ distribution, so $P(\bar{x} < 295) = 0.0072$.

5.31   **(a)** $P(X = 1) = 18/38 = 9/19$ and $P(X = -1) = 10/19$; $\mu_X = -1/19$ dollars; and $\sigma_x \doteq 0.9986$ dollars. **(b)** In the long run, the gambler's average losses will be close to $-\$0.0526$ per bet. **(c)** 95% of the time, the mean winnings will fall between $-\$0.3350$ and $\$0.2298$; his total winnings will be between $-\$16.75$ and $\$11.49$. **(d)** 0.6453. **(e)** 95% of the time, the mean winnings are between $-\$0.05882$ and $-\$0.04644$. The casino winnings are between $\$5882$ and $\$4644$.

5.33 $P(\bar{x} > 2) \doteq P(Z > 4.71) = 0$ (essentially).

5.35   **(a)** $N(0.9 \text{ g/mi}, 0.01342)$. **(b)** $P(Z > 2.326) = 0.01$ if Z is $N(0, 1)$, so $L = 0.9 + (2.326)(0.01342) = 0.9312$ g/mi.

5.37 $L = \mu - 1.645\sigma/\sqrt{n} = 12.513$.

5.39   **(a)** $\mu_{\bar{x}} = 360$ g and $\mu_{\bar{y}} = 385$ g, so $\mu_{\bar{y}-\bar{x}} = 25$ g. $\sigma_{\bar{x}} = 12.298$ g and $\sigma_{\bar{y}} = 11.180$ g, so $\sigma_{\bar{y}-\bar{x}} = 16.62$ g. **(b)** $\bar{x}$ is $N(360 \text{ g}, 12.298)$, $\bar{y}$ is $N(385 \text{ g}, 11.180)$, and $\bar{y} - \bar{x}$ is $N(25 \text{ g}, 16.62)$. **(c)** $P(\bar{y} - \bar{x} \geq 25) = P(Z \geq 0) = 0.5$.

5.41   **(a)** $\bar{y}$ is $N(\mu_Y, \sigma_Y/\sqrt{m})$, and $\bar{x}$ is $N(\mu_X, \sigma_X/\sqrt{n})$. **(b)** $\bar{y} - \bar{x}$ is $N(\mu_Y - \mu_X, \sqrt{(\sigma_Y^2/m) + (\sigma_X^2/n)})$.

5.43 $F - L$ is $N(0, 2\sqrt{2})$, so $P(|F - L| > 5) = P(|Z| > 1.7678) = 0.0771$.

5.45   **(a)** Yes. This is always true; it does not depend on independence. **(b)** No. It is not reasonable to believe that X and Y are independent.

5.47   The centerline is at $\mu = 75°$; the control limits should be at $75° \pm 3\sigma/\sqrt{4}$, which means $74.25°$ and $75.75°$.

5.49   **(a)** Centerline: 11.5; control limits: 11.2 and 11.8. **(c)** Set B is from the in-control process. The process mean shifted suddenly for Set A; it appears to have changed on about the 12th sample (which falls outside the control limits). The mean drifted gradually for the process in Set C; it has a "run of nine" ending on observation 18.

5.51   **(a)** Centerline: 10 psi; control limits: 7.922 and 12.078. **(b)** There are no runs that should concern us here. Lot 13 signals that the process is out of control. The two samples that follow the bad one are fine, so it may be that whatever caused the low average for the 13th sample was an isolated incident. The operator should investigate to see if there is such an explanation.

5.53   Control charts focus on ensuring that the *process* is consistent, not that the *product* is good. An in-control process may consistently produce some percentage of low-quality products.

5.55 $P(15 \text{ points within } 1\sigma \text{ level}) \doteq 0.68^{15} \doteq 0.0031$.

5.57   Centerline: 162 lbs; control limits: 159.4 and 164.6 lbs. The first five points and the eighth are above the upper control limit; the first nine points are a "run of nine" above the centerline. However, the overall impression is that Joe's weight returns to being "in control."

5.59   **(a)** Mean: 0.1; standard deviation: $\sqrt{p(1-p)/400} = 0.015$. **(b)** Approximately $N(0.1, 0.015)$. **(c)** Centerline: 0.1; control limits: 0.055 and 0.145. **(d)** This process is out of control. Points below the lower control limit would not be a problem here, but beginning with lot number 2, we see many points above the upper control limit, and every value of $\hat{p}$ is above the centerline (with the exception of two points that fall *on* the centerline). A failure rate above 0.1 is strongly indicated.

5.61 Centerline: 0.0225; control limits: -0.02724 and 0.07224. Since -0.02724 is a meaningless value for a proportion, the lower control limit may as well be set to 0, especially since we are only concerned with the failure proportion being too high.

5.63 (a) $P(Z > 1/3) = 0.3707$. (b) Mean: 100; standard deviation: 1.93649. (c) $P(Z > 2.5820) = 0.0049$. (d) The answer to (a) could be quite different; (b) would be the same (it does not depend on normality at all). The answer we gave for (c) would still be fairly reliable because of the central limit theorem.

5.65 No. There is no fixed number of trials.

5.67 $P(750/12) < \bar{x} < 825/12) = P(-1.732 < Z < 2.598) = 0.9535$.

5.69 (a) No. Among other reasons, one could never have $X = 0$. (b) No. A count assumes only whole-number values, so it cannot be normally distributed. (c) $N(1.5, 0.02835)$. (d) $700\bar{x}$ has (approximately) the $N(1050, 19.84)$ distribution; $P(700\bar{x} > 1075) = 0.1038$.

5.71 (a) The machine that makes the caps and the machine that applies the torque are not the same. (b) $T - S$ is $N(-3, 1.5)$, so $P(T > S) = P(T - S > 0) = 0.0228$.

5.73 (a) 0.4313. (b) Centerline: 3.0 μm. Control limits: 2.797 and 3.203 μm.

5.75 $X - Y$ is $N(0, 0.4243)$, so $P(|X - Y| \geq 0.8) = 0.0593$.

Chapter 6

6.1 (a) $\sigma_{\bar{x}} \doteq 0.9186$ kg. (b) $\bar{x} \doteq 61.7916$, so the interval is 59.99 to 63.59 kg. Since 65 kg is well above the upper confidence limit, we have good evidence that $\mu < 65$ kg.

6.3 (a) 1 kg is 2.2 lbs, so $\bar{x}^* \doteq 135.942$ lbs. (b) $\sigma_{\bar{x}^*} \doteq 2.021$ lbs. (c) 132.0 to 139.9 lbs.

6.5 $\sigma_{\bar{x}} \doteq 1.291$ bushels/acre. (a) 121.3 to 126.3 bushels/acre. (b) Smaller. With a larger sample comes more information, which gives less uncertainty about the value of μ. (c) Narrower.

6.7 11.78 ± 0.77, or 11.01 to 12.55 years.

6.9 35.091 ± 3.250, or 31.84 to 38.34.

6.11 (a) 2.352 points. (b) 7.438 points. (c) Take $n = 62$, which is under the maximum.

6.13 $n = \lfloor(1.645 \times 10)/4\rfloor^2 \doteq 16.91$. Take $n = 17$.

6.15 $\$23,453 \pm \$439$, or \$23,014 to \$23,892.

6.17 (a) No. We can only be 95% confident. (b) The interval was based on a method that gives correct results (i.e., includes the correct percentage) 95% of the time. (c) $\sigma_{\text{estimate}} \doteq 1.53\%$. (d) No. It only accounts for random fluctuation.

6.19 (a) $(0.95)^7 \doteq 0.698 = 69.8\%$. (b) $\binom{7}{6}(0.95)^6(0.05) + (0.95)^7 \doteq 0.956 = 95.6\%$.

6.21 Probably not, because the interval is so wide. Such a large margin of error ($\pm\$2000$) would suggest either a very small sample size or a large standard deviation, but neither of these seems very likely. It is more likely that this range was based on looking at the list of first-year salaries and observing that most were between \$20,000 and \$24,000.

6.23 No. The interval refers to the mean math score, not to individual scores, which will be much more variable.

6.25 (a) Now $\bar{x} = 63.012$ kg and $\sigma_{\bar{x}} = 0.9$, so the interval is 61.248 to 64.776 kg. (b) The interval from Exercise 6.1 may be better, since 92.3 kg is an obvious outlier and may need to be excluded.

6.27 (a) H_0: $\mu = 18$ seconds; H_a: $\mu < 18$ seconds. (b) H_0: $\mu = 50$; H_a: $\mu > 50$. (c) H_0: $\mu = 24$; H_a: $\mu \neq 24$.

6.29 (a) H_0: $\mu = \$52,500$; H_a: $\mu > \$52,500$. (b) H_0: $\mu = 2.6$ hours; H_a: $\mu \neq 2.6$ hours.

6.31 While we might expect *some* difference in the amount of ethnocentrism between church attenders and nonattenders, the observed difference was so large that it is unlikely to be due to chance.

6.33 There almost certainly was *some* difference between the sexes and between blacks and whites; the observed difference between men and women was so large that it is unlikely to be due to chance. For black and white students, however, the difference was small enough that it could be attributed to random variation.

6.35 **(a)** $z \doteq 3.01$, which gives $P = 0.0013$. We conclude that the older students do have a higher mean score. **(b)** We assume the 20 students were an SRS from a normal population. The assumption that we have an SRS is more important.

6.37 **(a)** H_0: $\mu = 20$; H_a: $\mu > 20$. $z \doteq 2.548$, so $P \doteq 0.0054$. This is strong evidence that $\mu > 20$—the students have a higher average than past students have. **(b)** Randomly assign some (25 to 30) students to take the course, and compare their ACT mean score with those who did not take the course.

6.39 **(a)** H_0: $\mu = 32$; H_a: $\mu > 32$. **(b)** $\bar{x} = 35.091$, so $z = 1.864$ and $P \doteq 0.0314$. This is fairly good evidence that children in this district have a mean score higher than the national average. Observations this extreme would only occur in about 3 out of 100 samples if H_0 were true.

6.41 **(a)** Yes, because $z > 1.645$. **(b)** Yes, because $z > 2.326$.

6.43 When a test is significant at the 1% level, it means that if the null hypothesis is true, outcomes similar to those seen are expected to occur less than once in 100 repetitions of the experiment or sampling. "Significant at the 5% level" means we have observed something which occurs in less than 5 out of 100 repetitions (when H_0 is true). Something that occurs "less than once in 100 repetitions" also occurs "less than 5 times in 100 repetitions," so significance at the 1% level implies significance at the 5% level (or any higher level).

6.45 Since $0.215 < 0.674$, $P > 0.25$. In fact, $P = P(Z > 0.215) = 0.4149$.

6.47 Since $1.282 < 1.37 < 1.645$, the P-value is between $2(0.05) = 0.10$ and $2(0.10) = 0.20$. From Table A, $P = 2(0.0853) = 0.1706$.

6.49 **(a)** 59.99 to 63.59 kg. **(b)** No, since 61.3 is inside the interval. **(c)** No, since 63 is inside the interval.

6.51 $P = 0.1292$. Although this sample showed *some* difference in market share between pioneers with patents or trade secrets and those without, the difference was small enough that it could have arisen merely by chance.

6.53 A test of significance answers question **(b)**.

6.55 **(a)** $P = 0.3821$. **(b)** $P = 0.1711$. **(c)** $P = 0.0013$.

6.57 **(a)** $z = 1.64 < 1.645$—not significant at 5% level ($P = 0.0505$). **(b)** $z = 1.65 > 1.645$—significant at 5% level ($P = 0.0495$).

6.59 **(a)** No. In a sample of size 500, we expect to see about 5 people who have a "P-value" of 0.01 or less. These four *might* have ESP, or they may simply be among the "lucky" ones we expect to see. **(b)** The researcher should repeat the procedure on these four to see if they again perform well.

6.61 Using $\alpha/12 = 0.0041\bar{6}$ as the cutoff, the fifth ($P = 0.001$), sixth ($P = 0.004$), and eleventh ($P = 0.002$) are significant.

6.63 $z \geq 2.326$ is equivalent to $\bar{x} \geq 460.4$, so $P(\text{reject } H_0 \text{ when } \mu = 460) = P(\bar{x} \geq 460.4 \text{ when } \mu = 460) = P(Z \geq 0.0894) = 0.4644$. This is quite a bit less than the "80% power"; this test is not very sensitive to a 10-point increase in the mean score.

6.65 Reject H_0 when $\bar{x} \leq 300 - 1.645(3/\sqrt{n})$. **(a)** $P(Z \leq 0.021\bar{6}) = 0.5086$. **(b)** $P(Z \leq 1.688\bar{3}) = 0.9543$.

6.67 **(a)** $P(\bar{x} > 0 \text{ when } \mu = 0) = P(Z > 0) = 0.50$. **(b)** $P(\bar{x} < 0 \text{ when } \mu = 0.3) = P(Z < -0.9) = 0.1841$. **(c)** $P(\bar{x} < 0 \text{ when } \mu = 1) = P(Z < -3) = 0.0013$.

6.69 $P(\text{Type I error}) = 0.01 = \alpha$. $P(\text{Type I error}) = 1 - 0.4644 = 0.5356$.

6.71 **(a)** H_0: the patient is ill (or "the patient should see a doctor"); H_a: the patient is healthy (or "the patient should not see a doctor"). A Type I error means a false negative—clearing a patient who should be referred to a doctor. A Type II error is a false positive—sending a healthy patient to the doctor. **(b)** Answers will vary.

6.73 $\bar{x} = -5.3\bar{6}$ mg/dl, so $\bar{x} \pm 1.645\sigma/\sqrt{6}$ is 4.76 to 5.97 mg/dl.

6.75 **(a)** The plot is reasonably symmetric for such a small sample. **(b)** $\bar{x} = 30.4$ μg/l; 26.06 to 34.74 μg/l. **(c)** H_0: $\mu = 25$; H_a: $\mu > 25$. $z = 2.44$; so $P = 0.007$. This is fairly strong evidence against H_0: the beginners' mean threshold is higher than 25 μg/l.

6.77   Divide everything by 52.14: $\$653.55 \pm \$6.21$, or $\$647.34$ to $\$659.76$.

6.79   12.3 to 13.5 g/100 ml. This assumes that the babies are an SRS from the population. The population should not be too nonnormal.

6.81   **(a)** $N(0, 5.3932\%)$. **(b)** $P = 0.1003$. **(c)** Not significant at $\alpha = 0.05$. The study gives *some* evidence of increased compensation, but it is not very strong; it would happen 10% of the time just by chance.

6.83 $H_0: p = 18/38$; $H_a: p \neq 18/38$.

6.85   Yes. Significance tests allow us to discriminate between random differences ("chance variation") that might occur when the null hypothesis is true and differences that are unlikely to occur when H_0 is true.

6.87   For each sample, find $\bar{x}$, then take $\bar{x} \pm (1.96)(5/\sqrt{5}) = \bar{x} \pm 4.383$. We "expect" to see that 95 of the 100 intervals will include 20 (the true value of μ); probability tells us that about 99% of the time, 90 or more intervals will include 20.

6.89   For each sample, find $\bar{x}$, then compute $z = (\bar{x} - 22.5)/(5/\sqrt{5})$. Choose a significance level α and the appropriate cutoff point (z^*). E.g., with $\alpha = 0.10$, reject H_0 if $|z| > 1.645$; with $\alpha = 0.05$, reject H_0 if $|z| > 1.96$. Since $Z = (\bar{x} - 20)/(5/\sqrt{5})$ has the $N(0, 1)$ distribution, the probability we will accept H_0 is $P(-z^* < (\bar{x} - 22.5)/(5/\sqrt{5}) < z^*) = P(-z^* < Z - 1.118 < z^*) = P(1.118 - z^* < Z < 1.118 + z^*)$. If $\alpha = 0.10$ ($z^* = 1.645$), this probability is 0.698; if $\alpha = 0.05$ ($z^* = 1.96$), this probability is 0.799. For smaller α, the probability will be larger. Thus, we "expect" to (wrongly) accept H_0 a majority of the time.

6.91   **(b)** Since $\sigma_{\bar{x}} = 10$, compute $z = (\bar{x} - 460)/10$. **(c)** Rarely would more than 4 of the 25 samples lead you to reject H_0. In the long run, about 5% of samples would wrongly reject H_0.

Chapter 7

7.1   **(a)** df $= 11$, $t^* = 1.796$. **(b)** df $= 29$, $t^* = 2.045$. **(c)** df $= 17$, $t^* = 1.333$.

7.3   **(a)** df $= 14$. **(b)** 1.761 and 2.145. **(c)** 0.05 and 0.025. **(d)** $0.025 < P < 0.05$. **(e)** Significant at 5%, but not at 1%. **(f)** $P = 0.0345$.

7.5   **(a)** df $= 11$. **(b)** $0.01 < P < 0.02$. **(c)** $P = 0.0161$.

7.7   **(a)** The data are right-skewed, with a high outlier of 2433 (and possibly 1933). The quantile plot shows these two outliers but otherwise is not strikingly different from a line. **(b)** $\bar{x} = 926$, $s = 427.2$, $SE_{\bar{x}} = 69.3$ (all in mg). **(c)** Using 30 df, we have 784.5 to 1067.5 mg; Minitab reports 785.6 to 1066.5 mg.

7.9   **(a)** The transformed data have $\bar{x} = 77.17\%$, $s = 35.6\%$, and $SE_{\bar{x}} = 5.78\%$; the confidence interval is 65.37% to 88.79% (using $t^* = 2.042$ with df $= 30$), or 65.46% to 88.87% (from Minitab). **(b)** After dividing the intervals from Exercise 7.7 by 12, the intervals are the same (up to rounding error).

7.11   **(a)** The distribution is right-skewed with high outliers of 63 and 79. (In fact, according to the $1.5 \times IQR$ outlier test, 48 is an outlier, too.) The quantile plot suggests that the distribution is not normal. **(b)** $\bar{x} = 23.56$, $s = 12.52$, $SE_{\bar{x}} = 1.77$ can openers. Using df $= 40$, we have $t^* = 2.021$ and the interval is 19.98 to 27.14 can openers; Minitab reports 20.00 to 27.12 can openers. **(c)** With such a large sample size, the t distribution is fairly good in spite of the skewness and outliers.

7.13   **(a)** Each store averaged $\$50.65$, giving a total of $(\$50.65)(3275) = \$165,879$. **(b)** Multiply the interval from Exercise 7.12 by 3275. Using df $= 40$, this gives $\$140,684$ to $\$191,100$. Using df $= 49$, this gives $\$140,825$ to $\$190,959$.

7.15   For large df, we use normal distribution critical values: 87.6 to 104.4 days.

7.17   **(a)** One possible display is a stemplot where the digits are the stems, and all leaves are "0"—this is essentially the same as a histogram. The scores are slightly left-skewed. **(b)** $\bar{x} = 3.618$, $s = 3.055$, $SE_{\bar{x}} = 0.524$. **(c)** Using df $= 30$, we have $t^* = 2.042$ and the interval is 2.548 to 4.688. Minitab reports 2.551 to 4.684.

7.19 $H_0: \mu = 0$; $H_a: \mu < 0$, where μ is the mean decrease in vitamin C content. Subtract "After" from "Before" to give -53, -52, -57, -52, and -61 mg/100 g. Then $\bar{x} = -55$, $s = 3.94$, and $SE_{\bar{x}} = 1.76$ mg/100 g, so $t = -31.24$, which is significant for any reasonable α (using df $= 4$). Cooking does decrease the vitamin C content.

7.21 (a) H_0; $\mu = 0$; H_a: $\mu > 0$. $t = 50.07$; since $P \doteq 0$, we conclude that the new policy would increase credit card usage. (b) Using $t^* = 1.984$ (df $= 100$): \$328 to \$356. Using $t^* = 1.9695$ (df $= 249$, from, e.g., Minitab): \$329 to \$355. (c) The sample size is very large, and we are told that we have an SRS. This means that outliers are the only potential snag, and there are none. (d) Make the offer to an SRS of 250 customers, and choose another SRS of 250 as a control group. Compare the mean increase for the two groups.

7.23 (a) $\bar{x} = 5.36$ mg/dl and $s \doteq 0.6653$, so $\mathrm{SE}_{\bar{x}} = 0.2716$ mg/dl. (b) df $= 5$, $t^* = 2.015$, and the interval is 4.819 to 5.914 mg/dl.

7.25 H_0: $\mu = 4.8$; H_a: $\mu > 4.8$. $t \doteq 2.086$. For df $= 5$, $0.025 < P < 0.05$. This is fairly strong, though not overwhelming, evidence that the patient's phosphate level is above normal.

7.27 (a) 111.2 to 118.6. (b) The essential assumption is that the 27 men tested can be regarded as an SRS from a population, such as all healthy white males in a stated age group. The assumption that blood pressure in this population is normally distributed is *not* essential, because $\bar{x}$ from a sample of 27 will be roughly normal in any event, as long as the population is not too greatly skewed and has no outliers.

7.29 (b) H_0: $\mu = 105$; H_a: $\mu \neq 105$. $\bar{x} = 104.13$ and $s = 9.40$ pCi/l, so $t \doteq -0.32$. With df $= 11$, $P > 2(0.25) = 0.50$, which gives us little reason to doubt that $\mu = 105$ pCi/l.

7.31 (a) SEM $=$ standard error of the mean ($\mathrm{SE}_{\bar{x}}$). (b) $s = (\sqrt{3})(\mathrm{SE}_{\bar{x}}) \doteq 0.0173$. (c) Using $t^* = 2.920$ (with df $= 2$): $0.84 \pm (2.920)(0.01)$, or about 0.81 to 0.87.

7.33 (a) For each subject, randomly select (e.g., by flipping a coin) which knob (right or left) that subject should use first. (b) H_0: $\mu = 0$ vs. H_a: $\mu < 0$, where μ is the mean of (right-thread time $-$ left-thread time). (c) $\bar{x} = -13.32$ seconds; $\mathrm{SE}_{\bar{x}} = 22.94/\sqrt{25} \doteq 4.59$ seconds, so $t = -2.90$. With df $= 24$, we see that $0.0025 < P < 0.005$. We have good evidence that the mean difference really is negative, i.e., that the mean time for right-threaded knobs is less than the mean time for left-threaded knobs.

7.35 (a) H_0: $\mu = 0$ vs. H_a: $\mu > 0$, where μ is the mean improvement in score (posttest $-$ pretest). (b) The differences are slightly left-skewed, with no outliers; the t test should be reliable. (c) $\bar{x} = 1.450$; $\mathrm{SE}_{\bar{x}} \doteq 0.716$, so $t = 2.02$. With df $= 19$, $0.025 < P < 0.05$. This is significant at 5% but not at 1%. We have some evidence that scores improve, but it is not overwhelming. (d) Minitab gives 0.211 to 2.689; using $t^* = 1.729$ and the values of $\bar{x}$ and $\mathrm{SE}_{\bar{x}}$ above, we obtain 0.212 to 2.688.

7.37 H_0: $\mu = 0$; H_a: $\mu > 0$, where μ is the mean of (Variety A $-$ Variety B). $t \doteq 1.295$; with df $= 9$, $0.10 < P < 0.15$. We do not have enough evidence to conclude that Variety A has a higher yield.

7.39 With all 50 states listed in the table, we have information about the entire population in question; no statistical procedures are needed (or meaningful).

7.41 (a) We reject H_0: $\mu - 0$ if $t > 1.833$, which translates to $\bar{x} > 0.4811$. The power against $\mu = 0.5$ lb/plant is $P(\bar{x} > 0.4811$ when $\mu = 0.5) = P(Z > -0.072) = 0.5279$. (b) We reject H_0 if $t > 1.711$, which translates to $\bar{x} > 0.2840$. The power against $\mu = 0.5$ lb/plant is $P(\bar{x} > 0.2840$ when $\mu = 0.5) = P(Z > -1.301) \doteq 0.9032$.

7.43 (a) H_0: population median $= 0$ vs. H_a: population media < 0, or H_0: $p = 1/2$ vs. H_a: $p < 1/2$, where p is the proportion of (right $-$ left) differences that are positive. (b) One pair of the 25 had no difference; of the remaining 24, only 5 differences were positive. If X (the number of positive differences) has a $B(24, 1/2)$ distribution, the P-value is $P(X \leq 5)$, for which the normal approximation gives 0.0021, or 0.0040 (with the continuity correction). In any case, this is strong evidence against H_0, indicating that the mean right-threaded knob time is shorter.

7.45 We cannot use the sign test, since we cannot determine the number of positive/negative differences in the original data.

7.47 $\bar{x} = 2.5552$, $s = 0.0653$, and $\mathrm{SE}_{\bar{x}} = 0.0292$, giving the interval 2.4929 to 2.6175.

7.49 (a) H_0: $\mu_1 = \mu_2$; H_a: $\mu_1 > \mu_2$. $\bar{x}_1 = 95.3$ and $s_1 = 0.990$ mg/100 g; $\bar{x}_2 = 95.85$ and $s_2 = 2.19$ mg/100 g. Since $\bar{x}_2 > \bar{x}_1$, we have no evidence against H_0; further analysis is not necessary. (b) -11.3 to 10.2 mg/100 g vitamin E lost (using df $= 1$), or -7.36 to 6.26 mg/100 g (using df $= 1.39$).

7.51   (a) Control scores are fairly symmetrical, while piano scores are slightly left-skewed. Scores in the piano group are generally higher than scores in the control group.
(b) Piano: $n_1 = 34, \bar{x}_1 = 3.618, s_1 = 3.055, SE_{\bar{x}_1} = 0.524$. Control: $n_2 = 44$, $\bar{x}_2 = 0.386, s_2 = 2.423, SE_{\bar{x}_2} = 0.365$. (c) $H_0: \mu_1 = \mu_2; H_a: \mu_1 > \mu_2$. $t = 5.06$. Whether df $= 33$ or df $= 61.7$, $P < 0.0001$, so we reject H_0 and conclude that piano lessons improved the test scores.

7.53   Having the control group in 7.51 and 7.52 makes our conclusions more reliable, since it accounts for increases in scores that may come about simply from the passage of time. Arguably, there is an advantage to the test since we have a one-sided alternative; the confidence interval by its nature is two-sided.

7.55   (a) $H_0: \mu_1 = \mu_2; H_a: \mu_1 \neq \mu_2$. For the low-fitness group, $\bar{x}_1 = 4.64$ and $s_1 = 0.69$. For the high-fitness group, $\bar{x}_2 = 6.43$ and $s_2 = 0.43$. $t = -8.23$, so $P < 0.0001$ (using either df $= 13$ or df $= 21.8$); this difference is significant at 5% and at 1% (and much lower). (b) All the subjects were college faculty members. Additionally, all the subjects volunteered for a fitness program, which could bring in some further confounding.

7.57   (a) $H_0: \mu_1 = \mu_2; H_a: \mu_1 < \mu_2$. $t = -7.34$, which gives $P < 0.0001$ whether df $= 133$ or df $= 140.6$. Cocaine use is associated with lower birth weights. (b) -488.7 to -281.3 g. (c) The "other" group may include drug users, since some in it were not tested. Among drug users, there may have been other (confounding) factors that affected birth weight.

7.59 $\$0.66 \pm t^*(\$0.037014)$, where t^* is chosen with either df $= 44$ or df $= 97.7$. Whatever choice of df is made, $t^* \doteq 2$, so the interval is about $\$0.59$ to $\$0.73$.

7.61   (a) $H_0: \mu_A = \mu_B; H_a: \mu_A \neq \mu_B$; $t = -1.484$. Using $t(149)$ and $t(297.2)$ distributions, P equals 0.1399 and 0.1388, respectively; not significant in either case. The bank might choose to implement Proposal A even though the difference is not significant, since it may have a *slight* advantage over Proposal B. Otherwise, the bank should choose whichever option costs them less. (b) Because the sample sizes are equal and large, the t procedure is reliable in spite of the skewness.

7.63   (a) $H_0: \mu_1 = \mu_2$ vs. $H_a: \mu_1 < \mu_2$, where μ_1 is the beta-blocker population mean pulse rate and μ_2 is the placebo mean pulse rate. We find $t = -2.4525$; with a $t(29)$ distribution, we have $0.01 < P < 0.02$, which is significant at 5% but not at 1%. With a $t(57.8)$ distribution, we have $P = 0.0086$, which is significant at 5% and at 1%. (b) With df $= 29$: -10.83 to 0.63 beats/minute. With df $= 57.8$: -10.64 to 0.44 beats/minute.

7.65 $H_0: \mu_{skilled} = \mu_{novice}$ vs. $H_a: \mu_s \neq \mu_n$. Use the "Unequal" values: $t = 0.5143$ with df $= 11.8$; its P-value is 0.6165. There is no reason to reject H_0; skilled and novice rowers seem to have (practically) the same mean weight.

7.67   (a) Using (e.g.) back-to-back stemplots, we see that both distributions are slightly skewed to the right and have one or two moderately high outliers. A t procedure may be (cautiously) used nonetheless, since the sum of the sample sizes is almost 40. (b) $H_0: \mu_w = \mu_m; H_a: \mu_w > \mu_m$. $\bar{x}_w = 141.056, s_w = 26.4363, \bar{x}_m = 121.250$, $s_m = 32.8519$, and $t = 2.0561$, so $P = 0.0277$ (with df $= 17$) or $P = 0.0235$ (with df $= 35.6$). This gives fairly strong evidence—significant at 5% but not 1%—that the women's mean is higher. (c) For $\mu_m - \mu_w$: -36.57 to -3.05 (with df $= 17$) or -36.07 to -3.54 (with df $= 35.6$).

7.69   (a) $t = 1.604$ with df $= 9$ or df $= 15.6$; the P-value is either 0.0716 or 0.0644, respectively. Both are similar to the P-value in Example 7.20, and the conclusion is essentially the same. (b) With df $= 9$: -0.76 to 11.30 (margin of error: 6.03). With df $= 15.6$: -0.48 to 11.02 (margin of error: 5.75). Both margins of errors are similar to (but slightly larger than) the margin of error in Example 7.21.

7.71   (a) $SE_{\bar{x}_2} \doteq 11.35$. $SE_{\bar{x}_1 - \bar{x}_2} \doteq 15.61$. (b) $H_0: \mu_1 = \mu_2$ vs. $H_a: \mu_1 \neq \mu_2$; $t = 1.249$. Using $t(9)$ and $t(25.4)$ distributions, P equals 0.2431 and 0.2229, respectively; the difference is not significant. (c) Using $t(9)$: -15.8 to 54.8. $t(25.4)$: -12.6 to 51.6. These intervals had to contain 0 because according to (b), the observed difference would occur in more than 22% of samples when the means are the same; thus 0 would appear in any confidence interval with a confidence level greater than about 78%.

7.73 (a) Using $t^* = 1.660$ (df = 100), the interval is \$412.68 to \$635.58. Using $t^* = 1.6473$ (df = 620), the interval is \$413.54 to \$634.72. Using $t^* = 1.6461$ (df = 1249.2), the interval is \$413.62 to \$634.64. (b) Because the sample sizes are so large (and the sample sizes are almost the same), deviations from the assumptions have little effect. (c) The sample is not *really* random, but there is no reason to expect that the method used should introduce any bias into the sample. (d) Students without employment were excluded, so the survey results can only (possibly) extend to *employed* undergraduates. Knowing the number of unreturned questionnaires would also be useful.

7.75 $s_p^2 = 27.75$, $s_p = 5.2679$, and $t = 0.6489$ with df = 293, so $P = 0.5169$—not significant. The conclusion is similar to that in Example 7.16.

7.77 With equal variances, $t = 0.5376$ (df = 16), which gives $P = 0.5982$. As before, there is no reason to reject H_0; skilled and novice rowers seem to have (practically) the same mean weight.

7.79 (a) From an $F(9, 20)$ distribution, $F^* = 2.39$. (b) P is between $2(0.025) = 0.05$ and $2(0.05) = 0.10$; $F = 2.45$ is significant at the 10% level but not at the 5% level.

7.81 $H_0: \sigma_1 = \sigma_2$; $H_a: \sigma_1 \neq \sigma_2$. $F = 1.59$; comparing with an $F(33, 43)$ distribution, we find $P = 0.1529$ (from the table, use an $F(30, 40)$ distribution and observe that $P > 0.1$). We do not have enough evidence to conclude that the standard deviations are different.

7.83 (a) An $F(1, 1)$ distribution; with a two-sided alternative, we need the critical value for $p = 0.025$: $F^* = 647.79$. This is a very low power test, since large differences between σ_1 and σ_2 would rarely be detected. (b) $H_0: \sigma_1 = \sigma_2$ vs. $H_a: \sigma_1 \neq \sigma_2$. $F = 3.963$. Not surprisingly, we do not reject H_0.

7.85 (a) $H_0: \sigma_1 = \sigma_2$; $H_a: \sigma_1 \neq \sigma_2$. (b) $F = 2.196$ from an $F(7, 9)$ distribution, so $P > 0.20$.

7.87 (a) $H_0: \sigma_m = \sigma_w$; $H_a: \sigma_m > \sigma_w$. (b) $F = 1.544$ from an $F(19, 17)$ distribution. (c) Using the $F(15, 17)$ entry in the table, we find $P > 0.10$ (in fact, $P = 0.1862$).

7.89 df = 198; we reject H_0 if $t > 1.660$ (from the table, with df = 100) or $t > 1.6526$ (using df = 198). The noncentrality parameter is $\delta \doteq 3.2636$; the power is about 95%, regardless of which t^*-value is used.

7.91 The standard error is $650\sqrt{2/n}$, and the t^*-values are 2.0106, 1.9845, 1.9761, 1.9720, and 1.9696 (using df = 48, 98, 148, 198, and 248). The margins of error are therefore 369.6, 258.0, 209.8, 181.3, and 161.9.

7.93 The distributions appear similar; the most striking difference is the relatively large number of boys with low GPAs. Testing the difference in GPAs, we obtain $t = -0.91$, which is not significant, regardless of whether we use df = 30 or 74.9. For the difference in IQs, we find $t = 1.64$, which is fairly strong evidence, although it is not quite significant at the 5% level: $0.05 < P < 0.10$ (df = 30), or $P = 0.0503$ (df = 56.8).

7.95 Both distributions have two high outliers. With those houses, $\bar{x}_3 = \$171{,}717$, $\bar{x}_4 = \$178{,}123$, and $t \doteq -0.5268$. For testing $H_0: \mu_3 = \mu_4$ vs. $H_a: \mu_3 < \mu_4$, $P \doteq 0.3$ whether we take df = 21 or df = 35.8, so we have little reason to reject H_0. The confidence interval for $\mu_3 - \mu_4$ is about $-\$31{,}694$ to \$18,882 (df = 21) or $-\$31{,}073$ to \$18,261 (df = 35.8). Without the outliers, $\bar{x}_3 = \$164{,}668$, $\bar{x}_4 = \$165{,}510$, and $t \doteq -0.1131$, so we have little reason to believe that $\mu_3 < \mu_4$ ($P \doteq 0.455$ with either df = 19 or df = 34.5). The interval for $\mu_3 - \mu_4$ is about $-\$16{,}432$ to \$14,747 (df = 19) or $-\$15{,}971$ to \$14,286 (df = 34.5).

7.97 It is reasonable to have a prior belief that people who evacuated their pets would score higher, so we test $H_0: \mu_1 = \mu_2$ vs. $H_a: \mu_1 > \mu_2$. We find $t = 3.65$, so $P < 0.0005$ no matter how we choose df. As one might suspect, people who evacuated their pets have a higher mean score. One might also compute a 95% confidence interval for the difference: 0.77 to 2.61 points.

7.99 $H_0: \mu_1 = \mu_2$; $H_a: \mu_1 < \mu_2$. $t = -0.7586$, which is not significant regardless of df. There is not enough evidence to conclude that nitrites decrease amino acid uptake.

7.101 (a) The stemplot (after truncating the decimal) shows that the data are left-skewed; there are some low observations but no particular outliers. (b) $\bar{x} = 59.58\%$, $SE_{\bar{x}} \doteq 6.255/\sqrt{9} \doteq 2.085$, and for df = 8, $t^* = 2.306$, so the interval is 54.8% to 64.4%.

7.103 The similarity of s_1 and s_2 suggests that the population standard deviations are likely to be similar. $t = -0.3533$, so $P = 0.3621$ (df $= 179$)—still not significant.

7.105 No. Counties in California could not be considered an SRS of Indiana counties.

7.107 **(a)** $H_0\colon \mu_1 = \mu_2$ vs. $H_a\colon \mu_1 > \mu_2$; $t = 1.1738$, so $P = 0.1265$ using $t(22)$, or 0.1235 using $t(43.3)$. Not enough evidence to reject H_0. **(b)** -14.57 to 52.57 (df $= 22$), or -13.64 to 51.64 (df $= 43.3$). **(c)** 165.53 to 220.47. **(d)** We are assuming that we have two SRSs from each population and that underlying distributions are normal. It is unlikely that we have random samples from either population.

7.109 We test H_0: population median $= 0$ vs. H_a: population median > 0, or $H_0\colon p = 1/2$ vs. $H_a\colon p > 1/2$, where p is the proportion of (city $-$ rural) differences that are positive. Ignoring missing values and the 2 "zero" differences, there are 6 negative differences and 18 positive differences. If X (the number of positive differences) has a $B(24, 1/2)$ distribution, the P-value is $P(X \le 6) = 0.0113$. This is strong evidence against H_0, indicating that the median city level is higher.

7.111 With 95% confidence intervals, e.g., the mean abdomen skinfold difference is between 11.62 and 15.38 mm (using df $= 103.6$). With the same df, the mean thigh skinfold difference is between 9.738 and 12.86 mm.

7.113 **(a)** With a pooled variance, $t = 3.9969$ with df $= 222$, so $P < 0.0001$. With unpooled variances, $t = 4.0124$ with df $= 162.2$, and again $P < 0.0001$. The test of $\sigma_1 = \sigma_2$ gives $F = 1.03$ with 144 and 78 df. The P-value is 0.9114, so the pooled procedure should be appropriate. In either case, we conclude that male mean SATM is higher than female mean SATM. A 99% confidence interval for the (male $-$ female) difference is 16.36 to 77.13. **(b)** With a pooled variance, $t = 0.9395$ with df $= 222$, so $P = 0.3485$. With unpooled variance, $t = 0.9547$ with df $= 167.9$, so $P = 0.3411$. The test of $\sigma_1 = \sigma_2$ gives $F = 1.11$ with 144 and 78 df. The P-value is 0.633, so the pooled procedure should be appropriate. In either case, we cannot see a difference between male and female mean SATV scores. A 99% confidence interval for the (male $-$ female) difference is -21.49 to 45.83. **(c)** The results may generalize fairly well to students in different years, less well to students at other schools, and probably not very well to college students in general.

7.115 **(a)** Pooling is appropriate; $t = 2.8679$ with df $= 42$, so $P = 0.0032$ (since our test is one-sided). We conclude that the mean score using DRTA is higher than the mean score with the Basal method; the difference in the average scores is 5.68. **(b)** Pooling is appropriate; $t = 1.8773$ with df $= 42$, so $P = 0.0337$. We conclude that the mean score using Strat is higher than the mean score with the Basal method; the difference in the average scores is 3.23.

7.117 E.g., for a difference of 0.05, we will reject H_0 if $|t| > 2.0244$ (assuming this is a two-sided alternative). The noncentrality parameter is $\delta \doteq 2.2588$; using the normal approximation, the power is $P(Z < -t^* - \delta \text{ or } Z > t^* + \delta) = 0.5927$, or about 60%.

7.119 The margin of error is $t^* s / \sqrt{n} = t^* / \sqrt{n}$, taking t^* from t distributions with 4, 9, 14, ..., 99 df. In order to get all the t^*-values, software is needed. For reference, $t^* = 2.7764$ with df $= 4$ and $t^* = 1.9842$ with df $= 99$; the margin of error gradually decreases from 1.2416 when $n = 5$ to 0.19842 when $n = 100$.

Chapter 8

8.1 **(a)** No: $n(1 - \hat{p}) = 3 < 10$. **(b)** Yes: $n\hat{p} = n(1 - \hat{p}) = 12.5$. **(c)** No: $n\hat{p} = 4 < 10$. **(d)** Yes: $n\hat{p} = 360$ and $n(1 - \hat{p}) = 240$.

8.3 No. Some of those who lied about having a degree may also have lied about their major. *At most* 24 applicants lied about having a degree or about their major.

8.5 $\hat{p} = 386/1711 \doteq 0.2256$, and $SE_{\hat{p}} = \sqrt{\hat{p}(1 - \hat{p})/1711} \doteq 0.0101$, so the 95% confidence interval (CI) is $0.2256 \pm (1.96)(0.0101)$, or 0.2058 to 0.2454.

8.7 $\hat{p} = 0.86$, and $SE_{\hat{p}} \doteq 0.0347$, so the 95% CI is 0.7920 to 0.9280.

8.9 $\hat{p} = 0.66$, and $SE_{\hat{p}} \doteq 0.0335$, so the 95% CI is 0.5943 to 0.7257.

8.11 $H_0\colon p = 0.36$ vs. $H_a\colon p \ne 0.36$. We have $\hat{p} = 0.38$ and $z = 0.9317$. This is clearly not significant (in fact, $P \doteq 0.35$).

8.13 **(a)** $\hat{p} \doteq 0.4202$, and $SE_{\hat{p}} \doteq 0.0117$, so the 99% CI is 0.3901 to 0.4503. **(b)** Yes. The 99% CI does not include 0.50 or more. **(c)** Use $n = 16,167$.

8.15 $\hat{p} \doteq 0.173$, and $\text{SE}_{\hat{p}} \doteq 0.0437$, so the 95% CI is 0.0877 to 0.2590.

8.17   **(a)** Testing $H_0: p = 0.5$ vs. $H_a: p \neq 0.5$, we have $z = 1.34$. This is not significant at $\alpha = 0.05$. **(b)** The 95% CI is 0.4969 to 0.5165.

8.19   **(a)** $H_0: p = 0.384$; $H_a: p > 0.384$. **(b)** $\hat{p} = 0.625$ and $z = 3.13$. **(c)** Reject H_0 since $z > 1.645$; $P = 0.0009$. **(d)** The 90% CI is 0.4991 to 0.7509. There is strong evidence that Leroy has improved. **(e)** We assume that the 40 free throws are an SRS; more specifically, each shot represents an independent trial with the same probability of success, so the number of free throws made has a binomial distribution. To use the normal approximation, we also need (for the test) $np_0 = 15.36 > 10$ and $n(1 - p_0) = 24.64 > 10$, and (for the confidence interval) $n\hat{p} = 25 > 10$ and $n(1 - \hat{p}) = 15 > 10$.

8.21 $n = (1.96/0.05)^2 (0.2)(0.8) = 245.9$. Use $n = 246$.

8.23 $n = (1.645/0.04)^2 (0.7)(0.3) = 355.2$. Use $n = 356$. With $\hat{p} = 0.5$, $\text{SE}_{\hat{p}} = 0.0265$, so the true margin of error is $(1.645)(0.0265) = 0.0436$.

8.25   **(a)** The margins of error are $1.96 \sqrt{\hat{p}(1 - \hat{p})/100} = 0.196 \sqrt{\hat{p}(1 - \hat{p})}$. This gives 0.0588 when $\hat{p} = 0.1$ or $\hat{p} = 0.9$, 0.0784 when $\hat{p} = 0.2$ or $\hat{p} = 0.8$, 0.0898 when $\hat{p} = 0.3$ or $\hat{p} = 0.7$, 0.0960 when $\hat{p} = 0.4$ or $\hat{p} = 0.6$, and 0.0980 when $\hat{p} = 0.5$. **(b)** No, because $n\hat{p} = (100)(0.04) = 4$ is less than 10.

8.27   **(a)** $\hat{p}_{\text{f}} = 0.8$, so $\text{SE}_{\hat{p}} \doteq 0.05164$ for females. $\hat{p}_{\text{m}} \doteq 0.39$, so $\text{SE}_{\hat{p}} \doteq 0.04253$ for males. **(b)** $\text{SE}_D \doteq 0.0669$, so the interval is 0.2749 to 0.5372. There is (with high confidence) a considerably higher percentage of juvenile references to females than to males.

8.29   **(a)** We have $\hat{p}_1 \doteq 0.1786$ and $\hat{p}_2 \doteq 0.1981$, which gives $\text{SE}_D \doteq 0.0570$, so the interval is -0.1132 to 0.0742. Since this interval includes 0, we have little evidence here to suggest that the two proportions are different.

8.31   Test $H_0: p_{\text{f}} = p_{\text{m}}$ vs. $H_a: p_{\text{f}} \neq p_{\text{m}}$ (assuming we have no belief, before seeing the data, that the difference will lie in a particular direction—e.g., that $p_{\text{f}} < p_{\text{m}}$). The pooled estimate of p is $\hat{p} \doteq 0.3168$, which gives $s_p \doteq 0.0357$, so $z \doteq -5.53$. This gives $P < 0.0001$; it is significant at any reasonable α, so we conclude (with near certainty) that there is a difference between the proportions.

8.33   **(a)** $H_0: p_1 = p_2$; $H_a: p_1 \neq p_2$. **(b)** $\hat{p} \doteq 0.3634$, which gives $s_p \doteq 0.0483$, so $z \doteq 1.22$. This gives $P = 2(0.1112) = 0.2224$; there is little evidence to suggest a difference between rural and urban households. **(c)** $\text{SE}_D \doteq 0.04859$, so the interval is -0.0209 to 0.1389.

8.35   **(a)** $\hat{p}_1 \doteq 0.5107$ and $\hat{p}_2 \doteq 0.4082$. **(b)** $\text{SE}_D \doteq 0.0294$. **(c)** 0.0268 to 0.1783. Since 0 is not in this interval, there appears to be a real difference in the proportions (though it might be fairly small).

8.37   **(a)** $\hat{p} \doteq 0.4540$. **(b)** $s \doteq 0.0295$. **(c)** $H_0: p_1 = p_2$; $H_a: p_1 \neq p_2$ (assuming we have no prior information about which might be higher). **(d)** $z \doteq 3.47$, which gives $P \doteq 2(0.0003) = 0.0006$. We reject H_0 and conclude that there is a real difference in the proportions.

8.39 $\hat{p}_1 \doteq 0.0930$ and $\hat{p}_2 \doteq 0.2857$; test $H_0: p_1 = p_2$ vs. $H_a: p_1 < p_2$. $\hat{p} \doteq 0.0943$ and $s \doteq 0.0554$; so $z \doteq -3.48$, which gives $P \doteq 0.0002$. We reject H_0 and conclude that there is a real difference in the proportions; abnormal chromosomes are associated with increased criminality.

8.41 $\hat{p}_1 \doteq 0.3895$ and $\hat{p}_2 \doteq 0.3261$, $\text{SE}_D \doteq 0.04297$, so the CI is -0.0208 to 0.1476.

8.43   **(a)** For $H_0: p_1 = p_2$ vs. $H_a: p_1 \neq p_2$, we have $\hat{p}_1 \doteq 0.4217$, $\hat{p}_2 \doteq 0.1103$, $\hat{p} \doteq 0.2283$, and $s \doteq 0.0585$. Then $z \doteq 5.33$, so $P < 0.0001$. We reject H_0 and conclude that there is a real difference in the proportions for the two shield types. **(b)** $\text{SE}_D = 0.0605$, so the interval is 0.2119 to 0.4109. The flip-up shields are much more likely to remain on the tractor.

8.45   **(a)** $\hat{p}_1 = 0.5$ and $\hat{p}_2 \doteq 0.72$. **(b)** -0.4825 to 0.0380. **(c)** $H_0: p_1 = p_2$; $H_a: p_1 < p_2$. $\hat{p} \doteq 0.61$, $s \doteq 0.1625$, and $z \doteq -1.37$; the P-value is 0.0853. There is some evidence that the proportions are different, but it is not significant at the 5% level; if the two proportions were equal, we would observe such a difference between $\hat{p}_1$ and $\hat{p}_2$ about 8.5% of the time.

8.47   **(a)** $\hat{p}_1 = 0.5$, $\hat{p}_2 \doteq 0.72$, $\hat{p} \doteq 0.61$, $s \doteq 0.1149$, and $z \doteq -1.93$; the P-value is 0.0268 (recall the alternative hypothesis is one-sided). This is significant evidence of a difference—specifically, that $p_1 < p_2$. **(b)** With the larger sample size, the difference $\hat{p}_2 - \hat{p}_2 \doteq -0.2$ is less likely to have happened by chance.

8.49 The proportions are (respectively) 0.872, 0.900, 0.537, 0.674, 0.935, 0.688, 0.643, 0.647, 0.710, and 0.876. The z-values are 4.64, 6.69, 0.82*, 5.31, 5.90, 5.20, 3.02*, 2.10*, 6.60, 9.05. Those marked with an asterisk were the only ones for which we might *not* reject H_0; all others had $P < 0.0005$. These three had $P = 0.413$ (for which we would definitely not reject H_0), $P = 0.002$ (pretty strong evidence against H_0, but perhaps not enough to reject H_0 if we are using, e.g., Bonferroni's procedure), and $P = 0.036$. The last three texts do not seem to be any different from the first seven; the gender of the author does not seem to affect the proportion.

8.51 No. The percentage was based on a voluntary response sample and so cannot be assumed to be a fair representation of the population. Such a poll is likely to draw a higher-than-actual proportion of people with a strong opinion, especially a strong negative opinion. A confidence statement like the one given is not reliable under these circumstances.

8.53 $\hat{p} \doteq 0.053$ and $SE_{\hat{p}} \doteq 0.0130$; the CI for p is about 0.0279 to 0.0788. The CI for $p - p_0$ is found by subtracting 0.11 from the previous interval: -0.0821 to -0.0312. In other words, we are 95% confident that the new process has a nonconformity rate that is 3.12% to 8.21% lower than the old process.

8.55 **(a)** $p_0 \doteq 0.2515$. **(b)** $\hat{p} = 0.5$. **(c)** $H_0: p = p_0$; $H_a: p > p_0$. $z \doteq 3.14$, so $P = 0.0008$; we reject H_0 and conclude that women are more likely to be among the top students than their proportion in the class suggests.

8.57 **(a)** $\hat{p}_1 \doteq 0.0165$ and $\hat{p}_2 \doteq 0.0078$; $SE_D \doteq 0.0028$, so the CI is 0.0032 to 0.0141. **(b)** H_0: $p_1 = p_2$; $H_a: p_1 > p_2$. $\hat{p} \doteq 0.0126$, and $s \doteq 0.0029$. Then $z \doteq 2.98$, so $P = 0.0014$. This difference is unlikely to occur by chance, so we conclude that high blood pressure is associated with a higher death rate.

8.59 **(a)** $H_0: p_1 = p_2$; $H_a: p_1 \neq p_2$. $z = -0.5675$ and $P = 0.5704$ (or 0.5686 using the tables). **(b)** Gastric freezing is not significantly more (or less) effective than a placebo treatment.

8.61 $SE_D = \sqrt{0.48/n}$. With $z^* = 1.96$, the 95% CI is $0.2 \pm 1.96\sqrt{0.48/n}$, and the margin of error is $1.96\sqrt{0.48/n}$. Then when $n = 15$, the CI is -0.151 to 0.551 (m.e. $= 0.351$); when $n = 25$, the CI is -0.072 to 0.472 (m.e. $= 0.272$); when $n = 50$, the CI is 0.008 to 0.392 (m.e. $= 0.192$); when $n = 75$, the CI is 0.043 to 0.357 (m.e. $= 0.157$); when $n = 100$, the CI is 0.064 to 0.336 (m.e. $= 0.136$); when $n = 500$, the CI is 0.139 to 0.261 (m.e. $= 0.061$). The interval narrows as n increases.

8.63 **(a)** The margin of error is $z^*SE_D = z^*\sqrt{0.5/n}$. With $z^* = 1.96$, this means we need to choose n so that $1.96\sqrt{0.5/n} \leq 0.05$. The smallest such n is 769. **(b)** Solving $z^*\sqrt{0.5/n} \leq m$ gives $n \geq 0.5(z^*/m)^2$.

8.65 It is likely that little or no useful information would come out of such an experiment; the proportion of people dying of cardiovascular disease is so small that out of a group of 200, we would expect very few to die in a five- or six-year period. This experiment would only detect differences between treatment and control if the treatment was *very* effective (or dangerous)—i.e., if it almost completely eliminated (or drastically increased) the risk of CV disease.

Chapter 9

9.1 **(a)** Under 40: 511 employees, $\hat{p} = 0.0137$ terminated, $SE_{\hat{p}} = 0.005142$. Over 40: 806 employees, $\hat{p} = 0.0509$ terminated, $SE_{\hat{p}} = 0.007740$. **(b)** The expected counts are $(48)(511)/1317 = 18.624$, $(48)(806)/1317 = 29.376$, $(1269)(511)/1317 = 492.376$, and $(1269)(806)/1317 = 776.624$, so $X^2 = 12.303$.
Comparing with a $\chi^2(1)$ distribution, we find $P < 0.0005$; we conclude that there is an association between age and whether or not the employee was terminated—specifically, older employees were more like to be terminated.

9.3 **(a)** Use column percents, because we suspect that "source" is explanatory. First column: 36.2% and 63.8%. Second column: 40.0% and 60.0%. Third column: 27.2% and 72.8%. Overall: 32.6% were "cases" and 67.4% were "control." **(b)** The expected counts are (top row) 111.92, 13.05, 91.03; (bottom row) 231.08, 26.95, 187.97. The test statistic is $X^2 = 6.611$. Comparing with a $\chi^2(2)$ distribution, we find $0.025 < P < 0.05$. The conclusion depends on the chosen value of α.

9.5   This is a 2×3 table, with each household classified by pet (cat or dog) and by source. If we view "source" as explanatory for pet type, then we should look at the conditional distribution of pet type, given the source. Private: 29.7% cats, 70.3% dogs. Pet store: 26.1% cats, 73.9% dogs. Other: 58.8% cats, 41.2% dogs. Overall, 38.0% were cats and 62.0% were dogs; it appears that cats are more likely to come from an "other" source. The test statistic bears this out: $X^2 = 90.624$ (df = 2), so that $P < 0.0005$. We conclude that there is a relationship between source and pet type.

9.7   **(a)** With a letter, 51.2% responded; without, the response rate was 52.6%. **(b)** The null hypothesis is that there is no relationship between whether or not a letter is sent and whether or not the subject responds, while the alternative is that there is a relationship. The test statistic is $X^2 = 1.914$; comparing with a $\chi^2(1)$ distribution, we find $0.15 < P < 0.20$. There is little reason to reject the null hypothesis.

9.9   **(a)** No. No treatment was imposed. **(b)** Among those who did not own a pet, 71.8% survived, while 94.3% of pet owners survived. Overall, 84.8% of the patients survived. **(c)** H_0 says that there is no relationship between patient status and pet ownership (i.e., that survival is independent of pet ownership). H_a says that there is a relationship between survival and pet ownership. **(d)** $X^2 = 8.851$ (df = 1), so $0.0025 < P < 0.005$. **(e)** Provided we believe that there are no confounding or lurking variables, we reject H_0 and conclude that owning a pet improves survival.

9.11   **(a)** Regular season: 2584 hits, 7280 "misses," totaling 9864 at-bats. World Series: 35 hits, 63 "misses," totaling 98 at-bats. All games: 2619 hits, 7343 "misses," totaling 9962 at-bats. **(b)** His batting average was .262 during the regular season and .357— much higher—during the World Series. **(c)** H_0 says that the regular season and World Series distributions (batting averages) are the same; the alternative is that the two distributions are different. $X^2 = 4.536$ (df = 1), so $0.025 < P < 0.05$. We have fairly strong (though not overwhelming) evidence that Jackson did better in the World Series.

9.13   69.7% of the second-year winners also had been winners in the first year, while only 29.7% of second-year losers had been winners in the first year. This suggests some persistence in performance. The test statistic supports this: $X^2 = 38.411$ (df = 1), so $P < 0.0005$. We have strong evidence to support persistence of fund performance.

9.15   With the retrospective approach, we have $\hat{p}_1 \doteq 0.7083$, $\hat{p}_2 \doteq 0.3083$, and "pooled" estimate $\hat{p} \doteq 0.5083$. The test statistic is $z \doteq 6.197$. This agrees with the previous result: $z^2 = 38.411$.

9.17   There is no reason to consider one of these as explanatory, but a conditional distribution is useful to determine the nature of the association. Among nonsmokers, 34.5% were nondrinkers, 19.1% drank 0.01 to 0.10 oz/day, 27.6% drank 0.11 to 0.99 oz/day, and 18.8% drank 1 oz or more per day. For light smokers (1–15 mg of nicotine per day), these percentages were 10.8%, 7.7%, 56.9%, and 24.6%. For heavy smokers, they were 13.3%, 15.7%, 50.6%, and 20.5%. Overall, the percentages were 27.2%, 16.8%, 36.1%, and 19.9%. $X^2 = 42.252$ (df = 6) so $P < 0.005$; we conclude that alcohol and nicotine consumption are not independent. The chief deviation from independence is that nondrinkers are more likely to be nonsmokers than we might expect, while those drinking 0.11 to 0.99 oz/day are less likely to be nonsmokers than we might expect.

9.19   **(a)** Blood pressure is explanatory. Of the low-blood-pressure group, 0.785% died from cardiovascular disease, compared to 1.648% of the high-blood-pressure group, suggesting that high blood pressure increases the risk of cardiovascular disease. **(b)** $H_0: p_1 = p_2$; $H_a: p_1 < p_2$. $\hat{p}_1 = 21/2676$, $\hat{p}_2 = 55/3338$, and the "pooled" estimate of p is $\hat{p} = 76/6014$. The test statistic is $z \doteq -2.98$, so that $P = 0.0014$; we conclude that the high-blood-pressure group has a greater risk. **(c)** The four entries in the two-way table are 21, 55, 2655, and 3283. The chi-square test is not appropriate since the alternative is one-sided. **(d)** -0.0141 to -0.0032.

9.21 25% of those with low antacid use, 62.5% of the medium-use group, and 80% of the high-use group had Alzheimer's, suggesting a connection. $X^2 = 7.118$ (df = 3), so $0.05 < P < 0.10$. There is some evidence for the connection, but it is not statistically significant.

9.23 71.3% of Irish, 76.0% of Portuguese, 69.5% of Norwegians, and 75.0% of Italians can taste PTC; there seems to be some variation in the percentages among the countries. $X^2 = 5.957$ (df = 3), so $0.10 < P < 0.15$. The observed differences between the percentages are not significant.

9.25 For the British study, $X^2 = 0.249$ (df = 1), which gives $P = 0.618$. There is very little evidence of an association. For the American study, $X^2 = 25.014$ (df = 1), which gives $P < 0.0005$. This is strong evidence of an association: aspirin reduced the risk of a fatal heart attack. The difference in the conclusions can be attributed to the larger sample size for the American study (important for something as rare as a heart attack), as well as the shorter duration of the study and the lower dosage (taking the aspirin every other day rather than every day).

9.27 **(a)** $X^2 = 2.186$ (df = 1), which gives $0.10 < P < 0.15$; we do not have enough evidence to conclude that the observed difference in death rates is due to something other than chance. **(b)** Good condition: $X^2 = 0.289$ (df = 1), which gives $P > 0.25$. Poor condition: $X^2 = 0.019$ (df = 1), which gives $P > 0.25$. In both cases, we cannot reject the hypothesis that there is no difference between the hospitals. **(c)** No.

9.29 The entries in the sex/social comparison two-way table are 21 (HSC/F), 49 (HSC/M), 46 (LSC/F), 18 (LSC/M). This gives $X^2 = 23.450$ (df = 1), so $P < 0.0005$; this is strong evidence of a link between gender and social comparison. The entries in the sex/mastery two-way table are 35 (HM/F), 36 (HM/M), 32 (LM/F), 31 (LM/M). This gives $X^2 = 0.030$ (df = 1), so $P > 0.25$; there is no evidence of a link between gender and mastery. It appears that the difference between male and female athletes observed in Example 9.4 is in social comparison, not in mastery.

9.31 For the table, $X^2 = 43.487$ (df = 12), so $P < 0.0005$, indicating that there is a relationship between PEOPLE score and field of study. Science has a large proportion of low-scoring students, while liberal arts/education has a large percentage of high-scoring students. (These two entries from the table make the largest contributions to the value of X^2.)

9.33 For example, there were 966 hypertensive hypokalemic patients, and therefore there were 128 nonhypertensive hypokalemic patients. The hypertension/potassium table thus has entries 966, 3662, 11, 128, 1027, 16. The respective X^2-values are 83.147, 48.761, 12.042, and 13.639, all with df = 2, which are all significant (the largest P-value is 0.003). If we drop the hyperkalemic group, the X^2-values are 57.764, 33.125, 11.678, and 8.288, all with df = 2, which are also all significant (the largest P-value is 0.004). Thus it appears that there is an association between potassium level and each of the four risk factors. Looking at the percentages in the table, the hyperkalemic group is generally (for all but diabetes) quite different from the other two groups; the large sample sizes for hypokalemic and normal groups make even small differences (like the difference for gender) statistically significant.

9.35 The percentage of women pharmacy students has gradually increased since 1987, from 60% to nearly 64%; by the end it seems to have nearly leveled out. The rate of increase is considerably less than that shown from 1970 to 1986. To summarize, women were a minority of pharmacy students in the early 70s, but the proportion of women steadily increased until the mid-80s and has increased less rapidly since then. Women became the majority in the early 80s.

9.37 For cats: $X^2 = 8.460$ (df = 4), which gives $P = 0.077$. We do not reject H_0 this time; with the 2×3 table, we had $P = 0.037$, so having more cells has "weakened" the evidence. For dogs: $X^2 = 33.208$ (df = 4), which gives $P < 0.0005$. The conclusion is the same as before: we reject H_0.

9.39 Before we had $X^2 = 7.118$; with the counts doubled, $X^2 = 14.235$ (df = 3), which gives $P = 0.003$. The proportions are the same, but the increased sample size makes the differences between the categories statistically significant.

Chapter 10

10.1 (a) Ignoring the outlier, there is a weak positive association. (b) $\hat{y} = 43.4 + 0.0733x$. The significance test for the slope yields $t = 2.85$ from a t distribution with df $= 59 - 2 = 57$. This is significant—$P < 2(0.005) = 0.01$. We conclude that linear regression on LOS is useful for predicting wages. (c) With $b_1 = 0.0733$, we can say that wages increase by 0.0733 per week of service. (Note: This is not $0.0733, since we don't know the units of "Wages.") (d) From software, $SE_{b_1} = 0.02571$; we compute $b_1 \pm t^* SE_{b_1}$. Using $t^* = 2.009$ (df $= 50$, from the table), the interval is 0.0216 to 0.1250. With $t^* = 2.0025$ (df $= 57$, from software), the interval is 0.0218 to 0.1248.

10.3 (a) There is a weak positive association. (b) $\hat{y} = 51{,}938 + 47.7x$. The significance test for the slope yields $t = 4.61$ (df $= 59 - 2 = 57$). This is significant: $P < 2(0.0005) = 0.001$. We conclude that linear regression on square footage is useful for predicting selling price.

10.5 (a) There is a fairly strong positive relationship. There are no particular outliers or unusual observations, but the spread seems to increase over time. (b) $\hat{y} = -3545 + 1.84x$. The slope is significantly different from 0 ($t = 13.06$, with df $= 38$). Yield has increased at an average rate of 1.84 bushels/acre each year.

10.7 (a) Powerboats registered is the explanatory variable, so it should be on the horizontal axis. The (positive) association appears to be a straight-line relationship. (b) $\hat{y} = -41.4 + 0.125x$. (c) $H_0: \beta_1 = 0$; $H_a: \beta_1 > 0$. The test statistic is $t = 14.24$, which is significant (df $= 12$, $P < 0.0005$); this is good evidence that manatee deaths increase with powerboat registrations. (d) Use $x = 716$: the equation gives $y = 48.1$, or about 48 manatee deaths. The mean number of manatee deaths for 1991–1993 is 42—less than the 48 predicted.

10.9 (a) $\hat{y} = -61.1 + 9.32(18) = 106.66$, for a prediction of 2.9107 m. (b) This is an example of extrapolation—trying to make a prediction outside the range of given x-values. Minitab reports $SE_{\hat{y}} = 19.56$, so that a 95% prediction interval for $\hat{y}$ when $x^* = 18$ is about 62.6 to 150.7. The width of the interval is an indication of how unreliable the prediction is.

10.11 (a) β_1 represents the increase in gas consumption (in hundreds of cubic feet) for each additional degree-day per day. 95% CI: 0.2517 to 0.2862. (b) The margin of error would be smaller here, since for a fixed confidence level, the critical value t^* decreases as df increases. (Additionally, the standard error is slightly smaller here.) The margin of error for Exercise 10.6 was $t^* SE_{b_1} = 0.027$, while it is 0.017 here.

10.13 (a) $t = 17.66$. (b) With df $= 7$, we have $t^* = 1.895$. We reject H_0 at this level (or any reasonable level). (c) From the table, we report $P < 0.0005$. This is probably more readily understandable than the software value: $P \doteq 2.3 \times 10^{-7} = 0.00000023$.

10.15 (a) $\bar{x} = 13.07$ and $\sum(x_i - \bar{x})^2 = 443.201$. (b) $H_0: \beta_1 = 0$; $H_a: \beta_1 > 0$. $SE_{b_1} = s / \sqrt{\sum(x_i - \bar{x})^2} = 0.0835$, so $t = 10.80$. For any reasonable α, this is significant; we conclude that the two variables are positively associated. (c) With df $= 8$, we have $t^* = 3.355$: 0.622 to 1.182. (d) $\hat{y} = 14.56$. $SE_{\hat{y}} = 1.850$ and $t^* = 1.860$, so the prediction interval is 11.12 to 18.00.

10.17 (a) The plot reveals no outliers or unusual points. (b) $\hat{y} = -0.06485 + 1.184x$, so we estimate $1/R = b_1 = 1.184$. $SE_{b_1} = 0.07790$ and $t^* = 3.182$, so the confidence interval is 0.936 to 1.432. (c) $R \doteq 1/b_1 = 0.8446$; the confidence interval is 0.698 to 1.068. (d) $SE_{b_0} = 0.1142$, so $t = -0.5679$. $P > 2(0.25) = 0.50$. We have little reason to doubt that $\beta_0 = 0$.

10.19 (a) The plot reveals no outliers or unusual points. (b) $\hat{y} = -2.80 + 0.0387x$. (c) $t = 16.10$ (with df $= 17$); we reject H_0 (since $P < 0.0005$). We conclude that linear regression on HR is useful for predicting VO2. (d) When $x = 95$, we have $\hat{y} = 0.8676$ and $SE_{\hat{y}} = 0.1269$, so the prediction interval is 0.5998 to 1.1354. When $x = 110$, we have $\hat{y} = 1.4474$ and $SE_{\hat{y}} = 0.1238$, so the interval is 1.1861 to 1.7086. (e) It depends on how accurately they need to know VO2; the regression equation only predicts the subject's *mean* VO2 for a given heart rate, and the intervals in (d) reveal that a particular *observation* may vary quite a bit from that mean.

10.21 (a) $H_0: \beta_1 = 0; H_a: \beta_1 > 0.$ $t = 3.654$; with df $= 16$, we have $0.001 < P < 0.0025$. We reject H_0 and conclude that greater airflow increases evaporation. (b) A 95% confidence interval for β_1 is 0.00279 to 0.01051.

10.23 (a) The prediction interval is 122.91 to 168.30 bushels/acre: $\hat{y} = 145.60$, and $SE_{\hat{y}} = 11.21$, so the 95% prediction interval is 122.71 to 168.49 (using the table value for df $= 30$), or 122.91 to 168.30 (using the software critical value for df $= 38$). (b) The centers are similar (150.25 vs. 145.60), but this interval is narrower. (c) The margin of error in Example 10.14 was 47 bushels/acre, compared to 23 here. It is smaller here because the sample size is larger, which decreases t^* (since df is larger) and decreases $SE_{\hat{y}}$ (since $1/n$ and $1/\sum(x_i - \bar{x})^2$ are smaller).

10.25 The log Yield model is $\hat{y} = -16.3 + 0.00925x$; for this regression, $r^2 = 81.7\%$ and $t = 13.02$. By comparison, for the original model we had $t = 13.06$ and $r^2 = 81.8\%$. The log model is not particularly an improvement over the original; the plot still suggests that the spread increases as "year" increases, though a plot of residuals vs. year shows some improvement in this respect, and the numerical measures are actually slightly smaller than those in the original.

10.27 (a) Regression: df $= 1$, SS $= 3.7619$, MS $= 3.7619$. Error: df $= 17$, SS $= 0.2467$, MS $= 0.0145$. Total: df $= 18$, SS $= 4.0085$. $F = 259.27$. (b) $H_0: \beta_1 = 0$; this says that VO2 is not linearly related to HR. (c) If H_0 is true, F has an $F(1, 17)$ distribution; $F = 259.27$ has $P < 0.001$. (d) We found $t = 16.10$ and $t^2 = 259.21$. (e) $r^2 = SSM/SST = 3.7619/4.0085 = 93.8\%$.

10.29 (a) $t = -5.160$. (b) We have df $= 711$, with $t = -5.16$, $P < 0.001$; this is significant (for any reasonable α), so we conclude that $\rho \neq 0$.

10.31 (a) Full set: $b_0 = -0.03328$, $b_1 = 1.01760$, $s = 0.01472$, $s_{b_1} = 0.01242$. Reduced set: $b_0 = -0.05814$, $b_1 = 1.03111$, $s = 0.01527$, $s_{b_1} = 0.01701$. b_0 and s_{b_1} have changed the most. (b) The most important difference is that F_{reduced} is about half as big as F_{full} (though both are quite significant). MSE is similar in both tables, reflecting the similarity in s in the full and reduced regressions. (c) Full set: $r = 0.996$. Reduced set: $r = 0.997$. (d) The relationship is still strong even with half as many data points; most values are similar in both regressions. (e) Since these values did not change markedly for $n = 25$ vs. $n = 50$, it seems likely that they will be similar when $n = 100$.

10.33 (a) $\hat{y} = 9.1 + 1.09x$. (b) $SE_{b_1} = 0.6529$. (c) $t = 1.66$, which gives $0.10 < P < 0.20$—not significant. With the original data, $t = 17.66$, which is strong evidence against H_0.

10.35 (a) There is a moderate positive relationship, with one outlier (Player 8). (b) $r = 0.687$, so $t = 2.99$. With df $= 10$, this gives $0.01 < P < 0.02$— fairly strong evidence that $\rho \neq 0$. (c) $r = 0.842$, so $t = 4.68$. With df $= 9$, this gives $0.001 < P < 0.002$ stronger evidence that $\rho \neq 0$. The outlier makes the plot less linear and so decreases the correlation.

10.37 With $n = 20$, $t = 2.45$ (df $= 18$, $0.02 < P < 0.04$), while with $n = 10$, $t = 1.63$ (df $= 8$, $0.1 < P < 0.2$). With the larger sample size, r should be a better estimate of ρ, so we are less likely to get $r = 0.5$ unless ρ is really not 0.

10.39 (a) There is a positive association between scores. The 47th pair of scores (SAT 420, ACT 21) is an outlier—the ACT score is higher than one would expect for the SAT score. Since this SAT score is so low, this point may be influential. No other points fall outside the pattern. (b) $\hat{y} = 1.63 + 0.0214x$; $t = 10.78$, so $P < 0.001$ (df $= 58$). (c) $r = 0.817$.

10.41 (a) SAT: $\bar{x} = 912.7$ and $s_x = 180.1$ points. ACT: $\bar{y} = 21.13$ and $s_y = 4.714$ points. So, $a_1 = 0.02617$ and $a_0 = -2.756$. (c) E.g., the first prediction is $-2.756 + (0.02617)(1000) = 23.42$. Up to rounding error, the mean and standard deviation are the same.

Chapter 11

11.1 (a) $H_0: \beta_1 = \beta_2 = \cdots = \beta_{13} = 0$ vs. H_a: not all of the β_j equal 0. The degrees of freedom are 13 and 2215, and $P < 0.001$ (referring to an $F(12, 1000)$ distribution). We have strong evidence that at least one of the β_j is not 0. (b) 29.7% of the variation is explained by the regression. **(c)** Each t statistic tests H_0: $\beta_j = 0$ vs. H_a: $\beta_j \neq 0$, and has df $= 2215$. The critical value is $t^* = 1.961$. **(d)** The only three coefficients that are *not* significantly different from 0 are those for "total payments," "male borrower," and "married." **(e)** Interest rates are lower: for larger loans, for longer terms, with larger down payments, when there is a cosigner, when the loan is secured, when the borrower has a higher income, when the credit report is not considered "bad," for older borrowers, when the borrower owns a home, and for borrowers who have lived for a long time at their present address.

11.3 In Exercise 11.1, we found that 10 factors have a significant effect on the interest rate for direct loans, while in Exercise 11.2, only 4 of the factors examined have a significant impact on the interest rate for indirect loans. Furthermore, a greater proportion of the variation in interest rates is explained by the regression for direct loans than for indirect.

11.5 **(a)** With the given values, $\mu_{GPA} = \beta_0 + 9\beta_1 + 8\beta_2 + 7\beta_3$. **(b)** We estimate $\widehat{GPA} = 2.697$. Among all computer science students with the given high school grades, we expect the mean college GPA after three semesters to be about 2.7.

11.7 **(a)** 0.09862 to 0.23852 ($t^* \doteq 1.9708$ for df $= 220$). This coefficient gives the average increase in college GPA for each 1-point increase in high school math grade.
(b) -0.0312 to 0.12136. This coefficient gives the average increase in college GPA for each 1-point increase in high school English grade.

11.9 **(a)** $\widehat{GPA} = 0.590 + 0.169HSM + 0.034HSS + 0.045HSE$. **(b)** $s = \sqrt{MSE} = 0.69984$.
(c) H_0: $\beta_1 = \beta_2 = \beta_3 = 0$; H_a: not all of the β_j equal 0. In words, H_0 says that none of the high school grade variables are predictors of college GPA (in the form given in the model); H_a says that at least one of them is. **(d)** Under H_0, F has an $F(3, 220)$ distribution. Since $P = 0.0001$, we reject H_0. **(e)** 20.46% of the variation in GPA is explained by the regression.

11.11 A 95% prediction interval is $2.11 to $2.16. The actual price falls in this interval (in fact, it is less than one standard error below the predicted value), so there is not enough evidence to reject H_0, which in this situation would be "there was no manipulation."

11.13 All of the plots display lots of scatter; high school grades seem to be poor predictors of college GPA. Of the three, the math plot seems to most strongly suggest a positive association, although the association appears to be quite weak, and almost nonexistent for HSM < 5. We also observe that scores below 5 are unusual for all three high school variables and could be considered outliers and influential.

11.15 $\widehat{GPA} = 1.289 + 0.002283\ SATM - 0.00002456\ SATV$. The residual plots show no striking patterns, but one should notice the similarity between the predicted GPA and SATM plots.

11.17 $\widehat{GPA} = 1.28 + 0.143\ HSE + 0.000394\ SATV$, and $R^2 = 8.6\%$. The t tests reveal that the coefficient of SATV is not significantly different from 0 ($t = 0.71$). For mathematics variables, we had $R^2 = 19.4\%$—not overwhelmingly large but considerably more than that for verbal variables.

11.19 $\widehat{GPA} = 0.648 + 0.205\ HSM + 0.0018\ HSS + 0.0324\ HSE$, with $R^2 = 25.1\%$. Only the coefficient of HSM is significantly different from 0 (even the constant 0.648 has $t = 1.17$ and $P = 0.247$). Regression with HSM and HSE gives $\widehat{GPA} = 0.648 + 0.206\ HSM + 0.0333\ HSE$, and $R^2 = 25.1\%$—but the P-values for the constant and coefficient of HSE have changed very little. With HSM alone, $\widehat{GPA} = 0.821 + 0.220\ HSM$, R^2 decreases only sightly to 24.9%, and both the constant and coefficient are significantly different from 0. Comparing the results to males, we see that both HSM and HSS were fairly useful for men, but based on R^2, HSM alone does a better job for women than all three variables for men.

11.21 **(a)** $\widehat{GPA} = -2.83 + 0.0822IQ + 0.163C3$. $R^2 = 45.9\%$, and the t statistic for the coefficient of C3 is 2.83, which has $P = 0.006$—significantly different from 0. C3 increases R^2 by 5.7%. **(b)** $\widehat{GPA} = -3.49 + 0.0761IQ + 0.0670C3 + 0.0369SC$. $R^2 = 47.5\%$, and the t statistic for the coefficient of C3 is 0.78, which has $P = 0.436$—*not* significantly different from 0. When self-concept (SC) is included in the model, C3 adds little. **(c)** The values change because coefficients are quite sensitive to changes in the model, especially when the explanatory variables are highly correlated.

11.23   Taste: $\bar{x} = 24.53$, $M = 20.95$, $s = 16.26$, and $IQR = 23.9$. Acetic: $\bar{x} = 5.498$, $M = 5.425$, $s = 0.571$, and $IQR = 0.656$. H2S: $\bar{x} = 5.942$, $M = 5.329$, $s = 2.127$, and $IQR = 3.689$. Lactic: $\bar{x} = 1.4420$, $M = 1.4500$, $s = 0.3035$, and $IQR = 0.430$. None of the variables show striking deviations from normality in the quantile plots. Taste and H2S are slightly right-skewed, and Acetic has two peaks. There are no outliers.

11.25 $\widehat{\text{Taste}} = -61.5 + 15.6$ Acetic; the slope is significantly different from 0 ($P = 3.48$, $P = 0.002$). The regression explains 30.2% of the variation in Taste. The residuals seem to have a normal distribution (based on stem- and quantile plots). Scatterplots reveal positive associations between residuals and both H2S and Lactic. Further analysis of the residuals shows a stronger positive association between residuals and Taste, while the plot of residuals vs. Acetic suggests greater scatter in the residuals for large Acetic values.

11.27 $\widehat{\text{Taste}} = -29.9 + 37.7$ Lactic. The slope is significantly different from 0 ($t = 5.25$, $P < 0.0005$). The regression explains 49.6% of the variation in Taste. The residuals seem to have a normal distribution (based on stem- and quantile plots). Scatterplots reveal a moderately strong positive association between residuals and Taste but no striking patterns for residuals vs. the other variables.

11.29 $\widehat{\text{Taste}} = -26.9 + 3.80$ Acetic $+ 5.15$ H2S; this model explains 58.2% of the variation in Taste. For the coefficient of Acetic, $t = 0.84$ ($P = 0.406$). It does not add significantly to the model when H2S is used, because Acetic and H2S are correlated. This model does a better job than any of the three simple linear regression models, but it is not much better than the model with H2S alone.

11.31 $\widehat{\text{Taste}} = -28.9 + 0.33$ Acetic $+ 3.91$ H2S $+ 19.7$ Lactic; this explains 65.2% of the variation in Taste. Residuals are positively associated with Taste, but appear to be normally distributed and show no patterns in other scatterplots. The coefficient of Acetic is not significantly different from 0 ($P = 0.942$); there is no gain in adding Acetic to the model with H2S and Lactic. It appears that the H2S/Lactic model of Exercise 11.30 is best.

11.33   **(a)** $\widehat{\text{Corn}} = -46.2 + 4.76$ Soybeans. The slope has $t = 16.04$ (df $= 38$), so $P < 0.0005$—it is significantly different from 0. $r^2 = 87.1\%$. **(b)** The quantile plot is fairly close to linear; the one high residual is not so high that we would call it an outlier. **(c)** Plotting residuals vs. year reveals a curved pattern: generally, the residuals are negative in the earlier and later years, and mostly positive from 1964 to 1990.

11.35   **(a)** $\widehat{\text{Corn}} = -964 + 0.480$ Year $- 0.0451$ Year2 $+ 3.90$ Soybeans. **(b)** H_0: All $\beta_j = 0$ vs. H_a: Not all $\beta_j = 0$. $F = 233.05$ (3 and 36 df) has $P < 0.0005$, so we conclude that at least one coefficient is not 0. **(c)** $R^2 = 95.1\%$ (without Year2: 90.5%). **(d)** All three coefficients are significantly different from 0: the t values are 2.97, -5.82, and 9.51, all with df $= 36$; the largest P-value (for the first of these) is 0.005, and the other two are less than 0.0005. **(e)** The residuals seem to be (close to) normal, and they have no apparent relationship with the explanatory or response variables.

11.37   For the simple linear regression, the predicted yield is 145.6; the 95% prediction interval is 122.91 to 168.30. For the multiple regression, the predicted yield is 130.98; the 95% prediction interval is 100.91 to 161.04. The second prediction is lower because with the multiple regression model, corn yield grows less rapidly in later years than it did in the earlier years.

11.39   The outlier is observation 15. Without it, $\widehat{\text{Wages}} = 43.4 + 0.0733$ LOS; the t statistic for the slope is 2.85 ($P = 0.006$). Although this is significant, the predictions are not too good: only 12.5% of the variation in Wages is explained by the regression.

11.41 $\widehat{\text{Wages}} = 37.6 + 0.0829$ LOS $+ 8.92$ Size. Both coefficients are significantly different from 0 ($t = 3.53$ and $t = 3.63$, respectively); regression explains 29.1% of the variation in Wages.

Chapter 12

12.1   **(a)** Immediate: $n = 2$, $\bar{x} = 48.705$, $s = 1.534$, $SE_{\bar{x}} = 1.085$ mg. One day: $n = 2$, $\bar{x} = 41.955$, $s = 2.128$, $SE_{\bar{x}} = 1.505$ mg. Three days: $n = 2$, $\bar{x} = 21.795$, $s = 0.771$, $SE_{\bar{x}} = 0.545$ mg. Five days: $n = 2$, $\bar{x} = 12.415$, $s = 1.082$, $SE_{\bar{x}} = 0.765$ mg. Seven days: $n = 2$, $\bar{x} = 8.320$, $s = 0.269$, $SE_{\bar{x}} = 0.190$ mg. **(b)** H_0: $\mu_1 = \mu_2 = \cdots = \mu_5$;

H_a: not all μ_i are equal. $F = 367.74$ with 4 and 5 df; $P < 0.0005$, so we reject the null hypothesis. **(c)** Vitamin C content decreases over time.

12.3   **(a)** All four data sets appear to be reasonably close to normal, although "3 promotions" seems to have a low outlier. **(b)** One promotion: $n = 40, \bar{x} = 4.2240$, $s = 0.2734$, $SE_{\bar{x}} = 0.0432$. Three promotions: $n = 40, \bar{x} = 4.0627, s = 0.1742$, $SE_{\bar{x}} = 0.0275$. Five promotions: $n = 40, \bar{x} = 3.7590, s = 0.2526, SE_{\bar{x}} = 0.0399$. Seven promotions: $n = 40, \bar{x} = 3.5487, s = 0.2750, SE_{\bar{x}} = 0.0435$. **(c)** The ratio of the largest to smallest standard deviations is about 1.58, so the assumption of equal standard deviations is reasonable. **(d)** H_0: $\mu_1 = \mu_2 = \mu_3 = \mu_4$; H_a: not all μ_i are equal. $F = 59.90$ with 3 and 156 df; $P < 0.0005$, so we reject H_0. With more promotions, expected price decreases.

12.5   **(a)** Piano: $n = 34, \bar{x} = 3.618, s = 3.055, SE_{\bar{x}} = 0.524$. Singing: $n = 10, \bar{x} = -0.300, s = 1.494, SE_{\bar{x}} = 0.473$. Computer: $n = 20, \bar{x} = 0.450, s = 2.212, SE_{\bar{x}} = 0.495$. None: $n = 14, \bar{x} = 0.786, s = 3.191, SE_{\bar{x}} = 0.853$. **(b)** H_0: $\mu_1 = \mu_2 = \mu_3 = \mu_4$; H_a: not all μ_i are equal. $F = 9.24$ with 3 and 74 df; $P < 0.0005$, so we reject the null hypothesis. The type of lesson affects the mean score change.

12.7   We test the hypothesis H_0: $\psi = \mu_1 - (1/3)(\mu_2 + \mu_3 + \mu_4) = 0$; the sample contrast is $c = 3.306$. The pooled standard deviation estimate is $s_p \doteq 2.735$, so $SE_c \doteq 0.6356$. Then $t \doteq 5.20$, with df $= 74$. This is enough evidence ($P < 0.001$) to reject H_0 in favor of H_a: $\psi > 0$, so we conclude that mean score changes for piano students are greater than the average of the means for the other three groups.

12.9   **(a)** Response: Typical number of hours of sleep. Populations: Nonsmokers, moderate smokers, heavy smokers. $I = 3, n_i = 100$ ($i = 1, 2, 3$), $n = 300$. **(b)** Response: Strength of the concrete. Populations: Mixtures A, B, C, and D. $I = 4, n_i = 5$ ($i = 1, 2, 3, 4$), $N = 20$. **(c)** Response: Scores on final exam. Populations: Students using Methods A, B, and C. $I = 3, n_i = 20$ ($i = 1, 2, 3$), $N = 60$.

12.11 **(a)** Lemon yellow: $\bar{x} = 47.17, s = 6.79$. White: $\bar{x} = 15.67, s = 3.33$. Green: $\bar{x} = 31.50, s = 9.91$. Blue: $\bar{x} = 14.83, s = 5.34$. **(b)** H_0: $\mu_1 = \mu_2 = \mu_3 = \mu_4$; H_a: not all μ_i are equal. ANOVA tests if there are differences in the mean number of insects attracted to each color. **(c)** $F = 30.55$ with 3 and 20 df; $P < 0.0005$, so we reject the null hypothesis. The color of the board does affect the number of insects attracted. The pooled standard deviation is $s_p = 6.784$, and $R^2 = 82.1\%$.

12.13 If doing computations by hand, note that $s_p \sqrt{(1/n_i) + (1/n_j)} \doteq 3.916$. The t statistics for the multiple comparisons are $t_{12} \doteq 8.04, t_{13} \doteq 4.00, t_{14} \doteq 8.26, t_{23} \doteq -4.04, t_{24} \doteq 0.21, t_{34} \doteq 4.26$. These indicate that the only nonsignificant difference is between white and blue boards; lemon yellow is the best.

12.15 **(a)** The low observation for 1977 is an outlier, as are the maximum and minimum in 1987 (using the $1.5 \times IQR$ criterion). There is no strong suggestion of a trend in the medians. **(b)** $F = 5.75$ is significant ($P = 0.0002$), indicating that the mean log-concentration does vary over the years. The t tests for individual differences suggest that the mean in 1987 is significantly lower than the others.

12.17 Yes. The ratio of the largest to smallest standard deviations is $12.2/9.2 \doteq 1.33 < 2$. The pooled variance is $s_p^2 \doteq 127.845$, so $s_p \doteq 11.3$.

12.19 **(a)** The sources of variation are "among groups" (i.e., among the means for nonsmokers, moderate smokers, and heavy smokers), with df $= 2$, and "within groups," with df $= 297$. The total variation has df $= 299$. **(b)** "Among groups" (i.e., variation between the means for each of the four concrete mixtures), with df $= 3$, and "within groups," with df $= 16$. The total variation has df $= 19$. **(c)** "Among groups" (i.e., variation between the means for each of the three teaching methods), with df $= 2$, and "within groups," with df $= 57$. The total variation has df $= 59$.

12.21 **(a)** H_0: $\mu_1 = \mu_2 = \mu_3 = \mu_4$; H_a: not all μ_i are equal. **(b)** "Among groups" (i.e., among the means for each of the four classes), with df $= 3$, and "within groups," with df $= 196$. The total variation has df $= 199$. **(c)** If H_0 is true, F has an $F(3, 196)$ distribution. **(d)** Referring to the $F(3, 100)$ distribution, the critical value is 2.70.

12.23 **(a)** First column: DFT $= 35$. Second column: SSG $= 476.88$, SSE $= 2009.92$, SST $= 2486.8$. Last column: $F \doteq 2.531$. **(b)** H_0: $\mu_1 = \mu_2 = \mu_3 = \mu_4$; H_a: not all μ_i are equal. **(c)** If H_0 is true, F has an $F(3, 32)$ distribution. Referring to the $F(3, 30)$ distribution, P is between 0.05 and 0.10 (there is *some* evidence of a difference, but not what we would usually call significant). **(d)** $s_p^2 = $ MSE $= 62.81$, so $s_p \doteq 7.925$.

12.25 (a) $s_p^2 = \text{MSE} \doteq 72{,}412$. (b) First column: DFG = 2, DFE = 230, DFT = 232. Second column: SSG = 6,572,551, SSE = 16,654,788, SST = 23,227,339. Third column: MSG = 3,286,275.5, MSE = 72,412. Last column: $F \doteq 45.38$. (c) $H_0: \mu_1 = \mu_2 = \mu_3$; H_a: not all μ_i are equal. (d) If H_0 is true, F has an $F(2, 230)$ distribution. Referring to the $F(2, 200)$ distribution, $P < 0.001$; we conclude that the means are not all the same. (e) $R^2 \doteq 0.283 = 28.3\%$.

12.27 (a) $\psi_1 = 0.5\mu_1 + 0.5\mu_2 - 0.5\mu_3 - 0.5\mu_4$ (or $\psi_1 = 0.5\mu_3 + 0.5\mu_4 - 0.5\mu_1 - 0.5\mu_2$). (b) $\psi_2 = \mu_1 - \mu_2$ (or $\psi_2 = \mu_2 - \mu_1$). (c) $\psi_3 = \mu_3 - \mu_4$ (or $\psi_3 = \mu_4 - \mu_3$).

12.29 (a) For $\psi_1 = 0.5\mu_1 + 0.5\mu_2 - \mu_3$: $H_0: \psi_1 = 0$ vs. $H_a: \psi_1 > 0$. For $\psi_2 = \mu_1 - \mu_2$: $H_0: \psi_2 = 0$ vs. $H_a: \psi_2 \neq 0$. (b) $c_1 = 0.93$ and $c_2 = 0.02$. (c) $\text{SE}_{c_1} \doteq 0.2299$ and $\text{SE}_{c_2} \doteq 0.3421$. (d) $t_1 \doteq 4.045$ (df = 221, $P < 0.0005$). We conclude that science majors have higher mean HS math grades than other majors. $t_2 \doteq 0.0585$ (df = 221, $P > 0.5$). The difference in mean HS math grades for computer science vs. other science students is not significant. (e) Use $t^* = 1.984$ (for df = 100, from the table), or $t^* = 1.9708$ (for df = 221). For ψ_1: 0.474 to 1.386, or 0.477 to 1.383. For ψ_2: -0.659 to 0.699, or -0.654 to 0.694.

12.31 (a) $\psi_1 = \mu_T - \mu_C$; $H_0: \psi_1 = 0$ vs. $H_a: \psi_1 < 0$. $\psi_2 = \mu_T - (1/2)(\mu_C + \mu_S)$; $H_0: \psi_2 = 0$ vs. $H_a: \psi_2 < 0$. $\psi_3 = \mu_J - (1/3)(\mu_T + \mu_C + \mu_S)$; $H_0: \psi_3 = 0$ vs. $H_a: \psi_3 < 0$. (b) First note $s_p = \sqrt{\text{MSE}} \doteq 7.925$ and df = 32. $c_1 = -5.5$, $\text{SE}_{c_1} \doteq 4.341$, and $t_1 \doteq -1.27$, which has $0.10 < P < 0.15$—not significant. $c_2 = -5.9$, $\text{SE}_{c_2} \doteq 3.315$, and $t_2 \doteq -1.78$, which has $0.025 < P < 0.05$—fairly strong evidence of a difference. $c_3 \doteq -6.103$, $\text{SE}_{c_3} \doteq 2.916$, and $t_3 \doteq -2.093$, which has $0.02 < P < 0.025$—fairly strong evidence of a difference. The contrasts allow us to determine which differences between sample means represent "true" differences in population means: T is not significantly better than C but is better than the average of C and S. Joggers have lower mean depression scores than the average of the other three groups. (c) No. The treatment imposed by the study (the T group) did not produce a significantly lower result than the control group. The contrasts that *were* significant involved the two groups that did not have treatments imposed on them. In these cases, causation cannot be determined, because of confounding or "common response" issues.

12.33 $\alpha = 0.05$ means that for all three comparisons, there is a probability no more than 0.05 that we will falsely conclude means are unequal. $s_p \doteq 269.095$; the t statistics (and standard errors) for the differences are $t_{12} \doteq 9.27$ ($\text{SE}_{12} \doteq 40.232$), $t_{13} \doteq 2.11$ ($\text{SE}_{13} \doteq 46.517$), $t_{23} \doteq -5.95$ ($\text{SE}_{23} \doteq 46.225$). The mean toddler food intake for Kenya is significantly less than the means for the other two countries.

12.35 $s_p = \sqrt{\text{MSE}} \doteq 7.925$; the t statistics for the differences are $t_{TC} \doteq -1.27$, $t_{TJ} \doteq 0.63$, $t_{TS} \doteq -1.78$, $t_{CJ} \doteq 1.79$, $t_{CS} \doteq -0.18$, $t_{JS} \doteq -2.44$. None of the differences are significant.

12.37 $\overline{\mu} = 3.0$ and $\lambda = 18n/529$; these values are 1.7013, 3.4026, 5.1040, 5.9546, and 6.8053. The values of F^* are 3.0576, 3.0261, 3.0158, 3.013, and 3.0108, and the df are 2 and $3n - 3$. Software gives the following values for the power (these may vary slightly from one package to another): 0.1940, 0.3566, 0.5096, 0.5780, and 0.6399. Choices of sample size might vary.

12.39 (a) The plot shows granularity (which varies between groups), but that should not make us question independence; it is due to the fact that the scores are all integers. (b) The ratio of the largest to the smallest standard deviation is less than 2. (c) Apart from the granularity, the quantile plots are reasonably straight. (d) Again, apart from the granularity, the quantile plots look pretty good.

12.41 (a) $\psi_1 = (1/3)\mu_1 + (1/3)\mu_2 + (1/3)\mu_3 - \mu_4$, $\psi_2 = (1/2)\mu_1 + (1/2)\mu_2 - \mu_3$, $\psi_3 = \mu_1 - \mu_2$. (b) $s_p \doteq 1.1958$. $\text{SE}_{c_1} \doteq 0.2355$, $\text{SE}_{c_2} \doteq 0.1413$, $\text{SE}_{c_3} \doteq 0.1609$. (c) Testing $H_0: \psi_i = 0$ vs. $H_a: \psi_i \neq 0$ for each contrast, we find $c_1 = -12.51$, $t_1 = -53.17$, $P_1 < 0.0005$; $c_2 = 1.269$, $t_2 = 8.98$, $P_2 < 0.0005$; $c_3 = 0.191$, $t_3 = 1.19$, $P_3 \doteq 0.2359$. The Placebo mean is significantly higher than the average of the other three, and the Keto mean is significantly lower than the average of the two Pyr means. The difference between the Pyr means is not significant.

12.43 For the vitamin A data, we have the following. Immediate: $n = 2$, $\overline{x} = 67.0\%$, $s = 0.28284\%$, $\text{SE}_{\overline{x}} = 0.2\%$. One day: $n = 2$, $\overline{x} = 64.8\%$, $s = 1.13137\%$, $\text{SE}_{\overline{x}} = 0.8\%$.

Three days: $n = 2, \bar{x} = 64.2\%, s = 1.41421\%, \mathrm{SE}_{\bar{x}} = 1.0\%$. Five days: $n = 2, \bar{x} = 66.1\%, s = 1.55563\%, \mathrm{SE}_{\bar{x}} = 1.1\%$. Seven days: $n = 2, \bar{x} = 59.3\%, s = 1.27279\%$, $\mathrm{SE}_{\bar{x}} = 0.9\%$. The transformation has no effect on vitamin E, since the number of milligrams remaining is also the percentage of the original 100 mg. Although the SS and MS entries are different for vitamin A from those of Exercise 12.2, everything else is the same: $F = 12.09$ with 4 and 5 df; $P = 0.009$. So we (again) conclude that vitamin A content decreases over time. Since the vitamin E numbers are unchanged, we again fail to reject H_0 ($F = 0.69$ with 4 and 5 df; $P = 0.630$). In summary, doing any linear transformation has no effect on the results of the ANOVA.

12.45 **(a)** Basal: $n = 22, \bar{x} = 5.\overline{27}, s = 2.7634$. DRTA: $n = 22, \bar{x} \doteq 5.09, s = 1.9978$. Strat: $n = 22, \bar{x} \doteq 4.954, s = 1.8639$. **(b)** No marked deviations from normality. **(c)** $2.7634/1.8639 = 1.4826$; ANOVA is reasonable. **(d)** $H_0: \mu_B = \mu_D = \mu_S$; H_a: at least one mean is different. $F = 0.11$ with 2 and 63 df, so $P = 0.895$; no evidence against H_0. **(e)** There is no reason to believe that mean PRE2 scores differ between methods.

12.47 **(a)** Basal: $n = 22, \bar{x} \doteq 6.681, s = 2.7669$. DRTA: $n = 22, \bar{x} \doteq 9.772, s = 2.7244$. Strat: $n = 22, \bar{x} \doteq 7.772, s = 3.9271$. **(b)** No marked deviations from normality. **(c)** $3.9271/2.7244 = 1.4415$; ANOVA is reasonable. **(d)** $H_0: \mu_B = \mu_D = \mu_S$; H_a: at least one mean is different. $F = 5.32$ with 2 and 63 df, so $P = 0.007$; this is strong evidence that the means differ. **(e)** $s_p \doteq 3.18852$. For $\psi = \mu_B - (1/2)\mu_D - (1/2)\mu_S$, we have $c \doteq -2.09$, $\mathrm{SE}_c = 0.8326$, and $t = -2.51$ with df $= 63$. The one-sided P-value (for the alternative $\psi < 0$) is 0.0073; this is strong evidence that the Basal mean is less than the average of the other two means. Confidence interval: -3.755 to -0.427. **(f)** For $\psi = \mu_D - \mu_S$, we have $c = 2$, $\mathrm{SE}_c = 0.9614$, and $t = 2.0504$ with df $= 63$. The two-sided P-value is 0.0415; this is fairly strong evidence that the DRTA and Strat means differ. Confidence interval: 0.079 to 3.921. **(g)** Among POST1 scores, the Basal mean is lowest, Strat scores are in the middle, and DRTA scores are highest. The differences are big enough that they are not likely to occur by chance.

12.49 **(a)** $F = 1.33$ with 3 and 12 df, giving $P = 0.310$; not enough evidence to stop believing that all four means are equal. **(b)** With the correct data, $F = 12.08$ with 3 and 12 df, giving $P = 0.001$. This is fairly strong evidence that the means are not all the same. Though the outlier made the means more different, it also increased the variability. **(c)** No nematodes: $\bar{x} = 34.95, s = 48.74$ cm. 1000 nematodes: $\bar{x} = 10.425, s = 1.48633$ cm. 5000 nematodes: $\bar{x} = 5.600, s = 1.24365$ cm. 10,000 nematodes: $\bar{x} = 5.450, s = 1.77106$ cm. The marked difference in the values for 0 nematodes would have caught our attention, especially the relatively large standard deviation.

12.51 No nematodes: $\bar{x} = 1.02165, s = 0.080039$. 1000 nematodes: $\bar{x} = 1.01438, s = 0.067147$. 5000 nematodes: $\bar{x} = 0.74084, s = 0.090196$. 10,000 nematodes: $\bar{x} = 0.71698, s = 0.154997$. The hypotheses are the same as before ($H_0: \mu_0 = \mu_{1000} = \mu_{5000} = \mu_{10,000}$ vs. H_a: at least one mean is different), except that now μ_i represents the mean *logarithm* of the growth for each group. The new F is 10.39, with 3 and 12 df; the P-value is 0.001. The conclusion is the same as with the original data (although the new F is slightly smaller than the old F, meaning P is slightly greater). For the original data, $s_p = 1.667$ and $R^2 = 75.1\%$. For the transformed data, $s_p = 0.1038$, and $R^2 = 72.2\%$.

12.53 **(a)** $\bar{\mu} = 590$ and $\lambda = 20n/81$. The degrees of freedom are 3 and $4n - 4$. Answers will vary with the choices of α and n. For example, with $\alpha = 0.05$ and $n = 25, 50, 75, 100$, the powers are 0.5128, 0.8437, 0.9618, and 0.9922. **(b)** The power rises to about 0.90 for $n = 60$; it continues rising (getting closer to 1) after that, but much more slowly. **(c)** Choice of sample size will vary.

12.55 $\hat{y} = 4.36 - 0.116x$. The regression is significant (i.e., the slope is significantly different from 0): $t = -13.31$ with df $= 158$, giving $P < 0.0005$. The regression on number of promotions explains 52.9% of the variation in expected price. (This is similar to the ANOVA value: $R^2 = 53.5\%$.) The granularity of the "number of promotions" observations makes interpreting the plot a bit tricky. For 5 promotions, the residuals seem to be more likely to be negative, while for 3 promotions, the residuals are weighted toward the positive side. This suggests that a linear model may not be appropriate.

Chapter 13

13.1 **(a)** Response variable: Yield (pounds of tomatoes/plant). Factors: Variety ($I = 5$) and fertilizer type ($J = 2$). $N = 40$. **(b)** Response variable: Attractiveness rating. Factors: Packaging type ($I = 6$) and city ($J = 6$). $N = 1800$. **(c)** Response variable: Weight loss. Factors: Weight-loss program ($I = 4$) and gender ($J = 2$). $N = 80$.

13.3 **(a)** Variety (df = 4), fertilizer type (df = 1), variety/fertilizer interaction (df = 4), and error (df = 30). Total df = 39. **(b)** Packaging type (df = 5), city (df = 5), packaging/city interaction (df = 25), and error (df = 1765). Total df = 1799. **(c)** Weight-loss program (df = 3), gender (df = 1), program/gender interaction (df = 3), and error (df = 72). Total df = 79.

13.5 **(b)** Nonwhite means are all slightly higher than white means. Mean systolic blood pressure rises with age. There does not seem to be any interaction; both plots rise in a similar fashion. **(c)** By race, the marginal means are 135.98 (white) and 137.82 (nonwhite). By age, they are 131.65, 133.25, 136.2, 140.35, 143.05. The nonwhite minus white means are 1.3, 1.9, 2, 1.9, and 2.1. The mean systolic blood pressure rises about 2 to 4 points from one age group to the next; the nonwhite means are generally about 2 points higher.

13.7 **(b)** There seems to be a fairly large difference between the means based on how much the rats were allowed to eat, but not very much difference based on the chromium level. There may be an interaction: the NM mean is lower than the LM mean, while the NR mean is higher than the LR mean. **(c)** L mean: 4.86. N mean: 4.871. M mean: 4.485. R mean: 5.246. LR minus LM: 0.63. NR minus NM: 0.892. Mean GITH levels are lower for M than for R; there is not much difference for L vs. N. The difference between M and R is greater among rats who had normal chromium levels in their diets (N).

13.9 The "Other" category had the lowest mean SATM score for both genders; the marginal means are CS: 605, EO: 624.5, O: 566. Males had higher mean scores in CS and O, while females are slightly higher in EO; this seems to be an interaction. Overall, the marginal means are 611.7 (males) and 585.3 (females).

13.11 **(a)** S1-Aspen: $\bar{x} = 387.333$, $s = 68.712$. S1-Birch: $\bar{x} = 292.667$, $s = 121.829$. S1-Maple: $\bar{x} = 323.333$, $s = 52.013$. S2-Aspen: $\bar{x} = 335.667$, $s = 60.136$. S2-Birch: $\bar{x} = 455.333$, $s = 117.717$. S2-Maple: $\bar{x} = 293.667$, $s = 183.919$. The species marginal means are 362 (aspen), 374 (birch), and 308 (maple). The flake size means are 334 (S1) and 362 (S2). **(b)** There appear to be differences between species and a smaller effect by flake size. There also seems to be an interaction (birch has the smallest mean for S1 but the largest mean for S2). **(c)** With A = species and B = flake size, $F_A = 0.59$ with 2 and 12 df; this has $P = 0.5687$. $F_B = 0.27$ with 1 and 12 df; this has $P = 0.6129$. $F_{AB} = 1.70$ with 2 and 12 df; this has $P = 0.2237$. None of these are significant; the differences we observed could easily be attributable to chance.

13.13 **(a)** In the order listed in the table: $\bar{x}_{11} = 25.0307$, $s_{11} = 0.0011541$; $\bar{x}_{12} = 25.0280$, $s_{12} = 0$; $\bar{x}_{13} = 25.0260$, $s_{13} = 0$; $\bar{x}_{21} = 25.0167$, $s_{21} = 0.0011541$; $\bar{x}_{22} = 25.0200$, $s_{22} = 0.0019999$; $\bar{x}_{23} = 25.1060$, $s_{23} = 0$; $\bar{x}_{31} = 25.0063$, $s_{31} = 0.0015275$; $\bar{x}_{32} = 25.0127$, $s_{32} = 0.0011552$; $\bar{x}_{33} = 25.0093$, $s_{33} = 0.0011552$; $\bar{x}_{41} = 25.0120$, $s_{41} = 0$; $\bar{x}_{42} = 25.0193$, $s_{42} = 0.0011552$; $\bar{x}_{43} = 25.0140$, $s_{43} = 0.0039997$; $\bar{x}_{51} = 25.9973$, $s_{51} = 0.0011541$; $\bar{x}_{52} = 25.0060$, $s_{52} = 0$; $\bar{x}_{53} = 25.0003$, $s_{53} = 0.0015277$. **(b)** Except for tool 1, mean diameter is highest at time 2. Tool 1 had the highest mean diameters, followed by tool 2, tool 4, tool 3, and tool 5. **(c)** With A = tool and B = time, $F_A = 412.98$ with 4 and 30 df. $F_B = 43.61$ with 2 and 30 df. $F_{AB} = 7.65$ with 8 and 30 df; all three have $P < 0.0005$. **(d)** There is strong evidence of a difference in mean diameter among the tools and among the times. There is also an interaction (tool 1's mean diameters changed differently over time compared to the other tools).

13.15 **(a)** In the order listed in the table: $\bar{x}_{1,10} = 4.4230$, $s_{1,10} = 0.1848$; $\bar{x}_{1,20} = 4.2250$, $s_{1,20} = 0.3856$; $\bar{x}_{1,30} = 4.6890$, $s_{1,30} = 0.2331$; $\bar{x}_{1,40} = 4.9200$, $s_{1,40} = 0.1520$; $\bar{x}_{3,10} = 4.2840$, $s_{3,10} = 0.2040$; $\bar{x}_{3,20} = 4.0970$, $s_{3,20} = 0.2346$; $\bar{x}_{3,30} = 4.5240$, $s_{3,30} = 0.2707$; $\bar{x}_{3,40} = 4.7560$, $s_{3,40} = 0.2429$; $\bar{x}_{5,10} = 4.0580$, $s_{5,10} = 0.1760$; $\bar{x}_{5,20} = 3.8900$, $s_{5,20} = 0.1629$; $\bar{x}_{5,30} = 4.2510$, $s_{5,30} = 0.2648$; $\bar{x}_{5,40} = 4.3930$, $s_{5,40} = 0.2685$;

$\bar{x}_{7,10} = 3.7800, s_{7,10} = 0.2144; \bar{x}_{7,20} = 3.7600, s_{7,20} = 0.2618; \bar{x}_{7,30} = 4.0940, s_{7,30} = 0.2407; \bar{x}_{7,40} = 4.2690, s_{7,40} = 0.2699$. Mean expected price decreases as percent discount increases, and also as the number of promotions increases. **(b)** With A = number of promotions and B = percent discount, $F_A = 47.73$ with 3 and 144 df; $P < 0.0005$. $F_B = 47.42$ with 3 and 144 df; $P < 0.0005$. $F_{AB} = 0.44$ with 9 and 144 df; $P = 0.912$. **(c)** There is strong evidence of a difference in mean expected price based on the number of promotions and the percent discount. Specifically, more promotions and higher discounts decrease the expected price. There is no evidence of an interaction.

13.17 **(a)** DFA = 1, DFB = 1, DFAB = 1, DFE = 36, DFT = 39. SST = 7.04487. MSA = 0.00121, MSB = 5.79121, MSAB = 0.17161, MSE = 0.03002. $F_A = 0.04030$, $F_B = 192.89$, $F_{AB} = 5.7159$. **(b)** $F_{AB} = 5.7159$. If there is no interaction, this comes from an $F(1, 36)$ distribution; 5.7159 gives $P = 0.0222$. **(c)** For Chromium, $F_A = 0.04030$. If there is no effect, this comes from an $F(1, 36)$ distribution; 0.04030 gives $P = 0.8420$. For Eat, $F_B = 192.89$. If there is no effect, this comes from an $F(1, 36)$ distribution; 192.89 gives $P < 0.001$. **(d)** $s_p^2 = $ MSE $= 0.03002$, so $s_p = 0.1733$. **(e)** The amount the rats were allowed to eat made a difference in mean GITH levels, but chromium levels had no (significant) effect by themselves, although there was a Chromium/Eat interaction.

13.19 **(a)** All three F-values have 1 and 945 df, the P-values are <0.001, <0.001, and 0.1477. Gender and handedness both have significant effects on mean lifetime, but there is no interaction. **(b)** Women live about 6 years longer than men (on the average), while right-handed people average 9 more years of life than left-handed people. Handedness affects both genders in the same way, and vice versa.

13.21 Men in CS: $n = 39, \bar{x} = 526.949, s = 100.937$. Men in EOS: $n = 39, \bar{x} = 507.846$, $s = 57.213$. Men in Other: $n = 39, \bar{x} = 487.564, s = 108.779$. Women in CS: $n = 39, \bar{x} = 543.385, s = 77.654$. Women in EOS: $n = 39, \bar{x} = 538.205, s = 102.209$. Women in Other: $n = 39, \bar{x} = 465.026, s = 82.184$. Students who stay in the sciences seem to have higher mean SATV scores than those who end up in the "Other" group. Female CS and EO students have higher scores than males in those majors, but males have the higher mean in the Other group. Quantile plots suggest some right-skewness in the "Women in CS" group and also some nonnormality in the tails of the "Women in EO" group. Other groups look reasonably normal. Only the effect of major is significant ($F = 93.2$, 2 and 228 df, $P < 0.0005$).

13.23 Men in CS: $n = 39, \bar{x} = 7.79487, s = 1.50752$. Men in EOS: $n = 39, \bar{x} = 7.48718$, $s = 2.15054$. Men in Other: $n = 39, \bar{x} = 7.41026, s = 1.56807$. Women in CS: $n = 39, \bar{x} = 8.84615, s = 1.13644$. Women in EOS: $n = 39, \bar{x} = 9.25641, s = 0.75107$. Women in Other: $n = 39, \bar{x} = 8.61539, s = 1.16111$. Females seem to have higher HSE grades than males. For a given gender, there is not too much difference among majors. Quantile plots show no great deviations from normality, apart from the granularity of the grades (most evident among Women in EO). Only the effect of gender is significant ($F = 50.32$, 1 and 228 df, $P < 0.0005$).

DATA TABLE INDEX

INDEX

TABLE B Random digits

Line

101	19223	95034	05756	28713	96409	12531	42544	82853
102	73676	47150	99400	01927	27754	42648	82425	36290
103	45467	71709	77558	00095	32863	29485	82226	90056
104	52711	38889	93074	60227	40011	85848	48767	52573
105	95592	94007	69971	91481	60779	53791	17297	59335
106	68417	35013	15529	72765	85089	57067	50211	47487
107	82739	57890	20807	47511	81676	55300	94383	14893
108	60940	72024	17868	24943	61790	90656	87964	18883
109	36009	19365	15412	39638	85453	46816	83485	41979
110	38448	48789	18338	24697	39364	42006	76688	08708
111	81486	69487	60513	09297	00412	71238	27649	39950
112	59636	88804	04634	71197	19352	73089	84898	45785
113	62568	70206	40325	03699	71080	22553	11486	11776
114	45149	32992	75730	66280	03819	56202	02938	70915
115	61041	77684	94322	24709	73698	14526	31893	32592
116	14459	26056	31424	80371	65103	62253	50490	61181
117	38167	98532	62183	70632	23417	26185	41448	75532
118	73190	32533	04470	29669	84407	90785	65956	86382
119	95857	07118	87664	92099	58806	66979	98624	84826
120	35476	55972	39421	65850	04266	35435	43742	11937
121	71487	09984	29077	14863	61683	47052	62224	51025
122	13873	81598	95052	90908	73592	75186	87136	95761
123	54580	81507	27102	56027	55892	33063	41842	81868
124	71035	09001	43367	49497	72719	96758	27611	91596
125	96746	12149	37823	71868	18442	35119	62103	39244
126	96927	19931	36089	74192	77567	88741	48409	41903
127	43909	99477	25330	64359	40085	16925	85117	36071
128	15689	14227	06565	14374	13352	49367	81982	87209
129	36759	58984	68288	22913	18638	54303	00795	08727
130	69051	64817	87174	09517	84534	06489	87201	97245
131	05007	16632	81194	14873	04197	85576	45195	96565
132	68732	55259	84292	08796	43165	93739	31685	97150
133	45740	41807	65561	33302	07051	93623	18132	09547
134	27816	78416	18329	21337	35213	37741	04312	68508
135	66925	55658	39100	78458	11206	19876	87151	31260
136	08421	44753	77377	28744	75592	08563	79140	92454
137	53645	66812	61421	47836	12609	15373	98481	14592
138	66831	68908	40772	21558	47781	33586	79177	06928
139	55588	99404	70708	41098	43563	56934	48394	51719
140	12975	13258	13048	45144	72321	81940	00360	02428
141	96767	35964	23822	96012	94591	65194	50842	53372
142	72829	50232	97892	63408	77919	44575	24870	04178
143	88565	42628	17797	49376	61762	16953	88604	12724
144	62964	88145	83083	69453	46109	59505	69680	00900
145	19687	12633	57857	95806	09931	02150	43163	58636
146	37609	59057	66967	83401	60705	02384	90597	93600
147	54973	86278	88737	74351	47500	84552	19909	67181
148	00694	05977	19664	65441	20903	62371	22725	53340
149	71546	05233	53946	68743	72460	27601	45403	88692
150	07511	88915	41267	16853	84569	79367	32337	03316

Table entry for p and C is the point t^* with probability p lying above it and probability C lying between $-t^*$ and t^*.

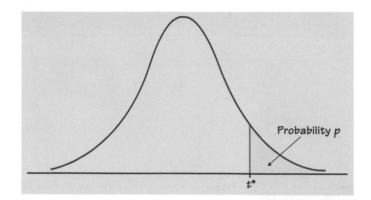

Probability p

t^*

TABLE D	t distribution critical values

df	Tail probability p											
	.25	.20	.15	.10	.05	.025	.02	.01	.005	.0025	.001	.0005
1	1.000	1.376	1.963	3.078	6.314	12.71	15.89	31.82	63.66	127.3	318.3	636.6
2	0.816	1.061	1.386	1.886	2.920	4.303	4.849	6.965	9.925	14.09	22.33	31.60
3	0.765	0.978	1.250	1.638	2.353	3.182	3.482	4.541	5.841	7.453	10.21	12.92
4	0.741	0.941	1.190	1.533	2.132	2.776	2.999	3.747	4.604	5.598	7.173	8.610
5	0.727	0.920	1.156	1.476	2.015	2.571	2.757	3.365	4.032	4.773	5.893	6.869
6	0.718	0.906	1.134	1.440	1.943	2.447	2.612	3.143	3.707	4.317	5.208	5.959
7	0.711	0.896	1.119	1.415	1.895	2.365	2.517	2.998	3.499	4.029	4.785	5.408
8	0.706	0.889	1.108	1.397	1.860	2.306	2.449	2.896	3.355	3.833	4.501	5.041
9	0.703	0.883	1.100	1.383	1.833	2.262	2.398	2.821	3.250	3.690	4.297	4.781
10	0.700	0.879	1.093	1.372	1.812	2.228	2.359	2.764	3.169	3.581	4.144	4.587
11	0.697	0.876	1.088	1.363	1.796	2.201	2.328	2.718	3.106	3.497	4.025	4.437
12	0.695	0.873	1.083	1.356	1.782	2.179	2.303	2.681	3.055	3.428	3.930	4.318
13	0.694	0.870	1.079	1.350	1.771	2.160	2.282	2.650	3.012	3.372	3.852	4.221
14	0.692	0.868	1.076	1.345	1.761	2.145	2.264	2.624	2.977	3.326	3.787	4.140
15	0.691	0.866	1.074	1.341	1.753	2.131	2.249	2.602	2.947	3.286	3.733	4.073
16	0.690	0.865	1.071	1.337	1.746	2.120	2.235	2.583	2.921	3.252	3.686	4.015
17	0.689	0.863	1.069	1.333	1.740	2.110	2.224	2.567	2.898	3.222	3.646	3.965
18	0.688	0.862	1.067	1.330	1.734	2.101	2.214	2.552	2.878	3.197	3.611	3.922
19	0.688	0.861	1.066	1.328	1.729	2.093	2.205	2.539	2.861	3.174	3.579	3.883
20	0.687	0.860	1.064	1.325	1.725	2.086	2.197	2.528	2.845	3.153	3.552	3.850
21	0.686	0.859	1.063	1.323	1.721	2.080	2.189	2.518	2.831	3.135	3.527	3.819
22	0.686	0.858	1.061	1.321	1.717	2.074	2.183	2.508	2.819	3.119	3.505	3.792
23	0.685	0.858	1.060	1.319	1.714	2.069	2.177	2.500	2.807	3.104	3.485	3.768
24	0.685	0.857	1.059	1.318	1.711	2.064	2.172	2.492	2.797	3.091	3.467	3.745
25	0.684	0.856	1.058	1.316	1.708	2.060	2.167	2.485	2.787	3.078	3.450	3.725
26	0.684	0.856	1.058	1.315	1.706	2.056	2.162	2.479	2.779	3.067	3.435	3.707
27	0.684	0.855	1.057	1.314	1.703	2.052	2.158	2.473	2.771	3.057	3.421	3.690
28	0.683	0.855	1.056	1.313	1.701	2.048	2.154	2.467	2.763	3.047	3.408	3.674
29	0.683	0.854	1.055	1.311	1.699	2.045	2.150	2.462	2.756	3.038	3.396	3.659
30	0.683	0.854	1.055	1.310	1.697	2.042	2.147	2.457	2.750	3.030	3.385	3.646
40	0.681	0.851	1.050	1.303	1.684	2.021	2.123	2.423	2.704	2.971	3.307	3.551
50	0.679	0.849	1.047	1.299	1.676	2.009	2.109	2.403	2.678	2.937	3.261	3.496
60	0.679	0.848	1.045	1.296	1.671	2.000	2.099	2.390	2.660	2.915	3.232	3.460
80	0.678	0.846	1.043	1.292	1.664	1.990	2.088	2.374	2.639	2.887	3.195	3.416
100	0.677	0.845	1.042	1.290	1.660	1.984	2.081	2.364	2.626	2.871	3.174	3.390
1000	0.675	0.842	1.037	1.282	1.646	1.962	2.056	2.330	2.581	2.813	3.098	3.300
z^*	0.674	0.841	1.036	1.282	1.645	1.960	2.054	2.326	2.576	2.807	3.091	3.291
	50%	60%	70%	80%	90%	95%	96%	98%	99%	99.5%	99.8%	99.9%

Confidence level C